SQL Server Analytical Toolkit

Using Windowing, Analytical, Ranking, and Aggregate Functions for Data and Statistical Analysis

Angelo Bobak

Apress®

SQL Server Analytical Toolkit: Using Windowing, Analytical, Ranking, and Aggregate Functions for Data and Statistical Analysis

Angelo Bobak
Hastings On Hudson, NY, USA

ISBN-13 (pbk): 978-1-4842-8666-1   ISBN-13 (electronic): 978-1-4842-8667-8
https://doi.org/10.1007/978-1-4842-8667-8

Managing Director, Apress Media LLC: Welmoed Spahr
Acquisitions Editor: Joan Murray
Development Editor: Laura Berendson
Coordinating Editor: Gryffin Winkler

Cover image designed by Freepik (www.freepik.com)

Distributed to the book trade worldwide by Springer Science+Business Media LLC, 1 New York Plaza, Suite 4600, New York, NY 10004. Phone 1-800-SPRINGER, fax (201) 348-4505, e-mail orders-ny@springer-sbm.com, or visit www.springeronline.com. Apress Media, LLC is a California LLC and the sole member (owner) is Springer Science + Business Media Finance Inc (SSBM Finance Inc). SSBM Finance Inc is a **Delaware** corporation.

For information on translations, please e-mail booktranslations@springernature.com; for reprint, paperback, or audio rights, please e-mail bookpermissions@springernature.com.

Apress titles may be purchased in bulk for academic, corporate, or promotional use. eBook versions and licenses are also available for most titles. For more information, reference our Print and eBook Bulk Sales web page at http://www.apress.com/bulk-sales.

Any source code or other supplementary material referenced by the author in this book is available to readers on GitHub.

Paper in this product is recyclable

I would like to dedicate this book to my wife, Cathy, for all her support and patience throughout my career and all my book projects. I would like to thank Apress for this opportunity, specifically Joan Murray, the acquisition editor, who gave me this chance and Laura Berendson and Gryffin Winkler for their valuable help and suggestions. Last but not least, I thank all the technical reviewers for their suggestions, tips, and corrections. What's good about this book is because of them. What's not so good is entirely due to me!

Table of Contents

About the Author

Angelo R. Bobak is a published author with more than three decades of experience and expertise in the areas of business intelligence, data architecture, data warehouse design, data modeling, master data management, and data quality using the Microsoft BI Stack across several industry sectors such as finance, telecommunications, engineering, publishing, and automotive.

About the Technical Reviewer

Alicia Moniz is a leader in Data & AI at Microsoft, an organizer for Global AI Bootcamp – Houston Edition, and a #KafkaOnAzure Evangelista and prior was a three-time Microsoft AI MVP. She is an active supporter of women in technology and volunteers her time at events that help make AI technology accessible to the masses. She is a co-author of the Apress publication *Beginning Azure Cognitive Services: Data-Driven Decision Making Through Artificial Intelligence* along with fellow Microsoft MVPs Matt Gordon, Ida Bergum, Mia Chang, and Ginger Grant. With over 14 years of experience in data warehousing and advanced analytics, Alicia is constantly upskilling and holds more than 12 in-demand IT certifications including AWS, Azure, and Kafka. She is active in the Microsoft User Group community and enjoys speaking on AI, SQL Server, #KafkaOnAzure, and personal branding for women in technology topics. Currently, she authors the blog HybridDataLakes.com, a blog focused on cloud data learning resources, and produces content for the YouTube channel #KafkaOnAzure.

Introduction

Welcome to my book, *SQL Server Analytical Toolkit.*

What's this book about?

This is a book on applying Microsoft SQL Server aggregate, analytical, and ranking functions across various industries for the purpose of statistical, reporting, analytical, and historical performance analysis using a series of built-in SQL Server functions affectionately known as the window functions!

No, not window functions like the ones used in the C# or other Microsoft Windows application programming. They are called window functions because they implement windows into the data set generated by a query. These windows allow you to control where the functions are applied in the data by creating partitions in the query data set.

"What's a partition?" you might ask. This is a key concept you need to understand to get the most out of this book. Suppose you have a data set that has six rows for product category A and six rows for product category B. Each row has a column that stores sales values that you wish to analyze. The data set can be divided into two sections, one for each product category. These are the partitions that the window functions use. You can analyze each partition by applying the window functions (more on this in Chapter 1).

We will see that the window in each partition can be further divided into smaller windows. The mechanism of a window frame allows you to control which rows in the partition are submitted to the window function relative to the current row being processed. For example, apply a function like the SUM() function to the current row being processed and any prior rows in the partition to calculate running totals by month. Move to the next row in the partition and it behaves the same.

The book focuses on applying these functions across four key industries: sales, finance, engineering, and inventory control. I did this so that readers in these industries can find something they are familiar with in their day-to-day job activities. Even if you are not working across these industries, you can still benefit by learning the window functions and seeing how they are applied.

Maybe you want to interview for a developer role in the finance sector? Or maybe you work in engineering or telecommunications or you are a manufacturer of retail products. This book will help you acquire some valuable skills that will help you pass the job interview.

Although you could perform these functions with tools like Power BI, performing these functions at the SQL level precalculates results and improves performance so that reporting tools use precalculated data.

By the way, there are many books out there on SQL Server and window (or windowing) functions. What's so different about this book?

Approach

This book takes a cookbook approach. Not only are you shown how to use the functions, but you are shown how to apply them across sales, finance, inventory control, and engineering scenarios.

These functions are grouped into three categories, so for each industry use case we look at, we will dedicate a chapter to each function category:

- Aggregate functions

- Analytical functions

- Ranking functions

For each function, a query is created and explained. Next, the results are examined and analyzed.

Where applicable the results are used to generate some interesting graphs with Microsoft Excel spreadsheets, like creating normal distribution charts for sales data.

Appendix A contains descriptions and syntax for these functions in case you are not familiar with them, so feel free to examine them before diving into the book.

Key to mastering the concepts in this book is understanding what the OVER() clause does. Chapter 1 starts off by defining what the OVER() clause is and how it is used with the window functions.

Several diagrams clearly explain what data sets, partitions, and window frames are and how they are key to using the window functions.

Each of the industries we identified earlier has three dedicated chapters, one for each of the window function categories. Each chapter provides a specification for the query to be written, the code to satisfy the specification, and then one or more figures to show the results.

The book is unique in that it goes beyond just showing how each function works; it presents use case scenarios related to statistical analysis, data analysis, and BI (BI stands for business intelligence by the way).

The book also makes available all code examples including code to create and load each of the four databases via the publisher's Google website.

Lastly, just enough theory is included to introduce you to statistical analysis in case you are not familiar with terms like standard deviation, mean, normal distribution, and variance. These are important as they will supply you with valuable skills to support your business users and enhance your skills portfolio. Hey, a little business theory can't hurt!

Appendix B has a brief primer on statistics, so make sure to check it out in case these topics are new to you. It discusses standard deviation, variance, normal distribution, other statistical calculations, and bell curves.

Back to the window functions. These functions generate a lot of numerical data. It's great to generate numbers with decimal points but even more interesting to graph them and understand what they mean. A picture is worth a thousand words. Seeing a graph that shows sales decreasing month by month is certainly worth looking at and should raise alarms!

You can also use the Excel spreadsheets to verify your results by using the spreadsheets' built-in functions to make sure they match the results of your queries. Always test your data against a set of results known to be correct (you might just learn a little bit about Microsoft Excel too!). The spreadsheets used in this book will also be available on the publisher's Google website.

The book includes tips and discussions that will take you through the process of learning the SQL Server aggregate, ranking, and analytical functions. These are delivered in a step-by-step approach so you can easily master the concepts. Data results are analyzed so that you can understand what the function does and how the windows are used to analyze the data work.

Expectations

Now that you know what you are in for, what do I expect from you?

Not much really, at a high level.

I expect you to be an intermediate to advanced SQL Server developer or data architect who needs to learn how to use window functions. You can write medium-complexity queries that use joins, understand what a CTE (common table expression) is, and be able to create and load database tables.

You could also be a tech-savvy business analyst who needs to apply sophisticated data analysis for your business users or clients.

Lastly, you could be a technology manager who wants to understand what your development team does in their roles as developers. All will benefit from this book.

In conclusion, you need

- A keen interest in learning data analysis skills

- A basic knowledge of relational database technology and SQL skills

- A basic understanding of mathematical calculations like calculating the average of values and performing basic arithmetic operations like addition, subtraction, division, and multiplication

- Lastly, a working knowledge of SQL Server Management Studio (SSMS), the tool we will use to create and execute the queries

In case you do not know how to use SSMS, there are many excellent YouTube videos and sites that can show you how to use this tool in a short amount of time.

You can also check out my podcast "GRUMPY PODCAST 01 NAVIGATING SSMS" at www.grumpyolditguy.com under the TSQL Podcasts menu selection in the menu bar.

What's in It for You?

You, my dear reader, will learn how to use the Microsoft SQL Server analytical, aggregate, and ranking functions. These are valuable skills that will enhance your capability to develop solutions for your company role and the business users you support or will support.

What Else Do You Get?

There are tons of code examples that you can access and play around with to become familiar with the workings of the window functions. Practice makes perfect. Break the code, modify it, and experiment. This is the best way to learn.

Note You can download the code from the publisher's Google website at https://github.com/Apress/SQL-Server-Analytical-Toolkit.

One Final Note

Besides the queries discussed in the chapters, DDL (data declaration language) code is included to create the four practice databases and load the tables in each. These scripts clearly identify the steps and the sequence you need to take to create and load all the data objects (some will generate millions of rows).

Once you create the databases and load the tables, you can practice by executing the supplied queries as you read along in each of the chapters. You can also practice performance tuning by creating indexes and examining the query plans that can be generated with SSMS.

Lastly, I will also touch upon performance tuning a bit to ensure that the queries you create run fast and efficient. The functions you will learn are resource intensive, so we need to write queries that will not tie down server resources, like CPU, memory, and disk! (Or have your users pull their hair when a query runs for hours or days....)

As added value, you will be shown how to apply good coding standards to improve code legibility. Code that is easy to read is easy to maintain, fix, and modify. We want to be good teammates in case your code is handed to another developer once you are promoted (or get a better job!). For these reasons, we will use the coding standards suggested by Microsoft Corporation for SQL Server.

These are easy to learn and adopt. All TSQL commands and keywords are in uppercase, and data objects referenced are named with one or more words that describe the data object. The first letter of the word that makes up the data object name is in uppercase, and the remaining letters are lowercase, for example, **DeptId** for department identifier.

A sample query that follows this coding standard is

```
SELECT DeptId, MgrId, MgrLastName
FROM Department
GO
```

Very simple and easy to read. It is clear by the names used that the query retrieves departmental information, specifically the department identifier, the manager identifier, and the manager's last name. Notice the name of the table. There is no doubt as to what the table contains.

That's it! Let's begin our journey into window functions.

CHAPTER 1

Partitions, Frames, and the OVER( ) Clause

The goal of this chapter is to describe the OVER() clause, its various configurations in terms of how partitions and window frames are created, and how data is sorted in the partition so that the window function can operate on the values. (The OVER() clause together with partitions and window frames is what gives power to the window functions.)

We will cover window frames and how they are defined to present a set of partition rows that are processed by the window functions.

Several diagrams illustrate the concepts, and then several examples are shown to illustrate the various ways partitions and window frames work on real data sets using window functions. We will discuss ROWS and RANGE clauses and how they operate on data in the partition.

We will also look at a case where we use subqueries in the ORDER BY clause instead of columns so as to eliminate expensive sort steps in a query plan.

A final short example is presented to show the new named window feature in SQL Server 2022.

What Are Partitions and Window Frames?

A TSQL query will generate an initial data result set. The result set can be divided into sections called partitions, like a partition for each set of rows by specific years. The window frame is like a smaller window into the partition defined by some row boundaries or ranges, for example, rows prior to and including the current row being processed or all rows after and including the current row within the partition. The entire

© Angelo Bobak 2023
A. Bobak, *SQL Server Analytical Toolkit*, https://doi.org/10.1007/978-1-4842-8667-8_1

data set can also be a partition by itself, and a window can be defined that uses all rows in the single partition. It all depends on how you include and define the PARTITION BY, ORDER BY, and ROWS/RANGE window frame clauses.

These conditions can be specified with the OVER() clause. Let's see how this works.

What Is an OVER() Clause?

As stated earlier, the OVER() clause allows you to create partitions and windows into a data set generated by a query. It allows you to divide the data set into sections called partitions. The windows allow you to control how the window functions are applied in terms of the data they will operate on.

Smaller windows called window frames can be created to further carve up the larger window defined by the PARTITION BY clause using ROWS and RANGE frame clauses. These windows grow or move (as rows are processed) based on the boundaries defined by ROWS and RANGE frame specifications included in the OVER() clause. You can also specify how the rows are sorted via an ORDER BY clause so that the window function can process them.

This capability allows you to create queries such as three-month rolling averages and year-to-date, quarter-to-date, and month-to-date totals. Each of the functions we discuss in this book can utilize the OVER() clause in this manner.

This is why they are called window functions!

History of the OVER() Clause and Window Functions

The following is a brief history of the window functions:

- Aggregate/ranking window functions without the ORDER BY clause support were introduced in 2005.

- Seven years later, aggregate functions with support for the ORDER BY clause were introduced in 2012.

- Support for window frames (which we will discuss shortly) was also introduced in 2012.

- Ranking functions and some window capability were introduced in 2015.

- The batch-mode window aggregate operator was introduced in 2016.

- The STRING_AGG function was introduced in 2017.

- The named WINDOW capability was introduced in SQL Server 2022.

The capability of the window functions has grown over the years and delivers a rich and powerful set of tools to analyze and solve complex data analysis problems.

The Window Functions

The window functions (which are the focus of this book) that can be used with the OVER() clause are assigned to three categories and are listed in Table 1-1.

Table 1-1. *Aggregate, Analytical, and Ranking Functions*

Aggregate Functions	Analytical Functions	Ranking Functions
COUNT()	CUME_DIST()	RANK()
COUNT_BIG()	FIRST_VALUE()	DENSE_RANK()
SUM()	LAST_VALUE()	NTILE()
MAX()	LAG()	ROW_NUMBER()
MIN()	LEAD()	
AVG()	PERCENT_RANK()	
GROUPING()	PERCENTILE_CONT()	
STRING_AGG()	PERCENTILE_DISC()	
STDEV()		
STDEVP()		
VAR()		
VARP()		

Each of the subsequent chapters will create and discuss queries for these categories for four industry-specific databases that are in scope for this book. Please refer to Appendix A for syntax and descriptions of what each of the preceding functions does if you are unfamiliar with them or need a refresher on how to use them in a query.

The OVER( ) Clause

The OVER() clause appears right after the window function in the SELECT clause. Our first example in Listing 1-1 uses the SUM( ) function to aggregate sales amounts by year and month.

Listing 1-1. SUM() Function with the OVER() Clause

```
SELECT OrderYear,OrderMonth,SalesAmount,
      SUM(SalesAmount) OVER(
            PARTITION BY OrderYear
            ORDER BY OrderMonth ASC
            ROWS BETWEEN UNBOUNDED PRECEDING AND CURRENT ROW
            ) AS AmountTotal
FROM OverExample
ORDER BY OrderYear,OrderMonth
GO
```

Between a set of parentheses after the OVER keyword, three other clauses can be included such as PARTITION BY, ORDER BY, and ROWS or RANGE clauses (to define the window frame that presents the rows to the function for processing).

Even if you have an ORDER BY clause in the OVER( ) clause, you can also include the usual ORDER BY clause at the end of the query to sort the final processed result set in any order you feel is appropriate to the business requirements the query solves.

Syntax

The following are three basic syntax templates that can be used with the window functions. Reading these syntax templates is easy. Just keep in mind keywords between square brackets mean they are optional. The following is the first syntax template available for the OVER() clause:

Syntax 1

```
<Window Function> OVER (
        [ <PARTITION BY expression> ]
        [ <ORDER BY expression ASC | DESC > ]
        [ <ROWS or RANGE expression > ]
        )
```

Most of the window functions use this first syntax, and it is composed of three main clauses, the PARTITION BY clause, the ORDER BY clause, and a ROWS or RANGE specification. You can include one or more of these clauses or none. These combinations will affect how the partition is defined. For example, if you do not include a PARTITION BY clause, the entire data set is considered a one large partition. The expression is usually one or more columns, but in the case of the PARTITION BY and ORDER BY clauses, it could also be a subquery (refer to Appendix A).

The **Window Function** is one of the functions identified in Table 1-1.

This first syntax is pretty much the same for all functions except for the PERCENTILE_DISC() and PERCENTILE_CONT() functions that use a slight variation:

Syntax 2

```
PERCENTILE_DISC (numeric literal ) WITHIN GROUP (
            ORDER BY expression [ ASC | DESC ]
    ) OVER ( [ <PARTITION BY expression> ] )
```

These functions are used to calculate the percentile discrete and percentile continuous values in a data set column. The numeric literal can be a value like .25, .50, or .75 that is used to specify the percentile you wish to calculate. Notice that the ORDER BY clause is inserted between the parentheses of the WITHIN GROUP command and the OVER() clause just includes the PARTITION BY clause.

Don't worry about what this does for now. Examples will be given that make the behavior of this code clear. For now, just understand that there are three basic syntax templates to be aware of.

In our chapter examples, the expression will usually be a column or columns separated by commas although you can use other data objects like queries. Please refer to the Microsoft SQL Server documentation to check out the detailed syntax specification or Appendix A.

Lastly, our third syntax template applies to SQL Server 2022 (release 16.x). The window capability has been enhanced that allows you to specify window options in a named window that appears at the end of the query:

Syntax 3

```
WINDOW <window name> AS (
      [ <PARTITION BY clause> ]
      [ <ORDER BY clause> ]
      [ <ROWS or RANGE clause> ]
      )
```

As of this writing, SQL Server 2022 is available for evaluation only. In Listing 1-2 is an example TSQL query that uses this new feature.

Listing 1-2. SQL Server 2022 Named Window Feature

```
SELECT OrderYear,OrderMonth,SalesAmount,
      SUM(SalesAmount) OVER SalesWindow AS SQPRangeUPCR
FROM OverExample
WINDOW SalesWindow AS (
            PARTITION BY OrderYear
            ORDER BY OrderMonth
            RANGE BETWEEN UNBOUNDED PRECEDING AND CURRENT ROW
            );
GO
```

The name of the window is **SalesWindow**, and it is used right after the OVER operator instead of the PARTITION BY, ORDER BY, and RANGE clauses as used in the first syntax template we discussed.

Probably a good feature in case you have multiple window functions in your SELECT clause that need to use this partition and window frame configuration. This would avoid repeating the partition code in each column of the SELECT clause.

The PARTITION BY, ORDER BY, and RANGE clauses are declared at the end of the query between parentheses after the WINDOW keyword instead of right after the OVER keyword.

If you want to play around with this, download and install the 2022 evaluation license and try it out on the example code available with the book or on your own queries. The setup and download are fast and simple. Make sure you get the latest version of SSMS. These are available on Microsoft's download website.

Partitions and Frames

Finally, we get to discuss what a partition looks like and what frames or window frames are.

Basically, a query will generate a data result set, which you can carve up into sections called partitions by including the PARTITION BY clause inside the OVER() clause. For each partition you can further define smaller windows to fine-tune how the window functions will be applied.

A picture is worth a thousand words, so let's look at one now. Please refer to Figure 1-1.

PARTITIONS & FRAMES

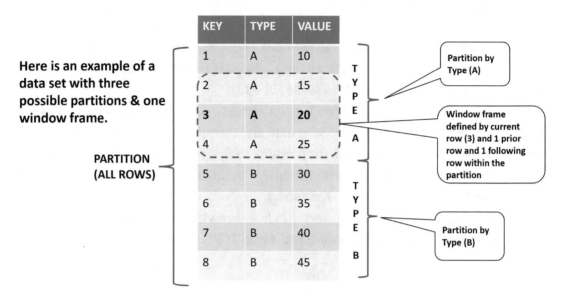

Figure 1-1. *A simple data set with three partitions and an example frame*

Here we have a simple data set composed of eight rows. There are three example partitions in this data set. One can include all eight rows of the data set; the other two include rows identified by the TYPE column. There are only two type values, type A and type B, so each of these partitions will have four rows. By the way, you cannot include multiple PARTITION BY clauses in an OVER() clause.

You can define only one partition per OVER() clause although you can have more than one column in the SELECT clause of the query that uses a partition. You can specify different column combinations to define the partitions.

Where the power of this architecture comes in is that we can create smaller window frames against the partition by using the ROWS or RANGE operator. These will allow you to specify how many rows before and/or after the current row being processed will be used by the window function.

In our preceding example snapshot, the current row is row 3, and the window frame is defined so it includes only the prior row, the current row, and the next row relative to the current row. If we apply the SUM() function to this window frame and add all the values, we get the result 60 (15 + 20 + 25). (Remember this is within the first partition, which contains only four rows.)

If processing continues on the next row, row 4, only rows 3 and 4 are available to the SUM() function, and the result is 45 (20 + 25). I neglected to mention that if we start at row 1, then only rows 1 and 2 are available to the SUM() function because there is no prior row. The function returns the value 25 (10 + 15).

How do we control this type of processing? All we need to do is add a ROWS or RANGE specification to the query if required. We could also include an ORDER BY clause to specify how to order the rows within the partition so that the window function is applied as needed. For example, generate rolling totals by month, starting of course at month 1 (January) and ending at month 12 (December).

Sounds easy, but we need to be aware of a few scenarios around default processing when we leave the ORDER BY clause and/or the PARTITION clause out. We will discuss these shortly.

ROWS Frame Definition

The ROWS clause operates on the physical set of rows belonging to the partition. Because the ROWS clause operates on the rows in the partition, it is considered a physical operation.

The ORDER BY clause allows you to specify the logical order of the rows in the partition so the window function can evaluate them. For example, if you have sales data by year and month, you would set the logical order of the rows in the partition by month so that the SUM() function can generate rolling year-to-date totals for each month. You can specify optional ascending and descending sort orders by using the keywords ASC and DESC.

The ROWS clause allows you to define the window frame into the partition. There are several variations of the ROWS clause that we need to be aware of. The following are two variations that generate the same window frame:

```
ROWS BETWEEN UNBOUNDED PRECEDING AND CURRENT ROW
ROWS UNBOUNDED PRECEDING
```

This clause tells the function to operate on the current row and all rows preceding the current row if there are any in the partition. A simple diagram in Figure 1-2 makes it all clear.

ROWS BETWEEN UNBOUNDED PRECEDING AND CURRENT ROW

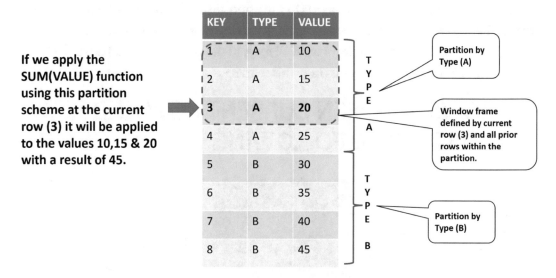

If we apply the SUM(VALUE) function using this partition scheme at the current row (3) it will be applied to the values 10,15 & 20 with a result of 45.

Figure 1-2. *Include the current row and all preceding rows*

If we start at row 1, since there are no prior rows in the partition before this row, the SUM() function returns the value 10.

Moving on to row 2, the SUM() function will include the only available prior row (row 1), so the result is 25 (10 + 15).

Next (shown in the preceding figure), the current row to be processed is row 3. The SUM() function will evaluate row 3 plus rows 1 and 2 in order to generate the total. The result is 45 (10 + 15 + 20).

Lastly, moving to row 4, the function will include the current row and all prior rows in its calculation and return 70 (10 + 15 + 20 + 25). Processing now concludes for this partition.

Moving to partition B, the processing repeats itself, and only rows from partition B are included. All rows in partition A are ignored.

Our next ROWS clause takes us in the opposite direction:

```
ROWS BETWEEN CURRENT ROW AND UNBOUNDED FOLLOWING
ROWS UNBOUNDED FOLLOWING (will not work, not supported)
```

The first clause is legal but the second is not. It is not supported. One would think that if there is a ROWS UNBOUNDED PRECEDING, there should be a ROWS UNBOUNDED FOLLOWING, but it is not supported at this time. Go figure!

As stated earlier, this clause takes us in the opposite direction than the prior scenario we just discussed. It will allow the aggregate or other window functions to include the current row and all succeeding rows until all rows in the partition are exhausted. Our next diagram in Figure 1-3 shows us how this works.

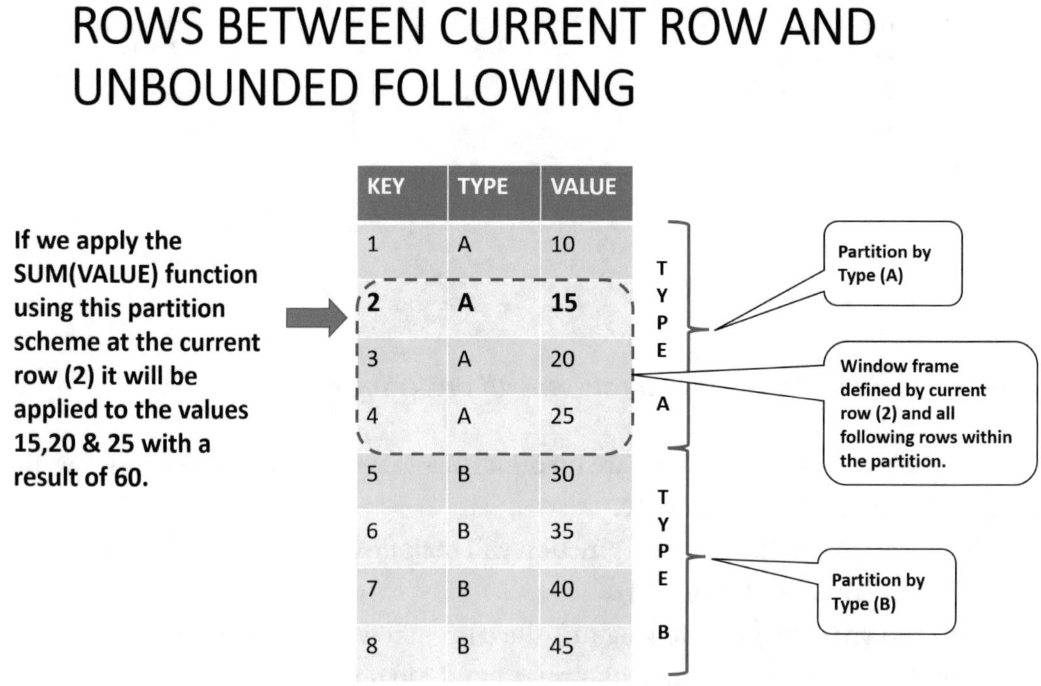

ROWS BETWEEN CURRENT ROW AND UNBOUNDED FOLLOWING

If we apply the SUM(VALUE) function using this partition scheme at the current row (2) it will be applied to the values 15,20 & 25 with a result of 60.

KEY	TYPE	VALUE
1	A	10
2	A	15
3	A	20
4	A	25
5	B	30
6	B	35
7	B	40
8	B	45

TYPE A

TYPE B

Partition by Type (A)

Window frame defined by current row (2) and all following rows within the partition.

Partition by Type (B)

Figure 1-3. *Process the current row and all succeeding rows*

Remember rows are processed one by one in the partition.

If processing starts at row 1, then all four values are included to calculate a sum of 70 (10 + 15 + 20 +25).

Next, if the current row being processed is row 2 as in the preceding example, then the SUM() function will include rows 2–4 to generate the total value. It will generate a result of 60 (15 + 20 + 25).

Moving on to row 3, it will only include rows 3 and 4 and generate a total value of 45 (20 + 25).

Once processing gets to row 4, only the current row is used as there are no more rows available in the partition. The SUM() function calculates a total of 25.

When processing resumes at the next partition, the entire scenario is repeated.

What if we do not want to include all prior or following rows but only a few before or after? The next window frame clause will accomplish the trick for limiting the number of following rows:

```
ROWS BETWEEN CURRENT ROW AND n FOLLOWING
```

Including this clause in the OVER() clause will allow us to control the number of rows to include relative to the current row, that is, how many rows after the current row are going to be used from the available partition rows. Let's examine another simple example where we want to include only one row following the current row in the calculation.

Please refer to Figure 1-4.

ROWS BETWEEN CURRENT ROW AND N = 1 FOLLOWING

If we apply the SUM(VALUE) function at the current row (2) using this partition scheme it will be applied to the values 15 & 20 with a result of 35.

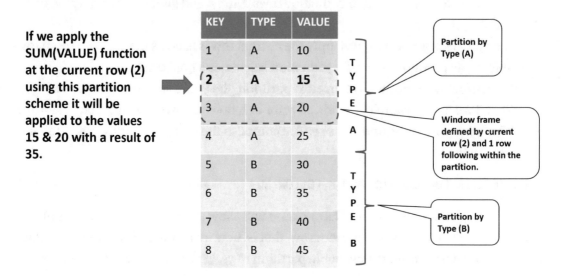

Figure 1-4. *Current row and one row following*

Here is the window frame clause to include in the OVER() clause:

ROWS BETWEEN CURRENT AND 1 FOLLOWING

Processing starts at row 1. I think by now you understand that only rows 1 and 2 are used and the result is 25 (10 + 15).

Next, in the preceding example, row 2 is the current row. If only the next row is included, then the SUM() function will return a total of 35 (15 + 20).

Moving to row 3 (the next current row), the SUM() function will return 45 as the sum (20 + 25).

Finally, when processing gets to the last row in the partition, then only row 4 is used and the SUM() function returns 25.

When processing continues to the next partition, the sequence repeats itself ignoring the values from the prior partition.

We can now see how this process creates windows into the partition that change as the rows are processed one by one by the window function. Remember, the order in which the rows are processed is controlled by the ORDER BY clause used in the OVER() clause.

The next example takes us in the opposite direction. We want to include the current row and the prior two rows (if there are any) in the calculation.

The window frame clause is

```
ROWS BETWEEN n PRECEDING AND CURRENT ROW
```

The letter n represents an unsigned integer value specifying the number of rows. The following example uses 2 as the number of rows.

Please refer to Figure 1-5.

ROWS BETWEEN N = 2 PRECEDING AND CURRENT ROW

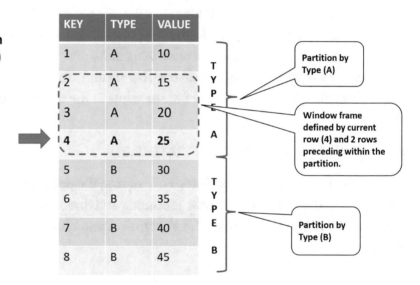

Figure 1-5. *Include two prior rows and the current row*

The window frame for this scenario is

```
ROWS BETWEEN 2 PRECEDING AND CURRENT ROW
```

Starting at row 1 (the current row), the SUM() function will only use row 1 as there are no prior rows. The value returned by the window function is 10.

Moving to row 2, there is only one prior row so the SUM() function returns 25 (10 + 15).

Next, the current row is row 3, so the SUM( ) function uses the two prior rows and returns a value of 45 (10 + 15 +20).

In the preceding figure, the current row being processed is row 4 and the preceding two rows are included; the SUM( ) function when applied will return a total value of 60 (15 + 20 + 25).

Since all rows of the first partition have been processed, we move to partition B, and this time only row 5 is used as there are no prior rows. We are at the beginning of the partition. The SUM( ) function returns a total value of 30.

Processing continues at row 6, so the SUM function processes rows 5 and 6 (the current and prior rows). The total value calculated is 65 (30 + 35).

Next, the current row is 7 so the window function will include rows 5, 6, and 7. The SUM( ) function returns 105 (30 + 35 + 40).

Finally, processing gets and ends at row 8 of the partition. The rows used by the SUM( ) function are rows 6, 7, and 8. The SUM( ) function() returns 120 (35 + 40 + 45). There are no more partitions so the processing ends.

What if we want to specify several rows before and after the current row? The next clause does the trick. This was one of the example in the chapter, but let's review it again and make one change:

ROWS BETWEEN n PRECEDING AND n FOLLOWING

In this case we want to include the current row, the prior row, and two following rows. We could well specify two rows preceding and three rows following or any combination within the number of rows in the partition if we had more rows in the partition.

Please refer to Figure 1-6.

ROWS BETWEEN N = 1 PRECEDING AND N = 2 FOLLOWING

If we apply the SUM(VALUE) function at the current row (2) using this partition scheme it will be applied to the values 10,15,20 & 25 with a result of 70.

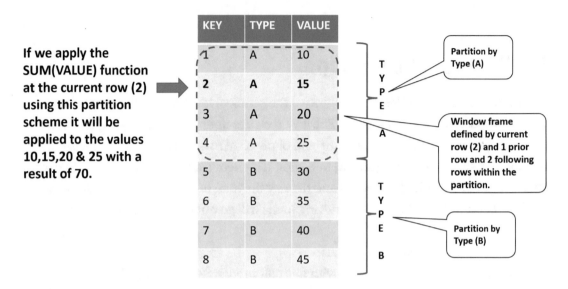

Figure 1-6. Rows between n = 1 preceding and n = 2 following

Here is the ROWS clause for this scenario:

```
ROWS BETWEEN 1 PRECEDING AND 2 FOLLOWING
```

Starting at row 1 of the partition, the SUM() function can only use rows 1, 2, and 3 as there are no rows in the partition prior to row 1. The result calculated by the SUM() function is 45 (10 + 15 + 20).

Moving to the next row, row 2 of the partition, the SUM() function can use rows 1, 2 (the current row), 3, and 4. The result calculated by the SUM() function is 70 (10 + 15 + 20 + 25).

Next, the current row is row 3. This window frame uses row 2, 3, and only 4 so the SUM() function will calculate a total sum of 60 (15 + 20 + 25).

Finally, moving to row 4 (last row of the partition), the SUM() function can only use rows 3 and 4 as there are no more available rows following in the partition. The result is 45 (20 + 25).

Processing at the next partition continues in the same fashion, ignoring all rows from the prior partition.

We have pretty much examined most of the configurations for the ROWS clause. Next, we look at window frames defined by the RANGE clause, which works at a logical level. It considers values instead of physical row position. Duplicate ORDER BY values will yield a strange behavior.

RANGE Frame Definition

We discussed how the ROWS clause works at the physical level by using values relative to the row positions they belong to. The RANGE clause operates at the logical level. It considers the values of the column instead of physical row position within the partition. It displays a strange (in my opinion) behavior. If it encounters duplicate values in the row sort column, it will add them all up and display the total in all rows with the duplicate ORDER BY column. Any values like moving totals will go out the window! (Pardon the pun.)

An example illustrates this scenario.

Suppose you have a small table with five columns: Row, Year, Month, Amount, and Running Total. The table is loaded with 12 rows for each year, let's say 2010 and 2011, plus a duplicate month row giving us a total of 13 rows. Each row represents a month in the year, but for the month of March, two rows are inserted for the first year (that's why we have 13 rows and not 12 rows). In other words, an extra amount value exists for the same month, resulting in duplicate month numbers.

The RANGE clause will use all the rows with duplicate current month values to apply the window function to (in addition to any other rows specified in the RANGE clause).

Check out the partial table, Table 1-2.

Table 1-2. *Partial RANGE Running Totals*

Row	Year	Month	Amount	Running Total
1	2010	1	100.00	100.00
2	2010	2	100.00	200.00
3	**2010**	**3**	**200.00**	**600.00**
4	**2010**	**3**	**200.00**	**600.00**

When we get to row 3, the first of two duplicate months, the window function (SUM( ) in this case) will include all prior rows and add the values in row 3 and the next row 4 to generate a total value of 600.00. This is displayed for both rows 3 and 4. Weird! One would expect a running total value of 400.00 for row 3.

Let's try another one. If we apply the following RANGE clause to calculate running totals

RANGE BETWEEN UNBOUNDED PRECEDING AND CURRENT ROW

or

RANGE UNBOUNDED PRECEDING

we get interesting results. Check out the partial results in Figure 1-7.

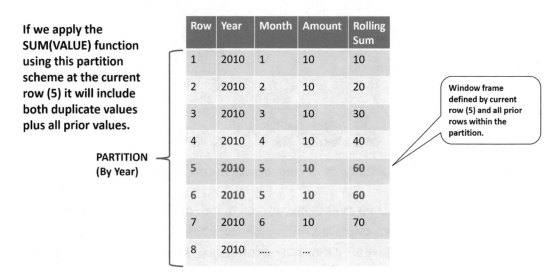

RANGE BETWEEN UNBOUNDED PRECEDING AND CURRENT ROW

If we apply the SUM(VALUE) function using this partition scheme at the current row (5) it will include both duplicate values plus all prior values.

PARTITION (By Year)

Row	Year	Month	Amount	Rolling Sum
1	2010	1	10	10
2	2010	2	10	20
3	2010	3	10	30
4	2010	4	10	40
5	2010	5	10	60
6	2010	5	10	60
7	2010	6	10	70
8	2010	....	...	

Window frame defined by current row (5) and all prior rows within the partition.

Figure 1-7. *Sales by year and month*

Everything works as expected until we get to the duplicate rows with a value of 5 (May) for the month. The window frame for this partition includes rows 1–4 and the current row 5, but we also have a duplicate month value in row 6, which calculates the rolling total value 60. This value is displayed in both rows 5 and 6. Aggregations continue for the rest of the months (not shown).

Let's process rows in the opposite direction. Here is our next RANGE clause:

```
RANGE BETWEEN CURRENT ROW AND UNBOUNDED FOLLOWING
RANGE UNBOUNDED FOLLOWING (will not work)
```

By the way, the second RANGE clause is not supported, but I included it so you are aware that it will not work and generate an error. Please refer to Figure 1-8.

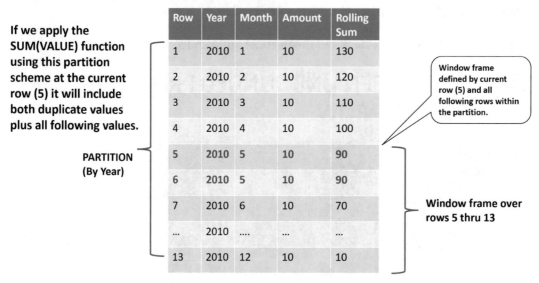

Figure 1-8. *RANGE between current row and unbounded following*

Rows 5 and 6 represent two sales figures for the month of May. Since these are considered duplicates, row 5 displays a rolling value of 90 and row 6 also shows 90 instead of 80 as one would expect if they did not understand the behavior of this window frame declaration.

Notice, by the way, how the rolling sum values decrease as the row processing advances from row 1 onward. We started with 130 and ended with 10 – moving totals in reverse.

Putting it all together, here is a conceptual view that illustrates the generation of window frames as rows are processed when the following ROWS clause is used:

```
ROWS BETWEEN 1 PRECEDING AND 1 FOLLOWING
```

This was one of the example in the chapter. Only the partition for type A rows is considered. Please refer to Figure 1-9.

WINDOW FRAMES - ROWS BETWEEN N = 1 PRECEDING AND N = 1 FOLLOWING

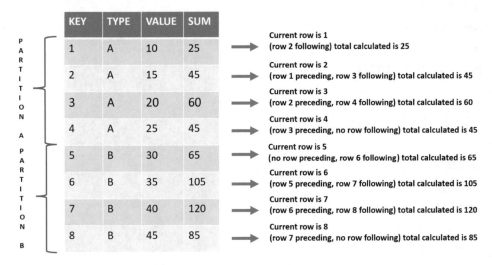

Figure 1-9. *Window frame and processed rows*

Both partitions are shown so you can see how the frame resets when it gets to the next partition.

The generated window frames, as processing goes row by row, show which values are included for the window function to process within the partition. The first window and last window only have two rows to process, but the second and third both have three rows to process, the prior, current, and next rows. The same pattern repeats itself for the second partition.

Let's put our knowledge to the test by looking at a simple query based on the figures we just examined.

Example 1

We will start by creating a small test temporary table for our analysis.

The following listing shows the test table CREATE TABLE DDL (data declaration language) statement created as a temporary table and one INSERT statement to load all the rows we wish to process.

Please refer to the partial listing in Listing 1-3.

Listing 1-3. Creating the Test Table

```
CREATE TABLE #TestTable (
     Row SMALLINT,
     [Year]SMALLINT,
     [Month]SMALLINT,
     Amount DECIMAL(10,2)
     );

     INSERT INTO #TestTable VALUES
     -- 2010
     (1,2010,1,10),
     (2,2010,2,10),
     (3,2010,3,10),
     (4,2010,4,10),
     (5,2010,5,10),
     (6,2010,5,10),
     (7,2010,6,10),
     (8,2010,7,10),
     (9,2010,8,10),
     (10,2010,9,10),
     (11,2010,10,10),
     (12,2010,11,10),
     (13,2010,12,10),
     -- 2011
     (14,2011,1,10),
     (15,2011,2,10),
     (16,2011,3,10),
     (17,2011,4,10),
     (18,2011,5,10),
     (19,2011,5,10),
     (20,2011,6,10),
     (21,2011,7,10),
     (22,2011,7,10),
     (23,2011,7,10),
```

```
(24,2011,8,10),
(25,2011,9,10),
(26,2011,10,10),
(27,2011,11,10),
(28,2011,12,10);
```

The CREATE statement is used to create a simple table with four columns: a Row column to identify the row numbers, a Year column and a Month column for the calendar information, and an Amount column to store a numerical value.

The column names Year and Month are not good names as they are reserved words in SQL Server, but since this is a simple test table, I did not get too rigorous with naming standards. Also, I used them in the figures for the RANGE examples, so we are consistent (at least for our simple examples).

We wish to load two years' worth of data with some duplicate months so we can see the difference between how the ROWS clause and the RANGE clause treat these situations.

Here is our first query. Let's calculate some amount totals by year and month. We go forward with a RANGE clause that considers the current row and all following rows.

Please refer to the partial listing in Listing 1-4.

Listing 1-4. RANGE, Current Row, and Unbounded Following

```
SELECT Row,[Year],[Month],Amount,
     SUM(Amount) OVER (
            PARTITION BY [Year]
            ORDER BY [Month]
            RANGE BETWEEN CURRENT ROW AND UNBOUNDED FOLLOWING
            ) AS RollingSalesTotal
     FROM #TestTable
ORDER BY [Year],[Month] ASC
GO
```

Notice the square brackets around the Year and Month column names. Using these will eliminate the colors SSMS uses to display reserved keywords. Leaving them out won't hurt, but the colors will be used in the query pane, which makes you aware you are using reserved keywords. Don't forget to drop the temporary table.

In Figure 1-10 are the partial results so you can see the entire range of values for one year.

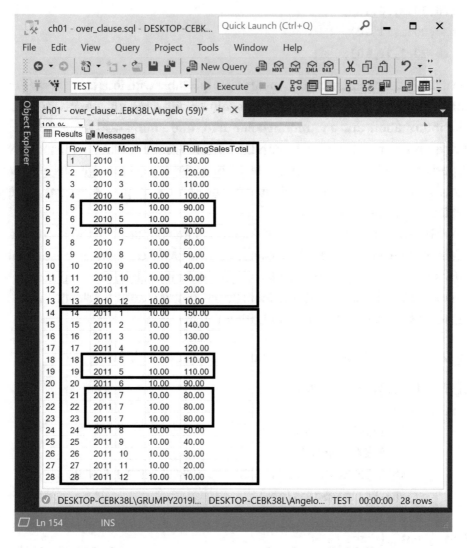

Figure 1-10. *RANGE between current row and unbounded following*

Looks like the totals are decreasing, which makes sense as we are going forward from the current row being processed by the window function. Look at the familiar duplicate behavior in rows 5 and 6 when both months are 5 (May). Both rows have the same value of 90.00.

When we get to December, in row 13, we have no more rows to process in the partition, so we only return the value 10.00 for the sum. This is the value for the current and last row in the partition.

In the second year partition, we have two clusters of duplicate rows, one for the month of May and one for the month of July. Again, we see the behavior that is characteristic of the RANGE clause.

By the way, the following window frame clauses are not supported:

```
RANGE BETWEEN CURRENT ROW AND n FOLLOWING (will not work)
RANGE BETWEEN n PRECEDING AND CURRENT ROW (will not work)
RANGE BETWEEN n PRECEDING AND n FOLLOWING (will not work)
```

These do not work, as the RANGE clause is a logical operation, so specifying a number for the rows to include in the window frame is not allowed. If you try to use these, you will get this wonderful error message:

RANGE is only supported with UNBOUNDED and CURRENT ROW window frame delimiters

Pretty clear, don't you think?

The last item we need to discuss in this chapter is the default behavior if we do not include window frame clauses in our OVER() clause. The behavior changes depending on how you include or not include the ORDER BY and PARTITION BY clauses.

ROWS and RANGE Default Behavior

By now we understand the syntax of the OVER() clause, and we know that it could include the following clauses:

- PARTITION BY clause

- ORDER BY clause

- ROWS or RANGE clause

- None of these (empty)

The first three clauses are optional. You can leave them out or include one or more as required. There are two scenarios to consider when applying (or not) these clauses.

Scenario 1

The default behavior of the window frames is dependent on whether the `ORDER BY` clause is included or not. There are two configurations to consider:

If the `ORDER BY` clause and the `PARTITION BY` clause are omitted and we do not include the window frame clause

If the `ORDER BY` clause is omitted but the `PARTITION BY` clause is included and we do not include the window frame clause

The default window frame behavior for both these cases is

```
ROWS BETWEEN UNBOUNDED PRECEDING AND UNBOUNDED FOLLOWING
```

Scenario 2

On the other hand, if we include an `ORDER BY` clause, the following two conditions also have a default window frame behavior:

If the `ORDER BY` clause is included but the `PARTITION BY` clause is omitted and we do not include the window frame clause

If the `ORDER BY` clause is included and the `PARTITION BY` clause is included and we do not include the window frame clause

The default window frame behavior for both these cases is

```
RANGE BETWEEN UNBOUNDED PRECEDING AND CURRENT ROW
```

Make sure you keep these default behaviors in mind as you start to develop queries that use the window functions. This is very important; otherwise, you will get some unexpected and possibly erroneous results. Your users will not be happy (is your resume up to date?).

Table 1-3 might help you remember these default behaviors.

Table 1-3. *Window Frame Default Behaviors*

ORDER BY	PARTITION BY	DEFAULT FRAME
No	No	ROWS BETWEEN UNBOUNDED PRECEDING AND UNBOUNDED FOLLOWING
No	Yes	ROWS BETWEEN UNBOUNDED PRECEDING AND UNBOUNDED FOLLOWING
Yes	No	RANGE BETWEEN UNBOUNDED PRECEDING AND CURRENT ROW
Yes	Yes	RANGE BETWEEN UNBOUNDED PRECEDING AND CURRENT ROW

With this new knowledge under our belts, let's check out the ROWS and RANGE clauses with some more simple examples. Remember you can override default behavior by including the ROWS or RANGE clause you need.

ROWS and RANGE Window Frame Examples

Let's start off with the ROWS clause. This clause you will use most of the time. The RANGE clause is a bit dodgy in my opinion and should be avoided.

Data Set

Prior to creating some queries, we need to create a small table and load it so we can practice. The table we create is called OverExample (clever name, I might add). It is made up of three columns: OrderYear, OrderMonth, and SalesAmount. Now we created this table in the beginning of the chapter, but I include it here for easy reference.

Please refer to the partial listing in Listing 1-5.

Listing 1-5. Practice Table OverExample

```
CREATE TABLE OverExample(
    OrderYear    SMALLINT,
    OrderMonth   SMALLINT,
    SalesAmount DECIMAL(10,2)
    );
```

```
INSERT INTO OverExample VALUES
(2010,1,10000.00),
(2010,2,10000.00),
(2010,2,10000.00),
--missing rows
(2010,8,10000.00),
(2010,8,10000.00),
(2010,9,10000.00),
(2010,10,10000.00),
(2010,11,10000.00),
(2010,12,10000.00),
-- 2011
(2011,1,10000.00),
(2011,2,10000.00),
(2011,2,10000.00),
 --missing rows
(2011,10,10000.00),
(2011,11,10000.00),
(2011,12,10000.00);
GO
```

The INSERT statements load two years' worth of data so we have enough rows to test out queries with the window functions and the OVER() clause. The SUM() aggregate function will be used to generate some rolling total calculations.

Please refer to Listing 1-6.

Listing 1-6. Scenario 1

```
SELECT OrderYear,OrderMonth,SalesAmount,
    SUM(SalesAmount) OVER (
            ) AS NPBNOB,
    SUM(SalesAmount) OVER (
            PARTITION BY OrderYear
            ) AS PBNOB,
    SUM(SalesAmount) OVER (
            PARTITION BY OrderYear
            ORDER BY OrderMonth
```

```
        ROWS BETWEEN UNBOUNDED PRECEDING AND UNBOUNDED FOLLOWING
        ) AS PBOBUPUF
FROM OverExample;
GO
```

The SUM() function is used three times with different ORDER BY/PARTITION BY configurations. I also used column names that are meant to indicate the structure of the OVER() clause in terms of whether the ORDER BY, PARTITION BY, and ROWS/RANGE clauses are included. I did this so when you examine the results, you can quickly remember what clauses were used. I know, weird names, but you can easily remember the frame clauses we are using.

The first SUM() function has an empty OVER() clause (the name NPBNOB = No PARTITION BY, No ORDER BY). Recall that the default window frame behavior for this combination is

```
ROWS BETWEEN UNBOUNDED PRECEDING AND UNBOUNDED FOLLOWING
```

The second SUM() function has an OVER() clause that includes a PARTITION BY but no ORDER BY, so the default window frame behavior is also

```
ROWS BETWEEN UNBOUNDED PRECEDING AND UNBOUNDED FOLLOWING
```

The column name is PBNOB (PARTITION BY, No ORDER BY).

This time two partitions are created, one for each year in the data set.

Last but not least, the third SUM() function uses an OVER() clause that has both a PARTITION BY and an ORDER BY clause. It also uses a window frame ROWS clause. The column name is PBOBUPUF (PARTITION BY, ORDER BY, UNBOUNDED PRECEDING UNBOUNDED FOLLOWING).

The default frame behavior for this last column configuration is

```
RANGE BETWEEN UNBOUNDED PRECEDING AND CURRENT ROW
```

Looks like we can override this behavior by replacing the RANGE clause and including a ROWS clause. It can be done!

Let's see what the results look like when we execute this query. Please refer to Figure 1-11.

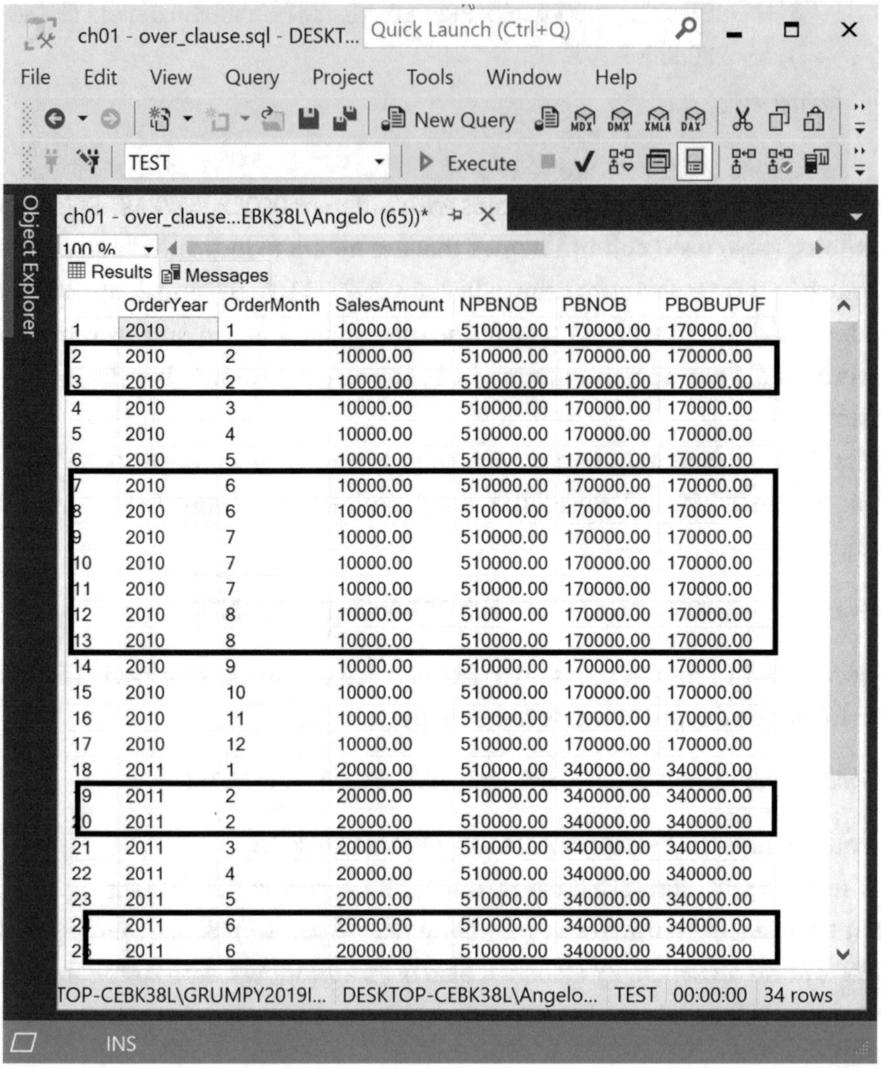

Figure 1-11. *Scenario 1 results*

Interesting results. The first SUM()/OVER() clause combination simply displays the grand total for all the rows in the table, just as the default window frame behavior enforces.

The second and third SUM()/OVER() clause combinations partition the data set by year, so the grand totals by year are displayed in each row. They appear to be identical. The window frame behavior is the same for both combinations:

```
ROWS BETWEEN UNBOUNDED PRECEDING AND UNBOUNDED FOLLOWING
```

That's because for the second combination, the default behavior kicked in and for the third combination, we overrode the RANGE behavior and included the ROWS clause. This behavior is enforced within each year partition. Let's try another example.

Example 2

This time we check out some combinations of the SUM()/OVER() clause that include default window frame behaviors vs. declared RANGE behaviors. (Keep in mind the default window frame behaviors we discussed earlier.)

Let's see the results these combinations deliver. Please refer to Listing 1-7.

Listing 1-7. Scenario 2 – Various Default Window Frame vs. RANGE Behaviors

```
SELECT OrderYear,OrderMonth,SalesAmount,
    SUM(SalesAmount) OVER (
            ORDER BY OrderYear,OrderMonth
            ) AS NPBOB,

    -- same as PBOBRangeUPCR
    SUM(SalesAmount) OVER (
        PARTITION BY OrderYear
            ORDER BY OrderMonth
            ) AS PBOB,

    SUM(SalesAmount) OVER (
            ORDER BY OrderYear,OrderMonth
            RANGE BETWEEN UNBOUNDED PRECEDING AND CURRENT ROW
            ) AS NPBOBRangeUPCR,

    -- same as PBOB
    SUM(SalesAmount) OVER (
            PARTITION BY OrderYear
            ORDER BY OrderMonth
            RANGE BETWEEN UNBOUNDED PRECEDING AND CURRENT ROW
            ) AS PBOBRangeUPCR
FROM OverExample;
GO
```

This time we added an ORDER BY clause to the first SUM()/OVER() combination and a PARTITION BY and an ORDER BY clause in the second SUM()/OVER() combination. The third SUM()/OVER() combination included an ORDER BY clause and no PARTITION BY clause, but we included a RANGE window frame clause. The fourth SUM()/OVER() combination included both a PARTITION BY and an ORDER BY clause plus a RANGE window frame clause. Results should be interesting.

Please refer to Figure 1-12.

	OrderYear	OrderMonth	SalesAmount	NPBOB	PBOB	NPBOBRangeUPCR	PBOBRangeUPCR
1	2010	1	10000.00	10000.00	10000.00	10000.00	10000.00
2	2010	2	10000.00	30000.00	30000.00	30000.00	30000.00
3	2010	2	10000.00	30000.00	30000.00	30000.00	30000.00
4	2010	3	10000.00	40000.00	40000.00	40000.00	40000.00
5	2010	4	10000.00	50000.00	50000.00	50000.00	50000.00
6	2010	5	10000.00	60000.00	60000.00	60000.00	60000.00
7	2010	6	10000.00	80000.00	80000.00	80000.00	80000.00
8	2010	6	10000.00	80000.00	80000.00	80000.00	80000.00
9	2010	7	10000.00	110000.00	110000.00	110000.00	110000.00
10	2010	7	10000.00	110000.00	110000.00	110000.00	110000.00
11	2010	7	10000.00	110000.00	110000.00	110000.00	110000.00
12	2010	8	10000.00	130000.00	130000.00	130000.00	130000.00
13	2010	8	10000.00	130000.00	130000.00	130000.00	130000.00
14	2010	9	10000.00	140000.00	140000.00	140000.00	140000.00
15	2010	10	10000.00	150000.00	150000.00	150000.00	150000.00
16	2010	11	10000.00	160000.00	160000.00	160000.00	160000.00
17	2010	12	10000.00	170000.00	170000.00	170000.00	170000.00
18	2011	1	20000.00	190000.00	20000.00	190000.00	20000.00
19	2011	2	20000.00	230000.00	60000.00	230000.00	60000.00
20	2011	2	20000.00	230000.00	60000.00	230000.00	60000.00
21	2011	3	20000.00	250000.00	80000.00	250000.00	80000.00
22	2011	4	20000.00	270000.00	100000.00	270000.00	100000.00
23	2011	5	20000.00	290000.00	120000.00	290000.00	120000.00
24	2011	6	20000.00	330000.00	160000.00	330000.00	160000.00
25	2011	6	20000.00	330000.00	160000.00	330000.00	160000.00
26	2011	7	20000.00	390000.00	220000.00	390000.00	220000.00
27	2011	7	20000.00	390000.00	220000.00	390000.00	220000.00
28	2011	7	20000.00	390000.00	220000.00	390000.00	220000.00
29	2011	8	20000.00	430000.00	260000.00	430000.00	260000.00
30	2011	8	20000.00	430000.00	260000.00	430000.00	260000.00
31	2011	9	20000.00	450000.00	280000.00	450000.00	280000.00
32	2011	10	20000.00	470000.00	300000.00	470000.00	300000.00
33	2011	11	20000.00	490000.00	320000.00	490000.00	320000.00
34	2011	12	20000.00	510000.00	340000.00	510000.00	340000.00

Figure 1-12. *More default vs. RANGE behaviors*

At first glance, all the results seem to be the same! This is true for the first year's partition, but once we get to the second year's partition, values for the first and third columns keep incrementing, while values for the second and fourth columns are reset due to the new year partition.

Values for duplicate month entries have been framed with boxes. Column NPBOB has no PARTITION BY clause, but it has an ORDER BY clause, so the default window frame behavior defined by RANGE BETWEEN UNBOUNDED PRECEDING and CURRENT ROW kicks in. Once again notice how the rows with the duplicate month numbers result in the totals being the same. Month 6 (June) for 2010 has totals of 8000.00, the same for months 7 (July) and 8 (August).

The second SUM()/OVER() combination does have a PARTITION BY and an ORDER BY clause, so the default window frame behavior is

RANGE BETWEEN UNBOUNDED PRECEDING AND CURRENT ROW

The PARTITION BY clause defines the partitions for each year, and the default frame behavior causes the window function to calculate values by considering the current row and prior row values until duplicate month entries are encountered. If this is a running total, we would expect 7000.00 and 8000.00, but both entries for these rows have the higher value: 8000.00.

The third SUM()/OVER() combination in the query includes an ORDER BY clause but no PARTITION BY clause. A window frame specification is included:

RANGE BETWEEN UNBOUNDED PRECEDING AND CURRENT ROW

The results are the same as the prior combination, but once we get to the partition for year 2011, the rolling sum just keeps going as a PARTITION BY clause was not included. Once again, notice the behavior when rows with the same month are encountered.

Last but not least, the fourth SUM()/OVER() combination has a PARTITION BY, an ORDER BY, and a RANGE window frame clause. This combination behaves just like the combination for the PBOB column.

We can now see how the default behaviors kick in for the various combinations of the ORDER BY and PARTITION BY clauses. When a window frame specification is included, it behaves just like the default behavior in the other columns. Keep these in mind when you start creating window function–based queries.

Example 3

The next example compares the ROWS window frame specifications vs. the RANGE window frame specification.

Please refer to Listing 1-8.

Listing 1-8. ROWS vs. RANGE Comparison

```
SELECT OrderYear,OrderMonth,SalesAmount,
     SUM(SalesAmount) OVER (
          PARTITION BY OrderYear
          ORDER BY OrderMonth
          ROWS BETWEEN UNBOUNDED PRECEDING AND CURRENT ROW
          ) AS POBRowsUPCR,

     SUM(SalesAmount) OVER (
          PARTITION BY OrderYear
          ORDER BY OrderMonth
          RANGE BETWEEN UNBOUNDED PRECEDING AND CURRENT ROW
          ) AS PBOBRangeUPCR
FROM OverExample;
GO
```

The SUM( ) function in both cases has identical PARTITION BY and ORDER BY clauses. The first has a ROWS BETWEEN UNBOUNDED PRECEDING AND CURRENT ROW clause, while the second combination has a RANGE BETWEEN UNBOUNDED PRECEDING AND CURRENT ROW clause.

What do you think the results will be?

Please refer to Figure 1-13.

	OrderYear	OrderMonth	SalesAmount	POBRowsUPCR	PBOBRangeUPCR
1	2010	1	10000.00	10000.00	10000.00
2	2010	2	10000.00	20000.00	30000.00
3	2010	2	10000.00	30000.00	30000.00
4	2010	3	10000.00	40000.00	40000.00
5	2010	4	10000.00	50000.00	50000.00
6	2010	5	10000.00	60000.00	60000.00
7	2010	6	10000.00	70000.00	80000.00
8	2010	6	10000.00	80000.00	80000.00
9	2010	7	10000.00	90000.00	110000.00
10	2010	7	10000.00	100000.00	110000.00
11	2010	7	10000.00	110000.00	110000.00
12	2010	8	10000.00	120000.00	130000.00
13	2010	8	10000.00	130000.00	130000.00
14	2010	9	10000.00	140000.00	140000.00
15	2010	10	10000.00	150000.00	150000.00
16	2010	11	10000.00	160000.00	160000.00
17	2010	12	10000.00	170000.00	170000.00
18	2011	1	20000.00	20000.00	20000.00
19	2011	2	20000.00	40000.00	60000.00
20	2011	2	20000.00	60000.00	60000.00
21	2011	3	20000.00	80000.00	80000.00
22	2011	4	20000.00	100000.00	100000.00
23	2011	5	20000.00	120000.00	120000.00
24	2011	6	20000.00	140000.00	160000.00
25	2011	6	20000.00	160000.00	160000.00
26	2011	7	20000.00	180000.00	220000.00
27	2011	7	20000.00	200000.00	220000.00
28	2011	7	20000.00	220000.00	220000.00
29	2011	8	20000.00	240000.00	260000.00
30	2011	8	20000.00	260000.00	260000.00
31	2011	9	20000.00	280000.00	280000.00
32	2011	10	20000.00	300000.00	300000.00
33	2011	11	20000.00	320000.00	320000.00
34	2011	12	20000.00	340000.00	340000.00

Figure 1-13. ROWS vs. RANGE, UNBOUNDED PRECEDING AND CURRENT ROW

The results are almost the same, but the SUM()/OVER() combination that uses the RANGE window frame specification will generate the same values for any rows having duplicate ORDER BY values. As we discussed earlier, this clause works at a logical level (vs. physical for ROWS); it will add up all the duplicate values and post the same results in each duplicate row.

Example 4

In this next example, we will look at two simple queries. The only difference is that the first query includes an ORDER BY clause that calls out column names, while the second query uses a simple SELECT query instead of the column names in the ORDER BY clause.

Why do we do this? Because using a simple query will eliminate a possibly expensive sort step in the estimated query plan. This technique is used when you have only one large partition. The query you need to use is a scalar query, which means it will only return one value, so you cannot process multiple partitions.

Let's check out the query with the column names first.

Please refer to Listing 1-9a.

Listing 1-9a. ORDER BY Without Subquery

```
WITH YearQtrSales (
      SalesYear,
      SalesQtr,
      SalesMonth,
      SalesTotal
)
AS
(
SELECT
    SalesYear,
      SalesQtr,
      SalesMonth,
      SalesTotal
FROM dbo.TestSales
)
```

```
SELECT
     SalesYear,
     SalesQtr,
     SalesMonth,
     SalesTotal,
     SUM(SalesTotal) OVER(
          ORDER BY SalesYear,SalesQtr,SalesMonth
          ROWS BETWEEN UNBOUNDED PRECEDING AND CURRENT ROW
          ) AS RollMonthlySales1

FROM YearQtrSales
ORDER BY
     SalesYear,
     SalesQtr,
     SalesMonth
GO
```

As can be seen, we have no PARTITION BY clause and an ORDER BY clause that sorts by three columns. This configuration will give us rolling totals by month for both years. In other words, when we get to the second year in the result set, the totals do not reset but keep on incrementing by month till the end of the data set.

The second query replaces the column names in the ORDER BY clause with a simple SELECT statement: SELECT (1).

Please refer to Listing 1-9b.

Listing 1-9b. ORDER BY with Subquery

```
WITH YearQtrSales (
     SalesYear,
     SalesQtr,
     SalesMonth,
     SalesTotal
)
AS
```

```
(
SELECT
      SalesYear,
      SalesQtr,
      SalesMonth,
      SalesTotal
FROM dbo.TestSales
)
SELECT
      SalesYear,
      SalesQtr,
      SalesMonth,
      SalesTotal,
      SUM(SalesTotal) OVER(
            ORDER BY (SELECT (1))
            ROWS BETWEEN UNBOUNDED PRECEDING AND CURRENT ROW
            ) AS RollMonthlySales2
FROM YearQtrSales
ORDER BY
      SalesYear,
      SalesQtr,
      SalesMonth
GO
```

This is the same query as the previous one. Let's compare execution plans. Please refer to Figure 1-14.

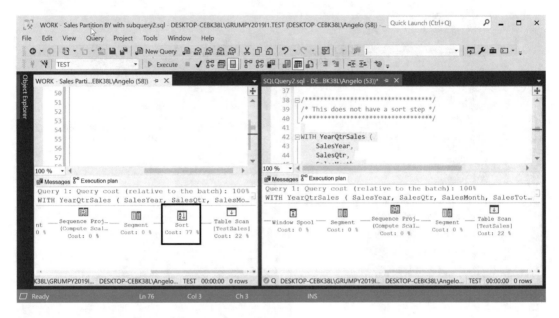

Figure 1-14. *Comparing execution plans*

As can be seen in the left-hand query plan, we have a pretty expensive sort task with a cost of 77%. The query plan on the right does not have this task, so it should be faster in terms of performance. Let's check out the results for both queries, side by side.

Please refer to Figure 1-15.

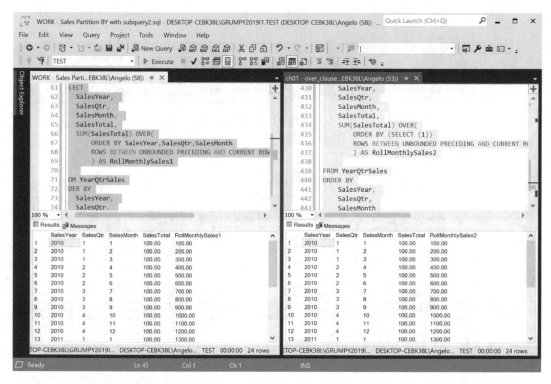

Figure 1-15. *Comparing results*

Results match. Notice how the rolling totals keep incrementing even when we start a new year. This type of configuration where we use a subquery is usually good when you only deal with one single partition with a large volume of rows. Eliminating a sort step on a query result set of a couple of hundred of thousand or more rows can increase performance.

Example 5

Our last example is a simple one that illustrates a new feature available with SQL Server 2022. As of this writing, it is available as an evaluation release. It allows you to specify the PARTITION BY, ORDER BY, and RANGE/ROWS clauses within a set of parentheses and give it a name. It's called the named window clause.

Please refer to Listing 1-10 for a simple example.

Listing 1-10. Named Windows

```
SELECT OrderYear
      ,OrderMonth
      ,SUM(SalesAmount) OVER SalesWindow AS TotalSales
FROM dbo.OverExample
WINDOW SalesWindow AS (
      PARTITION BY OrderYear
      ORDER BY OrderMonth
      RANGE BETWEEN UNBOUNDED PRECEDING AND CURRENT ROW
      )
GO
```

Notice how the named window code appears at the end of the query. The keyword WINDOW is followed by the name you wish to give this window, and then you can add the PARTITION BY, ORDER BY, and RANGE/ROWS clauses between the parentheses as we just discussed in this chapter.

The OVER() clause still appears after the window function, but it simply refers to the name. Is this value added? I do not know but I would have preferred something like

```
SUM(SalesAmount) OVER WINDOW (
      PARTITION BY OrderYear
      ORDER BY OrderMonth
      RANGE BETWEEN UNBOUNDED PRECEDING AND CURRENT ROW
      ) AS ...
```

At least it is clear that we are creating a window into the partition. But that's me.

On the other hand, if the OVER() clause is to be used in multiple columns, then the SQL Server 2022 version just discussed makes more sense. Yes, I think this is better.

Let's look at the results for this last query.

Please refer to Figure 1-16.

	OrderYear	OrderMonth	TotalSales
1	2010	1	10000.00
2	2010	2	30000.00
3	2010	2	30000.00
4	2010	3	40000.00
5	2010	4	50000.00
6	2010	5	60000.00
7	2010	6	80000.00
8	2010	6	80000.00
9	2010	7	110000.00
10	2010	7	110000.00
11	2010	7	110000.00
12	2010	8	130000.00
13	2010	8	130000.00
14	2010	9	140000.00
15	2010	10	150000.00
16	2010	11	160000.00
17	2010	12	170000.00
18	2011	1	10000.00
19	2011	2	30000.00
20	2011	2	30000.00
21	2011	3	40000.00

Figure 1-16. *Named WINDOW SQL Server 2022 enhancement*

No surprises here. Works just like the prior examples. Same behavior when duplicate ORDER BY column values exist. Go to the Microsoft SQL Server website to download the latest version. Make sure to check out what's new in this version.

Summary

We've covered the window functions we will use in this book. More importantly, we discussed the OVER() clause and how to set up partitions and window frames to control how the window functions are applied to the data sets, partitions, and window frames.

We discussed how window frames have default behavior depending on the combination of PARTITION BY and ORDER BY clauses within the OVER() clauses.

We also looked at some basic diagrams to illustrate how window frames affect the window function processing flow and under what conditions default window frames kick in.

Next, we briefly discussed how you can use subqueries with the `PARTITION BY` and `OVER()` clauses. This is of value as using subqueries eliminates a sort step and can make the overall query perform faster. This was demonstrated by looking at query plans for queries that used subqueries and queries that did not. Some tricks were tested to see how you can retrieve multiple values but package them as a single value with the `STRING_AGG()` function which we will discuss in future chapters.

Lastly, we looked at some simple examples to put our knowledge to work and briefly discussed a new SQL Server 2022 feature called named windows.

If you are unfamiliar with how window functions such as `RANK()` or other functions work, now is a good time to check out Appendix A, which gives you a brief description on what these functions do. These will be used many times in the remaining chapters in the book.

CHAPTER 2

Sales DW Use Case: Aggregate Functions

Our main goal in this chapter is to learn and apply the category of window functions called aggregate functions. We start off by describing the sales data warehouse (DW) called `APSales` that we will use.

A simple business conceptual model is presented together with some simple data dictionaries to help us understand the business contents of the database.

We include discussions on performance tuning and then proceed to take each window function for a test drive. We conclude the chapter with a brief discussion on the new SQL Server 2022 feature called `WINDOW`, which is used to define the `OVER()` clause for the window functions.

Note The DDL used to create this small data warehouse is found on the Google website for this book together with scripts to create all tables and views and to load the tables with test data. The code is easy to use. Each step is labeled, and where necessary comments help clarify what each step does.

Let's spend a little time understanding what our small sales data warehouse is all about.

Sales Data Warehouse

The `APSales` data warehouse stores transaction and historical information for a fictitious company that bakes and sells all sorts of cakes, pies, tarts, and chocolates with a traditional European style.

© Angelo Bobak 2023
A. Bobak, *SQL Server Analytical Toolkit*, https://doi.org/10.1007/978-1-4842-8667-8_2

The company has 100 customers, and we want to track and analyze their purchase patterns over several years' worth of data. The company sells 60 products that fall under several categories like cakes, pies, tarts, croissants, and chocolates.

Customers purchase their favorite confection at one of 16 stores situated throughout the United States.

Lastly, a staging table is used to simulate purchases by each customer of a few years. Your first assignment is to create the database and tables, load them, and write some simple queries to see the actual data you will analyze.

Figure 2-1 is a snapshot of SSMS Object Explorer that shows the database and the tables. Once you create the database and all table objects, your own Object Explorer panel should look something like this.

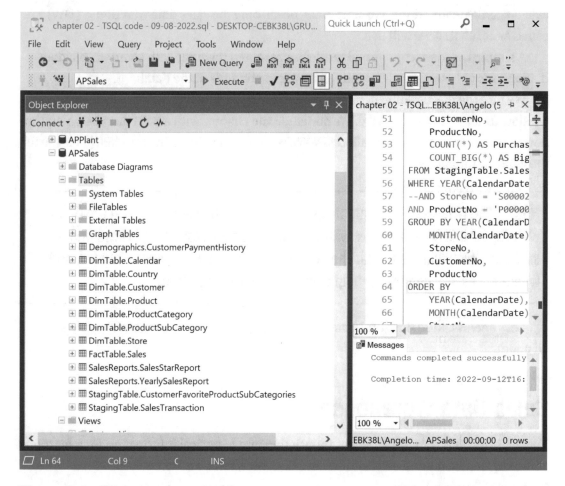

Figure 2-1. *Data warehouse tables*

Notice the tables in the `SalesReports` and `StagingTable` schemas. The tables assigned to the `SalesReports` schema represent denormalized views generated by querying the Sales fact table with a star query that joins all dimensions to the fact table so that we can present business analysts with some data that uses all the names, codes, etc.

The tables assigned to the `StagingTable` schema are used by TSQL batch scripts that generate some random transaction data so we can then load it into the Sales fact table. Let's check out a simple business conceptual model to understand how all this ties together.

Sales Data Warehouse Conceptual Model

The following is a simple conceptual business data model that identifies the main business objects of the sales data warehouse. The target database is a basic star schema model in that it has one fact table and several dimension tables (you can have multiple fact tables of course).

A couple of denormalized report tables are also included to facilitate running complex queries.

These tables are called `SalesStarReport` and `YearlySalesReport` and are assigned to the `SalesReports` schema. These are used to tie in the fact and dimension tables and replace the surrogate keys with the actual dimension table columns. They are preloaded each night so that they are available to users the following day. Preloading the tables provides performance benefits by allowing calculations on data to be performed prior to inserting into the table; simpler queries will run faster.

For those not familiar with data warehouses or data marts, columns assigned the role of surrogate keys are used to establish the associations between dimension and fact tables. These are numerical columns as they are faster to use in joins than alphanumeric columns.

Viewed in a diagram, using your imagination, they resemble a star and are therefore called star schemas.

The following model is intended to just show the basic business objects and the relationships. It is not a full physical data warehouse model. By the way, I assume you know what a fact table is and what a dimension is.

Note   There are many sources on the Web that describe what a data warehouse or data mart is. Here is a link that might be helpful if you are not familiar with this architecture:

`www.snowflake.com/data-cloud-glossary/data-warehousing/`

Another good resource is `www.kimballgroup.com/`.

Each dimension table has a surrogate key that plays the role of primary key. All these surrogate keys appear in the fact table as foreign keys. Technically, all the foreign keys in the fact table can be used as the primary key for the fact table. The fact table also contains columns that represent data you can count, in other words numerical data. In special cases a fact table can contain other types of columns, but this is a topic for another book. (Look up factless fact tables and junk dimensions.)

Please refer to Figure 2-2.

AP Sales Conceptual Model

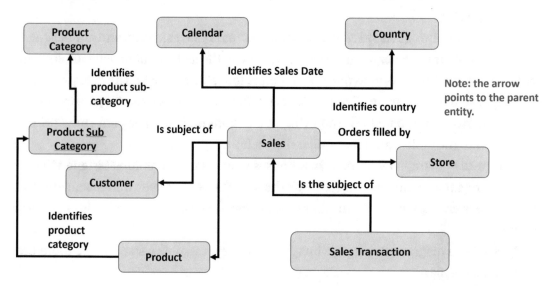

Figure 2-2. *The APSales conceptual business model*

This is called a conceptual business model as it is intended to clearly identify the main business data objects to nontechnical users (our friends, the business analysts, and managers). These types of models are used during the initial phases of the database design.

Data modelers, business analysts, and other stakeholders are gathered to identify the main data objects that will be in the database or data warehouse. At this stage of the design process, only basic metadata is identified, like entity names and their relationships between each other.

Subsequent sessions evolve the conceptual model in a physical denormalized star or snowflake schema design where all the surrogate keys, columns, and other objects are identified.

I present this simple model so that you can have a basic understanding of what the database that will implement the data warehouse looks like.

Next, we need to create a series of simple documents called data dictionaries that further describe the metadata.

Table 2-1 is a simple table that shows the associations between the entities that will become the fact table and the dimension tables. You need to understand these so you can build queries that join the tables together.

Table 2-1. *Fact-Dimension Associations*

Dimension	Business Rule	Fact/Dimension Table	Cardinality
Country	Identifies location for	Sales	One to zero, one, or many
Store	Fulfills sales for	Sales	One to zero, one, or many
Customer	Is the subject of	Sales	One to zero, one, or many
Product	Is subject of	Sales	One to zero, one, or many
Product Category	Category classification of a product	Product	One to zero, one, to many
Product Subcategory	Subcategory classification for	Product Category	One to one, many to one
Calendar	Identifies sales date of	Sales	One to zero, one to many

Typically, these types of tables appear in data dictionaries that contain information about the various data objects that make up a database. Looking at the associations in the preceding table, we also want to highlight the business rules that link these together.

A country identifies a location in zero, one, or more rows in the Sales fact table. A fact table row needs to contain a country code, but it is possible to have rows in the Country table that have no related row in the Sales table. In the database you will build, there are only sales transactions for the United States although the Country dimension contains most of the country information for all the ISO standard country codes.

Note   A data warehouse is a specialized type of table. In this chapter I use the terms interchangeably.

A store can be the subject of zero, one, or more records in the Sales fact table. It is possible to have a new store that has not sold any products yet, so there are no records in the Sales fact table.

The same rule applies for a customer. A customer can appear in zero, one, or more records in the Sales table. It is possible to have a new customer that has not bought anything yet, so the customer will appear in the Customer dimension but not in the Sales fact table.

A product can appear in zero, one, or more sales records in the Sales fact table. It is possible to have a product not yet sold appear in the Product table.

A product category can refer to one or more product subcategories, and each product category can refer to one or more products. It is possible to have product categories and product subcategories that have not been assigned to a product yet for various reasons, like it has not been manufactured yet. (For example, we plan to sell ice creams but have not manufactured them yet.)

Last but not least is a simple data dictionary that describes the tables themselves although the names are fairly self-explanatory. Please refer to Table 2-2.

Table 2-2. *Table Definition Data Dictionary*

Table Name	Description
Calendar	Basic calendar date objects used for time reporting (like date, quarter, month, and year objects).
Country	Contains ISO two- and three-character codes for most of the world's countries, for example, US or USA.
Customer	Basic customer table used to store names of the customers and identifiers.
Product	Basic product table that stores names and descriptions of products.
Product Category	High-level category codes and description for products: chocolates, cakes, pies, croissants, and tarts.
Product Subcategory	Further breaks down a product category. For example, chocolates can be broken down into dark chocolates – small, dark chocolates – medium, and dark chocolates – large.
Store	Contains basic information for 16 stores: the store number, name, and territory. Useful for sales performance analysis.
Sales	Stores all sales transactions by customers over multiple years. Who bought what, when, and where!

Keep in mind that this is just a basic set of design specifications for the simple data warehouse we will be using. I took great liberties in just presenting the most basic concepts of data warehouse and star schema. You would also need data dictionaries that describe the primary and foreign keys and surrogate keys and also data dictionaries that describe each of the columns and their data types in the fact and dimension tables.

To summarize, our goal is to have a simple data warehouse we can load and use as we study the capabilities of all the window functions that are discussed in this book. Hopefully this section gave you a basic understanding of the data warehouse you will build with the script that can be found on the Google website for this book.

A Word About Performance Tuning

Key to creating queries and batch scripts that utilize the window functions (and the OVER() clause) is performance. We want to create queries that are not only accurate but fast. No user will be happy with a query that runs more than one minute.

You will need to develop some skills in performance tuning, which entails using tools provided by SQL Server and how to create indexes based on the feedback these tools give you when you analyze the queries you are designing.

There are several tools available with SQL Server and tools available from different vendors. We will concentrate on the following SQL Server tools:

- Query plans

- Setting STATISTICS IO ON/OFF

- Setting STATISTICS TIME ON/OFF

- Client statistics

- Creating indexes (suggested or your own)

- Lots of curiosity and patience

There are also table denormalization, partitioned tables, physical table file placement, and hardware upgrades, but we will not cover these.

So what are query plans?

Visually, query plans are a graphical image that shows the steps SQL Server will take to execute a query. It shows you the flow and the cost of each task in terms of execution time. This tool can be accessed via the menu bar by clicking **Query** and then scrolling down to **Display Estimated Execution Plan**.

Please refer to Figure 2-3.

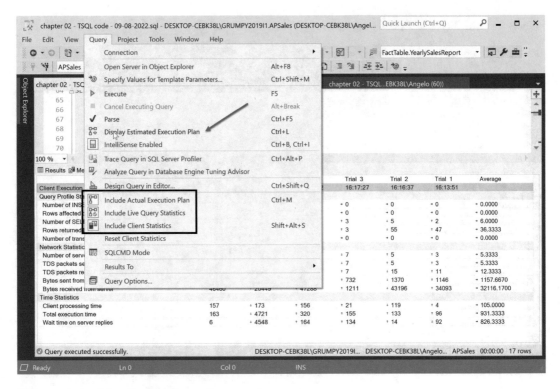

Figure 2-3. *The Query drop-down menu*

You can also include the execution plan, live query statistics, and client statistics by selecting the options called out in the box in the preceding figure.

Another way of accessing these tools is by clicking one or more of the buttons in the menu bar that are called out in Figure 2-4 by the boxes.

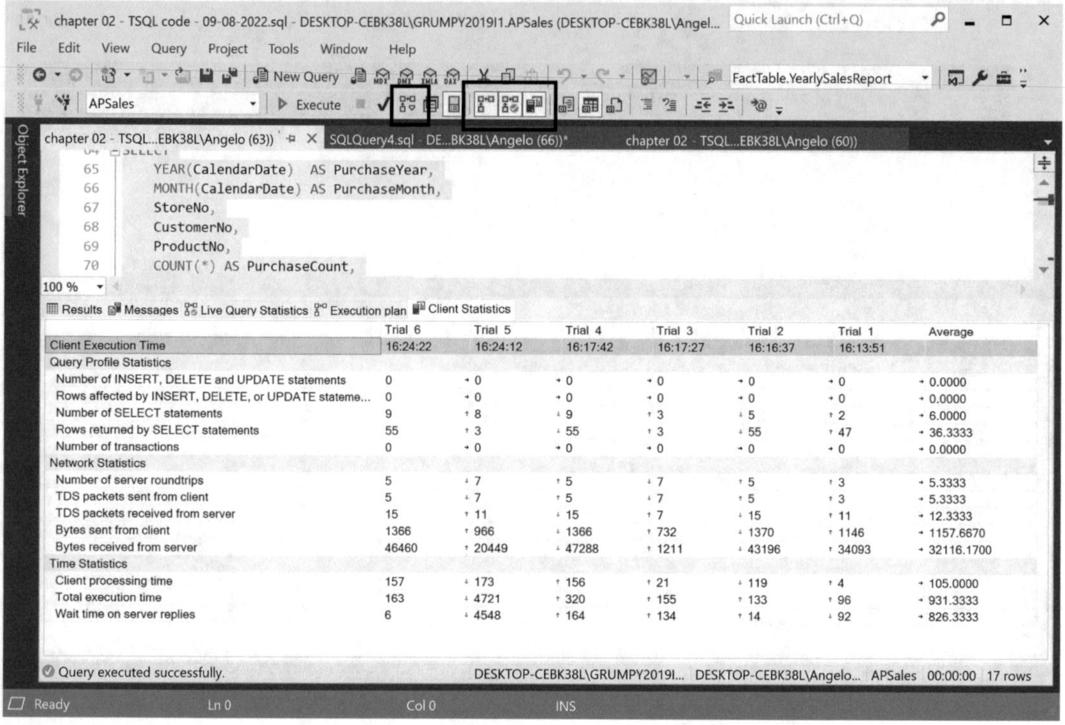

Figure 2-4. *Query performance tools*

The first button will turn on the **Display Estimated Execution Plan** tool. Referring to the next three buttons called out by the boxes in the preceding figure, you can also generate query plans by clicking the **Include Actual Query Plan** button or generate IO statistics by clicking the **Include Live Query Statistics** button. Lastly, you can include client statistics by clicking the last button called **Include Client Statistics**. Using your mouse, move the cursor over each button to generate a message that shows you what the button does.

Let's look at an example query plan. Please refer to Figure 2-5.

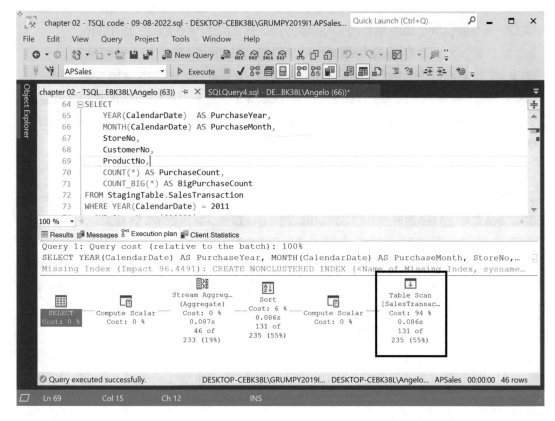

Figure 2-5. *Estimated query plan*

The query plan is read from right to left. It contains icons called tasks (or operations or plain old steps) that perform specific functions like sort data or merge data or use indexes. Associated with each task is a cost estimate displayed as a percentage of the total time spent to execute a query.

For example, looking at the first task on the right-hand side of the plan in the preceding figure, we see a table scan that has a cost of 94%. This tells us immediately that we have a problem. This task is taking up most of the execution time.

These types of tasks are the ones that we need to address with some sort of strategy, like adding an index or rewriting the query.

> **Note**   Sometimes, due to large volumes, all that can be done is to run the
> query overnight, so it loads a denormalized report table. Why do we want to
> do this? Because this will give the appearance of fast performance to any user
> accessing the denormalized table. Joins, calculations, and other complex logic are
> eliminated when processing the user report. The user is querying a preloaded and
> calculated table!
>
> Lastly, make sure to run these queries within a reporting tool, like SSRS, so you
> can add filters to limit what the user sees. (Create indexes based on how the users
> filter the query.) No one wants to scroll through millions of rows of data.

We also see a step that sorts the data, and it has a cost of 6% of the total execution time. Notice that 94% + 6% = 100%. Usually when you add up all the estimated costs, they will add up to 100%.

Next, we set IO and TIME statistics on with the following commands:

```
SET STATISTICS IO ON
GO

SET STATISTICS TIME ON
GO
```

By the way, at the end of the analysis, turn off statistics by replacing the keyword ON with OFF:

```
SET STATISTICS IO OFF
GO

SET STATISTICS TIME OFF
GO
```

Using these commands and executing the query, we are presented with some important performance statistics:

```
SQL Server parse and compile time:
   CPU time = 0 ms, elapsed time = 0 ms.

 SQL Server Execution Times:
   CPU time = 0 ms,  elapsed time = 0 ms.
```

```
SQL Server Execution Times:
   CPU time = 0 ms,  elapsed time = 0 ms.
SQL Server parse and compile time:
   CPU time = 15 ms, elapsed time = 70 ms.
```
(46 rows affected)
Table 'Worktable'. Scan count 0, **logical reads 0**, physical reads 0, page
server reads 0, read-ahead reads 0, page server read-ahead reads 0, lob
logical reads 0, lob physical reads 0, lob page server reads 0, lob read-
ahead reads 0, lob page server read-ahead reads 0.
Table 'SalesTransaction'. Scan count 1, logical reads 718, physical reads
0, page server reads 0, read-ahead reads 718, page server read-ahead reads
0, lob logical reads 0, lob physical reads 0, lob page server reads 0, lob
read-ahead reads 0, lob page server read-ahead reads 0.
(6 rows affected)

Some of the interesting values we want to pay attention to are scan counts, logical and physical reads, and whether a work table is used. Notice the logical read count of 718. This needs to go down. Logical reads are performed in memory if you have enough memory available or else on hard disk, which really slows things down. Aim for single-digit read counts.

The goal right now is to make you aware of these tools and what the output looks like. We will examine what they mean as we develop our TSQL queries that use the window functions and the OVER() clause. For now, just know where to find them, how to turn them on, and how to look at the results when you run the query.

Tip   You might want to look up the preceding counters in the Microsoft documentation to begin understanding what they mean. Yes, there are quite a few, but understanding them will pay dividends: `https://learn.microsoft.com/en-us/sql/relational-databases/showplan-logical-and-physical-operators-reference?view=sql-server-ver16`.

The preceding URL is fairly long to type. Just use your favorite search engine and search for "Showplan Logical and Physical Operators Reference."

Finally, the **Display Estimated Execution Plan** tool will also suggest a missing index that is needed to improve query performance. Let's look at an example. Please refer to Listing 2-1.

Listing 2-1. Suggested Index by Query Plan Tool

```
/*
Missing Index Details from chapter 02 - TSQL code - 09-08-2022.sql -
DESKTOP-CEBK38L\GRUMPY2019I1.APSales (DESKTOP-CEBK38L\Angelo (63))
The Query Processor estimates that implementing the following index could
improve the query cost by 96.4491%.
*/

/*
USE [APSales]
GO
CREATE NONCLUSTERED INDEX [<Name of Missing Index, sysname,>]
ON [StagingTable].[SalesTransaction] ([ProductNo])
INCLUDE ([CustomerNo],[StoreNo],[CalendarDate])
GO
*/
```

Lots of nice comments and a template for the index are generated. Notice the square brackets by the way. When SSMS generates the code for you, these are always included, probably just in case you used names with spaces or weird characters in your database table and columns!

The statement generated tells us a non-clustered index is needed and it is based on the table called SalesTransaction assigned to the StagingTable schema. The columns to include are the ProductNo column and the CustomerNo, StoreNo, and CalendarDate columns that appear after the INCLUDE keyword.

If we create this index and rerun the **Display Estimated Execution Plan** tool, we should see some improvements.

Please refer to Figure 2-6.

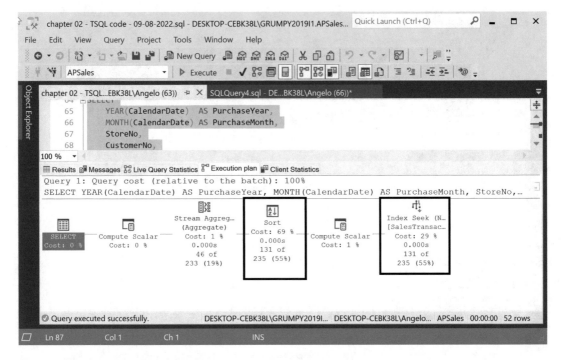

Figure 2-6. *Revised estimated query plan*

Success! The table scan step has been replaced with an index seek step at a cost of 29%. The sort step is still there with an increased cost of 69%. Is this better or worse?

Let's execute the query with the statistics settings turned on and see what we get. Please refer to Figure 2-7.

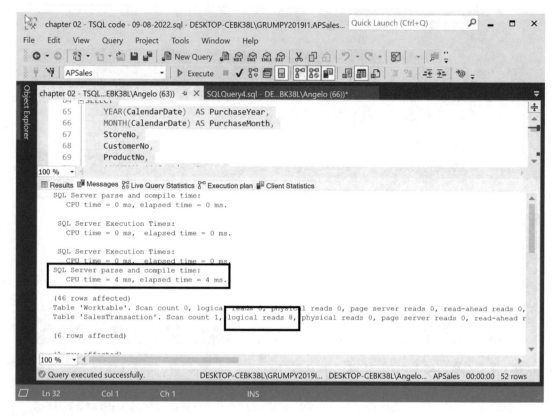

Figure 2-7. *Statistics IO report*

Wow! The prior logical read count of 718 is now reduced to 8. CPU time before was 15 ms, and now it is reduced to 4 ms. Dramatic improvement. Recall that logical reads indicate how many times the memory cache must be read. If enough memory is not available, then physical disk must be read.

This will be expensive when the tables are large and your disk is slow! Corporate production environments usually implement TEMPDB on a solid-state drive for increased performance.

I almost forgot client statistics. Here is the report that is generated when you click this selection. Please refer to Figure 2-8.

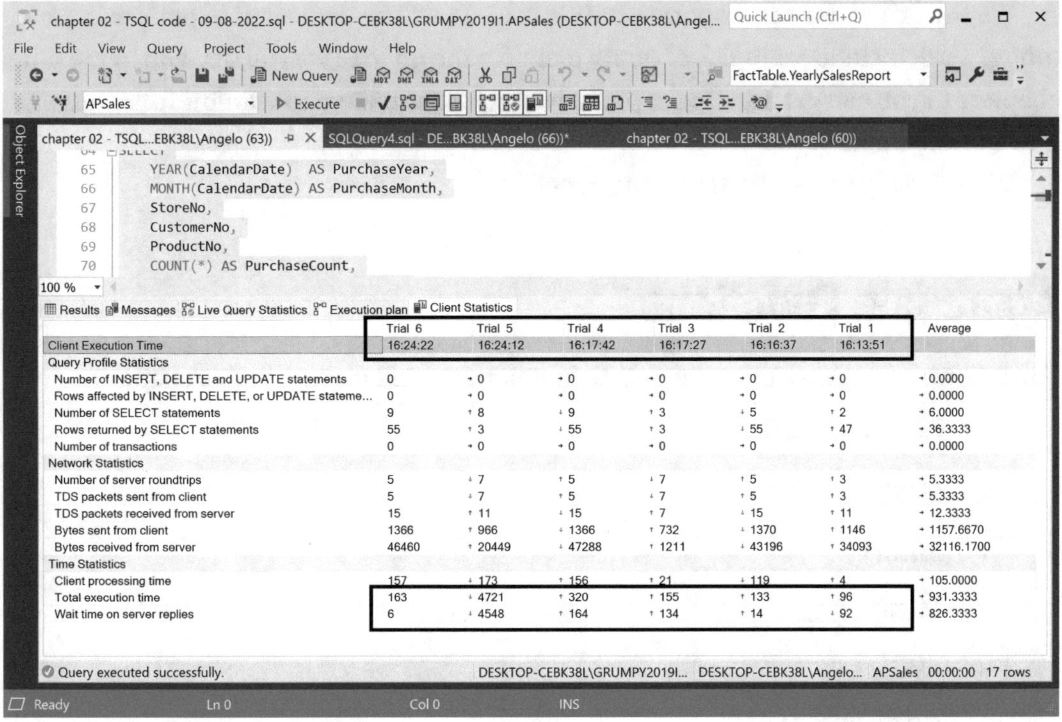

Figure 2-8. *Client statistics*

A very comprehensive set of statistics in this report. Notice all the trials. What this means is that the query was run six times and we obtained client execution times for each trial run. There is also average execution time for the statistics. Another important set of information is the total execution time and wait time on server replies. Trial 6 information seems to be the best, possibly because of the index.

This together with the output of the tools we just examined will allow you to diagnose and identify performance bottlenecks. Make sure you practice a bit with some simple queries and see if you can interpret the statistics that are returned.

By the way, I do not want to imply that performance tuning is a walk in the park. This is a difficult discipline to master and takes years of practice to achieve expert level. I am just touching the surface of this very interesting and important activity.

This section was long, and as stated earlier, it was meant to introduce you to the tools and look at the output. Lots of information, so do not worry if you don't understand what all the statistics mean yet. What's important at this stage is that you know where to find them and to try them out on some queries. As we proceed with the chapter, we will do a deep dive into what the important query plan steps tell us and how to use the information to create indexes to deliver better performance.

We will do this for a few queries, but Chapter 14 will be dedicated to performance tuning. I will include examples of query plans for the queries we will develop in the other chapters so you can get a feeling of performance issues related to window functions and how to solve them.

We are ready to start writing some code!

Aggregate Functions

The following are the window aggregate functions we will discuss in this chapter:

- COUNT() and COUNT_BIG()
- SUM()
- MAX()
- MIN()
- AVG()
- GROUPING()
- STRING_AGG()
- STDEV()
- STDEVP()
- VAR()
- VARP()

The approach we will take is to present the code for each function, discuss it, display the results when it is executed, and review the results. We might even generate a few graphs with Excel to better analyze the results.

For the GROUPING() and STDEV() functions, we generate IO and TIME statistics and an estimated query execution plan by following the steps discussed earlier. Usually, the query plan will also suggest a recommended index to create. If an index is suggested, we will copy the template and modify it by giving it a name and create it.

Once the index is created, we regenerate statistics and the query plan and compare the new set to the prior set of statistics to see what improvements in performance have been gained by creating the new index.

Sometimes creating indexes does not help and can even hurt performance, not to mention that the more indexes you create, the more performance decreases when you need to delete, update, or insert rows.

COUNT( ), MAX( ), MIN( ), AVG( ), and SUM( ) Functions

Let's start examining the workhorse functions you will use frequently in your queries and reports. The following is a query that reports the count of transactions, the largest number and smallest number of transactions, and the total quantities of a particular product purchased.

We want to create a crosstab report by year, month, store, and of course product to give us a sales performance profile. Let's not confuse transaction counts vs. quantities of items sold. A single transaction could have 10 products or 1 or maybe 20 or more. Note the difference.

Lastly, we want to limit the results by a single store, product, and year so we can analyze the results. Please refer to Listing 2-2 for the query.

Listing 2-2. Basic Sales Profile Report

```
SELECT YEAR(CalendarDate) AS PurchaseYear,
     MONTH(CalendarDate) AS PurchaseMonth,
     StoreNo,
     ProductNo,
     ProductName,
     COUNT(*) AS NumTransactions,
     MIN(TransactionQuantity) AS MinQuantity,
     MAX(TransactionQuantity) AS MaxQuantity,
     AVG(TransactionQuantity) AS AvgQuantity,
     SUM(TransactionQuantity) AS SumQuantity
FROM SalesReports.YearlySalesReport
WHERE StoreNo = 'S00001'
AND ProductNo = 'P0000001112'
AND YEAR(CalendarDate) = 2010
GROUP BY YEAR(CalendarDate),
      MONTH(CalendarDate),
     StoreNo,
```

```
        ProductNo,
        ProductName
ORDER BY YEAR(CalendarDate),
        MONTH(CalendarDate),
        StoreNo,
        ProductNo,
        ProductName
GO
```

Nothing fancy, no OVER() clause, no CTE, just a SELECT clause that retrieves rows from the YearlySalesReport table. We include the COUNT(), MIN(), MAX(), AVG(), and SUM() functions and pass the TransactionQuantity column as a parameter.

A WHERE clause filters the results, so we only get rows for store S00001, product P0000001112, and calendar year 2010.

The mandatory GROUP BY clause is included when you use aggregate functions, and we utilize an ORDER BY clause to sort the results. Please refer to the partial results in Figure 2-9.

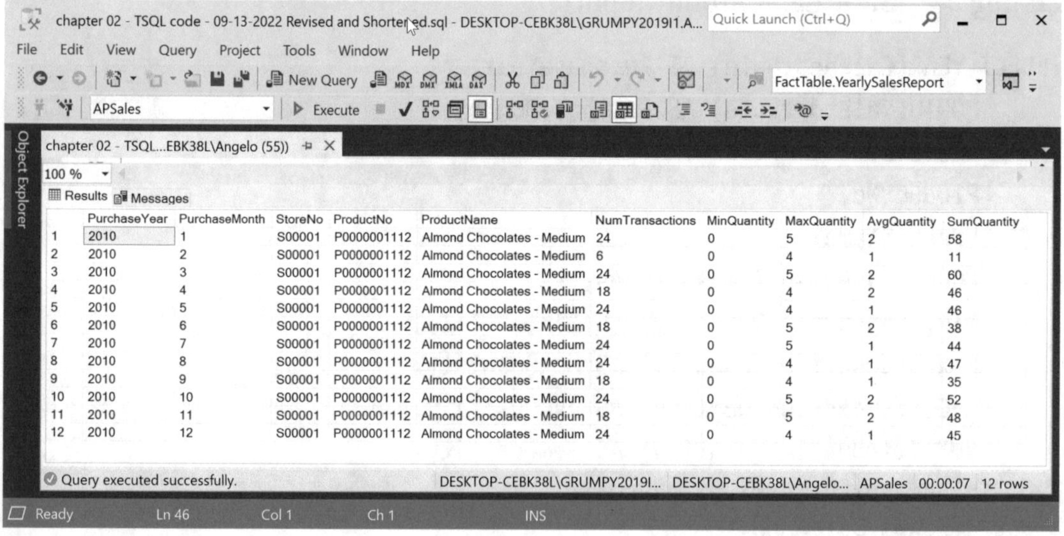

Figure 2-9. *Sales profile report*

Seems like February was a slow month, which is strange as you have the Valentine's Day holiday. You would think sales would be up.

With OVER()

Let's use the prior query and add some window capabilities. There are two parts to this query. First, we need to create a CTE (common table expression) called `ProductPurchaseAnalysis` that performs a little date manipulation and also includes the `COUNT()` function to generate the count of transactions by year, month, store number, and product number.

The second part to the solution is to create a query that references the CTE and uses all the basic aggregate functions, but this time we will include an `OVER()` clause to create and order rows for processing in the partitions. Let's examine the code.

Please refer to Listing 2-3.

Listing 2-3. Part 1, the CTE

```
WITH ProductPurchaseAnalysis (
      PurchaseYear,PurchaseMonth,CalendarDate,StoreNo,CustomerFullName,
      ProductNo,ItemsPurchased,NumTransactions
)
AS (
SELECT YEAR(CalendarDate) AS PurchaseYear,
      MONTH(CalendarDate) AS PurchaseMonth,
      CalendarDate,
      StoreNo,
      CustomerFullName,
      ProductNo,
      TransactionQuantity AS ItemsPurchased,
      COUNT(*)            AS NumTransactions
FROM SalesReports.YearlySalesReport
GROUP BY YEAR(CalendarDate) ,
      MONTH(CalendarDate),
      CalendarDate,
      StoreNo,
      CustomerFullName,
      ProductNo,
      ProductName,
      TransactionQuantity
)
```

Pretty straightforward. The calendar year and month are pulled out from the calendar date, which is included in the SELECT clause. The store number, customer's full name, and product number are also included. Lastly, the transaction quantity, meaning how many items were purchased, is included, and we count the number of transactions within each group as defined by the GROUP BY clause.

Now for the second part of this batch, the query that references the CTE. This query will give us a profile of the number of transactions and quantities purchased for each product by year, month, customer, and store.

Please refer to Listing 2-4.

Listing 2-4. Part 2 – Using Window Functions

```
SELECT PurchaseYear,PurchaseMonth,CalendarDate,StoreNo,
    CustomerFullName,ProductNo,NumTransactions,
    SUM(NumTransactions) OVER (
        PARTITION BY PurchaseYear,CustomerFullName
        ORDER BY CustomerFullName,PurchaseMonth
    ) AS SumTransactions,ItemsPurchased,
    SUM(ItemsPurchased) OVER (
        PARTITION BY PurchaseYear,CustomerFullName
        ORDER BY CustomerFullName,PurchaseMonth
    ) AS TotalItems,
    AVG(CONVERT(DECIMAL(10,2),ItemsPurchased)) OVER (
        PARTITION BY PurchaseYear,CustomerFullName
        ORDER BY CustomerFullName,PurchaseMonth
    ) AS AvgPurchases,
    MIN(ItemsPurchased) OVER (
        PARTITION BY PurchaseYear,CustomerFullName
        ORDER BY CustomerFullName,PurchaseMonth
    ) AS MinPurchases,
    MAX(ItemsPurchased) OVER (
        PARTITION BY PurchaseYear,CustomerFullName
        ORDER BY CustomerFullName,PurchaseMonth
    ) AS MaxPurchases
```

```
FROM ProductPurchaseAnalysis
WHERE StoreNo = 'S00001'
AND ProductNo = 'P0000001112'
AND PurchaseYear = 2010
AND PurchaseMonth = 1
AND ItemsPurchased > 0
GROUP BY PurchaseYear,PurchaseMonth,CalendarDate,StoreNo,
     CustomerFullName,ProductNo,NumTransactions,ItemsPurchased
ORDER BY CustomerFullName,PurchaseYear,PurchaseMonth,CalendarDate,StoreNo,
     ProductNo,ItemsPurchased
GO
```

Notice I included a filter to only report results for January of 2010. This was done in order to keep the result sets small so we can debug and analyze the query. Once it works correctly, you can remove these filters to get all months (when you download the code). You can also remove the PurchaseYear filter predicate to get more years. If you want this information for all stores and customers, you can modify the query so it inserts the results in a report table and then you can write queries to view any combination of customers, stores, and products that you wish.

Let's walk through the rest of the code.

An ORDER BY clause and a PARTITION BY clause are included in each OVER() clause. The data set results are partitioned by the PurchaseYear and CustomerFullName columns. The partition rows are sorted by CustomerFullName and PurchaseMonth.

As mentioned earlier, a WHERE clause is included to filter results by store "S00001," product "P0000001112," and purchase year 2010 and for the month of January. I also filter out any ItemsPurchased values equal to zero. (They somehow sneaked in during the table load process of the test data. This number is randomly generated, so zero values will sneak in!)

Let's see the results. Please refer to the Figure 2-10.

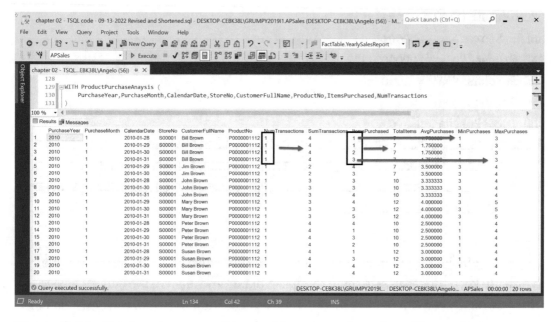

Figure 2-10. *Transaction quantities and items purchased profile*

Let's focus on purchases made by Bill Brown. There are four dates on which purchases were made. Assuming one purchase per day, this comes to a total of four transactions, which appear in the SumTransactions column.

Next, if we look at the ItemsPurchased column, we see the four values corresponding to each of the four days Bill made purchases. On January 28 and 29, he made one purchase each day. On January 30 he made two purchases, and on January 31 he made three purchases. If we add them all up, we get a total of seven purchases, which is reflected in the TotalItems column.

Notice the default behavior of the PARTITION BY and ORDER BY clauses if no window frame clause is included. The totals appear in each row of the result. This means include any previous and following rows (relative to the current row being processed) as each total value for the current row is calculated.

The same applies to the average calculation. Lastly, notice the minimum and maximum values. They can be clearly seen in the ItemsPurchased column. The minimum purchase Bill made was one item, and the largest purchase he made was three items.

This processing pattern repeats itself for each set of rows defined by the partition.

GROUPING( ) Function

The GROUPING() function allows you to create rollup values like sums for each level of a hierarchy defined by categories of data (think of rollups in Excel pivot tables). A picture is worth a thousand words and clarifies the results of this function. Please refer to the Figure 2-11.

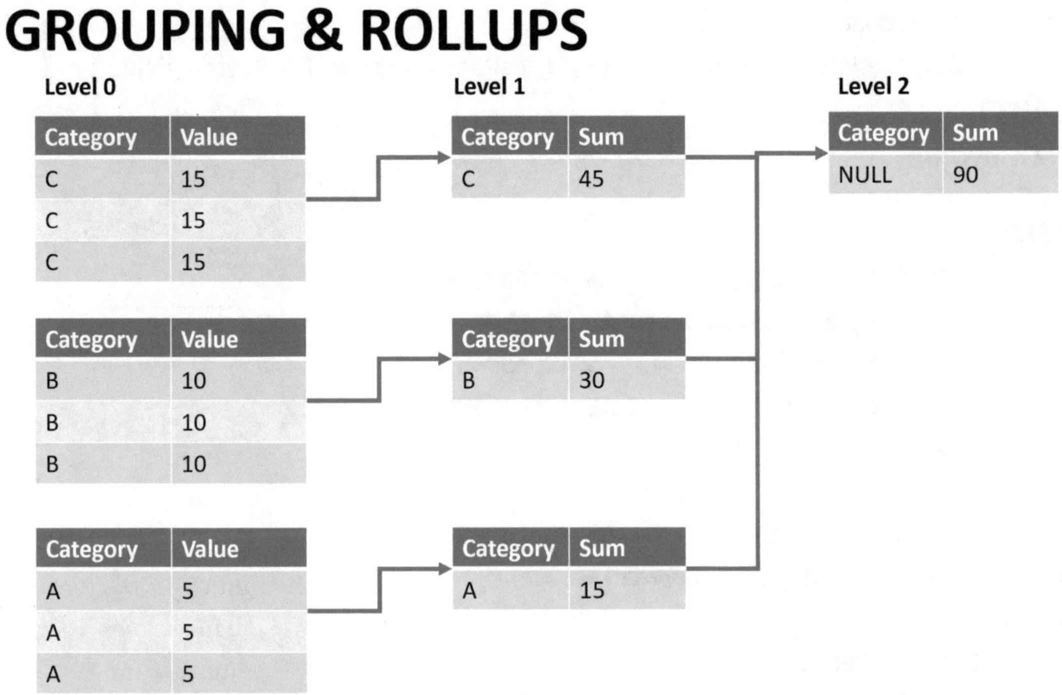

GROUPING & ROLLUPS

Figure 2-11. *Conceptual model of rollups within groups of data*

Starting with the three categories of rows to the left of the diagram, each row in the data set has a category and a value. The categories are A, B, and C. If we summarize the values for each category at level 0, we get rollups of 15 for category A, 30 for category B, and 45 for category C at level 1. If we summarize these three rollup totals, we get a final value of 90 at level 2.

This looks easy, but imagine the complexity if you must summarize data over five or more levels in the group hierarchy. For example, we can summarize by year, month, product category, product subcategory, and product. Let's throw in customer too so we can see totals by customer! This could make the output difficult to read and analyze, so make sure you provide the minimum required information that satisfies user requirements.

Armed with this knowledge, let's write a query that will create rollups for our sales data and try to use a clever trick in the ORDER BY clause to make the results easy to navigate.

Please refer to Listing 2-5.

Listing 2-5. Generating a Rollup Report

```
WITH StoreProductSalesAnalysis
(TransYear,TransQuarter,TransMonth,TransDate,StoreNo,ProductNo,Monthly
Sales)
AS
(
SELECT
      YEAR(CalendarDate)        AS TransYear,
      DATEPART(qq,CalendarDate) AS TransQuarter,
      MONTH(CalendarDate)       AS TransMonth,
      CalendarDate              AS TransDate,
      StoreNo,
      ProductNo,
      SUM(TotalSalesAmount)     AS MonthlySales
FROM FactTable.YearlySalesReport
GROUP BY
      CalendarDate,
      StoreNo,
      ProductNo
)

SELECT TransYear,
      TransQuarter,
      TransMonth,
      StoreNo,
      ProductNo,
      MonthlySales,
      SUM(MonthlySales)      AS SumMonthlySales,
      GROUPING(MonthlySales) AS RollupFlag
```

```
FROM StoreProductSalesAnalysis
WHERE TransYear = 2011
AND ProductNo = 'P0000001103'
AND StoreNo = 'S00001'
GROUP BY TransYear,
      TransQuarter,
      TransMonth,
      StoreNo,
      ProductNo,
      MonthlySales WITH ROLLUP
ORDER BY TransYear,
      TransQuarter,
      TransMonth,
      StoreNo,
      ProductNo,
          (
          CASE
                WHEN MonthlySales IS NULL THEN 0
          END
        ) DESC,
      GROUPING(MonthlySales) DESC
GO
```

We use the same CTE as the prior example, but the query has some complexity to it. We use the SUM() function to generate summaries for monthly sales and the GROUPING() function to create rollups by the columns included in the GROUP BY clause. The GROUPING() function appears right after the last column in the SELECT clause.

Why are we doing all this?

Specifically, we want to roll up sales summaries by year, quarter, month, store number, and product number. These represent the levels in our rollup hierarchy.

We need to include some clever code in the ORDER BY clause so that the display looks a little bit like the diagram we just discussed. Each time we roll up a level, a lot of NULLS appear in the rows for the values that no longer apply to the level in the hierarchy because we summarized the prior level's values. If you do not sort the results correctly, it will be impossible to understand.

What we need to do is introduce a CASE statement in the ORDER BY clause right after the ProductNo column that will tell the query what to do with the MonthlySales column if it is NULL:

```
ProductNo,
    (
    CASE
            WHEN MonthlySales IS NULL THEN 0
    END
    ) DESC,
```

If the value is NULL, then use 0 and sort in descending order. Notice the GROUPING() function at the end. For these values sort in descending order also. For all other columns in the ORDER BY clause, sort in ascending order. (Reminder: If you do not include the ASC keyword, then the default is ascending sort.)

Let's execute the query and see the results.

Please refer to Figure 2-12.

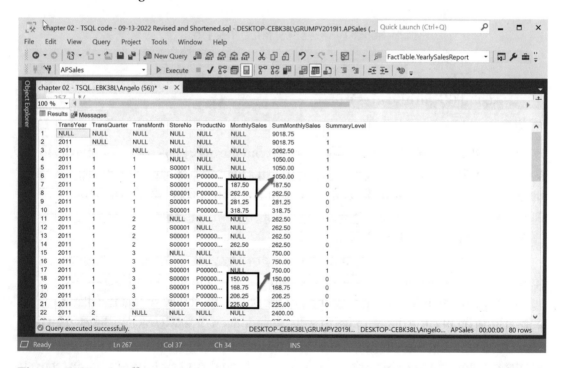

Figure 2-12. *A rollup report*

Not bad if I say so myself!

The results look like the ones in the conceptual diagram we discussed earlier (maybe I'm stretching it a bit).

As you navigate up the hierarchy, more NULLS appear. Also notice the SummaryLevel column. If the value is zero, then we are at a leaf row (or lowest-level node) in the hierarchy. If the value is 1, we are at a summary row somewhere in the hierarchy.

Checking out the two highlighted sets of rows, if we add $150.00, $168.75, $206.25, and $225.00, we get a total value of $750.00. If we examine the second group and add $187.50, $262.50, $281.25, and $318.75, we get $1050.00. This last value is the summary for the level group defined by year, quarter, month, store, and product.

This is only a partial screenshot, but the rollup pattern repeats for each set of rows in the hierarchy group.

Peeking at the $2062.50 total in row 3, this represents the rollup for the first quarter in the year. Row 2 contains the total for the year and row 1 the grand total. It is the same as the year total. If we had included other years, this total would of course be much larger.

Try this query on your own by downloading the code on the book's dedicated Google website and include two or three years' worth of data to see what the results look like.

Tip   With queries like this, always start off with small result sets by using the WHERE clause so you can make sure everything tallies up and of course sorts correctly so you can generate legible and understandable output. Once you are satisfied it works correctly, then you can remove the WHERE clause filters, so the original query specification is satisfied. (Test on small sets of data!)

GROUPING: Performance Tuning Considerations

Although Chapter 14 is dedicated to performance tuning, let me sneak in an example. In Figure 2-13 is the query plan for the query we just discussed.

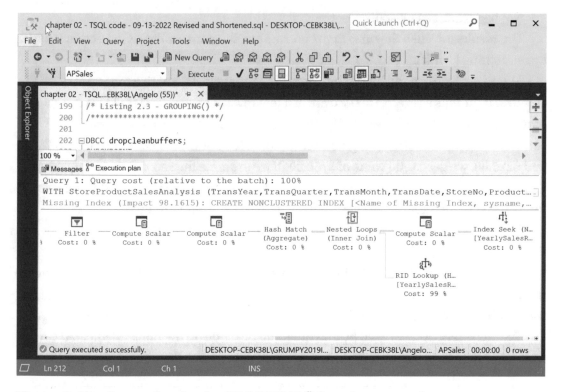

Figure 2-13. *Query plan for the GROUPING() function example*

Notice the index seek to the right of the plan. Even though we are using an index, the query plan estimator is suggesting we add a different index. The index seek has 0% cost, but the RID (row ID) lookup task cost 99%. The RID lookup is performed against the YearlySalesReport table based on the RID in the index. So this looks very expensive. The remaining tasks have 0% costs, so we will ignore these for now.

Let's copy the suggested index template by right-clicking it and selecting "Missing Index Details." It appears in a new query window in SSMS. All we really need to do is give the new index a name.

Please refer to Listing 2-6.

Listing 2-6. Suggested New Index

```
/*
Missing Index Details from chapter 02 - TSQL code - 09-13-2022.sql -
DESKTOP-CEBK38L\GRUMPY2019I1.APSales (DESKTOP-CEBK38L\Angelo (55))
The Query Processor estimates that implementing the following index could
improve the query cost by 98.1615%.
*/

/*
USE [APSales]
GO
CREATE NONCLUSTERED INDEX [<Name of Missing Index, sysname,>]
ON [SalesReports].[YearlySalesReport] ([ProductNo],[StoreNo])
INCLUDE ([CalendarDate],[TotalSalesAmount])
GO
*/

DROP INDEX IF EXISTS [ieProductNoStoreNoDateTotalSalesAmt]
ON [SalesReports].[YearlySalesReport]
GO

CREATE NONCLUSTERED INDEX [ieProductNoStoreNoDateTotalSalesAmt]
ON [SalesReports].[YearlySalesReport] ([ProductNo],[StoreNo])
INCLUDE ([CalendarDate],[TotalSalesAmount])
GO
```

Copy the suggested index template and paste it below the comments. All you need to do is give it a name: ieProductNoStoreNoDateTotalSalesAmt.

As we are learning about indexes and performance tuning, I decided to give verbose index names, so when you are working with them, you know what columns are involved. This is a non-clustered index. Notice how the date and sales amount columns are included right after the INCLUDE keyword. Let's create the index and generate a new query plan.

We need to split up the query plan as it is fairly large.

Please refer to Figure 2-14 for the first part.

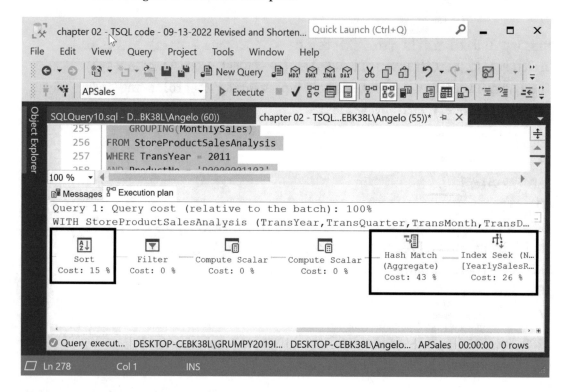

Figure 2-14. *Revised query plan, part A*

The RID lookup is gone, so is the nested loops join. Our index seek now has a 26% cost associated with it. There is also a sort step of the data stream at a cost of 15%. Let's look at the second part of the query plan. Please refer to Figure 2-15.

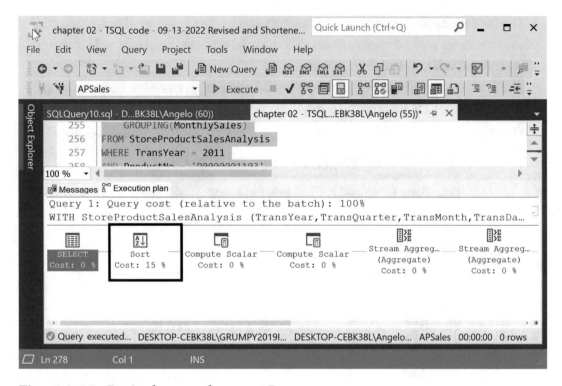

Figure 2-15. *Revised query plan, part B*

Some more compute scalar and stream aggregate steps appear but at a cost of 0%. We will ignore these. There is a second sort step with a cost of 15%. Is this better?

Let's rerun the query, but before we do, we need to clear the memory buffers and generate some IO statistics by executing the following commands:

```
DBCC dropcleanbuffers;
CHECKPOINT;
GO

-- turn set statistics io/time on

SET STATISTICS IO ON
GO

SET STATISTICS TIME ON
GO
```

The DBCC command will make sure the memory buffers are cleared out, so we start with a clean slate. If you do not perform this step, older query plans will be in memory cache, and you will get erroneous results.

The next two commands will generate our IO and timing statistics. Let's see what results we get when we execute the query.

Please refer to Table 2-3.

Table 2-3. *Comparing Statistics Between Plans*

SQL Server Parse and Compile Times	Existing Index	New Index
CPU time (ms)	15	16
Elapsed time (ms)	**4492**	**89**
Statistics (YearlySalesReport)	Existing Index	New Index
Scan count	1	1
Logical reads	**260**	**10**
Physical reads	1	1
Read-ahead reads	**51**	**7**
SQL Server Execution Times	Existing Index	New Index
CPU time (ms)	**83**	**74**
Elapsed time (ms)	0	0

Checking out the statistics, we see CPU time went up by 1 ms (millisecond), but elapsed time went down from 4492 ms to 89 ms. Scan count remained the same, but logical reads dropped to 10 from 260. This is good.

Physical reads remained the same at 1, and read-ahead reads dropped from 51 to 7. Finally, SQL Server execution CPU time dropped from 83 ms to 74 ms. Overall, it looks like this new index improved performance.

This was a quick example of the steps that you can take to analyze performance. As stated earlier, Chapter 14 will entirely be dedicated to performance tuning on selected queries we covered. Let's look at our next function, the STRING_AGG() function, to tie some strings together (pun intended!).

STRING_AGG Function

What does this do?

This function allows you to aggregate strings, just as the name implies. This comes in handy when you need to list items of interest, like when you want to show all stops in a flight itinerary or, in our case, items purchased on a specific date. Another interesting example is if you need to list stops in a logistics scenario, like all the stops a truck will make when delivering the chocolate candies and cakes sold by the stores in our examples.

This function is easy to use. You need to supply the name of the column and a delimiter, which is usually a comma.

Let's take a look at a simple example that shows what items customer C00000008 purchased. We want results by year and month.

Please refer to Listing 2-7.

Listing 2-7. Product Report Using STRING_AGG()

```
WITH CustomerPurchaseAnalysis(PurchaseYear,PurchaseMonth,CustomerNo,Product
No,PurchaseCount)
AS
(
SELECT DISTINCT
      YEAR(CalendarDate)  AS PurchaseYear,
      MONTH(CalendarDate) AS PurchaseMonth,
      CustomerNo,
      ProductNo,
      COUNT(*) AS PurchaseCount
FROM StagingTable.SalesTransaction
```

```
GROUP BY YEAR(CalendarDate),
     MONTH(CalendarDate),
     CustomerNo,
     ProductNo
)
SELECT
     PurchaseYear,
     PurchaseMonth,
     CustomerNo,
     STRING_AGG(ProductNo,',') AS ItemsPurchased,
     COUNT(PurchaseCount)     AS PurchaseCount
FROM CustomerPurchaseAnalysis
WHERE CustomerNo = 'C00000008'
GROUP BY
     PurchaseYear,
     PurchaseMonth,
     CustomerNo
ORDER BY CustomerNo,
     PurchaseYear,
     PurchaseMonth
GO
```

Notice how the STRING_AGG() function is used. The ProductNo column is passed together with a comma as the delimiter. A CTE is used so we can pull out the year and month from the CalendarDate column and count the number of purchases.

In this example, the count value will always be the value 1 in the CTE, so we need to count the occurrences to get the total number value of the purchases. The resulting purchase count and the aggregate product numbers should match up, as far as the number of items purchased.

Tip   Always test the query in the CTE to make sure it works as expected!

Let's take a look at the results in Figure 2-16.

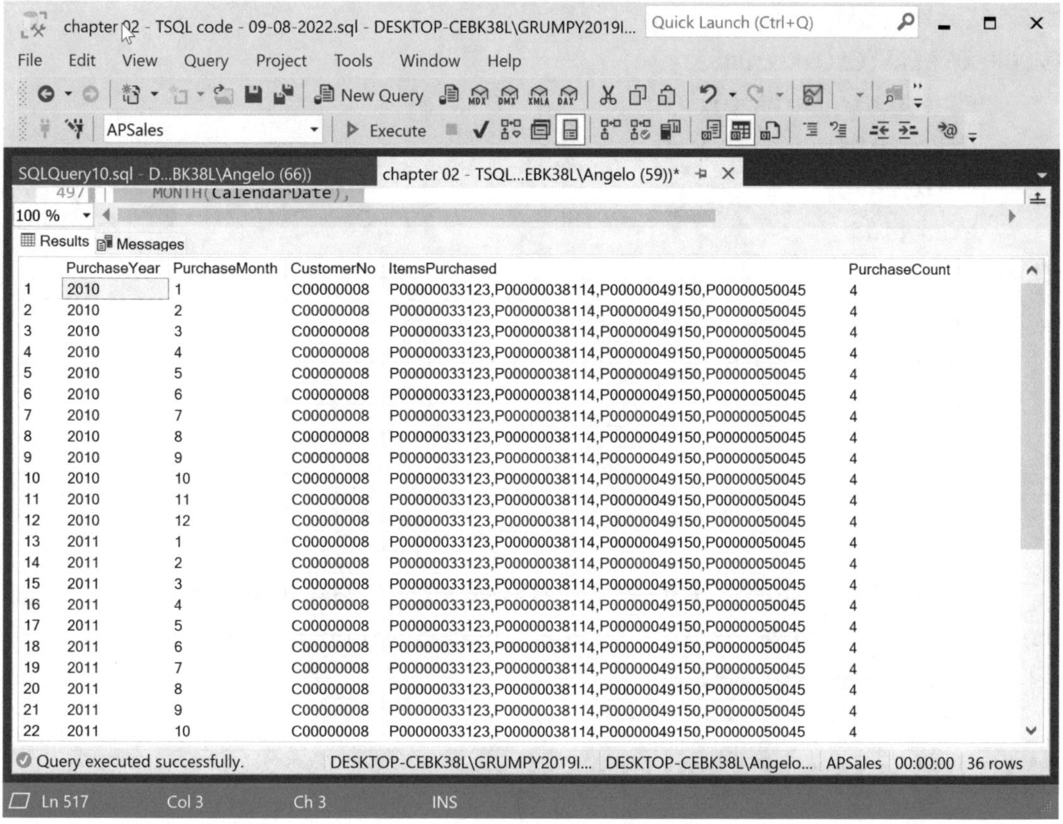

Figure 2-16. *Purchased product list report*

This looks correct. The `ItemsPurchased` column lists four items, and the `PurchaseCount` column has a value of 4. It seems this customer always buys the same items each month and the same quantity. When you like something, stick with it. What can I say?

You can validate the results by modifying the CTE query, so it filters by the customer:

```
SELECT DISTINCT
      YEAR(CalendarDate)  AS PurchaseYear,
      MONTH(CalendarDate) AS PurchaseMonth,
      CalendarDate,
      CustomerNo,
      ProductNo,
      COUNT(*) AS PurchaseCount
FROM StagingTable.SalesTransaction
```

```
WHERE CustomerNo = 'C00000008'
GROUP BY YEAR(CalendarDate),
      MONTH(CalendarDate),
      CalendarDate,
      CustomerNo,
      ProductNo
ORDER BY
      YEAR(CalendarDate),
      MONTH(CalendarDate),
      CalendarDate,
      CustomerNo,
      ProductNo
      GO
```

Something for you to try. To make things interesting, tweak the load script that loads the SalesTransaction table so that different products are bought each month.

It's time for some statistics.

STDEV( ) and STDEVP( ) Functions

The STDEV() function calculates the statistical standard deviation of a set of values when the entire population is not known or available.

The STDEVP() function calculates the statistical standard deviation when the entire population of the data set is known.

But what does this mean?

What this means is how close or how far apart the numbers are in the data set from the arithmetic **MEAN** (which is another word for average). In our context the average or **MEAN** is the sum of all data values in question divided by the number of data values.

There are other types of **MEAN** like geometric, harmonic, or weighted **MEAN**, which I will briefly cover in Appendix B.

So how is the standard deviation calculated?

You could use the following algorithm to calculate the standard deviation with TSQL:

- Calculate the average (**MEAN**) of all values in question and store it in a variable.

- For each value in the data set

- Subtract the calculated average.

- Square the result.

- Take the sum of all the results and divide the value by the number of data elements minus 1. (This step calculates the variance; to calculate for the entire population, do not subtract 1.)

- Now take the square root of the variance, and that's your standard deviation.

Or you could just use the STDEV() function! There is a STDEVP() function as mentioned earlier. As stated earlier, the only difference is that STDEV() works on part of the population of the data. This is when the entire data set is not known. The STDEVP() function works on the data set that represents the entire population.

Here is the syntax:

```
SELECT STDEV( ALL | DISTINCT ) AS [Column Alias]
FROM [Table Name]
GO
```

You need to supply a column name that contains the data you want to use. You can optionally supply the ALL or DISTINCT keyword if you wish. If you leave these out, just like the other functions, the behavior will default to ALL. Using the DISTINCT keyword will ignore duplicate values in the column.

Let's look at a TSQL query that uses this window function. We want to calculate the standard deviation plus the average by year, customer, store number, and product number.

As usual we start off with a CTE to do some preprocessing and return our data set.

Please refer to Listing 2-8.

Listing 2-8. Standard Deviation Sales Analysis

```
WITH CustomerPurchaseAnalysis
(PurchaseYear,PurchaseMonth,StoreNo,ProductNo,CustomerNo,TotalSalesAmount)
AS
(
SELECT
      YEAR(CalendarDate)  AS PurchaseYear,
      MONTH(CalendarDate) AS PurchaseMonth,
      StoreNo,
```

```
      ProductNo,
      CustomerNo,
      SUM(TransactionQuantity * UnitRetailPrice) AS TotalSalesAmount
FROM StagingTable.SalesTransaction
GROUP BY
YEAR(CalendarDate),MONTH(CalendarDate),ProductNo,CustomerNo,StoreNo
)

SELECT
      cpa.PurchaseYear,
      cpa.PurchaseMonth,
      cpa.StoreNo,
      cpa.ProductNo,
      c.CustomerNo,
      CONVERT(DECIMAL(10,2),cpa.TotalSalesAmount) AS TotalSalesAmount,
      AVG(CONVERT(DECIMAL(10,2),cpa.TotalSalesAmount)) OVER(
            --PARTITION BY cpa.PurchaseYear,c.CustomerNo
            ORDER BY cpa.PurchaseYear,c.CustomerNo
            ) AS AvgPurchaseCount,
      STDEV(CONVERT(DECIMAL(10,2),cpa.TotalSalesAmount)) OVER(
            ORDER BY cpa.PurchaseMonth
            ) AS StdevTotalSales,
      STDEVP(CONVERT(DECIMAL(10,2),cpa.TotalSalesAmount)) OVER(
            ORDER BY cpa.PurchaseMonth
            ) AS StdevpTotalSales,
      STDEV(CONVERT(DECIMAL(10,2),cpa.TotalSalesAmount)) OVER(
            ) AS StdevTotalSales,
      STDEVP(CONVERT(DECIMAL(10,2),cpa.TotalSalesAmount)) OVER(
            ) AS StdevpYearTotalSales
FROM CustomerPurchaseAnalysis cpa
JOIN DimTable.Customer c
ON cpa.CustomerNo = c.CustomerNo
WHERE cpa.CustomerNo = 'C00000008'
AND PurchaseYear = 2011
AND ProductNo = 'P00000038114';
GO
```

The CTE pulls out the year and month parts from the date. We also calculate the total sales amount by multiplying the transaction quantity by the unit retail price.

The average is calculated for the entire year as we wish to use that value in an Excel spreadsheet to calculate and graph the normal distribution bell curve. We will see it shortly.

Next, the standard deviation values are calculated by using the STDEV() and STDEVP() functions. We perform the calculations two different ways. The first pair uses an ORDER BY clause in the OVER() clause so that we get rolling monthly results. The second pair does not use a PARTITION BY or ORDER BY clause, so the default frame behavior kicks in: to process results, including all prior and following rows relative to the current row.

Let's see the results in Figure 2-17.

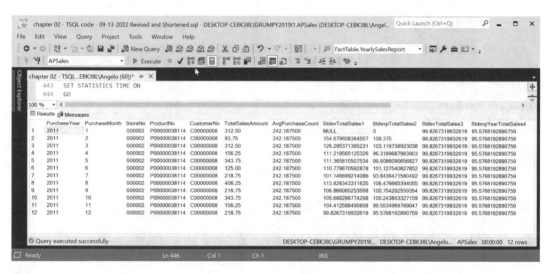

Figure 2-17. *Calculating standard deviation*

Short and sweet. Results for one customer, one year by month. Notice the NULL value for the first standard deviation results in column StdevTotalSales1. This happens because it is the first value in the partition, and it cannot calculate the standard deviation for one value, so NULL is the result. The STDEVP() function though is happy. It simply returns 0. For the remaining months, the rolling standard deviation is calculated month by month.

If you do not like the NULL, just use a CASE block that tests the value returned and substitute it with a zero!

Column `StdevTotalSales2` uses `STDEVP()`, and the values as can be seen are a bit lower since the entire population of the data set is used.

Finally, columns `StdevTotalSales3` and `StdevTotalSales4` calculate the standard deviation for the entire year. We will use the `STDEV()` year result in our Excel spreadsheet to generate normal distribution values and generate a graph called a bell curve based on each normal distribution value.

Let's check out the graph now in Figure 2-18.

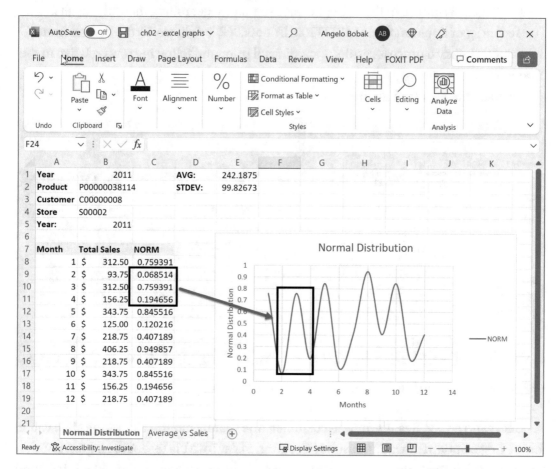

Figure 2-18. *Standard deviation bell curve(s)*

What we have here are a series of small bell curves. For example, notice the curves between months 2 and 4 and 4 and 6. In this case each curve is generated from three values, a low one, then a high one, and back to a low value. The curve generated for months 2–4 has values 0.068514, 0.759391, and 0.194656. These are great values for the bell curve.

To generate the normal distribution for each month, the Excel NORM.DIST function was used. This function uses these parameters: the value, the average (MEAN), and the standard deviation generated by our query. We then insert a suggested graph for these values.

STDEV: Performance Tuning Considerations

Let's do some performance analysis. By highlighting the query we just discussed and generating a query plan in one of the usual methods, we are presented with a plan that appears to already use an index.

Please refer to Figure 2-19.

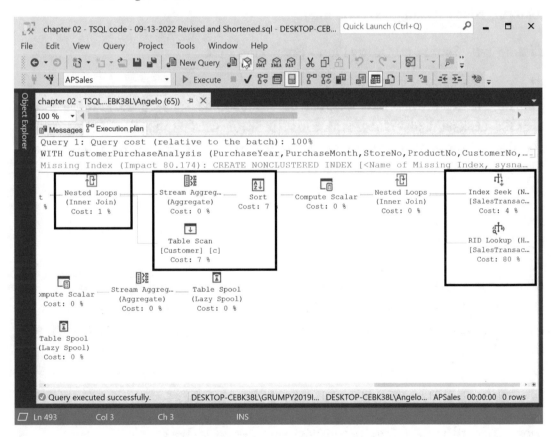

Figure 2-19. *First analysis query plan*

Starting from right to left, we see an index seek with a cost of 4% and a RID (row ID) lookup task with an estimated cost of 80%. This is where most of the work seems to be happening.

Ignoring the tasks with 0% cost, we see a table scan on the Customer table with a cost of 7% and a sort step with a cost of 7%. Finally, we see a nested loops inner join step with a cost of 1%.

This is a partial screenshot. I ignored the steps with zero cost so we can concentrate on the high-cost steps (in your queries, take all steps into account).

Notice that a suggested index was generated. Hold the mouse pointer over the index (it will be in green font) and right-click the index. When the pop-up menu appears, click "Missing Index Details" to generate the TSQL code in a new query window. The following code in Listing 2-9 is presented.

Listing 2-9. Estimated Query Plan – Suggested Index

```
/*
Missing Index Details from chapter 02 - TSQL code - 09-13-2022.sql -
DESKTOP-CEBK38L\GRUMPY2019I1.APSales (DESKTOP-CEBK38L\Angelo (65))
The Query Processor estimates that implementing the following index could
improve the query cost by 80.174%.
*/

/*
USE [APSales]
GO
CREATE NONCLUSTERED INDEX [<Name of Missing Index, sysname,>]
ON [StagingTable].[SalesTransaction] ([CustomerNo],[ProductNo])
INCLUDE ([StoreNo],[CalendarDate],[TransactionQuantity],[UnitRetailPrice])
GO
*/

/* Copy code from above and paste and supply name */

DROP INDEX IF EXISTS [CustNoProdNoStoreNoDateQtyPrice]
ON [StagingTable].[SalesTransaction]
GO
```

```
CREATE NONCLUSTERED INDEX [CustNoProdNoStoreNoDateQtyPrice]
ON [StagingTable].[SalesTransaction] ([CustomerNo],[ProductNo])
INCLUDE ([StoreNo],[CalendarDate],[TransactionQuantity],[UnitRetailPrice])
GO
```

All I needed to do was copy and paste the suggested index template and supply a name for the index. I used all the columns related to the index in the name, so it is clear what the index covers.

In a real production environment, you would most likely just give it a shorter name, but as we are in learning mode, I decided to be a bit verbose as far as the names are concerned. I also added a DROP INDEX command so that it is available to you when you start analyzing this query on your own.

Let's create the index, generate a new query plan, and then execute the query so we generate some statistics with the SET STATISTICS commands.

Please refer to Figure 2-20 for the new plan.

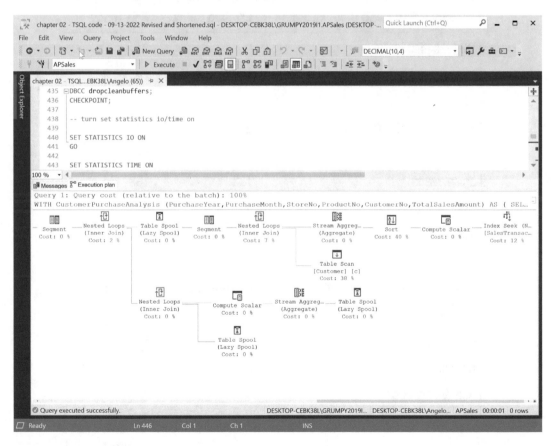

Figure 2-20. *Revised query plan*

Some definite improvements. Starting from right to left in the query plan, we now see an index seek on the SalesTransaction table at a cost of 12%. A fairly expensive sort step at 40% appears next. The Customer table uses a table scan at a cost of 38%, which cannot be helped because we did not create an index for this table. This table has only 100 rows, so an index would most likely not be used, but on your own, you can create one and experiment to see if there are any benefits.

Right in the middle, there is a nested loops inner join at 7% cost. Not too expensive. There is also a final nested loops inner join at 2%. For our discussion we will ignore the tasks at 0% cost.

So is this an improvement? Looks like it because of the index seek. The IO and TIME statistics will give us a clearer picture though, so let's look at them.

Please refer to Table 2-4.

Table 2-4. *Comparing IO Statistics*

SQL Server Parse and Compile Times	Existing Index	New Index
CPU time (ms)	0	31
Elapsed time (ms)	11	79
Statistics (Work Table)	**Existing Index**	**New Index**
Scan count	18	18
Logical reads	133	133
Physical Reads	0	0
Read-ahead reads	0	0
Statistics (SalesTransaction)	**Existing Index**	**New Index**
Scan count	1	1
Logical reads	**43**	**5**
Physical Reads	0	0
Read-ahead reads	0	0

(continued)

Table 2-4. (*continued*)

Statistics (Customer)	Existing Index	New Index
Scan count	1	1
Logical reads	216	216
Physical reads	0	0
Read-ahead reads	0	0

SQL Server Execution Times	Existing Index	New Index
CPU time (ms)	16	0
Elapsed time (ms)	4	30

This table shows statistics before and after the index was created.

Some interesting results, I think. Starting with SQL Server parse and compile times, time statistics actually went up with the new index, not down.

Statistics for the work table remained the same, while statistics for the SalesTransaction table went down significantly. Logical reads went from 43 to 5, which is very good. Physical reads and read-ahead reads remained at 0.

Finally, at the end of processing, for SQL Server execution times, CPU time went from 16 to zero, so that's a significant improvement, while elapsed time increased from 4 to 30 ms.

In conclusion, adding the index reduced logical reads, but elapsed time for parsing and execution increased.

The SalesTransaction table has about 24,886 rows, which is not a lot. Real performance improvements or changes would be noticeable if we had a couple of million rows.

Table 2-5 represents before and after statistics after I loaded about 1.75 million rows by removing any WHERE clause filters in the load script. (This is available in the load script for this chapter on the publisher's Google site.)

Table 2-5. *Comparing Statistics on 2 Million Rows*

SQL Server Parse and Compile Times	Existing Index	New Index
CPU time (ms)	**32**	**0**
Elapsed time (ms)	**62**	**0**

Statistics (Work Table)	Existing Index	New Index
Scan count	18	18
Logical reads	133	134
Physical reads	0	0
Read-ahead reads	0	0

Statistics (SalesTransaction)	Existing Index	New Index
Scan count	**5**	**1**
Logical reads	**41840**	**76**
Physical reads	0	3
Read-ahead reads	**38135**	**71**

Statistics (Customer)	Existing Index	New Index
Scan count	1	1
Logical reads	18	18
Physical reads	0	0
Read-ahead reads	0	0

SQL Server Execution Times	Existing Index	New Index
CPU time (ms)	485	0
Elapsed time (ms)	**2926**	**48**

Now we really have some interesting statistics. For the SQL Server parse and compile time step, CPU and elapsed time statistics went down from 32 and 62, respectively, to 0.

Statistics for the work table remained the same.

For the SalesTransaction we really see significant improvement in the logical read category, from 41,840 reads before the index was created to 76. Physical reads went up a bit from 0 to 3, while read-ahead reads went down from 38,135 to 71.

Statistics for the Customer table remained the same, both before the index was created and after the index was created. (This makes sense as the index was created on a different table.)

Wrapping it up with the SQL Server execution times, the CPU time went from 485 ms to 0 ms, and the elapsed execution time went from 2926 ms to 48 ms.

Looks like we really need a significant amount of data rows to work with to see how indexes improve or in some case decrease performance.

One last comment, the statistics generated will vary from time to time as you keep performing analysis on your queries, so make sure you use the DBCC command to clear all memory buffers.

Also, run the DBCC and SET STATISTICS ON steps separately from the query; otherwise, if you run them together, you will get statistics for both these little steps and the query being evaluated. This could be confusing.

VAR() and VARP() Functions

The VAR() and VARP() functions are used to generate the variance for a set of values in a data sample. The same discussion related to the data sets used in the STDEV()/STDEVP() examples applies to the VAR() and VARP() functions relative to population of the data.

The VAR() function works on a partial set of the data when the entire data set is not known or available, while the VARP() function works on the whole data set population when it is known.

So, if your data set consists of ten rows, these function names ending with P will include all ten rows when processing the values; the ones not ending in the letter P will look at N – 1 or 10 – 1 = 9 rows.

Okay, so what does *variance* mean anyway?

Informally, if you take all the differences of each value from the mean (average) and square the results and then add all the results and finally divide them by the number of data samples, you get the variance.

For our scenario, let's say we had some data that forecasted sales amounts for one of our products during any particular year. When sales data was finally generated for the product, we wanted to see how close or far the sales figures were from the forecast. It answers the questions:

- Did we miss our targets?

- Did we meet our targets?

- Did we exceed our targets?

This statistical formula is also discussed in Appendix B by the way. We will develop TSQL code to duplicate the STDEV() and VAR() functions. Let's see if they get the same results!

Let's check out our query that uses these functions in Listing 2-10.

Listing 2-10. Calculating Sales Variance

```
WITH CustomerPurchaseAnalysis
(PurchaseYear,PurchaseMonth,StoreNo,ProductNo,CustomerNo,TotalSalesAmount)
AS
(
SELECT
      YEAR(CalendarDate)  AS PurchaseYear,
      MONTH(CalendarDate) AS PurchaseMonth,
      StoreNo,
      ProductNo,
      CustomerNo,
      SUM(TransactionQuantity * UnitRetailPrice) AS TotalSalesAmount
FROM StagingTable.SalesTransaction
GROUP BY YEAR(CalendarDate),MONTH(CalendarDate),
      ProductNo,CustomerNo,StoreNo
)
SELECT
      cpa.PurchaseYear,
      cpa.PurchaseMonth,
      cpa.StoreNo,
      cpa.ProductNo,
```

```
      c.CustomerNo,
      c.CustomerFullName,
      CONVERT(DECIMAL(10,2),cpa.TotalSalesAmount) AS TotalSalesAmount,
      AVG(CONVERT(DECIMAL(10,2),cpa.TotalSalesAmount)) OVER(
            ORDER BY cpa.PurchaseMonth) AS AvgPurchaseCount,
      VAR(CONVERT(DECIMAL(10,2),cpa.TotalSalesAmount)) OVER(
            ORDER BY cpa.PurchaseMonth
            ) AS VarTotalSales,
      VARP(CONVERT(DECIMAL(10,2),cpa.TotalSalesAmount)) OVER(
            ORDER BY cpa.PurchaseMonth
            ) AS VarpTotalSales,
      VAR(CONVERT(DECIMAL(10,2),cpa.TotalSalesAmount)) OVER(
            ) AS VarTotalSales,
      VARP(CONVERT(DECIMAL(10,2),cpa.TotalSalesAmount)) OVER(
            ) AS VarpYearTotalSales
FROM CustomerPurchaseAnalysis cpa
JOIN DimTable.Customer c
ON cpa.CustomerNo = c.CustomerNo
WHERE cpa.CustomerNo = 'C00000008'
AND PurchaseYear = 2011
AND ProductNo = 'P00000038114';
GO
```

Basically I did a copy and paste of the standard deviation query and just replaced the STDEV( ) functions with the VAR( ) functions. The usual CTE is used. I also commented out the Product and Customer columns as they are for one customer and one product as can be identified in the WHERE clause. I did this as the results had too many columns and they did not display well.

When you test out these queries, please feel free to uncomment them and get results for more than one customer. Also display the customers and product.

Let's see the results in Figure 2-21.

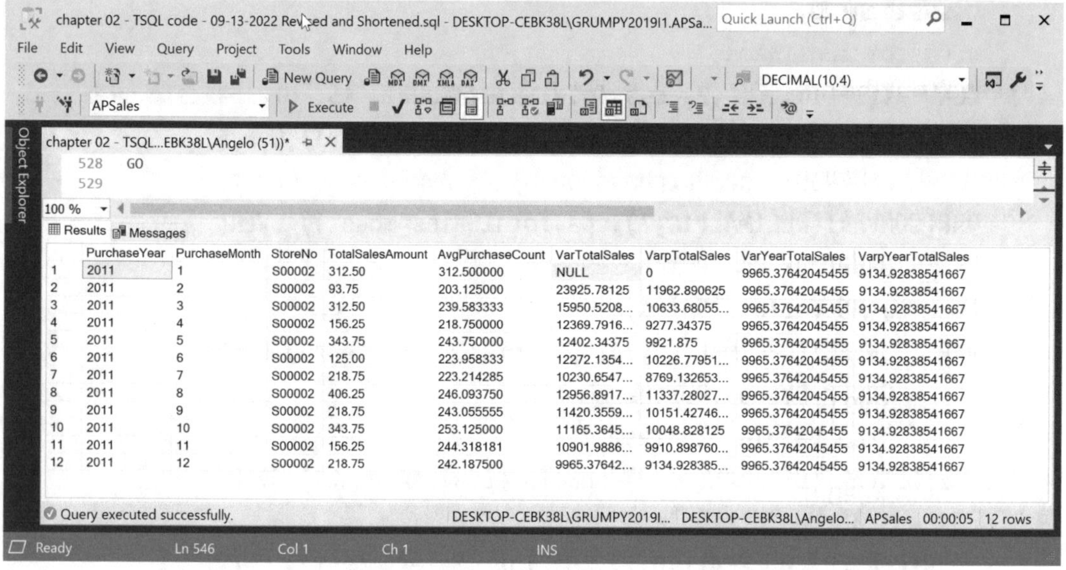

Figure 2-21. *Comparing variance vs. variance for the whole population*

The average and the first two variance columns are rolling monthly values as can be seen by the different values for each month. These only have an `ORDER BY` clause, so the default window frame behavior is

`RANGE BETWEEN UNBOUNDED PRECEDING AND CURRENT ROW`

The last two columns do not specify either a `PARTITION BY` or an `ORDER BY` clause, so the default behavior is

`ROWS BETWEEN UNBOUNDED PRECEDING AND UNBOUNDED FOLLOWING`

If you have not done so already, download the code for this chapter and try these examples out. For this example, uncomment the customer- and product-related columns. When you are comfortable with the code, also try out the query for multiple years and then multiple customers and products. Lastly, try playing around with the `RANGE` and `ROWS` clauses to see how they affect window frame behavior.

One comment on report generation: What you do not want to do is generate a report that has so much data that it cannot be read. You can create a query that SSRS (SQL Server Reporting Services) can use that includes data for all customers, products, and years. All you need to do is add some filters in the SSRS report so the user can pick and choose what combinations of the filters they want to view from a drop-down list.

SQL Server 2022: Named Window Example

Let's check out the new WINDOW feature in SQL Server 2022 by creating a query that uses the AVG() function with the OVER() clause. We will gather the IO and TIME statistics plus generate a query plan in the usual manner to see how this new feature affects performance.

If the query plan suggests an index, which it will as there are none, we will create the index and then run the query again and look at the new plan and statistics.

Note   I ran this on an evaluation copy of SQL Server 2022. Feel free to download and install it. Also do some research on the other new features of this version. Very important to keep up with the changes of the latest editions.

Please refer to Listing 2-11 for the query code.

Listing 2-11. Average by Year, Month, and Customer

```
WITH CustomerPurchaseAnalysis
(PurchaseYear,PurchaseMonth,CustomerNo,TotalSalesAmount)
AS
(
SELECT
     YEAR(CalendarDate)  AS PurchaseYear,
     MONTH(CalendarDate) AS PurchaseMonth,
     CustomerNo,
     SUM(TransactionQuantity * UnitRetailPrice) AS TotalSalesAmount
FROM StagingTable.SalesTransaction
GROUP BY YEAR(CalendarDate),MONTH(CalendarDate),CustomerNo
)
SELECT
     cpa.PurchaseYear,
     cpa.PurchaseMonth,
     c.CustomerNo,
     c.CustomerFullName,
     cpa.TotalSalesAmount,
```

```
        AVG(cpa.TotalSalesAmount) OVER SalesWindow AS AvgTotalSales
FROM CustomerPurchaseAnalysis cpa
JOIN DimTable.Customer c
ON cpa.CustomerNo = c.CustomerNo
WHERE cpa.CustomerNo = 'C00000008'
WINDOW SalesWindow AS (
            PARTITION BY cpa.PurchaseYear
            ORDER BY cpa.PurchaseYear ASC,cpa.PurchaseMonth ASC
            )
GO
```

As usual we use our CTE code structure to set things up. No changes to the CTE syntax of course. What we want to do is calculate the total sales amount by year, month, and customer. We use the SUM() function in the CTE so we need to include a GROUP BY clause.

The query refers to the CTE and calculates the average of all the sums. This time the OVER() clause simply refers to a name of the block that defines the PARTITION and ORDER BY clauses. This is called SalesWindow and is declared at the end of the query.

The new WINDOW command is used followed by the name SalesWindow, and then between parentheses the PARTITION BY and ORDER BY clauses are declared:

```
WINDOW SalesWindow AS (
            PARTITION BY cpa.PurchaseYear
            ORDER BY cpa.PurchaseYear ASC,cpa.PurchaseMonth ASC
            )
```

Executing the query, we are presented with the results in Figure 2-22.

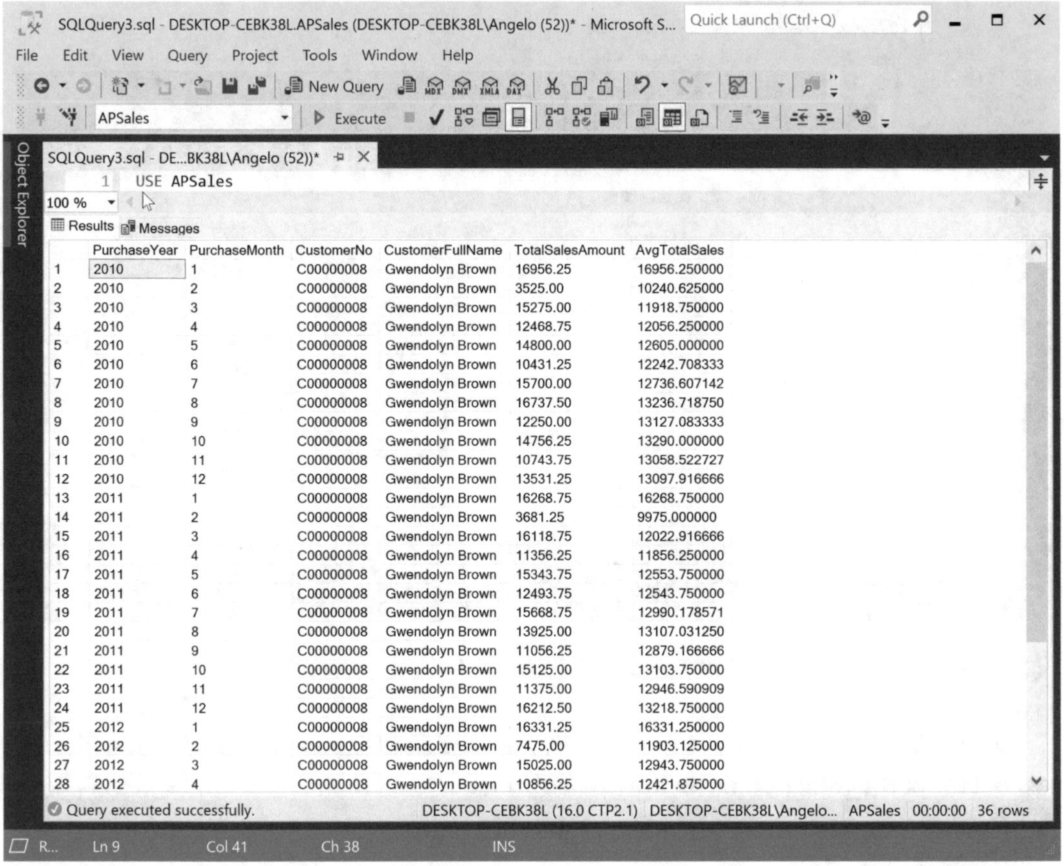

Figure 2-22. *Query results*

This query generated 36 rows. Large enough to evaluate the results. No surprises here. Works as expected.

The rolling averages seem to work. Let's generate a query plan and see if there are any suggestions for indexes or more importantly if this new feature generated more steps or less steps in the query plan.

Please refer to Figure 2-23.

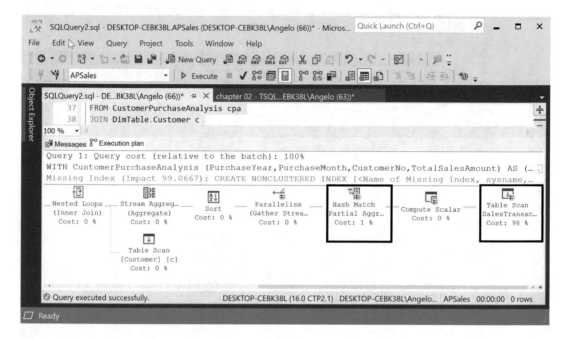

Figure 2-23.   *Pre-index query plan*

As expected, since there are no indexes, we see a nice expensive table scan with a cost of 98% of the total execution time. We also see a hash match task with a small cost of 1%. There is a suggestion for an index. Right-click it and extract the index to a new query window in the usual manner.

Please refer to Listing 2-12 for what we get.

Listing 2-12.   Suggested Index

```
/*
Missing Index Details from SQLQuery2.sql - DESKTOP-CEBK38L.APSales
(DESKTOP-CEBK38L\Angelo (66))
The Query Processor estimates that implementing the following index could
improve the query cost by 99.0667%.
*/

/*
USE [APSales]
GO
CREATE NONCLUSTERED INDEX [<Name of Missing Index, sysname,>]
```

```
ON [StagingTable].[SalesTransaction] ([CustomerNo])
INCLUDE ([CalendarDate],[TransactionQuantity],[UnitRetailPrice])
GO
*/

DROP INDEX IF EXISTS [CustomerNoieDateQuantityRetailPrice]
ON [StagingTable].[SalesTransaction]
GO

CREATE NONCLUSTERED INDEX [CustomerNoieDateQuantityRetailPrice]
ON [StagingTable].[SalesTransaction] ([CustomerNo])
INCLUDE ([CalendarDate],[TransactionQuantity],[UnitRetailPrice])
GO
```

A little copy-and-paste action so we can give the index a name. I left the square brackets alone as they were generated by SQL Server (and I am too lazy to remove them). Let's create the index and rerun statistics and generate a new query plan.

Please refer to Figure 2-24.

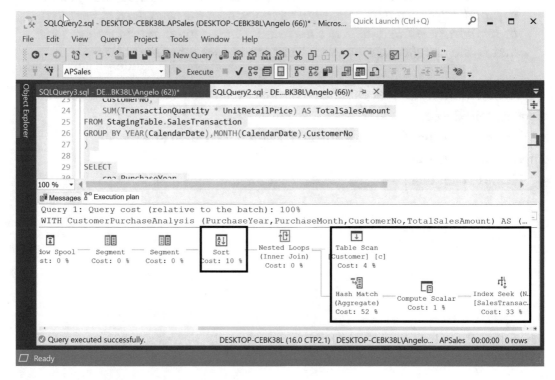

Figure 2-24. *Post-index query plan*

We see a bunch of steps, but at least there is an index seek now on the SalesTransaction table. This task costs 33% of the total execution time. There's a small compute scalar task at 1% and a hash match at a cost of 52%. Finally, there's a sort task that costs 10% (of the total execution time).

There is also a window spool task with a cost of 0%, which is important. This task is associated with the OVER() clause and ROWS and RANGE clauses as it loads the required rows in memory so they can be retrieved as often as required. If we do not have enough memory, then the rows need to be stored in temporary storage, which could slow things down considerably. This task is one you should always look out for and see what the cost is!

As usual I am ignoring the tasks with 0% cost (except for the window spool) to keep things simple, but on your own, look them up in the Microsoft documentation so you know what they do.

This is important. Remember what the window spool does:

> *This task is associated with the OVER() clause and ROWS and RANGE clauses as it loads the required rows in memory so they can be retrieved as often as required. If we do not have enough memory, then the rows need to be stored in temporary storage, which could slow things down considerably. This task is one you should always look out for and see what the cost is!*

Seems like the index seek eliminated the first expensive table scan, but several new steps are introduced.

Let's compare the IO statistics pre- and post-index creation to see what improvements there are.

Please refer to Table 2-6.

Table 2-6. *IO Statistics Comparison, Before and After Index*

SQL Server Parse and Compile Times	No Index	Index
CPU time (ms)	31	0
Elapsed time (ms)	290	49

Statistics (Work Table)	No Index	Index
Scan count	39	39
Logical reads	217	217
Physical reads	0	0
Read-ahead reads	0	0

Statistics (Work File)	No Index	Index
Scan count	0	0
Logical reads	0	0
Physical reads	0	0
Read-ahead reads	0	0

Statistics (SalesTransaction)	No Index	Index
Scan count	5	1
Logical reads	17244	42
Physical reads	0	1
Read-ahead reads	17230	10

(continued)

101

Table 2-6. (*continued*)

Statistics (Customer)	No Index	Index
Scan count	**1**	**1**
Logical reads	**648**	**18**
Physical reads	1	1
Read-ahead reads	0	0
SQL Server Execution Times	**No Index**	**Index**
CPU time (ms)	156	0
Elapsed time (ms)	1210	105

Looking through the statistics, there seems to be significant improvement. What we are looking for is how significant the improvements are. Does the new WINDOW feature cause execution to be more expensive, faster, or the same?

Statistics for the work table and the work file remained the same, while statistics for the SalesTransaction table have improved significantly. Our logical reads went from 17,244 to 42 and read-ahead reads went down from 17,230 to 10. This is great!

Customer table logical reads went from 648 to 18, and our CPU time went from 156 to 0 ms. Lastly, elapsed execution time went from 1210 to 105 ms.

In conclusion, the good news is that the new WINDOW feature does not decrease performance. Improvements in the query plans are similar to the prior OVER() clause syntax.

Seems like the benefit is that we can place the OVER() clause logic in one place and call it from multiple columns, eliminating redundant typing if the window frame specifications are the same.

There's more. You can define more than one window and even have a window reference another window definition as illustrated by the partial code in Listing 2-13.

Listing 2-13. Defining Multiple Windows

```
SELECT
     cpa.PurchaseYear,
     cpa.PurchaseMonth,
     c.CustomerNo,
     c.CustomerFullName,
     cpa.TotalSalesAmount,

     AVG(cpa.TotalSalesAmount) OVER AvgSalesWindow AS AvgTotalSales,
     STDEV(cpa.TotalSalesAmount) OVER StdevSalesWindow AS StdevTotalSales,
     SUM(cpa.TotalSalesAmount) OVER SumSalesWindow AS SumTotalSales
FROM CustomerPurchaseAnalysis cpa
JOIN DimTable.Customer c
ON cpa.CustomerNo = c.CustomerNo
WHERE cpa.CustomerNo = 'C00000008'
WINDOW
     StdevSalesWindow AS (AvgSalesWindow),
     AvgSalesWindow AS (
          PARTITION BY cpa.PurchaseYear
          ORDER BY cpa.PurchaseYear ASC,cpa.PurchaseMonth ASC
          ),
       SumSalesWindow AS (
          );
GO
```

Interesting feature. Let's wrap up the chapter. Feel free to go back and review some areas you might be fuzzy on.

Summary

Did we meet our objectives?

We introduced the APSales data warehouse and studied the conceptual model for this database.

We also became familiar with the tables, columns, and relationships between the tables by introducing some simple data dictionaries.

We've covered the window functions in the aggregate function category by creating examples that used the OVER() clause to define window frames for the data sets generated by the queries.

One important objective was that we needed to understand performance analysis and tuning associated with queries that use the window functions and OVER() clauses. This was accomplished by generating and studying query plans and IO statistics. This information was used to arrive at conclusions on how to improve performance.

We also identified the window spool statistics as an important value to monitor. Make sure you understand what this means.

Last but not least, we took a peek at the new WINDOW feature in SQL Server 2022 that shows a lot of promise in the areas of code readability and implementation. Our quick analysis showed that it did not degrade performance but also it did not improve performance over the old syntax, but further testing needs to be done with larger data sets.

By the way, now is a good time, if you have not done so already, to download the example script for this chapter and practice what you just learned.

In our next chapter, we cover the analytical functions, which are a very powerful set of tools used in data analysis and reporting. We will also start adding ROWS and RANGE clauses to the queries we discuss that override the default behavior of window frames when we do not specify them.

CHAPTER 3

Sales Use Case: Analytical Functions

Now that we have created and familiarized ourselves with the sales data warehouse and created queries that use aggregate window functions, we can move to our next step, learning and creating queries using the analytical functions available with SQL Server.

In the last chapter, we touched upon performance tuning the queries. We will use the same approach in this chapter, but we will look at other performance tools such as client statistics to give us a more complete picture of how the queries run.

We will make heavy use of execution statistics, that is, we compare statistics before recommended indexes are created and after, to see if they improve performance or not.

Lastly, we will look at setting up preloaded reporting tables and memory-optimized tables to see if they can yield higher query performance.

The functions we are studying are processing intensive, so we want to make sure we write efficient, accurate, and fast-executing queries and scripts. Query tuning, index creation, and performance analysis are all part of the process for delivering useful and actionable data to your business analysts.

Analytical Functions

The following are the eight functions for this category sorted by name. They can be used in queries together with the OVER() clause to generate valuable reports for our business analysts:

- CUME_DIST()
- FIRST_VALUE()
- LAST_VALUE()

© Angelo Bobak 2023
A. Bobak, *SQL Server Analytical Toolkit*, https://doi.org/10.1007/978-1-4842-8667-8_3

- LAG()

- LEAD()

- PERCENT_RANK()

- PERCENTILE_CONT()

- PERCENTILE_DISC()

The same approach as the prior chapter will be taken. I describe what the function does and present the code and the results. Next, some performance analysis is performed so we can see what improvements, if any, are required.

As we saw in the examples in Chapter 2, improvements can be adding indexes or even creating a denormalized report table so that when the query using the analytical window functions is executed, it goes against preloaded data avoiding joins, calculations, and complex logic.

CUME_DIST( ) Function

This function calculates the relative position of a value within a data set like a table, partition, or table variable loaded with test data.

How does it work?

First, calculate how many values come before it or are equal to it (call this value C1). Next, calculate the number of values or rows in the data set (call this C2). C1 is then divided by C2 to deliver the cumulative distribution result. The values returned are a float data type, so you need to use the FORMAT( ) function to convert it to a percentage.

By the way, this function is similar to the PERCENT_RANK( ) function, which works as follows: Given a set of values in a data set, this function calculates the rank of each individual value relative to the entire data set. This function also returns a percentage.

Back to the CUME_DIST( ) function. We will look at two examples, an easy one and then one that will query our sales data warehouse.

Let's look at a simple example first so we can see how the prior description works on a small data set. Please refer to Listing 3-1a.

Listing 3-1a. A Simple Example

```
USE TEST
GO

DECLARE @CumDistDemo TABLE (
     Col1 VARCHAR(8),
     ColValue INTEGER
     );

INSERT INTO @CumDistDemo VALUES
('AAA',1),
('BBB',2),
('CCC',3),
('DDD',4),
('EEE',5),
('FFF',6),
('GGG',7),
('HHH',8),
('III',9),
('JJJ',10)

SELECT Col1,ColValue,
     CUME_DIST() OVER(
          ORDER BY ColValue
     ) AS CumeDistValue,
     A.RowCountLE,
     B.TotalRows,
     CONVERT(DECIMAL(10,2),A.RowCountLE)
          / CONVERT(DECIMAL(10,2),B.TotalRows) AS MyCumeDist
FROM @CumDistDemo CDD
CROSS APPLY (
     SELECT COUNT(*) AS RowCountLE FROM @CumDistDemo
     WHERE ColValue <= CDD.ColValue
     ) A
```

```
CROSS APPLY (
     SELECT COUNT(*) AS TotalRows FROM @CumDistDemo
     ) B
GO
```

A simple table variable called @CumeDistDemo is declared and loaded with ten rows of data. Only two columns are created. The first is called Col1 and acts as a key column, while the ColValue column holds the data. In our case these are the numbers 1–10. Really simple.

The query utilizes the CUME_DIST() function and an OVER( ) clause. No PARTITION BY clause is included, but the ORDER BY clause is used to order the partition rows by the ColValue column in ascending order.

We need to include the formula for calculating cumulative distribution so we can see how it works and compare it to the SQL Server version. Here is the formula:

Count of rows with values less than or equal to the current column value divided by the total number of rows

To get the values for this formula, two CROSS APPLY operators are used. The first one calculates the number of rows with values less than or equal to the current column value, and the second calculates the total number of rows in the data set.

This query can probably be optimized, but for our purposes, it should illustrate how the CUME_DIST() function works.

The values returned by the queries in the CROSS APPLY block are used in the formula in the SELECT clause:

```
CONVERT(DECIMAL(10,2),A.RowCountLE)
        / CONVERT(DECIMAL(10,2),B.TotalRows) AS MyCumeDist
```

Let's execute the query and see what we get. Please refer to Figure 3-1.

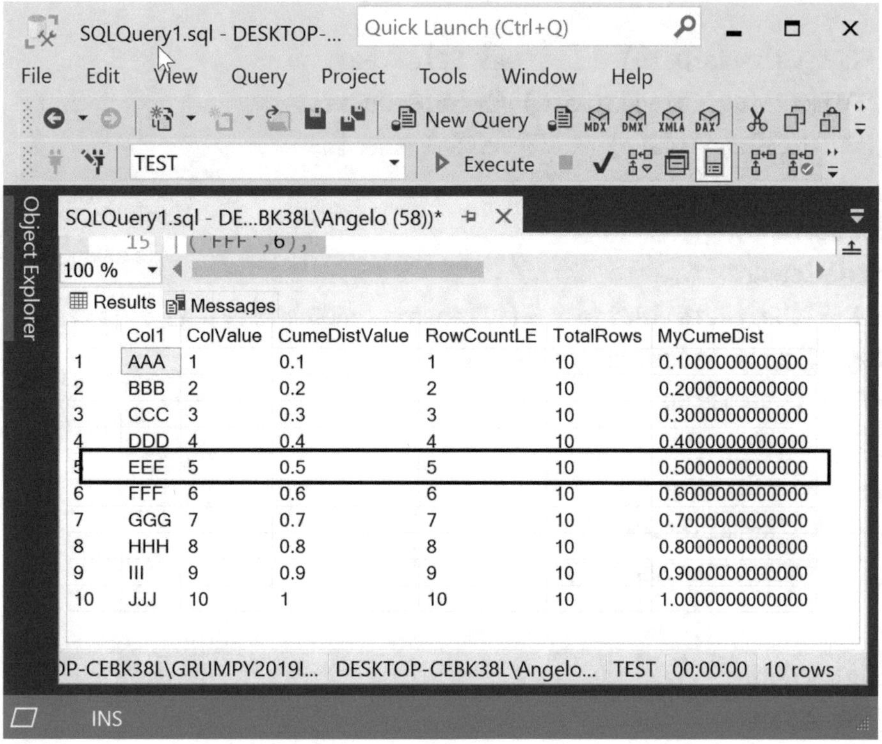

Figure 3-1. *Two ways of calculating cumulative distribution*

Looks good and predictable, I think! The MyCumeDist column has the values calculated by the formula we created, and the column called CumeDistValue contains the values the SQL Server function calculated. Maybe a little formatting is needed so they look the same, but the values match.

Now that we understand how this function works, let's create a query that goes against our sales data. We will use the strategy we used in the prior chapter. That is, we will use a CTE to set up the data, and then the functions we are learning are in the query that accesses the CTE.

Please refer to Listing 3-1b.

Listing 3-1b. The CUME_DIST() Function in Action

```
WITH CustSales (
SalesYear,SalesQuarter,SalesMonth,CustomerNo,StoreNo,CalendarDate,
SalesTotal
)
AS
```

```
(
SELECT YEAR(CalendarDate)         AS SalesYear
      ,DATEPART(qq,CalendarDate) AS SalesQuarter
      ,MONTH(CalendarDate)        AS SalesMonth
      ,ST.CustomerNo,
      ST.StoreNo,
      ST.CalendarDate,
      SUM(ST.UnitRetailPrice * ST.TransactionQuantity) AS SalesTotal
FROM StagingTable.SalesTransaction ST
GROUP BY ST.CustomerNo
      ,ST.StoreNo
      ,ST.CalendarDate
      ,ST.UnitRetailPrice
      ,ST.TransactionQuantity
)

SELECT SalesYear
      ,SalesQuarter
      ,SalesMonth
      ,CustomerNo
      ,SUM(SalesTotal) AS MonthlySalesTotal
      ,CUME_DIST() OVER (
            PARTITION BY SalesYear
            ORDER BY SUM(SalesTotal)
      ) AS CumeDist
FROM CustSales
WHERE SalesYear IN(2010,2011)
AND CustomerNo = 'C00000001'
GROUP BY SalesYear
      ,SalesQuarter
      ,SalesMonth
      ,CustomerNo
GO
```

Our usual CTE is set up to pull transaction data from the SalesTransaction table so we can pull out the individual date objects: year, quarter, and month. The sales total is also calculated by summing the results of multiplying the UnitRetailPrice by the

TransactionQuantity column. (Not efficient I have to admit. I could have used the TotalSalesAmount column and avoided the formula. Oh well.)

The CUME_DIST() function uses an OVER() clause that includes a PARTITION BY clause and an ORDER BY clause. A partition is set up over the SalesYear column, and the row set of the partition is sorted by the SalesTotal column.

Results are summed up again because we eliminated the store number and only want to see totals by year, quarter, month, and customer. A WHERE clause is included to filter out the results for only one customer and for only two years. I did this to keep the result set small, but you can remove them to see what you get.

Please refer to the results in Figure 3-2.

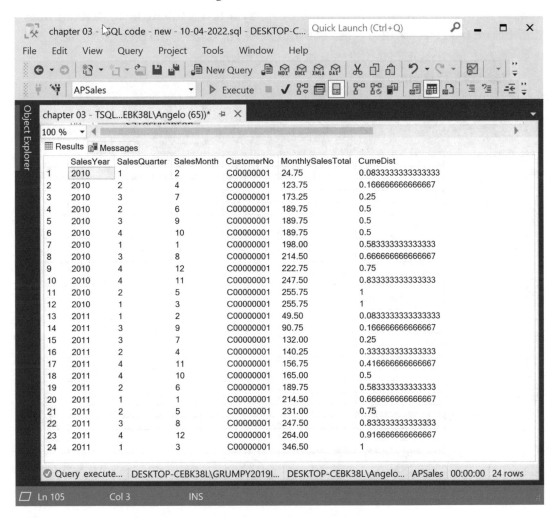

Figure 3-2. *Query results for the test script*

The results were sorted by the SalesTotal column in the OVER() clause in order to see the ascending values of the sales totals and cumulative distribution results so you can see how the formula was applied. Sorting by the sales amounts in ascending order lines things up nicely.

Try out the example by sorting results by sales year, quarter, and month, which is what a business analyst would normally want to see.

Performance Considerations

Let's generate an estimated query execution plan in the usual manner. The results are long, so we only look at the right half of the display as this is where the most interesting statistics for the tasks are displayed. The other tasks or steps in the plan are mostly 0% cost, so we ignore them in the interest of space constraints.

Please refer to Figure 3-3.

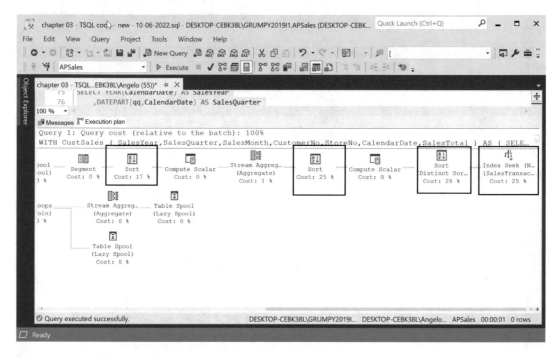

Figure 3-3. *Estimated query plan for the script*

Yikes! Look at the three sort steps. At least there is an index seek, and no other indexes were suggested by the estimated performance tool.

The index seek task costs 25% of the total execution plan, and the sort step next to it costs 28%. A second sort step at 25% can be seen, so a lot of sorting seems to occur with this query. A third sort step costs 17%!

This is an indication that we could improve performance by staging the initial data generated by the CTE in some sort of report or staging table and building indexes to assist when the main query is executed. A large benefit would be that you load this table once a day, and assuming that users are happy with 24-hour data refresh, the processing of the loading is not included when the users execute the main query that has the window function.

By the way, if you position your mouse over each sort step, you will see that the following columns are sorted:

- First sort step: Store and calendar date

- Second sort step: Sort by customer

- Third sort step: Sort by customer again

Now we are second-guessing the query plan estimator. On your own, experiment by creating one or two indexes to support the required sorting based on the preceding information. See if the sort steps are eliminated or at least reduced. Keep in mind that the more indexes you create on a table, the slower performance will be when loading or editing the table rows.

I have included a script based on the CTE logic in the listing that stages the data in a table. An index is also created to support the sorting (the index is suggested by the estimated query plan tool). The results are that the index seek costs a little less, but the sorting still exists and is a bit higher.

Where the improvement appears is in the query statistics. In the first query, when accessing the sales transaction table, the logical reads are 23, but this is reduced to 10 when using the staging table. The physical reads in the first query are 1, while they are 0 in the query accessing the staging table.

This strategy seems to work. Once again, try it on your own before looking at my example.

Please refer to Table 3-1 for some statistics.

Table 3-1. *Execution Statistics – No Staging Table vs. Staging Table*

SQL Server Parse and Compile Times	No Staging Table	Staging Table
CPU time (ms)	0	0
Elapsed time (ms)	**77**	**6**
Statistics (Work Table)	**Existing Index**	**New Index**
Scan count	26	26
Logical reads	**198**	**196**
Physical reads	0	
Read-ahead reads	0	
Statistics (SalesTransaction)	**Existing Index**	**New Index**
Scan count	**1**	**2**
Logical reads	**23**	**10**
Physical reads	**0**	**1**
Read-ahead reads	**0**	**3**
SQL Server Execution Times	**Existing Index**	**New Index**
CPU time (ms)	0	0
Elapsed time (ms)	35	2

I reran the original query and the query that accesses the staging table. No changes were made to the original query (that contains the CUME_DIST() function) except that it now accesses the staging table instead of the CTE. Also, some new indexes were created.

The CTE in the original query was modified so it loads the staging table once. Assume in a real production environment it would be run nightly at off hours.

As can be seen, the first set of elapsed time went from 77 to 6 ms. Good start. The scan count remained the same and the logical reads went down a bit (198 vs. 196) for the work table.

Looking at the `SalesTransaction` table statistics, the scan count went from 1 to 2, while the logical reads went down to 10. This is also good!

Now we get to physical reads and read-ahead reads; they went from 0 to 1 and 3, respectively. Not so good. Lastly, CPU time was zero in both cases, but it went from 35 to 2 ms for elapsed time, so some major improvement here.

Some final comments: When you perform this analysis, you will get different results due to your hardware configuration. You need to run these tests several times to get a complete picture of what works and does not work. Do not forget to run DBCC so you always start off with clean buffers. Not executing this command before your run each test will give you misleading results!

Note   By the way, DBCC stands for database consistency check. A very powerful and useful tool. I suggest you research the Microsoft SQL Server documentation on this.

Let's check out the client statistics for the first query and the query that uses the staging table. The client statistics button can be found in the menu bar to the right of the button with the little green checkbox. Click it to turn it on or off.

Please refer to Figure 3-4.

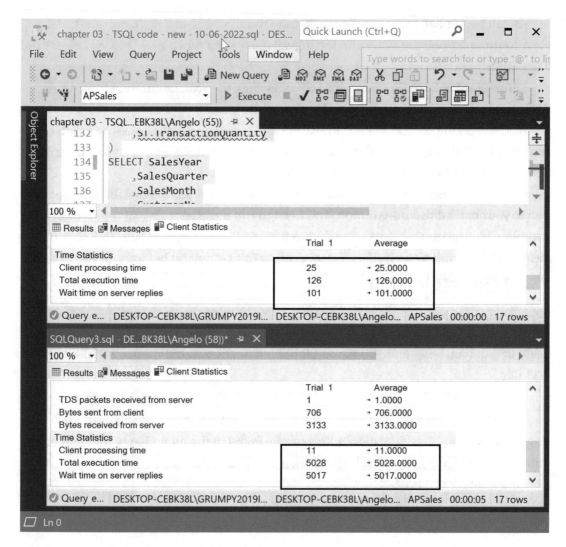

Figure 3-4. *Client statistics for the queries*

This is interesting. The top report is for the first query. I am only showing client processing time, total execution time, and wait time on server replies. They are respectively 25, 126, and 101.

The second client statistics report shows client processing time at 11 (good), but check out the total execution time at 5028 and wait time on server replies at 5017. These went up significantly.

So there are a few takeaways from our discussion. The first is you have a set of statistics and plans generated by some valuable tools that you should use when performing query design and tuning.

Second, you need to run the tests several times to get an "average" view of the results. Creating indexes seems to be a good idea, but most of the time, they will give you the same results as or worse results than the indexes suggested by the query estimated execution plan tool. If you forget to run the DBCC command to clear the buffers, you will get misleading results that make you think performance is improving.

Lastly, working on small sets of data like less than 200,000 rows will not show much difference when you are running tests. You need to see how everything works when dealing with millions of rows or more! (Look over the load scripts for the sales data warehouse and see if you can load 5–10 million rows in the fact tables we are using by modifying my load script. Make sure you make a backup.)

Our next function in our bag of tricks is the PERCENT_RANK() function.

PERCENT_RANK() Function

Next on deck is the PERCENT_RANK() function. What does it do?

Using a data set like results from a query, partition, or table variable, this function calculates the relative rank of each individual value relative to the entire data set (as a percentage). You need to multiply the results (the value returned is a float) by 100.00 or use the FORMAT() function to convert the results to a percentage.

Also, it works a lot like the CUME_DIST() function; at least that's what the Microsoft documentation says. We discussed this function at the beginning of the chapter, so go back and take a look at what it does if you forgot.

You will also see that it works a lot like the RANK() function (in theory anyway). The RANK() function returns the position of the value, while PERCENT_RANK() returns a percentage value. We will use this function together with the CUME_DIST() function just discussed so we can compare the results and also see what effect it will have on the query plan we will generate shortly.

But first let's examine another simple example, like the prior discussion, so we can clearly see the results generated from a simple data set.

I will also include a homegrown version of the percent rank calculation based on the following formula:

(current row column value - 1) / (total row count of the data set - 1)

Simple! We will use the same table variable data as the CUME_DIST() function.

For our homegrown code, we need to simulate the RANK() function, which will be introduced in the next chapter. I also include the RANK() function in the SELECT clause so we can compare its results with our custom code.

Please refer to Listing 3-2.

Listing 3-2. A Simple Percent Rank Example

```
DECLARE @CumeDistDemo TABLE (
      Col1      VARCHAR(8),
      ColValue DECIMAL(10,2)
      );

INSERT INTO @CumeDistDemo VALUES
('AAA',1.0),
('BBB',2.0),
('CCC',3.0),
('DDD',4.0),
('EEE',5.0),
('FFF',6.0),
('GGG',7.0),
('HHH',8.0),
('III',9.0),
('JJJ',10.0)

SELECT Col1,ColValue,A.RowCountLTE AS MyRank,
      RANK() OVER(
            ORDER BY ColValue
          ) AS SQLRank,

      PERCENT_RANK() OVER(
            ORDER BY ColValue
      ) AS PCTRank,

      /* current value rank - 1 /data sample total row count - 1 */
      (RANK() OVER(
            ORDER BY ColValue
          ) - 1.0) / CONVERT(DECIMAL(10,2),(
```

```
        SELECT COUNT(*) AS SampleRowCount
        FROM @CumeDistDemo) - 1.0
        ) AS MyPctRank
FROM @CumeDistDemo CDD
CROSS APPLY (
    SELECT COUNT(*) AS RowCountLTE FROM @CumeDistDemo
    WHERE ColValue <= CDD.ColValue
    ) A
GO
```

Notice I repurposed the @CumeDistDemo table variable (too lazy to rename it!).

As before, the data values are incremented from 1.0 to 10.0 sequentially. This will make it easy to correlate the values to the rank results. A single CROSS APPLY block is used in the FROM clause so we can calculate the total row count of the data set less than or equal to the current row:

```
CROSS APPLY (
    SELECT COUNT(*) AS RowCountLTE
    FROM @CumeDistDemo WHERE ColValue <= CDD.ColValue
    ) A
```

Applying the simple formula, the following code delivers the percent rank:

```
(RANK() OVER(
        ORDER BY ColValue
        ) - 1.0) / CONVERT(DECIMAL(10,2),(SELECT COUNT(*) AS
SampleRowCount FROM @CumeDistDemo) - 1.0) AS MyPctRank
```

Notice that we are using the RANK() function to calculate the ranking assignments (we will discuss it in the next chapter). If we sort the data being ranked, namely, the ColValue column in ascending order, as a trick, we could also use the ROW_NUMBER() function as shown in the following (we will also discuss this in the next chapter):

```
    ROW_NUMBER() OVER(
        ORDER BY ColValue
        ) AS RowNumberAsRank,
```

One final alternative is to just use the column RowCountLTE as the rank:

```
/* current value rank - 1 /sample total row count - 1 */
    FORMAT(
    (RowCountLTE - 1.0) /
                CONVERT(DECIMAL(10,2),(
                        SELECT COUNT(*) AS SampleRowCount
                FROM @CumeDistDemo
                ) - 1.0),'P'
    ) AS MyPctRank
```

Try to modify the code by using all versions of percentile rank so you can verify the results are the same and the formula works. Try it on your own before looking at the bonus script.

Let's look at the results. Please refer to Figure 3-5.

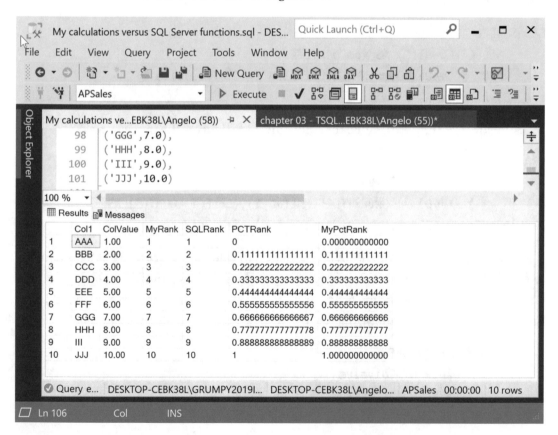

Figure 3-5. *Calculating percent rank with the SQL Server function and homegrown script*

Here are the sorted results. Both percent rank results match, and so do both the rank calculations used with the SQL Server function and the homegrown function. You can use the FORMAT() function to convert the results to percentages if you like, for example:

```
FORMAT(
       PERCENT_RANK() OVER(
       ORDER BY ColValue
       ),'P'
   ) AS PCTRank,
```

Just wrap the PERCENT_RANK() code block with parentheses, use the FORMAT() function, and include the 'P' parameter before the last parenthesis. Very easy to use.

Tip   In my opinion, creating your own version of a function based on the formula is a great way to understand what the function does.

Let's try this function out on our sales data warehouse.

Please refer to Listing 3-3.

Listing 3-3. The PERCENT_RANK() Function in Action

```
WITH CustSales (
      SalesYear,SalesQuarter,SalesMonth,CustomerNo,
      StoreNo,CalendarDate,SalesTotal
)
AS
(
SELECT YEAR(CalendarDate)        AS SalesYear
      ,DATEPART(qq,CalendarDate) AS SalesQuarter
      ,MONTH(CalendarDate)       AS SalesMonth
      ,ST.CustomerNo
      ,ST.StoreNo
      ,ST.CalendarDate
      ,SUM(ST.TotalSalesAmount) AS SalesTotal
FROM StagingTable.SalesTransaction ST
GROUP BY ST.CustomerNo
      ,ST.StoreNo
```

```
        ,ST.CalendarDate
)
SELECT SalesYear
        ,SalesQuarter
        ,SalesMonth
        ,CustomerNo
        ,SUM(SalesTotal) AS MonthlySalesTotal
        ,CUME_DIST() OVER (
                PARTITION BY SalesYear
                ORDER BY SUM(SalesTotal)
        ) AS CumeDist
        ,PERCENT_RANK() OVER (
                PARTITION BY SalesYear
                ORDER BY SUM(SalesTotal)
        ) AS PctRank
FROM CustSales
WHERE SalesYear IN(2010,2011)
AND CustomerNo = 'C00000001'
GROUP BY SalesYear
        ,SalesQuarter
        ,SalesMonth
        ,CustomerNo
GO
```

I do not think we need to go over the CTE again except to highlight that we are summing up sales by store. We could use the staging table we created for the prior example, but let's work with the CTE again. Also, I included the CUME_DIST() function in the SELECT clause so we can compare the results between the two functions.

The OVER() clause uses the same ORDER BY and PARTITION BY clauses. We want to sum up values by year, quarter, month, and customer. We are not using the store number column, so we need to apply the SUM() function across the year, quarter, month, and customer number columns. This means we need to use a GROUP BY clause.

Once again, we use a WHERE clause to filter by years 2010 and 2011 and for a single customer ("C0000001") in order to keep the result set small. Feel free to download the script and modify the WHERE clause once you are comfortable with how the query works so you load more years (make sure your laptop or desktop is powerful).

Let's check out the results. Please refer to the partial results in Figure 3-6.

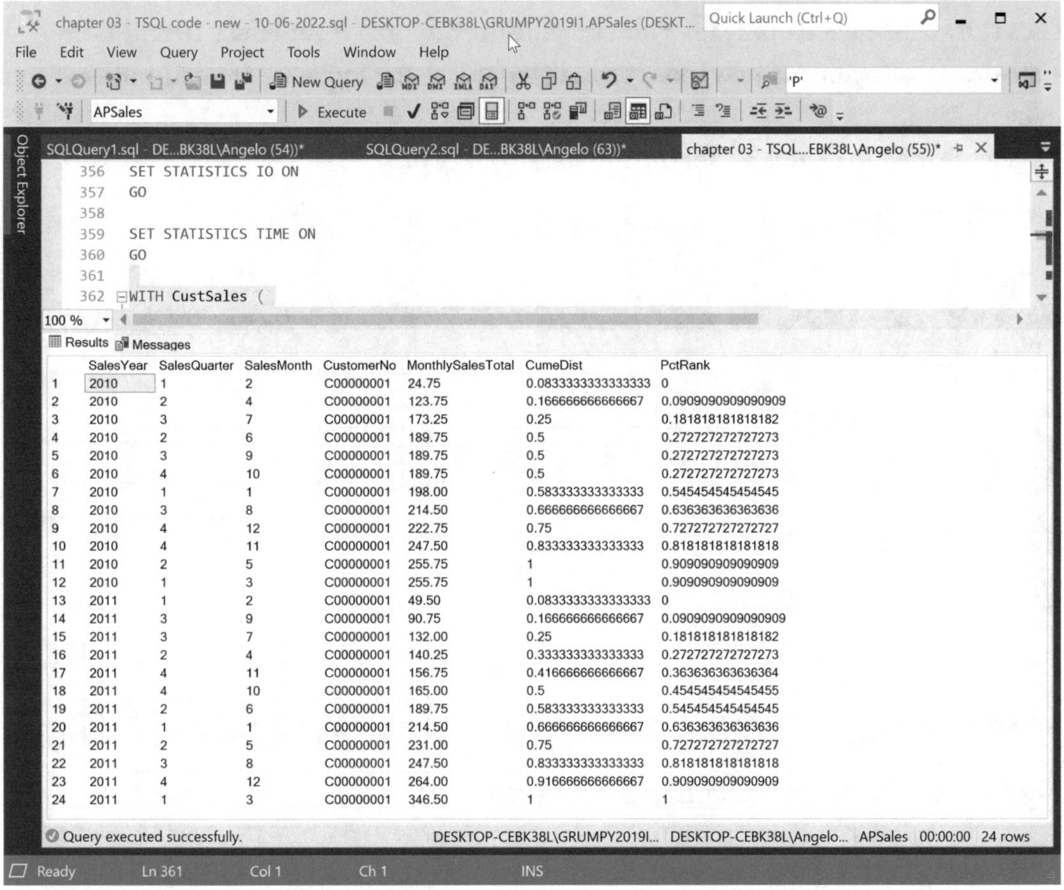

Figure 3-6. *Query results for the test script*

I changed the query a bit by eliminating the WHERE clause filter and including an ORDER BY clause. This change resulted in 16,380 rows being returned once the query was executed.

Lastly, there is a bonus query after this example in the script that uses the homegrown version of the percent rank logic and compares it to the SQL Server function using the formula we discussed.

The performance was still under one second. Let's see what the estimated query plan and statistics look like next.

Performance Considerations

Let's run an estimated query plan and execute the query so it generates IO and TIME statistics. Once again, no new index was suggested, so the index we created earlier seems to be utilized.

Please refer to Figure 3-7 for the estimated query plan.

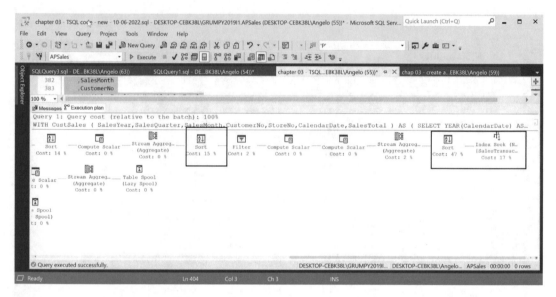

Figure 3-7. *Estimated execution plan results*

Working from right to left, we encounter an index seek at a cost of 17% and a sort task at a cost of 47%. The only other expensive step is a second sort task at 15%. The remaining tasks are either low value or zero, so we ignore them. Usually, in a real production environment, you want to at least be aware of them but mainly focus on the expensive steps to determine if you eliminate them or at least reduce the cost.

Sometimes creating indexes based on the columns of the dimensions might help. Check out the sort steps and see if an index based on the columns referenced in the sort steps is needed.

There does not seem much we can do for now, so let's look at the performance statistics to see what they tell us.

Please refer to Table 3-2.

Table 3-2. *Performance Analysis*

SQL Server Parse and Compile Times	Existing Index	New Index
CPU time (ms)	0	
Elapsed time (ms)	37	

Statistics (Work Table)	Existing Index	New Index
Scan count	26	
Logical reads	196	
Physical reads	0	
Read-ahead reads	0	

Statistics (SalesTransaction)	Existing Index	New Index
Scan count	1	
Logical reads	17	
Physical reads	1	
Read-ahead reads	14	

SQL Server Execution Times	Existing Index	New Index
CPU time (ms)	0	
Elapsed time (ms)	30	

The scan count step was high on the work table at 26. The logical read statistic also seems a bit high for both the work table and the SalesTransaction table, at 196 and 17, respectively. Lastly, the SalesTransaction table has a read-ahead at 14. Looks like some improvement might be called for. Let's look at the client statistics next.

Please refer to Figure 3-8.

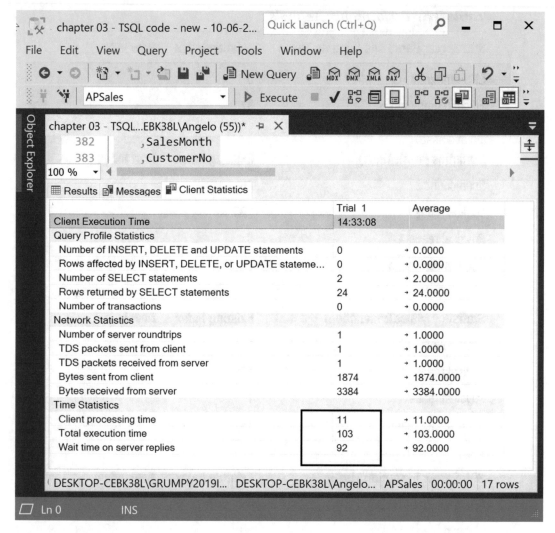

Figure 3-8. *Client statistics analysis*

Let's just look at the time statistics at the bottom. The others are important also, but these tell us some performance stats we need to know. Client processing time was 11, total execution time was a bit high at 103, and wait time on server replies was at 92.

Sometimes creating indexes is not enough. Other solutions could include modifying the query, utilizing temporary or report tables, or even creating memory-optimized tables. This last option loads all table rows to memory if you have enough, that is. We will discuss this next.

High-Performance Strategy

In this section we create a memory-optimized table and load it with the base data (this replaces the CTE query in the prior query). This strategy assumes that the table is loaded nightly once a day and the users query the memory-optimized table the next day.

This strategy also replaces the strategy of the staging or report table we discussed earlier as we are assuming tables that exist in memory will give faster performance than tables that exist on physical disk. Granted, if one runs out of memory, SQL Server will have to access data on physical disk anyway. That is why the steps to create a memory-optimized table include creating a dedicated file group and a physical file (or two).

The following are the steps to take to create a memory-optimized table.

Step 1: Check the compatibility level.

First things first. Make sure your computer or laptop supports a memory-optimized table by checking the compatibility level of SQL Server by running the simple query in Listing 3-4.

Listing 3-4. Check the Compatibility Level

```
SELECT d.compatibility_level
FROM sys.databases as d
WHERE d.name = Db_Name();
GO
```

You should be at level 130 or greater. If you are, we need to set a system parameter. If not, read this section anyway so you know what it is all about!

Step 2: Set parameter `MEMORY_OPTIMIZED_ELEVATE_TO_SNAPSHOT` to `ON`.

Please see Listing 3-5.

Listing 3-5. Set Memory Optimized Elevate to Snapshot

```
ALTER DATABASE [APSAles]
SET MEMORY_OPTIMIZED_ELEVATE_TO_SNAPSHOT = ON;
GO
```

Step 3: Next, add a dedicated file group for memory-optimized data by running the following command.

Please see Listing 3-6.

Listing 3-6. Add a File Group for Memory-Optimized Data

```
ALTER DATABASE APSales
ADD FILEGROUP APSalesMemOptimized CONTAINS MEMORY_OPTIMIZED_DATA;
GO
```

Notice that a directive needs to be included that specifies the file group will contain memory-optimized data.

Step 4: Create a dedicated file for memory-optimized tables.

Next, we add a file to the file group in the usual manner. Please see Listing 3-7.

Listing 3-7. Add a File to the File Group

```
ALTER DATABASE APSales
ADD FILE (
name='APSalesMemoOptData',
filename=N'D:\APRESS_DATABASES\AP_SALES\MEMORYOPT\AP_SALES_MEMOPT.mdf'
)
TO FILEGROUP APSalesMemOptimized
GO
```

Step 5: Create the memory-optimized table.

Next, we create a table in the usual manner but need to add a bit of code that defines it as a memory-optimized table. Please see Listing 3-8.

Listing 3-8. Create a Memory-Optimized Table

```
CREATE TABLE [SalesReports].[MemorySalesTotals](
    [SalesTotalKey] INTEGER NOT NULL IDENTITY PRIMARY KEY NONCLUSTERED,
    [SalesYear]     [int] NOT NULL,
    [SalesQuarter]  [int] NOT NULL,
    [SalesMonth]    [int] NOT NULL,
    [CustomerNo]    [nvarchar](32)   NOT NULL,
    [StoreNo]       [nvarchar](32)   NULL,
    [CalendarDate]  [date]           NOT NULL,
    [SalesTotal]    [decimal](21, 2) NULL
)
WITH (
```

```
    MEMORY_OPTIMIZED = ON,
    DURABILITY = SCHEMA_AND_DATA
    );
GO
```

We need to make sure to identify a primary key and set two parameters at the end of the CREATE TABLE command:

```
    MEMORY_OPTIMIZED = ON,
    DURABILITY = SCHEMA_AND_DATA
```

Step 6: Check that it was created.

Next, we need to query two system tables to make sure the table is registered as a memory-optimized table. Please see Listing 3-9.

Listing 3-9. Check the Filegroups and Database_files Tables

```
SELECT g.name, g.type_desc, f.physical_name
 FROM sys.filegroups g JOIN sys.database_files f ON g.data_space_id =
 f.data_space_id
 WHERE g.type = 'FX' AND f.type = 2
 GO
```

The sys.filegroups and sys.database_files tables have a lot of interesting columns, and I suggest you check them out. For our little discussion, I will only list the basic columns to save space.

Step 7: Load the Memory-Optimized Table.

Now it's time to load the table. Please see Listing 3-10.

Listing 3-10. Load the Memory-Optimized Table

```
INSERT INTO [SalesReports].[MemorySalesTotals]
SELECT YEAR(CalendarDate)         AS SalesYear
      ,DATEPART(qq,CalendarDate) AS SalesQuarter
      ,MONTH(CalendarDate)        AS SalesMonth
      ,ST.CustomerNo
      ,ST.StoreNo
      ,ST.CalendarDate
      ,SUM(ST.UnitRetailPrice * ST.TransactionQuantity) AS SalesTotal
```

129

```
FROM StagingTable.SalesTransaction ST
GROUP BY ST.CustomerNo
      ,ST.StoreNo
      ,ST.CalendarDate
      ,ST.UnitRetailPrice
      ,ST.TransactionQuantity
GO
```

Remember, this was the query used in the CTE from the prior example we discussed. Now we are ready for our performance analysis.

Step 8: Check the estimated query plan.

Before we run the query, we need to check the estimated query plan in the usual manner. See if an index is suggested. If it is, create the index and rerun the query plan estimator tool. Once the index is created, set the parameters to display performance statistics.

Step 9: Run the query.

Clear buffers with DBCC, make sure IO and TIME statistics are set, and run the query. Please see Listing 3-11.

Listing 3-11. Query the Memory-Optimized Table

```
DBCC dropcleanbuffers;
CHECKPOINT;
GO
SET STATISTICS IO ON
GO
SET STATISTICS TIME ON
GO

SELECT SalesYear
      ,SalesQuarter
      ,SalesMonth
      ,CustomerNo
      ,SUM(SalesTotal) AS MonthlySalesTotal
      ,CUME_DIST() OVER (
            PARTITION BY SalesYear
            ORDER BY SUM(SalesTotal)
```

```
    ) AS CumeDist
    ,PERCENT_RANK() OVER (
          PARTITION BY SalesYear
          ORDER BY SUM(SalesTotal)
    ) AS PctRank
FROM [SalesReports].[MemorySalesTotals]
WHERE SalesYear IN(2010,2011)
AND CustomerNo = 'C00000001'
GROUP BY SalesYear
    ,SalesQuarter
    ,SalesMonth
    ,CustomerNo
GO
```

Always clear the buffers by running the DBCC command. Next, run the two commands to set the IO and TIME statistics on and then run the query.

Step 10: Create the suggested index.

Here is the code for the suggested index. I placed it here, but you should run it after you check out the first estimated query plan. Come to think of it, running the query prior to creating the index so you can record the performance statistics.

Please see Listing 3-12.

Listing 3-12. Create the Suggested Index

```
ALTER TABLE SalesReports.MemorySalesTotals
ADD INDEX ieCustNoSaleYearMemTable
NONCLUSTERED (CustomerNo,SalesYear)
GO
```

Step 11: Create a Second Estimated Query Plan.

After the index is created, create a second estimated query plan. Maybe you can do this in two horizontally placed query panes. The top one would contain the query plan without the suggested index, and the bottom one would contain the query plan after the suggested index is created so that you can compare the two.

Step 12: Rerun the query and make sure all statistics are turned on.

Make sure you clear buffers with **DBCC** when checking estimated query plans and running queries with statistics on.

I ran the estimated query plan twice. The first time was to see what index the estimator suggested. The second time was after I created the suggested index.

Please refer to Figure 3-9 for the estimated query plan after the index was created.

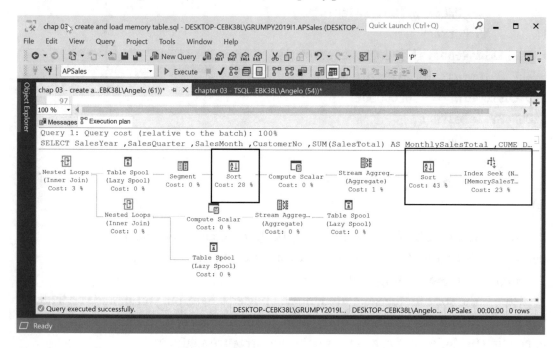

Figure 3-9. *The estimated query plan*

Reading the query plan right to left, we notice an index seek at a cost of 23%. Following it is a sort step at 43%. Ignoring the tasks with zero or minimal costs, we see a second sort step at 28%. There is a nested loops join at 3%. Not shown are more tasks to the left of the nested loops join, but they are all zero.

Download the code yourself and check out the plans and statistics. It is worthwhile learning how to do this performance analysis activity.

Let's wrap things up by comparing statistics generated for the CTE version of the query and the query that utilizes a memory-optimized table.

Please refer to Table 3-3 for the CTE vs. memory-optimized table statistics analysis.

Table 3-3. *Comparing CTE vs. Memory-Optimized Table Statistics*

SQL Server Parse and Compile Times	Using a CTE	Memory-Optimized Table
CPU time (ms)	**46**	**0**
Elapsed time (ms)	**83**	**0**

Statistics (Work Table)	Using a CTE	Memory-Optimized Table
Scan count	26	26
Logical reads	196	198
Physical reads	0	0
Read-ahead reads	0	0

Statistics (SalesTransaction) and MemorySalesTotals	Using a CTE	Memory-Optimized Table
Scan count	1	N/A
Logical reads	**23**	N/A
Physical reads	1	N/A
Read-ahead reads	20	N/A

SQL Server Execution Times	Using a CTE	Memory-Optimized Table
CPU time (ms)	**0**	**0**
Elapsed time (ms)	**34**	**0**

This is interesting. For the memory-optimized table strategy, we only have statistics for the work table – no statistics for the MemorySalesTotal table, which is now loaded in memory.

We see significant improvement in both sets of CPU time and elapsed time for parse and compile times and for execution times. With the memory-optimized table, they are all zero. Lastly, for the work table the results are about the same.

Looks like the memory-optimized table yields significant performance results.

One last comment: While performing this type of analysis, it might be a good idea to store statistics in a spreadsheet so you can analyze and compare. Run several tests that

- Generate query plans.

- Generate IO and TIME statistics.

- Generate client statistics.

And another last comment: As stated earlier, sometimes you need to go beyond index creation and look at other modifications like report and memory-enhanced tables (or even the query itself).

The last resort is getting management to upgrade your servers with more and/or faster memory and fast disks or SAN solutions. (A nice solid-state drive for TEMPDB would also help!)

Let's forge ahead with our next two analytical functions.

LAST_VALUE() and FIRST_VALUE()

So what do this related pair of functions do for us? The names give us a hint.

Given a set of sorted values, the FIRST_VALUE() function will return the first value.

Given a set of sorted values, the LAST_VALUE() function will return the last value.

Pretty simple, let's dive right into an example. This time we eliminated the CTE and utilized the memory-enhanced table we created earlier.

Please refer to Listing 3-13.

Listing 3-13. The FIRST_VALUE() and LAST_VALUE() Functions in Action

```
SELECT SalesYear
     ,SalesQuarter
     ,SalesMonth
     ,CustomerNo
     ,SUM(SalesTotal) AS MonthlySalesTotal
     ,FIRST_VALUE(SUM(SalesTotal)) OVER (
          PARTITION BY SalesYear
          ORDER BY SalesMonth
     ) AS SalesTotalFirstValue
     ,LAST_VALUE(SUM(SalesTotal)) OVER (
          PARTITION BY SalesYear
```

```
        ORDER BY SalesMonth
    ) AS SalesTotalLastValue
    ,FIRST_VALUE(SUM(SalesTotal)) OVER (
        PARTITION BY SalesYear
        ORDER BY SalesMonth
    ) -
    LAST_VALUE(SUM(SalesTotal)) OVER (
        PARTITION BY SalesYear
        ORDER BY SalesMonth
    ) AS Change
    ,CASE
        WHEN (
            FIRST_VALUE(SUM(SalesTotal)) OVER (
            PARTITION BY SalesYear
            ORDER BY SalesMonth
            ) - -
            LAST_VALUE(SUM(SalesTotal)) OVER (
                PARTITION BY SalesYear
                ORDER BY SalesMonth
                )
            ) > 0 THEN 'Sales Increase'
        WHEN (
            FIRST_VALUE(SUM(SalesTotal)) OVER (
            PARTITION BY SalesYear
            ORDER BY SalesMonth
            ) -
            LAST_VALUE(SUM(SalesTotal)) OVER (
                PARTITION BY SalesYear
                ORDER BY SalesMonth
                )
            ) < 0 THEN 'Sales Decrease'
        ELSE 'No change'
    END AS [Sales Performance]
FROM SaleReports.MemorySalesTotals
WHERE SalesYear IN(2010,2011)
```

```
AND CustomerNo = 'C00000001'
GROUP BY SalesYear
      ,SalesQuarter
      ,SalesMonth
      ,CustomerNo
GO
```

The usual OVER() structure for these functions is used. A PARTITION BY clause sets up partitions by sales year. An ORDER BY clause sorts the partition rows by the SalesMonth column.

We sum a second time as we want the results by customer only as the memory-enhanced table rows are summed by store and customer plus the usual year, quarter, and month columns.

Please refer to the partial results in Figure 3-10.

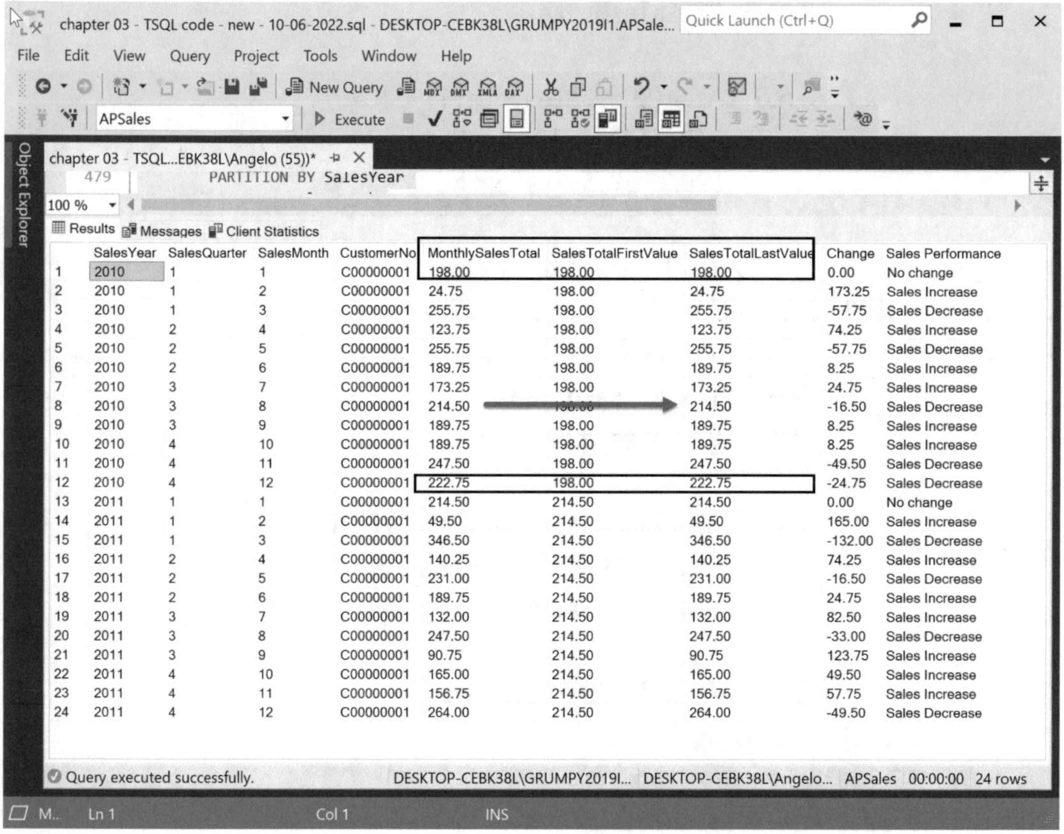

	SalesYear	SalesQuarter	SalesMonth	CustomerNo	MonthlySalesTotal	SalesTotalFirstValue	SalesTotalLastValue	Change	Sales Performance
1	2010	1	1	C00000001	198.00	198.00	198.00	0.00	No change
2	2010	1	2	C00000001	24.75	198.00	24.75	173.25	Sales Increase
3	2010	1	3	C00000001	255.75	198.00	255.75	-57.75	Sales Decrease
4	2010	2	4	C00000001	123.75	198.00	123.75	74.25	Sales Increase
5	2010	2	5	C00000001	255.75	198.00	255.75	-57.75	Sales Decrease
6	2010	2	6	C00000001	189.75	198.00	189.75	8.25	Sales Increase
7	2010	3	7	C00000001	173.25	198.00	173.25	24.75	Sales Increase
8	2010	3	8	C00000001	214.50	198.00	214.50	-16.50	Sales Decrease
9	2010	3	9	C00000001	189.75	198.00	189.75	8.25	Sales Increase
10	2010	4	10	C00000001	189.75	198.00	189.75	8.25	Sales Increase
11	2010	4	11	C00000001	247.50	198.00	247.50	-49.50	Sales Decrease
12	2010	4	12	C00000001	222.75	198.00	222.75	-24.75	Sales Decrease
13	2011	1	1	C00000001	214.50	214.50	214.50	0.00	No change
14	2011	1	2	C00000001	49.50	214.50	49.50	165.00	Sales Increase
15	2011	1	3	C00000001	346.50	214.50	346.50	-132.00	Sales Decrease
16	2011	2	4	C00000001	140.25	214.50	140.25	74.25	Sales Increase
17	2011	2	5	C00000001	231.00	214.50	231.00	-16.50	Sales Decrease
18	2011	2	6	C00000001	189.75	214.50	189.75	24.75	Sales Increase
19	2011	3	7	C00000001	132.00	214.50	132.00	82.50	Sales Increase
20	2011	3	8	C00000001	247.50	214.50	247.50	-33.00	Sales Decrease
21	2011	3	9	C00000001	90.75	214.50	90.75	123.75	Sales Increase
22	2011	4	10	C00000001	165.00	214.50	165.00	49.50	Sales Increase
23	2011	4	11	C00000001	156.75	214.50	156.75	57.75	Sales Increase
24	2011	4	12	C00000001	264.00	214.50	264.00	-49.50	Sales Decrease

Figure 3-10. *Query results for the test script*

Results are returned for two years. Notice the first and last values for the first row. They are equal as the window frame consists of only one row. When we move to the second row, the first value remains the same, but the second value now reflects the monthly sales total for February.

As the window frames get larger, the last value is adjusted to reflect the last month in the partitions. I also included a formula that calculates the change between the first and last total sales values and a little message that shows you if sales increased or decreased or if no change occurred at all.

This might make a pretty good report for our analysts if we include all years. A tool like SQL Server Reporting Services can use this query and include a drop-down list so analysts can select one, two, or more years.

Performance Considerations

Let's generate an estimated query plan in the usual manner and see if the memory-optimized table plus the index created to support it yields positive results.

Please refer to Figure 3-11.

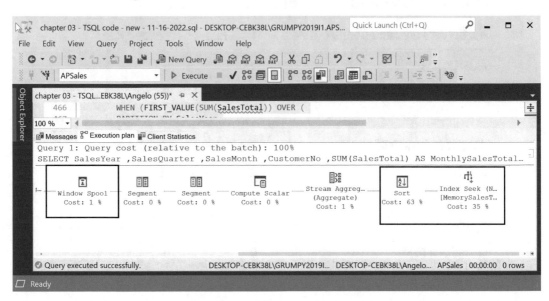

Figure 3-11. *Estimated query plan using a memory-optimized table*

Looks good. Reading the steps right to left, we see an index seek at a cost of 35%, a sort at a cost of 63%, and then a stream aggregate step at 1%. Ignoring the 0% cost steps, we have a window spool task at 1%. The rest of the steps not shown are all zeroes, so we can ignore them for the purpose of our discussion. (In a real production environment, look at all tasks!)

Side comment: I added an index based on the following command:

```
ALTER TABLE [SalesReports].[MemorySalesTotals]
ADD INDEX ieSalesYearQuarterMonth(SalesYear,SalesQuarter,SalesMonth)
GO
```

I also added an ORDER BY clause that sorted rows by the year, quarter, and month columns. I generated another estimated query plan, and the first step, the index seek task, went down to 25% from 35% and the sort step went down to 46% from 63%.

So sometimes you need to create other indexes not suggested by the query plan estimator.

Now we check our IO and TIME statistics that were generated when the query was run. Please refer to Table 3-4.

Table 3-4. *TIME and IO Statistics Analysis Report*

SQL Server Parse and Compile Times	First Run	Second Run
CPU time (ms)	15	**5**
Elapsed time (ms)	47	**5**

Statistics (Work Table)	First Run	Second Run
Scan count	26	26
Logical reads	145	145
Physical reads	0	0
Read-ahead reads	0	0

SQL Server Execution Times	First Run	Second Run
CPU Time (ms)	0	0
elapsed time (ms)	**1**	**0**

I ran the query twice. The first result had CPU and elapsed time values of 15 and 47 in the parse step. The second run had 5 for these two values. Most likely the query plan was cached, so some reduction in time for parsing occurred. This illustrates the importance of running DBCC to clear the buffers each time you run the query with statistics turned on (like moving old plans out of cache).

Please refer to Figure 3-12 for client statistics.

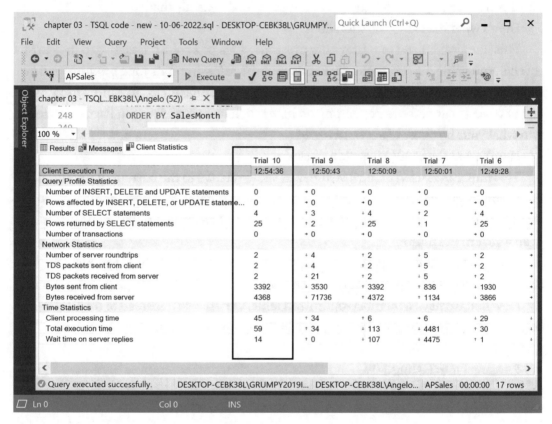

Figure 3-12.   *Client statistics performance report*

Keep in mind that running these performance tools will yield different results each time, not by much, but they will most likely be different than the ones displayed in this book due to the difference in the computing platforms you are using. Let's move on to our next set of functions.

LAG( ) and LEAD( )

These are very important functions to include in your programming toolkit to analyze data. One of the activities your business analysts perform is to look back in time or forward in time within a set of data that is historical in nature. Of course, the data could be a snapshot that includes current and past data but no future data. These functions allow you to create powerful queries and reports to do exactly that.

Here's what the LAG( ) function does:

This function is used to retrieve the previous value of a column relative to the current row column value. For example, given this month's sales amounts for a product, pull last month's sales amounts for the same product or specify some offset, like three months ago or even a year or more depending on how much historical data you have collected and what your business users want to see.

Here's what the LEAD( ) function does:

This function is used to retrieve the next value relative to the current row column. For example, given this month's sales amounts for a product, pull next month's sales amounts for the same product or specify some offset, like three months in the future or even a year in the future.

These functions come particularly handy in a sales scenario where you need to analyze historical sales performance or current performance against past performance.

Now that we have a memory-optimized table, we will use it instead of the CTE scheme we used for most of the examples in this chapter. Performance should be very fast!

Please refer to Listing 3-14.

Listing 3-14. The LAG() and LEAD() Functions in Action

```
SELECT
        SalesYear,
        SalesQuarter,
        SalesMonth,
        StoreNo,
        ProductNo,
        CustomerNo,
        CalendarDate AS SalesDate,
        SalesTotal,
```

```
LAG(SalesTotal) OVER (
        PARTITION BY SalesYear,CustomerNo
        ORDER BY SalesMonth,CustomerNo,CalendarDate
) AS LastMonthlySales,

LEAD(SalesTotal) OVER (
        PARTITION BY SalesYear,CustomerNo
        ORDER BY SalesMonth,CustomerNo,CalendarDate
) AS NextMonthylSales

FROM [SalesReports].[MemorySalesTotals]
WHERE StoreNO = 'S00005'
AND SalesYear = 2002
GO
```

This time we let the query rip with filters for the store number and year. The query generated 902 rows (still not a lot of rows as far as SQL Server is concerned).

The time to execute was less than one second. Please refer to the partial results in Figure 3-13.

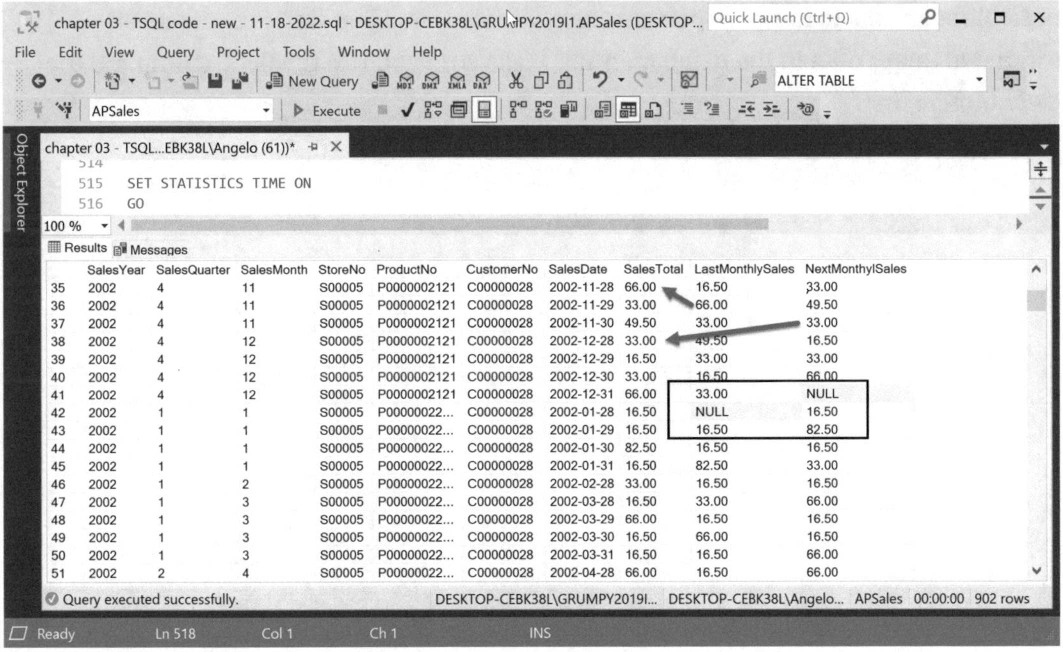

Figure 3-13. *Query results for the test script*

I scrolled down a bit so you can see the data patterns when the window frame moves to the next partition. Notice the red arrows. We can see that the LastMonthlySales column points to the prior month's value and that the NextMonthlySales column points to next month's value in the SalesTotal column.

We can see the window frame action in rows 41 and 42. The next month's sales amount is set to NULL when the partition changes, and the prior month's sales value in row 42 is also NULL as we are starting a new partition (keep in mind the difference between a partition and a window frame within a partition).

Download the script and remove the filter predicate for the year so you get all years for this store. If you want not only the years but all the stores, you might need to fiddle a bit with the sort orders, so everything lines up correctly.

That's why I always recommend starting with a small data set when creating a complex script or query. When you are sure it works correctly, you can expand the scope by removing any filters.

Performance Considerations

Next, we see what kind of performance issues are related to these functions and some possible solutions to improve performance. As usual our first step is to create an estimated query plan in the usual manner. Make sure you run the DBCC command to clear buffers first.

Please refer to Figure 3-14 for the estimated execution plan.

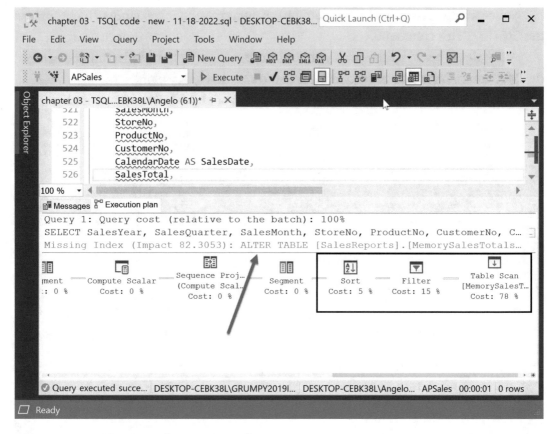

Figure 3-14. *Estimated execution plan*

This estimated query plan shows that the query needs attention.

Reading from right to left, we see a table scan at a cost of 78% followed by a filter at a cost of 15% and a sort step at a small cost of 5%. An index was suggested, so let's pull it out and give it a name.

Please refer to Listing 3-15 for the suggested index.

Listing 3-15. Estimate Query Plan Tool Suggested Index

```
ALTER TABLE [SalesReports].[MemorySalesTotals]
ADD INDEX ieSalesYearStoreNo
NONCLUSTERED ([SalesYear],[StoreNo])
GO
```

Notice we must use the ALTER command as memory-enhanced tables do not support the CREATE INDEX command!

This index is based on the sales year and store number columns. Let's create it, run the DBCC command again, and generate a new estimated query plan.

Please refer to Figure 3-15 for the estimated plan after the index was created.

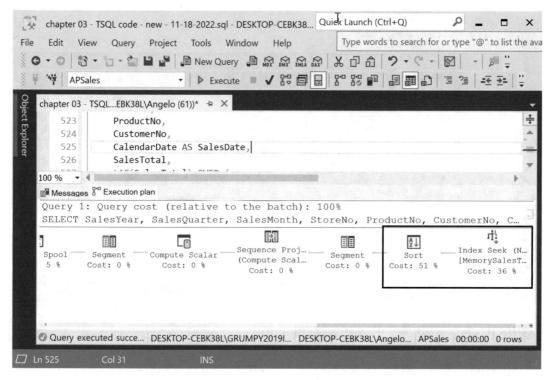

Figure 3-15. *Estimated query plan after the index was created*

An index scan at a cost of 36% replaces the table scan that had a cost of 78%, so we are off to a good start. The filter task is gone, but the sort step increased from 5% to 51%. Once again it seems when we use an index seek, the sort step cost goes up. Let's look at the second part of the plan by scrolling to the left.

Please refer to Figure 3-16.

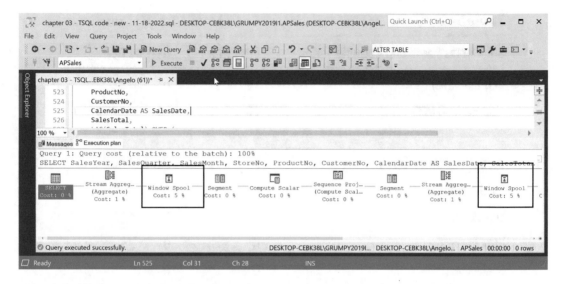

Figure 3-16. *The left-hand side of the estimated query plan*

Working from right to left again, notice the first window spool task. This has a cost of 5% and is used to reference data in the partition (which is in memory) multiple times. There are some other low-cost tasks, and we see a second window spool task. This also costs 5%. One thing to keep in mind: We should aim for 0% for window spool tasks. This means the action is in memory. If the value is greater than zero, then the spool activity is on disk. (Will more memory help?)

Here are the final IO and TIME statistics once the index is created. The query was run two times. Each time the DBCC command was executed to clear the memory buffers (a little reminder ...).

Please refer to Table 3-5.

Table 3-5. *IO and TIME Statistics for LAG() and LEAD() Functions*

SQL Server Parse and Compile Times	First Run	Second Run
CPU time (ms)	**16**	**0**
Elapsed time (ms)	**76**	**39**

Statistics (Work Table)	First Run	Second Run
Scan count	0	0
Logical reads	0	0
Physical reads	0	0
Read-ahead reads	0	0

Memory Table	First Run	Second Run
Scan count	N/A	N/A
Logical reads	N/A	N/A
Physical reads	N/A	N/A
Read-ahead reads	N/A	N/A
	N/A	N/A

SQL Server Execution Times	First Run	Second Run
CPU time (ms)	31	31
Elapsed time (ms)	**174**	**121**

Seems like this query is very fast. Double-digit times in milliseconds. Parse time went down in the second run, there are all zeros for the work table, and there are no statistics for the memory table (makes sense as the table is in memory). Finally for the execution times, the CPU time remained the same at 31 ms, and the elapsed time statistic went down from 174 ms to 121 ms.

PERCENTILE_CONT( ) and PERCENTILE_DISC( )

What do these two functions do?

The PERCENTILE_CONT() function works on a continuous set of data based on a required percentile so as to return a value within the data set that satisfies the percentile (you need to supply a percentile as a parameter, like .25, .5, .75, etc.).

With this function the value is interpolated. It usually does not exist, so it is introduced. Or by coincidence it could use one of the values in the data set if the numbers line up correctly.

What is continuous data anyway? Simply put, it is data that changes over specific periods of time, like sales amounts of time periods over time, for example, total sales in a one-month period for a specific product.

The PERCENTILE_DISC() function works on a discrete set of data based on a required percentile so as to return a value within the data set that satisfies the percentile. With this function the value exists in the data set; it is not interpolated like in the PERCENTILE_ CONT() function.

What is discrete data anyway?

Simply put, discrete data values are numbers that have values that you can count over time, like sales amounts for a product over days, months, and years. You can add these up to get a grand total sales amount.

Continuous data is data measured over a period of time, like boiler temperatures in equipment over a 24-hour period. Summing up each boiler temperature makes no sense. Summing up the boiler temperatures and then dividing them by the time period to get an average does.

Let's start off with a simple example before we dive into some analysis of our sales data.

Please refer to Listing 3-16.

Listing 3-16. The PERCENTILE_CONT() and PERCENTILE()_DISC Functions in Action

```
DECLARE @ExampleValues TABLE (
     TestKey VARCHAR(8) NOT NULL,
     TheValue SMALLINT NOT NULL
     );
```

```
INSERT INTO @ExampleValues VALUES
('ONE',1),('TWO',2),('THREE',3),('FOUR',4),('SIX',6),('SEVEN',7),('EIGHT',8
),('NINE',9),('TEN',10),('TWELVE',12);

SELECT
     TestKey,TheValue,
     PERCENTILE_CONT(.5)
          WITHIN GROUP (ORDER BY TheValue)
          OVER() AS PctCont, -- continuous
     PERCENTILE_DISC(.5)
          WITHIN GROUP (ORDER BY TheValue)
          OVER() AS PctDisc -- discrete
FROM @ExampleValues
GO
```

A very simple example. The data set consists of ten rows with two columns, a text value and a numerical value.

The PERCENTILE_CONT() function is passed .5 as a parameter to interpolate a value that will fit into the 50th percentile position in the data set. It treats the data set as a continuous distribution of values.

The PERCENTILE_DISC() function is also passed .5 as a parameter to identify an existing value that falls into the 50th percentile position in the data set. It treats the data set as a discrete distribution of values. (You can pass any value less than or equal to 1, like .75 or .25).

Let's see what the results are. Please refer to Figure 3-17.

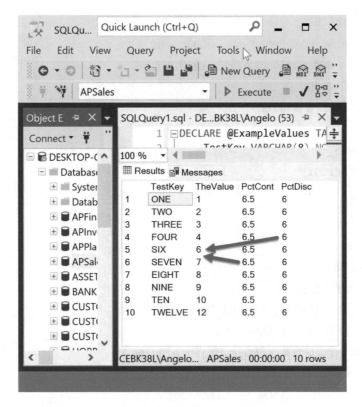

Figure 3-17. *Percentile continuous vs. discrete results*

The percentile continuous value is interpolated, while the percentile discrete is not. Notice that the 6.5 value falls between the values 6 and 7 in the TheValue column. The percentile discrete value is 6, and it exists in the TheValue column. Walk in the park!

With this powerful knowledge, let's try these functions out on our sales data. Please refer to Listing 3-17, and we will go back to our friend, the CTE.

Listing 3-17. Percentile Continuous and Discrete Analysis

```
WITH StoreSalesAnalysis (
     SalesYear,SalesMonth,StoreNo,StoreName ,StoreTerritory,TotalSales
)
AS (
SELECT YEAR(CalendarDate) AS SalesYear
       ,MONTH(CalendarDate) AS SalesMonth
     ,StoreNo
     ,StoreName
```

```
      ,StoreTerritory
      ,SUM(TotalSalesAmount) AS TotalSales
FROM APSales.SalesReports.YearlySalesReport
GROUP BY YEAR(CalendarDate)
        ,MONTH(CalendarDate)
      ,StoreNo
      ,StoreName
      ,StoreTerritory

SELECT  SalesYear
      ,SalesMonth
      ,StoreNo
      ,StoreName
      ,StoreTerritory
      ,FORMAT(TotalSales,'C') AS TotalSales
      ,FORMAT(PERCENTILE_CONT(.5)
            WITHIN GROUP (ORDER BY TotalSales)
            OVER (
                PARTITION BY SalesYear
              ),'C') AS PctCont
      ,FORMAT(PERCENTILE_DISC(.5)
            WITHIN GROUP (ORDER BY TotalSales)
            OVER (
                PARTITION BY SalesYear
              ),'C') AS PctDisc
FROM StoreSalesAnalysis
WHERE SalesYear IN(2010,2011)
AND StoreNo = 'S00004'
GO
```

I decided to format the percentiles to currency, so they look nice in the reports. Notice that the OVER() clause can only contain the PARTITION BY clause; no ORDER BY is allowed, and you will see why. (In this case we partition by SalesYear.)

Also notice right above the OVER() clause we need to include a WITHIN GROUP (ORDER BY TotalSales) directive. So, as you can see, the syntax for these functions is slightly different.

Let's check out the results. Please refer to the partial results in Figure 3-18.

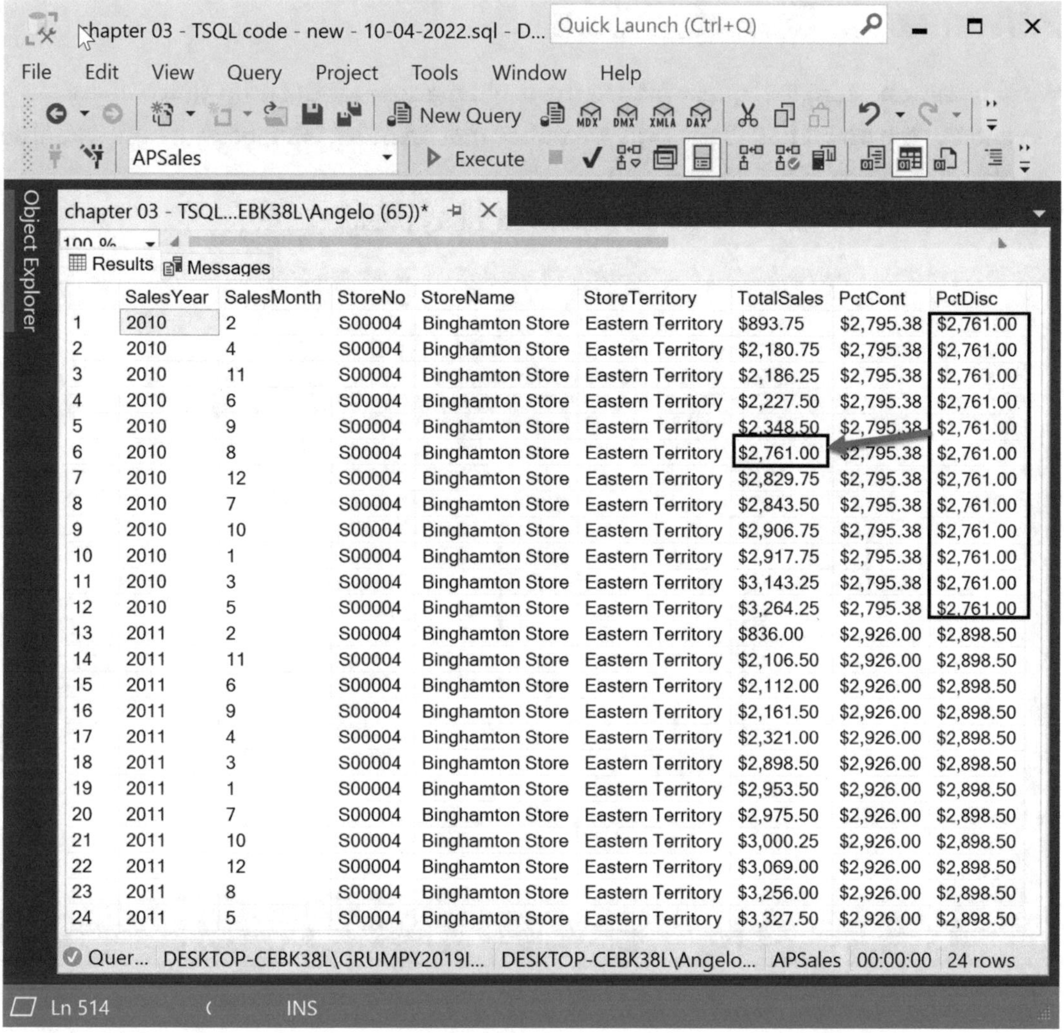

	SalesYear	SalesMonth	StoreNo	StoreName	StoreTerritory	TotalSales	PctCont	PctDisc
1	2010	2	S00004	Binghamton Store	Eastern Territory	$893.75	$2,795.38	$2,761.00
2	2010	4	S00004	Binghamton Store	Eastern Territory	$2,180.75	$2,795.38	$2,761.00
3	2010	11	S00004	Binghamton Store	Eastern Territory	$2,186.25	$2,795.38	$2,761.00
4	2010	6	S00004	Binghamton Store	Eastern Territory	$2,227.50	$2,795.38	$2,761.00
5	2010	9	S00004	Binghamton Store	Eastern Territory	$2,348.50	$2,795.38	$2,761.00
6	2010	8	S00004	Binghamton Store	Eastern Territory	$2,761.00	$2,795.38	$2,761.00
7	2010	12	S00004	Binghamton Store	Eastern Territory	$2,829.75	$2,795.38	$2,761.00
8	2010	7	S00004	Binghamton Store	Eastern Territory	$2,843.50	$2,795.38	$2,761.00
9	2010	10	S00004	Binghamton Store	Eastern Territory	$2,906.75	$2,795.38	$2,761.00
10	2010	1	S00004	Binghamton Store	Eastern Territory	$2,917.75	$2,795.38	$2,761.00
11	2010	3	S00004	Binghamton Store	Eastern Territory	$3,143.25	$2,795.38	$2,761.00
12	2010	5	S00004	Binghamton Store	Eastern Territory	$3,264.25	$2,795.38	$2,761.00
13	2011	2	S00004	Binghamton Store	Eastern Territory	$836.00	$2,926.00	$2,898.50
14	2011	11	S00004	Binghamton Store	Eastern Territory	$2,106.50	$2,926.00	$2,898.50
15	2011	6	S00004	Binghamton Store	Eastern Territory	$2,112.00	$2,926.00	$2,898.50
16	2011	9	S00004	Binghamton Store	Eastern Territory	$2,161.50	$2,926.00	$2,898.50
17	2011	4	S00004	Binghamton Store	Eastern Territory	$2,321.00	$2,926.00	$2,898.50
18	2011	3	S00004	Binghamton Store	Eastern Territory	$2,898.50	$2,926.00	$2,898.50
19	2011	1	S00004	Binghamton Store	Eastern Territory	$2,953.50	$2,926.00	$2,898.50
20	2011	7	S00004	Binghamton Store	Eastern Territory	$2,975.50	$2,926.00	$2,898.50
21	2011	10	S00004	Binghamton Store	Eastern Territory	$3,000.25	$2,926.00	$2,898.50
22	2011	12	S00004	Binghamton Store	Eastern Territory	$3,069.00	$2,926.00	$2,898.50
23	2011	8	S00004	Binghamton Store	Eastern Territory	$3,256.00	$2,926.00	$2,898.50
24	2011	5	S00004	Binghamton Store	Eastern Territory	$3,327.50	$2,926.00	$2,898.50

Figure 3-18. *Query results for the test script*

Notice how the results for percentile discreet appear in the TotalSales column. In other words, the results are not interpolated. This is not the case for percentile continuous; the results may appear in the TotalSales column, but usually they are interpolated. The value $2,795.38 does not appear as an actual value in the TotalSales column.

Note SQL Server 2022 includes two new functions, `APPROX_PERCENTILE_DIS()` and `APPROC_PERCENTILE_CONT()`, which we will cover in another chapter.

Let's see what kind of estimated query plan these functions generate.

Performance Considerations

We generate an estimated query plan in the usual manner for the drop-down query selection in the menu bar. Right away we notice that an index is suggested. Let's examine the plan.

Please refer to Figure 3-19 for the estimated query plan.

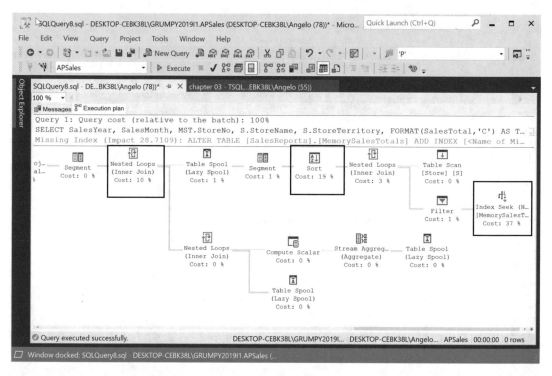

Figure 3-19. *Estimated query plan for percentile functions*

Reading from right to left, we see an index seek task at a cost of 37%, but apparently this index does not fulfill all performance requirements. Despite having an index lying around that was used, a new index was suggested, so we will have to address this requirement.

The next expensive tasks are a sort task at a cost of 19% and a nested loops join task at a cost of 10%. All other tasks have zero or low costs, so we will not discuss them at this time. You should always examine each task, even the low-cost ones, so you understand how the query will be processed.

In Table 3-6 are the statistics when this query is executed with the existing index.

Table 3-6. *Statistics with the Existing Index*

SQL Server Parse and Compile Times	Existing Index	New Index
CPU time (ms)	0	TBD
Elapsed time (ms)	98	TBD

Statistics (Work Table)	Existing Index	New Index
Scan count	9	TBD
Logical reads	171	TBD
Physical reads	0	TBD
Read-ahead reads	0	TBD

Statistics (Work File)	Existing Index	New Index
Scan count	0	TBD
Logical reads	**0**	TBD
Physical reads	0	TBD
Read-ahead reads	0	TBD

Statistics (YearlySalesReport)	Existing Index	New Index
Scan count	1	TBD
Logical reads	4033	TBD
Physical reads	1	TBD
Read-ahead reads	4041	TBD

SQL Server Execution Times	Existing Index	New Index
CPU time (ms)	47	TBD
Elapsed time (ms)	1025	TBD

Starting at the top, the elapsed time for parse and compile is 98 ms. A work file and a work table appear in the statistics, and the YearlySalesReport table is also included. For the work table, we see a scan count of 9 and 171 logical reads. These can be improved.

Looking at the work file, all values are 0, so we do not need to worry about this set of statistics.

Now the statistics for the YearlySalesReport need to be paid attention to. We see a scan count of 1. Logical reads come in at 4,033, physical reads at 1, and finally read-ahead reads at 4,041.

Lastly CPU time is 47 ms and elapsed time is 1025 ms. A bit high. Let's create the suggested index and see if it helps lower these statistics.

Please refer to Listing 3-18 that shows the SQL Server–generated index and comment.

Listing 3-18. Suggested Index

```
/*
Missing Index Details from chapter 03 - TSQL code - new - 10-06-2022.sql -
DESKTOP-CEBK38L\GRUMPY2019I1.APSales (DESKTOP-CEBK38L\Angelo (52))
The Query Processor estimates that implementing the following index could
improve the query cost by 87.5919%.
*/

/*
USE [APSales]
GO
CREATE NONCLUSTERED INDEX [<Name of Missing Index, sysname,>]
ON [SalesReports].[YearlySalesReport] ([StoreNo])
INCLUDE ([StoreName],[StoreTerritory],[CalendarDate],[TotalSalesAmount])
GO
*/
```

SQL Server tells us that if we create this index, after we give it a name, it will improve performance by 87.5%.

Okay, sounds good, but we will see.

Let's call this index ieStoreTerritoryDateTotalSales. A little copy-and-paste action to include the command with the name results in the following DDL query:

```
CREATE NONCLUSTERED INDEX ieStoreTerritoryDateTotalSales
ON [SalesReports].[YearlySalesReport] ([StoreNo])
INCLUDE ([StoreName],[StoreTerritory],[CalendarDate],[TotalSalesAmount])
GO
```

We execute it and then generate a new estimated query plan.

Please refer to Figure 3-20.

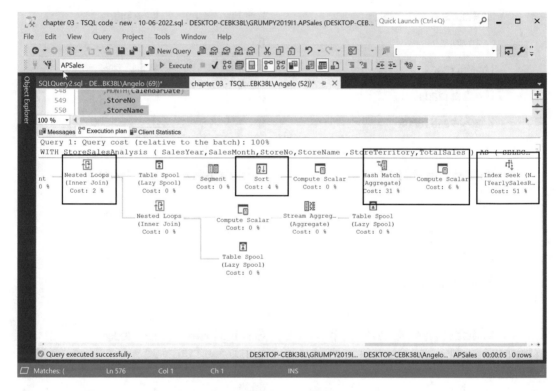

Figure 3-20. *Estimated query plan after the index was created*

Right off the bat, we see an improvement. Looks like the new index is being used. The index seek tasks chimes in at 51% cost. We do have a hash match aggregate at a cost of 31%. The sort is now at 4% cost and the nested loops join is at 2%.

In Table 3-7 are the statistics once the query is executed with the new index.

Table 3-7. *Statistics After the Suggested Index Was Created*

SQL Server Parse and Compile Times	Existing Index	New Index
CPU time (ms)	0	31
Elapsed time (ms)	98	113

Statistics (Work Table)	Existing Index	New Index
Scan count	9	9
Logical reads	171	171
Physical reads	0	0
Read-ahead reads	0	0

Statistics (Work File)	Existing Index	New Index
Scan count	0	0
Logical reads	0	0
Physical reads	0	0
Read-ahead reads	0	0

Statistics (YearlySalesReport)	Existing Index	New Index
Scan count	1	1
Logical reads	**4033**	**343**
Physical reads	1	0
Read-ahead reads	**4041**	**332**

SQL Server Execution Times	Existing Index	New Index
CPU time (ms)	**47**	**16**
Elapsed time (ms)	**1025**	**159**

For the parse step, the CPU time went up to 31 ms, and the elapsed time went up to 113 ms. We are going in the wrong direction!

No change for the work table and work file, but if we look at statistics for processing the YearlySalesReport table, we see that

- The scan count remained the same – no news is good news.

- Logical reads went from 4,033 to 343 – a major improvement.

- Physical reads went from 1 to 0 – we will take it.

- Read-ahead reads went from 4,041 to 332 – another major improvement.

Finally, for the SQL execution times, the CPU time went from 47 to 16 ms, and the elapsed time went from 1025 to 159 ms. Creating the suggested index helped a lot.

Will a denormalized report table that replaces the CTE help even more? Or what about a memory-optimized table?

Using a Report Table

Sometimes creating one or more indexes to improve performance for queries that utilize window functions or complex calculations or both is just not enough. The next step to consider is creating what is called a denormalized table or a report table that is loaded at off hours, assuming the users can tolerate data that is 24 hours old.

Key to this step is understanding business analyst requirements. Some users will want or need current, live data, while other users can accept aged data, like when they need to run month-end totals or some other type of report that summarizes past performance.

As a SQL developer or architect, you need to work with your business counterpart and collect the data and performance reports the business users expect. Hopefully, after your requirements gathering sessions, you will arrive at the scenario where 80% of users do not need live data; instead, day-old data or older is sufficient. The other 20% of the users that need live data require complex solutions that might involve hardware, complex replication schemes, or querying transaction tables directly, which is not recommended.

Let's take a look at a possible solution for the query we just examined that uses a report table that is loaded off hours. The CTE portion of the query is replaced with the report table. We will preload it and then check out performance on the query that contains our window functions.

Please refer to Listing 3-19 for the script to load the report table and the revised query.

Listing 3-19. YearlySummaryReport Table for Preloading

```
DROP TABLE IF EXISTS [SalesReports].[YearlySummaryReport]
GO

CREATE TABLE [SalesReports].[YearlySummaryReport](
      [SalesYear]        [int] NULL,
      [SalesMonth]       [int] NULL,
      [StoreNo]          [nvarchar](32) NOT NULL,
      [StoreName]        [nvarchar](64) NOT NULL,
      [StoreTerritory]   [nvarchar](64) NOT NULL,
      [TotalSales]       [decimal](10, 2) NULL
) ON [AP_SALES_FG]
GO

INSERT INTO APSales.SalesReports.YearlySummaryReport
SELECT YEAR(CalendarDate) AS SalesYear
       ,MONTH(CalendarDate) AS SalesMonth
      ,StoreNo
      ,StoreName
      ,StoreTerritory
      ,SUM(TotalSalesAmount) AS TotalSales
FROM APSales.SalesReports.YearlySalesReport
GROUP BY YEAR(CalendarDate)
       ,MONTH(CalendarDate)
      ,StoreNo
      ,StoreName
      ,StoreTerritory
GO
```

```
CREATE NONCLUSTERED INDEX [ieYearlySalesStoreTerritorySummary]
ON [SalesReports].[YearlySummaryReport] ([StoreNo],[SalesYear])
INCLUDE ([SalesMonth],[StoreName],[StoreTerritory],[TotalSales])
GO
```

This script uses the output columns defined for the CTE and uses them to create a physical table called YearlySummaryReport. Once the table is created, the query that was originally used for the CTE is used in an INSERT statement to load the table. This step was easy. Now for the query that contains the window functions and the analysis we need to perform to see if we have an improvement on the estimated query plan and the IO and TIME statistics, please refer to Listing 3-20.

Listing 3-20. Querying the Report Table

```
DBCC dropcleanbuffers
CHECKPOINT;
GO
SET STATISTICS TIME ON
GO
SET STATISTICS IO ON
GO

SELECT      SalesYear
      ,SalesMonth
      ,StoreNo
      ,StoreName
      ,StoreTerritory
      ,FORMAT(TotalSales,'C') AS TotalSales
      ,FORMAT(PERCENTILE_CONT(.5)
            WITHIN GROUP (ORDER BY TotalSales)
            OVER (
            PARTITION BY SalesYear
         ),'C') AS PctCont
      ,FORMAT(PERCENTILE_DISC(.5)
            WITHIN GROUP (ORDER BY TotalSales)
            OVER (
            PARTITION BY SalesYear
```

```
        ),'C') AS PctDisc
FROM APSales.SalesReports.YearlySummaryReport
WHERE SalesYear IN(2010,2011)
AND StoreNo = 'S00004
GO

SET STATISTICS TIME OFF
GO

SET STATISTICS IO OFF
GO
```

We have covered this query earlier, so let's get right to it and generate the estimated query plan in the usual manner. I did create an index on the new report table. This index was suggested by the estimated query plan tool. Please refer to the Listing 3-21.

Listing 3-21. Index to Support the Report Table

```
CREATE NONCLUSTERED INDEX [ieYearlySalesStoreTerritorySummary]
ON [SalesReports].[YearlySummaryReport] ([StoreNo],[SalesYear])
INCLUDE ([SalesMonth],[StoreName],[StoreTerritory],[TotalSales])
GO
```

We need to compare the new plan with the prior plan to see what improvements, if any, have occurred.

Please refer to Figure 3-21 for the estimated query plans for both approaches.

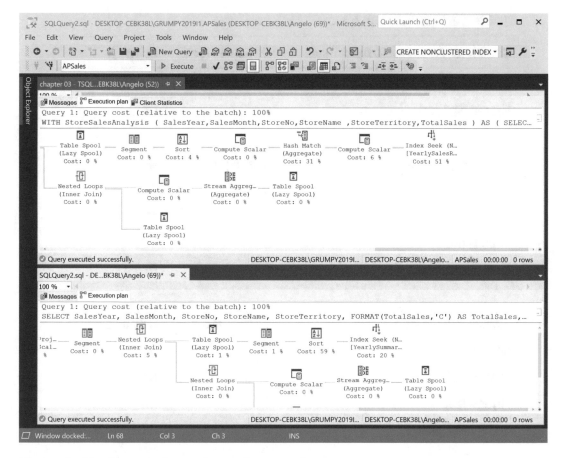

Figure 3-21. *Comparing estimated query plans*

The top estimated query plan represents the original query with the CTE. The bottom query plan is for the revised script that queries the reporting table that was preloaded.

What we see is that the index seek for the reporting table query is at a cost of 20% compared with the index seek of 51% for the original query. The new query also has a sort step significantly higher than the sort step in the original query that uses the CTE. Seems like when index seek cost goes down, sort cost goes up. Something to keep an eye out for.

Is this an improvement? Not sure, so let's look at the IO and TIME statistics to see what they tell us.

Please refer to Table 3-8.

Table 3-8. *Comparing IO and TIME Statistics*

SQL Server Parse and Compile Times	No Report Table	Report Table
CPU time (ms)	15	0
Elapsed time (ms)	**20**	**58**
Statistics (Work Table)	**Existing Index**	**New Index**
Scan count	9	9
Logical reads	171	171
Physical reads	0	0
Read-ahead reads	0	0
Statistics (Work File)	**Existing Index**	**New Index**
Scan count	0	NA
Logical reads	0	NA
Physical reads	0	NA
Read-ahead reads	0	NA
Statistics (YearlySalesReport) vs. YearlySummaryReport	**Existing Index**	**New Index**
Scan count	**1**	**2**
Logical reads	**343**	**6**
Physical reads	**2**	**1**
Read-ahead reads	**345**	**8**
SQL Server Execution Times	**Existing Index**	**New Index**
CPU time (ms)	**16**	**0**
Elapsed time (ms)	**132**	**67**

Looking at the statistics that were generated, we see significant improvements when we use the reporting table called YearlySummaryReport. The scan count went up from 1 to 2, but the logical reads went from 343 to 6. Physical reads decreased by 1 in the new reporting table, and best of all, read-ahead reads went down from 345 to 8. So, yes, there seems to be a significant improvement when using preloaded and precalculated reporting tables in queries that use window functions. Come to think of it – eliminating the CTE or load step from the analysis definitely will improve performance of the query.

Last but not least, CPU time for execution times went down from 16 to 0 ms, and elapsed time went from 132 to 67 ms.

One last comment when performing this type of analysis: Always run several tests, clearing the buffers with DBCC for each trial. Record the results in a spreadsheet for later analysis and evaluation as to which performance improvement modification works best.

By the way, isolating the loading of a report table does not mean we completely ignore performance analysis on the load steps. If we are loading millions of rows nightly, then we need to come up with a load strategy that will not take so long it creeps into work hours!

Summary

We just covered our second set of window functions, analytical functions. I think some interesting discoveries were made not only in analytical techniques to use to make our business analysts happy but also in performance tuning.

We did the usual analysis with the estimated query plans and IO and TIME statistics, and we also looked at client statistics.

Next, we introduced denormalized staging or reporting tables in order to simulate a production environment that loads data overnight so as to perform all calculations and joins. We performed the same performance analysis as we did for the CTE-based queries.

Finally, we introduced memory-optimized tables, which are loaded into memory. We discovered that these could be very fast.

In conclusion, you now know how to use the analytical functions within partitions that use the OVER() clause. In the next chapter, we will take a look at the ranking functions and play around with window frame definitions to make things interesting.

One more comment: We are only touching on performance analysis and tuning techniques. The scope of this book is on the window functions, but a bare minimum of learning how to use some of the performance tuning tools and techniques is warranted as it will help you write fast and effective scripts and queries.

For a great read on performance tuning, please refer to *SQL Server 2022 Query Performance Tuning: Troubleshoot and Optimize Query Performance* (sixth edition, Apress) by Grant Fritchey.

A great read and in my opinion a must in your library if you are serious about learning performance analysis, associated tools, and performance improvement.

CHAPTER 4

Sales Use Case: Ranking/ Window Functions

This is our last chapter dealing with the sales data warehouse. We will look at the third category of functions called window or ranking functions. We will take the same approach as prior chapters, that is, provide a brief explanation of the function, present the code and the query results, and do some performance analysis and tuning by adding indexes and report tables. Lastly, we will present a data analysis problem called gaps and islands where we use some of the window system functions we studied in Chapter 3 to identify gaps and sequences in dates for sales data.

Ranking/Window Functions

There are only four functions in this category, but we will revisit the `PERCENT_RANK()` function from the last chapter as it has a lot of similarities with some of these functions:

- RANK()
- PERCENT_RANK()
- DENSE_RANK()
- NTILE()
- ROW_NUMBER()

The `RANK()` function shows the rank of the value in each row column of a data set relative to the other row values in the data set. In case of ties, the rank is the same for the ties, but the next value after the duplicate values is assigned a rank equal to the number of ties plus the current rank value.

© Angelo Bobak 2023
A. Bobak, *SQL Server Analytical Toolkit*, https://doi.org/10.1007/978-1-4842-8667-8_4

So, if you have a three-way tie for a value like 4 and it is assigned a rank of 4, the next rank for a value greater than 4 will be 4 + 3 (the number of ties) = 7 (yikes!).

The DENSE_RANK() function works almost the same as the RANK() function. In case of ties, the rank is the same for the ties, but the next value after the duplicate values simply is assigned the next rank number. So, if the rank for the ties is 4, the next rank assigned to the next higher value is 5 (makes sense to me …).

The PERCENT_RANK() function assigns a ranking as a percentage. Ties get the same rank percentage values. Using a data set like results from a query, partition, or table variable, this function calculates the relative rank of each individual value relative to the entire data set (as a percentage). You need to multiply the results (the value returned is a float data type) by 100.00 or use the FORMAT() function to convert the results to a percentage.

The ROW_NUMBER() function simply assigns the next highest number to the row regardless of ties.

So, if the current row is the fourth row in the data set, the row number 4 will be assigned. For row number 5, then, wait for it, the number 5 is assigned and so on. This function does not care if there are ties in the column that contains values. It only cares about the position of the row within the data set.

The NTILE() function allows you to divide a set of rows in a data set into tiles or buckets. If you have a data set of 12 rows and you want to assign 4 tiles, each tile will have 3 rows. This function comes in handy when you want to evaluate salesperson performance and grant them bonuses based on sales performance. (We will see how to do this. We will generate several tiles to implement performance buckets and assign a salesperson to them.)

Okay, I am sure that all this was a bit confusing, and you are scratching your head. Let's use the simple example from the prior chapter, modify it a bit, and then look at some simple data so we can explain what each function does (except for NTILE()) and how they are alike to (and differ from) each other.

Please refer to Listing 4-1.

Listing 4-1. Ranking Functions in Action

```
DECLARE @ExampleValues TABLE (
     TestKey VARCHAR(8) NOT NULL,
     TheValue SMALLINT NOT NULL
     );
```

```
INSERT INTO @ExampleValues VALUES
('ONE',1),('TWO',2),('THREE',3),('FOUR',4),
('FOUR',4),('SIX',6),('SEVEN',7),
('EIGHT',8),('NINE',9),('TEN',10);

SELECT
     TestKey,
     TheValue,
     ROW_NUMBER()          OVER(ORDER BY TheValue) AS RowNo,
     RANK()                OVER(ORDER BY TheValue) AS ValueRank,
     DENSE_RANK()          OVER(ORDER BY TheValue) AS DenseRank,
     PERCENT_RANK()        OVER(ORDER BY TheValue) AS ValueRank,
     FORMAT(PERCENT_RANK() OVER(ORDER BY TheValue),'P') AS ValueRankAsPct
FROM @ExampleValues;
GO
```

We declare our simple table variable and insert ten rows into it. Notice the duplicate values of ('FOUR',4) in the INSERT statement. This was done on purpose, so we will see how duplicate values affect the behavior of each function.

Each function is used with an OVER() clause. We include only an ORDER BY clause that sorts the partition data set by the values in the "TheValue" column. Notice the use of the FORMAT() function to display the results of the PERCENT_RANK() function as a percentage. Let's see the results.

Note   In this example the partition is the entire data set, as no PARTITION BY clause was included. After all, we only have ten rows.

Please refer to Figure 4-1.

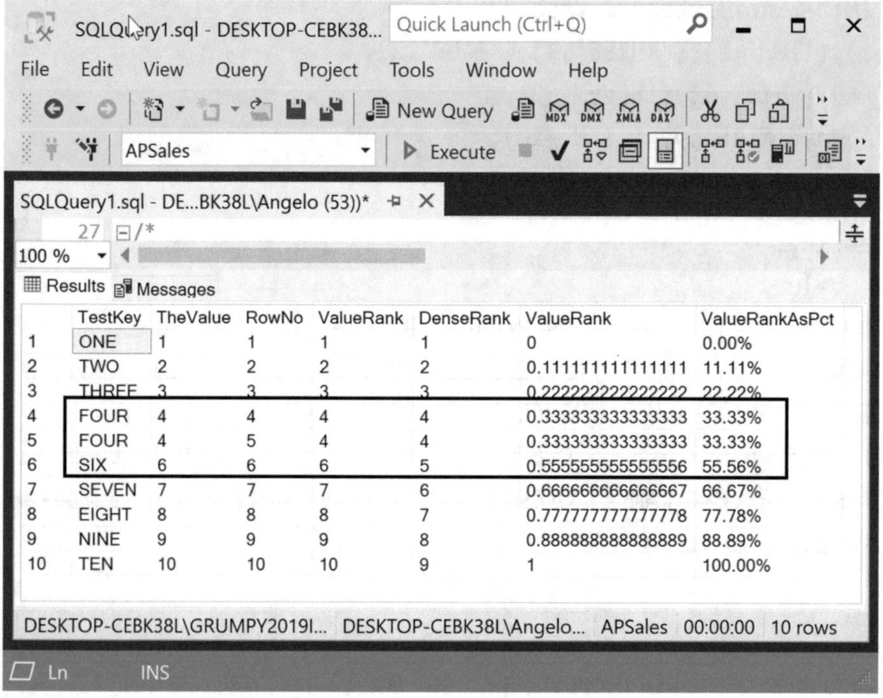

	TestKey	TheValue	RowNo	ValueRank	DenseRank	ValueRank	ValueRankAsPct
1	ONE	1	1	1	1	0	0.00%
2	TWO	2	2	2	2	0.111111111111111	11.11%
3	THREE	3	3	3	3	0.222222222222222	22.22%
4	FOUR	4	4	4	4	0.333333333333333	33.33%
5	FOUR	4	5	4	4	0.333333333333333	33.33%
6	SIX	6	6	6	5	0.555555555555556	55.56%
7	SEVEN	7	7	7	6	0.666666666666667	66.67%
8	EIGHT	8	8	8	7	0.777777777777778	77.78%
9	NINE	9	9	9	8	0.888888888888889	88.89%
10	TEN	10	10	10	9	1	100.00%

Figure 4-1. *Ranking functions in action results*

Let's start with the `ROW_NUMBER()` function results. There are ten rows, and the row numbers are 1–10 in sequential order. It does just what the name implies.

Next are the RANK() function results. Notice what happens with the two ties represented by the value 4. Each is assigned the rank 4 because they are the same. But look at what happens to the next row value. It skips the number 5 and is assigned a value of 6. Just like the formula we discussed earlier, it takes the number of ties (2) and adds it to the current rank 4 to assign a rank of 6.

Now look at the `DENSE_RANK()` function results. Just like the RANK() function, ties are assigned the same rank value, but the next row value is assigned the next higher rank, which is 5.

Last but not least, the `PERCENT_RANK()` function assigns a rank as a percentage between 0 and 1. I added a column so as to use the `FORMAT()` function to format the results as a percentage. This way we can compare the formats and decide which one makes sense to use.

NTILE() Example

Let's create some tiles. No, not kitchen floor tiles but tiles that are buckets of data (I know, bad pun). We will start off with another simple example so we can clearly see what this function does. We will use a table variable loaded with ten rows of year-to-date sales amounts for our sales team.

What we want to do is create three tiles or buckets of data so we can use them as criteria to award bonuses to the sales team members according to the tile they fall into.

Please refer to Listing 4-2.

Listing 4-2. Assigning Performance Buckets for Bonuses

```
DECLARE @SalesPersonBonusStructure TABLE (
    SalesPersonNo VARCHAR(4) NOT NULL,
    SalesYtd       MONEY       NOT NULL
    );

INSERT INTO @SalesPersonBonusStructure VALUES
('S001',2500.00),
('S002',2250.00),
('S003',2000.00),
('S004',1950.00),
('S005',1800.00),
('S006',1750.00),
('S007',1700.00),
('S008',1500.00),
('S009',1250.00),
('S010',1000.00);

-- Care must be taken how you sort (ASC or DESC)

SELECT SalesPersonNo
    ,SalesYtd
    ,NTILE(3) OVER(ORDER BY SalesYtd DESC) AS BonusBucket
    ,CASE
            WHEN (NTILE(3) OVER(ORDER BY SalesYtd DESC)) = 1
                THEN 'Award $500.00 Bonus'
            WHEN (NTILE(3) OVER(ORDER BY SalesYtd DESC)) = 2
                THEN 'Award $250.00 Bonus'
            WHEN (NTILE(3) OVER(ORDER BY SalesYtd DESC)) = 3
```

```
                THEN 'Award $150.00 Bonus'
    END AS BonusAward
FROM @SalesPersonBonusStructure
GO
```

The solution is simple. The NTILE() function is used with an OVER() clause to create the three tiles. There is no PARTITION BY clause due to the small data set, but we include an ORDER BY clause so we can sort the year-to-date sales amounts in descending order.

Next, a series of three CASE blocks determine the tile assignment by using the NTILE() function again so as to print out a message of the amount the salesperson is awarded based on the tile they are in.

Maybe a little brute force as we could have used a CTE to determine the bucket and then write the query so it just tests the tile value instead of using the NTILE() function again. See if you can download the script for this chapter and code a query using the CTE approach. I did provide the solution just in case you get stuck in the script for this section.

Let's see what kind of awards each salesperson received. Please refer to Figure 4-2.

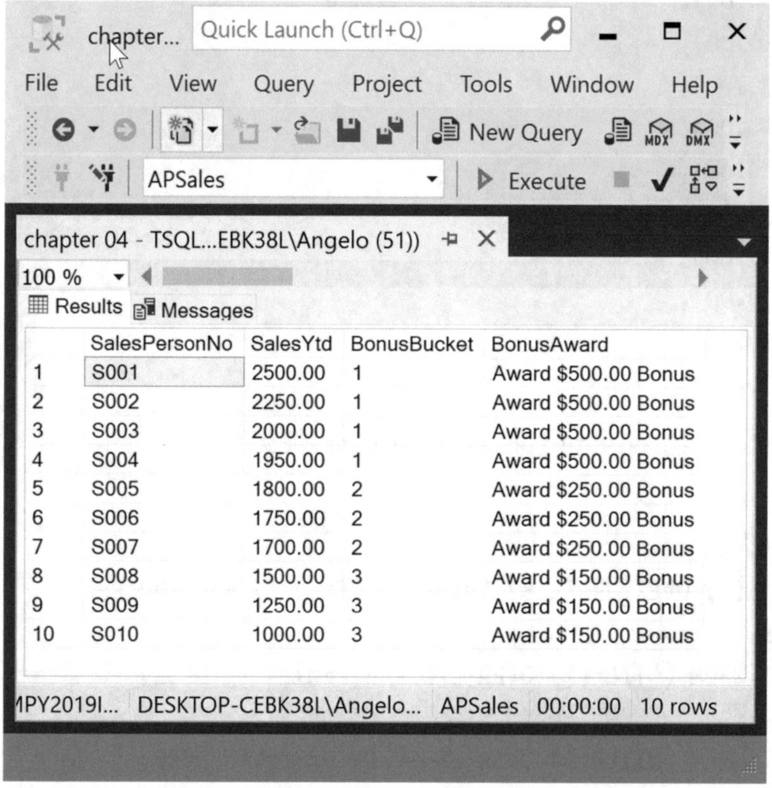

Figure 4-2. *Bonus performance buckets*

There are ten rows in the result set, and we specified only three buckets, so the bucket assignment is uneven. The first two buckets get three rows each, and the last bucket gets four rows. If we had 12 rows, then each bucket would get 4 rows. Either case these are rather cheesy bonuses. Our sales staff needs to sell more if they want larger bonuses!

Armed with our understanding of how these functions work, let's use the functions against our sales data warehouse. We will start with RANK() and PERCENT_RANK() and revisit NTILE() later in the chapter.

RANK() vs. PERCENT_RANK()

Let's see how these two functions work and compare the results to see any similarities. The PERCENT_RANK() falls under the analytical function category, which we covered in the last chapter, but I include it in this chapter so we can compare it to the RANK() function.

We will use our usual structure to put these functions through their paces. Please refer to Listing 4-3.

Listing 4-3. Rank vs. Percent Rank

```
WITH CustomerRanking (
      CalendarYear,CalendarMonth,CustomerFullName,TotalSales
      )
AS
(
SELECT CalendarYear
      ,CalendarMonth
      ,CustomerFullName
      ,SUM(TotalSalesAmount) AS TotalSales
FROM SalesReports.YearlySalesReport YSR
JOIN DimTable.Calendar C
ON YSR.CalendarDate = C.CalendarDate
GROUP BY C.CalendarYear
      ,C.CalendarMonth
      ,CustomerFullName
)
```

```
SELECT
     CalendarYear
     ,CalendarMonth
     ,CustomerFullName
     ,FORMAT(TotalSales,'C') AS TotalSales
     ,RANK()
          OVER (
--            PARTITION BY CalendarYear
          ORDER BY TotalSales
     ) AS Rank
     ,PERCENT_RANK()
          OVER (
--            PARTITION BY CalendarYear
          ORDER BY TotalSales
     ) AS PctRank
FROM CustomerRanking
WHERE CalendarYear = 2011
AND CalendarMonth = 1
ORDER BY
     RANK() OVER (
          PARTITION BY CalendarYear
          ORDER BY TotalSales
     ) DESC
GO
```

Our CTE simply assembles some columns we will need to report on and also includes the SUM() function to calculate total sales by year, month, and customer name. To limit the number of rows returned, a WHERE clause is used to filter the results by year and for one month only (2011, January).

Reminder – start with small data sets. Develop the solution and test it. Once you are satisfied it works correctly, apply the solution to larger data sets.

Both the RANK() and PERCENT_RANK() functions use an OVER() clause that includes an ORDER BY clause so the partition rows are sorted by the TotalSales column. The PARTITION BY clause is commented out as we are only retrieving one-year worth of data. Once you are comfortable with the logic, feel free to retrieve multiple rows and uncomment the PARTITION BY clause so we can process multiple years.

172

Lastly, here is something interesting. You can include a window function in the query ORDER BY clause (as opposed to the OVER() clause). We do so with the same code used for the RANK() function. Leaving it out will give you the same results but in reverse order. But at least you know it can be done. Just understand the differences. When the ORDER BY clause is in the OVER() clause, it sorts the rows in the partition. When the ORDER BY clause is at the end of the query, it sorts the final results of the query.

Please refer to the partial results in Figure 4-3.

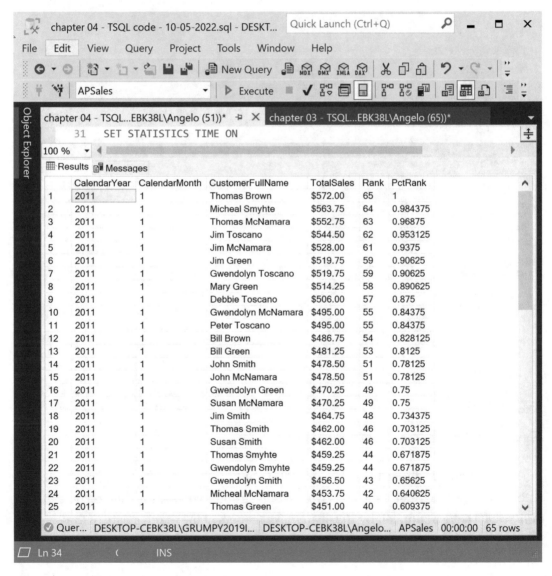

Figure 4-3. *Rank vs. percent rank*

Results are in descending order. Notice the behavior of the duplicates in rows 6 and 7 and rows 21 and 22. As mentioned earlier, multiply percent rank by 100.00 so you see values as percentages or use the FORMAT() function instead. This way you can graph the results in Microsoft Excel as shown in the following.

I copied and pasted the results and multiplied the percent rank results by 100.00, so we see a nice graph instead of small values less than or equal to 1.0.

Please refer to Figure 4-4.

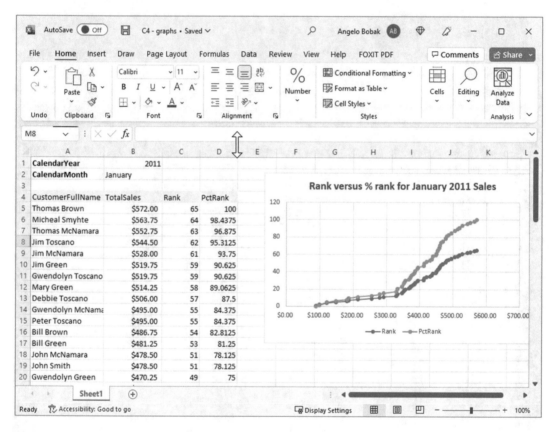

Figure 4-4. *Rank vs. percent rank analysis*

Not too bad. We do see the graph results are almost the same for both functions with a nice upward trend. I should have included a comment stating that the percent rank results were multiplied by 100.00 to make readers of the graph aware of the manipulation.

The data is displayed to the left for easy reference. It is always a good idea to generate some nice graphs and charts with Microsoft Excel for your users so they can interpret the results. A picture is worth a thousand words! Microsoft Power BI is also a great tool for visualizing data.

Let's see the performance characteristics for this query.

Tip   As a SQL developer or architect, you want to also master some other skills, like having a good working knowledge of Microsoft Excel or Power BI so you can generate powerful visuals for your users and management. Some ETL skills with SSIS (SQL Server Integration Services) would also be valuable.

Performance Considerations

Let's start off by generating an estimated query plan in the usual manner (menu bar ➤ Query ➤ Display Estimated Execution Plan). You will see some rather interesting and puzzling results.

Please refer to Figure 4-5.

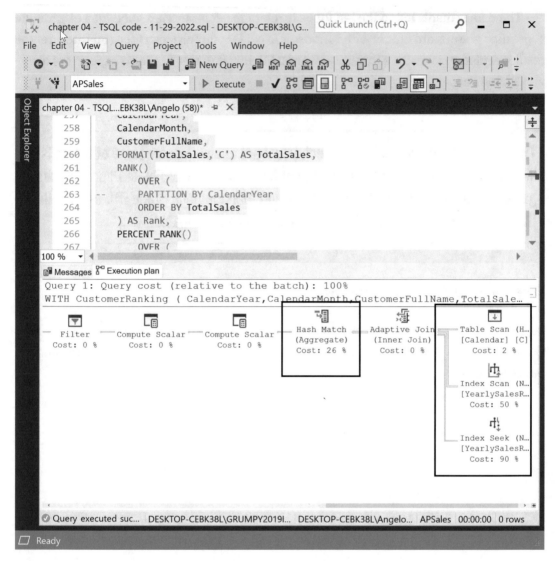

Figure 4-5. Estimated query plan – rank vs. percent rank

We had some indexes lying around, so the estimator did not suggest any, but wait a minute – the tasks add up to more than 100%:

- Table scan: 2%

- Index scan: 50%

- Index seek: 90%

- Hash match: 26%

These values add up to 168%.

This is not possible. Costs cannot be over 100%. Why is this?

You will not like the answer.

This is a bug with the client-side software, and occasionally you will get surprising results like this. I recommend you research this on the Microsoft website to see if newer versions of SSMS will clear this up. Fortunately, it looks like it does not happen often.

Back to the query plan, as you can see some indexes were lying around from the prior chapter, so they were used. This is what you want; you want your indexes to be used by multiple queries and to not have to create a new index every time a new request comes in.

Let's examine the IO and TIME statistics generated by this query.

Please refer to Table 4-1.

Table 4-1. *Query IO and Time Statistics*

SQL Server Parse and Compile Times	Existing Indexes
CPU time (ms)	0
Elapsed time (ms)	127

Statistics (Work Table 1)	Existing Indexes
Scan count	0
Logical reads	0
Physical reads	0
Read-ahead reads	0

Statistics (Work Table 2)	Existing Indexes
Scan count	0
Logical reads	0
Physical reads	0
Read-ahead reads	0

(*continued*)

Table 4-1. (*continued*)

Statistics (YearlySalesReport)	Existing Indexes
Scan count	1
Logical reads	**1384**
Physical reads	1
Read-ahead reads	**1401**

Statistics (Calendar)	Existing Indexes
Scan count	1
Logical reads	48
Physical reads	0
Read-ahead reads	0

SQL Server Execution Times	Existing Indexes
CPU time (ms)	125
Elapsed time (ms)	334

Seems like we have issues with logical reads and read-ahead reads. These exist on the CTE side with the YearlySalesReport table. Logical reads might not be an issue though as they are performed in memory (i.e., if you have enough!). Stay tuned. We need to look at IO and TIME statistics to see if issues exist related to logical reads.

The query in the CTE might be a candidate for a denormalized report or staging table or, better yet, memory-enhanced table. The indexes helped, but there is just so much you can do by creating indexes. The more indexes you add, the more performance will degenerate when you are loading, modifying, or deleting rows in this table.

Based on the code examples from the last chapter, see if you can modify this table by replacing the CTE with a script that loads a report or memory-enhanced table and then try this analysis on your own.

RANK( ) vs. DENSE_RANK( )

Let's look at the RANK( ) function again, but now we will compare it to the DENSE_RANK( ) function.

As a reminder, so you do not have to flip back to the start of the chapter, the DENSE_RANK( ) function works almost the same as the RANK( ) function. In case of ties, the rank is the same for the ties, but the next value after the duplicate values simply is assigned the next rank number.

Please refer to Listing 4-4.

Listing 4-4. Rank vs. Dense Rank

```
WITH CustomerRanking (
      CalendarYear,CalendarMonth,CustomerFullName,TotalSales
      )
AS
(
SELECT YEAR(CalendarDate)
      ,MONTH(CalendarDate)
      ,CustomerFullName

      -- add one duplicate value on the fly
      ,CASE
            WHEN CustomerFullName = 'Jim OConnel' THEN 17018.75
            ELSE SUM(TotalSalesAmount)
      END AS TotalSales
FROM SalesReports.YearlySalesReport
GROUP BY YEAR(CalendarDate)
      ,MONTH(CalendarDate)
      ,CustomerFullName
)
SELECT
      CalendarYear
      ,CalendarMonth
      ,CustomerFullName
```

```
        ,FORMAT(TotalSales,'C') AS TotalSales
        ,RANK()
                OVER (
                ORDER BY TotalSales
        ) AS Rank
        ,DENSE_RANK()
                OVER (
                ORDER BY TotalSales
        ) AS DenseRank
FROM CustomerRanking
WHERE CalendarYear = 2011
AND CalendarMonth = 1
ORDER BY
        DENSE_RANK() OVER (
                PARTITION BY CalendarYear
                ORDER BY TotalSales
        ) DESC
GO
```

I included a CASE statement in the CTE to hard-code a duplicate value, so we see the results. This is just a trick to use when prototyping. You can always introduce the duplicate in your data if it is test data. Never mess with production data. If you do, I hope your resume is up to date!

Basically, this is a carbon copy of the prior query, but I included the DENSE_RANK() function so we can compare results. We did say that these functions are similar. The numerical results together with the chart in Microsoft Excel will verify this statement.

Once again, to keep the result set small, the query uses a WHERE clause filter to pull data only for one year, 2011, and for the month of January. Let's see the query results.

Please refer to Figure 4-6.

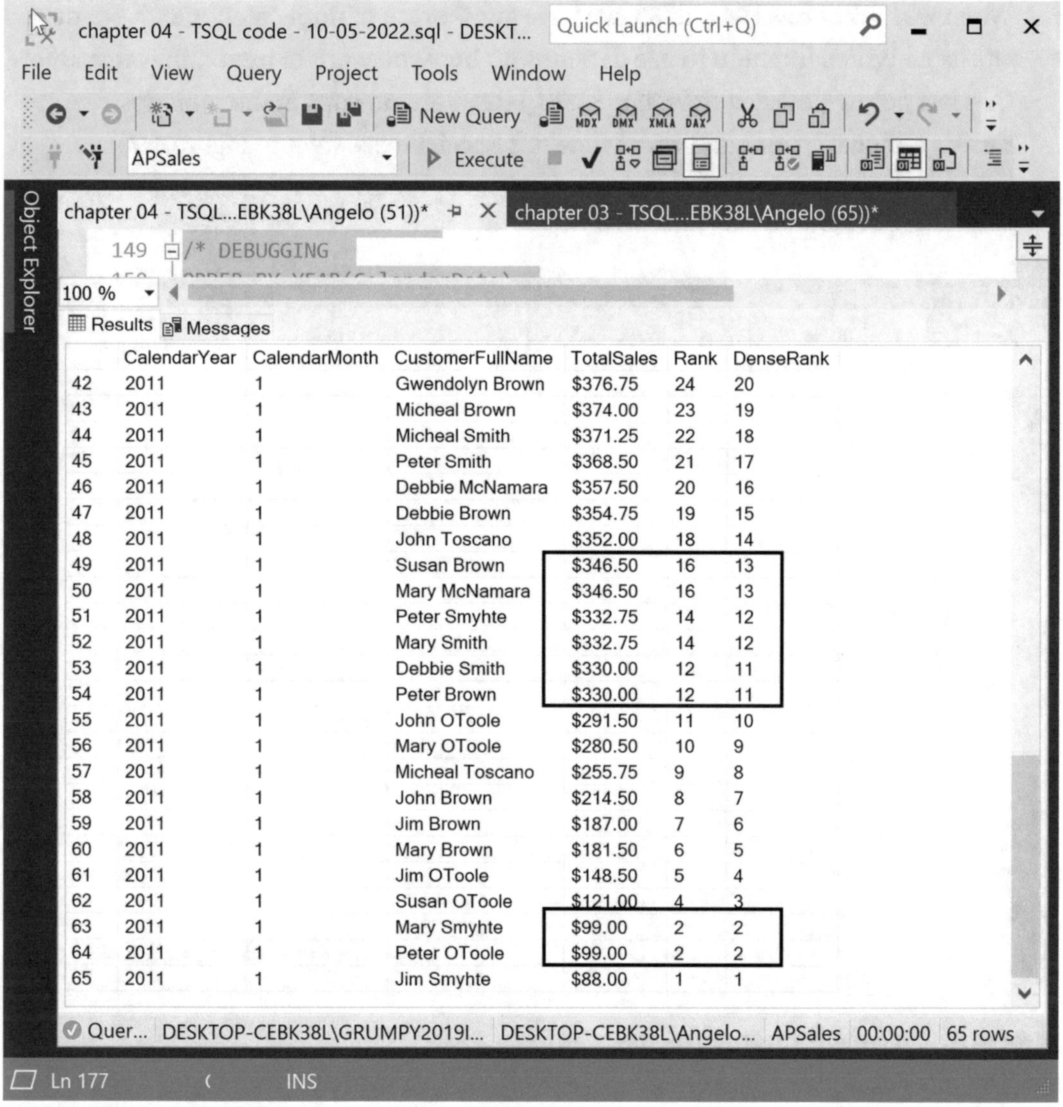

Figure 4-6. *Rank vs. dense rank analysis*

What we want to look at here is how the results differ between these functions if duplicate values are present. Starting from the bottom with rows 63 and 64, both sales totals are $99.00, so this is where the fun begins. Both get a rank and dense rank of 2, but look at what happens when we move to row 62. The DENSE_RANK() value starts at the next value 3, while the RANK() value skips a number and goes to 4. Value 3 is ignored.

When we look at rows 54 and 53, we have another set of duplicates, and the pattern repeats itself. Which function to use depends on how you want to treat duplicate values.

One last point: Notice with DENSE_RANK() there are no gaps in the rankings.

Here is our chart generated with Microsoft Excel. Please refer to Figure 4-7.

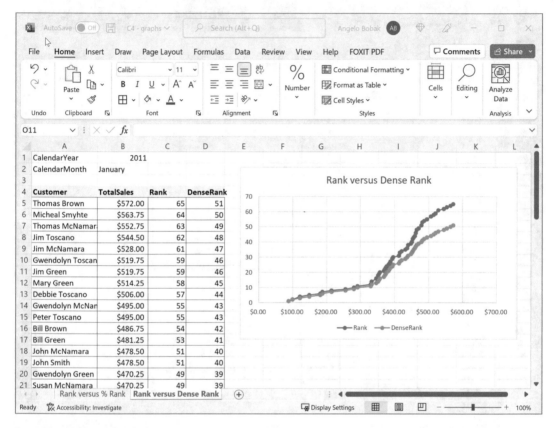

Figure 4-7. *Graphing rank vs. dense rank*

Sure enough, the results are similar once again. A complete set of data and graphs like these for your users will guarantee productive analysis and decision making.

Generating these graphs is easy by the way. Copy and paste the results to a spreadsheet. Tighten up the data and then highlight it so you can generate the graph. Click "Insert" in the menu bar and then select "Recommended Charts." Pick a chart, and you're done except for some formatting to make it look professional.

Performance Considerations

I think there will be no surprises here due to all the indexes we created in prior chapters. Let's create an estimated query plan in the usual manner (click "Query" in the menu bar and then click "Display Estimated Execution Plan" in case you forgot).

Please refer to Figure 4-8.

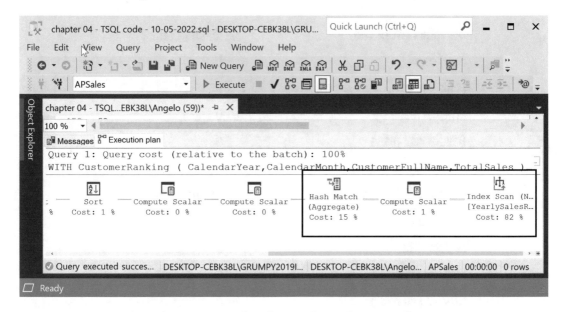

Figure 4-8. *Estimated execution plan for rank vs. dense rank*

Once again, an existing index results in an index scan task with a cost of 82%. If we add up all the costs, we get 82% + 1% + 15% = 98%. Not shown to the left are two sort tasks at 1% each, so the total cost adds up to 100% here. No bug to contend with this time. Let's look at IO and TIME statistics next.

Please refer to Table 4-2.

Table 4-2. *Query IO and Time Statistics*

SQL Server Parse and Compile Times	Existing Index
CPU time (ms)	4
Elapsed time (ms)	4

Statistics (Work File)	Existing Index
Scan count	0
Logical reads	0
Physical reads	0
Read-ahead reads	0

Statistics (Work Table)	Existing Index
Scan count	0
Logical read	0
Physical reads	0
Read-ahead reads	0

Statistics (YearlySalesReport)	Existing Index
Scan count	1
Logical reads	1384
Physical reads	1
Read-ahead reads	1401

SQL Server Execution Times	Existing Index
CPU time (ms)	62
Elapsed time (ms)	58

Statistics on the work file and the work table are all good. Logical reads seem high, but consider that logical reads usually go against memory cache so most likely this value is nothing to worry about. There is one physical read, which means SQL Server had to go to the physical disk to retrieve the pages it required. The read-ahead read statistic indicates how many pages anticipated were required and loaded into memory (recall that if you have a lot of memory, this is usually not a problem). Just be on the lookout for table spool tasks, which means TEMPDB is used.

Lastly, the scan count of 1 and the physical read of 1 are low values that we shouldn't worry about. Don't forget to run DBCC to clear cache and to set statistics on and off each time you perform some analysis, so you do not get incorrect results. Turn statistics on at the end, so they are only collected for the query and not for when you run an estimated execution plan – in other words, right before you execute the query.

NTILE( ) Function Revisited

We covered this function in the beginning of this chapter, but let's take another look. Here is a little credit rating and payment delinquency example that places customers into buckets so they can be assigned to collection agents based on how many days they are delinquent in their payments. Code is provided in the script folder for this chapter to load the credit-related tables.

Please refer to Listing 4-5.

Listing 4-5. Assigning Credit Analysts to Delinquent Accounts

```
DECLARE @NumTiles INT;

SELECT @NumTiles = COUNT(DISTINCT [90DaysLatePaymentCount])
FROM Demographics.CustomerPaymentHistory
WHERE [90DaysLatePaymentCount] > 0;

SELECT CreditYear
    ,CreditQtr
    ,CustomerNo
    ,CustomerFullName
    ,SUM([90DaysLatePaymentCount]) AS Total90DayDelinquent
    ,NTILE(@NumTiles) OVER (
        PARTITION BY CreditYear,CreditQtr
```

```
                ORDER BY CreditQtr
                ) AS CreditAnaystBucket

        ,CASE NTILE(@NumTiles) OVER (
                PARTITION BY CreditYear,CreditQtr
                ORDER BY CreditQtr
                )
                WHEN 1 THEN 'Assign to Collection Analyst 1'
                WHEN 2 THEN 'Assign to Collection Analyst 2'
                WHEN 3 THEN 'Assign to Collection Analyst 3'
                WHEN 4 THEN 'Assign to Collection Analyst 4'
                WHEN 5 THEN 'Assign to Collection Analyst 5'
        END AS CreditAnalystAssignment
FROM Demographics.CustomerPaymentHistory
WHERE [90DaysLatePaymentCount] > 0
GROUP BY CreditYear
        ,CreditQtr
        ,CustomerNo
        ,CustomerFullName
ORDER BY CreditYear
        ,CreditQtr
        ,SUM([90DaysLatePaymentCount]) DESC
        GO
```

We make use of a variable called @NumTiles to count the number of instances of the account being delinquent 90 days. The following query initializes it:

```
SELECT @NumTiles = COUNT(DISTINCT [90DaysLatePaymentCount])
FROM Demographics.CustomerPaymentHistory
WHERE [90DaysLatePaymentCount] > 0;
```

The main query pulls out the year, quarter, customer number, and customer full name together with the total days the customer is delinquent. Next, the CreditAnalystBucket column is assigned the bucket number the customer falls in by using the NTILE() function and an OVER() clause that is partitioned by year and quarter. The partition rows are sorted by quarter and customer. As we have results loaded for only one year, we do not need to partition by year.

Next, a case block is set up to print a message that states

"Assign to Collection Analyst N"

N is a value from 1 to 5. If the credit analyst bucket value is 1, the customer account is assigned to credit analyst 1; if the value is 2, the customer account is assigned to credit analyst 2; and so on. Try to load multiple years' worth of credit data by modifying the load script for this table. You will need to add a `PARTITION BY` clause.

Let's see the results. Please refer to Figure 4-9.

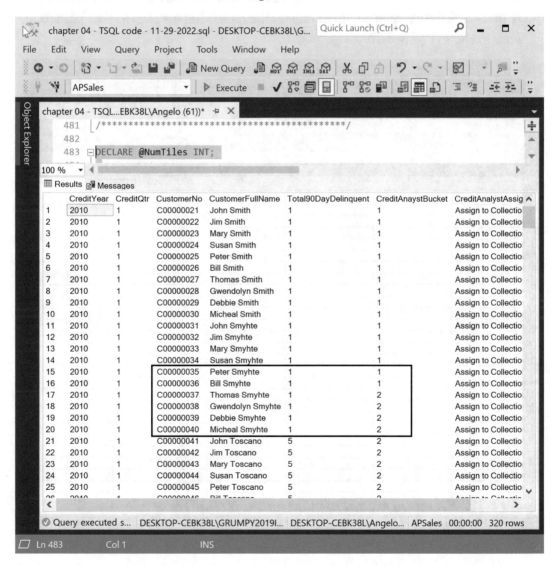

Figure 4-9. *Credit analyst assignments for 90-day accounts*

It works well, but as can be seen there are some overlaps. Some customers that are 90 days late are assigned to either credit analyst 1 or 2. So this strategy tries to balance out the assignments by volume as opposed to the number of times a customer was 90 days late.

Let's see how the NTILE() function works as far as performance is concerned.

Performance Considerations

Let's generate an estimated execution plan, which is created in the usual manner. (Remember, there are a few ways to create it.) This time there are two plans, one for the query that loads the variable and then one for the query that uses the NTILE() function.

Please refer to Figure 4-10.

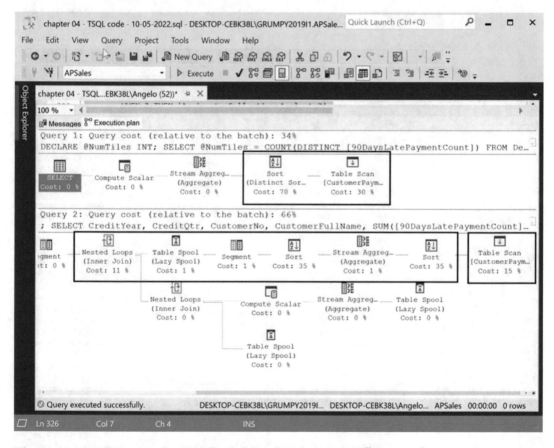

Figure 4-10. *Estimated execution plans for the NTILE() example*

Looking at the first execution plan for the query that sets the @NumTiles variable, we get the sense that it is expensive. We see (going right to left) the table scan task at a cost of 30% and a very expensive sort step at 70%. The rest of the steps cost 0%, so we will not worry about them.

An index was not suggested, but I wonder if an index created on the column used in the sort step would help. The answer is no. In this case the table has only 400 rows, so indexes would not help or be used. The only alternative is to sort it when it is loaded or pop the table into memory. That is, create it as a memory-optimized table.

Let's look at the second plan. No indexes were suggested, and we see a table scan, two sort tasks and some table spools, and two nested loops joins. Nothing to do here.

As I stated earlier, the table has only 400 rows, so let's leave well enough alone. Might as well look at some statistics though for the sake of a complete analysis exercise.

Please refer to Table 4-3.

Table 4-3. *Query IO and Time Statistics*

SQL Server Parse and Compile Times	First Query	Second Query
CPU time (ms)	0	16
Elapsed time (ms)	43	41
Statistics (Work table)	**First Query**	**Second Query**
Scan count	0	3
Logical reads	0	657
Physical reads	0	0
Read-ahead reads	0	0

(continued)

Table 4-3. (*continued*)

Statistics (CustomerPaymentHistory)	First Query	Second Query
Scan count	1	1
Logical reads	5	5
Physical reads	0	0
Read-ahead reads	0	0
SQL Server Execution Times	**First Query**	**Second Query**
CPU time (ms)	16	0
Elapsed time (ms)	41	78

Also, nothing to worry about here. The second query has 657 logical reads, which means it went to memory cache to retrieve data it needed 657 times, but that's not a significant value.

As I stated earlier, when you have a query that works on a small volume of data and the query executes under one second, move on to more troublesome queries (choose your battles).

Note   At the end of the day, if your query ran under one second, leave it alone. Ultimately, it is long-running queries that need to be addressed. Also, what are your users willing to tolerate? Times up to one minute? Take this into account.

ROW_NUMBER() Function

Let's look at the ROW_NUMBER() function again. In this next example, we want to keep a running total by month for products that were sold during a two-year time (2011, 2012) period for two specific stores. Also, we want to track one product only: dark chocolates – medium-size box.

The SUM() aggregate function is used, and the ROW_NUMBER() function is used in a trivial manner to assign an entry number for each month. This function will be used three times so we can see the behavior when we use different PARTITION BY and ORDER BY clause combinations in the query.

Our last example in this chapter will show you how to use the ROW_NUMBER() function to solve the islands and gaps problem related to data ranges. This will be interesting and practical. Estimated query plans should be interesting also. Let's see how to calculate rolling totals next.

Please refer to Listing 4-6.

Listing 4-6. Rolling Sales Total by Month

```
WITH StoreProductAnalysis
(TransYear,TransMonth,TransQtr,StoreNo,ProductNo,ProductsBought)
AS
(
SELECT
      YEAR(CalendarDate)        AS TransYear
      ,MONTH(CalendarDate)      AS TransMonth
      ,DATEPART(qq,CalendarDate) AS TransQtr
      ,StoreNo
      ,ProductNo
      ,SUM(TransactionQuantity)  AS ProductsBought
FROM StagingTable.SalesTransaction
GROUP BY YEAR(CalendarDate,
      ,MONTH(CalendarDate)
      ,DATEPART(qq,CalendarDate)
      ,StoreNo
      ,ProductNo
)

SELECT
      spa.TransYear
      ,spa.TransMonth
      ,spa.StoreNo
      ,spa.ProductNo
      ,p.ProductName
```

```
        ,spa.ProductsBought
        ,SUM(spa.ProductsBought) OVER(
                PARTITION BY spa.StoreNo,spa.TransYear
                ORDER BY spa.TransMonth
                ) AS RunningTotal
        ,ROW_NUMBER() OVER(
                PARTITION BY spa.StoreNo,spa.TransYear
                ORDER BY spa.TransMonth
                ) AS EntryNoByMonth
        ,ROW_NUMBER() OVER(
                PARTITION BY spa.StoreNo,spa.TransYear,TransQtr
                ORDER BY spa.TransMonth
                ) AS EntryNoByQtr
        ,ROW_NUMBER() OVER(
                ORDER BY spa.TransYear,spa.StoreNo
                ) AS EntryNoByYear
FROM StoreProductAnalysis spa
JOIN DimTable.Product p
ON spa.ProductNo = p.ProductNo
WHERE spa.TransYear IN(2011,2012)
AND spa.StoreNo IN ('S00009','S00010')
AND spa.ProductNo = 'P00000011129'
GO
```

Starting with the CTE, the query generates the sum of transactions by year, month, store, and product. The sum is generated for each month.

The query that uses the CTE needs to generate a rolling total of the sums by month per year. For this purpose the SUM() function is used with an OVER() clause that contains a partition by store number and transaction year and includes an ORDER BY clause by transaction month.

The same OVER() clause is used for the ROW_NUMBER() function to generate the entry numbers. This is used three times so we can see how the behavior is if we modify the PARTITION BY clause, so it reflects months, quarters, and years.

Let's check the results. Please refer to Figure 4-11.

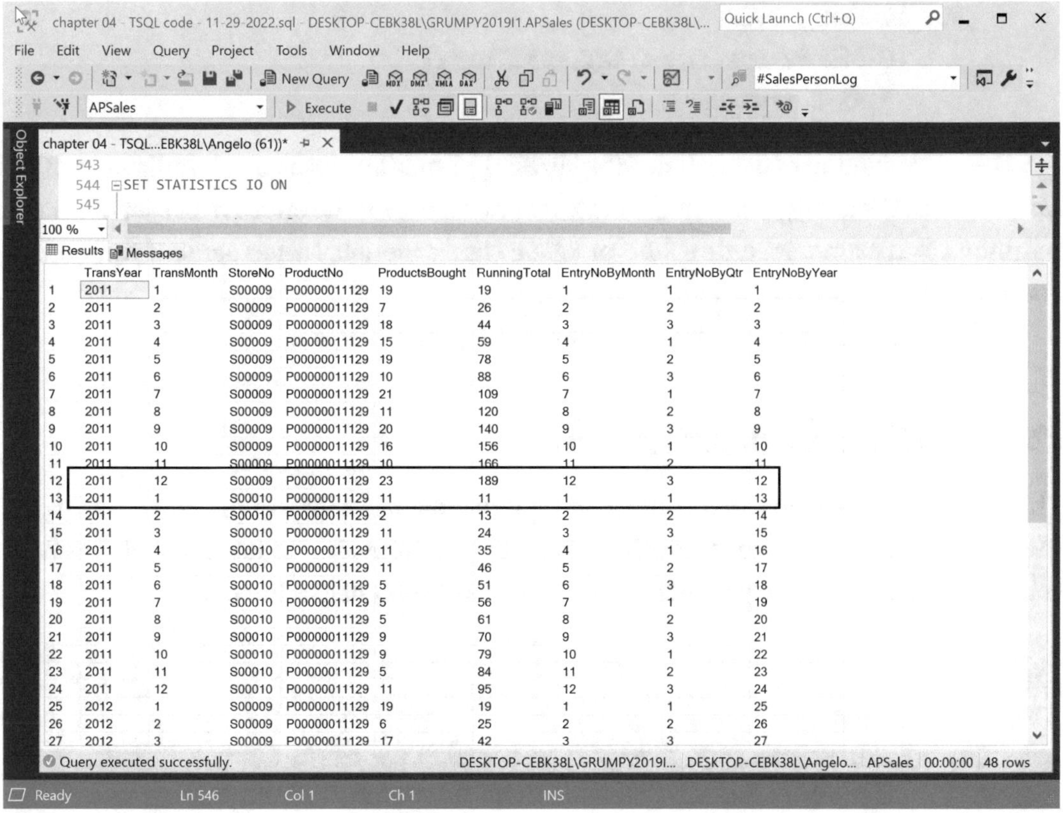

	TransYear	TransMonth	StoreNo	ProductNo	ProductsBought	RunningTotal	EntryNoByMonth	EntryNoByQtr	EntryNoByYear
1	2011	1	S00009	P00000011129	19	19	1	1	1
2	2011	2	S00009	P00000011129	7	26	2	2	2
3	2011	3	S00009	P00000011129	18	44	3	3	3
4	2011	4	S00009	P00000011129	15	59	4	1	4
5	2011	5	S00009	P00000011129	19	78	5	2	5
6	2011	6	S00009	P00000011129	10	88	6	3	6
7	2011	7	S00009	P00000011129	21	109	7	1	7
8	2011	8	S00009	P00000011129	11	120	8	2	8
9	2011	9	S00009	P00000011129	20	140	9	3	9
10	2011	10	S00009	P00000011129	16	156	10	1	10
11	2011	11	S00009	P00000011129	10	166	11	2	11
12	2011	12	S00009	P00000011129	23	189	12	3	12
13	2011	1	S00010	P00000011129	11	11	1	1	13
14	2011	2	S00010	P00000011129	2	13	2	2	14
15	2011	3	S00010	P00000011129	11	24	3	3	15
16	2011	4	S00010	P00000011129	11	35	4	1	16
17	2011	5	S00010	P00000011129	11	46	5	2	17
18	2011	6	S00010	P00000011129	5	51	6	3	18
19	2011	7	S00010	P00000011129	5	56	7	1	19
20	2011	8	S00010	P00000011129	5	61	8	2	20
21	2011	9	S00010	P00000011129	9	70	9	3	21
22	2011	10	S00010	P00000011129	9	79	10	1	22
23	2011	11	S00010	P00000011129	5	84	11	2	23
24	2011	12	S00010	P00000011129	11	95	12	3	24
25	2012	1	S00009	P00000011129	19	19	1	1	25
26	2012	2	S00009	P00000011129	6	25	2	2	26
27	2012	3	S00009	P00000011129	17	42	3	3	27

***Figure 4-11.** Rolling total sales by month*

The report works well enough in that it calculates running totals by month. Notice where the new year starts or the store changes. The rolling total is reset, and so are the entry number levels. We can generate row numbers by month or quarter or year depending on how the PARTITION BY clause is defined:

```
,ROW_NUMBER() OVER(
     PARTITION BY spa.StoreNo,spa.TransYear
     ORDER BY spa.TransMonth
     ) AS EntryNoByMonth
,ROW_NUMBER() OVER(
     PARTITION BY spa.StoreNo,spa.TransYear,TransQtr
     ORDER BY spa.TransMonth
     ) AS EntryNoByQtr
```

```
,ROW_NUMBER() OVER(
     ORDER BY spa.TransYear,spa.StoreNo
     ) AS EntryNoByYear
```

This combination of uses of the ROW_NUMBER() function gives you an indication of how it works. Generally, you can generate row numbers for the entire result set or for the partitions. You cannot include a ROWS or RANGE clause though. Makes sense if you think about it. It must work on the entire partition or result set!

Let's see the performance impact when this function is used.

Performance Considerations

This time I will show most of the estimated query plan; it is a long one. I need to split the execution plan into two screenshots. We will start from right to left as usual. The first screenshot in Figure 4-12 shows the first (right-hand side) half of the plan.

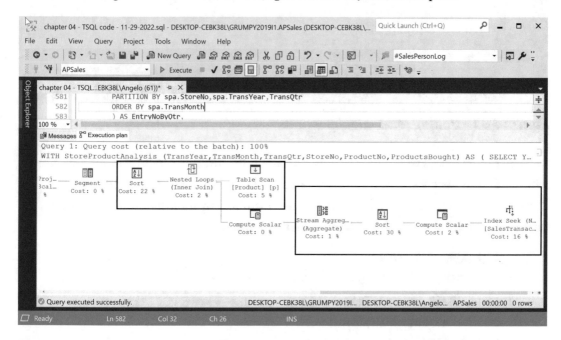

Figure 4-12. *Estimated index plan for rolling monthly sales analysis*

An index seek appears as the first step with a cost of 16% of the total estimated execution time. Ignoring the low-cost steps, a sort step clocks in at 30%. Next, since we are linking (with a `JOIN`) the `SalesTransaction` table to the `Product` table, a table scan step appears at a cost of 5%. Not too expensive as this is a small table, so no index is needed. A nested loops join task joins both streams of data. Following is a sort step at 22% and finally a window spool at 1%.

We always want to pay attention to spool tasks as this is an indicator that cached data is spooled so it can be repeatedly accessed. Depending on the type of spool tasks (table, index, lazy, eager), temporary storage is used, and then things could get really expensive in terms of execution time (having `TEMPDB` on a solid-state drive helps …).

Research Microsoft documentation to see which spool tasks are physical (`TEMPDB`), logical (memory cache), or both. Here is Table 4-4 with selected tasks that identifies if they are logical or physical or both.

Table 4-4. *Logical and Physical Plan Tasks*

Task	Physical	logical
Sort	X	X
Eager spool		X
Lazy spool		X
Spool	X	
Table spool		X
Window spool	X	X
Table scan	X	X
Index scan	X	X
Index seek	X	X
Index spool		X

Lastly, physical tasks are tasks that go against physical disk, and logical tasks are tasks that go against memory cache. As can be seen in the preceding table, some tasks can be both depending on the query. Keep this information in mind together with the `IO` and `TIME` statistics when you are analyzing performance.

Note   I have seen inexpensive solid-state USB drives at around $90 for those of you that want to experiment and have some cash lying around. Install SQL Server 2022 and make sure TEMPDB is created on the solid-state drive. Thinking of getting one myself for Christmas!

Let's look at the left-hand side of the estimated execution plan. Please refer to Figure 4-13.

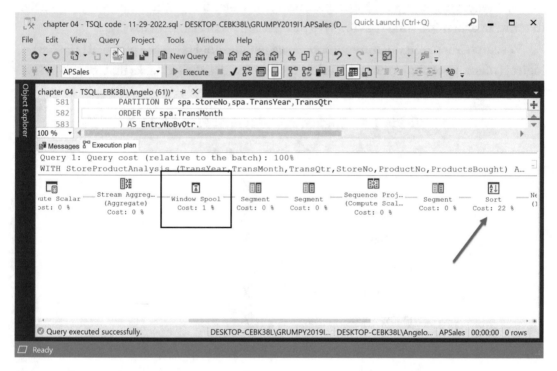

Figure 4-13.   *Left-hand side of the plan*

There's the sort step that costs 22% as a point of reference from the prior screenshot. All we see here is the window spool step we just discussed. This appears because we are using window functions. A 1% cost is a nice low value, so we do not need to worry about it. Or do we?

If logical reads are not zero and the window spool tasks are greater than zero, then the window spool is not using memory.

Here come our statistics. Please refer to Table 4-5.

Table 4-5. *Query IO and Time Statistics*

SQL Server Parse and Compile Times	Existing Index
CPU time (ms)	15
Elapsed time (ms)	70

Statistics (Work Table)	Existing Index
Scan count	52
Logical reads	289
Physical reads	0
Read-ahead reads	0

Statistics (SalesTransaction)	Existing Index
Scan count	2
Logical reads	28
Physical reads	1
Read-ahead reads	21

Statistics (Product)	Existing Index
Scan count	1
Logical reads	1
Physical reads	1
Read-ahead reads	0

SQL Server Execution Times	Existing Index
CPU time (ms)	0
Elapsed time (ms)	44

The only high cost is the logical reads on the work table, but remember, logical reads go against cache in memory (i.e., if you have enough), so this is not a value that should cause concern.

The rest of the values are low, and the estimated query plan analyzer did not suggest a new index, so the query works well. Recall our window spool task. It is not zero, so that means it is performing spooling to TEMPDB on physical disk.

All this information is great, but what does it all mean? How can we use this information to improve performance?

The rule of thumb is to create indexes based on the columns in the PARTITION BY clause and then the ORDER BY clause. In this prior example, we had the following columns in the PARTITION BY clause: StoreNo, TransYear, TransQtr. And the ORDER BY clause used the TransMonth column.

Okay, but we have a poser for you. The date columns were all generated in the CTE by using the YEAR(), MONTH(), and DATEPART() functions. These used the CalendarDate column from the SalesTransaction table. These are all derived columns, so our index would need to be based on the StoreNo column and the CalendarDate column. Would a JOIN to the Calendar dimension table be more effective so we can pull out these date parts? Would this strategy be more efficient than deriving them with the date functions? Try it out!

Let's see what our index seek task is doing in the estimated query plan. Please refer to Figure 4-14.

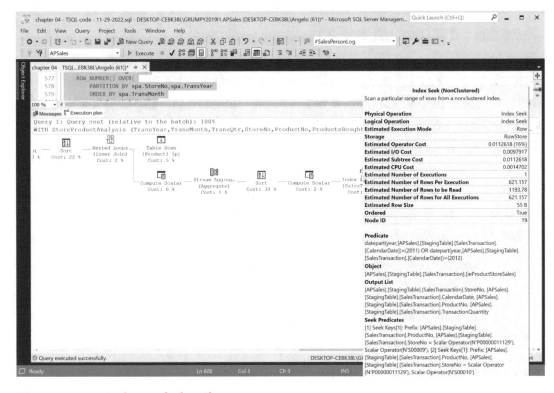

Figure 4-14. *Index seek details*

By placing the mouse pointer over any task, a nice yellow pop-up panel appears with all the details. This is a lot of information and might be hard to read, but let me highlight the columns identified in two of the four sections:

Predicate uses column `CalendarDate` belonging to the `SalesTransaction` table.

Seek Predicates uses seek keys `ProductNo` and `StoreNo` belonging to the `SalesTransaction` table.

So this indicates that we need to look at not only the columns in the `PARTITION BY` and `ORDER BY` clauses of the `OVER()` clause but also the columns in the `WHERE` clause, specifically in the `WHERE` clause `PREDICATES`.

The index used in this query is called `ieProductStoreSales`. It is based on the columns `ProductNo` and `StoreNo`, and the `INCLUDE` columns are `CalendarDate` and `TransactionQuantity`.

Reminder   INCLUDE columns are columns you include in the CREATE INDEX command by using the INCLUDE keyword. The prefix "ie" in the name stands for inversion entry index. This means that we are presenting the user with a non-unique access path to the data. If you see "pk," it means primary key; "ak" means alternate key, which is an alternate to the primary key; and "fk" stands for foreign key. I think it is good practice to include these prefixes in the index name.

So now we have taken our performance analysis to a lower level. By looking at the details behind the high-cost tasks, we can further see if a strategy is working or not.

So let's summarize what we need to base our performance analysis and strategy on:

- Generate estimated and actual query plans and identify high-cost steps.

- Be aware of which tasks are logical or physical operators or both.

- Generate IO and TIME performance statistics.

- Always run DBCC for each test run to make sure memory cache is clear.

- Base index columns on the columns used in the PARTITION BY clause (in the OVER() clause).

- Base index columns on the columns used in the ORDER BY clause (in the OVER()clause).

- Base index columns on the columns used in the WHERE clause (if one is used in the CTE).

- Base index columns on columns used in derived fields if we use functions like YEAR(), MONTH(), DATEPART(), etc. to pull out the year, month, quarter, and day parts of a date.

- Leverage prior indexes by modifying them so they support multiple queries.

- As a last resort, consider memory-enhanced tables or report tables that are preloaded.

- Solid-state drives for TEMPDB will help!

Denormalization Strategies

The preceding second-to-last step said to consider building staging or reporting tables so preprocessing data like data generated by a CTE can be loaded once in a staging table and used by the queries containing the window functions (assuming users can work with 24-hour data or older).

As part of the preceding effort, perform an analysis of all your queries to see if you can design a common index set and staging table for multiple queries. Indexes are powerful, but too many will slow down the table load and modification processing time to a dangerous level.

Consider building memory-optimized tables for fast performance.

If all else fails, we need to consider query or table redesign or even consider hardware upgrades like more memory or more and faster CPUs and storage, like solid-state disk.

Finally, test, test, and did I say test your queries? Come up with a simple test strategy and document results for analysis.

Suggestion   Once you are comfortable reading estimated execution plans and IO and TIME statistics, modify the load scripts that were used to load the databases and try loading a large number of rows, like millions of rows, and see how the performance is impacted.

Islands and Gaps Example

A classic data challenge is the concept of gaps and islands. For example, do we have gaps in the days that sales staff sell products? Or are there clusters of days that a product is selling consistently? This last example shows you how to solve this challenge by using some of the window functions we covered in this chapter.

The solution requires a few steps. The trick to this puzzle is to find out a way to generate some sort of category name we can use in a GROUP BY clause so we can extract the minimum start dates and maximum end dates of the islands and gaps using the MIN() and MAX() aggregate functions.

We will use the ROW_NUMBER() function and the LAG() function to identify when a gap or island starts but to also generate a number that will be used in the value that will be used in the GROUP BY clause mentioned earlier (like ISLAND1, GAP1, etc.).

In our test scenario, we are tracking whether a salesperson generated sales over a 31-day period only. Some sales amounts are set to zero for one or more days, so these gaps in days can be one day, two days, or more (these are the gaps).

Tip   Always start with a small data set. Apply your logic, test it, and, once you are satisfied it works, apply it to your larger production data set.

The same can be said for the groups of days that sales were generated (these are the islands).

Here are some steps that need to occur:

Step 1: Generate some test data. For simplicity include a column that identifies rows as islands or gaps. These will be used to create the groups used in the GROUP BY clause.

Step 2: Add a numerical column to identify the start days of islands or gaps. This will be used to identify the GROUP BY text value.

Step 3: Add a column that will contain a unique number, so for each set of islands and gaps, we can generate category names for the GROUP BY clause that look something like this: ISLAND1, ISLAND2, GAP1, GAP2, etc. (we will use the ROW_NUMBER() function for this).

Once these categories have been correctly assigned to each row, the start and stop days are easily pulled out with the MIN() and MAX() aggregate functions.

Here we go. Let's start by generating some test data.

Please refer to Listing 4-7a.

Listing 4-7a. Loading the SalesPersonLog

```
USE TEST
GO

DROP TABLE IF EXISTS SalesPersonLog
GO

CREATE TABLE SalesPersonLog (
    SalesPersonId VARCHAR(8),
    SalesDate     DATE,
```

```
        SalesAmount          DECIMAL(10,2),
        IslandGapGroup VARCHAR(8)
        );
TRUNCATE TABLE SalesPersonLog
GO

INSERT INTO SalesPersonLog
SELECT 'SP001'
        ,[CalendarDate]
        ,UPPER (
                CONVERT(INT,CRYPT_GEN_RANDOM(1)
        )) AS SalesAmount
        ,'ISLAND'
FROM APSales.[DimTable].[Calendar]
WHERE [CalendarYear] = 2010
AND [CalendarMonth] = 10
GO

/*******************/
/* Set up some gaps */
/*******************/

UPDATE SalesPersonLog
SET SalesAmount = 0,
        IslandGapGroup = 'GAP'
WHERE SalesDate BETWEEN '2010-10-5' AND '2010-10-6'
GO

UPDATE SalesPersonLog
SET SalesAmount = 0,
        IslandGapGroup = 'GAP'
WHERE SalesDate BETWEEN '2010-10-11' AND '2010-10-16'
GO

UPDATE SalesPersonLog
SET SalesAmount = 0,
        IslandGapGroup = 'GAP'
```

```
WHERE SalesDate BETWEEN '2010-10-22' AND '2010-10-23'
GO

-- Just in case the random sales value generator
-- set sales to 0 but the update labelled it as an ISLAND

UPDATE SalesPersonLog
SET IslandGapGroup = 'GAP'
WHERE SalesAmount = 0
GO
```

The gap and island text strings could have been generated in the following query, but I wanted to simplify things. Notice all the update statements. These set up the test scenario for us.

Please refer to Listing 4-7b.

Listing 4-7b. Generating the Gaps and Islands Report

```
SELECT SalesPersonId,GroupName,SUM(SalesAmount) AS TotalSales
     ,MIN(StartDate) AS StartDate,MAX(StartDate) AS EndDate
     ,CASE
          WHEN SUM(SalesAmount) <> 0 THEN 'Working, finally!'
          ELSE 'Goofing off again!'
     END AS Reason
FROM (
     SELECT SalesPersonId,SalesAmount,
          ,IslandGapGroup + CONVERT(VARCHAR,(SUM(IslandGapGroupId)
               OVER(ORDER BY StartDate) )) AS GroupName
          ,StartDate
          ,PreviousSalesDate AS EndDate
     FROM
     (
          SELECT ROW_NUMBER() OVER(ORDER BY SalesDate) AS RowNumber
               ,SalesPersonId
               ,SalesAmount
               ,IslandGapGroup
               ,SalesDate AS StartDate
               ,LAG(SalesDate)
```

```
                    OVER(ORDER BY SalesDate) AS PreviousSalesDate
            ,CASE
                    WHEN LAG(SalesDate) OVER(ORDER BY
                    SalesDate) IS NULL
                        OR
                        (
                        LAG(SalesAmount) OVER(ORDER BY
                        SalesDate) <> 0
                        AND SalesAmount = 0
                        ) THEN ROW_NUMBER() OVER(ORDER BY SalesDate)
                    WHEN (LAG(SalesAmount) OVER(ORDER BY SalesDate) = 0
                        AND SalesAmount <> 0)
                    THEN ROW_NUMBER() OVER(ORDER BY SalesDate)
                    ELSE 0
            END AS IslandGapGroupId
        FROM SalesPersonLog
    ) T1
)T2
GROUP BY SalesPersonId,GroupName
ORDER BY StartDate
GO
```

This query has three levels or three nested queries acting like tables to solve the problem. The in-line query used like a table that is labeled T1 produces the following values in Figure 4-15.

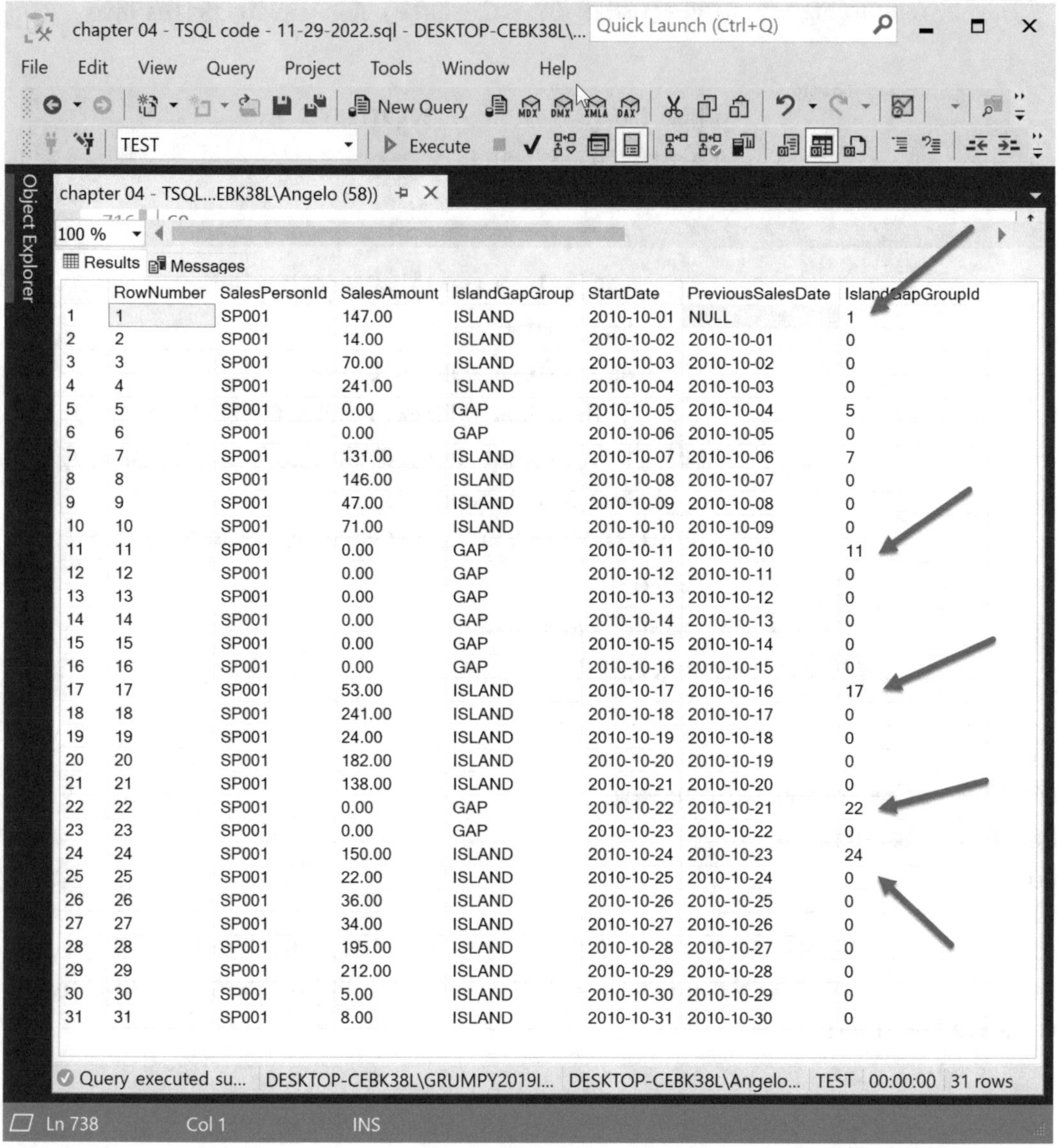

Figure 4-15. *Gaps and islands interim results*

Notice how the values that are used to identify the start of a gap or island are generated by the ROW_NUMBER() function. This guarantees they are unique. Let's look at the next level query.

Please refer to Figure 4-16.

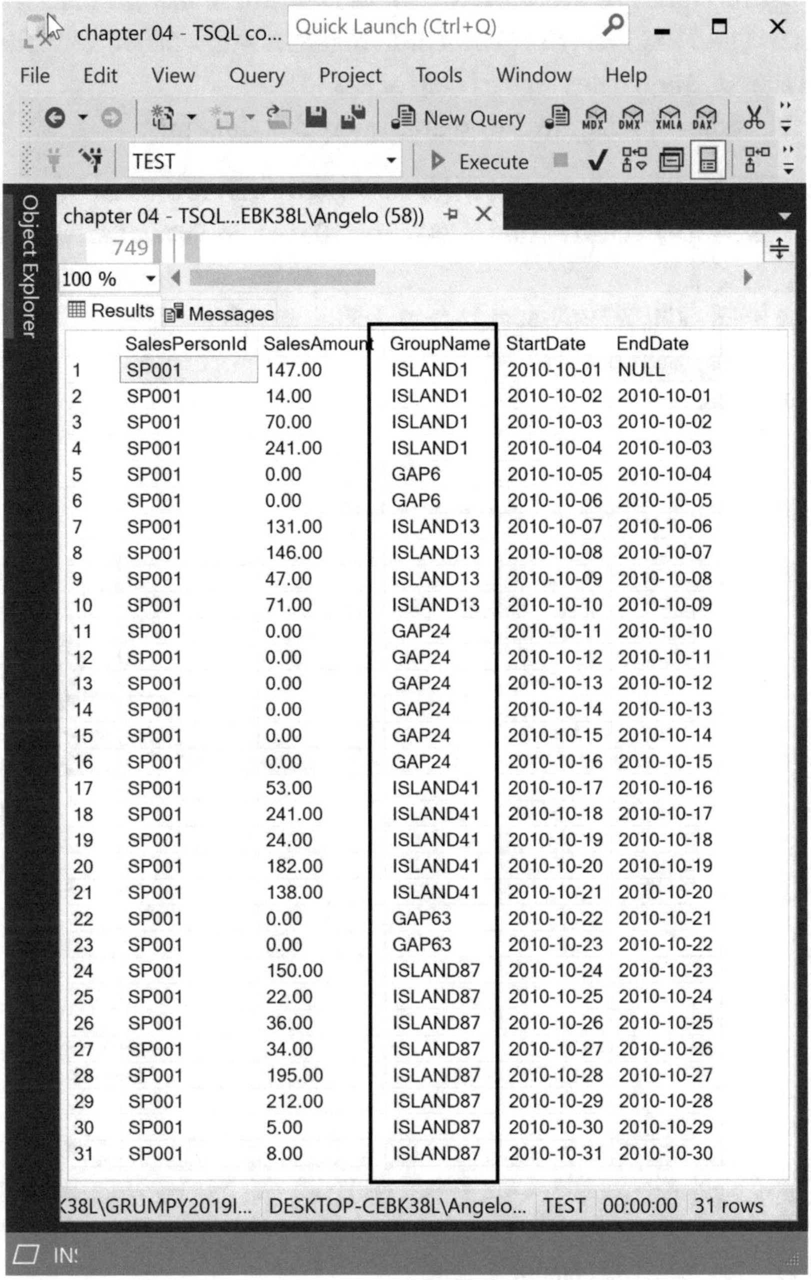

	SalesPersonId	SalesAmount	GroupName	StartDate	EndDate
1	SP001	147.00	ISLAND1	2010-10-01	NULL
2	SP001	14.00	ISLAND1	2010-10-02	2010-10-01
3	SP001	70.00	ISLAND1	2010-10-03	2010-10-02
4	SP001	241.00	ISLAND1	2010-10-04	2010-10-03
5	SP001	0.00	GAP6	2010-10-05	2010-10-04
6	SP001	0.00	GAP6	2010-10-06	2010-10-05
7	SP001	131.00	ISLAND13	2010-10-07	2010-10-06
8	SP001	146.00	ISLAND13	2010-10-08	2010-10-07
9	SP001	47.00	ISLAND13	2010-10-09	2010-10-08
10	SP001	71.00	ISLAND13	2010-10-10	2010-10-09
11	SP001	0.00	GAP24	2010-10-11	2010-10-10
12	SP001	0.00	GAP24	2010-10-12	2010-10-11
13	SP001	0.00	GAP24	2010-10-13	2010-10-12
14	SP001	0.00	GAP24	2010-10-14	2010-10-13
15	SP001	0.00	GAP24	2010-10-15	2010-10-14
16	SP001	0.00	GAP24	2010-10-16	2010-10-15
17	SP001	53.00	ISLAND41	2010-10-17	2010-10-16
18	SP001	241.00	ISLAND41	2010-10-18	2010-10-17
19	SP001	24.00	ISLAND41	2010-10-19	2010-10-18
20	SP001	182.00	ISLAND41	2010-10-20	2010-10-19
21	SP001	138.00	ISLAND41	2010-10-21	2010-10-20
22	SP001	0.00	GAP63	2010-10-22	2010-10-21
23	SP001	0.00	GAP63	2010-10-23	2010-10-22
24	SP001	150.00	ISLAND87	2010-10-24	2010-10-23
25	SP001	22.00	ISLAND87	2010-10-25	2010-10-24
26	SP001	36.00	ISLAND87	2010-10-26	2010-10-25
27	SP001	34.00	ISLAND87	2010-10-27	2010-10-26
28	SP001	195.00	ISLAND87	2010-10-28	2010-10-27
29	SP001	212.00	ISLAND87	2010-10-29	2010-10-28
30	SP001	5.00	ISLAND87	2010-10-30	2010-10-29
31	SP001	8.00	ISLAND87	2010-10-31	2010-10-30

***Figure 4-16.** Generating values for final GROUP BY clause*

At this level we have all the category values generated, and we used the LAG() function to set the end dates. We still have results in linear, sequential dates. All we need to do now is use the MAX() and MIN() functions together with a GROUP BY clause so we pull out the start and end dates for each gap or island.

Here is the code snippet from the original query that does this:

```
SELECT SalesPersonId,GroupName,SUM(SalesAmount) AS TotalSales,
    MIN(StartDate) AS StartDate,MAX(StartDate) AS EndDate,
    CASE
        WHEN SUM(SalesAmount) <> 0 THEN 'Working, finally!'
        ELSE 'Goofing off again!'
    END AS Reason
FROM ...
```

Let's see the results. Please refer to Figure 4-17.

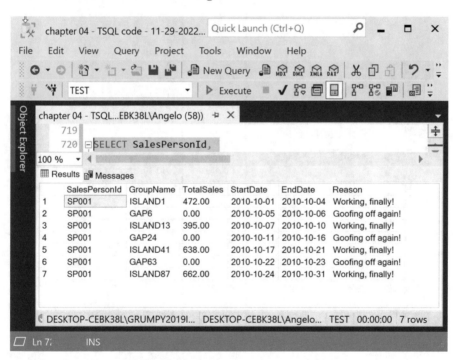

Figure 4-17. *The gaps and islands report*

Looks good. A column called Reason tells us the cause of gaps in sales days. This salesperson needs a good talking too!

In conclusion, come up with logic to identify the start dates of the islands and gaps. Next, come up with a strategy for generating values that can be used in the GROUP BY clause to categorize the islands and gaps. Finally, use the MAX() and MIN() functions to generate the start and stop dates of the gaps and islands.

Summary

We covered a lot of ground. What did we learn?

- We learned how to apply the window or ranking functions to our sales data warehouse.

- We took a deeper dive into performance analysis by understanding how data is either cached in memory or on physical disk.

- We started to analyze and come up with conclusions on performance based on the nature of estimated plan tasks and IO and TIME statistics.

- We had a brief discussion about and summarized some steps to consider when doing performance analysis.

- Lastly, we looked at a real-world data problem called gaps and islands.

Armed with all this knowledge, we continue to see how all our functions work against a financial database.

CHAPTER 5

Finance Use Case: Aggregate Functions

We now turn our focus to the financial industry. As was the case with the past three chapters, we will use the three categories of SQL Server functions available to us to analyze a stock trading scenario. We will dedicate one chapter to each category of functions. Each function will be used in a query. We will examine the results by looking at the output and, then in most cases, copying the output to a Microsoft Excel spreadsheet so we can generate a graph or two to visualize the results.

We will also get deeper into performance tuning. A powerful command called SET PROFILE STATISTICS ON can be used to get valuable insight into exactly what SQL Server is doing when executing a query. This will be a long chapter, so brew a lot of coffee!

Tip Keep in mind that estimated query plan costs are not the time it takes to complete a step, but a weighted value assigned by the query optimizer that takes into account IO, memory, and CPU costs.

Aggregate Functions

Aggregate functions coupled with the OVER() clause (so we can partition result sets) deliver a powerful set of tools. Recall that this category includes the following functions:

- COUNT() and COUNT_BIG()
- SUM()
- MAX()

© Angelo Bobak 2023
A. Bobak, *SQL Server Analytical Toolkit*, https://doi.org/10.1007/978-1-4842-8667-8_5

- MIN()

- AVG()

- GROUPING()

- STRING_AGG()

- STDEV() and STDEVP()

- VAR() and VARP()

But one thing to be aware of: If you use these with the OVER() clause, DISTINCT is not allowed as a keyword between parentheses! This will generate an error: AVG(DISTINCT ...) OVER(...).

Additionally, we will not review the COUNT_BIG() function again as it works just like the COUNT() function. The only difference is that it returns a BIGINT data type.

By the way, the range for this data type is between –9,223,372,036,854,775,808 and 9,223,372,036,854,775,807. It requires 8 bytes of storage compared with the integer data type we have been using in our examples. This data type takes only 4 bytes of storage, and the range of values is between –2,147,483,648 and 2,147,483,647. Impress your friends with these statistics (not).

Back to our aggregate functions. These functions can be part of a toolkit that lets you analyze financial data and set up powerful scripts and reports for your users, analysts, financial advisors, and managers. Include them in stored procedures, and they can be called from SSRS (SQL Server Reporting Services) to create valuable reports for financial stakeholders.

One thing to keep in mind as you go through this chapter is that financial data is characterized by large volumes of daily trade transactions and even larger volumes of historical trades. We are talking about hundreds of millions of trades a day for millions of customers depending on how large the financial institution is. For this reason, we need to keep performance analysis and tuning in mind so as to create powerful and fast queries. You might deal with billions of rows a year. I had to!

In order to do this, we must be very familiar with the database we are working with – the size, the table contents, and the relationships between tables.

Let's start by taking some time to understand the data model of the database we will be working with. We will do this with a simple conceptual business model.

Please refer to Figure 5-1.

AP Finance Conceptual Model (Trade)

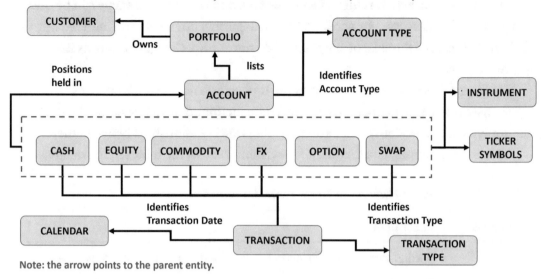

Figure 5-1. *Financial database physical data model*

This is called a conceptual model because it only identifies the high-level entities and relationships from a business perspective. This diagram can be understood by nontechnical analysts and management. They will have great input during the design phase of a database.

Let's walk through the entities and the relationships between them.

A customer owns a portfolio, which is made up of accounts. Accounts are identified by their types. An account can be one of the following:

- Cash account

- Equity account

- Commodity account

- FX (foreign exchange) account

- Option account (a type of derivative)

- Swap account (another type of derivative)

Each of these account types holds positions (balances) for financial instruments identified by their ticker symbol, like IBM for International Business Machines.

Finally, one or more transactions occur daily and update the positions (balances) of the accounts. A transaction is identified by its type and category. The type can be one of the types we just discussed for accounts. The category in our scenario is either a buy or sell transaction. (For the sake of simplicity, I do not show a Category entity as it will have only two values.)

Finally, the Calendar entity identifies the transaction date in a transaction.

Next, we need a set of documents called data dictionaries that describe the key components in the model, in our case the entities and relationships between them.

Table 5-1 is a simple data dictionary describing each of the tables in the database.

Table 5-1. *Financial Database Table Data Dictionary*

Table Name	Description
Account	Used to track trading balances in a customer's account. Links to Portfolio.
Account Versus Target	This entity is used to track account actual balances vs. target balances to see if an account performed or underperformed.
Account Type	Identifies the account type: equity, cash, FX (foreign exchange), commodity, or derivative.
Transaction	Stores daily buy and sell transactions for all customers and transaction types.
Transaction Type	Identifies transaction type: equity, cash, FX (foreign exchange), commodity, or derivative.
Transaction Account Type	Links account types to transaction types. For example, TT00001 (cash transaction) links to account type AT00001, which is a cash balance account.
Portfolio	Records monthly positions for the following portfolio types: FX - FINANCIAL PORTFOLIO EQUITY - FINANCIAL PORTFOLIO COMMODITY - FINANCIAL PORTFOLIO OPTION - FINANCIAL PORTFOLIO SWAP - FINANCIAL PORTFOLIO CASH - FINANCIAL PORTFOLIO
Calendar	Basic calendar date objects used for time reporting (like date, quarter, month, and year objects).

(*continued*)

Table 5-1. (*continued*)

Table Name	Description
Country	Contains ISO two- and three-character codes for most of the world's countries, for example, US or USA.
Customer	Basic customer table used to store names of the customers and identifiers.
Customer Ticker Symbols	Used to identify the ticker symbols a customer trades in.
Ticker Symbols	Identifies name of a financial trading instrument ticker, for example, AA for Alcoa and SWAADY10.RT for I/R (interest rate) Swap 10-Year.
Ticker Price Range History	Stores historical lows, highs, and spreads for 37 sample ticker symbols by date. For example: Ticker BSBR Company: Banco Santander Brasil Ticker Date: 2015-03-19 Low: 123.10 High: 139.90 Spread: 16.80
Ticker History	Stores history of trades for one fictitious ticker symbol only (GOITGUY). Used for an example.

Next, Table 5-2 is a simple data dictionary describing each of the relationships between the tables in the database.

Table 5-2. *Financial Table Data Dictionary*

Parent Table	Business Rule	Child Table	Cardinality
Customer	Owns a	Portfolio	One to zero, one, or many
Portfolio	Is made up of	Account	One to zero, one, or many
Account Type	Identifies type of	Account	One to zero, one, or many
Account	May be a	Cash Account	One, zero to one
	May be a	Equity Account	One, zero to one
	May be a	Commodity Account	One, zero to one
	May be a	FX Account	One, zero to one
	May be a	Option Account	One, zero to one
	May be a	Swap Account	One, zero to one
Account	Refers to an	Instrument	One to zero, one, or many
Instrument	Is identified by	Ticker Symbol	One to One
Transaction	Updates	Account	Zero, one, many to one
Transaction Type	Identifies type of	Transaction	One to zero, one, or many
Calendar	Identifies date of	Transaction	One to zero, one, or many

This business conceptual model is meant to convey that a customer holds one or more portfolios. A portfolio is made up of accounts that include cash, equity, commodities (gold, silver, etc.), FX (foreign currency exchange), options, and derivatives. I could have created a subtype called derivatives and have all the swap, option, and exotic derivatives like swaptions appear as subtypes, but I want to keep the model simple. You can try it out. Research what these instruments are. You can modify the data model and database if you are inspired to do so and if you are an expert financial type.

Let's discuss transactions next. Transactions deal with the instruments (not the ones you play) mentioned earlier, and there can be one or more transactions a day per customer. Lastly, the Calendar entity identifies the data objects for each date in the Transaction entity, that is, not only dates for days but the year, month, quarter, and week parts of a date.

I did not include a data dictionary for the primary and foreign key attributes, but you can identify these in the physical tables themselves by downloading the code for this chapter used to create the database and tables and also populate the tables. An example of a primary key attribute for a Customer entity is "Customer Number." This can be used to uniquely identify a customer. It acts as a foreign key when it appears in another entity like the Transaction entity. This attribute is used to link both entities together.

Note   The DDL commands used to create the APFinance financial database can be found on the Google website dedicated to this book for this chapter together with scripts to create all tables and views and scripts to load the tables with test data. The code is easy to use. Each step is labeled, and where necessary comments help clarify what each step does. (Spreadsheets used in the graph examples are available also.)

Make sure you review the model and data dictionaries we just discussed so that you are comfortable understanding the queries we will discuss, especially when we refer to and join multiple tables in a query.

In case you are not reading the chapters sequentially, I will include a brief description of what each function does again. I always find it a bit cumbersome to flip back 20, 30, or more pages to review a concept that is new to me. If you are new to these, a little review will not hurt. If you know what they do, just skip the descriptions.

COUNT( ) and SUM( ) Functions

The COUNT() function allows you to count the number of times a data value appears in a row column. I include the SUM() function, which adds up data values in a column. Both of these will be used in our example to see how that can be applied to our financial banking scenario.

The syntax is easy:

```
COUNT(*) or COUNT(ALL|DISTINCT <column name>)
SUM(<column name>) or SUM(ALL|DISTINCT <column name>)
```

Back to the COUNT() function. Use the wildcard character symbol "*" to count all rows that satisfy the query requirements; by this I mean if the query includes or does not include a GROUP BY clause. If you call out a column, you can optionally include the keyword ALL or DISTINCT or leave it out entirely.

For example, assume you have ten rows in a table and the table includes a column called ProductType. For six of the rows in the table, the column contains "Type A," and for four of the rows, the value for the column is "Type B." If we count the number of types, we get 2. If we count the number of values in the first type, we get six. If we count the number of values in the second type, we get four. If we count all values regardless of type, we get ten. Walk in the park!

Let's see the results if we use the syntax variations of the function in some simple queries.

Please refer to Listing 5-1.

Listing 5-1. COUNT() Function in All Its Flavors

```
DECLARE @CountExample TABLE (
     ProductType VARCHAR(64)
     );

INSERT INTO @CountExample VALUES
('Type A'),
('Type A'),
('Type A'),
('Type A'),
('Type A'),
('Type A'),
('Type B'),
('Type B'),
('Type B'),
('Type B');

SELECT COUNT(*) FROM @CountExample;
SELECT COUNT(DISTINCT ProductType) FROM @CountExample;
SELECT COUNT(ALL ProductType) FROM @CountExample;
SELECT COUNT(ProductType) FROM @CountExample;
GO
```

The example is simple. But there are subtle variations you need to be aware of.

Using DISTINCT, for example, will only count distinct values, and this might be what you want or not what you want depending on the business specification given to you by your business users.

Make sure you count what you think you want to count. For example, if you have 2, 2, 5, 7, 7, and 2 trades in a day (at different times for the same or different instruments), you want to count them all, not just the distinct values. Here we have six trade events that add up to 25 trades, so be aware of the difference.

Let's execute the simple queries and check out the results. Please refer to Figure 5-2.

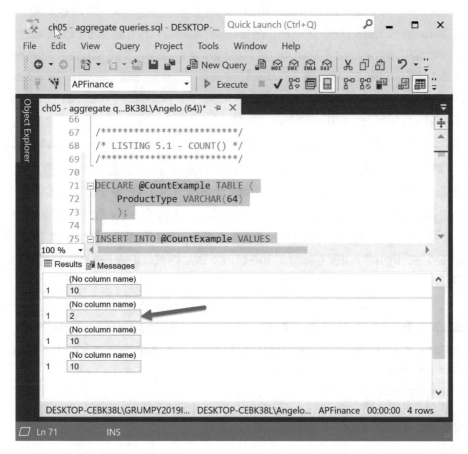

Figure 5-2. *Using ALL vs. DISTINCT*

Back to the query example, we see results for queries 1, 3, and 4 are the same, the value of 10. The second query that uses the DISTINCT keyword generates a value of 2!

So be aware that if you leave out the ALL keyword, you will get the same result as when you include it in the COUNT() function. Using the wildcard character, the asterisk will count everything.

Keep this in mind when you use all the aggregate and other functions as this syntax is common to all of them.

If at this stage you are a little bit confused, why not download the code and try it out. Modify the query a bit until you feel comfortable.

Let's try this function out in a more complex query with our financial data. Let's start with a simple business specification for our first example.

Our business user wishes to perform an analysis on customer C0000001, Mr. John Smith. (Remember him?) Specifically, it's an analysis on buy trades for the year 2012, for the financial instrument AA, which is the company Alcoa. The account number of interest is A02C1, which is an equity (stock) account. Lastly, this needs to be a four-day rolling count, so we need to specify a ROWS() frame for our OVER() clause. The data can be 24 hours old (hint: we can load it overnight).

Note   Microsoft SQL Server defines a CTE as having three parts: the CTE **expression**, the CTE **query**, and the **outer query**. The CTE expression follows the WITH keyword and names the CTE; this is followed by the column list that the CTE returns. The CTE query is the query between the parentheses; it follows the AS ( ) block. And finally the outer query is the query that processes the rows returned by the CTE.

This solution will use both the COUNT() and SUM() functions to meet the business requirements. It is in what I call our usual CTE inner and outer query format, but let's look at the query used for the CTE first.

Please refer to Listing 5-2a.

Listing 5-2a. Customer Analysis CTE Query

```
SELECT YEAR(T.TransDate)           AS TransYear
      ,DATEPART(qq,(T.TransDate)) AS TransQtr
      ,MONTH(T.TransDate)          AS TransMonth
      ,DATEPART(ww,T.TransDate)    AS TransWeek
      ,T.TransDate
      ,C.CustId
```

```
        ,C.CustFname
        ,C.CustLname
        ,T.AcctNo
        ,T.BuySell
        ,T.Symbol
        ,COUNT(*) AS DailyCount
FROM [Financial].[Transaction] T
JOIN [MasterData].[Customer] C
ON T.CustId = C.CustId
GROUP BY YEAR(T.TransDate)
        ,DATEPART(qq,(T.TransDate))
        ,MONTH(T.TransDate)
        ,T.TransDate
        ,T.BuySell
        ,T.Symbol
        ,C.CustId
        ,C.CustFname
        ,C.CustLname
        ,AcctNo
ORDER BY C.CustId
        ,YEAR(T.TransDate)
        ,DATEPART(qq,(T.TransDate))
        ,MONTH(T.TransDate)
        ,T.TransDate
        ,T.BuySell
        ,T.Symbol
        ,C.CustFname
        ,C.CustLname
        ,AcctNo
GO
```

This is a straightforward query. Notice the COUNT() function and the supporting attributes together with a mandatory GROUP BY clause because we are using an aggregate function. I included the ORDER BY clause so we can get a better idea of what the data looks like. This needs to be commented out when used as the CTE query. If you do not, you will get an error message.

Since this is a query used in historical analysis, we might want to consider using it to load a report table in order to improve the performance and eliminate the CTE. The query returns 444,144 rows and is a snapshot in time, so it only needs to be loaded once. Something to consider.

Back to our discussion. Please refer to the partial results in Figure 5-3.

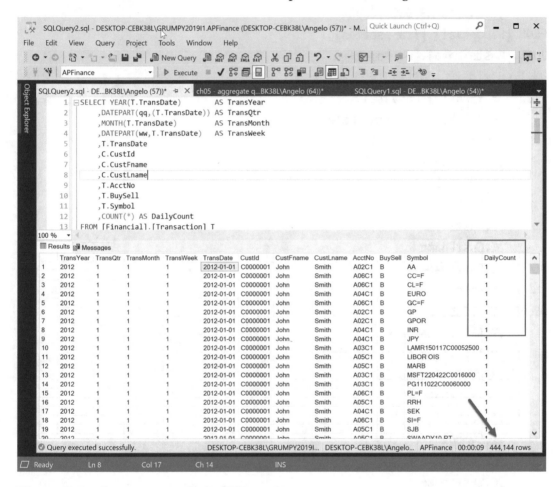

Figure 5-3. *Customer analysis CTE results*

There are two items to focus our attention on. Notice that all the values returned by the COUNT() function are 1. That's because the customer trades in each instrument only once a day. That's how I loaded the test data. Try modifying the code that loads this table so that a customer has more than one trade in a day for one or more instruments. Try to mix it up so there are two or three buy transactions and one, two, or three sell transactions. It is perfectly reasonable to assume a customer will perform multiple buy or sell transactions a day on the same instrument. One warning: The Transaction table will get big!

Note   The financial industry refers to stocks, contracts, products, etc. as financial instruments. Lots of interesting jargon to learn if you want to work in this sector.

Now let's look at the remaining part of the query.

Please refer to Listing 5-2b.

Listing 5-2b. Customer Analysis Outer Query

```
SELECT TransYear
      ,TransQtr
      ,TransMonth
      ,TransWeek
      ,TransDate
      ,CustId
      ,CustFName + ' ' + CustLname AS [Customer Name]
      ,AcctNo
      ,BuySell
      ,Symbol
      ,DailyCount
      ,SUM(DailyCount) OVER (
            PARTITION BY TransQtr
            ORDER BY TransQtr,TransMonth,TransDate
            ROWS BETWEEN 3 PRECEDING AND CURRENT ROW
      )AS RollingTransCount
FROM CustomerTransactions
WHERE CustId  = 'C0000001'
AND TransYear = 2012
AND BuySell   = 'B'
AND Symbol    = 'AA'
AND AcctNo    = 'A02C1'
ORDER BY TransYear
      ,TransQtr
```

```
    ,TransMonth
    ,TransDate
    ,CustId
    ,BuySell
    ,Symbol
GO
```

Let's start by analyzing the OVER() clause. It includes a PARTITION BY and an ORDER BY clause together with a ROWS clause. We partition by calendar quarter and order the partition results by quarter, month, and transaction date. The ROWS clause sets up the four-day rolling sum (3 prior days + current day = 4 days).

Notice I sneaked in the SUM() function, which will be covered in the next section. We want to add up all the transaction counts returned by the CTE query. Let's see what the results look like.

Please refer to the partial results in Figure 5-4.

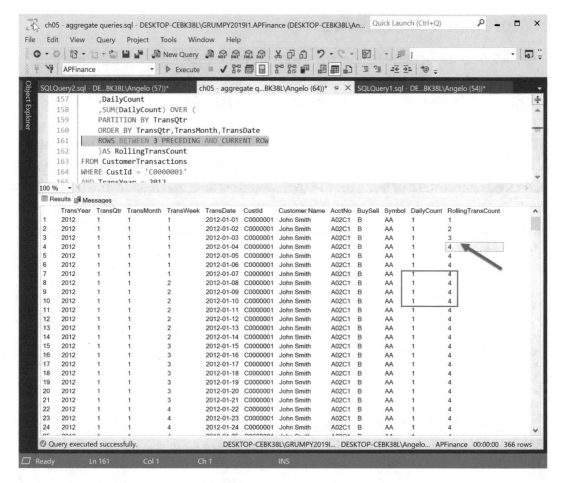

Figure 5-4. *Transaction count four-day rolling totals*

The first four rows look good as the rolling count sums increase by one each day. Past day 4 though, they are all the same. Each day has only one buy transaction, so we keep getting the value 4 as the total for the last four days. In a real-world scenario, this could be an interesting pattern to look for: days a customer executes only one trade vs. days a customer executes multiple trades.

This was expected as we discussed how the customer only places a trade a day. Try modifying this query so we take into account both buy and sell trades. Also load some more transactions for select customers and instruments so the results are a little more interesting.

Performance Considerations

Next, let's do some performance analysis with the tools we learned so far. Keep in mind what the results are if you decide to load multiple trades each day. The Transaction table will get very large fast, and your performance will deteriorate.

The approach we will take is to look at the original CTE-based query first. Then we will pull out the CTE query part and check out the query plan for it. We want to use the CTE query to load a report table so that the query with the OVER( ) clause runs against a preloaded table instead of using the CTE (one would assume that this will run faster). If any index is suggested, we will build it, use the query to load the table, and then create an estimated query plan on the original query that has been modified to use the report table instead of the CTE.

Let's look at the original estimated query plan. Please refer to Figure 5-5a.

Figure 5-5a. *Original query, first part*

Lots of expensive tasks. Starting from right to left, we see a table scan on the Customer table at 5%. No worries as the table has only five customers, so there is not much we need to do with it. Indexes will not help as SQL Server loads the small table into memory.

We do see an index seek at a cost of 14%.

Index seeks are good. The seek means that the required row is retrieved via a structure called a b-tree (binary tree). Index scans are not desirable because each row in a table needs to be checked one at a time, in sequential order, until the desired row is found. Keep on the lookout for these.

Another aspect of query plans to keep in mind as you progress in your performance analysis activities.

Remember: "Index **seek** good, index **scan** bad," most of the time anyway.

Next, we see an expensive sort at 50% cost and finally another sort at 19%. But wait. This is only half the plan. Let's see what the second half at the left-hand side contains.

Please refer to Figure 5-5b.

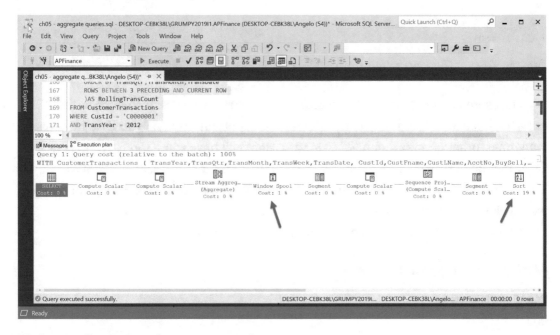

Figure 5-5b.   *Original query, second part*

There's the sort we just discussed at 19% as a placeholder so we know where we are relative to the whole plan. The only other task to consider is the window spool due to the ROWS clause. This is only 1%, but it does indicate that the spooling is performed against physical storage vs. in memory.

We now have a performance baseline. We want to consider the CTE query for use to load a report table. Remember the data can be up to 24 hours old.

Running an estimated query plan on the CTE query, we get the following plan in Figure 5-5c.

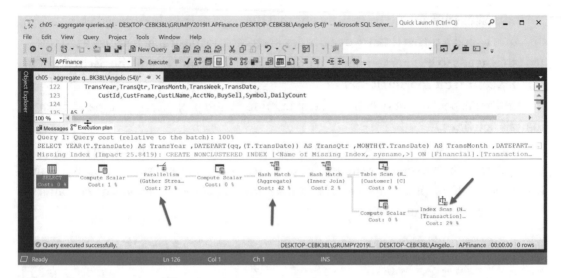

Figure 5-5c. *CTE estimated query plan*

The estimated query plan did suggest an index to improve performance by 25.8%; not a lot, but still it would help when loading the report table overnight. If we minimize load times, other queries that need to load at night will have some room. Notice the Customer table scan costs 0%. Also notice that the index task is a scan and not a seek. This costs 29%. That is why the index was suggested by the query plan estimator tool. We would like to see an index seek.

Listing 5-3 is the index DDL command that was suggested.

Listing 5-3. Suggested Index

```
USE [APFinance]
GO
CREATE NONCLUSTERED INDEX [<Name of Missing Index, sysname,>]
ON [Financial].[Transaction] ([CustId])
INCLUDE ([TransDate],[Symbol],[AcctNo],[BuySell])
GO
```

The name for the index I came up with is

```
ieTransDateSymbolAcctNoBuySell
```

Copying this name in the preceding code and then executing the command creates the index. Let's run a second estimated query plan and see what improvements there are.

Please refer to Figure 5-6 for the new estimated query plan.

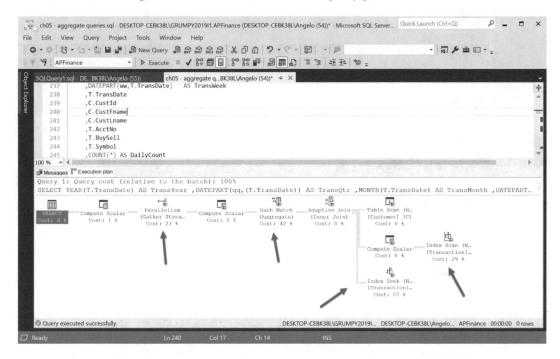

Figure 5-6.   *Revised estimated query plan*

This is interesting. The index scan is still there, but we now have an index seek operation. Both are based on the same index. An adaptive join follows at 0% cost with a hash match at 42% cost. Finally, all the streams from the tasks are gathered for the final row output.

Running this query to load the report table took ten seconds for 444,144 rows. Seems like a lot of time. Listing 5-4 is the INSERT command using the table we just discussed.

Listing 5-4.   Loading the Customer Analysis Table

```
TRUNCATE TABLE Report.CustomerC0000001Analysis
GO

INSERT INTO Report.CustomerC0000001Analysis
SELECT YEAR(T.TransDate)         AS TransYear
     ,DATEPART(qq,(T.TransDate)) AS TransQtr
```

```
        ,MONTH(T.TransDate)           AS TransMonth
        ,DATEPART(ww,T.TransDate)     AS TransWeek
        ,T.TransDate
        ,C.CustId
        ,C.CustFname
        ,C.CustLname
        ,T.AcctNo
        ,T.BuySell
        ,T.Symbol
        ,COUNT(*) AS DailyCount
FROM [Financial].[Transaction] T
JOIN [MasterData].[Customer] C
ON T.CustId = C.CustId
GROUP BY YEAR(T.TransDate)
        ,DATEPART(qq,(T.TransDate))
        ,MONTH(T.TransDate)
        ,T.TransDate
        ,T.BuySell
        ,T.Symbol
        ,C.CustId
        ,C.CustFname
        ,C.CustLname
        ,AcctNo
GO
```

This took 15 seconds to run. This took even longer than when we tested the preceding query. We might want to consider a memory-optimized table for the customer data as we are dealing with a small number of rows. (I include code in the script for this chapter to create the memory-optimized table, so try it out.)

Let's run our original query and modify it so it uses the report table. Figure 5-7 is the revised estimated execution plan when we use the preloaded report table instead of the CTE.

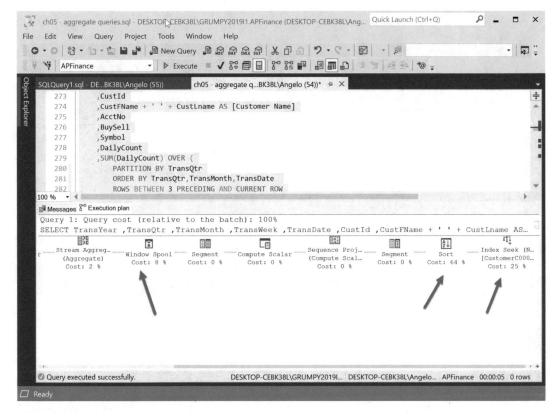

Figure 5-7. *Revised query estimated execution plan*

This seems to be a more streamlined plan. We are seeing an index seek at a cost of 25% but an expensive sort at 64%, and check out the window spool! This comes in at 8%, which means we are using physical storage. We need to dive deeper into this strategy and look at the profile statistics. We should look at the IO and TIME statistics also, but in the interest of space, we will look at the PROFILE statistics only.

Let's examine the PROFILE statistics before and after the index was created. Make sure you execute the following commands before executing the query:

```
DBCC dropcleanbuffers;
CHECKPOINT;
GO

SET STATISTICS PROFILE ON
GO
```

Let's execute the query and check out the PROFILE statistics. Please refer to Figure 5-8.

231

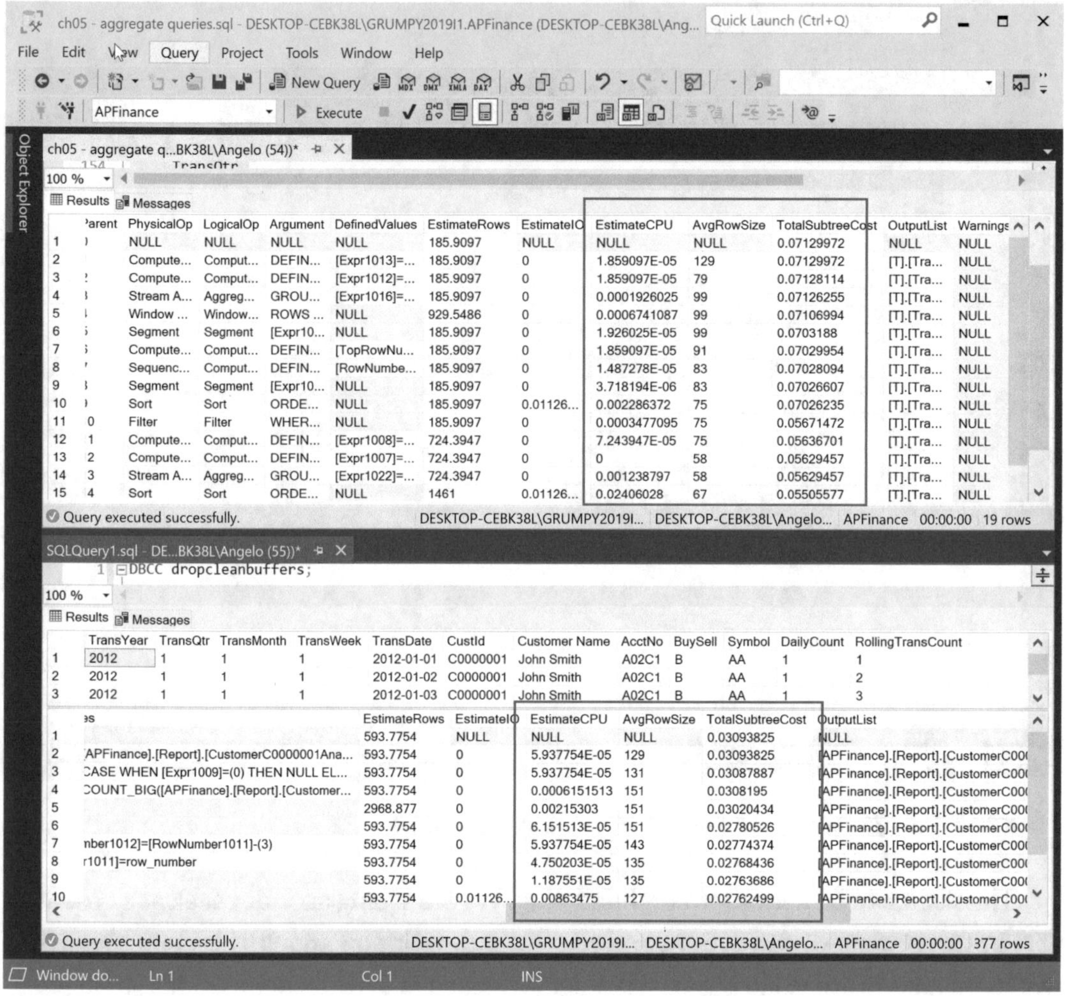

Figure 5-8. *Comparing partial performance statistics*

First of all, let's look at the steps required to process this query. We have 19 steps using the CTE vs. 11 steps with the report table strategy. The EstimateIO statistic is a little bit less with the reporting table that was preloaded. But of course, we have fewer steps. EstimateCPU is lower with the original CTE-based query, but TotalSubTreeCost is less in the second report table–based query.

I would say we have an improvement as we have fewer steps with the second option. Also, the second query ran in zero seconds vs. two seconds with the first query.

Please refer to Figure 5-9.

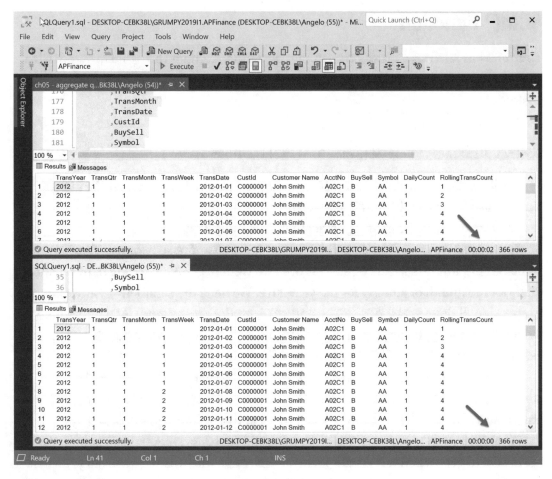

Figure 5-9. *Comparing subtree cost statistics*

Clearly, the CTE approach generated 19 steps, and the report table approach generated 11 steps.

In conclusion, after a rather exhausting performance analysis session, we see that creating a report table to preload data that originally was generated by a CTE yields performance enhancements. The suggested index did not hurt either.

One other conclusion: Setting the performance statistics on will yield valuable performance analysis data. Another tool to use!

SUM( ) Function

This time our analysis focuses on customer accounts. Accounts are used to summarize the positions of the trades daily and also monthly so we can issue reports to the customer.

Here is the syntax:

```
SUM(ALL|DISTINCT *) or SUM(ALL|DISTINCT <column name>)
```

I will not show any simple examples for this as I think by now you get the idea of when to use the ALL vs. DISTINCT keyword.

Let's examine the business specifications our analysts submitted.

Once again, we need some analysis for the same customer, customer C0000001 (John Smith), for the cash account. This time we want to generate rolling daily balance totals for the month of January for the year 2012. We need the capability to also generate a report for all three years, 2012, 2013, and 2014, for the same month. At a future date, we would like to compare performance for the current date vs. the same date in the prior year (this will involve the LAG( ) function).

Last but not least, the report needs to be sorted by year, month, posting date, customer, account name, and number. The output of the report needs to be copied and pasted into a Microsoft Excel spreadsheet in order to generate a graph for analysis.

A big thank you to our business analysts for this easy-to-follow specification. Always make sure to get written vs. verbal specifications; otherwise, the user will tell you got it wrong and will give you more verbal specifications and add-on requirements.

This report has many possibilities and can be repurposed for other account types like equity and FX. Let's review it.

Please refer to Listing 5-5.

Listing 5-5. Cash Account Totals Summary Analysis

```
SELECT YEAR(PostDate) AS AcctYear
    ,MONTH(PostDate)  AS AcctMonth
    ,PostDate
    ,CustId
    ,PrtfNo
    ,AcctNo
    ,AcctName
```

```
        ,AcctTypeCode
        ,AcctBalance
        ,SUM(AcctBalance) OVER(
                PARTITION BY YEAR(PostDate),MONTH(PostDate)
                ORDER BY PostDate
        ) AS RollingDailyBalance
FROM APFinance.Financial.Account
WHERE YEAR(PostDate) = 2012

/* Uncomment the line below, to report on more than 1 year,
    and comment out the line above*/
--WHERE YEAR(PostDate) IN(2012,2013,2014)

AND CustId = 'C0000001'
AND AcctName = 'CASH'
AND MONTH(PostDate) = 1
ORDER BY YEAR(PostDate)
        ,MONTH(PostDate)
        ,PostDate
        ,CustId
        ,AcctName
        ,AcctNo
GO
```

This is a basic report that does not use a CTE. The SELECT clause uses the YEAR() and MONTH() functions, which might prove costly. The SUM() function is used with an OVER() clause to generate the rolling daily balances. We partition by year and month although technically you do not need the year column, but it is there in case you want to expand the scope of the data processed over multiple years (by changing the WHERE clause).

The WHERE clause implements the specification requirements as far as filtering the output, and finally an ORDER BY clause sorts the report for the user presentation. Let's see what it looks like.

Please refer to the partial results in Figure 5-10.

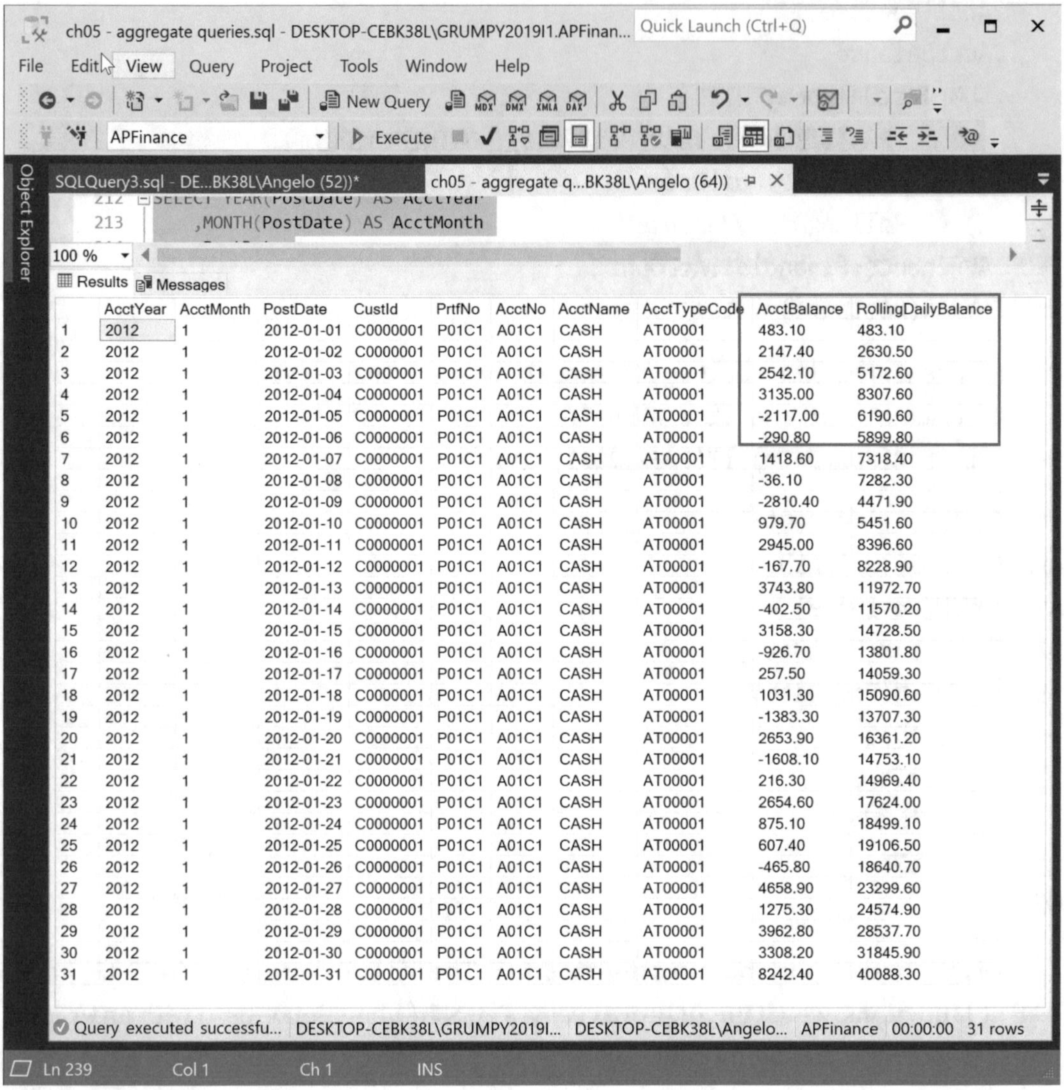

	AcctYear	AcctMonth	PostDate	CustId	PrtfNo	AcctNo	AcctName	AcctTypeCode	AcctBalance	RollingDailyBalance
1	2012	1	2012-01-01	C0000001	P01C1	A01C1	CASH	AT00001	483.10	483.10
2	2012	1	2012-01-02	C0000001	P01C1	A01C1	CASH	AT00001	2147.40	2630.50
3	2012	1	2012-01-03	C0000001	P01C1	A01C1	CASH	AT00001	2542.10	5172.60
4	2012	1	2012-01-04	C0000001	P01C1	A01C1	CASH	AT00001	3135.00	8307.60
5	2012	1	2012-01-05	C0000001	P01C1	A01C1	CASH	AT00001	-2117.00	6190.60
6	2012	1	2012-01-06	C0000001	P01C1	A01C1	CASH	AT00001	-290.80	5899.80
7	2012	1	2012-01-07	C0000001	P01C1	A01C1	CASH	AT00001	1418.60	7318.40
8	2012	1	2012-01-08	C0000001	P01C1	A01C1	CASH	AT00001	-36.10	7282.30
9	2012	1	2012-01-09	C0000001	P01C1	A01C1	CASH	AT00001	-2810.40	4471.90
10	2012	1	2012-01-10	C0000001	P01C1	A01C1	CASH	AT00001	979.70	5451.60
11	2012	1	2012-01-11	C0000001	P01C1	A01C1	CASH	AT00001	2945.00	8396.60
12	2012	1	2012-01-12	C0000001	P01C1	A01C1	CASH	AT00001	-167.70	8228.90
13	2012	1	2012-01-13	C0000001	P01C1	A01C1	CASH	AT00001	3743.80	11972.70
14	2012	1	2012-01-14	C0000001	P01C1	A01C1	CASH	AT00001	-402.50	11570.20
15	2012	1	2012-01-15	C0000001	P01C1	A01C1	CASH	AT00001	3158.30	14728.50
16	2012	1	2012-01-16	C0000001	P01C1	A01C1	CASH	AT00001	-926.70	13801.80
17	2012	1	2012-01-17	C0000001	P01C1	A01C1	CASH	AT00001	257.50	14059.30
18	2012	1	2012-01-18	C0000001	P01C1	A01C1	CASH	AT00001	1031.30	15090.60
19	2012	1	2012-01-19	C0000001	P01C1	A01C1	CASH	AT00001	-1383.30	13707.30
20	2012	1	2012-01-20	C0000001	P01C1	A01C1	CASH	AT00001	2653.90	16361.20
21	2012	1	2012-01-21	C0000001	P01C1	A01C1	CASH	AT00001	-1608.10	14753.10
22	2012	1	2012-01-22	C0000001	P01C1	A01C1	CASH	AT00001	216.30	14969.40
23	2012	1	2012-01-23	C0000001	P01C1	A01C1	CASH	AT00001	2654.60	17624.00
24	2012	1	2012-01-24	C0000001	P01C1	A01C1	CASH	AT00001	875.10	18499.10
25	2012	1	2012-01-25	C0000001	P01C1	A01C1	CASH	AT00001	607.40	19106.50
26	2012	1	2012-01-26	C0000001	P01C1	A01C1	CASH	AT00001	-465.80	18640.70
27	2012	1	2012-01-27	C0000001	P01C1	A01C1	CASH	AT00001	4658.90	23299.60
28	2012	1	2012-01-28	C0000001	P01C1	A01C1	CASH	AT00001	1275.30	24574.90
29	2012	1	2012-01-29	C0000001	P01C1	A01C1	CASH	AT00001	3962.80	28537.70
30	2012	1	2012-01-30	C0000001	P01C1	A01C1	CASH	AT00001	3308.20	31845.90
31	2012	1	2012-01-31	C0000001	P01C1	A01C1	CASH	AT00001	8242.40	40088.30

***Figure 5-10.** Cash account totals summary report*

Always do some data verification so that no surprises pop up. In this case check the rolling totals. Adding the account balances for the first two days, $483.10 + $2147.40, we do indeed get $2630.50. Since this is a small report, there are only 31 rows. Pop the results in a spreadsheet and validate the rolling totals.

Now we copy and paste the results in another Excel spreadsheet and create a nice line graph to see how the account is performing.

Please refer to the partial results in Figure 5-11.

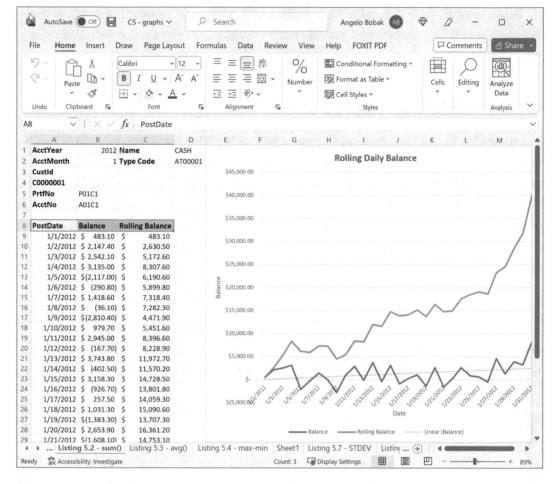

Figure 5-11. *Cash account summary graph*

Pasting the results to an Excel spreadsheet can give us a good visual perspective on financial performance – in this case, cash account positions, which involve deposits, withdrawals, and also money market fund positions.

Here we see a volatile line chart for balances, but the trend is up, so I guess that's good. The dotted line is a trend line, so it looks like it is steadily going up. Remember also that the cash accounts are used to fund buy trades. When instruments are sold, the profit goes back into the cash account.

The rolling balances are trending up also, so this is an indicator that the other financial investments are profitable.

In conclusion, data generated by our window functions in a query is important and valuable, but adding a graph will provide extra insights that might not stand out in raw numbers!

Performance Considerations

As a reminder, for the purposes and space constraints of this book, we only check out and analyze query plan tasks with costs greater than zero. For example, if a window spool task (which usually appears for ROWS or RANGE frame clauses in an OVER() clause) has a value of 0%, it means it is occurring in memory, so we will not worry about it. If it is greater than zero, we consider it in the analysis as it is most likely using physical storage.

Let's start and look at the initial query plan. Our Account table has 43,860 rows. Will an index be recommended?

Please refer to Figure 5-12.

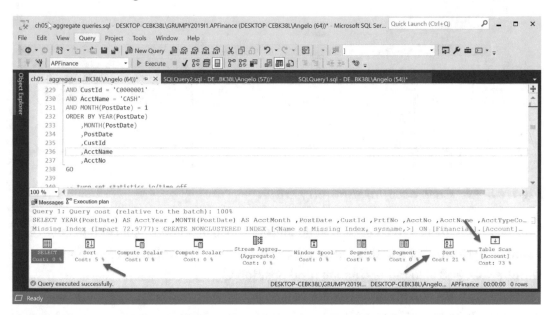

Figure 5-12. *The estimated execution plan suggests an index*

Yes, an index is recommended. Right away we see a table scan on the Account table with a cost of 73%. This is followed by a sort at 21% and a window spool task at 0% (performed in memory). Another sort task at 5% appears on the left-hand side.

Clearly, we need to accept the recommendation and create the suggested index. Listing 5-6 is the index after I supplied a verbose name.

Listing 5-6. Suggested Index

```
CREATE NONCLUSTERED INDEX [iePrtNoAcctNoTypeBalPostDate]
ON [Financial].[Account] ([CustId],[AcctName])
INCLUDE ([PrtfNo],[AcctNo],[AcctTypeCode],[AcctBalance],[PostDate])
GO
```

Creating the index, we generate and check out the new execution plan. Please refer to Figure 5-13.

Figure 5-13. *Revised execution plan*

Things have improved. We see an index seek at 20% and a sort task at 39%, but look at the window spool task. It now costs 3%, which means data went to disk, so this is not an improvement. Finally, we have a final sort task at 39% at the end of the query plan on the left-hand side.

Let's check out the profile statistics:

```
SET STATISTICS PROFILE ON
GO
```

Figure 5-14 is a comparison of the profile statistics before and after the creation of the index.

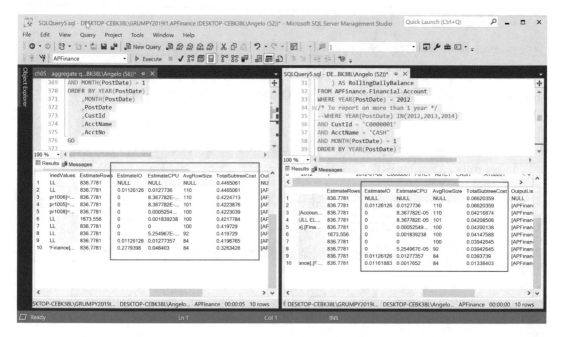

Figure 5-14. *Profile statistics with and without index*

We can see that both scenarios, without index and with index, have the same number of steps. We focus on the `EstimateIO`, `EstimateCPU`, and `TotalSubtreeCost` in both sets of statistics and see a slight improvement with the index version but not a significant amount. Let's set up some graphs that show us this information.

Please refer to Figure 5-15.

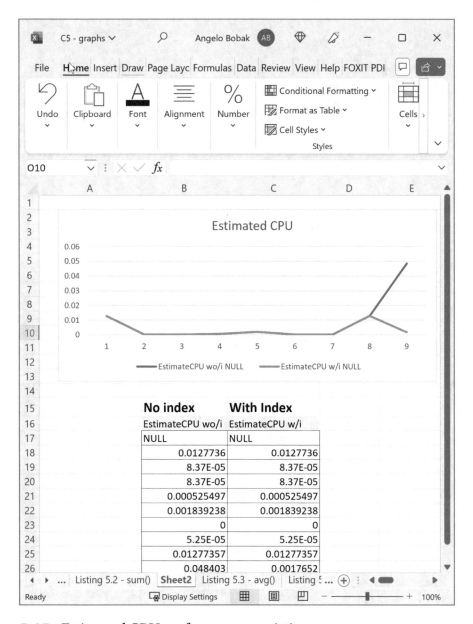

Figure 5-15. *Estimated CPU performance statistics*

Clearly, we see CPU statistics are the same until the last steps where they go down with the index but go up without the index. So this is another indicator we can use to see if indexes we create are valuable or not.

Please refer to Figure 5-16.

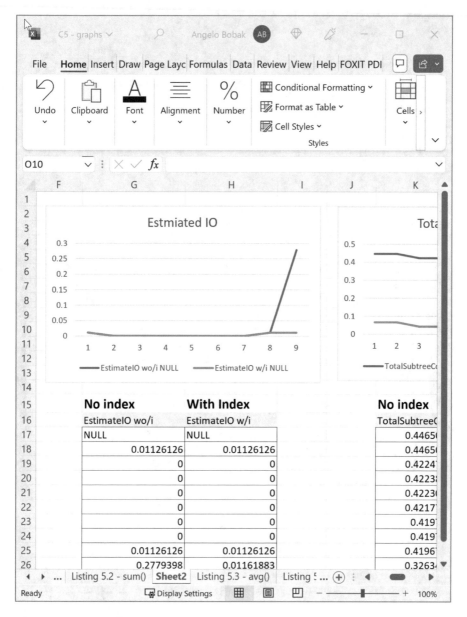

Figure 5-16. *Estimated IO statistics*

The same can be said for IO statistics. With the index IO goes down; without the index IO goes up. We want to keep CPU utilization and IO down. Lastly, let's look at subtree costs.

Please refer to Figure 5-17.

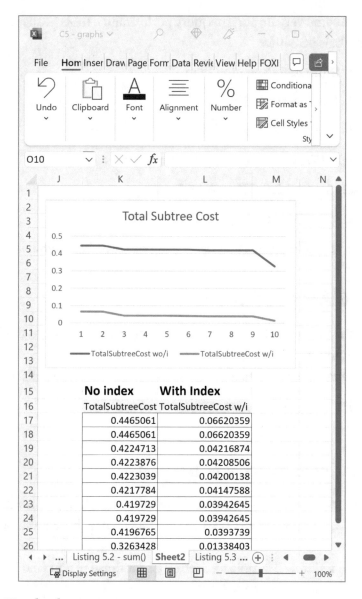

Figure 5-17. *Total subtree costs*

The subtree cost statistics are also considerably down with the index. So now we have looked at several performance statistics categories and tools we can use to analyze, evaluate, and improve the performance of queries that use the aggregate functions with the OVER() clauses to process data partitions.

All this analysis proves that the SQL Server Query Estimation Tool is usually correct!

You now have some very important query performance tools available for your performance analysis activities:

- Display Estimated Execution Plan

- Include Live Query Statistics

- Include Client Statistics

- The SET STATISTICS IO ON/OFF setting

- The SET STATISTICS TIME ON/OFF setting

- The SET STATISTICS PROFILE ON/OFF setting

Next, we look at two functions that usually appear together in analytical reports.

MIN() and MAX() Functions

A quick review, the MIN() and MAX() functions return the smallest and the largest value, respectively, in a column.

Here is the syntax, the same as the other aggregate functions we discussed:

```
MIN(ALL|DISTINCT *) or MIN(ALL|DISTINCT <column name>)
```

and

```
MAX(ALL|DISTINCT *) or MAX(ALL|DISTINCT <column name>)
```

The power of these functions is enhanced when we combine them with the OVER() clause. Worth repeating, you cannot use DISTINCT if you use these functions with the OVER() clause.

Let's start with a simple example as a refresher for those of you who have not seen or used these before. Please refer to Listing 5-7.

Listing 5-7. MIN() and MAX() in a Simple Example

```
DECLARE @MinMaxExample TABLE (
     ExampleValue SMALLINT
     );

INSERT INTO @MinMaxExample VALUES(20),(20),(30),(40),(60),(60);
```

```
SELECT MIN(ExampleValue)    AS MinExampleValue
      ,COUNT(ExampleValue) AS MinCount
FROM @MinMaxExample;
SELECT MIN(ALL ExampleValue)    AS MinExampleValueALL
      ,COUNT( ALL ExampleValue) AS MinCountALL
FROM @MinMaxExample;
SELECT MIN(DISTINCT ExampleValue)  AS MinExampleValueDISTINCT
      ,COUNT(DISTINCT ExampleValue) AS MinCountDISTINCT
FROM @MinMaxExample;

SELECT MAX(ExampleValue)    AS MaxExampleValue
      ,COUNT(ExampleValue) AS MAXCount
FROM @MinMaxExample;
SELECT MAX(ALL ExampleValue)    AS MaxExampleValueALL
      ,COUNT( ALL ExampleValue) AS MAXCountALL
FROM @MinMaxExample;
SELECT MAX(DISTINCT ExampleValue)    AS MaxExampleValueDISTINCT
      ,COUNT( DISTINCT ExampleValue) AS MAXCountDISTINCT
FROM @MinMaxExample;
GO
```

I included all variations of using the ALL or DISTINCT keyword, and the results are in Figure 5-18.

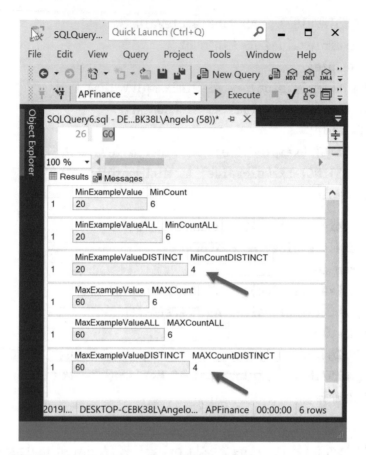

Figure 5-18. *MIN() and MAX(), ALL, DISTINCT*

Although you can use either `ALL` or `DISTINCT` with the `MIN()` and `MAX()` functions, the outputs are all the same. After all, minimum is minimum and maximum is maximum no matter how many duplicate values are present!

Let's turn our attention to our financial database example and see what the business specification for this query is.

For customer C000001, Mr. John Smith, we want to see the daily minimum and maximum cash account values for every month in 2012. The query needs to be written so that if we wish to see other years, we include a commented-out `WHERE` clause so we can switch back and forth. We are anticipating our business analyst will want other years in the analysis.

Please refer to Listing 5-8.

Listing 5-8. Cash Analysis for Customer C000001 Using MIN() and MAX()

```
SELECT YEAR(PostDate) AS AcctYear
      ,MONTH(PostDate) AS AcctMonth
      ,PostDate
      ,CustId
      ,PrtfNo
      ,AcctNo
      ,AcctName
      ,AcctTypeCode
      ,AcctBalance
      ,MIN(AcctBalance) OVER(
            PARTITION BY MONTH(PostDate)
            /* To report on more than 1 year */
            --PARTITION BY YEAR(PostDate),MONTH(PostDate)
            ORDER BY PostDate
      ) AS MonthlyAcctMin
      ,MAX(AcctBalance) OVER(
            PARTITION BY MONTH(PostDate)
            /* To report on more than 1 year */
            --PARTITION BY YEAR(PostDate),MONTH(PostDate)
            ORDER BY PostDate
      ) AS MonthlyAcctMax
FROM APFinance.Financial.Account
WHERE YEAR(PostDate) = 2012

/* To report on more than 1 year */
--WHERE YEAR(PostDate) IN(2012,2013)

AND CustId = 'C0000001'
AND AcctName = 'CASH'
ORDER BY YEAR(PostDate)
      ,MONTH(PostDate)
      ,PostDate
      ,CustId
      ,AcctName
      ,AcctNo
GO
```

Since we are initially only looking at year 2012, our `OVER()` clause for both the `MIN()` and `MAX()` functions has a `PARTITION BY` clause that uses the month part of the postdate (each day is a postdate). Both `ORDER BY` clauses sort by `PostDate`.

As mentioned earlier, if you want to see results for more than one year, uncomment the `PARTITION BY` clauses that partition by year and month. We do not need to sort the partition by month and postdate as postdate alone will keep things in their correct order.

Lastly, the `WHERE` clause filters out results for the customer C0000001 and the cash account. The same applies as far as reporting for multiple years. Use the commented out `WHERE` clause if you want to generate a report for multiple years.

To see some interesting results, like three-day rolling minimum and maximum balances, just replace the `OVER()` clauses with this code snippet:

```
,MIN(AcctBalance) OVER(
        PARTITION BY MONTH(PostDate)
        ORDER BY PostDate
        ROWS BETWEEN 2 PRECEDING AND CURRENT ROW
  ) AS MonthlyAcctMin
,MAX(AcctBalance) OVER(
        PARTITION BY MONTH(PostDate)
        ORDER BY PostDate
        ROWS BETWEEN 2 PRECEDING AND CURRENT ROW
  ) AS MonthlyAcctMax
```

Here we use the `ROWS` frame definition to define the three-day window frames. Notice two rows preceding plus the current row add up to a window frame that has three rows. But you knew that!

Let's see the results from the original query. Please refer to the partial results in Figure 5-19.

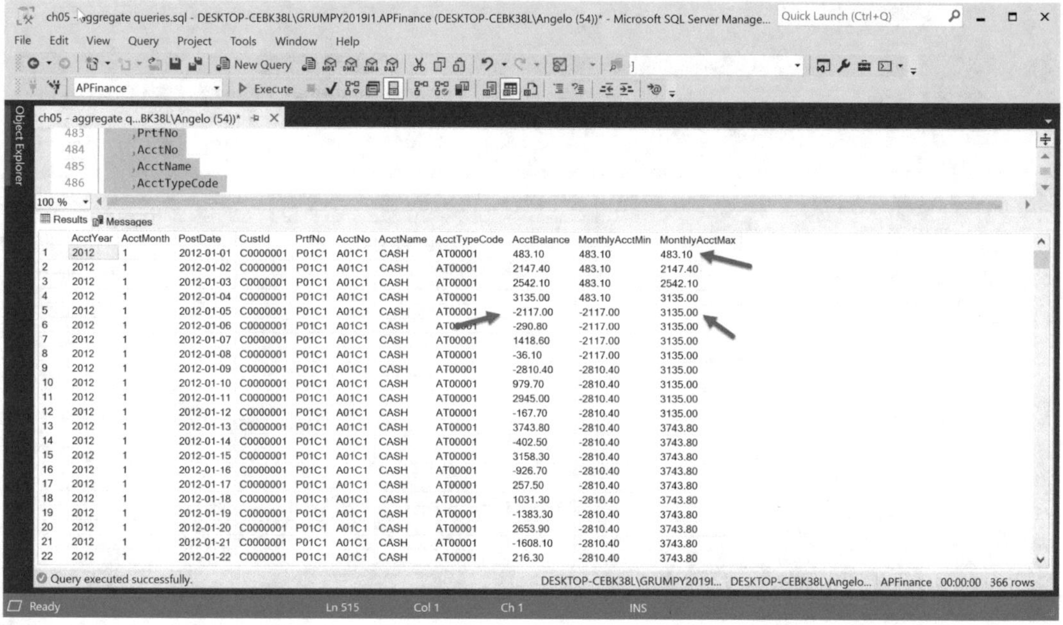

Figure 5-19. *Rolling daily minimum and maximum cash balances*

As can be seen, we start off with the same minimum and maximum balances. Each time a new minimum or maximum balance appears, the values are updated accordingly. These values result in an interesting graph when we use them in a Microsoft Excel spreadsheet.

Please refer to the partial results in Figure 5-20.

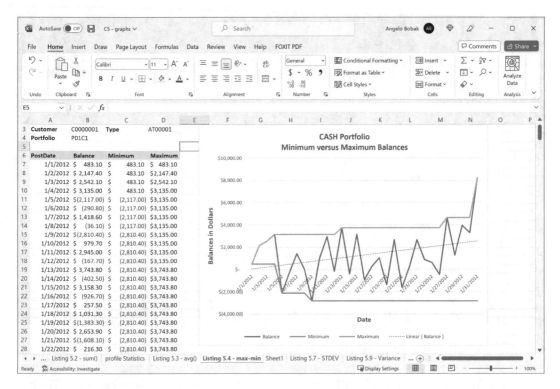

Figure 5-20. *Minimum and maximum cash account analysis*

The good news is that the cash balances are trending up. What we see though are fluctuations in the minimum and maximum balances, which indicate some heavy trading activity. Remember, the customer uses this account to fund buy and sell trades against other instruments like FX and equity transactions.

Performance Considerations

Using the MIN() and MAX() functions with partitions could also be performance intensive. Let's run through our usual analysis to see how we can evaluate and improve performance for this query.

Please refer to Figure 5-21.

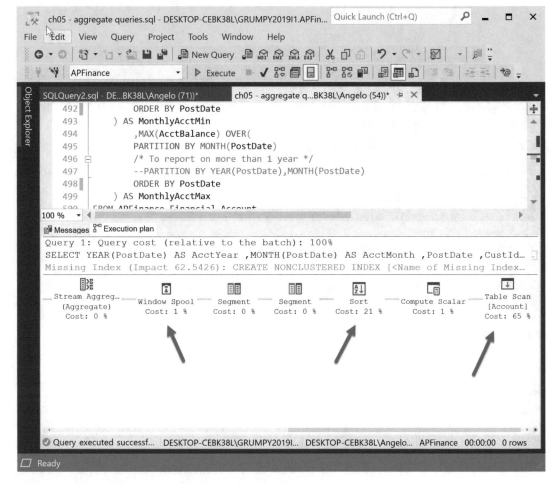

Figure 5-21. *Estimated query plan for the MIN/MAX query*

Right away we see a table scan task with a cost of 65% and a missing index message. A sort follows at 21%, with a window spool at 1%, which indicates we are using physical storage vs. memory. Let's look at part 2 of the estimated query plan in Figure 5-22.

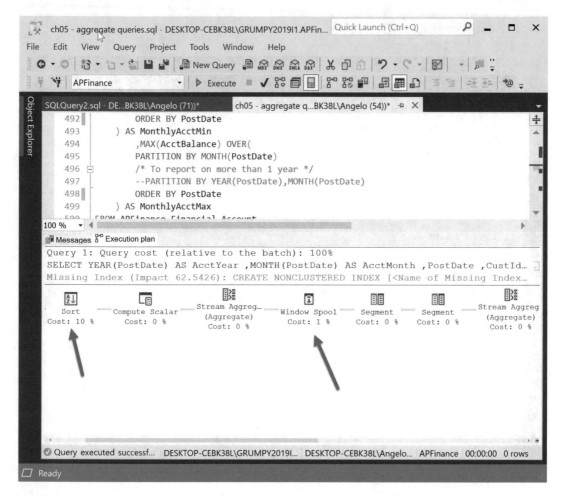

Figure 5-22. *Estimated query plan part 2*

We see a second window spool at 1%. Two window spool tasks make sense as we are using the MIN() and MAX() functions, so each function generates a spool task. Finally, we have a sort task at 10%. Remember that I am not discussing the 0% tasks because for the scope of performance tuning discussions in this chapter, we only want to address the more costly tasks. A good performance tuning approach though is to look at high-cost tasks first and then tasks that you do not know what they are doing. What you don't know can hurt you a whole lot!

Also make sure you look at any tasks that have a warning icon. Let's examine the suggested index and then create it using the suggested name.

Please refer to Listing 5-9.

Listing 5-9. Suggested Index

```
/*
Missing Index Details from ch05 - aggregate queries.sql - DESKTOP-CEBK38L\
GRUMPY2019I1.APFinance (DESKTOP-CEBK38L\Angelo (54))
The Query Processor estimates that implementing the following index could
improve the query cost by 62.5426%.
*/

/*
USE [APFinance]
GO
CREATE NONCLUSTERED INDEX [<Name of Missing Index, sysname,>]
ON [Financial].[Account] ([CustId],[AcctName])
INCLUDE ([PrtfNo],[AcctNo],[AcctTypeCode],[AcctBalance],[PostDate])
GO
*/
```

The query estimator says creating this index can improve performance by 62.54%. Sounds good!

Let's build it. We will call this index iePrtfNoAcctNoAcctTypeCodeAcctBalancePostDate.

Yes, I know the index name is long, but for the purpose of this book, I decided to create index names that contain the names of the columns for easy identification. This way, as you learn the topic, you can quickly identify what the index is for. So as you progress you can rename the indexes to shorter but still descriptive names. Your company has or should have some good naming standards for database objects.

The preceding code is copied to our script for this chapter with the suggested name, and the index is created. We also execute the usual DBCC command to clear the memory plan cache and generate a new query plan. Let's see how we do. Please refer to Figure 5-23.

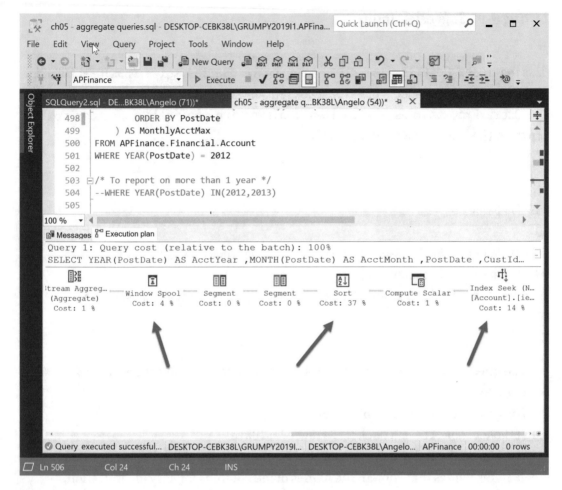

***Figure 5-23.** Post–index creation estimated query plan*

The table scan has been replaced by an index seek with a cost of 14%. So far, so good!

We now see an expensive sort task at 37% and a window spool at 4%. These went up. Not so good. Let's look at the left-hand side of the estimated query plan to see if things improved there.

Please refer to Figure 5-24.

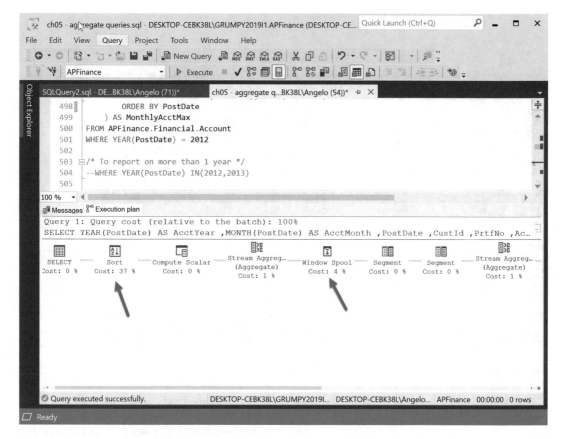

Figure 5-24. *Left-hand side of the query plan*

We find the same issue here; the window spool went up to 4%, and the last sort task went up to 37%. So we need to ask the question: Did the index help or make things worse? To answer this question, we need to look at the statistics that are generated when we execute the following command:

```
SET STATISTICS PROFILE OFF
GO
```

We need to do this pre-index creation and post-index creation to see if the IO and other statistics went up or down. Please refer to Figure 5-25 for the results.

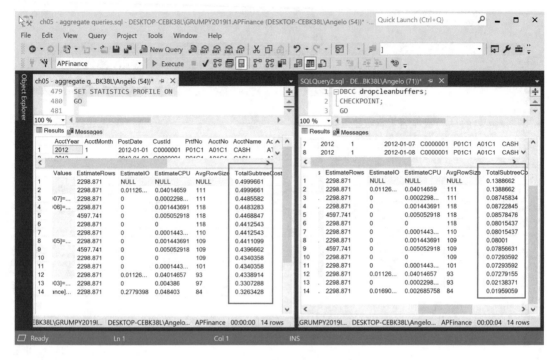

Figure 5-25. *Pre- and post-index PROFILE statistics*

This takes a bit of work to set up. First, you need to execute the DBCC command in a query pane for executing the query before the index is created. Set the profile statistics on and execute the query.

Open a second query pane and copy and paste the DBCC, SET PROFILE ON, and query plus the code to create the index. Follow these steps:

- Step 1: Create the index.

- Step 2: Execute the SET PROFILE ON command.

- Step 3: Execute the DBCC command.

- Step 4: Execute the query.

Make sure the query pane windows are side by side, vertically. As can be seen, the only improvement seems to be in the subtree costs. There is a minimal improvement in the estimated CPU statistic but not much.

Let's compare the IO and TIME statistics pre-index and post-index creation to see what they tell us. Please refer to Figure 5-26.

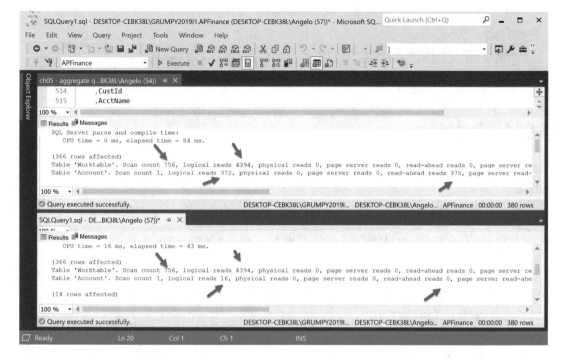

Figure 5-26. *IO and TIME statistics analysis*

What does this set of statistics tell us?

We can see that the scan count and logical reads for the work table remained the same.

We can see that the scan count for the Account table remained the same, but the logical reads went down from 372 to 16 and the read-ahead reads went from 375 to 0.

So, you can see, all our performance analysis tools are necessary so we can really see if an index helped or not. In this case it helped somewhat, and the creation of the index is justified. You do need to consider whether creating an index that helps performance improvement by 50% or less is justified. Will the index cause performance to degrade when a table is loaded? Does 50% increase mean the query that ran four seconds now runs in two? Take all these factors into consideration. If your users are happy with four-second wait times for the query to run, leave it alone. Don't fix what's not broken.

Next, we quickly look at the AVG( ) function. We will not perform the rigorous performance analysis that we did in this section. I leave that up to you as an exercise as now you have seen several examples of the tools and techniques you need to use to measure performance.

Pop Quiz   What does cost mean? Is it the time a step takes to execute or a weighted artificial value based on IO, memory usage, and CPU cost?

AVG( ) Function

The AVG( ) function calculates the average of a data set represented by a column in a table.

Here is the syntax:

```
AVG(ALL|DISTINCT *) or AVG(ALL|DISTINCT <column name>)
```

As usual, let's look at a simple example so you can see the effect on the results that occur when you include the ALL or DISTINCT keyword in the function call.

Please refer to Listing 5-10.

Listing 5-10. Test Query

```
USE TEST
GO

-- easy example

DECLARE @AVGMaxExample TABLE (
      ValueType VARCHAR(32),
      ExampleValue SMALLINT
      );

INSERT INTO @AVGMaxExample VALUES
('Type 1',20),
('Type 1',20),
('Type 1',30),
('Type 2',40),
('Type 2',60),
('Type 3',60);
```

```
SELECT AVG(ExampleValue)                    AS AVGExampleValue
FROM @AVGMaxExample;
SELECT AVG(ALL ExampleValue)                AS AVGExampleValueALL
FROM @AVGMaxExample;
SELECT AVG(DISTINCT ExampleValue)           AS AVGExampleValueDISTINCT
FROM @AVGMaxExample;

SELECT ValueType,AVG(ExampleValue)          AS AVGExampleValue
FROM @AVGMaxExample
GROUP BY ValueType;
SELECT ValueType,AVG(ALL ExampleValue)      AS AVGExampleValueALL
FROM @AVGMaxExample
GROUP BY ValueType;
SELECT ValueType,AVG(DISTINCT ExampleValue) AS AVGExampleValueDISTINCT
FROM @AVGMaxExample
GROUP BY ValueType;
GO
```

Using a table variable loaded with six rows of data, we want to see the results of averages by type. This is key to understand with simple data because if you are unsure of the results generated for duplicate data vs. distinct data, you will create serious errors in your results when querying hundreds of thousands or millions of rows. This could get serious!

For example, calculating the average number of trades per hour in a day and including duplicates will generate different values than if you only include distinct values. You want to include all trades, even duplicates, to get an accurate average (like two trades at 10 AM and two more trades at 1 PM, total of four trades that day)!

Let's see the results of our simple query. Please refer to Figure 5-27.

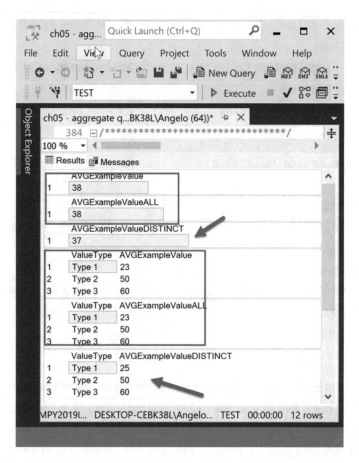

Figure 5-27. *Testing ALL vs. DISTINCT when calculating averages*

Starting from the top, using `ALL` or nothing at all in the `AVG()` function yields a result of 38. Using `DISTINCT` yields a result of 37. When we introduce the `ValueType` column, we see similar behavior.

Armed with this knowledge, let's try a query against our financial data.

Our business analyst needs a query that will generate a report that shows rolling three-day averages for the customer account. As usual we pick customer "C0000001" (John Smith). If you look at the Customer table, you see Mr. Smith earns $250,000 a year. Nice!

The query needs to show results for the cash account and for only the first month of the year 2012. We want the capability of showing other years, so we need to include commented-out code that we can enable as per the analyst's request (you can do this on your own).

Please refer to Listing 5-11.

Listing 5-11. Rolling Three-Day Averages

```
SELECT YEAR(PostDate)  AS AcctYear
      ,MONTH(PostDate) AS AcctMonth
      ,PostDate
      ,CustId
      ,PrtfNo
      ,AcctNo
      ,AcctName
      ,AcctTypeCode
      ,AcctBalance
      ,AVG(AcctBalance) OVER(
            PARTITION BY MONTH(PostDate)

      /* uncomment to report on more than 1 year */
      --PARTITION BY YEAR(PostDate),MONTH(PostDate)

            ORDER BY PostDate
      ROWS BETWEEN 2 PRECEDING AND CURRENT ROW
      ) AS [3 DayRollingAcctAvg]
FROM APFinance.Financial.Account
WHERE YEAR(PostDate) = 2012

/******************************************/
/* Uncomment to report on more than 1 year */
/******************************************/
--WHERE YEAR(PostDate) IN(2012,2013)

AND CustId = 'C0000001'
AND AcctName = 'CASH'
AND MONTH(PostDate) = 1
ORDER BY YEAR(PostDate)
      ,MONTH(PostDate)
      ,PostDate
      ,CustId
      ,AcctName
      ,AcctNo
GO
```

No CTE is used by the way. The AVG() function is used with an OVER() clause that partitions the data set by month and orders the partition result set by posting date. The ROWS clause is included to create a window frame that includes the prior two rows and the current row as the results are processed.

Remember this is for the cash account only. It would be interesting to compare the cash account averages vs. the equity account averages, for example, to see if Mr. Smith made or lost money. Did a lot of cash go out of the account to buy equities? Did the equities go up, and were they then sold at a profit so that cash deposits were made back to the cash account?

You can do this by modifying the query to include equity accounts.

Let's check out the results in Figure 5-28, and then we can graph them.

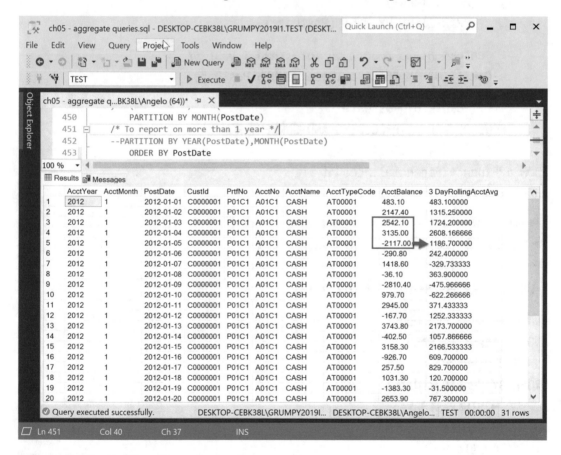

Figure 5-28. *Rolling three-day cash account averages*

The output is sorted by day so you can see how the averages are calculated on a daily basis. Notice that there are a lot of negative account balances, so this might imply a lot of volatile trading. In actual trading scenarios, checks are in place to prevent a trade in case funding account balances are negative, or maybe some sort of mechanism is put in place where funds can be moved from other accounts to cover the short falls just in case.

A picture is worth a thousand words, so let's copy this data to Excel and generate a nice graph.

Please refer to the Excel graph in Figure 5-29.

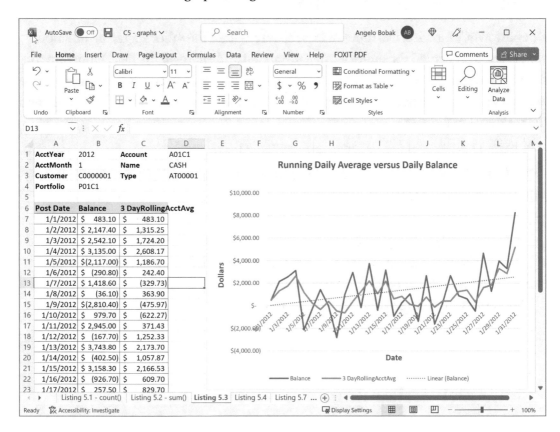

Figure 5-29. *Rolling three-day averages for the cash account*

Account balances and averages are generally trending upward, but there are a lot of up and down swings. The dotted linear balance line is used to show trends, so at least it is pointing up.

We are looking at these functions on a one-by-one basis, but valuable reports would give a profile of account performance. By this I mean use all the functions we discussed in a single report so the analyst can see all performance aspects of the account.

Try this on your own by repurposing the code from prior queries and creating a profile report for either the cash account or other account types like equity or FX (foreign exchange).

One last comment: Use Excel to generate the three-day rolling averages next to the values generated by the SQL Server functions to validate the results. Always validate data generated by your queries.

Performance Considerations

Does the AVG() function use more or less resources than the other aggregate functions we discussed? I copied the query we just discussed and changed the AVG() function to the SUM() function. I then generated an estimated query plan for each, and they were the same. Same steps, costs, etc. I also tried the STDEV() function vs. the AVG() function, and the results were the same.

Pop Quiz Answer   Costs are derived from weighted calculation that measures CPU, IO, and memory usage costs.

A deeper analysis is called for to answer this question.

Let's perform our usual analysis with the query plan estimator, this time paying attention to the ROWS clause in the OVER() clause.

Please refer to Figure 5-30.

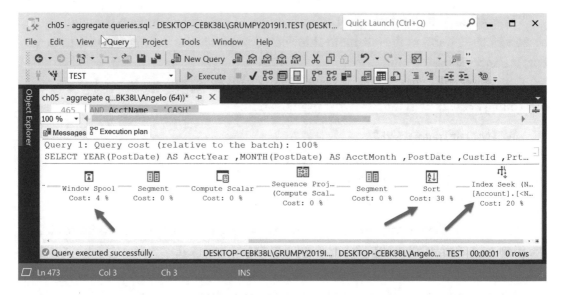

Figure 5-30. *Estimated query plan for the three-day rolling averages (part A)*

An existing index is used (20% cost), then followed by a sort step (38%), and there is our window spool at a cost 4%. This means spool activity is performed on physical storage. Let's look at the left-hand side of the estimated query plan. Please refer to Figure 5-31.

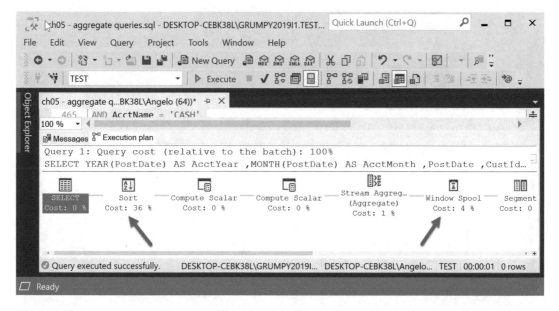

Figure 5-31. *Estimated query plan for the three-day rolling averages (part B)*

There's the same window spool task as a point of reference between the two screenshots. The sort at the left-hand side costs 36% and is the last highest task as far as cost is concerned.

Let's change the ROWS clause so we get a four-day rolling average instead of a three-day rolling average with the following modification:

```
,AVG(AcctBalance) OVER(
        PARTITION BY MONTH(PostDate)
        ORDER BY PostDate
    ROWS BETWEEN 3 PRECEDING AND CURRENT ROW
```

Running another estimated query plan after we execute the DBCC function (to clear the plan cache), the following plan in Figure 5-32 is generated.

Pop Quiz   Why do we use DBCC to clear the plan cache?

Changing the ROWS clause to generate a four-day rolling average calculation increased the window spool. Please refer to Figure 5-32.

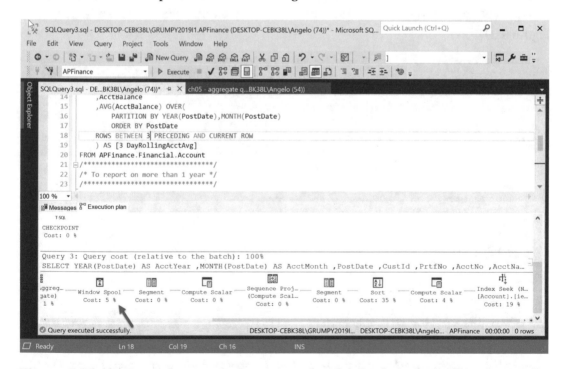

Figure 5-32. *Estimated query performance plan for the four-day rolling average query*

The index seek went down to 19% cost, the sort went down to 35%, and the window spool task is 5%. Seems that if we increase the number value for preceding rows in the ROWS clause, the window spool cost goes up. Sadly, this is not true all of the time. I changed the values to 4, 5, and 8, and the window spool cost went back down to three. Interesting.

I modified the query so we had three calls to the AVG() function with the preceding row values set to 3, 4, and 5, and the window spool costs were 3%, 3%, and 4%, respectively. So my initial assumption did not hold.

The sort costs went down a bit though. Try this analysis on your own to see what results you get.

Let's do some multidimensional analysis next by rolling up totals.

GROUPING Function

Recall from Chapter 2 that the GROUPING() function can be compared to a Microsoft Excel spreadsheet pivot table but implemented with TSQL. The results are powerful but confusing to view and complicated to analyze. Pivot tables are three dimensional; the function results are two dimensional.

We will create a query and look at the results. We will then modify the query so it only generates raw data, and we will copy it to a Microsoft Excel spreadsheet so we can create a pivot table. We will compare the two results to see if we get the same values.

As a refresher, recall from Chapter 2 how grouping works. Please refer to Figure 5-33.

GROUPING & ROLLUPS

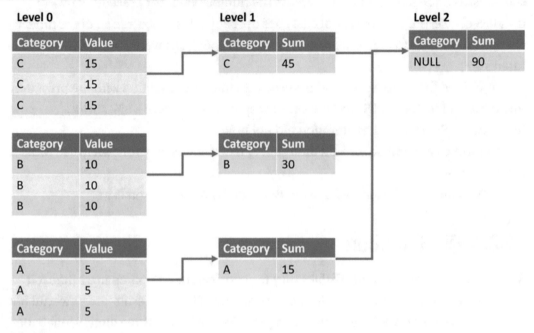

Figure 5-33. *How grouping and rollups work*

Starting with the three categories of rows to the left of the diagram, each row in the data set has a category and a value. The categories are A, B, and C. If we summarize the values for each category at level 0, we get rollups of 15 for category A, 30 for category B, and 45 for category C positioned at level 1. If we summarize these three rollup totals, we get a final value of 90 at level 2.

Imagine how powerful but confusing this structure is if we are grouping and rolling up totals for tens of thousands of rows!

Back to finance. Here is our business analyst requirement:

Generate a report for another customer C0000003 (Sherlock Carruthers) that shows transaction totals for EURO (European Union Currency) and GBP (Great Britain Pound Sterling) foreign currency trades. Show only for the month of January for the 2012 calendar year.

The results need to be grouped and rolled up by year, quarter, month, transaction date, and financial instrument symbol.

Listing 5-12 is the resulting query based on this specification.

Listing 5-12. Transaction Rollups for EURO and GBP

```
SELECT YEAR(TransDate)          AS TransYear
      ,DATEPART(qq,TransDate) AS TransQtr
      ,MONTH(TransDate)         AS TransMonth
      ,TransDate
      ,Symbol
      ,CustId
      ,TransAmount
      ,SUM(TransAmount)        AS SumOfTransAmt
      ,GROUPING(TransAmount) AS TransAmtGroup
FROM Financial.[Transaction]
WHERE Symbol IN ('EURO','GBP')
AND CustId = 'C0000003'
AND YEAR(TransDate) = 2012
AND MONTH(TransDate) = 1
GROUP BY YEAR(TransDate)
      ,DATEPART(qq,TransDate)
      ,MONTH(TransDate)
      ,TransDate
      ,Symbol
      ,CustId
      ,TransAmount WITH ROLLUP
ORDER BY YEAR(TransDate)
      ,DATEPART(qq,TransDate)
      ,MONTH(TransDate)
      ,TransDate
      ,Symbol
      ,CustId
      ,(CASE WHEN TransAmount IS NULL THEN 0 END)DESC
      ,SUM(TransAmount) DESC
      ,GROUPING(TransAmount) DESC
GO
```

The GROUPING() function appears right after the SUM() function. There are two important coding sections we need to be aware of.

First, we need to add the `WITH ROLLUP` directive to the `TransAmount` column in the `GROUP BY` clause so we can generate the rollups.

Second, in the `ORDER BY` clause, we need to include the conditional `CASE` code block, so we know how to sort `NULL` values for the `TransAmount` column. Notice we also include the `SUM()` and `GROUPING()` functions in the `ORDER BY` clause using a small `CASE` block so we generate a report that lines up all `NULL` values correctly as you navigate up the `ROLLUP` of the transaction amounts.

Please refer to the partial results in Figure 5-34.

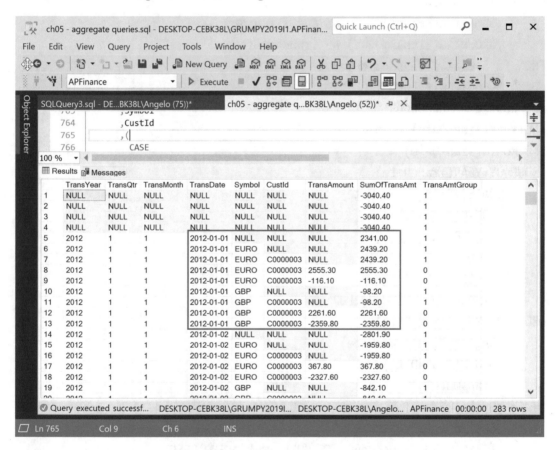

Figure 5-34. *Transaction rollup report for EURO and GBP*

Let's look at a small portion of the rollups. For January 1, 2012, if we add up the two GBP currency values, we see they roll up to –£98.20. Bad show! If we add up the EURO entries, we end up with a positive €2439.20. If we add up –£98.20 (converted to €98.20) + €2439.20, we get €2341.

Let's assume that GBP and EURO are at a one-to-one exchange rate for this example. Otherwise, values must be converted to US dollars or one or the other currency we are adding. Otherwise, the additions do not make sense.

Please refer to the Excel pivot table in Figure 5-35.

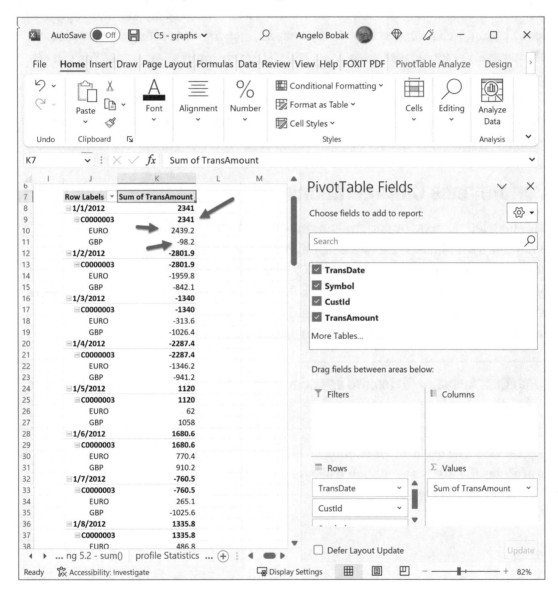

Figure 5-35. *Currency pivot table for EURO and GBP*

Looks good! Values match. Notice at the top for January 1, adding the EURO and GBP values assuming a one-to-one exchange rate yields the final total of €2341.

If you get inspired, make some modifications to the transaction table so foreign currency values are converted to USD with some make-believe conversion rates so you can enhance the reporting.

Back to our coding example. What does this tell us about the value of the GROUPING() function? It does work, but it needs some sorting tricks in the query to get the output to look like comprehensible rollups. In my opinion, copying and pasting the raw data to a Microsoft Excel spreadsheet pivot table will yield more impressive visuals. You can also create pivot charts to accompany your pivot table.

Last but not least, you can also load the raw data into SQL Server Analysis Services, creating and loading a multidimensional cube and performing the multidimensional analysis there.

Performance Considerations

This function seems to be performance intensive. We have several tools now for analyzing performance, from creating estimated and actual query plans to analyzing statistics such as IO, TIME, and STATISTICS PROFILE. Our process is to look at the estimated plan and then dive deeper into the performance statistics of the query being analyzed in order to have a complete picture of what is going on and if there are any indexes or reporting tables we can add to improve performance.

Pop Quiz Answer   To remove any prior query plans that could skew performance statistics.

Let's start, as usual, by generating an estimated query plan in the usual manner. This is probably the first step you should always take when starting to analyze performance.

Please refer to Figure 5-36.

Figure 5-36. *The estimated query plan shows an expensive table scan*

We now know that when we see a table scan on a large table, it means trouble. We also see that the estimator has suggested an index, so this is our first remedial course of action to take. If we create the index, SQL Server tells us that we will see an improvement of 93.72%

We copy and paste the suggested index into our script, give it a name, and execute the DDL command to create the index. Keep your fingers crossed!

Please refer to Listing 5-13.

Listing 5-13. Suggested Index for the GROUPING() Example

```
CREATE NONCLUSTERED INDEX [ieTranSymCustBuySell]
ON [Financial].[Transaction] ([Symbol],[CustId],[BuySell])
INCLUDE ([TransDate],[TransAmount])
GO
```

The name is not too long, and it does convey the purpose of the index. I could have added the words *Date* and *Amount* as an afterthought, but you get the picture. An index name of ieX-1234 is probably not a good idea.

Forging ahead we execute the command to create the index and generate another estimated query plan. Please refer to Figure 5-37.

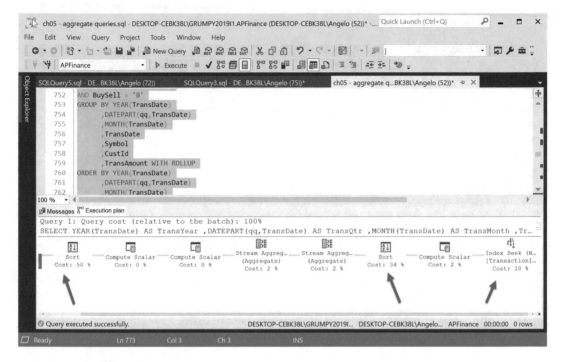

Figure 5-37. *Revised estimated query plan*

The table scan is gone and replaced by an index seek with a cost of 10%. Notice, though, we have two expensive sorts; the first costs 34% and the second one costs 50%. Seems each time we add what we think is a high-performing index, we also incur some expensive sort tasks!

For this reason, we need to take a deep dive into the query. The following are the steps you can follow:

- Copy the query to a new query pane (including the DBCC command, the command to create the query, and the SET STATISTICS PROFILE command).

- Move it to a new horizontal tab.

- Drop the new table index.

- Set the STATISTICS PROFILE to ON.

- Run the query in the original query pane.

- In the second query pane, create the index, run DBCC, and set the STATISTICS PROFILE to ON.

Run the query now that the index is created and compare the profile statistics as shown in Figure 5-38.

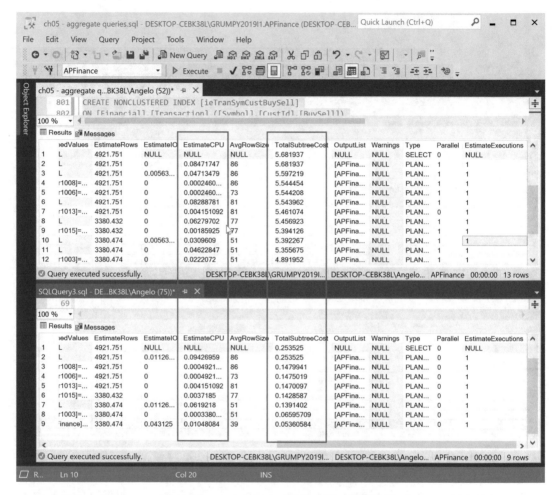

Figure 5-38. *Pre- and post–index creation profile statistics*

What do we see? Right off the bat, we see that the second results, the ones representing the statistics when the index was created, have fewer steps than the original query without the index. Good start. Fewer steps, less time to execute the query.

The Estimated IO column shows the values actually went up, but there are fewer steps. The Estimated CPU values trended up a bit, and a few went down. Again, there are fewer steps. Finally, the TotalSubTreeCost values went down considerably, so it looks like the creation of the index was the correct action to take. Fewer steps, lower subtree costs, and mixed CPU costs.

How about our old friends, the IO and TIME statistics?

Please refer to Figure 5-39.

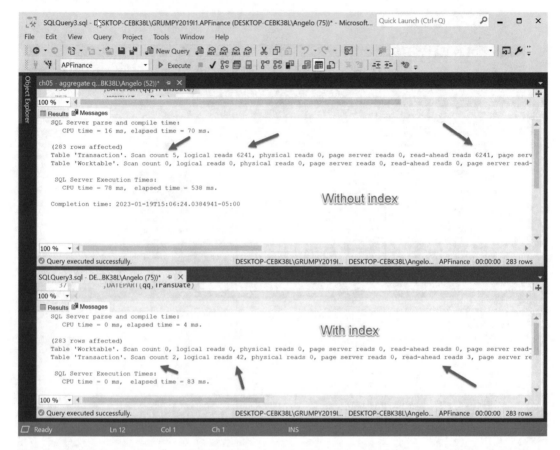

Figure 5-39. *Comparing IO and TIME statistics – pre- and post-index creation*

The last piece of the performance puzzle is to compare the scan counts, logical reads, and read-ahead reads for work tables and database tables used in the query. As can be seen by these statistical values, everything went down after the index was created. Values for logical reads and read-ahead reads went down in the scale of thousands. This is what you want to see.

To set it up, use the same steps as discussed earlier, view both sets of statistics in horizontal query panes.

STRING_AGG( ) Function

Here's the function used to aggregate text strings again. Let's look at a simple example that shows how to create message packets related to ticker price that could be used to communicate the information between trading platforms.

Here is the syntax:

```
STRING_AGG(<string>,<delimiter>)
```

By cleverly creating a text string that combines labels, messages, and data from a table, we can create data packets that can be used by an application for tracking ticker symbol prices by days and hours.

In this case, the analyst requirement is to assemble the ticker symbol, company name, trade date, hour, and price. This information is needed by traders as they execute buy and sell transactions. Some sort of packet start/stop token needs to be added at each end of the packet so the data stream can be processed at the receiving end.

Please refer to Listing 5-14.

Listing 5-14. Assemble 24-Hour Ticker Prices

```
SELECT STRING_AGG('msg start->ticker:' + Ticker
      + ',company:' + Company
      + ',trade date:' + convert(VARCHAR, TickerDate)
      + ',hour:' + convert(VARCHAR, QuoteHour)
      + ',price:' + convert(VARCHAR, Quote) + '<-msg stop', '!') + CHAR(10)
FROM MasterData.TickerHistory
WHERE TickerDate = '2015-01-01'
GO
```

This is just a simple example, and you need to use your imagination a bit, but it illustrates the applicability of even this function in a financial trading scenario.

Please refer to the results in Figure 5-40.

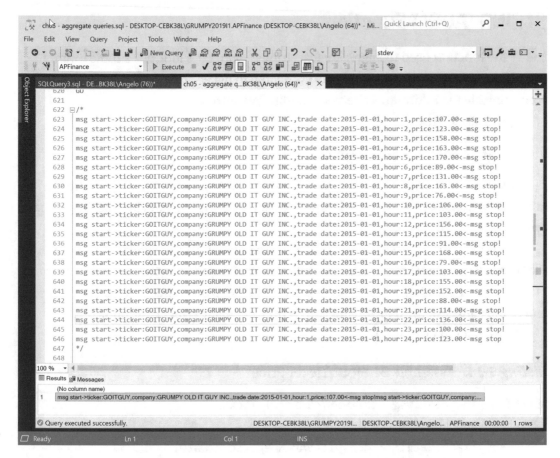

Figure 5-40. Twenty-four-hour ticker prices for symbol GOITGUY

The results are one long string with embedded cursor/return characters. I copied the results into a query pane and placed some comments around so you can see the results.

We will not look at any performance analysis as this will basically be one large table scan. Since we have a WHERE clause that filters by date, we can easily come to the conclusion an index would be needed on that column.

If the table stores thousands of ticker symbols for periods that span decades, then a clustered index on symbol and trade date would be required.

STDEV() and STDEVP() Functions

Recall from Chapter 2 that the STDEV() function calculates the statistical standard deviation of a set of values when the entire population is not known or available.

Also recall that the STDEVP() function calculates the statistical standard deviation when the entire population of the data set is known. Simple to remember; it has a capital "P" at the end of the name.

What this means is how close or how far apart the numbers are in the data set from the arithmetic **MEAN** (which is another word for average). In our context the average or **MEAN** is the sum of all data values in question divided by the number of data values. There are other types of **MEAN** like geometric, harmonic, and weighted **MEAN**, which are discussed in Appendix B.

The syntax is just like the other functions we covered so far:

```
STDEV(ALL|DISTINCT <column name>) and STDEVP(ALL|DISTINCT <column name>)
```

The fun begins when you start using an OVER() clause to generate results. We will look at two examples, one with our test data that exists in the APFinance database and one where the data is also test data but it is a small example with only 12 rows so we can see how to interpret the information and create an interesting graph called a bell curve.

We will also graph our financial data results, but because of the wild fluctuations in financial transactions, we will see a rather interesting and weird graph!

Here is the business specification our analyst gave us:

Create a report that will show monthly trade totals and the standard deviations both for partial data and full data set availability (hint: STDEVP). Results should be for, you guessed it, customer C0000001, John Smith, and for the 2012 calendar trading year. Show the year, quarter, and month. The report needs to identify the customer portfolio. I am only interested in the commodity portfolio at this time, but the query you build can be expanded to report on other portfolios. Lastly, generate a Microsoft Excel spreadsheet graph from the results.

After studying the requirements, Listing 5-15 is the code we came up with.

Listing 5-15. Portfolio Standard Deviation Analysis

```
WITH PortfolioAnalysis (
TradeYear,TradeQtr,TradeMonth,CustId,PortfolioNo,Portfolio,MonthlyValue
)
AS (
SELECT Year                    AS TradeYear
     ,DATEPART(qq,SweepDate)    AS TradeQtr
     ,Month                     AS TradeMonth
```

```
      ,CustId
      ,PortfolioNo
      ,Portfolio
      ,SUM(Value)                          AS MonthlyValue
FROM Financial.Portfolio
WHERE Year > 2011
GROUP BY CustId
      ,PortfolioNo
      ,Year
      ,Month
      ,SweepDate
      ,Portfolio
)

SELECT TradeYear
      ,TradeQtr
      ,TradeMonth
      ,CustId
      ,PortfolioNo
      ,Portfolio
      ,MonthlyValue
      ,SUM(MonthlyValue) OVER(
      PARTITION BY Portfolio
      ORDER BY Portfolio,TradeMonth
      ) AS RollingMonthlyValue
      ,STDEV(MonthlyValue) OVER(
      PARTITION BY Portfolio
      ORDER BY Portfolio,TradeMonth
      ) AS RollingMonthlyStdev
      ,STDEVP(MonthlyValue) OVER(
      PARTITION BY Portfolio
      ORDER BY Portfolio,TradeMonth
      ) AS RollingMonthlyStdevp
FROM PortfolioAnalysis
WHERE TradeYear = 2012
AND CustId = 'C0000001'
```

```
AND Portfolio = 'CASH - FINANCIAL PORTFOLIO'
ORDER BY CustId,Portfolio,TradeYear,TradeQtr,TradeMonth
GO
```

We are back to using our old friend, the CTE. The year, quarter, and month parts of the date are pulled out together with the customer and portfolio information. The value is summed up by year, month, and sweep date.

Note that results are for the cash portfolio, so technically, you do not need to call out the portfolio in the PARTITION BY clause. Download the script, try commenting it out, and then add it back in. Also add another portfolio name in the WHERE clause predicate. That's when the PARTITION BY portfolio kicks in. Let's see the report generated by this query.

Please refer to the results in Figure 5-41.

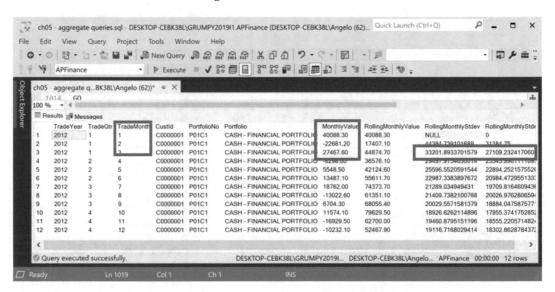

Figure 5-41. *Portfolio standard deviation analysis*

Looking at the first three months, we see the monthly value, rolling total value, and both standard deviation values. The results are updated as each month rolls by. Notice that the first set of standard deviation values are NULL and 0, respectively (STDEV() and STDEVP()).

At least STDEVP() gives us a zero!

Try modifying the WHERE BY clause as follows:

```
WHERE TradeYear IN(2012,2013)
AND CustId = 'C0000001'
AND Portfolio IN (
       'CASH - FINANCIAL PORTFOLIO',
       'EQUITY - FINANCIAL PORTFOLIO'
       )
```

This time the first row values have actual amounts. Interesting. More data will do this?

Let's copy and paste the original results into Microsoft Excel and generate the graph in Figure 5-42.

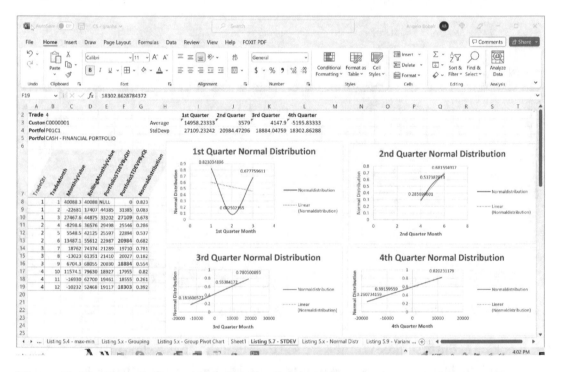

Figure 5-42. *Quarterly normal distribution curves*

To generate a graph for the standard deviation values, you need to calculate the normal distribution. I got a little creative and generated four graphs, one for each quarter. I took the average and standard deviation values for the last month in each quarter to use with the normal distribution function, that is, the 3rd, 6th, 9th, and 12th month set of values.

The first graph for the first quarter looks like an inverted bell curve – the remaining graphs not so much. This spreadsheet is available on the website where the code is stored for this chapter. Download the spreadsheet, analyze how it works, and then create your own queries, like generating bell curves for six-month intervals or the entire year.

Listing 5-16 is another listing that generates a more traditional bell curve for normal distribution using some staged data.

Listing 5-16. Setting Up the Example - Part 1

```
DECLARE @TradeYearStdev TABLE (
     TradeYear        SMALLINT     NOT NULL,
     TradeQuarter     SMALLINT     NOT NULL,
     TradeMonth       SMALLINT     NOT NULL,
     CustId           VARCHAR(32)  NOT NULL,
     PortfolioNo      VARCHAR(32)  NOT NULL,
     Portfolio        VARCHAR(64)  NOT NULL,
     MonthlyValue     DECIMAL(10,2) NOT NULL
     );

INSERT INTO @TradeYearStdev VALUES
('2012','1','1','C0000001','P01C1','COMMODITY - FINANCIAL
PORTFOLIO','13862.80'),
('2012','1','2','C0000001','P01C1','COMMODITY - FINANCIAL
PORTFOLIO','14629.50'),
('2012','1','3','C0000001','P01C1','COMMODITY - FINANCIAL
PORTFOLIO','15568.90'),
('2012','2','4','C0000001','P01C1','COMMODITY - FINANCIAL
PORTFOLIO','17004.80'),
('2012','2','5','C0000001','P01C1','COMMODITY - FINANCIAL
PORTFOLIO','18064.90'),
('2012','2','6','C0000001','P01C1','COMMODITY - FINANCIAL
PORTFOLIO','18500.30'),
('2012','3','7','C0000001','P01C1','COMMODITY - FINANCIAL
PORTFOLIO','17515.00'),
('2012','3','8','C0000001','P01C1','COMMODITY - FINANCIAL
PORTFOLIO','16779.50'),
```

```
('2012','3','9','C0000001','P01C1','COMMODITY - FINANCIAL
PORTFOLIO','15576.00'),
('2012','4','10','C0000001','P01C1','COMMODITY - FINANCIAL
PORTFOLIO','15941.60'),
('2012','4','11','C0000001','P01C1','COMMODITY - FINANCIAL
PORTFOLIO','14208.80'),
('2012','4','12','C0000001','P01C1','COMMODITY - FINANCIAL
PORTFOLIO','13804.30');

SELECT TradeYear
      ,TradeQuarter
      ,TradeMonth
      ,CustId
      ,PortfolioNo
      ,Portfolio
      ,MonthlyValue
      ,AVG(MonthlyValue) OVER(
      ) AS PortfolioAvg
      ,STDEVP(MonthlyValue) OVER(
      ) AS PortfolioStdevp
FROM @TradeYearStdev
ORDER BY CustId
      ,TradeYear
      ,TradeMonth
      ,PortfolioNo
GO
```

As I mentioned, the data (MonthlyValue) is staged so it will produce a nice bell curve
for the normal distribution. I set up the test data by loading it into a table variable with
some hard-coded INSERT statements.

Notice the OVER() clause for both the AVG() and STDEVP() functions. I want to
generate a single value for each so we can generate a graph for 12 months' worth of
sample data. That means each normal distribution uses the same average and standard
deviation values. All that changes is the monthly value. Here is the Excel formula: =NORM.
DIST(C8,$I$3,$I$4,TRUE).

Cell BC8 represents the `MonthlyValue` for January, the $I$3 reference is for the average for the entire year data sample, and the $I$4 cell reference is for the `STDEVP()` value for the entire year.

Note   The dollar ($) sign in front of the row and column coordinates for the cell reference indicates that the reference never changes as you use it for the other months. The cell reference for the total amounts does change to reflect each new row.

Let's see what this bell curve looks like. Please refer to the graphed normal distribution results in Figure 5-43.

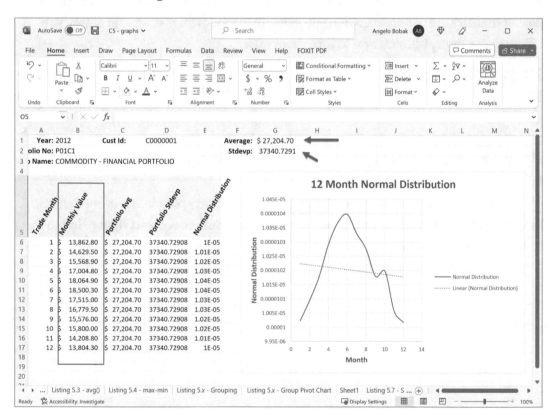

Figure 5-43. *Twelve-month normal distribution curve*

Looks more like the Alps in Italy, but this is a more recognizable bell curve generated by the data. This is of course staged data, but it looks like investments made went up until the middle of the year and then went down each month after till the end of the year.

A good analyst will ask why this is happening. I am certain our customer John Smith wants to know! Performance analysis time.

Performance Considerations

This time, in our performance analysis, we will take a deep dive into the PROFILE statistics and compare them with the tasks that appear in the graphical estimated query plan. This could get a bit into the weeds, but it is worth the effort as you will have a lot of information at hand that will assist you in your performance analysis efforts. Remember, profile statistics are generated when you execute the following command before you execute your query (don't forget to turn them off):

```
SET STATISTICS PROFILE ON
GO
```

Note   For this discussion, refer to Figures 5-44a through 5-44h.

As usual, let's start off by generating an estimated query plan in the usual manner. First, we examine the nodes that have high costs. Remember, we start right to left in the estimated execution plan.

Please refer to Figure 5-44a.

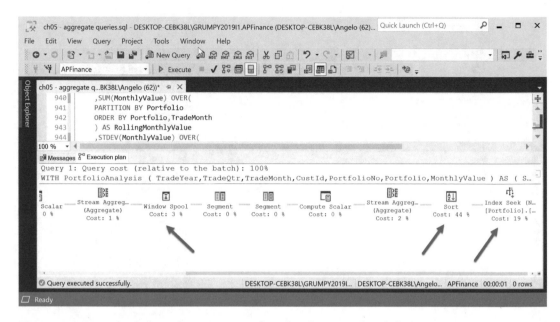

Figure 5-44a. *Estimated execution plan for the standard deviation query*

Not shown in the interest of space is one sort task at the far left of the plan with a cost of 33%, but we will discuss it later.

There are four high-cost tasks:

- An index seek, cost 19%

- First sort task, cost 44%

- Window spool task, cost 3%

- Second sort task, cost 33%

We can look at the PROFILE statistics for each node by placing the mouse over the node and allowing a pop-up tip panel to appear with the statistics.

Please refer to Figure 5-44b.

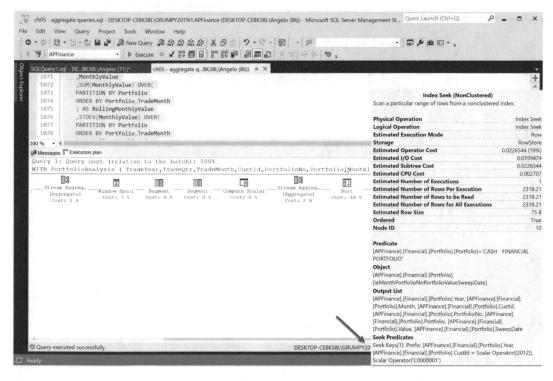

Figure 5-44b. *Index seek node profile statistics*

This index seek is a result of the WHERE clause predicate. If you look at the columns in the node, they match the columns in the WHERE clause predicate. This is called a covering index, and as we said earlier, an index seek is generally preferred to an index scan.

As a validation, if we look at the PROFILE statistics output column called statement text, we see

```
--Index Seek(OBJECT:(
[APFinance].[Financial].[Portfolio].
    [ieMonthPortfolioNoPortfolioValueSweepDate]),
   SEEK:(
   [APFinance].[Financial].[Portfolio].[Year]=(2012)
   AND [APFinance].[Financial].[Portfolio].[CustId]='C0000001'
   ),
   WHERE:([APFinance].[Financial].[Portfolio].[Portfolio]
        ='CASH - FINANCIAL PORTFOLIO')
ORDERED FORWARD
   )
```

A little cryptic to read, but we can clearly see the WHERE clause driving the index seek. Next, let's check out the expensive 44% sort.

Please refer to Figure 5-44c.

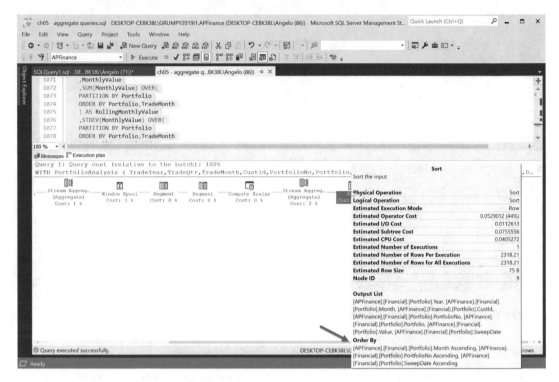

Figure 5-44c. *Sort step profile statistics*

Looking at the profile statistics output, it tells us the sort is by month, portfolio number, and sweep date:

```
--Sort(ORDER BY:(
     [APFinance].[Financial].[Portfolio].[Month] ASC,
     [APFinance].[Financial].[Portfolio].[PortfolioNo] ASC,
     [APFinance].[Financial].[Portfolio].[SweepDate] ASC
     )
)
```

This sort is for the output of the index seek.

If I remove the WHERE clause completely, this is what the sort parameters look like. Please refer to Figure 5-44d.

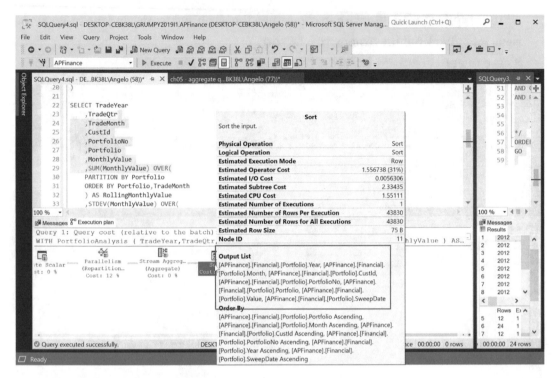

Figure 5-44d. *Sort tip if the WHERE clause is removed*

Notice the extra columns that are now included in the sort.

Next is the stream aggregate step. I did not generate a screenshot of this step, but it is used to generate sums via the SUM( ) function in the CTE query:

```
--Stream Aggregate(
     GROUP BY:([APFinance].[Financial].[Portfolio].[Month],
     [APFinance].[Financial].[Portfolio].[PortfolioNo],
      [APFinance].[Financial].[Portfolio].[SweepDate])

     DEFINE:([Expr1003]=SUM([APFinance].[Financial].[Portfolio].[Value]),
     [APFinance].[Financial].[Portfolio].[CustId]
          =ANY([APFinance].[Financial].[Portfolio].[CustId]),
       [APFinance].[Financial].[Portfolio].[Year]=
          ANY([APFinance].[Financial].[Portfolio].[Year]),
     [APFinance].[Financial].[Portfolio].[Portfolio]
          =ANY([APFinance].[Financial].[Portfolio].[Portfolio])
     )
)
```

Even though the CTE `GROUP BY` includes the following columns

```
GROUP BY CustId,PortfolioNo,Year,Month,SweepDate,Portfolio
```

it only sorts by month, portfolio, and sort date because we are filtering by year and customer in the `WHERE` clause.

Same as before, if I remove the `GROUP BY` clause in the CTE, this step is modified to include the missing columns:

```
|--Stream Aggregate(
GROUP BY:(
      [APFinance].[Financial].[Portfolio].[Portfolio],
      [APFinance].[Financial].[Portfolio].[Month],
      [APFinance].[Financial].[Portfolio].[CustId],
      [APFinance].[Financial].[Portfolio].[PortfolioNo],
      [APFinance].[Financial].[Portfolio].[Year],
      [APFinance].[Financial].[Portfolio].[SweepDate]
      )
      DEFINE:([Expr1003]=SUM([APFinance].[Financial].[Portfolio].[Value])))
```

Tip   Make sure you include only the columns you need in the GROUP BY, ORDER BY, OVER(), and WHERE clauses to improve performance and to make sure the optimizer does not have to do extra work to ignore columns that are not needed.

Next on deck is the window spool task at 3% generated by the default RANGE clause. Please refer to Figure 5-44e.

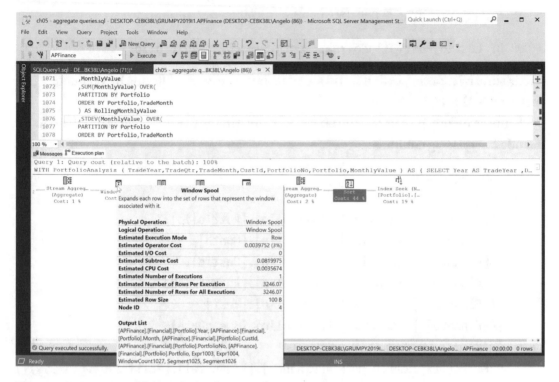

Figure 5-44e. *Profile statistics for window spool*

Recall from our discussions that the window spool task is used to load rows into memory or temporary storage that need to be processed repeatedly. In our case, this is generated by the OVER() clause, specifically the default RANGE clause.

This is used to calculate the sum and standard deviation values. If we look at the query, the ORDER BY clause in the OVER() clause sorts by portfolio and month. Look at the profile statistics details for the window spool in Figure 5-44f.

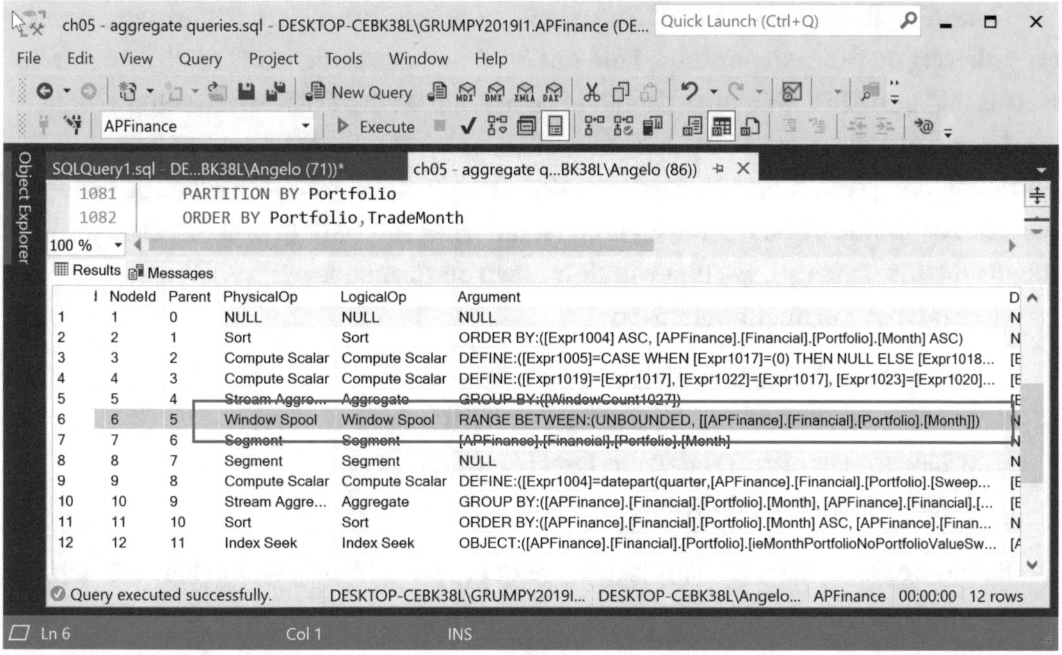

Figure 5-44f. *Window spool profile statistics details*

The window spool displays the default behavior when a PARTITION BY and a ORDER BY clause are included in the OVER() clause:

RANGE BETWEEN:([UNBOUNDED,[[APFinance].[Financial].[Portfolio].[Month]])

Wait a minute. I stated earlier that the ORDER BY clause was sorted by portfolio and month. Well, since the WHERE clause is filtering out rows for only one portfolio, the CASH - FINANCIAL PORTFOLIO, then this column is not needed. Once again, I removed the WHERE clause from the query, and this time the argument profile statistics value was

```
|--Window Spool(RANGE BETWEEN:(UNBOUNDED, [
     [APFinance].[Financial].[Portfolio].[CustId],
     [APFinance].[Financial].[Portfolio].[Year],
     [APFinance].[Financial].[Portfolio].[Portfolio],
     [APFinance].[Financial].[Portfolio].[Month]
     ])
     )
```

Now this tells us that we need to be careful how we set up our partition definition. By only filtering on the cash portfolio, I did not need to include the PARTITION BY portfolio clause, and the ORDER BY clause should have referenced the TradeMonth column only.

Since you might want to try this query so that it works on multiple customers, portfolios, and years, then you need the customer, portfolio, and year columns in the partition definitions. Make sure you understand what you want to process; else, you might add unnecessary code that will slow down performance.

Here is what you might want to try:

```
,STDEV(MonthlyValue) OVER(
PARTITION BY CustId, TradeYear,Portfolio
ORDER BY CustId, TradeYear,Portfolio,TradeMonth
) AS RollingMonthlyStdev
```

Finally, we look at the last sort task in the query plan. Please refer to Figure 5-44g.

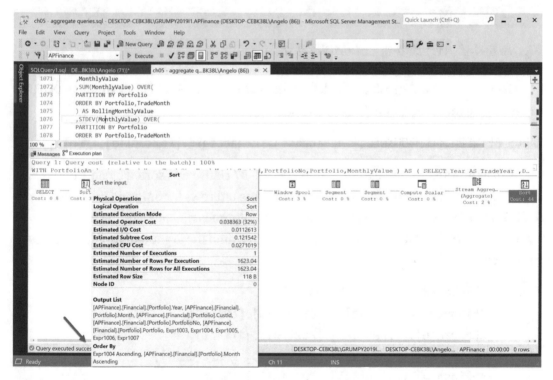

Figure 5-44g. *Final sort task in the query plan*

This last task orders the data stream by the following columns and corresponds to the ORDER BY clause that appears in the partition. The portfolio is not included as we have only one portfolio:

```
--Sort(
    ORDER BY:([Expr1004] ASC,
            [APFinance].[Financial].[Portfolio].[Month] ASC
            )
    )
```

If I modify the WHERE clause in the query so I filter for two years, two customers, and two portfolios as follows

```
FROM PortfolioAnalysis
WHERE TradeYear IN(2012,2013)
AND CustId IN('C0000001','C00000012')
AND Portfolio IN (
    'CASH - FINANCIAL PORTFOLIO',
    'EQUITY - FINANCIAL PORTFOLIO'
    )
ORDER BY CustId,Portfolio,TradeYear,TradeQtr,TradeMonth
GO
```

the sort string in the PROFILE statistics column StmtText is modified to represent the new columns:

```
--Sort(
    ORDER BY:(
    [APFinance].[Financial].[Portfolio].[CustId] ASC,
    [APFinance].[Financial].[Portfolio].[Portfolio] ASC,
    [APFinance].[Financial].[Portfolio].[Year] ASC,
    [Expr1004] ASC,
    [APFinance].[Financial].[Portfolio].[Month] ASC)
    )
```

Let's wrap it up for this function. We took our analysis a few steps deeper, but we see that we can gain valuable insight as to what is going on. Please also notice two important factors:

- Graphical estimated query plans are read right to left (we knew this).

- PROFILE statistics are read top down, so the last statement in the profile statistics corresponds to the right task in the estimated query plan.

- Node IDs between the estimated query plan and PROFILE statistics do not correspond. For example, in the PROFILE statistics output, the index seek has a NodeId of 12, but in the estimated query plan tip, it has a node ID of 10!

This is shown in Figure 5-44h.

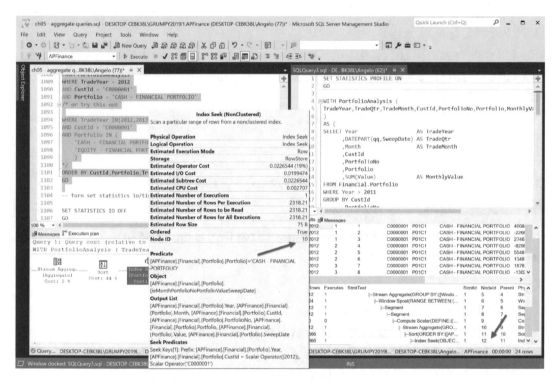

Figure 5-44h. *How to read the PROFILE statistics compared with the estimated query plan*

As can be seen, reading top to bottom of the PROFILE statistics corresponds to reading the tasks in the estimated query plan left to right (if this was not confusing enough).

So what did all this analysis get us?

The SQL Server estimated query plan is pretty good at recommending the correct index to use. But if performance is still an issue, we need to dive into these details and determine other actions to take, like denormalized reporting tables or rewriting the query. We won't even get into hardware considerations like more memory or faster disks. This is a performance tuning topic that can be part of a whole book on performance tuning.

Practice and get familiar with what each task does and how it corresponds to the entries in the PROFILE statistics so you can determine what additional steps you might need to take to improve performance.

There's one more aggregate function to discuss in this chapter.

VAR() and VARP() Functions

Recall from Chapter 2 that the VAR() and VARP() functions are used to generate the variance for a set of values in a data sample. The same discussion related to the data sets used in the STDEV()/STDEVP() examples applies to the VAR() and VARP() functions relative to population of the data:

The VAR() function works on a partial set of the data when the entire data set is not known or available, while the STDEVP() function works on the whole data set population when it is known.

For us developers here's a simpler explanation:

Informally, if you take all the differences of each value from the mean (average) and square the results and then add all the results and finally divide them by the number of data samples, you get the variance.

For our scenario, let's say we had some data that forecasted stock performance targets for one of our portfolios during a particular year. We want to see how close or far the gains were to or from the forecast. It answers the following questions:

- Did we **miss** our target investment goals?

- Did we **meet** our target investment goals?

- Did we **exceed** our target investment goals?

Data generated by these functions can be used in statistical reports and Microsoft Excel graphs to answer these questions.

The syntax is the same as the other functions:

```
VAR(ALL|DISTINCT <column name>) and VARP(ALL|DISTINCT <column name>)
```

But one thing to be aware of: If you use these with the OVER() clause, DISTINCT is not allowed!

Please refer to Listing 5-17.

Listing 5-17. Setting Up the Example

```
DECLARE @TradeYearStdev TABLE (
        TradeYear            SMALLINT       NOT NULL,
        TradeQuarter         SMALLINT       NOT NULL,
        TradeMonth           SMALLINT       NOT NULL,
        CustId               VARCHAR(32)    NOT NULL,
        PortfolioNo          VARCHAR(32)    NOT NULL,
        Portfolio            VARCHAR(64)    NOT NULL,
        MonthlyValue         DECIMAL(10,2)  NOT NULL
        );

INSERT INTO @TradeYearStdev VALUES
('2012','1','1','C0000001','P01C1','COMMODITY - FINANCIAL
PORTFOLIO','13862.80'),
('2012','1','2','C0000001','P01C1','COMMODITY - FINANCIAL
PORTFOLIO','14629.50'),
('2012','1','3','C0000001','P01C1','COMMODITY - FINANCIAL
PORTFOLIO','15568.90'),
('2012','2','4','C0000001','P01C1','COMMODITY - FINANCIAL
PORTFOLIO','17004.80'),
('2012','2','5','C0000001','P01C1','COMMODITY - FINANCIAL
PORTFOLIO','18064.90'),
('2012','2','6','C0000001','P01C1','COMMODITY - FINANCIAL
PORTFOLIO','18500.30'),
('2012','3','7','C0000001','P01C1','COMMODITY - FINANCIAL
PORTFOLIO','17515.00'),
('2012','3','8','C0000001','P01C1','COMMODITY - FINANCIAL
PORTFOLIO','16779.50'),
```

```
('2012','3','9','C0000001','P01C1','COMMODITY - FINANCIAL
PORTFOLIO','15576.00'),
('2012','4','10','C0000001','P01C1','COMMODITY - FINANCIAL
PORTFOLIO','15941.60'),
('2012','4','11','C0000001','P01C1','COMMODITY - FINANCIAL
PORTFOLIO','14208.80'),
('2012','4','12','C0000001','P01C1','COMMODITY - FINANCIAL
PORTFOLIO','13804.30');

SELECT TradeYear
     ,TradeQuarter
     ,TradeMonth
     ,CustId
     ,PortfolioNo
     ,Portfolio
     ,MonthlyValue
     -- generate values when there are 3 rows of values
     ,CASE
          WHEN TradeMonth % 3 = 0 THEN
               VAR(MonthlyValue) OVER(
                    PARTITION BY TradeQuarter
                    ORDER BY TradeMonth
                    ROWS BETWEEN 2 PRECEDING AND CURRENT ROW
               )
          ELSE 0
          END AS PortfolioquarterlyVar
     ,CASE
          WHEN TradeMonth % 3 = 0 THEN
               VARP(MonthlyValue) OVER(
                    PARTITION BY TradeQuarter
                    ORDER BY TradeMonth
                    ROWS BETWEEN 2 PRECEDING AND CURRENT ROW
               )
          ELSE 0
          END AS PortfolioquarterlyVarp
FROM @TradeYearStdev
```

```
ORDER BY CustId
      ,TradeYear
      ,TradeMonth
      ,PortfolioNo
GO
```

We are going to first declare a table variable to store some monthly portfolio positions. We include the trade year, quarter, and month and also the customer ID, portfolio number and name, and monthly account positions.

The table variable is loaded with 12 rows, one for each month, and I purposely assigned amounts that increase each month until the middle of the year and then after that start going down.

The query that generates the variance report based on this value is interesting. It uses CASE blocks to generate the variances for only the last month in each quarter. The modulo operator is used to test the value for the numerical month to see if it is divisible by 3. This way we report variances for the months of March, June, September, and December.

Next, in the OVER() clause, a partition is defined using the TradeQuarter column, and the partition is ordered by the TradeMonth column.

Finally, the entire query results are sorted by customer ID, trade year, trade month, and portfolio number. Let's see what the report looks like when we execute this query.

Please refer to Figure 5-45.

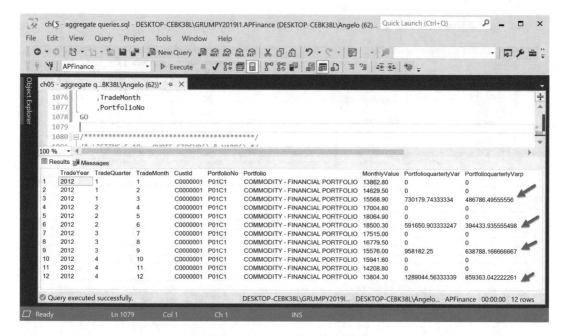

Figure 5-45.   *Quarterly portfolio variances for 2012*

We can see the variance values for every third month. Notice all the other variances are 0. Using this data, we copy and paste it into a Microsoft Excel spreadsheet for graphing.

Please refer to Figure 5-46.

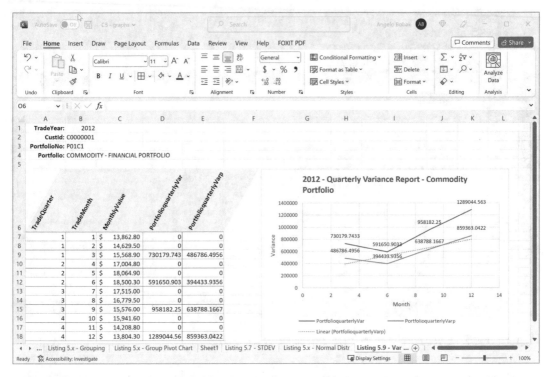

Figure 5-46. *Quarterly variance report and graph*

After a little fancy copying and pasting and some reformatting in Microsoft Excel, we have a nice variance report that only shows quarterly variance values. VAR and VARP are both trending up.

Ticker Analysis

Let's perform some statistical analysis on one financial instrument called Alcoa. The ticker symbol is AA.

The specification is to report the minimum low, the maximum low, the minimum high, and the maximum high for the year 2012. Please refer to Listing 5-18.

Listing 5-18. Alcoa Ticker Symbol Analysis

```
SELECT QuoteYear
      ,QuoteQtr
      ,QuoteMonth
      ,QuoteWeek
```

```sql
      ,Ticker
      ,Company
      ,TickerDate
      ,[Low]
      ,[High]
      ,MIN([Low]) OVER(
            PARTITION BY QuoteYear,QuoteQtr,QuoteMonth,[QuoteWeek],[Ticker]
            ORDER BY [Low]
            ROWS BETWEEN UNBOUNDED PRECEDING AND UNBOUNDED FOLLOWING
            ) AS MinLow
      ,MAX([Low]) OVER(
            PARTITION BY QuoteYear,QuoteQtr,QuoteMonth,[QuoteWeek],[Ticker]
            ORDER BY [Low]
            ROWS BETWEEN UNBOUNDED PRECEDING AND UNBOUNDED FOLLOWING
            ) AS MaxLow
      ,MIN([High]) OVER(
            PARTITION BY QuoteYear,QuoteQtr,QuoteMonth,[QuoteWeek],[Ticker]
            ORDER BY [High]
            ROWS BETWEEN UNBOUNDED PRECEDING AND UNBOUNDED FOLLOWING
            ) AS MinHigh
      ,MAX([High]) OVER(
            PARTITION BY QuoteYear,QuoteQtr,QuoteMonth,[QuoteWeek],[Ticker]
            ORDER BY [High]
            ROWS BETWEEN UNBOUNDED PRECEDING AND UNBOUNDED FOLLOWING
            ) AS MaxHigh
FROM [MasterData].[TickerPriceRangeHistoryDetail]
WHERE Ticker = 'AA'
AND QuoteYear = 2012
ORDER BY Ticker
      ,QuoteYear
      ,QuoteQtr
      ,QuoteMonth
      ,QuoteWeek
      ,Company
      ,TickerDate
GO
```

We want to set up a partition by year, quarter, month, week, and ticker symbol. For the low strike price, we order the partition by the Low column, and for the high strike price, we order the partition by the High column.

A window frame is defined by

```
ROWS BETWEEN UNBOUNDED PRECEDING AND UNBOUNDED FOLLOWING
```

This logic will allow us to see the lows and highs for each day but also the minimums and maximums for each week.

Copying and pasting the results into a Microsoft Excel spreadsheet yields a very interesting graph. Please refer to Figure 5-47.

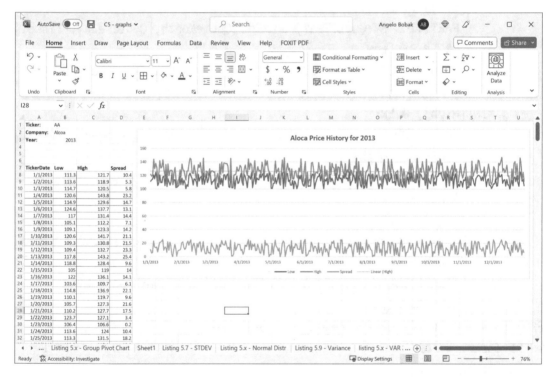

Figure 5-47. *Alcoa low, high, and spread analysis for 2012*

The spread line at the bottom also shows the volatility of the prices for this financial instrument. By the way, I made up all these prices, so they do not represent actual performance for Alcoa.

More Non-statistical Variance

What if we want to see the distance between two specific points of values, like if we want to compare current account balances vs. target account performance balances? Did we meet our targets or fall short? Nothing like a visual to point out this information. The following examples will not use the VAR() and VARP(), which calculate statistical variance. We want to use the results to graph the distance between each pair of actual vs. projected values the old-fashioned way.

This technique together with the use of the VAR() and VARP() functions from earlier examples will deliver to the business analyst a powerful set of reports and graphs to measure portfolio and account performance.

First, we need to set up the test. We will create a table called AccountVersusTarget and populate it with some test actual vs. forecast balances (also made up).

Please refer to Listing 5-19.

Listing 5-19. Create an Account Forecast Table

```
TRUNCATE TABLE [Financial].[AccountVersusTarget]
GO

INSERT INTO [Financial].[AccountVersusTarget]
SELECT [CustId]
      ,[PrtfNo]
      ,[AcctNo]
      ,[AcctName]
      ,[AcctTypeCode]
      ,[AcctBalance]
      ,[PostDate]

      ,CASE
      WHEN [AcctBalance] < 0 THEN (ABS([AcctBalance]) * 1.05)
      WHEN [AcctBalance] BETWEEN 500 AND 1000 THEN [AcctBalance] * 1.25
      WHEN [AcctBalance] BETWEEN 1001 AND 2000 THEN [AcctBalance] * 1.20
      WHEN [AcctBalance] BETWEEN 2001 AND 3000 THEN [AcctBalance] * 1.15
      WHEN [AcctBalance] BETWEEN 3001 AND 4000 THEN [AcctBalance] * 1.10
      ELSE [AcctBalance] * .90
      END AS TargetBalance,
```

```
        (
        CASE
        WHEN [AcctBalance] < 0 THEN  (ABS([AcctBalance]) * 1.05)
        WHEN [AcctBalance] BETWEEN 500 AND 1000 THEN [AcctBalance] * 1.25
        WHEN [AcctBalance] BETWEEN 1001 AND 2000 THEN [AcctBalance] * 1.20
        WHEN [AcctBalance] BETWEEN 2001 AND 3000 THEN [AcctBalance] * 1.15
        WHEN [AcctBalance] BETWEEN 3001 AND 4000 THEN [AcctBalance] * 1.10
        ELSE [AcctBalance] * .90
        END
        ) - [AcctBalance] AS Delta
FROM [APFinance].[Financial].[Account]
GO
```

A couple of CASE blocks are used to create some random forecast values and the difference between the actual balances and the forecast balances. The account information is retrieved from the Account table and inserted into our new table.

Listing 5-20 is the query we will use for our analysis.

Listing 5-20. Actual vs. Forecast Account Balance Query

```
SELECT [CustId]
        ,YEAR([PostDate])  AS PostYear
        ,MONTH([PostDate]) AS PostMonth
        ,[PostDate]
        ,[PrtfNo]
        ,[AcctNo]
        ,[AcctName]
        ,[AcctTypeCode]
        ,[AcctBalance]
        ,[TargetBalance]
        ,[Delta]
FROM [Financial].[AccountVersusTarget]
WHERE [CustId] = 'C0000001'
AND [PrtfNo] = 'P01C1'
AND [AcctName] = 'EQUITY'
```

```
AND YEAR([PostDate]) = '2012'
AND MONTH([PostDate]) = 1
GO
```

Very simple query. We pick our old friend, customer C0000001 (John Smith), and use some filters to check out the equity account in his portfolio for the month of January of 2012. No other bells or whistles in the query. What you see is what you get.

Please refer to Figure 5-48 for the results.

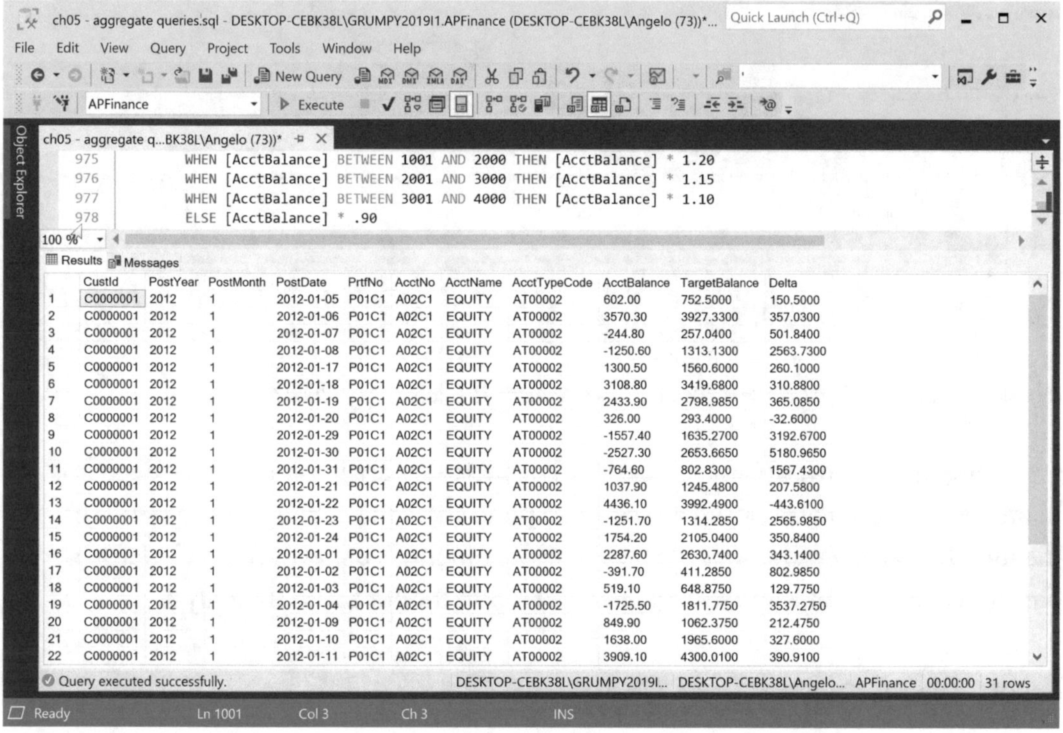

Figure 5-48. *Actual vs. forecast balances for John Smith*

Everything looks correct. We copy and paste the results into a Microsoft Excel spreadsheet so we can create a nice graph. Please refer to Figure 5-49.

Figure 5-49. *Actual vs. forecast delta report for January 2012*

This graph is powerful in my opinion. The increases are in blue and the decreases in orange bars. I superimposed a second graph to show the actual balance trends of the month. So even without powerful aggregate functions using the OVER() clause and partitions, we were able to create a powerful report for our business analyst.

Even More Statistical Variance

Let's try one more example. Let's go back and look at some ticker symbol performance. Let's use a couple of aggregate functions to take a snapshot of the GOITGUY ticker symbol again. This time we want to look at the price swings over a 24-hour period for January 5, 2015, and generate the total average, the standard deviation for the entire population of the data, and also the variance for the entire population of the data (assume the stock trades 24 by 7 around the world).

Please refer to Listing 5-21.

Listing 5-21. More Nice Bell Curves

```
SELECT [Ticker]
     ,[Company]
     ,[TickerDate]
     ,[QuoteHour]
     ,[Quote]
     ,AVG(Quote) OVER(
          ) AS AvgQuote
     ,STDEVP(Quote) OVER(
     ) AS TickerSTDEVP
     ,VARP(Quote) OVER(
     ) AS TickerVARP
FROM [APFinance].[MasterData].[TickerHistory]
WHERE TickerDate = '2015-01-05'
GO
```

Very simple query this time. Since we are only looking at 24 hours' worth of data, the OVER() clause for all functions is empty, no PARTITION BY or ORDER BY clauses.

Pop Quiz   What is the default window frame behavior when no PARTITION BY or ORDER BY clause is used in the OVER() clause?

Executing the query produces the output in Figure 5-50, which we use to populate an Excel spreadsheet so we can create our usual masterpiece graphs!

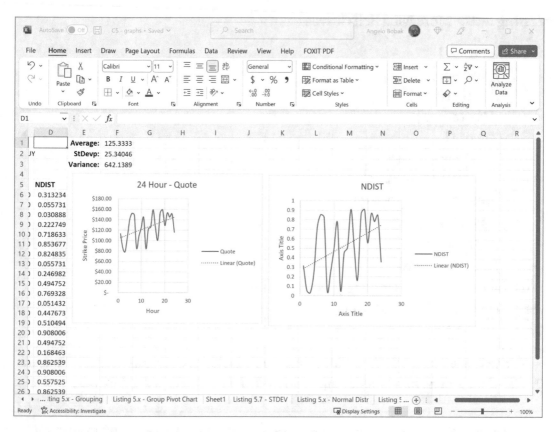

Figure 5-50. *More nice bell curves*

Very nice. We have two graphs. One shows the 24-hour price swings of the stock, which is trending up for that day, so that is a good thing. The other shows the bell curve or should I say bell curves created by using the normal distribution values for the 24-hour prices.

We can see that sometimes we need to use our aggregate window functions and sometimes we use plain queries. Together we can generate valuable reports and queries that show us the performance of a financial portfolio and also individual stocks or other types of financial instruments.

These reports and graphs provide the financial advisor with the tools they need to not only analyze past historical performance but to also forecast future performance. All that is missing from our discussion are the events that occurred during the ups and downs of stock prices, like did companies introduce a new product to market or did companies have issues with some of their products or was the economy facing an upturn or downturn. If current news and indicators tell us these events are about to occur again, we can take a good guess as to whether to sell, buy, or stay put!

Lots of factors to managing a financial portfolio, but with the right tools, decisions can be made to avoid severe loss and maybe make a couple of dollars!

Sadly, my strategy always seems to be buy high and sell low!

Summary

We accomplished a lot in this chapter. We went through the aggregate window functions as we did in Chapter 2 but this time with financial data. We also took a more detailed look at performance analysis and introduced some new tools like the SET PERFORMANCE ON command that generates detailed information on the steps the estimated query plan takes to produce the desired report.

Lastly, we generated several different graphs with Microsoft Excel that used the data generated by our queries and gave us some interesting visual information and patterns that gave us insight to the performance of accounts, portfolios, ticker symbols, and also financial transactions.

Next, we apply this methodology to the ranking window functions.

CHAPTER 6

Finance Use Case: Ranking Functions

Ranking functions, when applied to financial data, can produce interesting and valuable information for analysts or portfolio managers. This information is used to predict future trading patterns based on historical patterns. A properly managed portfolio based on solid analytical data and analysis will generate profits (hopefully).

This chapter will cover the application of the functions in this category to several interesting examples ranging from transaction to account, portfolio, and ticker performance analysis. Excel graphs will be generated, and we will dive deeper into performance tuning by using several of the performance tuning tools available with SQL Server.

Ranking Functions

The following are the four functions that belong to the ranking function category:

- RANK()

- DENSE_RANK()

- NTILE()

- ROW_NUMBER()

We first saw these functions in Chapter 4 when we applied them to a sales scenario. This time we apply them to a financial database called APFinance. The approach we take will be similar to other chapters, but we will dive deeper into performance tuning by introducing new tools like the SET PERFORMANCE STATISTIC ON setting that generates vital statistics related to each step of the execution plan. Additionally we will compare

313

© Angelo Bobak 2023
A. Bobak, *SQL Server Analytical Toolkit*, https://doi.org/10.1007/978-1-4842-8667-8_6

estimated query performance plans to actual query performance plans that show real-time timing and row movement statistics.

We will also generate some interesting graphs with Microsoft Excel that use the data generated by our scripts and queries. These, coupled with the SQL reports, will give us valuable insight into the trading patterns we have simulated for our five clients. This is called data visualization.

The chapter concludes with some interesting applications of the ROW_NUMBER() function and other functions we covered to solve the traditional gaps and islands problems that appear in many analysis projects.

RANK() Function

Recall from Chapter 4 that the RANK() function shows the rank of the value in each row column of a data set relative to the other row values in the data set. In case of ties, the rank is the same for the ties, but the next value after the duplicate values skips the next rank and a number equal to the number of ties plus the current rank value is assigned. We will demonstrate this behavior in our first example.

Example 1

Our first objective is to rank portfolio performance for our clients. Our business analyst has supplied a business specification for our report. It needs to show the total value of customer portfolios by trade year, customer identifier, portfolio number, and portfolio name. These results need to be ranked by year for each customer. The data can be 24 hours old as the only time updates occur is at night when the accounts are swept. This fact alone should guide us in how we structure the solutions (hint: CTE or reporting table approach).

This report will give the analyst some insight as to how each customer is performing relative to the other customers.

Please refer to Listing 6-1.

Listing 6-1. Ranking Customer Cash Portfolio

```
WITH PortfolioAnalysis (
TradeYear,CustId,PortfolioNo,Portfolio,YearlyValue
)
AS (
```

```
SELECT Year AS TradeYear
      ,CustId
      ,PortfolioNo
      ,Portfolio
      ,SUM(Value) AS YearlyValue
FROM Financial.Portfolio
WHERE Year > 2011
GROUP BY CustId,Year,PortfolioNo
      ,Portfolio
)

SELECT TradeYear
      ,CustId
      ,PortfolioNo
      ,Portfolio
      ,YearlyValue
      ,RANK() OVER(
            PARTITION BY Portfolio,TradeYear
            ORDER BY YearlyValue DESC
      ) AS PortfolioRankByYear
FROM PortfolioAnalysis
GO
```

The CTE is used to generate results that include the total value of the portfolio by year, customer ID, portfolio number, and the name of the portfolio.

The query that accesses the CTE uses the RANK() function, which is partitioned by portfolio number and trade year to generate the rankings.

Please refer to the partial results in Figure 6-1.

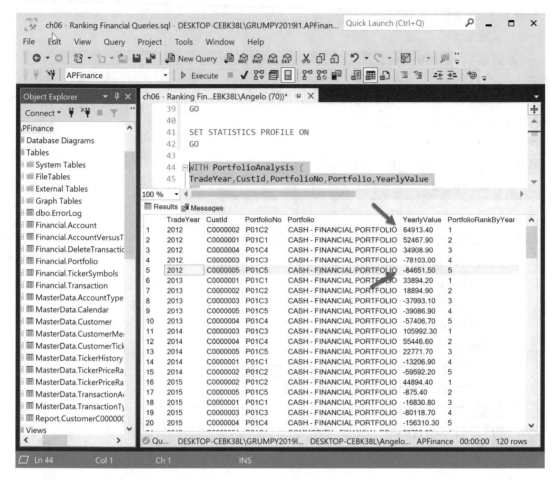

Figure 6-1. *Customer cash portfolio performance ranking*

For 2012, the top cash portfolio belongs to customer C0000002 and has a value of $64,913.40. This seems good, but our analyst would like to dig deep and understand why it is so high.

Did the customer sell off a lot of instruments?

Did the customer make frequent deposits into the cash accounts like checking or money market?

How about customer C0000005 that shows a –$84,651.50 balance and is ranked at the bottom of the list? Is this a result of bad trades? Our analyst would use this report as a starting point to study the trading patterns for these customers to understand what is going on.

What would help is also a nice graph based on this data to aid in the analysis. And here it is in Figure 6-2!

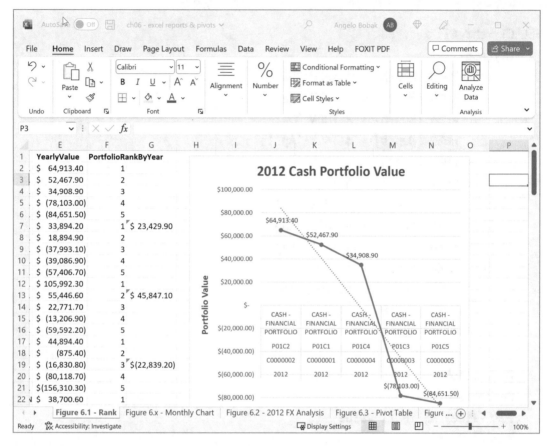

Figure 6-2. *Customer cash portfolio performance graph*

There is definitely a downward trend, but not only customer C0000005 but customer C0000003 seems to be in trouble. Adding a trend line also shows which direction portfolio values are going and can be used to grab the immediate attention of the analyst.

Let's turn our attention to customer C0000001 and look at their FX (foreign exchange) portfolio. Their cash portfolio seems on solid ground. Let's see how they are doing with their foreign currency portfolio.

Simple business specification: Our analyst requested a report that shows portfolio totals by year, month and then the rank of each month within the year for FX balances. Of course, they want an Excel graph, so we will create one for them. Lastly, months should be spelled out with three-character abbreviations and not numbers.

Please refer to Listing 6-2.

Listing 6-2. Customer C0000001 2012 Monthly Balance

```
SELECT Year AS TradeYear
     ,CASE
            WHEN [Month] = 1 THEN 'Jan'
            WHEN [Month] = 2 THEN 'Feb'
            WHEN [Month] = 3 THEN 'Mar'
            WHEN [Month] = 4 THEN 'Apr'
            WHEN [Month] = 5 THEN 'May'
            WHEN [Month] = 6 THEN 'Jun'
            WHEN [Month] = 7 THEN 'Jul'
            WHEN [Month] = 8 THEN 'Aug'
            WHEN [Month] = 9 THEN 'Sep'
            WHEN [Month] = 10 THEN 'Oct'
            WHEN [Month] = 11 THEN 'Nov'
            WHEN [Month] = 12 THEN 'Dec'
      END AS Trademonth
   ,CustId
   ,PortfolioNo
   ,Portfolio
   ,SUM(Value) AS YearlyValue
   ,RANK() OVER(
            ORDER BY SUM(Value) DESC
      ) AS PortfolioRankByMonth
FROM Financial.Portfolio
WHERE Portfolio = 'FX - FINANCIAL PORTFOLIO'
AND CustId = 'C0000001'
AND Year = 2012
GROUP BY CustId,Year,Month,PortfolioNo
      ,Portfolio
ORDER BY SUM(Value) DESC
GO
```

This is a straightforward query, without a CTE, that uses a case block to display the month abbreviations instead of just numbers. The SUM() function is used without an OVER() clause because we want to see the straight totals, not rolling totals by month. The RANK() function is used to assign a rank to each month to see what the best- and worst-performing months were.

Lastly, the mandatory GROUP BY clause is included, and we sort the results by the portfolio total in descending order.

Notice the ORDER BY clause just mentioned. Do we really need it? I commented out the line, re-executed the query, and got the same results. Conclusion: We do not need it. Steps like this could affect performance, so make sure you understand in the query plan what is going on. Run them side by side, in two horizontal query panes, to make sure there are no extra steps.

In this particular example, I did the analysis, and the query plan and profile statistics were the same for both queries, one with the extra ORDER BY and one without, so performance was not affected.

I made a slight alteration, and I changed the sort order in the OVER() clause to ascending and kept the query ORDER BY clause that sorted the data set in descending order, while the first query did not include the last ORDER BY clause. Here is where there was a slight degradation of performance as can be seen in the resulting performance statistics. Please refer to Figure 6-3.

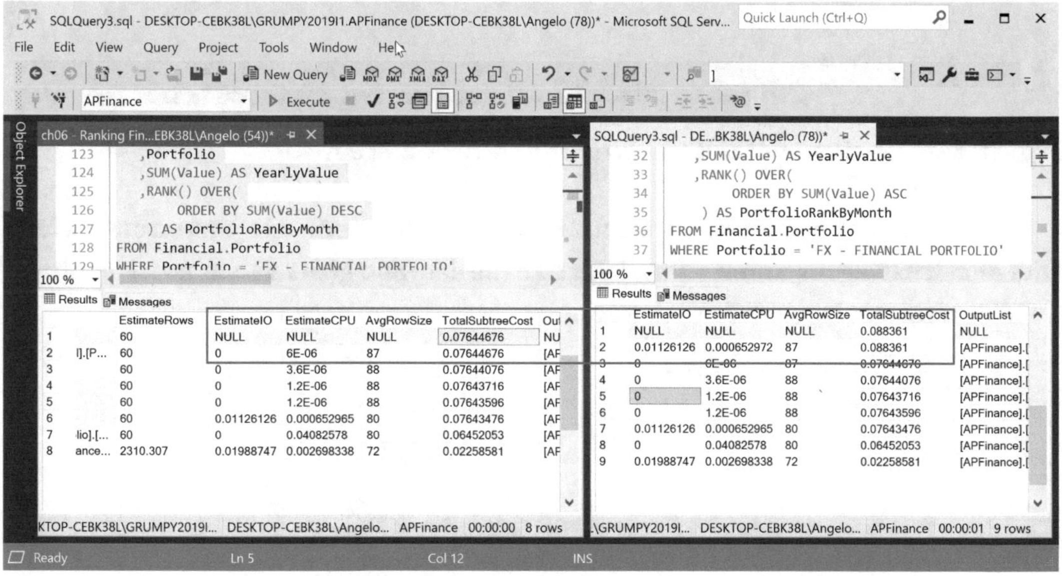

Figure 6-3. *Comparing performance statistics*

Does not seem like a substantial degradation in performance, but if your tables have millions of rows, this could be an issue. Look out for the small stuff. It can hurt you a lot!

Let's get back to the query and see what the results look like. Please refer to Figure 6-4.

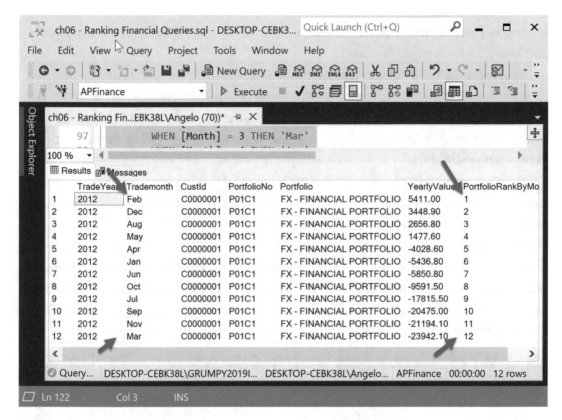

Figure 6-4. *Customer C0000001 2012 monthly balance report*

We see that February results were the best, but values seemed to jump around a lot with March having the lowest value. Let's take the data and generate our usual graph.

Please refer to Figure 6-5.

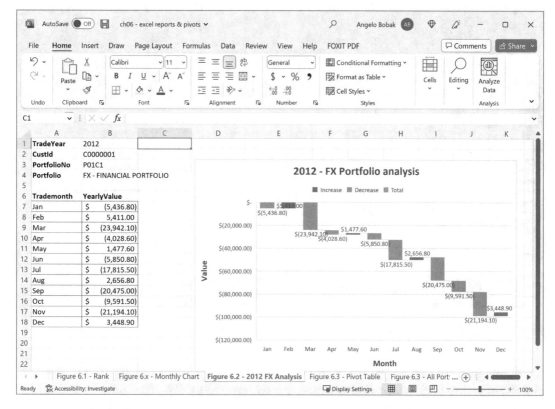

Figure 6-5. *Customer C0000001 2012 monthly balance chart*

This time the results are sorted by month in ascending order, and we see a nice downward spiral. This needs to be investigated. Lots of buying and selling in this portfolio. Are these panic sell-offs or sell-offs to cash in on profits?

Performance Considerations

Let's get back to our first query and do some performance analysis with the estimated query execution plan.

Please refer to Figure 6-6.

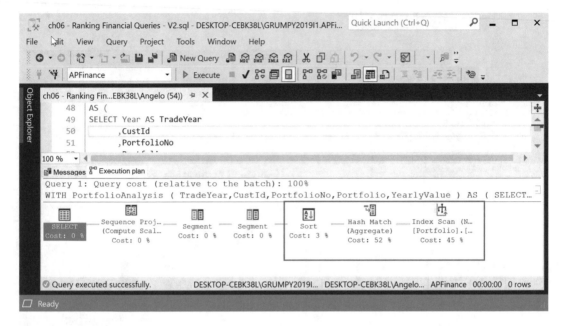

Figure 6-6. *Estimated query plan for the first query*

Reading from right to left, we see an expensive index scan and a subsequent hash match aggregate task, which is also expensive with a cost of 52%. The index is scanned, and the output goes into a hash table so it can be matched to rows from the Portfolio table itself. These rows are also placed into a hash bucket, and the matching rows are passed on to the sort task.

Note   Place your mouse pointer over this and other tasks to display a hint panel that includes a short explanation of the task. Do this for tasks you do not know what they do!

To improve the performance of this query, we would need to at least eliminate the hash match task. Recall that the data can be 24 hours old, so a possible solution could be to replace the CTE with a table that is loaded using the same query.

This is exactly what we will try out. Please refer to Listing 6-3.

Listing 6-3. Loading the Portfolio Sweep Table

```
TRUNCATE TABLE Financial.DailyPortfolioAnalysis
GO

INSERT INTO Financial.DailyPortfolioAnalysis
SELECT Year AS TradeYear
      ,CustId
      ,PortfolioNo
      ,Portfolio
      ,SUM(Value) AS YearlyValue
FROM Financial.Portfolio
GROUP BY CustId,Year,PortfolioNo
      ,Portfolio
ORDER BY Portfolio ASC,Year ASC, SUM(Value) DESC
GO

SELECT TradeYear,
      CustId,
      PortfolioNo,
      Portfolio,
      YearlyValue,
      RANK() OVER(
      PARTITION BY Portfolio,TradeYear
      ORDER BY YearlyValue DESC
      ) AS PortfolioRankByYear
FROM Financial.DailyPortfolioAnalysis
WHERE TradeYear > 2011
GO
```

Assume the INSERT command runs once a day, so the analyst only needs to execute the query against the loaded report table. We see that the query is historical in nature, so it does not make sense regenerating the entire historical data set each time the query is run.

If we now run the estimated query plans for each version of the query, we see the following results. Please refer to Figure 6-7.

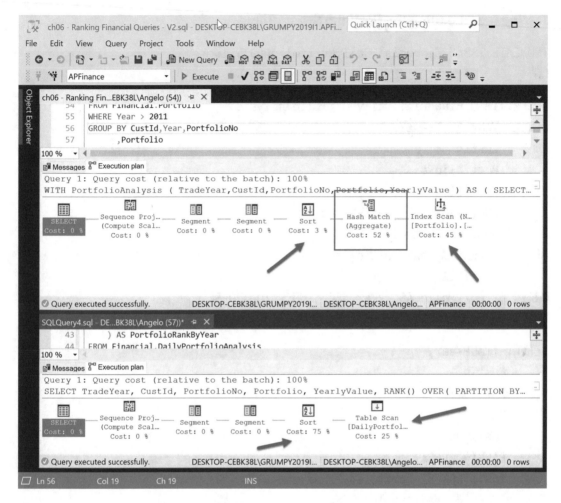

Figure 6-7. Comparing query plans

Notice the expensive table hash match task is gone. We do see a table scan task with a cost of 25%, but this is a small table, only 120 rows, so an index is not required. The sort is expensive at 75%, but look at the lines between the table scan and sort step. They are very thin, and if you click them, you will see that 120 rows were passed to be sorted.

Note   Running a live query performance plan will display the rows that were passed between tasks and the time each task took to complete. Pressing the button with the green checkmark will generate the live plan while the query is running.

With the first version of the estimated execution plan, the index scan passed 43,830 rows to the hash match task, and this task passed 600 rows to the sort task. Looks like this strategy worked!

Let's compare IO and TIME statistics. Please refer to Figure 6-8.

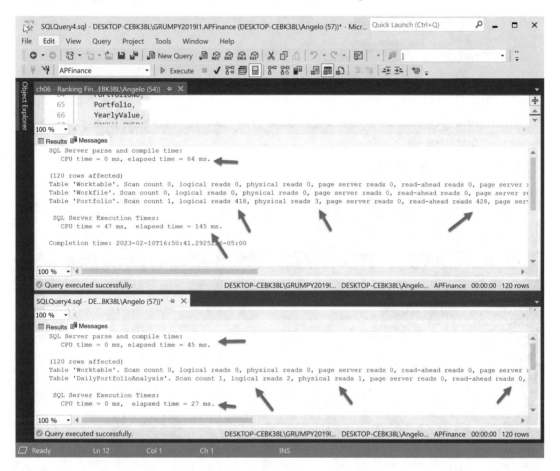

Figure 6-8. *IO and TIME statistics*

Let's start by comparing parse and execution times:

For SQL Server parse and compile times, the CTE query had 84 ms elapsed time vs. 45 ms for the report table strategy.

For SQL Server execution times, the CTE query had 145 ms elapsed time vs. 27 ms for the report table strategy.

Let's look at the IO statistics. Please refer to Table 6-1.

Table 6-1. *Comparing IO Statistics*

Step	CTE Query	Report Query
Scan count	1	1
Logical reads	418	2
Physical reads	3	1
Read-ahead reads	428	0

The results speak for themselves. Significant improvement for all statistics. What about profile statistics?

Please refer to Figure 6-9.

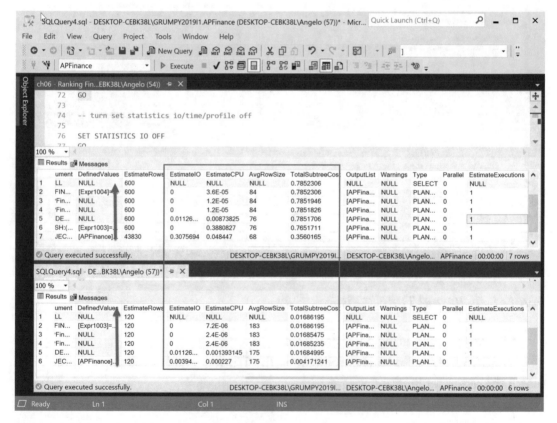

Figure 6-9. *Comparing profile statistics*

The profile statistics also speak for themselves. First, look at the number of rows passed between each step of the plans. The improved performance strategy has to work with fewer rows. If we examine the `EstimateIO`, `EstimateCPU`, and `TotalSubTreeCost`, we see significant improvements in the report table strategy. What are the takeaways?

The first takeaway is that understanding business requirements is important not only in delivering the report the analyst wants, but it may also provide hints on how we can improve performance. In our case we ended up creating the report table, which is loaded once a day, and making our base report query more efficient and faster.

Who would have thought?

One last point: Make sure the query results are equal. Might be a problem if a solution is faster but the results are wrong! Please refer to Figure 6-10.

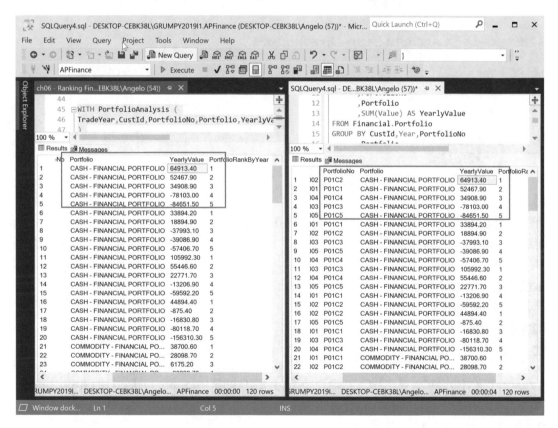

Figure 6-10. *Comparing query results*

Looks good! But look at the time parameter on the lower right-hand side. Our new and improved strategy took four seconds to run. When we see this, we should run the tests a couple of times and make sure we execute the DBCC command each time we run the queries – once per query. If the original query seems to run faster, then maybe we should drop the new and improved strategy!

Let's turn our attention back to analyzing portfolios!

Example 2

This time we look at the FX (foreign exchange) portfolio positions. We will look at the year 2012 and monthly positions for customer C0000001 again.

Please refer to Listing 6-4.

Listing 6-4. FX Account Deep-Dive Analysis

```
SELECT Year        AS TradeYear
      ,Month       AS Trademonth
      ,CustId
      ,PortfolioNo
      ,Portfolio
      ,SUM(Value) AS YearlyValue
FROM Financial.Portfolio
WHERE Portfolio = 'FX - FINANCIAL PORTFOLIO'
AND CustId = 'C0000001'
AND Year = 2012
GROUP BY CustId,Year,Month,PortfolioNo
      ,Portfolio
ORDER BY Month
GO
```

This is a straightforward query. The SUM() aggregate function is used to report the monthly positions. We want to see how things stand. Is the portfolio performing well, or is it losing money? Let's see the results.

Please refer to Figure 6-11.

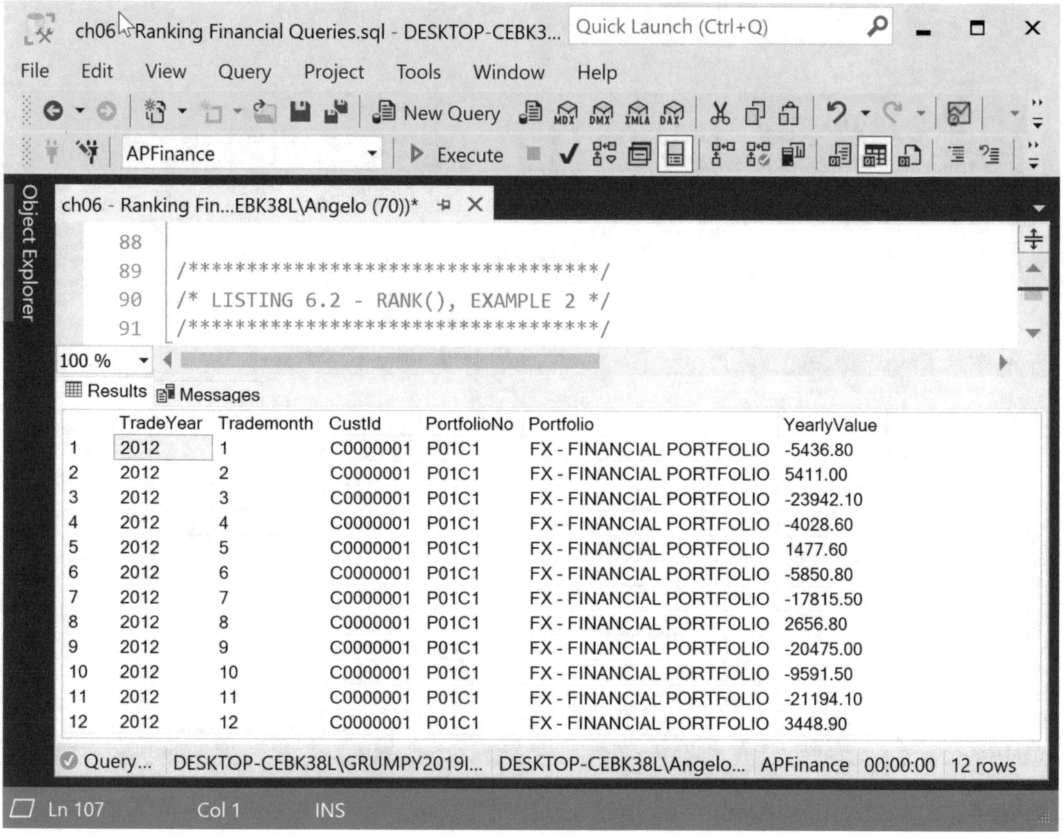

Figure 6-11. *Monthly FX positions for 2012*

Seems like a very volatile trading pattern. Lots of buys and sells. A graph will give us a better visual indicator of what is going on. Please refer to Figure 6-12.

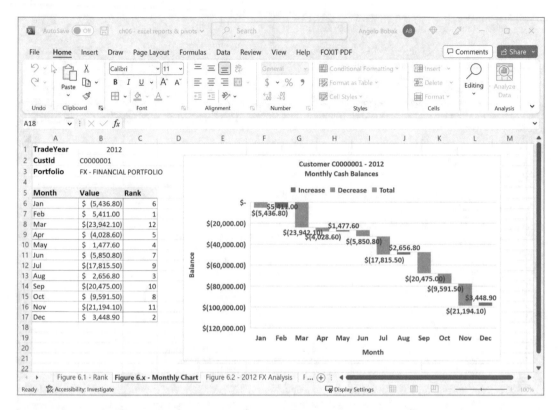

Figure 6-12. *2012 monthly FX positions and trends*

Looking at this graph, we see there are more decreases than increases. This means the customer is selling off foreign currency. Is this because there are profits to be made or the instrument is performing so badly the customer is selling off?

We need to compare performance across all portfolios to get an idea of what is going on. Sometimes trading strategies involve selling one instrument to purchase another. Let's put all this data into a Microsoft Excel pivot table so we can really do some analysis.

Please refer to the query in Listing 6-5.

Listing 6-5. 2012 Portfolio Value Pivot Table Query

```
SELECT [Year] AS TradeYear
     ,CASE
            WHEN [Month] = 1 THEN 'Jan'
            WHEN [Month] = 2 THEN 'Feb'
            WHEN [Month] = 3 THEN 'Mar'
            WHEN [Month] = 4 THEN 'Apr'
```

```
            WHEN [Month] = 5 THEN 'May'
            WHEN [Month] = 6 THEN 'Jun'
            WHEN [Month] = 7 THEN 'Jul'
            WHEN [Month] = 8 THEN 'Aug'
            WHEN [Month] = 9 THEN 'Sep'
            WHEN [Month] = 10 THEN 'Oct'
            WHEN [Month] = 11 THEN 'Nov'
            WHEN [Month] = 12 THEN 'Dec'
        END AS Trademonth
     ,CustId
     ,PortfolioNo
     ,Portfolio
     ,SUM(Value) AS [MonthTotal]
FROM Financial.Portfolio
WHERE CustId = 'C0000001'
AND Year = 2012
GROUP BY Year,Month,Custid,PortfolioNo,Portfolio
ORDER BY Custid,Year,Month,PortfolioNo,Portfolio
GO
```

Let's run the query and copy and paste the results to a Microsoft Excel spreadsheet and create a pivot table. This way we can perform some slicing and dicing to look at the various trading patterns. Please refer to Figure 6-13.

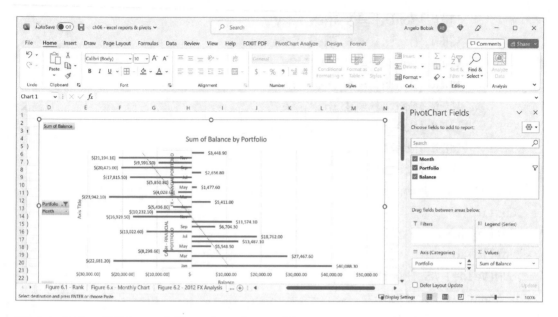

Figure 6-13. *2012 portfolio value pivot table*

We now can perform a deep dive into the transaction details to see what is going on, on a daily basis. This usually gets us to the root cause of any questionable trading patterns that lead to losses. Keep in mind that what we are looking at is the totals of buy and sell transactions by month, not the rolling totals as the months go by. These would give us a true picture of portfolio performance and what is left in the portfolio each month after all the trades have been summed up. Let's check out the query. Try to download the code prior to each chapter and follow along.

Please refer to Listing 6-6.

Listing 6-6. Detailed Transaction Analysis

```
SELECT MONTH(T.[TransDate]) AS TransMonth
    ,CASE
            WHEN MONTH(T.[TransDate]) = 1 THEN 'Jan'
            WHEN MONTH(T.[TransDate]) = 2 THEN 'Feb'
            WHEN MONTH(T.[TransDate]) = 3 THEN 'Mar'
            WHEN MONTH(T.[TransDate]) = 4 THEN 'Apr'
            WHEN MONTH(T.[TransDate]) = 5 THEN 'May'
            WHEN MONTH(T.[TransDate]) = 6 THEN 'Jun'
            WHEN MONTH(T.[TransDate]) = 7 THEN 'Jul'
```

```
            WHEN MONTH(T.[TransDate]) = 8 THEN 'Aug'
            WHEN MONTH(T.[TransDate]) = 9 THEN 'Sep'
            WHEN MONTH(T.[TransDate]) = 10 THEN 'Oct'
            WHEN MONTH(T.[TransDate]) = 11 THEN 'Nov'
            WHEN MONTH(T.[TransDate]) = 12 THEN 'Dec'
        END AS Trademonth
        ,T.[TransDate]
        ,T.[Symbol]
        ,T.[Price]
        ,T.[Quantity]
        ,T.[TransAmount]
        ,T.[CustId]
        ,T.[PortfolioNo]
        ,PAT.PortfolioAccountTypeCode
        ,PAT.PortfolioAccountTypeName
        ,T.[AcctNo]
        ,T.[BuySell]
    FROM [APFinance].[Financial].[Transaction] T
    JOIN [MasterData].[PortfolioAccountType] PAT
    ON T.PortfolioAccountTypeCode = PAT.PortfolioAccountTypeCode
    WHERE YEAR(T.TransDate) = 2012
    AND T.CustId = 'C0000001'
    GO
```

This query gives us a detailed view of the transaction level for all instruments and buy/sell categories. A CASE block prints out the month abbreviations for a more meaningful report. We use the Portfolio Account Type table (alias = PAT) and pull out the unique portfolio code and name for the client.

Let's look at some of the transaction details. Please refer to Figure 6-14.

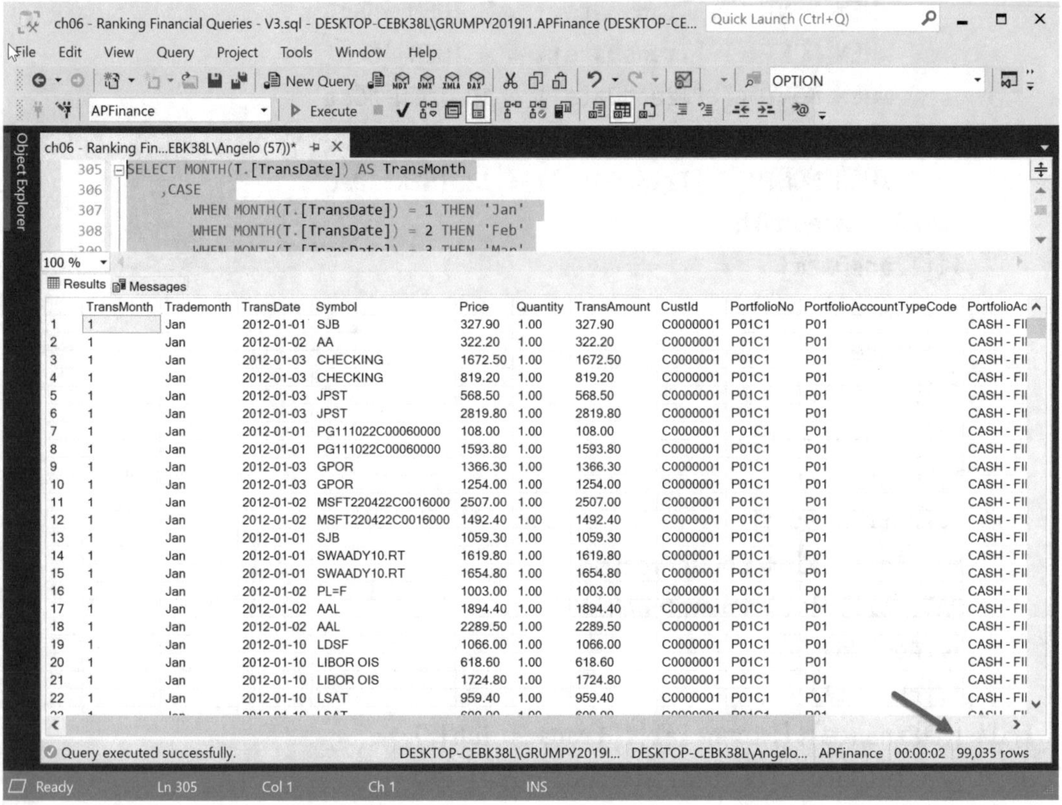

Figure 6-14. *Transaction details for customer C0000001*

This gives us a lot of information like transaction dates, price and quantity of each trade, the symbol, and whether the transaction was a buy, sell, deposit, or withdrawal (in case of cash accounts like checking, savings, or money market accounts). This level of detail allows the analyst to understand trading patterns at a daily level.

Let's generate another pivot table with Microsoft Excel, this time at the transaction level. Please refer to Figure 6-15.

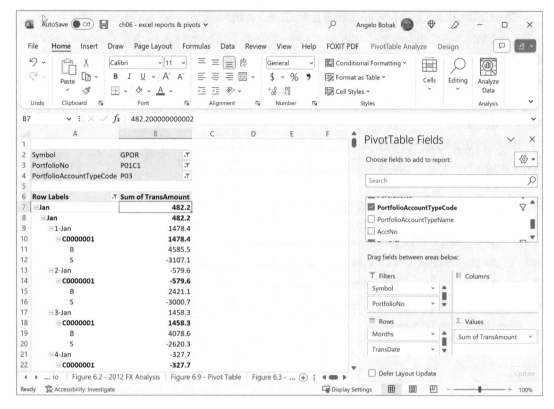

Figure 6-15. *Transaction detail pivot table*

The prior report was valuable because it gave us a lot of detail, but to be honest, it was hard to navigate through all the data. Taking the results and generating a pivot table with Microsoft Excel allows the analyst to slice and dice the data and switch around portfolios, ticker symbols, months, and even customers.

Here we see values not only for the month of January plus also values at the lowest granularity, the day level, for the ticker symbol GPOR (Gulfport Energy) and the totals for buy and sell transaction amounts. Imagine if we had multiple trades a day!

Note   For a quick reminder, the trade amounts are all randomly generated and do not in any way represent actual historical value for our instruments. Also, I kept the trades at one a day to keep things simple. Once you are comfortable with the techniques and how the function works, you can load more trades into the Transaction table to see what effect larger volumes of data have on query performance. Also make sure you update the Account and Portfolio tables after each load.

Hopefully, these last few sections have demonstrated how to perform analysis by using the various ranking and aggregate functions. The power is to drill down one or more levels to get a picture of the trading patterns that generated the results. This will tell us the conditions that generated the profits and losses so we can refine our trading strategy. Our tools are not only the ranking functions but also the performance tools and Microsoft Excel.

Beyond the scope of this book is loading our data into Power BI. We could create some impressive scorecards and dashboards with that tool!

Performance Considerations

Let's perform some analysis on the detailed transaction query as it retrieves data from the Transaction table, which has around 600,000 rows. Not a lot in the scheme of things but enough to challenge performance when you are running the query on a laptop. Let's see the estimated query execution plan generated by SSMS.

Please refer to Figure 6-16.

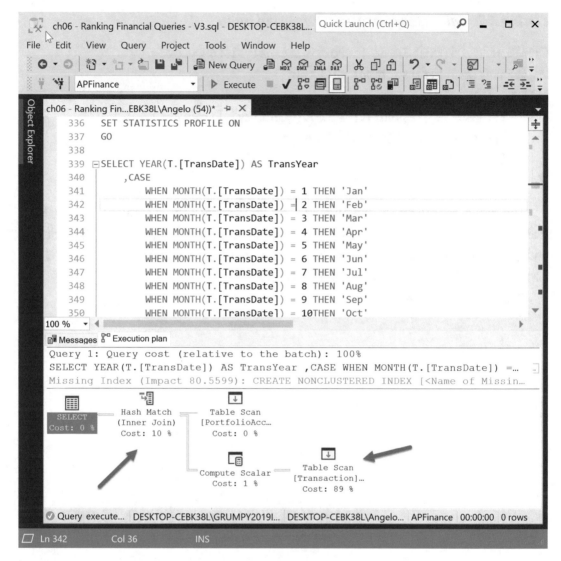

Figure 6-16. *Estimated query plan – first attempt*

We can see an expensive table scan for the Transaction table (89% cost), a table scan for the Portfolio table (0% cost), and our old friend, the hash match inner join task (cost 10% – use hashing to join table row streams). A suggested index can be seen, so let's create it with the following batch command. We make sure to give it a good name.

Please refer to Listing 6-7.

Listing 6-7. Estimated query plan suggested index

```
CREATE NONCLUSTERED INDEX [iePortfolioAnalysisC0000001]
ON [Financial].[Transaction] ([CustId])
INCLUDE (
    [TransDate],[Symbol],[Price],[Quantity],[TransAmount],
    [PortfolioNo],[AcctNo],[BuySell]
    )
GO
```

The suggested index is based on the customer identifier column called CustId, but it also includes a whole list of columns that appear in the query. Let's create the index and run another estimated execution plan. Running update statistics and our DBCC command would not hurt after we create the index.

Please refer to Figure 6-17.

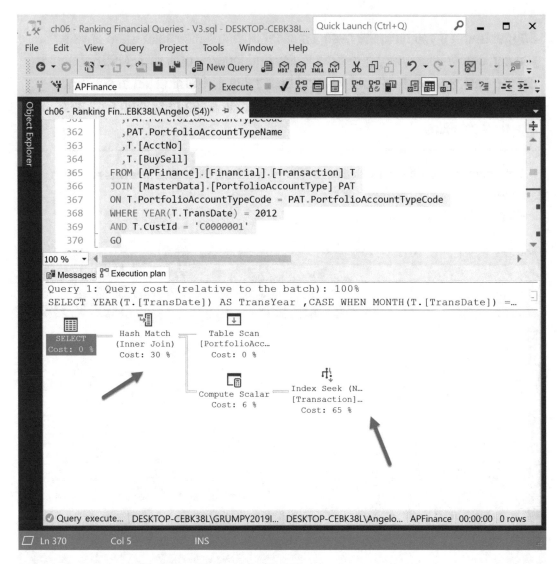

Figure 6-17.   *Second estimated query plan*

We see an improvement with the index seek against the Transaction table. This incurs a cost of 65%. The table scan on the Portfolio table is still 0%, so this is good. The hash match task cost went up to 30%. Some improvement can be seen.

Suggestion   If you have the cash, consider buying a nice solid-state drive so you can create your query-intensive tables and indexes on it or TEMPDB. Should set you back around $200, but it would be a valuable learning aid to see how hardware also needs to be taken into consideration when designing databases and reporting solutions.

Let's compare profile statistics for both approaches, without the index and with the index.

Please refer to Figure 6-18.

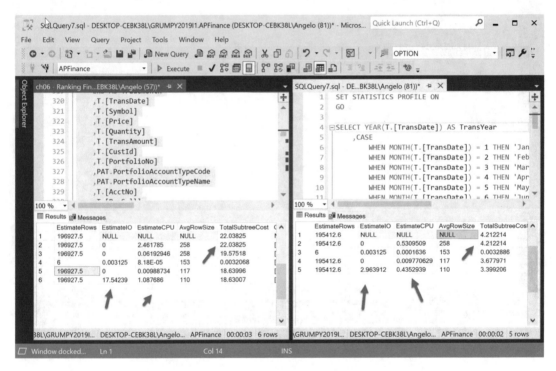

Figure 6-18. *Comparing profile statistics*

Once again, SQL Server has suggested a good index. Notice the EstimateIO, EstimateCPU, and TotalSubtreeCost have gone down considerably. Also notice we eliminated a step with the index solution. At this stage we can agree that running the estimated query plan tool by itself is okay but not enough to give us enough information to analyze performance. We also need to examine IO, TIME, and PERFORMANCE statistics.

> **Note**   Sorry but another note. When analyzing performance on a server scenario as opposed to a laptop or workstation setup, you need to take into account network performance. Your server could execute the query fast enough, but if your network is slow, it could give you misleading impressions regarding the quality of the query or script you just created. Hey, just blame the network administrators. Only kidding …

DENSE_RANK( ) Function

Next function. Recall from Chapter 4 that the DENSE_RANK( ) function works almost the same as the RANK( ) function. In case of ties, the dense rank is the same for the ties, but the next value after the duplicate values is simply assigned the next rank number. So, if the dense rank for the ties is 4, the next rank assigned to the next higher value is 5.

Not so for the RANK( ) function results. If you have five tie values of 6 and the current rank for the ties is 4, then the next rank is 9 (the 5 tie values + the current rank value 4 = 9). Strange indeed.

Example 1

Recall the behavior of RANK( ) vs. DENSE_RANK( ) when we look at the results of this query that will analyze portfolio performance rankings.

Please refer to Listing 6-8.

Listing 6-8. Portfolio Analysis with Dense Rank

```
WITH PortfolioAnalysis (
TradeYear,CustId,PortfolioNo,Portfolio,YearlyValue
)
AS (
SELECT [Year] AS TradeYear
     ,CustId
     ,PortfolioNo
     ,Portfolio
        -- introduce artificial duplicate value
```

```
            -- to see how rank vs dense_rank behave
        ,CASE
            WHEN Portfolio = 'CASH - FINANCIAL PORTFOLIO'
                AND [Year] = 2012 THEN 33894.20
                ELSE SUM(Value)
        END AS YearlyValue
FROM Financial.Portfolio
WHERE Year > 2011
GROUP BY CustId,Year,PortfolioNo
        ,Portfolio
)

SELECT TradeYear,
        ,CustId
        ,PortfolioNo
        ,Portfolio
        ,YearlyValue
        ,RANK() OVER(
            ORDER BY YearlyValue DESC
        ) AS PortfolioRank
        ,DENSE_RANK() OVER(
            ORDER BY YearlyValue DESC
        ) AS PortfolioDenseRank
FROM PortfolioAnalysis
WHERE Portfolio = 'CASH - FINANCIAL PORTFOLIO'
GO
```

We will use the CTE query construct and include a CASE statement block so we can introduce some artificial duplicates so we can put the RANK() and DENSE_RANK() functions through their paces (this way we can compare results).

When the portfolio is the cash portfolio and the year equals 2012, then we just hard-code the value $33,894.20 to generate some duplicates, triplicates, etc.

The RANK() and DENSE_RANK() functions include an ORDER BY clause that sorts the partition results by the YearlyValue transaction totals. The results should be interesting.

Please refer to the partial results in Figure 6-19.

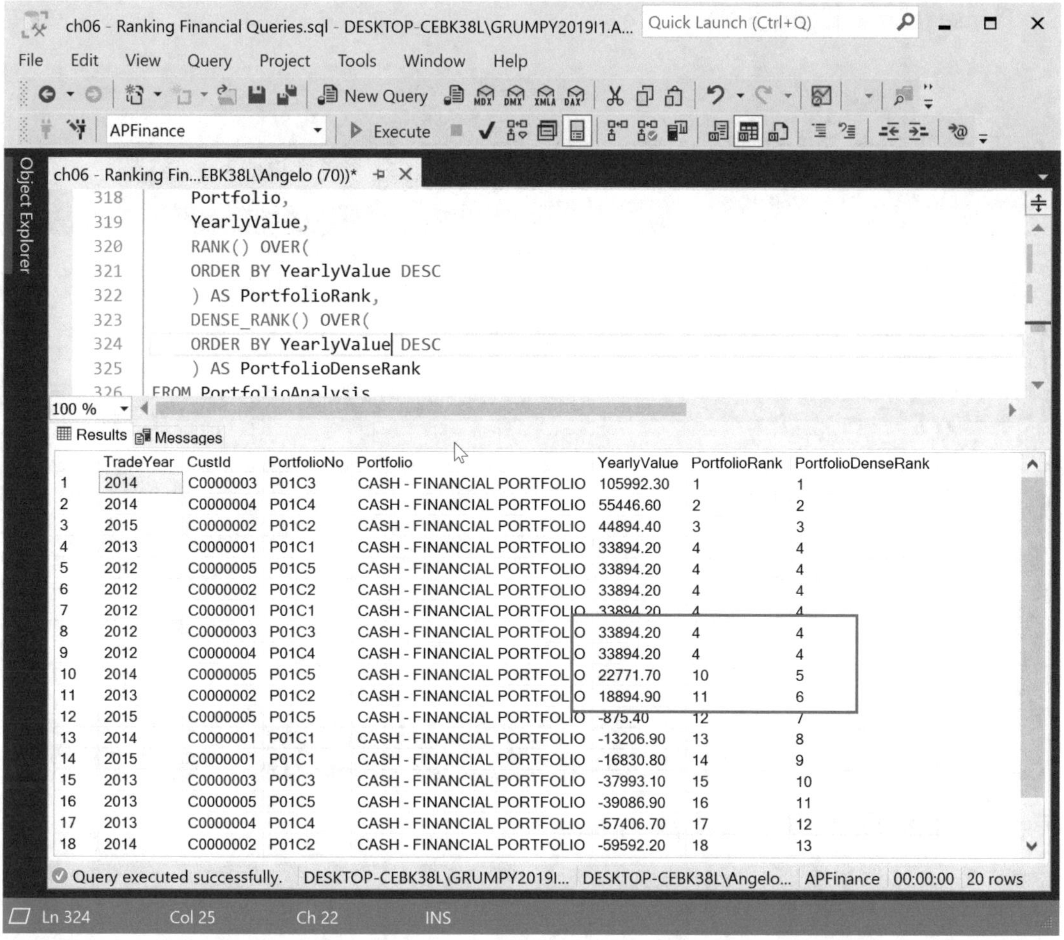

Figure 6-19. *RANK() vs. DENSE_RANK() against portfolio performance*

There it is. We have a large six-way tie across all customer portfolios for 2012 and one customer in 2013. The DENSE_RANK() function simply increases the rank to 5 for the next non-tie value, but the RANK() function adds the tie number count value of 6 to the rank of 4 to deliver the next ranking of 10. Then it proceeds along its merry way.

My personal opinion is that this is confusing. The DENSE_RANK() function gives more intuitive results, but that's just me.

Performance Considerations

Let's check out how this query performs. The following is the first estimated query plan for this query that will give us a baseline for our performance tuning efforts.

Please refer to Figure 6-20.

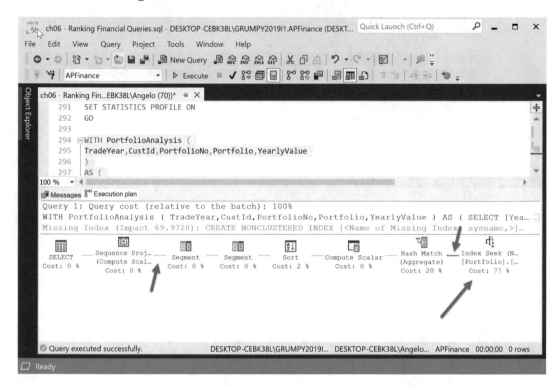

Figure 6-20.   *First estimated query plan*

Despite an index seek (better than an index scan), a new index is suggested. We see a hash match at 20% cost right after the index seek, and the only other task to be aware of is the sort at 2%. Let's extract the code for the suggested index and come up with a name for it so we can create it.

Please refer to Listing 6-9.

Listing 6-9.   Suggested Index

```
/*
Missing Index Details from ch06 - Ranking Financial Queries.sql - DESKTOP-
CEBK38L\GRUMPY2019I1.APFinance (DESKTOP-CEBK38L\Angelo (70))
The Query Processor estimates that implementing the following index could
improve the query cost by 69.9728%.
*/
```

```
/*
USE [APFinance]
GO
CREATE NONCLUSTERED INDEX [<Name of Missing Index, sysname,>]
ON [Financial].[Portfolio] ([Portfolio],[Year])
INCLUDE ([CustId],[PortfolioNo],[Value])
GO
*/

DROP INDEX IF EXISTS [ieCustIdPortfolioNoValue]
ON [Financial].[Portfolio]
GO

CREATE NONCLUSTERED INDEX [ieCustIdPortfolioNoValue]
ON [Financial].[Portfolio] ([Portfolio],[Year])
INCLUDE ([CustId],[PortfolioNo],[Value])
GO
```

The index is based on the portfolio and year columns, and we include the customer ID, portfolio number, and transaction total value. Reminder, I include the DROP command so that when you are trying the code out and practicing, you can drop it and recreate it as necessary.

Let's create the index and generate a new estimated execution plan.

Please refer to Figure 6-21.

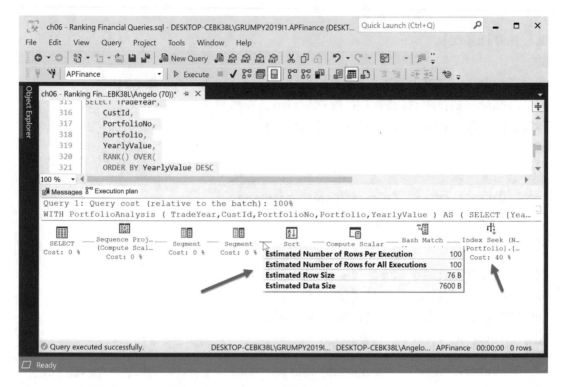

Figure 6-21. *Estimated execution plan after index creation*

Let's take a look at the before and after values of the costs between both plans. Please refer to Table 6-2.

Table 6-2. *Comparing Costs*

Step	Before	After
SELECT	0%	0%
Sequence projection	0%	0%
Segment	0%	0%
Segment	0%	0%
Sort	2%	8%
Compute scalar	0%	0%
Hash match	20%	52%
Index seek	77%	40%

The index seek went down from 77% to 40%, but of course the hash match went up to 52% from 20% and so did the sort step, from 2% to 8%. This was our experience with other queries when we created a suggested index.

I tried moving the WHERE clause predicate to the calling query and removed it from the CTE, but it did not help. I generated a plan that looked the same as the original. The only alternative is to stage the CTE-generated data in a temporary or report table and then query that so that we do not incur the cost of the CTE processing.

Reminder   Remember that the transaction values are generated randomly, so when you try these scripts, you will get different results from the figures in this book.

Let's try another example.

Example 2

For our next example, let's consider to calculate the spread of stock prices, that is, the value between the minimum price the stock had in a year and the maximum value in a year as a measure of performance. So, if the value is large, that means the stock is a high-performer and worth investing in. If the value is negative, then it is a low-performer!

What do you do? Dump it or go along for the ride hoping it increases sooner or later? Don't do like I said earlier what I do: buy high and sell low.

Please refer to Listing 6-10.

Listing 6-10. Top 20 Performing Ticker Symbols

```
SELECT TOP 20 [QuoteYear]
      ,[Ticker]
      ,[Company]
      ,MIN([Low])               AS MinLow
      ,MAX([High])              AS MaxHigh
      ,MAX([High]) - MIN([Low]) AS MaxSpread
      ,RANK() OVER(
                ORDER BY MAX([High]) DESC
          ) AS PerformanceRank
```

```
    ,DENSE_RANK() OVER(
            ORDER BY MAX([High]) DESC
        ) AS PerformanceDenseRank
FROM [APFinance].[MasterData].[TickerPriceRangeHistoryDetail]
WHERE [QuoteYear] = 2015
GROUP BY[QuoteYear]
    ,[Ticker]
    ,[Company]
GO
```

This is a slightly interesting query. It reports the minimum and maximum price quotes for the ticker symbols by year (2015 only). It also calculates the spread between the minimum and maximum values by using the MIN() and MAX() functions a second time. This is probably inefficient, and we should consider rewriting this query so it uses a report table instead of the CTE.

Let's see the results in Figure 6-22.

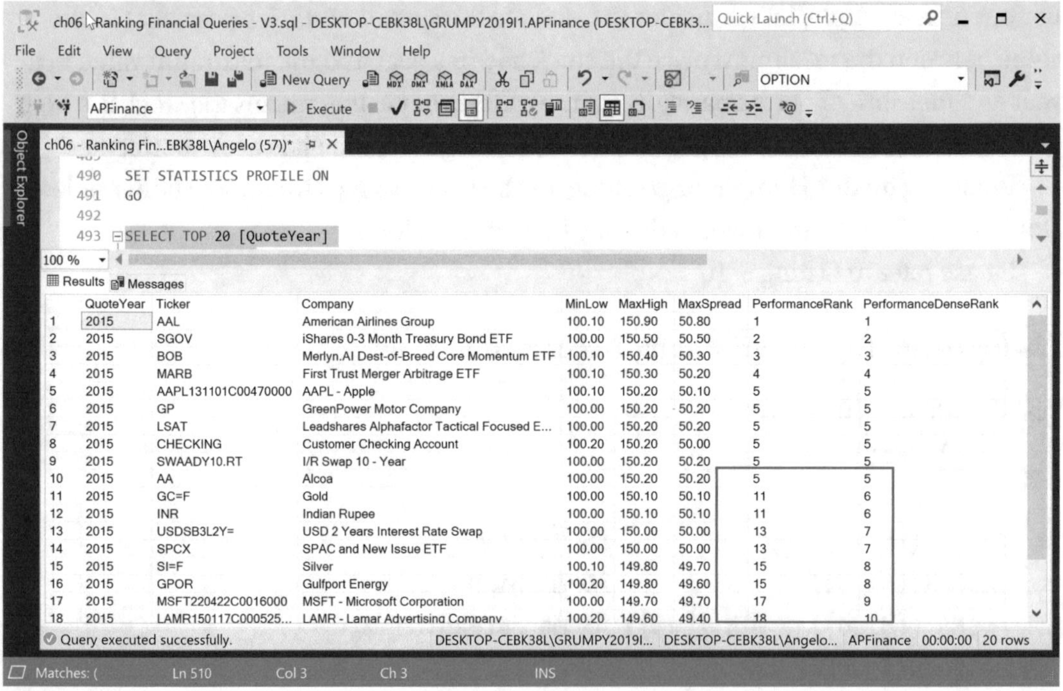

Figure 6-22. *Ticker low, high, and spread for 2015*

As can be seen, we have a couple of ties, and the same behavior is displayed as far as skipping ranking values vs. just going to the next value. Still, we get a sense of how the stocks perform.

Reminder   All stock and other instrument values, transactions, etc. are simulated and do not reflect the actual performance of these instruments, both historically and at present time. And, yes, do not base your investing strategy on mine because I always buy high and sell low.

Time for some performance analysis.

Performance Considerations

Let's run our baseline estimated execution plan. When you do this on your own, take the time to check out each of the help hint panels by placing the mouse pointer over each task. Record what parameters are confusing to you so you can look them up on the Microsoft documentation website. Now is the time to increase your knowledge of these important tasks and statistics.

Note   You will never be able to memorize them all; just understand the most common ones like index seek and scan.

Please refer to Figure 6-23.

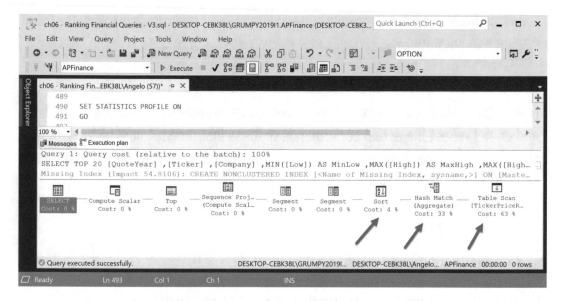

Figure 6-23. *Estimated query plan for the ticker analysis query*

There's an expensive table scan cost of 63% against a table that has 54,057 rows, so an index is suggested. Next, we see a hash match aggregate at 33% followed by a sort at 4%. Let's see what the suggested index looks like before we consider replacing the CTE with a report table.

Please refer to Listing 6-11.

Listing 6-11. Suggested Index

```
DROP INDEX IF EXISTS eTickerCompanyLowHigh
ON [MasterData].[TickerPriceRangeHistoryDetail]
GO

CREATE NONCLUSTERED INDEX [eTickerCompanyLowHigh]
ON [MasterData].[TickerPriceRangeHistoryDetail] ([QuoteYear])
INCLUDE ([Ticker],[Company],[Low],[High])
GO
```

As a force of habit, I always include the DROP INDEX command in case I need to experiment and rebuild the index. Let's build the index and generate another estimated query plan.

Please refer to Figure 6-24.

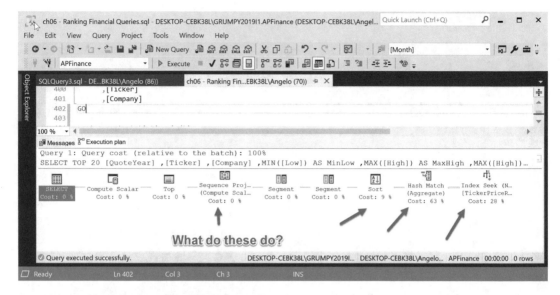

Figure 6-24. *Revised query plan for the ticker analysis query*

As usual, an index seek replaced the table scan. As usual, the hash match and sort went up.

Please refer to Table 6-3.

Table 6-3. *Before and After Index Costs*

Before			After		
Index Seek	Hash Match	Sort	Index Seek	Hash Match	Sort
63%	33%	4%	28%	63%	9%

We are seeing the usual pattern here. There are just so many indexes and report tables one can build. The trick is when you gather user specification reports to identify common requirements and create report tables to satisfy most of them, so you only end up with between two and five tables. Otherwise, things will get hard to manage, and loading will be a nightmare due to all the indexes that need to be updated.

Back to the plan. What does the sequence project task do? Additionally, you might be asking yourself, "What steps process the MIN(), MAX(),RANK(), and DENSE_RANK() functions?"

These questions can be answered by setting the PROFILE STATISTICS parameter on and rerunning the query. Let's see the results.

Please refer to Figure 6-25.

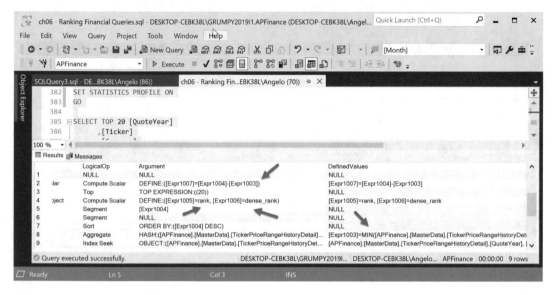

Figure 6-25. *Profile statistics report*

The first order of business is to identify what those **ExprXXXX** expressions mean:

Expr1003: `MIN()` function

Expr1004: `MAX()` function

Expr1005: `RANK()` function (in compute scalar tasks)

Expr1006: `DENSE_RANK()` function (in compute scalar tasks)

Exper1007: Expr1004 – Expr1003

Next, let's answer the question: ***What is the sequence project task used for?***

The answer is that it is used to include columns that participated in calculations, in this case the two rank functions.

So this is a step you cannot eliminate by performance tuning unless maybe if we rewrite the query so it uses the old report table query scheme. In other words, precalculate the values in these columns.

First, we need to compare the before and after `PROFILE STATISTICS` to see if the new index helped. Please refer to Figure 6-26.

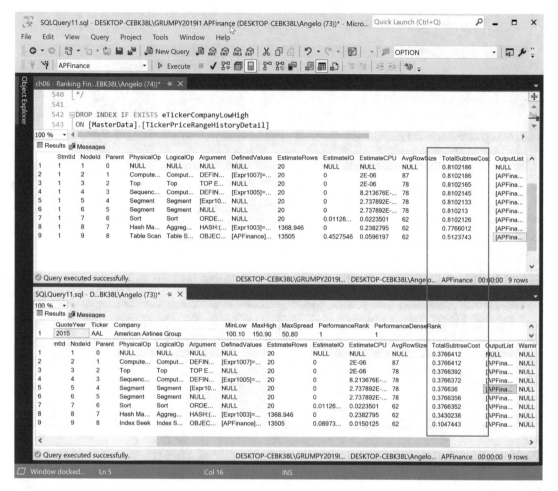

Figure 6-26. *Comparing profile statistics*

Looks like both pre- and post-index scenarios generated the same number of steps, and the only category of values that showed improvements were the total subtree costs highlighted in the figure that went down by more than 50%. At this stage we might ask the question: Was it worth another index?

Next, we need to complete the performance analysis profile by looking at the IO and TIME statistics. Please refer to Figure 6-27.

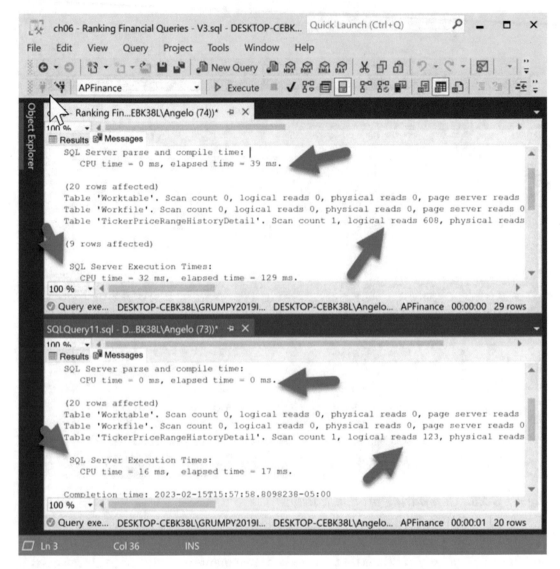

Figure 6-27. *Comparing IO and TIME statistics between both strategies*

Yes, I know, you love the nice big red arrows! All kidding aside, we see important statistics values went down once the index was created. And this is to be expected most of the time. The logical reads before and after are 608 and 123, so this is impressive. Physical reads are 0 in both cases, and the scan count is 1 in both cases. Looks like memory was utilized to process both query approaches.

CPU and elapsed time are a mixed bag. CPU time went up, but elapsed time went down.

Conclusion: We will keep the index! Next topic: We create some tiles to store financial information.

NTILE( ) Function

Remember the definition of this function? The NTILE( ) function allows you to divide a set of rows in a data set into tiles or buckets (I like this term better). If you have a data set of 12 rows and you want to assign 4 tiles, each tile will have 3 rows. This function comes in handy when you want to create sets of data like when you need to evaluate salesperson performance and grant them bonuses based on sales performance.

Since we are discussing finance, another example is setting up bucket categories for how stocks perform in a portfolio: best-performing in the first bucket, slightly worst-performing stocks in the next bucket, and so on. This way you can rate them and base investment strategies according to which bucket a stock falls into.

Note   If you have an odd number of rows, like 15, and you decide to create 4 tiles, you will get tiles with 4 rows, 4 rows, 4 rows, and 3 rows. So there is no way to control what goes into a tile; it just divides them up.

For our first example, we will start off easily. We create tiles so we can use them to categorize whether an investment is good, bad, or junk! I am breaking up the listing into two parts, so it is easy to explain.

Example 1

First, we set up a data set consisting of some fictitious stocks and assign them ratings of A+, A, B+, B, and JUNK!

Please refer to Listing 6-12a.

Listing 6-12a. Set Up Ticker Data

```
DECLARE @TickerRating TABLE
(
Ticker VARCHAR(8) NOT NULL,
Rating VARCHAR(8) NOT NULL
);
```

```
INSERT INTO @TickerRating VALUES
('AAA','A+'),
('AAB','A+'),
('AAC','A+'),
('AAD','A+'),

('BBB','A'),
('BBC','A'),
('BBD','A'),
('BBE','A'),

('CCC','B+'),
('CCD','B+'),
('CCE','B+'),
('CCE','B+'),

('DDD','B'),
('DDE','B'),
('DDF','B'),
('DDG','B'),

('ZZZ','JUNK'),
('ZZA','JUNK'),
('ZZB','JUNK'),
('ZZB','JUNK');
```

As usual, let's use a table variable called @TickerRating and insert four stocks (or ticker symbols if you like) for each category. We assign ratings to each one and load the rows into the table variable. Next, we create a query that will assign bucket numbers so we can evaluate them. Please refer to Listing 6-12b.

Listing 6-12b. Generate Recommendation Tiles

```
DECLARE @RatingBuckets INT;

SELECT @RatingBuckets = COUNT(DISTINCT Rating)
FROM @TickerRating;

SELECT Ticker
      ,Rating
```

```
,CASE
        WHEN NTILE(@RatingBuckets) OVER (ORDER BY Ticker) = 1
                THEN 'Highly Recommend'
        WHEN NTILE(@RatingBuckets) OVER (ORDER BY Ticker) = 2
                THEN 'Recommend'
        WHEN NTILE(@RatingBuckets) OVER (ORDER BY Ticker) = 3
                THEN 'Worth a shot'
        WHEN NTILE(@RatingBuckets) OVER (ORDER BY Ticker) = 4
                THEN 'Are you feeling lucky?'
        WHEN NTILE(@RatingBuckets) OVER (ORDER BY Ticker) = 5
                THEN 'Run away!'
END AS AnalystRecommends
FROM @TickerRating;
GO
```

This time we generate the number we will use to define the tiles by counting the distinct rating categories. In our case we will need to set up five tiles. Next, we create a query that contains a CASE block so we can decode the tile number and generate a message based on the value, for example, 1 for the first tile means that stocks in this bucket are "Highly Recommended." Very simple concept, but it illustrates we can use tiles to categorize things.

Let's see the results in Figure 6-28.

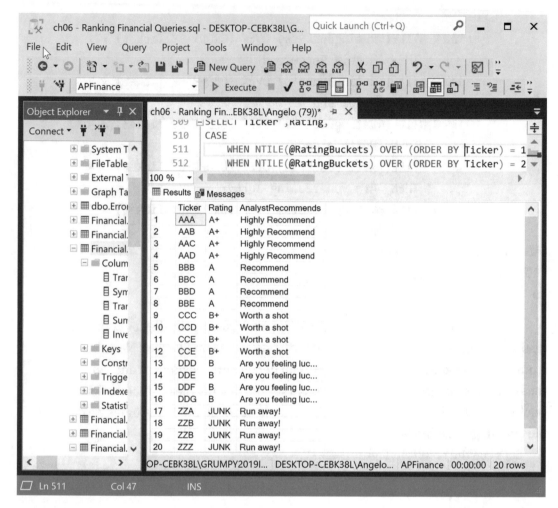

Figure 6-28. *Financial adviser investment recommendation*

As can be seen, the recommendations are very professional and sophisticated. My favorite recommendations are for ratings B and JUNK.

Performance Considerations

Although this is a very small table, it has several steps, so I thought it would be interesting to see what its estimated performance plan looks like. Please refer to Figure 6-29.

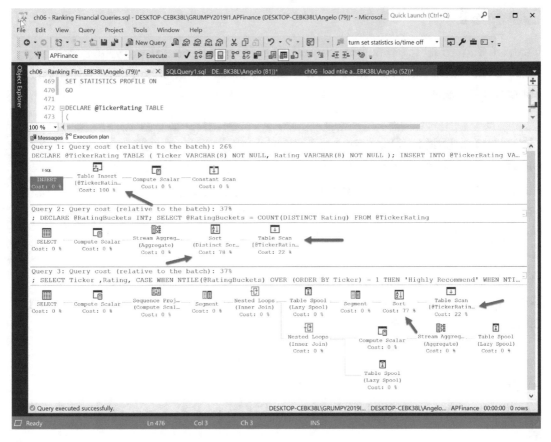

Figure 6-29. *NTILE query estimated execution plan*

We actually have three plans, the first for setting up and inserting rows into the table variable. Notice the cost for this query is 26% relative to the three-query batch script.

The second plan is for the query that calculates the number of tiles based on how many rating categories there are. The cost for this query is 37%.

Last but not least is the query that has the CASE block used to generate a recommendation based on the tile number of the stock. Lots of steps, and the cost for this query is 37% relative to the entire batch.

So, if we add up all three costs, 26% + 37% + 37%, we should get 100%, which is what we get.

Since the table is so small, we will not bother doing some performance tuning, but look at all the steps: table scan, table spooling, nested joins, etc. If our table had thousands or more rows, performance might become an issue.

Example 2

Next, we want to set up some buckets with NTILE() and then rank the values within the buckets with the RANK() function. We will work with the Transaction table, which has about 651,934 rows.

Recommendation   Once you are comfortable with this and other functions, try loading the Transaction table two or three times so as to simulate two or more trades a day. The volume will increase to about two million rows. Load the Account and Portfolio tables based on the new rows to see how performance is affected when you run the queries we just discussed.

This is another simple example that uses both the NTILE() and RANK() functions. These two functions process the data in different ways: one categorizes the data, and the other ranks it. We will see if the estimated query plan is complicated.

Please refer to Listing 6-13.

Listing 6-13. Rank Values Within Buckets

```
;WITH SymbolCTE (TransYear,Symbol,SumTransactions,InvestmentBucket)
AS
(
SELECT YEAR([TransDate])  AS TransYear
     ,[Symbol]
     ,SUM([TransAmount]) AS SumTransactions
     ,NTILE(9) OVER (
          PARTITION BY YEAR([TransDate])
          ORDER BY SUM([TransAmount]) DESC
          ) AS InvestmentBucket
FROM [APFinance].[Financial].[Transaction]
GROUP BY YEAR([TransDate]),Symbol
)

SELECT TransYear
     ,Symbol
     ,SumTransactions
```

```
      ,InvestmentBucket
      ,RANK() OVER(
                  PARTITION BY TransYear,InvestmentBucket
                  ORDER BY InvestmentBucket,SumTransactions DESC
            ) AS InvestmentRank
FROM SymbolCTE
GO
```

We are only interested in **BUY** transactions, so we set up a WHERE clause to filter out any transaction that is not a "**BUY**" transaction. The CTE sums up the transaction amounts, so the CTE needs a GROUP BY clause because we did not include an OVER() clause. The NTILE() function does use an OVER() clause with both a PARTITION BY and an ORDER BY clause. The partition is set up by year, and the result set in the partition is sorted by the total transaction amount.

Please refer to Figure 6-30.

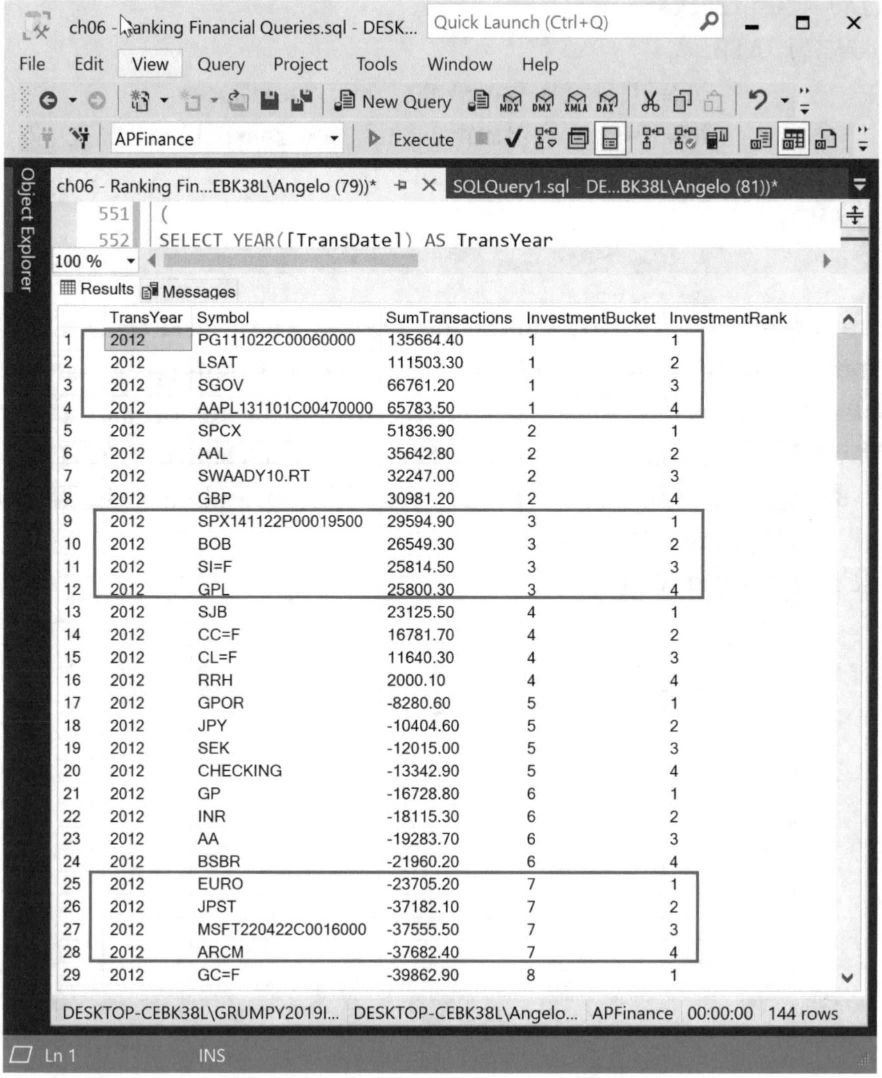

Figure 6-30. *2012 ticker bucket ratings*

Looks like we have nine buckets and four ticker symbols assigned to each bucket. No ties, so we do not see the peculiar rank pattern that this function produces. The financial instruments are rated 1 to 4 in each bucket. This could be valuable information when planning an investment portfolio.

Homework   As a small exercise, use the logic in the first example so you can rate the ticker symbols based on the rank within the partition. Hint: You need to use a CASE block.

I think this result set will make for some interesting graphs, so let's copy and paste the results into an Excel spreadsheet and generate some graphs.

Please refer to Figure 6-31.

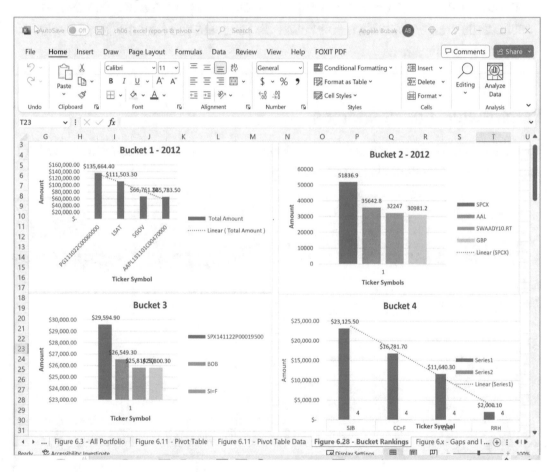

Figure 6-31. *Ticker bucket ratings for 2012*

This time we generated four smaller graphs, one for each bucket, but visually we can see the rank positions and the trend lines. Including the values also helps with understanding the information. Looking at the bucket 3 graph, which is the top-ranking instrument? Which is the lowest? There's almost a tie for the lowest, but this visual representation will help the analyst decide on what to invest in for their client's portfolio.

Thoughts and ideas   At this stage we can see how the SQL Server functions we have been studying are the first-line tools we use to generate analytical data from raw data. Our tools range from the reports generated via queries to graphs and pivot tables generated by the query results with Microsoft Excel. An extra tool is to use the analytical data generated with the functions to populate scorecards and dashboards created with Microsoft Power BI. I suggest you become familiar with both Excel and Power BI.

Performance Considerations

Let's see what an estimated query plan looks like for a query that contains both a RANK() and an NTILE() function. Please refer to Figure 6-32.

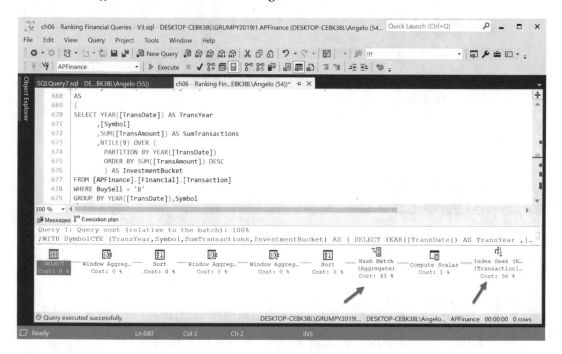

Figure 6-32. *NTILE() and RANK() estimated query plan*

As usual we start from right to left. We see an index seek at a cost of 56% followed by a compute scalar step at 1% and a hash match aggregate task at 43%. Let's take a detailed view of these three tasks by placing the mouse pointer over each one.

Please refer to Figure 6-33.

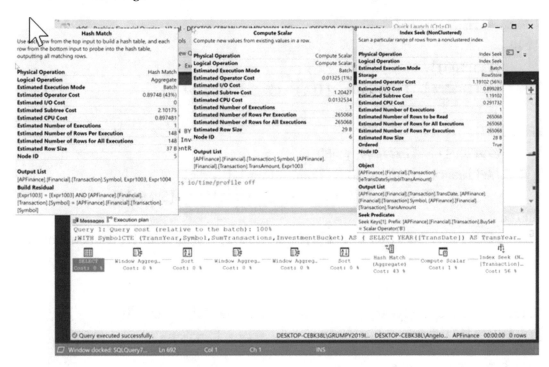

Figure 6-33. *Detailed task information*

Starting with the index scan, we see that the top and bottom objects are the `Transaction` table (top object) and the index `ieTransDateSymbolTransAmount` (bottom object). Next, the compute scalar task outputs expression Expr1003.

What is this? The following is the expression retrieved from the profile statistics:

```
[Expr1003]=datepart(year,[APFinance].[Financial].[Transaction].
[TransDate])))
```

Mystery solved. It is the results of the `DATEPART()` function, which is used to extract the year.

Finally, recall what the hash match tasks do; they match the rows from the top object in the index scan to the bottom object in the index scan and match them by building a hash table. I encourage you to examine the details for these nodes in a query plan as they clearly identify what is going on, especially for tasks you have not seen before. Knowing what they do can help you come up with performance improvement actions like modifying a query or creating staging or temporary tables.

Printing out the statement text for this step produces the following logic:

```
--Hash Match(Aggregate,
HASH:(
        [Expr1003],
        [APFinance].[Financial].[Transaction].[Symbol]
        ),
RESIDUAL:[
        Expr1003] = [Expr1003] AND
        [APFinance].[Financial].[Transaction].[Symbol]
                = [APFinance].[Financial].[Transaction].[Symbol])
DEFINE:(
        [Expr1004]=SUM([APFinance].[Financial].[Transaction].[TransAmount]))
        )

--Compute Scalar(DEFINE:
[Expr1003]=datepart(year,[APFinance].[Financial].[Transaction].
[TransDate])))
```

A little cryptic, I agree.

This identifies the two expressions that perform the SUM() values and pull out the year date part.

Looking back at the query plan, notice the width of the lines as we use the index scan and then how they get thinner as the flow goes from right to left. This feature can help us identify high-volume tasks and opportunities for performance improvement. Always be aware of these.

Last but not least are the three window aggregate tasks. Although they chime in at 0% cost, it is worth mentioning that these are the tasks associated with the two ranking functions that use the OVER() clause. If we look at this step in the profile statistics, we see that the EstimateIO value is 0, so this processing is performed in memory. That's what we want to see!

These are the statements for these tasks when copied from the profile statistics column StmtText:

```
--Window Aggregate(
  DEFINE:(
      [Expr1006]=rank),
      PARTITION COLUMNS:([Expr1003], [Expr1005]),
      ORDER BY:([Expr1003], [Expr1005]),
      ROWS BETWEEN:(UNBOUNDED, CURRENT ROW)
      )
--Window Aggregate(
DEFINE:(
      [Expr1005]=ntile),
      PARTITION COLUMNS:([Expr1003]),
      ROWS BETWEEN:(UNBOUNDED, CURRENT ROW)
--Window Aggregate(
  DEFINE:(
  [AggResult1009]=Count(*)),
      PARTITION COLUMNS:([Expr1003]),
        ORDER BY:([Expr1003]),
        RANGE BETWEEN:(UNBOUNDED, UNBOUNDED)
)
```

Now we can associate query plan steps to the actual query components. But where does the COUNT(*) step come in? This was not included in the query. Looks like it had to count the rows so the tiles could be defined and processed in the next step.

Might as well check out the PROFILE statistics for this script as we just referenced it.

Please refer to Figure 6-34.

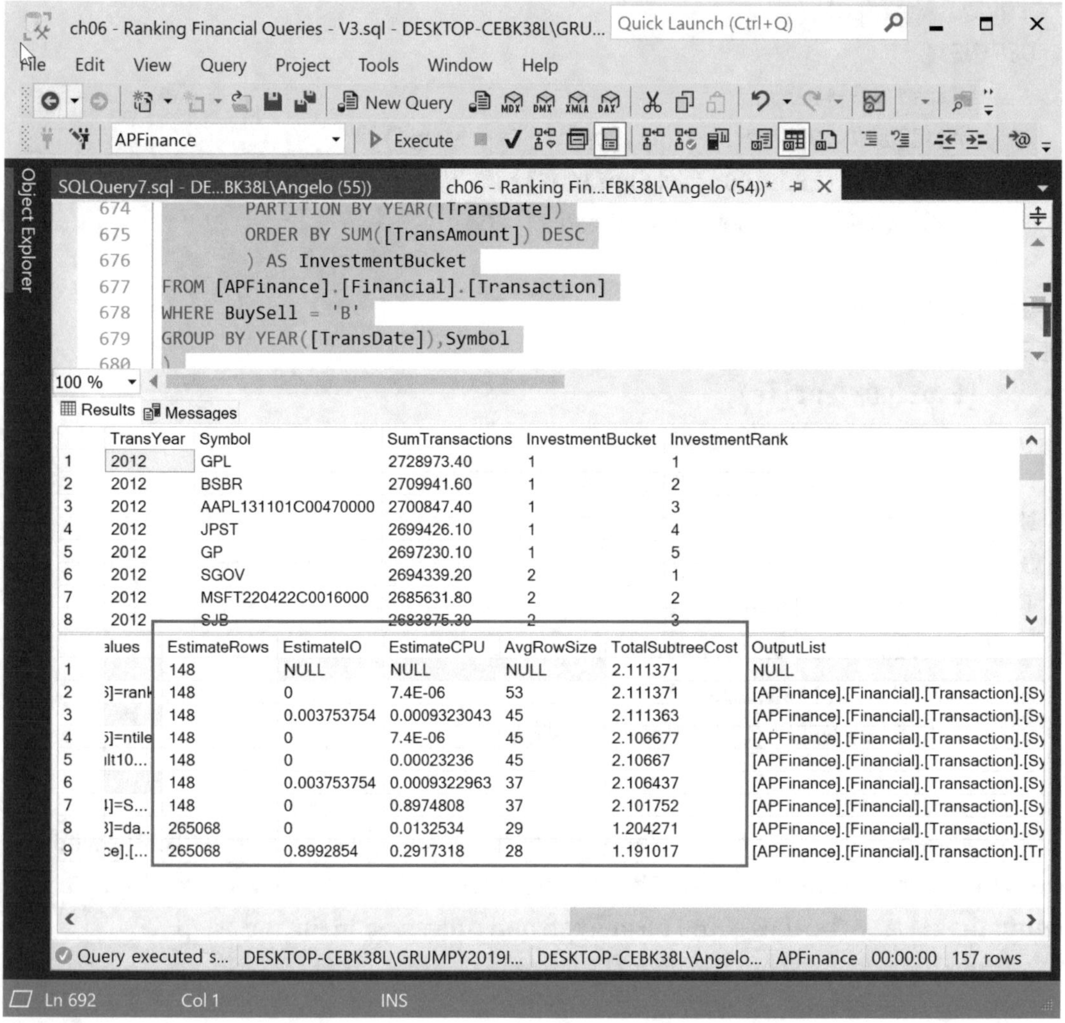

Figure 6-34. *PROFILE statistics for NTILE() and RANK()*

The total subtree costs seem high. One possible solution is to, you guessed it, replace the CTE with a reporting table, which is loaded once a day. This should improve things considerably. Notice that six out of nine steps have EstimateIO values of 0, so processing is using memory (the faster memory you have, the better).

You have code for a reporting table approach in the script for this chapter so you can try it on your own. I will compare the statistics between both scenarios shortly.

Let's take a peek at the other statistics. The following are the IO and TIME statistics for both approaches.

Please refer to Figure 6-35.

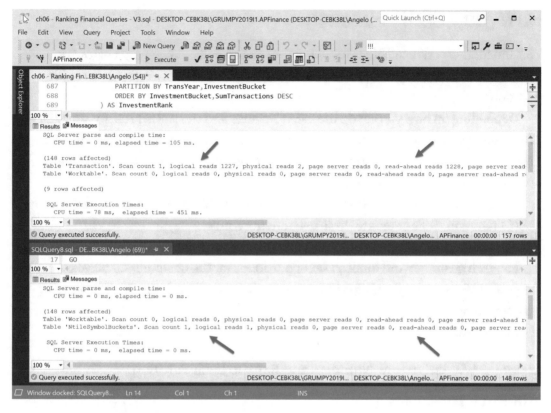

Figure 6-35. *Comparing IO and TIME statistics*

Here we look at the statistics generated by the CTE and report table approaches.

Looking at the logical reads and read-ahead reads for both approaches, we see a drastic improvement with the reporting table approach. Logical reads went down from 1227 to 1, and read-ahead reads went from 1228 to 0. So, if the business requirements allow for data that can be 24 hours old, you can load the computationally intensive data in a report table and then run the query that generates the results from the preloaded report table.

By the way, the expensive hash match aggregate task was eliminated when I ran an estimated query plan on the new query that uses the report table. Simple strategy that works!

Note   To save space I did not show the code for the report table strategy, but it is included in the code for this chapter, which you can download. Try it out, check it out, break it, and then improve it! That's how you learn.

ROW_NUMBER( ) Function

This function, if you recall, does just what the name implies. Given a set of rows in the data result set, it will assign a sequential number to each row in the data set (or partition).

Depending on how you set up the OVER( ) clause and the PARTITION BY and ORDER BY clauses, it will assign row numbers to each partition in the result set. That is, each time you move to the next partition, the row number starts at 1 and goes on sequentially until the last row in the partition.

Lastly, this function behaves like some of the other functions we discussed, like the RANK( ) function. Let's compare these two functions now.

The business requirements for this next script are as follows:

For each customer and for each set of portfolios for the customer for all years, rank the performance of the portfolio accounts in the customer portfolio. Include the row numbers for each set of results.

Please refer to Listing 6-14.

Listing 6-14. Assign Row Numbers and Rank to Each Partition

```
WITH PortfolioAnalysis (
TradeYear,CustId,PortfolioNo,PortfolioAccountType,Portfolio,YearlyValue
)
AS (
SELECT Year AS TradeYear
      ,CustId
      ,PortfolioNo
      ,PortfolioAccountTypeCode
      ,Portfolio
      ,SUM(Value) AS YearlyValue
FROM Financial.Portfolio
WHERE Year > 2011
GROUP BY CustId,Year,PortfolioNo,PortfolioAccountTypeCode,Portfolio
)

SELECT TradeYear
      ,CustId
      ,PortfolioNo
```

```
        ,PortfolioAccountType
        ,Portfolio
        ,YearlyValue
        ,RANK() OVER(
        PARTITION BY CustId,TradeYear
                ORDER BY CustId,TradeYear,YearlyValue DESC
        ) AS PortfolioRankByYear
        ,ROW_NUMBER() OVER(
                PARTITION BY CustId,TradeYear
        ORDER BY CustId,TradeYear,YearlyValue DESC
        ) AS PortfolioRowNumByYear
FROM PortfolioAnalysis
GO
```

The CTE strategy is used again so we can generate the transaction sum values and then let the outer query perform the ranking and row assignment. Both the ROW_ NUMBER() and RANK() functions use the same partition definition and sort order. The partitions are defined by customer ID and trade year, and the partition result set rows are sorted by customer ID, trade year, and the yearly value.

Please refer to the partial results in Figure 6-36.

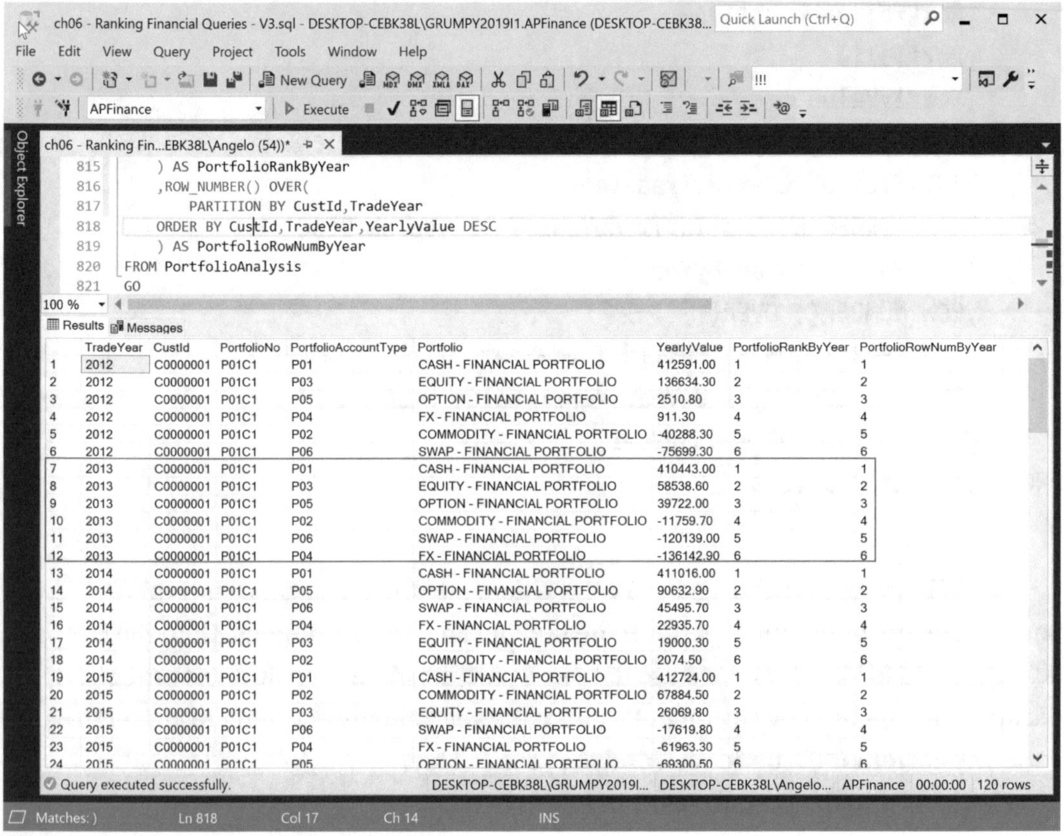

Figure 6-36. *Portfolio analysis report*

Works as expected. The cash portfolio seems to always be the top-performer. Does this mean the customers are always depositing extra cash to cover the trades, or are they selling stocks and making a profit?

The script that loads the transaction does deposit extra cash just to make sure our customers do not run out!

One last graph before we do some performance analysis. Please refer to Figure 6-37.

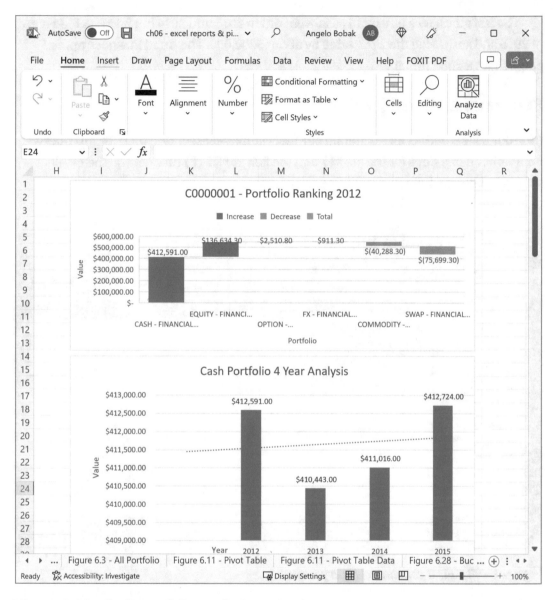

Figure 6-37. *Cash portfolio analysis – two views*

For customer C0000001 (our old friend, whom we always pick), we see the portfolio value across all four years and for all portfolios. These represent the sum of all buy and all sell transactions plus deposits (credits) and withdrawals (debits) for cash accounts. This is an example of how you can track performance across all portfolios but also for each portfolio across all years.

The last graph tells us a lot of cash went out in 2013 and 2014 but it was gained back in 2015 and surpassed the 2012 value by about $200,000. The trend line goes up, so I guess that's a good thing!

Performance Considerations

As is our usual routine, let's generate our first estimated execution plan for this query. Remember it has steps for the RANK() and the ROW_NUMBER() function, so the results should not be too different from our prior query.

Please refer to Figure 6-38.

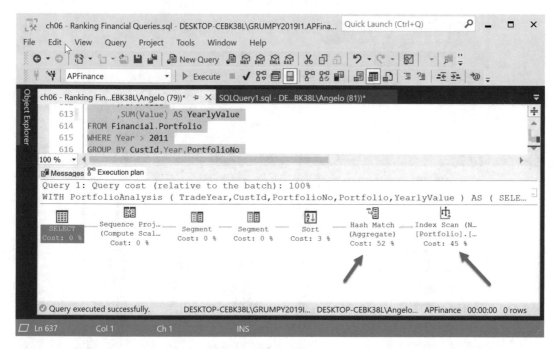

Figure 6-38. *First estimated query plan*

No surprises here, but we have an index scan, which is more costly than an index seek. This one chimes in at 45% cost. And there is our old friend, the expensive hash match aggregate step at 52% cost. The sort step at 3% is not expensive. Let's generate the execution plan again but this time with live statistics.

Please refer to Figure 6-39.

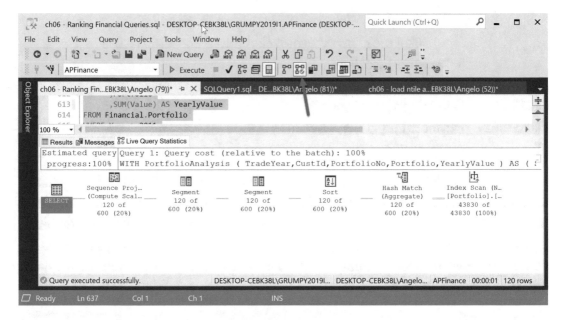

Figure 6-39. *Estimated query plan live statistics*

Clicking the icon in the menu bar with the little green checkmark will generate the live statistics. By the way, it is a good idea to run both the estimated and the live statistics.

Usually, they should not be too far apart, but if they are, some analysis and query tuning are warranted to find out the reason, like more or fewer steps between the plans. Notice that we see the row counts as the flow of data proceeds from right to left. The index scan retrieves 43,830 rows, and as processing begins, the row set is reduced to 120, which is the count for each of the remaining steps.

Wait a minute. What's up with the index scan showing 100% and the others showing 20%?

These are actually time progress indicators and not costs. If a query ran long enough, you could see the progress indicators increment for each step until the processing was finished. Let's see an example. Please refer to Listing 6-15.

Listing 6-15. Query Example to Generate Dynamic Statistics

```
SELECT TransYear,
       N.Symbol,
       SumTransactions,
       InvestmentBucket,
       RANK() OVER(
```

```
            PARTITION BY TransYear,InvestmentBucket
            ORDER BY InvestmentBucket,SumTransactions DESC
        ) AS InvestmentRank
FROM NtileSymbolBuckets N
CROSS JOIN [MasterData].[Calendar]
CROSS JOIN [Financial].[TickerSymbols]
GO
```

The query itself is meaningless. It just performs a three-table CROSS JOIN to generate a lot of rows and make the query run a long time. If you execute it, you can actually see the row number and time spent counters change as the query runs. This is another tool that lets you see tasks that take a long time to run and volumes of data so you can see if you can improve the query to minimize these intensive tasks. Check out the following output in Figure 6-40.

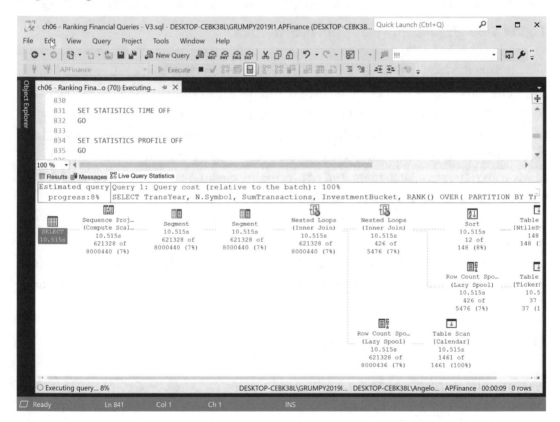

Figure 6-40. *Live query plan statistics*

We can see the time spent in seconds, like the sort step at the right-hand top of the plan that shows 10.515 seconds, and it is only 8% completed. We see the row counts also, so this output has a lot of valuable information to help you with your performance tuning.

Last but not least, notice the nested loops tasks with the big red **X** symbol. These are warnings. In this case a CROSS JOIN is occurring, and there are no ON predicates to link the tables. If you click these tasks, the message you get is "No join predicate."

Congratulations! Here is another tool you can add to your toolkit!

The Data Gaps and Islands Problem

This last set of scripts addresses the classic gaps and islands problem. Recall that gaps in our scenario are missing dates in a sequence of sequential dates, while islands are clusters of sequential dates that the customer executed trades for a particular instrument.

This should be interesting as there are a lot of steps to take and we will see how the data is transformed in each step until we generate the final report.

We will start off with gaps in trading days for the **LIBOR** (London Interbank Offered Rate) financial instrument. The script will be divided into five steps, the first being to load the test data and the remaining four to create the chained CTEs and the final query that generates the desired report. The script for identifying islands will have four steps, three to create the chained CTEs and one for the report.

Note   By chained CTEs I mean that the CTE queries the prior CTE until the last step where the query references the last CTE in the chain or sequence.

Our business analyst wants us to create a query that will identify the gaps and islands in trading dates for customer C0000005 (we give customer C0000001 a break). As mentioned earlier we will only concentrate on the **LIBOR** interest rate and for the year 2012. The report should include the start and stop dates of the gaps and islands but also a column that lists the dates for each category of data patterns.

Figure 6-41 is a sneak peek of the results for the solution.

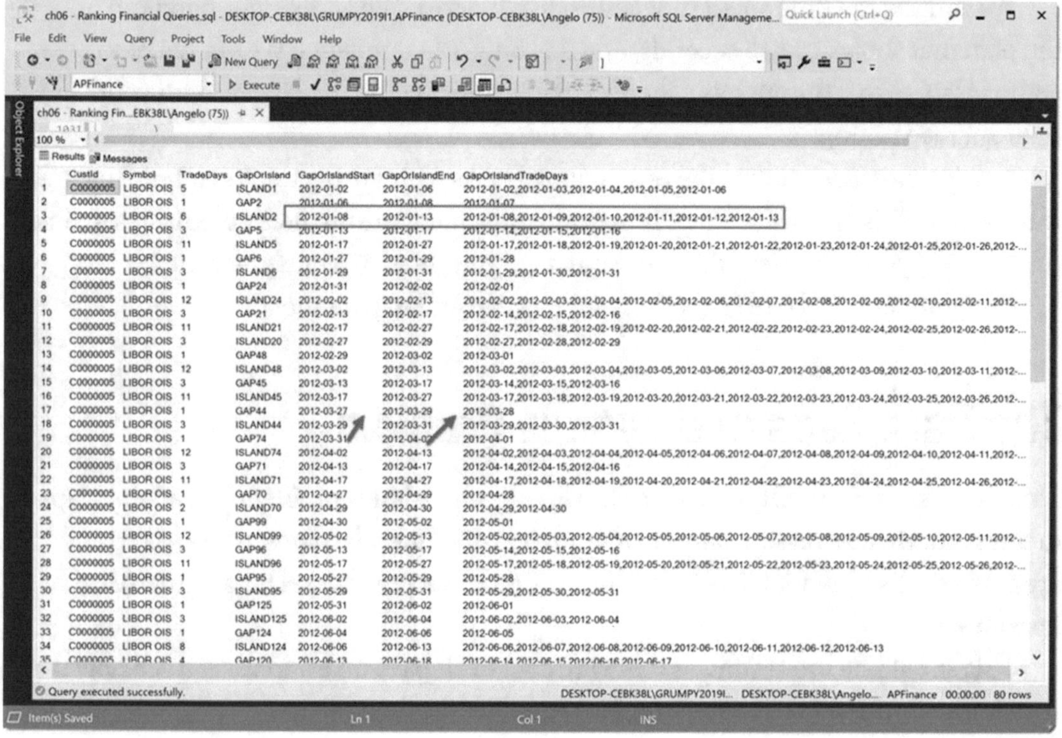

Figure 6-41. *Final gaps and islands report*

Notice the boxed area in the screenshot. We can see that the start and end dates of the island are followed by the list of dates that make up the island. These include the start and stop dates. The gap range identified by the arrows does not include the start and stop dates, only the dates in between. Lastly, we see the string that identifies the date pattern as either a gap or island. The numbers at the end make this code unique as we will see. We have a big job ahead of us.

Let's start by creating the table that will store our test data.

Please refer to Listing 6-16.

Listing 6-16. Set Up Data, Gaps, and Islands

```
CREATE TABLE Financial.CustomerBuyTransaction(
    TransDate              date            NOT NULL,
    CustId                 varchar(8)      NOT NULL,
    Symbol                 varchar(64)     NOT NULL,
    TradeNo                smallint        NOT NULL,
```

```
    TransAmount          decimal(10,2)  NOT NULL,
    TransTypeCode        varchar(8)     NOT NULL
) ON AP_FINANCE_FG
GO
```

All the column names are self-explanatory. Technically, the primary key is made up of the TransDate, CustId, Symbol, and TradeNo columns. We need the TradeNo column in case a customer places multiple trades on a single day for the same instrument.

Please refer to Listing 6-17.

Listing 6-17. Load Trades for the Customer

```
INSERT INTO Financial.CustomerBuyTransaction
SELECT
    CAL.[CalendarDate]
    ,FT.CustId
    ,FT.Symbol
    ,ROW_NUMBER() OVER (
        PARTITION BY CAL.[CalendarDate],FT.CustId,FT.Symbol,
            FT.TransAmount,FT.TransTypeCode
        ORDER BY CAL.[CalendarDate],FT.CustId,FT.Symbol,
            FT.TransAmount,FT.TransTypeCode
        ) AS TradeNo
    ,FT.TransAmount
    ,FT.TransTypeCode
FROM [MasterData].[Calendar] CAL
JOIN [Financial].[Transaction] FT
ON CAL.[CalendarDate] = FT.TransDate
WHERE [BuySell] = 'B'
ORDER BY CAL.[CalendarDate],FT.CustId,FT.Symbol
GO

INSERT INTO Financial.CustomerBuyTransaction
VALUES('2012-01-01','C0000001','AA',2,312.40,'TT00002')
GO
```

This is a standard INSERT/SELECT command used to load the table. What is interesting is that we use the ROW_NUMBER() function with an OVER() clause to generate a trade sequence number so we can compensate for the event of multiple trades being placed by the same customer on the same date for the same instrument.

The data we need to create this test data set comes from the Calendar and Transaction tables in our database. The second INSERT command simply inserts a second trade for January 1, 2012, for customer C0000001 and for instrument AA (Alcoa), so we have two trades in one day.

Let's see what the data we just inserted looks like by executing the following query:

```
SELECT *
FROM [Financial].[CustomerBuyTransaction]
ORDER BY 1,2,3
GO
```

The results are shown in Figure 6-42.

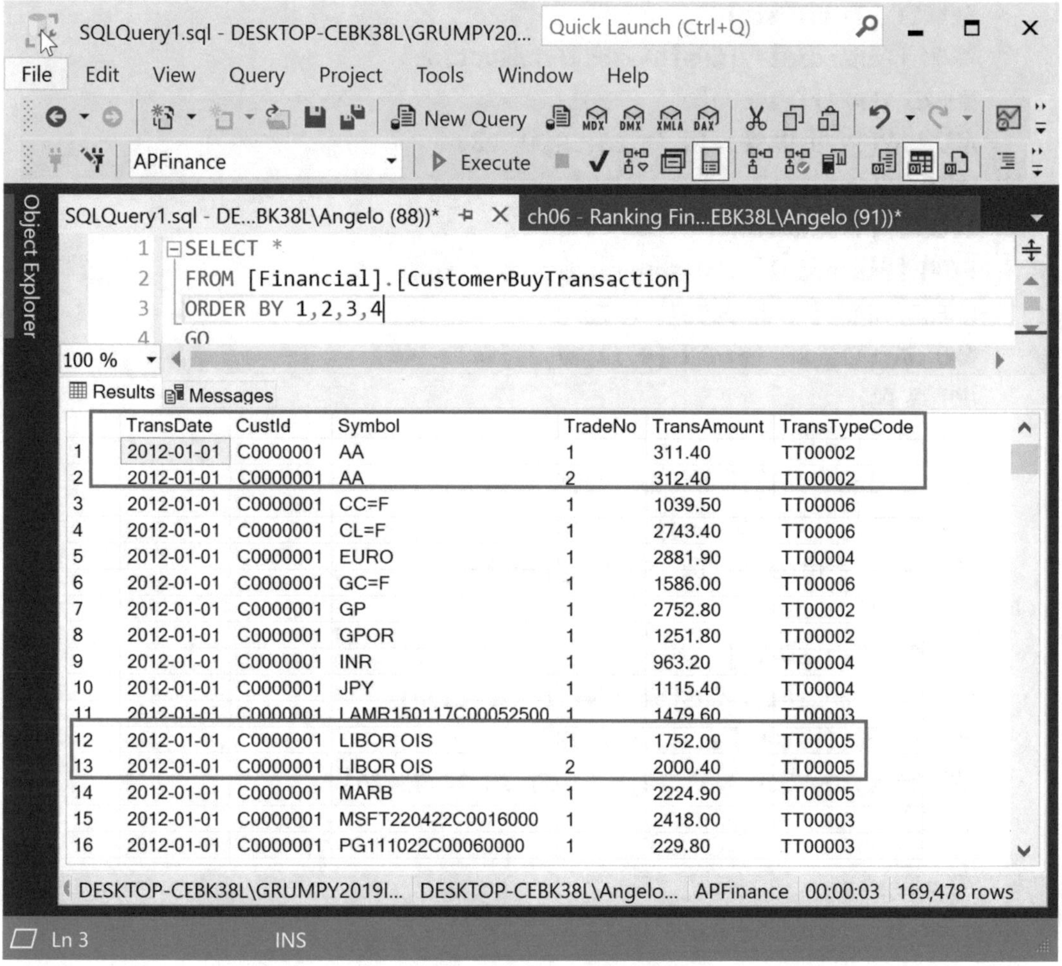

Figure 6-42. *Test data and instance of multiple trades in a day*

Looks like the ROW_NUMBER() function did its job. The first two rows show the multiple trades in the same day scenario. So do rows 12 and 13. I basically went through this to ensure we had a viable primary key for the table, but it will be interesting to see how it affects our gaps and islands logic.

Next, we need to create some artificial gaps for our little exercise.

Please refer to Listing 6-18.

Listing 6-18. Create Gaps in Trade Activity

```
DELETE FROM Financial.CustomerBuyTransaction
WHERE [TransDate] IN(
```

```
        SELECT [TransDate]
        FROM [Financial].[CustomerBuyTransaction]
        WHERE YEAR([TransDate]) = 2012
        AND DAY(TransDate) IN (1,14,15,16,28)
        UNION ALL
        SELECT [TransDate]
        FROM [Financial].[CustomerBuyTransaction]
        WHERE YEAR([TransDate]) = 2013
        AND DAY(TransDate) IN (5,17,22,23,24)
        UNION ALL
        SELECT [TransDate]
        FROM [Financial].[CustomerBuyTransaction]
        WHERE YEAR([TransDate]) = 2014
        AND DAY(TransDate) IN (9,19,25,26,27,28)
        UNION ALL
        SELECT [TransDate]
        FROM [Financial].[CustomerBuyTransaction]
        WHERE YEAR([TransDate]) = 2015
        AND DAY(TransDate) IN (2,8,11,17,23,24,25,26)
        )
        GO

-- need to insert these back in

INSERT INTO Financial.CustomerBuyTransaction
VALUES('2012-01-01','C0000001','AA',2,312.40,'TT00002')
GO

INSERT INTO Financial.CustomerBuyTransaction
VALUES('2012-01-01','C0000001','LIBOR OIS',2,2000.40,'TT00005')
GO
```

The DELETE command to remove some random dates is basically a series of UNION ALL commands that select the TransDate value for days listed in the IN() operator. This list is returned to the DELETE command, and the rows are gone.

Yes, I need to reinsert the multiple trades for the two days as we discussed as we loaded the table for the first time. This is optional; you do not need to do it if you do not want to.

Let's begin to build our solution to the gaps problem by creating our first CTE.

Step 1: Create the First CTE

This first CTE pulls out rows from our test table for the LIBOR instrument, for the year 2012, and for customer C0000005.

Please refer to Listing 6-19.

Listing 6-19. Create the First CTE

```
WITH TradeGapsCTE(
     TransDate,NextTradeDate,GapTradeDays,
     CustId,Symbol,TransAmount,TransTypeCode
)
AS
(
SELECT
     FT.[TransDate] AS TradeDate
     ,LEAD(FT.[TransDate]) OVER(
          ORDER BY FT.[TransDate]
          ) AS NextTradeDate
     ,CASE
          WHEN (DATEDIFF(dd,FT.[TransDate],LEAD(FT.[TransDate])
               OVER(ORDER BY FT.[TransDate])) - 1) = -1
          THEN 0
          ELSE (DATEDIFF(dd,FT.[TransDate],LEAD(FT.[TransDate])
               OVER(ORDER BY FT.[TransDate])) - 1)
      END AS GapTradeDays
     ,FT.CustId
     ,FT.Symbol
     ,FT.TransAmount
     ,FT.TransTypeCode
FROM Financial.CustomerBuyTransaction FT
```

```
WHERE CustId = 'C0000005'
AND Symbol = 'LIBOR OIS'
AND YEAR(FT.TransDate) = 2012
),
```

The first step is to retrieve the next trade date relative to the current trade date, so the LEAD() function is used. Inside the OVER() clause, an ORDER BY clause is used to sort the partition data set by the TransDate column.

The second step is to calculate the number of days between the two dates by subtracting the results of the same LEAD() function just discussed from the current trade date. We subtract the value 1 from the results because if one trade date follows the other, there are technically 0 gaps between the two dates.

Notice the use of the CASE block to determine if the number of days is –1. This occurs when you subtract the same day from itself and also deduct a 1 from it as can be seen in the formula that was used. This logic traps this value and simply reports it as 0. Yes, I admit this is a workaround.

Note   When we build the gap logic, we need to identify the number of days in the gap as the days between the start of the gap and end of the gap only. The start and end days are days on which trades occur, so we do not want to include them.

Let's see the data rows that the query in the first CTE produced.

Please refer to Figure 6-43.

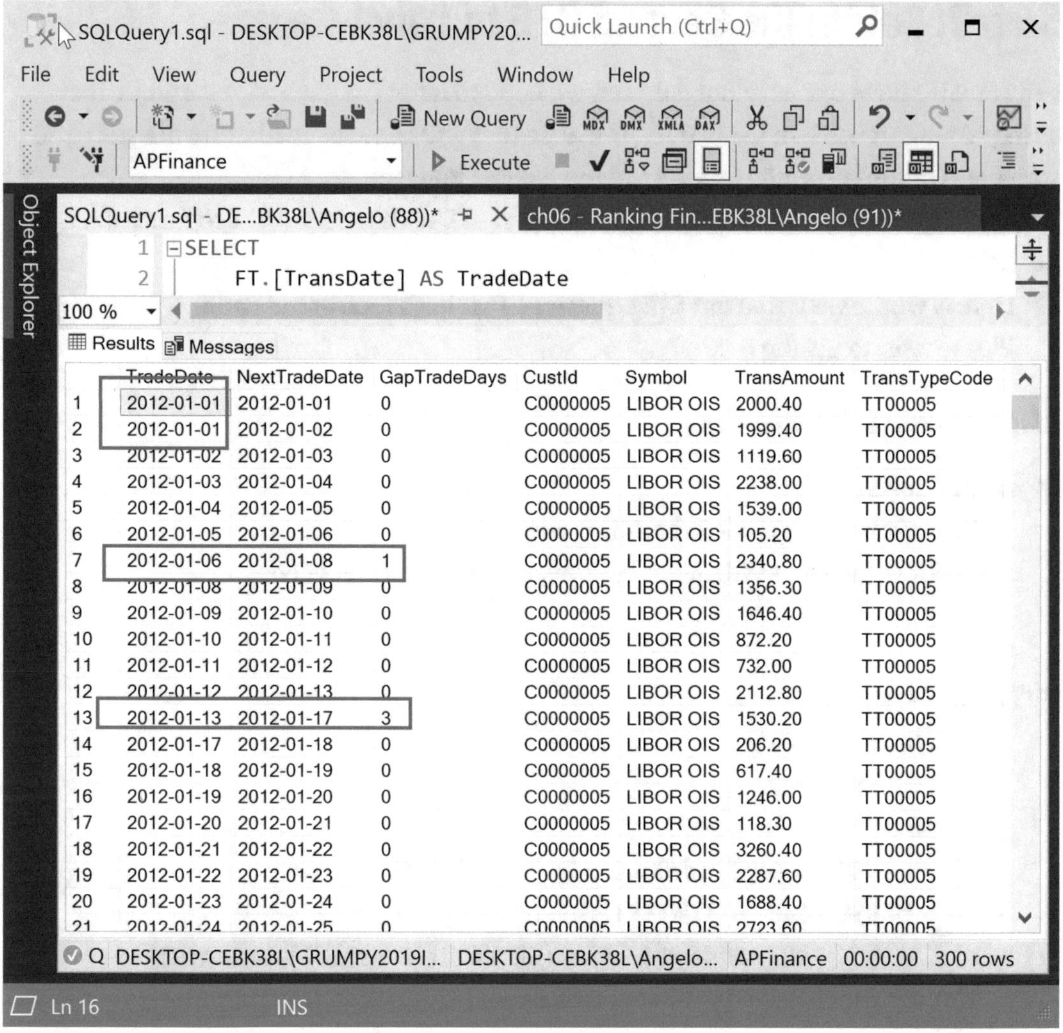

Figure 6-43. *First CTE, gaps and islands identified*

Notice our two trades for January 1. The gap is 0 days, so that is correct. Notice rows 7 and 13 in the result set. The gap in trade days was also calculated correctly. There is one day between January 6, 2012, and January 8, 2012 (January 7, 2012). Also, there are three days between January 13, 2012, and January 17, 2012 (January 14, 15, and 16, 2012).

So far, the logic seems to work. On to step 2 where we define our second CTE.

Step 2: Set Up the Second CTE to Label Gaps

Now that we have our baseline data set, we need to label the gaps with a unique text string so we can use it in the GROUP BY clause that will allow us to identify the start and stop dates of the gap while leaving out the days in between.

This next section of the code chains the first CTE to the second CTE to further set up the data.

Here is where we are so far: CTE2 queries CTE1. Let's look at the code.

Please refer to Listing 6-20.

Listing 6-20. Label the Gaps

```
GapsAndIslands(
      TransDate,NextTradeDate,GapTradeDays,
      GapOrIsland,CustId,Symbol,TransAmount,TransTypeCode
)
AS (
SELECT TG.TransDate
      ,TG.NextTradeDate
      ,TG.GapTradeDays
      ,CASE
            WHEN TG.GapTradeDays = 0 THEN 'ISLAND'
            ELSE 'GAP' + CONVERT(VARCHAR,
                  ROW_NUMBER() OVER (ORDER BY YEAR(TG.TransDate)))
      END AS GapOrIsland
      ,TG.CustId
      ,TG.Symbol
      ,TG.TransAmount
      ,TG.TransTypeCode
FROM TradeGapsCTE TG
),
```

Here comes the ROW_NUMBER() function in the CASE block back to the rescue. Notice how it is used to create a unique "GAP" identifier string by appending the numerical value to the end of the string. Also notice that no PARTITION BY or ORDER BY clause is used or needed.

Pop Quiz   What is the default behavior of the partition window frame when the `ORDER BY` and `PARTITION BY` clauses are not included?

The "ISLAND" string identifiers are all the same across all the islands. We will address how to make these unique for each island in the second example, but I promise you that it will be a little bit trickier than what we just used for the GAP identifier strings. Let's check out our interim results.

Please refer to Figure 6-44.

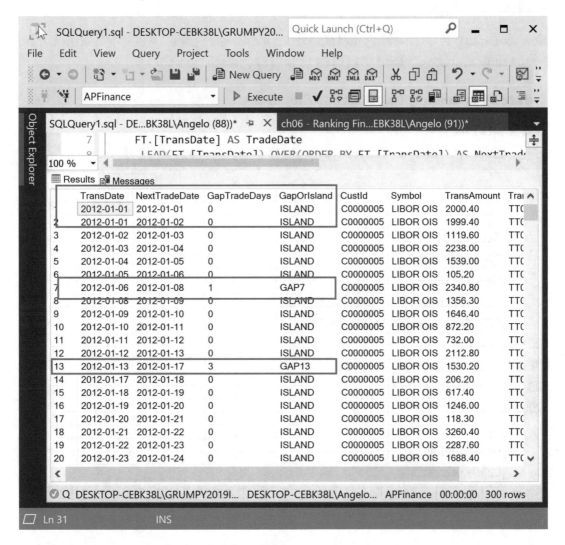

Figure 6-44. *Gaps and islands identified*

All the gaps and islands are labeled correctly (need to work on the ISLAND string later).

We can see how the value returned by the ROW_NUMBER() function is attached to the string "GAP" to produce a unique identifier. This is meaningless for the analyst reading the report but a very valuable mechanism for the GROUP BY clause that will finalize the report.

Step 3: Set Up the Third CTE and Identify Start/Stop Dates of Gaps

Now we get to pull out the start and stop dates for the date gaps. Listing 6-21 is the listing for the third and last chained CTE (CTE3 calls CTE2, which calls CTE1).

Listing 6-21. Identify Gap Start and Stop Dates

```
FinalGapIslanCTE (
      CustId,Symbol,GapTradeDays,GapOrIsland,GapStart,GapEnd
)
AS (
SELECT ,CustId
      ,Symbol
      ,GapTradeDays
      ,GapOrIsland
      ,MIN(TransDate)    AS GapStart
      ,MAX(NextTradeDate) AS GapEnd
FROM GapsAndIslands
WHERE NextTradeDate IS NOT NULL
AND GapTradeDays > 0
GROUP BY CustId
      ,Symbol
      ,GapTradeDays
      ,GapOrIsland
)
```

This CTE is simple. It is based on a standard query that uses the MIN() and MAX() functions to pull out the gap start and stop dates. An OVER() clause is not required. We filter out gap trade days of 0 (the islands) and any row where the next trade date is NULL, just to keep things tidy.

Notice the mandatory GROUP BY clause. Let's see the results of this step.

Please refer to Figure 6-45.

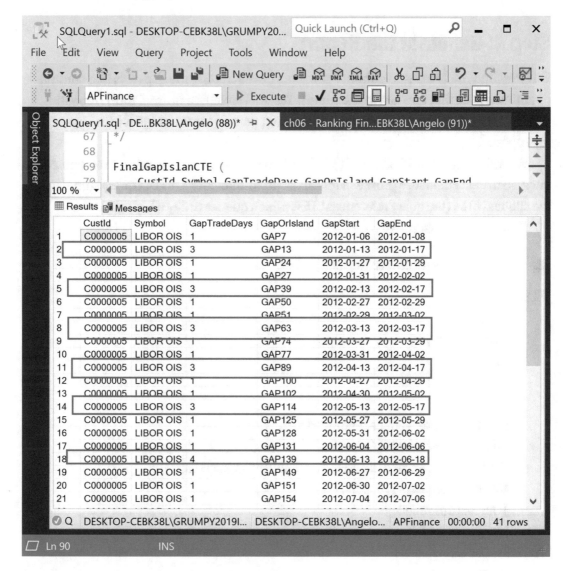

Figure 6-45. *Gaps with start and stop dates*

This is what we want to see: all the gaps have unique codes. I highlighted the gaps with values greater than one so we can spot-check if the start and stop dates make sense. Looking at row 18, **GAP139**, there are indeed four days between June 13, 2012, and June 18, 2012 (June 14, 15, 16, and 17, 2012). Wouldn't it be nice if we could somehow list them in a comma-delimited string next to these two dates?

That's just what we will do in the last step!

Step 4: Generate the Report

Our next and final step is to enhance the report by including the gap days in a comma-delimited string that appears in the report. Yes, this violates a normal form (repeating groups, **1NF**), but we are denormalizing the data pattern for the value it brings to the report.

We will bring in the STRING_AGG() function to the rescue and include it in an embedded query in the SELECT clause (how will this affect performance?). This last query completes the chained CTEs (the query references CTE3, which queries CTE2, which queries CTE1).

Please refer to Listing 6-22.

Listing 6-22. Assemble the Report

```
SELECT CustId
      ,Symbol
      ,GapTradeDays
      ,GapStart
      ,GapEnd
      (
      SELECT STRING_AGG(CalendarDate,',')
      FROM [MasterData].[Calendar]
      WHERE CalendarDate BETWEEN DATEADD(dd,1,GapStart)
      AND DATEADD(dd,-1,GapEnd)
      ) AS GapInTradeDays
FROM FinalGapIslanCTE
ORDER BY CustId
      ,Symbol
      ,GapStart
GO
```

The inline query that appears in the SELECT clause is a bit tricky in that it needs to skip the start and stop dates of the gap by using the DATEADD() function in the BETWEEN predicate. We add one day to the gap start day and deduct one day from the gap stop date. Remember that trades did occur on the start and stop dates, so they are not included in the gap days.

These days are retrieved from the Calendar table and fed to the STRING_AGG() function so it can create a comma-delimited list for us. Very powerful! Let's see the results.

Please refer to Figure 6-46.

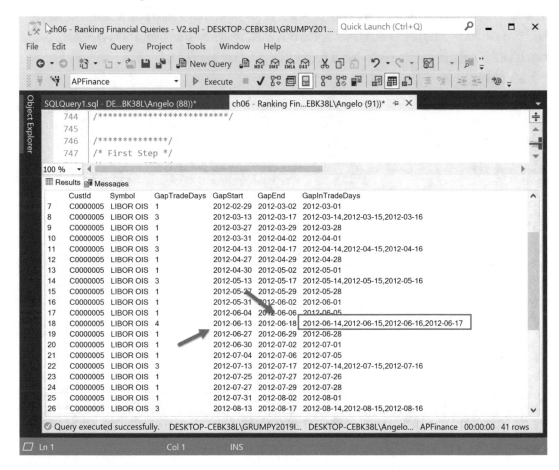

Figure 6-46. *Final gap report*

Looks good. Reminder, for the first gap (and all other gaps), notice we do not include the first and last days in the GapInTradeDays column as trades occurred on these days. It is the day (or days) in the middle where trades did not occur.

Looking at row 18, trades occurred on June 13, 2012, and June 18, 2012, but no trades occurred between these two dates, and that is correctly displayed in the date list in the GapInTradeDays column. It correctly lists four dates as identified in the GapTradeDays column.

That was easy. Let's check out the estimated query plan for this large script that contains three chained CTEs and a query.

Performance Considerations

The estimated query plan for this script is large, so I split it up into two screenshots. We will not perform a rigorous analysis as we did with the previous examples we discussed, but it will be of interest to see what it looks like for a large script.

Please refer to Figure 6-47.

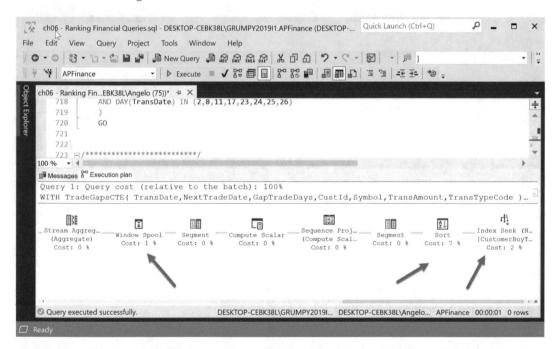

Figure 6-47. *First part of the estimated query plan (right side)*

Starting from right to left, we see a low-cost index seek at 2%. This is very efficient. Next is a sort step at 7% and a window spool at 1% (indicates spooling to storage and not memory). There are also a number of tasks with 0% costs, but we will ignore them as usual.

This is not to say that when you perform your own analysis you should also ignore them. Analyze each step, not only the high-cost steps but also steps you do not understand what they do. Look up in the Microsoft documentation what these steps mean and then determine what impact, if any, they have. This way you expand your knowledge base related to performance tuning.

Let's take a peek at the second half of the estimated query plan. Please refer to Figure 6-48.

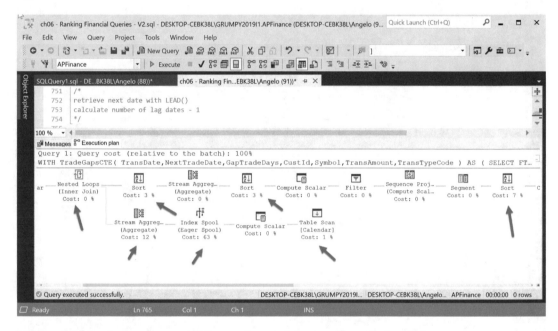

Figure 6-48. *Second part of the estimated query plan (left side)*

There is a lot more action going on here. At the first level we see a sort step at 7% cost followed by a second sort step at 3% cost. And, yes, one more sort step at 3% again.

In the lower branch, we see a small table scan at 1% cost, an index spool at 63%, and a stream aggregate at 12%.

Recall that the index spool means that search values get stored to temporary storage so they can be accessed more than once while the query is processing a partition.

This task feeds a stream aggregate task (12% cost) and finally the nested loops join task. This type of join is what causes the index spool. Any type of spooling to storage can be expensive and is something we should always examine to see if we can improve it. As a last resort, implement TEMPDB on a solid-state disk so spooling is fast!

Might as well set the PROFILE STATISTICS setting on and see what they look like for a large script. Please refer to Figure 6-49.

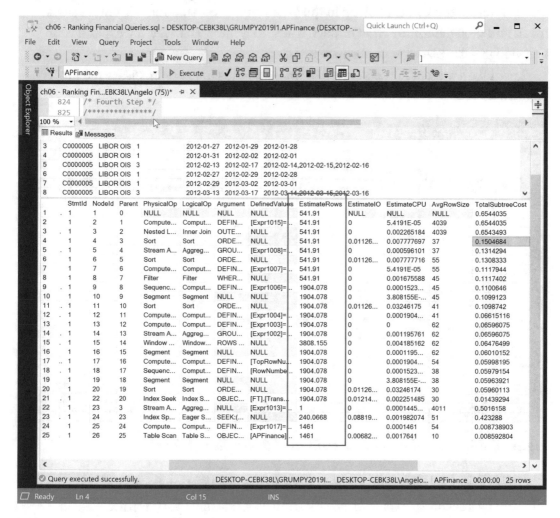

Figure 6-49. *Query profile statistics*

The estimated IO statistics (EstimateIO) are low, and so are the estimated CPU statistics (EstimateCPU), for the most part. The very first CPU statistic is high at 5.419, which is for the first CTE. Row 7 of the results also has a high CPU statistic (EstimateCPU), which happens to be for a CASE block. There are two remaining segment steps with CPU (EstimateCPU) values at 3.8.

Lastly, look at the estimated number of rows (EstimateRows) returned by each step. Here we see what is going on in those lines of various thickness between each step (as far as row volume). Interesting how the rows go from low to high and back to low as each step in the plan processes the data.

Homework   See if you can analyze this query faster and improve performance by modifying the query or adding indexes you feel might help.

Time for a cup of coffee before we dive into modifying our last and final report so we can identify islands and also include gaps in the same report.

Islands Next

Our last objective for this chapter is to enhance our gap report to include islands. Recall islands are clusters of data that appear in sequence, like our trades that occur multiple days in a row.

Recall that customers that trade every day for ten or more days, stop for three or four days, and then start trading again are an example of islands with a gap in between.

The approach we take will be like the one we followed for the gap report, but generating the unique island text strings that will be used in the GROUP BY clause will be a bit of a challenge.

The reason for this is that the ROW_NUMBER() function will return a unique value for each row in the island and it cannot be used by itself to generate the string. We will have to come up with a trick that involves the day of the month of a row and the difference between it and the row number value.

Step 1: Create the First CTE Using LAG( ) and LEAD( )

The first CTE in the CTE chain is like the one we used for the gap report script except for a couple of minor additions, which I will explain. Please refer to Listing 6-23.

Listing 6-23. Create Island CTE1

```
WITH TradeGapsCTE(
     TransDate,NextTradeDate,GapTradeDays,CustId,Symbol,
     TransAmount,TransTypeCode
```

```
)
AS
(
SELECT
        FT.[TransDate] AS TradeDate
        ,LEAD(FT.[TransDate]) OVER(ORDER BY FT.[TransDate]) AS NextTradeDate
        ,CASE
                WHEN LEAD(FT.[TransDate])
                    OVER(ORDER BY FT.[TransDate]) = FT.TransDate
                THEN 0
                ELSE DATEDIFF(dd,FT.[TransDate],LEAD(FT.[TransDate])
                    OVER(ORDER BY FT.[TransDate])) - 1
        END AS GapTradeDays
        ,FT.CustId
        ,FT.Symbol
        ,FT.TransAmount
        ,FT.TransTypeCode
FROM Financial.CustomerBuyTransaction FT
WHERE CustId = 'C0000005'
AND Symbol = 'LIBOR OIS'
AND YEAR(FT.TransDate) = 2012
--ORDER BY FT.TransDate
)
-- uncomment below to test CTE
SELECT * FROM TradeGapsCTE
ORDER BY 1
```

The only difference between this first CTE and the first CTE in the prior gap example is the code in bold and italic font. The logical tests for the condition when the transaction date is equal to the next transaction date. If it does, it sets the gap trade number to 0; otherwise, it takes the difference between these two dates and subtracts 1 from the difference to identify the number of gap days.

Makes sense in that under the condition of sequential trade days, there are 0 gaps. Let's just execute this CTE with a small query that shows the values generated.

Please refer to Figure 6-50.

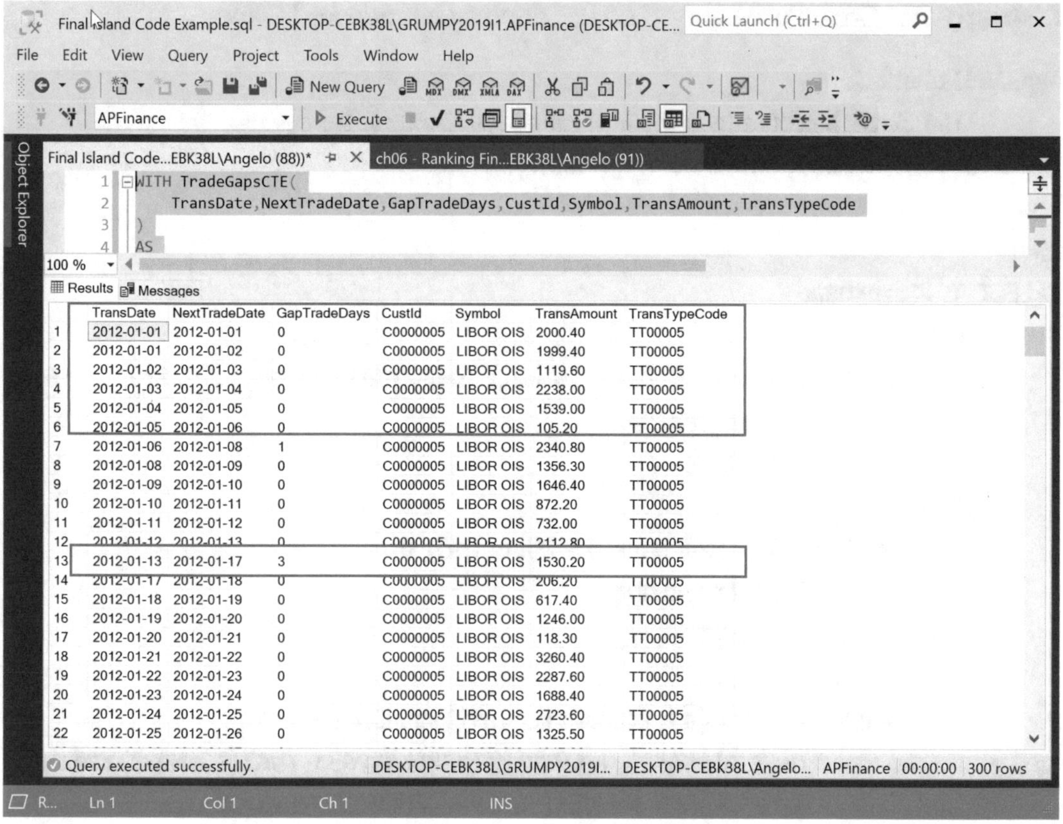

Figure 6-50. *Island CTE1 report*

Looks good. The gaps appear with the same values as the prior report. The members of the island groups appear sequentially though. We want to pull out the start and end days of these islands. We need to assign each island group a unique text string to identify the group of rows belonging to an island.

Step 2: Create the Second CTE That Labels Islands and Gaps

Our second CTE in the chained CTE sequence will start to label each row with its membership in either a gap or island. There are a few conditions we need to check for with CASE blocks, so the code will be interesting if not complex.

Please refer to Listing 6-24.

Listing 6-24. Create the Second CTE, Labeling Islands and Gaps

```
GapsAndIslands(
      TransDate,NextTradeDate,GapTradeDays,GapOrIsland,
      CustId,Symbol,TransAmount,TransTypeCode
)
AS (
SELECT TG.TransDate
      ,CASE
            WHEN TG.NextTradeDate IS NULL THEN DATEADD(dd,1,TG.TransDate)
            ELSE TG.NextTradeDate
       END AS NextTradeDate
      ,CASE
            WHEN TG.GapTradeDays IS NULL THEN 0
            ELSE TG.GapTradeDays
       END AS GapTradeDays
      ,CASE
            WHEN DATEDIFF(dd,TransDate,NextTradeDate) = 0
                 THEN 'ISLAND' + CONVERT(VARCHAR,ABS((DAY(TG.NextTrade
                 Date) - ROW_NUMBER()OVER (ORDER BY YEAR(TG.TransDate)))))
            WHEN DATEDIFF(dd,TransDate,NextTradeDate) = 1
                 THEN 'ISLAND' + CONVERT(VARCHAR,ABS((DAY(TG.NextTrade
                 Date) - ROW_NUMBER()OVER(ORDER BY YEAR(TG.TransDate)))))
            WHEN NextTradeDate IS NULL
                 THEN 'ISLAND' + CONVERT(VARCHAR,(ROW_NUMBER()
                      OVER(ORDER BY YEAR(TG.TransDate))))
                 ELSE 'GAP' + CONVERT(VARCHAR,ABS(DAY(TG.NextTradeDate) -
                 ROW_NUMBER()OVER(ORDER BY YEAR(TG.TransDate))))
       END AS GapOrIsland
      ,TG.CustId
      ,TG.Symbol
      ,TG.TransAmount
      ,TG.TransTypeCode
FROM TradeGapsCTE TG
)
```

```
-- uncomment below to test CTE
--SELECT TransDate,NextTradeDate,GapTradeDays,GapOrIsland,
--CustId,Symbol,TransAmount,TransTypeCode
--FROM GapsAndIslands
--ORDER BY 1
--GO
```

There are three CASE blocks. The first block tests for the condition when the next trade is NULL. This happens when there is no more data in the result set. If it is, it adds one day to the current trade date and reports it as the next trade day. Otherwise, it returns the valid next trade date:

```
,CASE
     WHEN TG.NextTradeDate IS NULL THEN DATEADD(dd,1,TG.TransDate)
     ELSE TG.NextTradeDate
  END AS NextTradeDate
```

The next CASE block tests for the condition that the number of gap or island days is NULL. If this is the case, it simply returns a value of 0 instead of NULL. Otherwise, the value for the number of days in the island or gap is returned:

```
,CASE
     WHEN TG.GapTradeDays IS NULL THEN O
     ELSE TG.GapTradeDays
  END AS GapTradeDays
```

The third CASE block is where we generate the island or gap group identifier string, like ISLAND10 or GAP 15:

```
,CASE
     WHEN DATEDIFF(dd,TransDate,NextTradeDate) = 0
          THEN 'ISLAND' + CONVERT(VARCHAR,(DAY(TG.NextTradeDate)
          - ROW_NUMBER() OVER (ORDER BY YEAR(TG.TransDate))))
     WHEN DATEDIFF(dd,TransDate,NextTradeDate) = 1
          THEN 'ISLAND' +  CONVERT(VARCHAR,(DAY(TG.NextTradeDate)
          - ROW_NUMBER() OVER (ORDER BY YEAR(TG.TransDate))))
     WHEN NextTradeDate IS NULL
          THEN 'ISLAND' + CONVERT(VARCHAR,(ROW_NUMBER()
```

```
          OVER (ORDER BY YEAR(TG.TransDate))))
     ELSE 'GAP' + CONVERT(VARCHAR,ABS(DAY(TG.NextTradeDate)
          - ROW_NUMBER() OVER (ORDER BY YEAR(TG.TransDate))))
END AS GapOrIsland
```

This is a bit complex, three conditions to test and one default condition.

The first condition checks the number of days between the transaction date and the next trade date. If the value is zero, we generate the text string for the island by concatenating the text "ISLAND" with the numerical value of subtracting the row number from the day part of the next trade date.

The second condition also checks the number of days between the transaction date and the next trade date. If the value is 1, we generate the text string for the island by concatenating the text "ISLAND" with the numerical value of subtracting the row number from the day part of the next trade date.

The third condition checks whether the next trade date is NULL. If the value is NULL, we generate the text string for the island by concatenating the text "ISLAND" with the numerical value of the row number only.

The default condition is to generate a string with the value "GAP" plus the absolute value of subtracting the row number from the next trade date. We want only positive numbers, so we do not see a dash (–) symbol separating the two parts of the text that is used to name the grouping of rows.

Let's execute the query that uses the two chained CTEs.

Please refer to Figure 6-51.

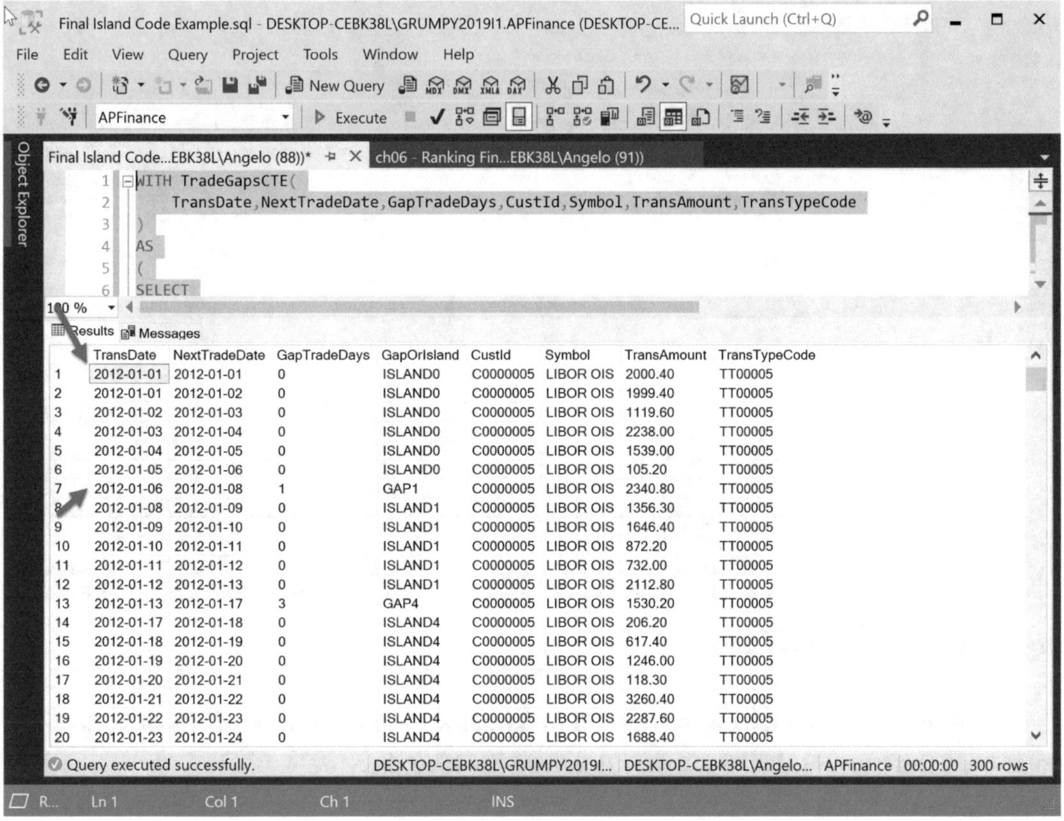

Figure 6-51. *Naming unique groups of gaps and islands*

Looks like it worked. As can be seen, each of the groups of rows has a unique value for the GapOrIsland column. This is what we will use in our GROUP BY clause when we extract the smallest and largest trade dates in each group of rows. That's what we do next by adding a third CTE to help us out with the MIN() and MAX() functions.

If we take the transaction date and gap trade days columns and pop them in an Excel spreadsheet, we can get a good visual of the gaps and islands. Please refer to Figure 6-52.

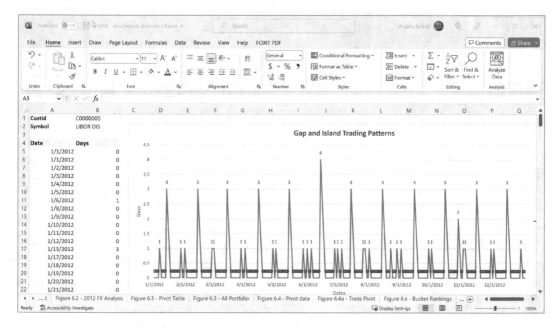

Figure 6-52.   *Gaps and islands chart*

Sometimes a good chart or visual will make sense of the raw data! The peaks of course are the gaps, and all the zero values are the islands.

Step 3: Identify Island Start/Stop Dates

This last CTE is rather simple. We basically use it to extract the start and end dates of each island and gap row set. Please refer to Listing 6-25.

Listing 6-25.   Identifying Start and End Dates for Gaps and Islands

```
FinalGapIslandCTE (
        CustId,Symbol,GapTradeDays,GapOrIsland,GapIslandStart,GapIslandEnd
)
AS (
SELECT CustId
        ,Symbol
        ,GapTradeDays
        ,GapOrIsland
        ,MIN(TransDate) AS GapIslandStart
        ,MAX(NextTradeDate) AS GapIslandEnd
```

```
FROM GapsAndIslands
GROUP BY CustId
       ,Symbol
       ,GapTradeDays
       ,GapOrIsland
)

--uncomment below to test CTE
SELECT * FROM FinalGapIslandCTE
ORDER BY CustId,GapIslandStart
GO
```

As stated earlier, we can now use the MIN() and MAX() functions because we have unique string names that help us identify each set of island or gap rows. Luckily for us the dates are all sequential, so this scheme works for this scenario. Depending on the data you will work with, you will have to come up with another scheme to generate those unique gap and island string identifiers. This data is in the GapOrIsland column, which appears in the GROUP BY clause.

Also notice the uncommented small query. This is used to check out the results at each step of the CTE chain so you can make sure things are working as expected. You can comment out the query once you tested all the parts of this script on your own. Let's see the results.

Please refer to Figure 6-53.

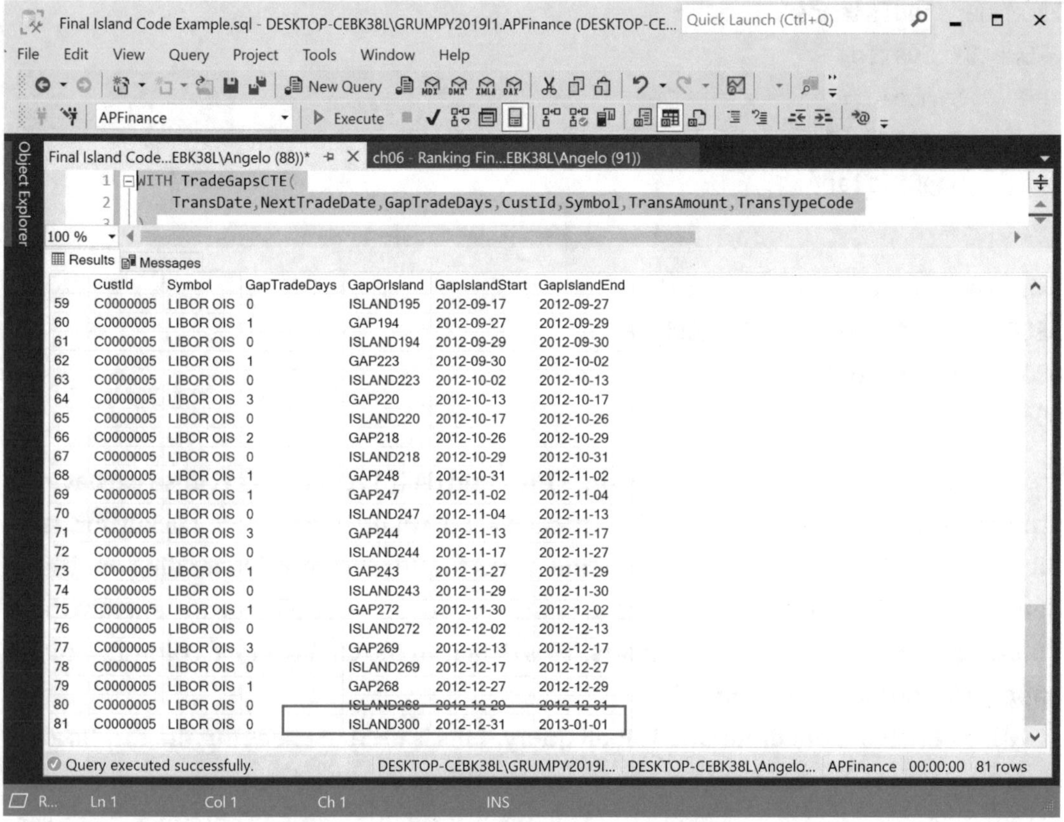

Figure 6-53. *Gaps and islands report with start and end dates*

This also works. We are almost finished. Notice the highlighted last row, which shows the island name and start and end dates. This is the last row of the data set as there are no more test trades after December 31, 2012, although the logic in the script adds a day to the start end date and uses it as the gap or island end date. We could have left it alone and displayed a NULL value, but I think this is neater.

All that is left is to write the master query that also includes all the dates in the islands or gaps in a comma-separated string. As mentioned before, this violates **1NF** (first normal form) design rules, but in this case, I think it is appropriate.

Step 4: Create the Final Report

This is the last step of the script. This script is a bit complex in that it also includes two CASE blocks to correctly calculate the number of days in the gaps and islands and also to create the string that displays all the dates in the gaps and islands but as one value.

Please refer to Listing 6-26.

Listing 6-26. The Final Gaps and Islands Report

```
SELECT CustId
    ,Symbol
    ,CASE
        WHEN GapOrIsland LIKE 'ISLAND%' THEN DATEDIFF(dd,GapIsland
        Start,GapIslandEnd) + 1
        ELSE GapTradeDays
    END AS TradeDays,
    ,GapOrIsland
    ,GapIslandStart AS GapOrIslandStart
    ,GapIslandEnd AS GapOrIslandEnd,
    ,CASE
        WHEN GapOrIsland LIKE 'GAP%' THEN
        (
            SELECT STRING_AGG(CalendarDate,',')
            FROM [MasterData].[Calendar]
            WHERE CalendarDate BETWEEN DATEADD(dd,1,GapIslandStart)
            AND DATEADD(dd,-1,GapIslandEnd)
        )
        ELSE (
            SELECT STRING_AGG(CalendarDate,',')
            FROM [MasterData].[Calendar]
            WHERE CalendarDate BETWEEN GapIslandStart AND GapIslandEnd
        )
        END AS GapOrIslandTradeDays
FROM FinalGapIslandCTE
ORDER BY CustId,Symbol,GapIslandStart
GO
```

The first CASE block checks for the string "ISLAND%" in the GapOrIsland column for each row. If it finds it, it generates the difference between the start and end dates using the DATEDIFF() function and adds 1 to the start date and subtracts 1 from the stop date.

If the value is not "ISLAND%", then it just accepts the current results in the TradeDays column. Please recall the percent (%) character means that some sort of numerical value is there, but right now for this conditional test, we do not care what it is.

The second CASE block contains the logic for generating the comma-separated list of dates in the islands or gaps. Remember that for islands we want to include the start and end dates but for gaps we only want the date or dates in between.

The first test looks for the string "GAP%". If this value is found, a sub-query pulls out the dates between the start and end dates, which have, respectively, been incremented and decremented by 1 day, so the query will pull out any dates between these modified dates from the Calendar table:

```
SELECT STRING_AGG(CalendarDate,',')
FROM [MasterData].[Calendar]
WHERE CalendarDate BETWEEN DATEADD(dd,1,GapIslandStart)
AND DATEADD(dd,-1,GapIslandEnd)
```

The trusty BETWEEN predicate is used for this job! *This could be an expensive strategy as we have a nested query in the SELECT clause and the Calendar table it queries has 1,461 rows.*

Finally, the second and last test is actually the default condition. It pulls out the list of dates with the same query, but this time there is no need to use the DATEADD() function:

```
SELECT STRING_AGG(CalendarDate,',')
FROM [MasterData].[Calendar]
WHERE CalendarDate BETWEEN GapIslandStart AND GapIslandEnd
```

The data is sorted by an ORDER BY clause, and we are finished. Let's see the results. Hope it worked. Please refer to Figure 6-54.

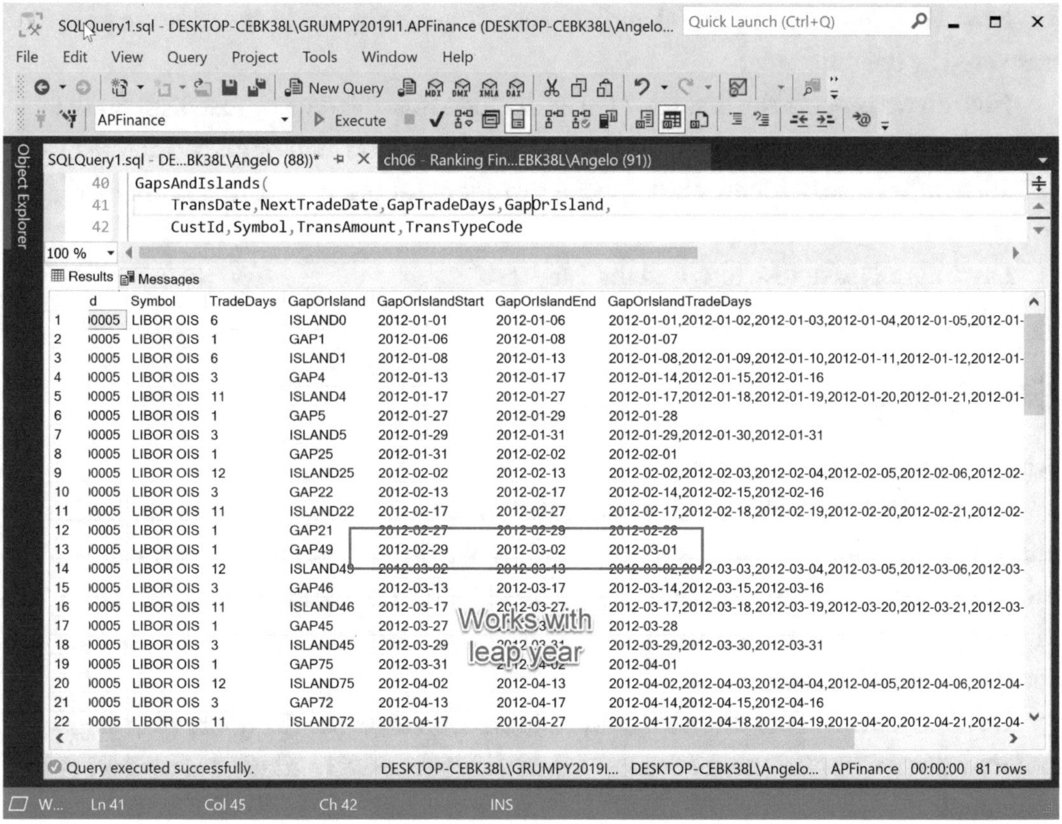

Figure 6-54. *Final islands and gaps report*

Looks good!

Notice that this time the start and stop dates are included for the islands. That is what we want as they are dates where trading occurred. Once again, for gaps, we do not include the start and stop dates in the long string that contains all the dates separated by commas.

It also works correctly with leap years, so I think we can congratulate ourselves for a job well done.

One final comment: When you receive a verbal or written business requirement from an analyst, you should take time out to write a simple technical specification that includes the following three steps:

Step 1: Define the initial condition.

Step 2: Write out the logic that will fulfill the business requirement.

Step 3: Define the final condition or conditions that will generate the report correctly.

Now what do I mean by initial and final conditions? For our little script, they are answered by the following:

How do we process the first row? Do we include missing data or just display NULL?

How do we process the last row? Do we include missing data or just display NULL?

So, in other words, identify any special treatment for processing that is not included in the main logic of the script.

Last but not least, test, test, and did I say test?

Summary

What did we learn in this chapter?

For one thing we realized our toolkit consists not only of the aggregate, ranking, and analytical functions but also of supporting tools like the estimated query plan generator; the various performance and runtime statistics like IO, TIME, and PERFORMANCE; and Microsoft Excel. We use Excel to visually interpret the data results to give our analyst a complete view of the results.

We also expanded our knowledge by doing a deep dive into the query plan steps and understanding how the sections of a query are processed and what statistics are generated.

Lastly, we did a quick discussion of the live query plan generator so we could see how much time is spent on each task and what volume of rows are passed between tasks in order to give us a complete picture of the processing behind the scenes. This helps identify high-performance steps that need to be analyzed.

Hopefully, at this stage of the book, you see how you can follow a structured approach to performance tuning and which tools and also solutions you can apply to fix performance problems. These include adding indexes, using report tables, and in some cases upgrading hardware for faster memory and solid-state disks for TEMPDB. We did not discuss memory-enhanced tables, but this is also a tool you can consider.

The next chapter puts analytical window functions through their paces against our financial database.

CHAPTER 7

Finance Use Case: Analytical Functions

This is the last chapter dealing with our financial database. We will put the analytical functions through their paces and perform the usual performance analysis, but we will also take a look at a few strategies that involve variations of the queries, like from CTE- to reporting table–based queries to using memory-enhanced tables to see which scheme yields better performance.

Analytical Functions

Just in case you did not read Chapter 3, the following are the eight functions for this category that we will discuss (they are sorted by name). They are used with the OVER() clause to generate valuable reports for our business analysts:

- CUME_DIST()
- FIRST_VALUE()
- LAST_VALUE()
- LAG()
- LEAD()
- PERCENT_RANK()
- PERCENTILE_CONT()
- PERCENTILE_DISC()

© Angelo Bobak 2023
A. Bobak, *SQL Server Analytical Toolkit*, https://doi.org/10.1007/978-1-4842-8667-8_7

The same approach as the prior chapter will be taken. I describe what the function does and present the code and the results. Next, some performance analysis is performed so we can see what improvements, if any, are required.

As we saw in the examples in Chapter 3, improvements can be adding indexes or even creating a denormalized report table so that when the query using the analytical window functions is executed, it goes against preloaded data avoiding JOINs, calculations, and complex logic. Rarely will you be delivered requirements that need window functions with real-time data. If these come up, the reporting table strategy will do you no good.

CUME_DIST() Function

This function calculates the relative position of a value within a data set like a table, partition, or table variable loaded with test data. The keyword here is relative.

How does it work?

First, calculate how many values come before it or are equal to it (call this value **C1**). Next, calculate the number of values or rows in the data set (call this **C2**). **C1** is then divided by **C2** to deliver the cumulative distribution result. Check the Microsoft documentation for the return data types of all the functions as there is a lot of good information from the descriptions in their documentation. In this case the data type is FLOAT(53). You may want to use the FORMAT() function to convert it to a percentage so it looks good in the resulting report.

Recall that this function is similar to the PERCENT_RANK() function, which works as follows:

Given a set of values in a data set, this function calculates the rank of each individual value relative to the entire data set. This function also returns a percentage, and the data type is also FLOAT(53).

Back to the CUME_DIST() function. We will look at two examples, an easy one and then one that will query our finance transactional database.

Our analyst wants us to create a report that will display the cumulative distribution by month and quarter in the Portfolio table grouped by year, quarter, month, customer ID, and of course portfolio number and name. The customer of course is our friend customer C0000001 (John Smith), and the year is 2012. John Smith's income bracket is between $150K and $250K, not bad.

Once the query is validated, it can be modified for more than one customer and year. Remember, generate small results that satisfy the requirements and then modify the data scope if required (after testing).

Let's look at a simple example first so we can see how the prior description works on a small data set. Please refer to Listing 7-1.

Listing 7-1. A Simple Example

```
USE TEST
GO

DECLARE @CumDistDemo TABLE (
     Col1 VARCHAR(8),
     ColValue INTEGER
     );

INSERT INTO @CumDistDemo VALUES
('AAA',1),
('BBB',2),
('CCC',3),
('DDD',4),
('EEE',5),
('FFF',6),
('GGG',7),
('HHH',8),
('III',9),
('JJJ',10)

SELECT Col1,ColValue,
     CUME_DIST() OVER(
           ORDER BY ColValue
     ) AS CumeDistValue,
     A.RowCountLE,
     B.TotalRows,
     CONVERT(DECIMAL(10,2),A.RowCountLE)
           / CONVERT(DECIMAL(10,2),B.TotalRows) AS MyCumeDist
FROM @CumDistDemo CDD
CROSS APPLY (
```

```
    SELECT COUNT(*) AS RowCountLE FROM @CumDistDemo
    WHERE ColValue <= CDD.ColValue
    ) A
CROSS APPLY (
    SELECT COUNT(*) AS TotalRows FROM @CumDistDemo
    ) B
GO
```

A simple table variable called @CumeDistDemo is declared and loaded with ten rows of data. Only two columns are created. The first is called Col1 and acts as a key column, while the ColValue column holds the data. In our case these are the numbers 1–10. Really simple.

The query utilizes the CUME_DIST() function and an OVER( ) clause. No PARTITION BY clause is included, but the ORDER BY clause is used to order the partition rows by the ColValue column in ascending order.

We need to include the formula for calculating cumulative distribution so we can see how it works and compare it to the SQL Server version. Here is the formula:

Count of rows with values less than or equal to the current column value divided by the total number of rows

To get the values for this formula, two CROSS APPLY operators are used. The first one calculates the number of rows with values less than or equal to the current column value, and the second calculates the total number of rows in the data set.

This query can probably be optimized, but for our purposes it should illustrate how the CUME_DIST() function works.

The values returned by the queries in the CROSS APPLY block are used in the formula in the SELECT clause:

```
    CONVERT(DECIMAL(10,2),A.RowCountLE)
            / CONVERT(DECIMAL(10,2),B.TotalRows) AS MyCumeDist
```

Let's execute the query and see what we get. Please refer to Figure 7-1.

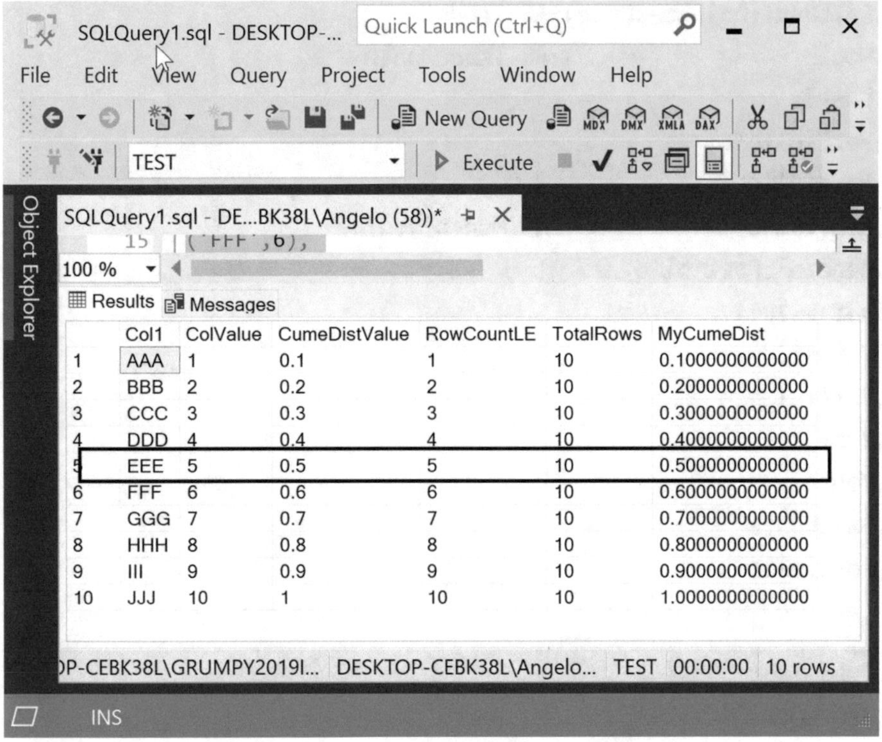

Figure 7-1. *Two ways of calculating cumulative distribution*

Looks good and predictable, I think! The MyCumeDist column has the values calculated by the formula we created, and the column called CumeDistValue contains the values the SQL Server function calculated. Maybe a little formatting is needed so they look the same, but the values match.

Now for a financial example. Our analyst supplies the following business requirements:

For customer John Smith (C0000001), generate a report that shows the cumulative distribution by month and quarter. Results need to be for the year 2012, and the year, quarter, month, customer ID, portfolio, and monthly values also need to be displayed.

Please refer to Listing 7-2.

Listing 7-2. Monthly Portfolio Cumulative Distribution Analysis

```
WITH PortfolioAnalysis (
TradeYear,TradeQtr,TradeMonth,CustId,PortfolioNo,Portfolio,MonthlyValue
)
AS (
SELECT Year AS TradeYear
```

413

```
        ,DATEPART(qq,SweepDate) AS TradeQtr
        ,Month                  AS TradeMonth
        ,CustId
        ,PortfolioNo
        ,Portfolio
        ,SUM(Value)             AS MonthlyValue
FROM Financial.Portfolio
WHERE Year > 2011
GROUP BY CustId
        ,PortfolioNo
        ,Year
        ,Month
        ,SweepDate
        ,Portfolio
)

SELECT TradeYear
        ,TradeQtr
        ,TradeMonth
        ,CustId
        ,PortfolioNo
        ,Portfolio
        ,MonthlyValue
        ,CUME_DIST() OVER(
            PARTITION BY Portfolio
            ORDER BY TradeMonth,Portfolio
        ) AS CumeDistByMonth
        ,CUME_DIST() OVER(
            PARTITION BY Portfolio
            ORDER BY TradeQtr,Portfolio
        ) AS CumeDistByMonth
FROM PortfolioAnalysis
WHERE TradeYear = 2012
AND CustId = 'C0000001'
GO
```

Our usual CTE template is used. The CTE calculates the monthly sum of the portfolios, and the query that references the CTE in order to calculate the cumulative distribution uses two OVER() clauses. The CTE returns results for years greater than 2011. We skip 2011 because this is when the account was opened, and we know the initial balance was $100,000 (all the accounts were initialized with this value).

Looking at the base query, the first OVER() clause contains a PARTITION BY clause that sets up partitions by month and portfolio name. The partition result set is sorted by trade month and portfolio name by an ORDER BY clause.

The second OVER() clause also contains a PARTITION BY clause, but this time the partition is set up by quarter and portfolio name. Additionally, the partition result set is sorted by trade quarter and portfolio name in the ORDER BY clause.

If you include multiple years, you can include an OVER() clause to calculate cumulative distributions by year. Good exercise for you to try out. Just add the following code to the SELECT clause:

```
,CUME_DIST() OVER(
        PARTITION BY Portfolio
        ORDER BY TradeYear,Portfolio
) AS CumeDistByMonth
```

Don't forget to modify the WHERE clause so it does not have a predicate that filters for one year only. This will give you a larger result, but it presents a more complete analysis as far as cumulative distribution goes.

No final query ORDER BY clause is required. Let's check out the results in Figure 7-2.

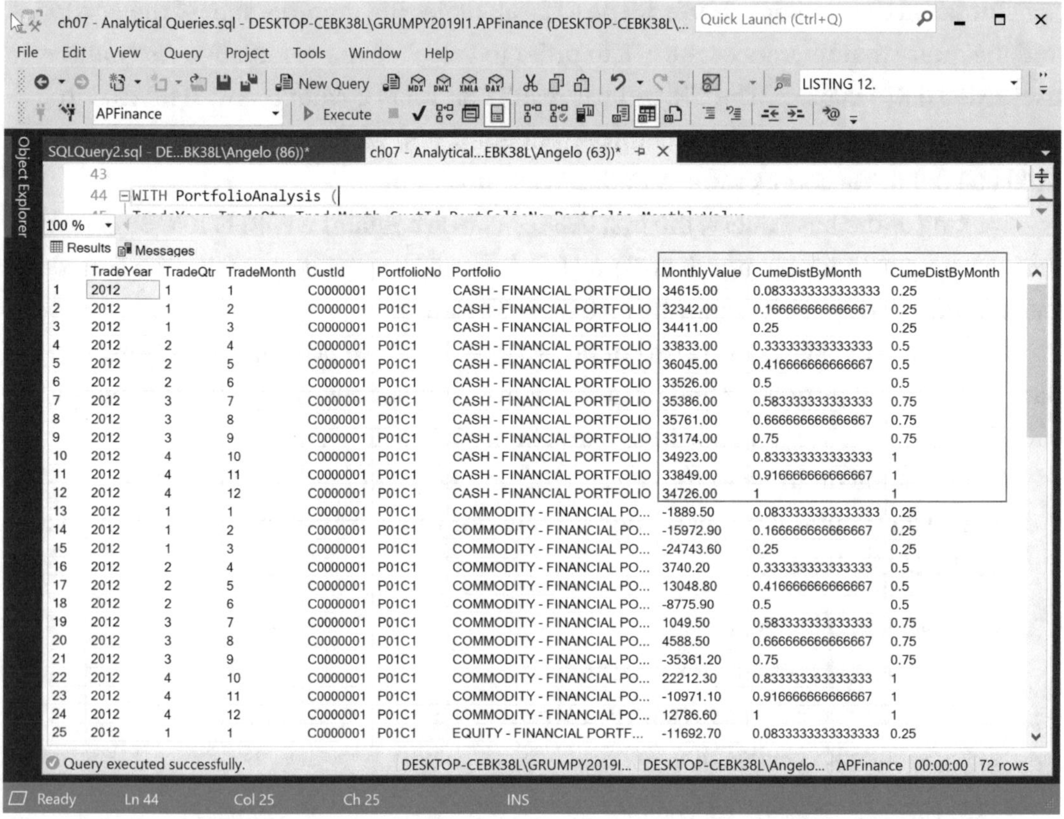

Figure 7-2. *Portfolio cumulative distribution analysis for customer C0000001*

As can be seen, we get cumulative distributions by month and quarter across portfolios. Very interesting query. The pattern repeats itself for each portfolio.

Performance Considerations

How does this query perform? Let's run our usual estimated query plan to establish a baseline for our analysis, after which we will analyze it and see if we can improve performance by adding an index or creating a report table in place of the CTE.

Please refer to Figure 7-3.

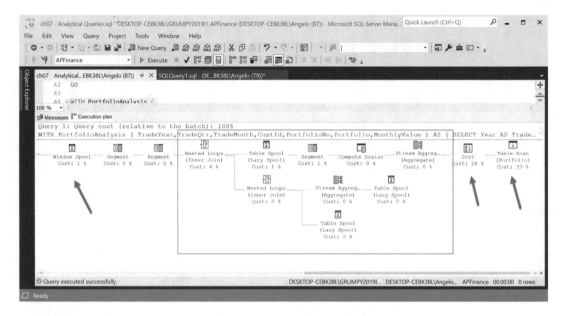

Figure 7-3. *Cumulative distribution baseline query plan – part 1*

This is the right-hand side of the estimated query plan. Notice the table scan; no index is required as this is a small table, around 1,470 rows. The table scan has a cost of 33% and is followed by a sort step with a cost of 28%. (Note the ID of the table scan by placing the mouse over the icon and viewing the ID. Most of the time, the table scan is used for other tasks, the temporary table in TEMPDB that is created, that is.)

The area boxed in red is for the first cumulative distribution function. It will be repeated in the next section of the query plan for the second cumulative distribution function call. It contains some zero-cost tasks and some table spool tasks that make the data available in temporary physical storage so that calculations can be applied. Two of these spool tasks chime in at 0% cost, while the last one comes in at 1% cost. Two nested loops join tasks complete this section, one at 0% and one at 4%. So far, nothing to be concerned about.

Let's look at the second section. Please refer to Figure 7-4.

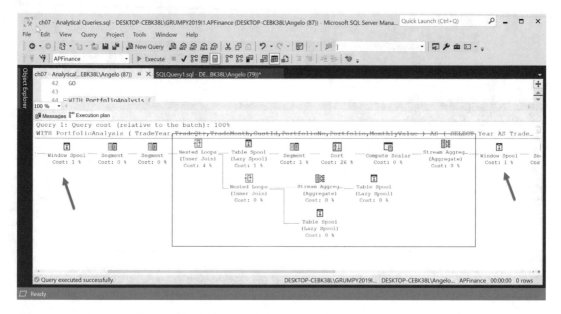

Figure 7-4. *Cumulative distribution baseline query plan – part 2*

Notice the window spool task at the right hand side of the plan. This comes in at a cost of 1%, so spooling action is performed on disk as opposed to memory. This window spool task is associated with the first call to the cumulative distribution function.

There's the boxed area, and it is the same as the first area we discussed. It also is followed by a window spool task. Looks like we have three tasks with low cost that we should be aware of, the table spool at 1%, the nested loops for the inner join at 4%, and finally the window spool task at 1%.

I tried the reporting table approach where the query in the CTE was used to populate a report table and the base query was executed against the reporting table. The improvements were marginal as far as the query plan was concerned (the code is included in the script for this chapter).

The TIME and IO statistics showed marginal improvements with the report table approach, so I include a screen print here. Please refer to Figure 7-5.

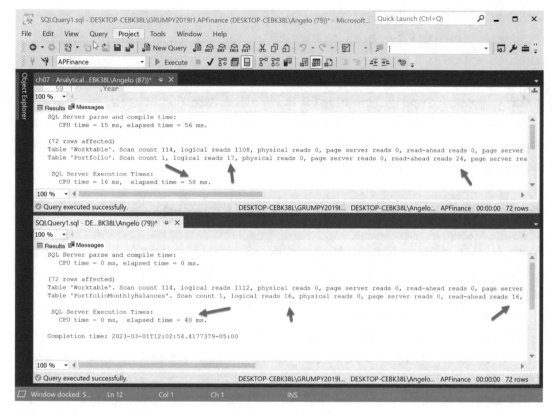

Figure 7-5. *Comparing IO and TIME statistics*

There's modest improvement in SQL Server execution time and read-ahead reads. Since this is historical data, the report table approach is probably more efficient than the CTE approach, especially if multiple users run the same query, like from a SQL Server Reporting Services (SSRS) report that allows analysts to select different years via a drop-down list in the report.

There is a cost to the report table strategy as it must be periodically loaded to update the data to keep it current. This could be performed daily or weekly and is usually performed overnight by a scheduled load process so users are not affected.

Last but not least, I modified the original CTE query so it does not use a CTE but goes directly against the Portfolio table and calculates the SUM( ) right before the two calls to the cumulative distribution functions. Results almost matched the CTE version, and SQL Server execution time was slightly less than the CTE version, 230 ms (CTE) vs. 171 ms (non-CTE). There were fewer steps in the non-CTE version (72 steps vs. 32 steps) as far as STATISTICS PROFILE values, but all in all, performance was the same except for the slightly faster execution time. So maybe CTE blocks are not the first line choice for these types of queries? If we need to query millions of rows, then the CTE might not be considered at all.

Always consider how the query will be used and run, like in a web report by multiple users as opposed to one user running the query alone. Also, do not use a feature like a CTE just because you feel it is slick and impresses your manager if you use it. The simplest, fastest, and of course accurate solution is the best solution. Also consider the labor cost of maintaining these queries.

FIRST_VALUE( ) and LAST_VALUE( ) Functions

So what do this related pair of functions do for us? If you did not read Chapter 3 or need a refresher, please read this next discussion; otherwise, feel free to skip ahead to the example.

The names give us a hint. They answer the following question:

Who comes first and who comes last in the partition?

Given a set of sorted values, the FIRST_VALUE( ) function will return the first value in the partition.

Given a set of sorted values, the LAST_VALUE( ) function will return the last value in the partition.

Both functions return whatever data type was used as an argument; it could be an integer, could be a string, etc. Pretty simple. Let's dive right into the example.

Here is our business specification:

This time the analyst asks us to turn our attention to all customer accounts. We are asked to create a report that will return the last and first account values in a partition defined by year and customer ID and sorted by month. The analyst wants to see these values for data sets (partitions) that include the current month and also all future months and also the current month and all past months. In other words, look forward and look back (hint: need to define a window frame).

Lastly, the report will be for cash accounts only.

Please refer to Listing 7-3.

Listing 7-3. First and Last Account Balances by Year and Customer, Sorted by Month

```
WITH MonthlyAccountBalances (
AcctYear,AcctMonth,CustId, PrtfNo, AcctNo, AcctName, AcctBalance
)
AS
(
SELECT YEAR(PostDate)  AS AcctYear
      ,MONTH(PostDate) AS AcctMonth
      ,CustId
      ,PrtfNo
      ,AcctNo
      ,AcctName
      ,SUM(AcctBalance)
FROM Financial.Account
GROUP BY YEAR(PostDate)
      ,MONTH(PostDate)
      ,CustId
      ,PrtfNo
      ,AcctNo
      ,AcctName
)

SELECT CustId
      ,AcctYear
      ,AcctMonth
      ,AcctNo
      ,AcctName
      ,AcctBalance

      ,FIRST_VALUE(AcctBalance) OVER (
            PARTITION BY AcctYear,CustId
            ORDER BY AcctMonth
            ROWS BETWEEN CURRENT ROW AND UNBOUNDED FOLLOWING
      ) AS FirstValueBalance

      ,LAST_VALUE(AcctBalance) OVER (
```

```
            PARTITION BY AcctYear,CustId
            ORDER BY AcctMonth
            ROWS BETWEEN CURRENT ROW AND UNBOUNDED FOLLOWING
    ) AS LastValueBalanceCRUF

    ,FIRST_VALUE(AcctBalance) OVER (
            PARTITION BY AcctYear,CustId
            ORDER BY AcctMonth
            ROWS BETWEEN UNBOUNDED PRECEDING AND CURRENT ROW
    ) AS FirstValueBalanceUPCR

    ,LAST_VALUE(AcctBalance) OVER (
            PARTITION BY AcctYear,CustId
            ORDER BY AcctMonth
            ROWS BETWEEN UNBOUNDED PRECEDING AND CURRENT ROW
    ) AS LastValueBalanceUPCR
FROM MonthlyAccountBalances
WHERE AcctYear >= 2013
AND AcctNAme = 'CASH'
ORDER BY CustId
        ,AcctYear
        ,AcctName
        ,AcctMonth
        GO
```

The CTE looks like the one we discussed in the prior example, but this time it references the Account table, which is at a lower granularity than the Portfolio table.

What is interesting is that we include two sets of FIRST_VALUE()OVER() and LAST_ VALUE()OVER() clauses with different ROWS frame clauses.

The first set uses the following ROWS window frame definition in order to define the window frame within the partition by the current row and all rows following (look ahead):

```
ROWS BETWEEN CURRENT ROW AND UNBOUNDED FOLLOWING
```

The second set uses the following ROWS window frame definition in order to define the window frame that processes the current row and all preceding rows (look back):

```
ROWS BETWEEN UNBOUNDED PRECEDING AND CURRENT ROW
```

Remember, the window frame is relative to the current row being processed as processing moves from row to row within the partition. If this is a bit fuzzy, review the topic in Chapter 1 and then try out a simple example using ten values, a table variable, and the functions so you can see them work on a simple data set. Check out the diagrams in Chapter 1 on this topic; they will make the concept clear.

The query was executed. This one took four seconds to run! Can't wait to see the estimated execution plan. Please refer to the partial results in Figure 7-6.

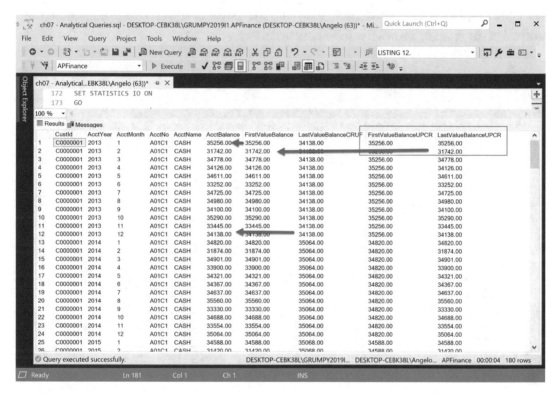

Figure 7-6. *First and last rows within a defined partition*

As can be seen in the first set of first and last balances, the first value is the value in the current row, while the last value always points to the last value in the partition, $34,138.00. Here we look forward.

In the second set of first and last balances, the first value always points to the first value in the partition, while the last value is always the value in the current row of the window frame. Here we look backward.

I guess this is of value as we can see where we started off, where we are at, and where we are going relative to the balances. This scheme tells you if the account is increasing in value or decreasing in value as the months go by.

Better check the estimated query plan as we saw this query took four seconds to run.

Performance Considerations

We need to split the estimated query plan into two parts. Starting right to left, about the middle of the screen, we see the following tasks in Figure 7-7.

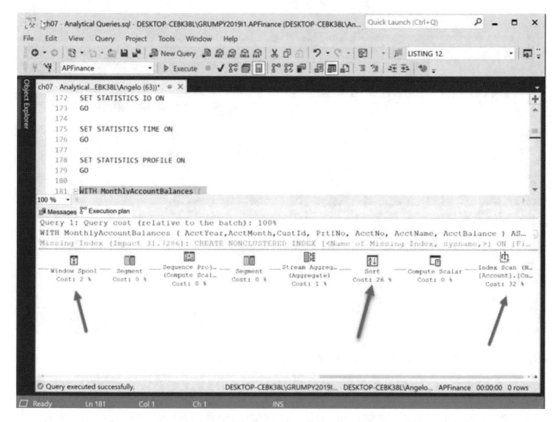

Figure 7-7. *Estimated query plan for FIRST_VALUE() and LAST_VALUE(), part 1*

Right away, we see we need to create an index as the estimated query plan tool suggested an index. We do see an existing index with a cost of 32%, but apparently, we need another one. We see a sort task at 26% cost and a window spool task at 2%.

This is only half of the estimated query plan, but the pattern repeats itself for all pairs of function calls, so any improvements we apply here can be applied to the other patterns in the query plan (by patterns I mean the combination of operators (tasks) and connecting flows).

Please refer to Figure 7-8.

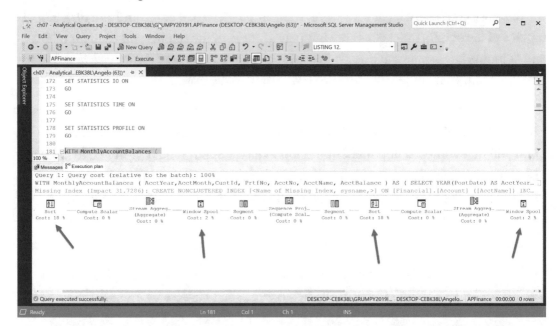

Figure 7-8. *Estimated query plan for FIRST_VALUE() and LAST_VALUE(), part 2*

As can be seen, the same pattern of compute scalar, sort, segment, sequence project, segment, and window spool tasks repeats itself. Costs are about the same but with the sort steps showing slightly different costs. The estimated query planner tells us the suggested index will improve performance by 31.7286%. I'll take it!

Let's copy and paste the index template, give it a name, and execute the command to build the index.

Listing 7-4 is the completed code for the suggested index.

Listing 7-4. Suggested Estimated Query Plan Index

```
CREATE NONCLUSTERED INDEX [ieAcctPrtBalancePostDate]
ON [Financial].[Account] ([AcctName])
INCLUDE ([CustId],[PrtfNo],[AcctNo],[AcctBalance],[PostDate])
GO
```

This is a non-clustered index based on the account name column, but we include the customer ID, portfolio number, account number, account balance, and postdate columns as suggested by the query plan tool. Notice those pesky square brackets. Since SQL Server generated this TSQL code, it included them. You might want to remove them, but is it worth it?

Suppose you have a large script with thousands of lines of code, and you do a global replace of the square brackets. What happens if some of the names you used for tables and columns have spaces in them? If you do the global replace of the square brackets to remove them, your script will blow up. That's right, blow up!

Note   Programming naming standards are great when applied realistically, but one can go overboard and most of the time will be spent typing pretty code instead of coming up with good working solutions. Tools are available to "pretty up" the code if your department has the cash for them. Well, that's my two cents anyway.

Let's create the index and see if the query plan improves. Please refer to Figure 7-9a.

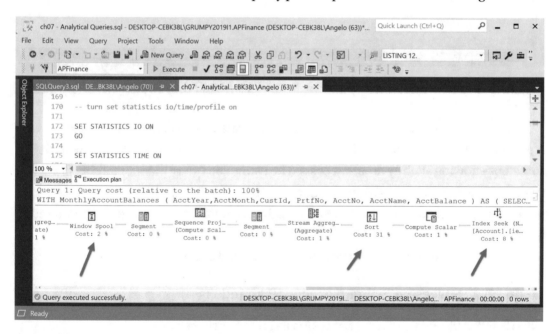

Figure 7-9a.   *Revised estimated query plan, first half*

As usual, the index was used but the sort went up. Our friend, the window spool, remained at 2% cost. Let's look at the second half of the query plan.

Please refer to Figure 7-9b.

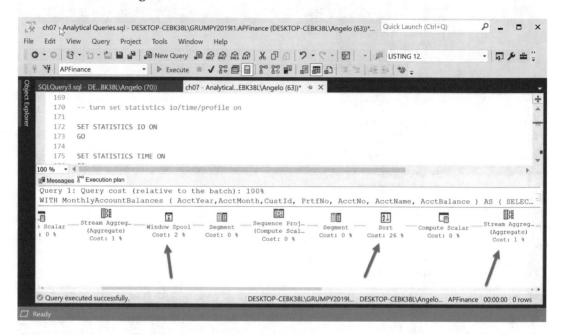

Figure 7-9b.   *Revised query plan, second half*

Same pattern in this section: a sort at 26% and a window spool at 2%.

Reminder   As this is a book on applying the window functions and not an exhaustive book on query plans and performance tuning, the tasks with 0% cost are not discussed. Again, in your real world, do not ignore them but understand what is going on.

We need to generate some more statistics to better understand what is going on as far as performance. Please refer to Figure 7-10 for IO and TIME statistics.

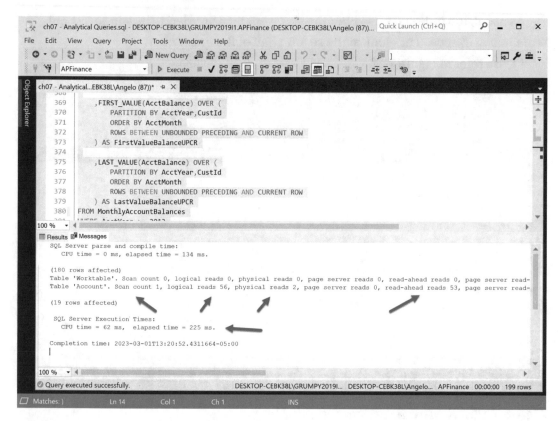

Figure 7-10. *IO and TIME statistics*

How about `STATISTICS PROFILE` statistics? These usually have a lot to say related to steps, costs of each step, and `IO` and `CPU` utilization. You can also take a deep dive and understand which logic is related to the tasks by looking at the `StmtText` (statement text) column. Please refer to Figure 7-11.

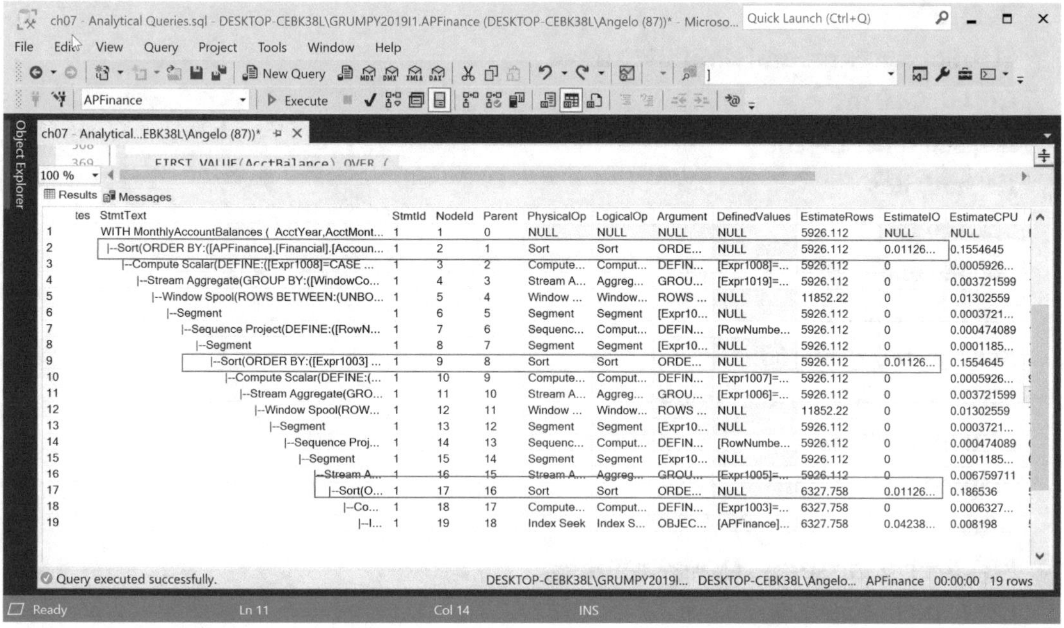

Figure 7-11. *Profile statistics for repeating execution patterns*

This is a good set of statistics in that it contains a lot of information. Clearly, the sort steps are doing most of the IO although the values are not high. The last step, the index seek step, has the highest IO coming in at 0.04238. Right now, the query runs under one second. The Account table has 43,860 rows. What would the IO be for these sort steps if we had 100,000 or 1,000,000 accounts (as in large international banks)? These IO statistics could become a problem.

This final step started me thinking. I had a final ORDER BY clause in the query, which ended up being redundant. I removed it, checked the results, and reran the query to generate the PROFILE STATISTICS. One expensive sort was removed, so that's an improvement. One less step means less time executing the query.

Pop Quiz   How do you generate PROFILE statistics? The answer will appear at the end of this section.

Lastly, comparing the IO and TIME statistics after I removed the last ORDER BY clause yielded the following results:

Before

SQL Server Parse and Compile Times

CPU time: 0 ms

Elapsed time: 134 ms

IO Statistics

Scan count: 1

Logical reads: 56

Physical reads: 2

Read-ahead reads: 53

SQL Server Execution Times

CPU time: 62 ms

Elapsed time: 225 ms

After

SQL Server Parse and Compile Times

CPU time: 16 ms

Elapsed time: 21 ms

IO Statistics

Scan count: 1

Logical reads: 56

Physical reads: 0

Read-ahead reads: 0

SQL Server Execution Times

CPU time: 31 ms

Elapsed time: 128 ms

There's noticeable improvement in IO, and execution time was cut down by 50% after the redundant ORDER BY clause was removed.

One more lesson learned: Review your code, and make sure you do not have redundant or unnecessary logic. Usually, in working environments, a good peer review session will help identify little glitches that you cannot see but that a pair of external eyes can help you find.

It is always uncomfortable to have your code reviewed and critiqued, but it is part of the game.

Pop Quiz Answer   By executing the SET STATISTICS PROFILE ON command.

LAG( ) and LEAD( ) Functions

This pair of functions are very important tools to include in your programming toolkit to analyze data. One of the activities your business analysts perform is to look back in time or forward in time within a set of historical data. Of course, the data could be a snapshot that includes current and past data but no future data (does not exist yet depending on the point of view).

These functions allow you to create powerful queries and reports to do exactly that.

Recall what the LAG( ) function does:

This function is used to retrieve the previous value of a column relative to the current row column value (usually within a time dimension). For example, given this month's sales amounts for a product, pull last month's sales amounts for the same product or specify some offset, like three months ago or even a year or more depending on how much historical data you have collected and what your business users want to see. Use this logic to calculate the difference. Did sales go up or down? Critical measurement to monitor and improve sales performance, not to mention awarding bonuses!

Here's what the LEAD( ) function does:

This function is used to retrieve the next value relative to the current row column. For example, given this month's sales amounts for a product, pull next month's sales amounts for the same product or specify some offset, like three months in the future or even a year in the future (historical data of course; one cannot see into the future from the present, and neither can your database if the data does not exist).

One exception, though, is if you have future projections for sales data. Makes for interesting analysis for the sales team.

Reminder   Both functions return the data type of the value used as an argument.

LAG( ) Function

Let's start with the LAG( ) function first. Our analyst sends us the following specifications for a report:

I would like to see this month's and last month's account balances in the same report. The results need to be grouped by year, month, customer identifier, portfolio number, and account number and name. Results need to be for the years greater than or equal to 2013. Lastly, I would like to see the difference between the current month's and last month's values in order to see if balances increased or decreased.

We are using the same CTE as the prior example so I will not repeat it in the listing. This could be a good candidate for converting the CTE to a report table loaded, let's say, once a month. We will see. Please refer to Listing 7-5.

Listing 7-5. Using LAG() to Calculate Last Month's Balances

```
SELECT CustId
      ,AcctYear
      ,AcctMonth
      ,PrtfNo
      ,AcctNo
      ,AcctName
      ,AcctBalance
      ,LAG(AcctBalance) OVER (
            PARTITION BY CustId,AcctYear,AcctName
            ORDER BY AcctMonth
      ) AS LastMonthBalance
      ,AcctBalance -
            (
            LAG(AcctBalance) OVER (
            PARTITION BY CustId,AcctYear,AcctName
            ORDER BY AcctMonth
            )
      ) AS Change
FROM MonthlyAccountBalances
WHERE AcctYear >= 2013
GO
```

The LAG() function uses an OVER() clause that contains a PARTITION BY clause and an ORDER BY clause. Partitions are defined by customer, year, and account name. The result set of the partition is sorted by account month in ascending order. There's the WHERE clause that filters the results by year greater than or equal to 2013.

Please refer to the partial results in Figure 7-12.

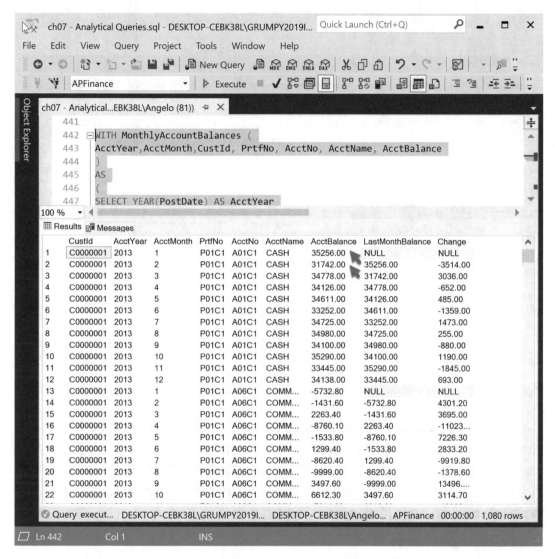

Figure 7-12. *Account balances for current-month and last-month analysis*

The report works correctly. We see the current and prior month's balances and the difference between the balances. Let's place these results, for cash accounts only and for the year 2013, into a Microsoft Excel spreadsheet chart.

Please refer to the Excel chart screenshot in Figure 7-13.

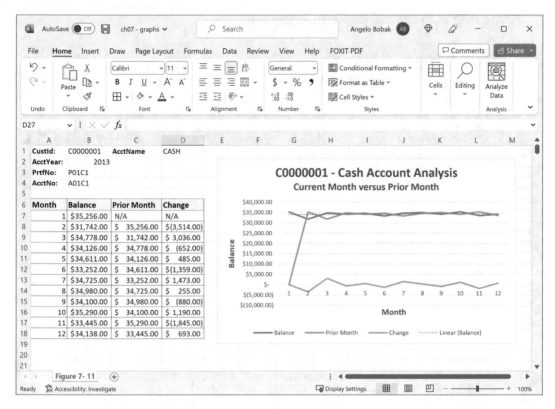

Figure 7-13. *Account balances for current-month and last-month analysis chart*

Visually we can see that this account remained constant at around $35,000 each month after an initial jump. There seems to be a slight trend upward in the balances, and the changes each month were between minus $5000 and plus $5000 (approximately).

For one last item, to see the values as rolling values across all years, modify the OVER() clause with the following code snippet:

```
,LAG(AcctBalance) OVER (
        PARTITION BY CustId,AcctName
        ORDER BY CustId,AcctYear,AcctMonth,AcctName
    ) AS LastMonthBalance
    ,AcctBalance -
        (
        LAG(AcctBalance) OVER (
        PARTITION BY CustId,AcctName
        ORDER BY CustId,AcctYear,AcctMonth,AcctName
```

```
    )
) AS Change
```

There are slight modifications to the PARTITION BY and ORDER BY clauses you need to make.

Here comes a question:

Referring to the preceding code snippet, is the way that the account balance difference between the current account balance and the last month's account balance is calculated efficient, or will it negatively affect performance?

Let's see.

Performance Considerations

Let's generate the baseline estimated execution plan in the usual manner, from the query selection in the menu bar or the estimated execution plan icon, whichever method you like.

Please refer to Figure 7-14.

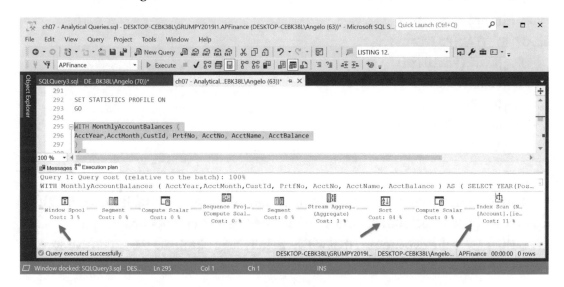

Figure 7-14. *Baseline estimated execution plan for the LAG() query*

Reading right to left, we are off to a good start. We see an index scan at 11% cost, but then we see an expensive sort at 84%. The compute scalar task between the two is for pulling out the year and month from the PostDate column. The sorting task is for the GROUP BY clause in the CTE.

Let's try the reporting table approach. I used the CTE query to create a reporting table in a new schema called `FinancialReports`. Also, we will create two indexes: one is a clustered index, and one is a non-clustered index.

Please refer to Listing 7-6.

Listing 7-6. Creating the Account Monthly Balances Report Table

```
CREATE SCHEMA FinancialReports
GO

SELECT YEAR(PostDate) AS AcctYear
      ,MONTH(PostDate) AS AcctMonth
      ,CustId
      ,PrtfNo
      ,AcctNo
      ,AcctName
      ,SUM(AcctBalance) AS AcctBalance
INTO FinancialReports.AccountClustered
FROM Financial.Account
GROUP BY YEAR(PostDate)
      ,MONTH(PostDate)
      ,CustId
      ,PrtfNo
      ,AcctNo
      ,AcctName
GO
```

This batch script uses the `SELECT/INTO` command to create the table on the fly. The chapter script code includes the truncate/load table with the `INSERT` command.

Listing 7-7 is the two indexes that will hopefully help eliminate the expensive sort step in the plan.

Listing 7-7. Creating the Clustered and Non-clustered Indexes

```
CREATE CLUSTERED INDEX ieYearMonthCustIdPrtfNoAcctNoAcctName
ON FinancialReports.AccountClustered (
[AcctYear],[AcctMonth],[CustId],[PrtfNo],[AcctNo],[AcctBalance]
)
```

```
GO

CREATE INDEX ieCustIdAcctYearAcctMonthAcctNameAcctName
ON FinancialReports.AccountClustered (
[CustId],[AcctYear],[AcctName],[AcctMonth]
)
GO
```

Listing 7-8 is the modified query so it accesses the new report table. Results are the same, so I will not show the output screenshot (always check when you do modifications – you never know).

Listing 7-8. Using the Account Report Table with Clustered and Non-clustered Indexes

```
SELECT CustId
      ,AcctYear
      ,AcctMonth
      ,PrtfNo
      ,AcctNo
      ,AcctName
      ,AcctBalance
      ,LAG(AcctBalance) OVER (
            PARTITION BY CustId,AcctYear,AcctName
            ORDER BY AcctMonth
      ) AS LastMonthBalance
      ,AcctBalance -
            (
            LAG(AcctBalance) OVER (
            PARTITION BY CustId,AcctYear,AcctName
            ORDER BY AcctMonth
            )
      ) AS Change
FROM FinancialReports.AccountClustered
WHERE AcctYear >= 2013
GO
```

Let's compare the query plans for the original query and the report table version.

Please refer to Figure 7-15.

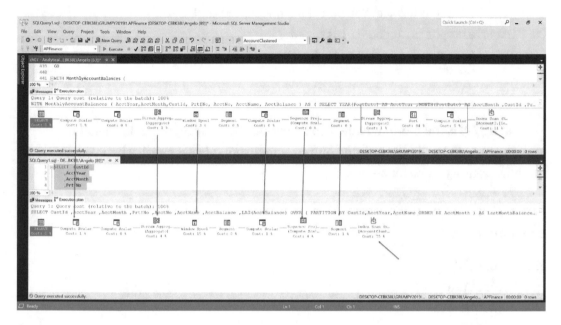

Figure 7-15. *CTE query vs. report table query plans*

The compute scalar (0%), sort (84%), and stream aggregate (1%) tasks in the top original estimated query plan were eliminated in the new plan with the new report table. Having fewer steps is good!

In the new plan though, the index scan went up to 75% from 11%, the segment task went up to 1% from 0%, the sequence project went up to 4% from 0%, the compute scalar went up to 1% from 0%, and the window spool went up to 15% from 3%. Lastly, the stream aggregate went up to 4% from 1%, and the last compute scalar went up to 1% from 0%.

Although tasks were eliminated, the improvement does not look promising.

Let's compare the IO and TIME statistics between both versions of the query.

Please refer to Figure 7-16.

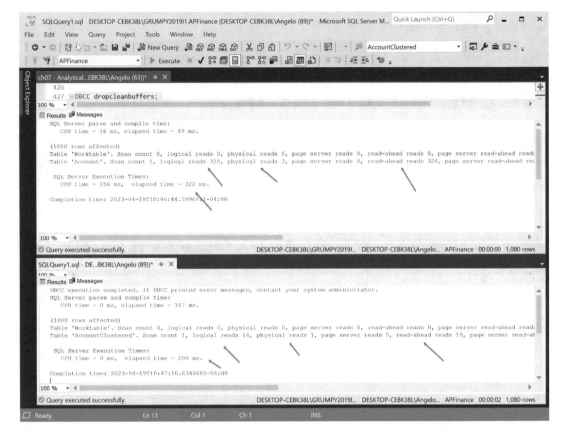

Figure 7-16.   *Comparing IO and TIME statistics*

I marked the statistics we should pay attention to with arrows:

> SQL Server parse and compile times went up from 49 ms to 167 ms.

> SQL Server execution times went down from 322 ms to 200 ms.

The following are the high-value statistics:

> Account table logical reads went from 328 to 14.

> Account table physical reads went from 3 to 1.

> Account table read-ahead reads went from 324 to 19.

Over all these statistics do show an improvement in performance even though the query plans were a bit inconclusive.

Let's use the same logic with the LEAD() function.

439

LEAD( ) Function

Let's now look at LAG( )'s counterpart, the LEAD( ) function. Our analyst has supplied the following business requirements for this next report:

Modify the prior query so that this time the report shows the current month's account balance and the next year's account balance for the same month for all customers. Start at year 2013 and include all years after that. The results should be for the cash account only, and the report needs to be grouped by customer ID, year, month, and account number and name.

Please refer to Listing 7-9.

Listing 7-9. Monthly Account Balance Analysis, Current Year vs. Next Year

```
SELECT CustId
      ,AcctYear
      ,AcctMonth
      ,AcctNo
      ,AcctName
      ,AcctBalance
      ,LEAD(AcctBalance,12,0) OVER (
            PARTITION BY CustId
            ORDER BY AcctName,AcctYear,AcctMonth
      ) AS NextYearBalance
FROM MonthlyAccountBalances
WHERE AcctYear >= 2013
AND AcctNAme = 'CASH'
GO
```

This time we need to pass some parameters to the LEAD( ) function besides the column name. We also pass the value 12 as the number of months to look forward, and the last parameter defines what the function returns when there are no more rows. In this case the function will return 0 instead of NULL.

The partition is set to customer ID, that is, we will have one partition per customer, and the rows in the partition are sorted by account name, account year, and account month. The WHERE clause filter pulls rows for years 2013 onward, and we are only interested in the cash account.

Pop Quiz   How do we modify the query if we want to see all the account types? Any changes required in the OVER() clause?

Let's look at the partial results in Figure 7-17.

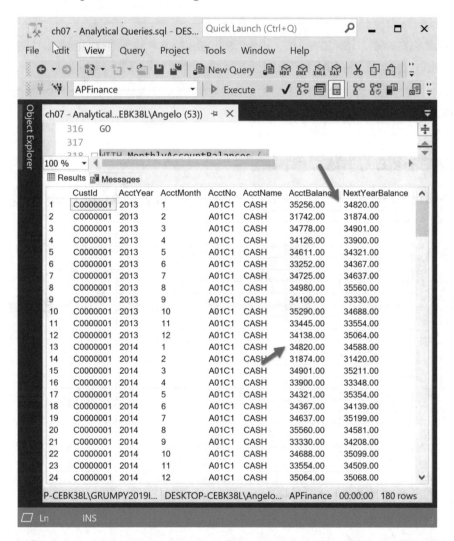

Figure 7-17. *Current and next year's monthly account balance analysis report*

Looks like it works. The next year's balance for the first row points to row 13, which is for January 2014. So we looked forward a year. The remaining row values point to the next year's values correctly. Try to download this query in the chapter script and modify it so we see the month abbreviations instead of numbers. Also include a calculation for the difference between this month's balance and next year's balance for the same month.

Pop Quiz Answer   Remove the filter predicate in the WHERE clause and include the account type in the PARTITION BY clause in the OVER() clause.

Performance tuning time.

Performance Considerations

Figure 7-18 is our baseline estimated query performance plan for the query we just discussed.

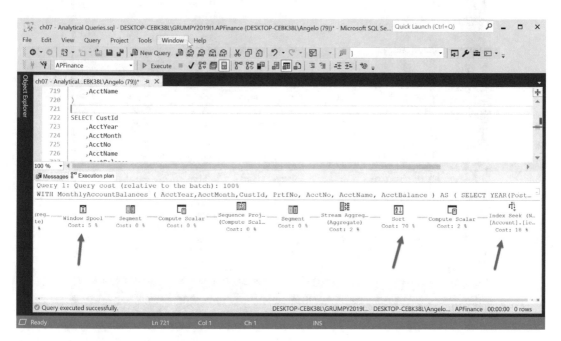

Figure 7-18. *Estimated query plan for the LEAD() function query*

We have seen this pattern many times now, the index seek or table scan followed by a sort task. In this case the index seek (caused by the WHERE clause) is at 18% cost, and the sort comes in at 70%. What stands out is the window spool task at 5%. Usually, we see 1%, but seems like this function does a lot of work in this particular query. Let's collect some more statistics.

Please refer to Figure 7-19.

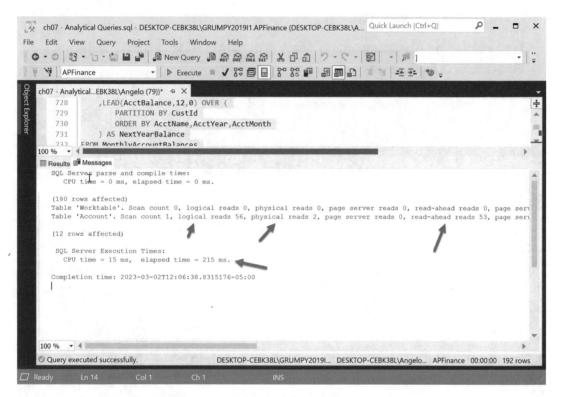

Figure 7-19. *IO and TIME statistics for the LEAD() function query*

Starting at the bottom, this query has the following statistics for SQL Server execution times:

CPU time: 15 ms

Elapsed time: 215 ms

There are all 0 values for the work table, but the Account table has the following statistics:

- Scan count: 1

- Logical reads: 56

- Physical reads: 2

- Read-ahead reads: 53

Looks like some IO is going on. Let's look at the STATISTICS PROFILE statistics. Please refer to Figure 7-20.

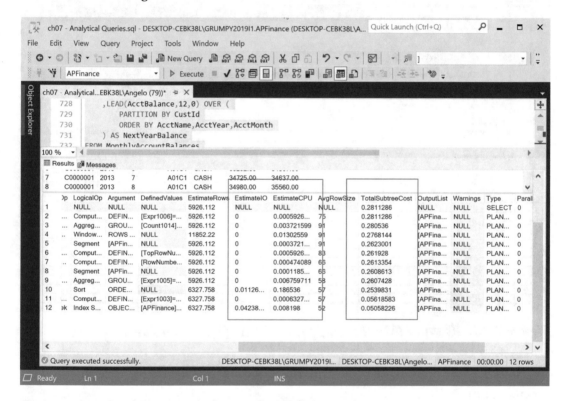

Figure 7-20. *Profile statistics for the LEAD() function query*

Looking at these statistics, we see almost all zero values for the EstimateIO statistics, which include the window spool. The window spool cost is 5% although the EstimateIO statistic is zero. The window spool cost is for the following logic that explains the default ROWS frame clause for the partition:

```
--Window Spool(ROWS BETWEEN:([TopRowNumber1016], [BottomRowNumber1017]))
```

Finally, the `EstimateCPU` costs are low, and the `TotalSubtreeCost` values are approximately .2 each. Looks like there is not much that can be done as an index was available and used. Maybe we can improve performance by modifying the logic so that the account data can be loaded in memory via a memory-enhanced table.

Memory-Optimized Strategy

Let's create a memory-optimized table and load the account information and run our query against it. We will check out the estimated query plan to see if there is improvement. Listing 7-10 is the SQL DDL command to create the table.

Listing 7-10. Create a Memory-Optimized Table

```
CREATE TABLE [FinancialReports].[AccountMonthlyBalancesMem]
(
       [AcctYear]    [int] NULL,
       [AcctMonth]   [int] NULL,
       [CustId]      [varchar](8) NOT NULL,
       [PrtfNo]      [varchar](8) NOT NULL,
       [AcctNo]      [varchar](64) NOT NULL,
       [AcctName]    [varchar](24) NOT NULL,
       [AcctBalance] [decimal](10, 2) NULL,

       INDEX [ieMonthlyAcctBalanceMemory] NONCLUSTERED
       (
       [AcctYear],
       [AcctMonth],
       [CustId],
       [PrtfNo],
       [AcctNo]
       )
)WITH ( MEMORY_OPTIMIZED = ON , DURABILITY = SCHEMA_ONLY )
GO
```

The table is called `AccountMonthlyBalancesMem`. Notice that we need to include the index declaration, and also notice the `WITH` directive that contains the settings to make this a memory-optimized table. Listing 7-11 is the simple `INSERT` command to load the table.

Listing 7-11. Load the Memory-Optimized Table

```
INSERT INTO [FinancialReports].[AccountMonthlyBalancesMem]
SELECT YEAR(PostDate)  AS AcctYear
      ,MONTH(PostDate) AS AcctMonth
      ,CustId
      ,PrtfNo
      ,AcctNo
      ,AcctName
      ,CONVERT(DECIMAL(10,2),SUM(AcctBalance)) AS AcctBalance
FROM Financial.Account
GROUP BY YEAR(PostDate)
      ,MONTH(PostDate)
      ,CustId
      ,PrtfNo
      ,AcctNo
      ,AcctName
GO
```

Nothing out of the ordinary with this INSERT command, so let's check out the query we will analyze. Please refer to Listing 7-12.

Listing 7-12. Using LEAD() and LAG() Against a Memory-Optimized Table

```
SELECT CustId
      ,AcctYear
      ,AcctMonth
      ,AcctNo
      ,AcctName
      ,AcctBalance
      ,LEAD(AcctBalance,12,0) OVER (
            PARTITION BY CustId
            ORDER BY AcctName,AcctYear,AcctMonth
      ) AS NextYearBalance
      ,LAG(AcctBalance,12,0) OVER (
            PARTITION BY CustId
            ORDER BY AcctName,AcctYear,AcctMonth
```

```
    ) AS LastYearBalance
FROM [FinancialReports].[AccountMonthlyBalancesMem]
WHERE AcctYear >= 2013
AND AcctNAme = 'CASH'
GO
```

Same query as before, but we use the memory-optimized table and added a LAG() section for the last year's balance for the same month. Recall that we partition by customer and sort the partition result set by account name, year, and month. Final requirements are satisfied by the WHERE clause filter by cash accounts only and for years equal to or greater than 2013.

No need to look at the results as we saw them in the prior non-memory-enhanced table example as they match except for the added LAG() function results in the output. (I want to save some space in this chapter).

Alert   Whenever you are performing these kinds of major strategy or code changes, always test results against a known set of correct data to make sure things are working as expected. Never take things for granted. Bugs can and will creep in. Don't forget to save a copy of the old script or query!

Let's look at the query plans for both. **This time the original plan is the one that has no indexes, not the one shown in Figure 7-18.** I dropped the index on the physical table for this test. Please refer to Figure 7-21.

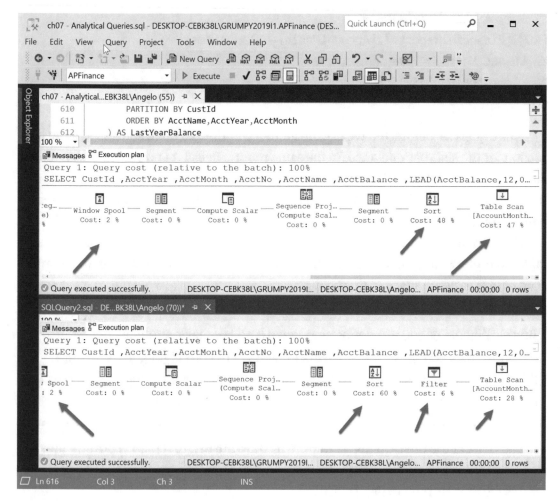

Figure 7-21. *Comparing estimated query plans*

Reading right to left, we see that table scan went down from 47% cost to 28% cost, a filter task with a cost of 6% was added in the memory table version (a little strange), and the sort went up from 48% cost to 60% cost, which seems to be standard with all the "enhanced" strategies we tried. The window spool tasks – there were two of them – stayed the same.

Does not seem to be a significant improvement. The memory tables has a default index but it was not used in the plan. Let's see the IO and TIME statistics.

Please refer to Figure 7-22.

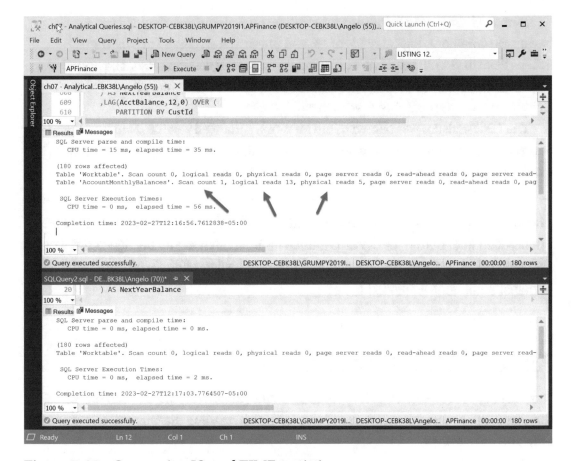

Figure 7-22. *Comparing IO and TIME statistics*

Here is where we see the improvement. The non-memory-enhanced version had a scan count of 1, logical reads of 13, and physical read counts of 5 for the AccountMonthlyBalances table. The memory-enhanced version did not have statistics on this table because it is in memory. The work table statistics were all zero, so there is significant improvement shown with these statistics.

The TIME statistics went down in the memory-enhanced version. Notice that SQL Server execution time went from 56 ms to 2 ms in the memory-enhanced version, so I would say this strategy was worth the development time.

Note   These examples were run on a laptop with 16 GB of memory and seventh-generation Intel Core i7 CPU. Performance improvement would be significant when running on large servers with multiple CPUs and having multiple queries launched against a memory-enhanced table.

PERCENT_RANK( ) Function

Next on deck is the PERCENT_RANK( ) function. What does it do? We covered this in Chapter 3, but in case you need a reminder or are not reading the chapters in sequential order, here is a review.

Using a data set like results from a query, partition, or table variable, this function calculates the relative rank of each individual value relative to the entire data set (as a percentage). You need to multiply the results (the value returned is a float) by 100.00 or use the FORMAT( ) function to convert the results to a percentage (this function also returns a FLOAT(53) data type).

Also, it works a lot like the CUME_DIST( ) function; at least that's what the Microsoft documentation says.

You will also see that it works a lot like the RANK( ) function (in theory anyway). The RANK( ) function returns the position of the value, while PERCENT_RANK( ) returns a percentage value. We will use this function together with the CUME_DIST( ) function just discussed so we can compare the results and also see what effect it will have on the query plan we will generate shortly.

Our analyst has submitted a simple business requirement for this next report. We need to create a query that shows the percent rank, dense rank, and rank values for account balances by customer ID, year, account number and name, and month. Results need to be for years greater than or equal to 2013.

Luckily for us, we have the memory-enhanced account balance table loaded. Let's take a quick look at the query that will satisfy this request.

Please refer to Listing 7-13.

Listing 7-13. Customer Yearly Account Balance Monthly Ranking

```
SELECT CustId
      ,AcctYear
      ,[AcctMonth]
      ,AcctNo
      ,AcctName
      ,AcctBalance
      ,PERCENT_RANK() OVER (
            PARTITION BY CustId,AcctYear,AcctMonth
            ORDER BY AcctBalance DESC
      ) AS PercentRank
      ,DENSE_RANK() OVER (
            PARTITION BY CustId,AcctYear,AcctMonth
            ORDER BY AcctBalance DESC
      ) AS DenseRank
      ,RANK() OVER (
            PARTITION BY CustId,AcctYear,AcctMonth
            ORDER BY AcctBalance DESC
      ) AS Rank
FROM [FinancialReports].[AccountMonthlyBalancesMem]
WHERE AcctYear >= 2013
GO
```

The query puts the PERCENT_RANK() function through its paces, but we also include the DENSE_RANK() and RANK() functions so we can compare results. The OVER() clause in all cases uses a PARTITION BY clause that creates partitions by customer ID, account year, and month. Partition results are sorted by the AcctBalance column. Let's see the results.

Please refer to Figure 7-23.

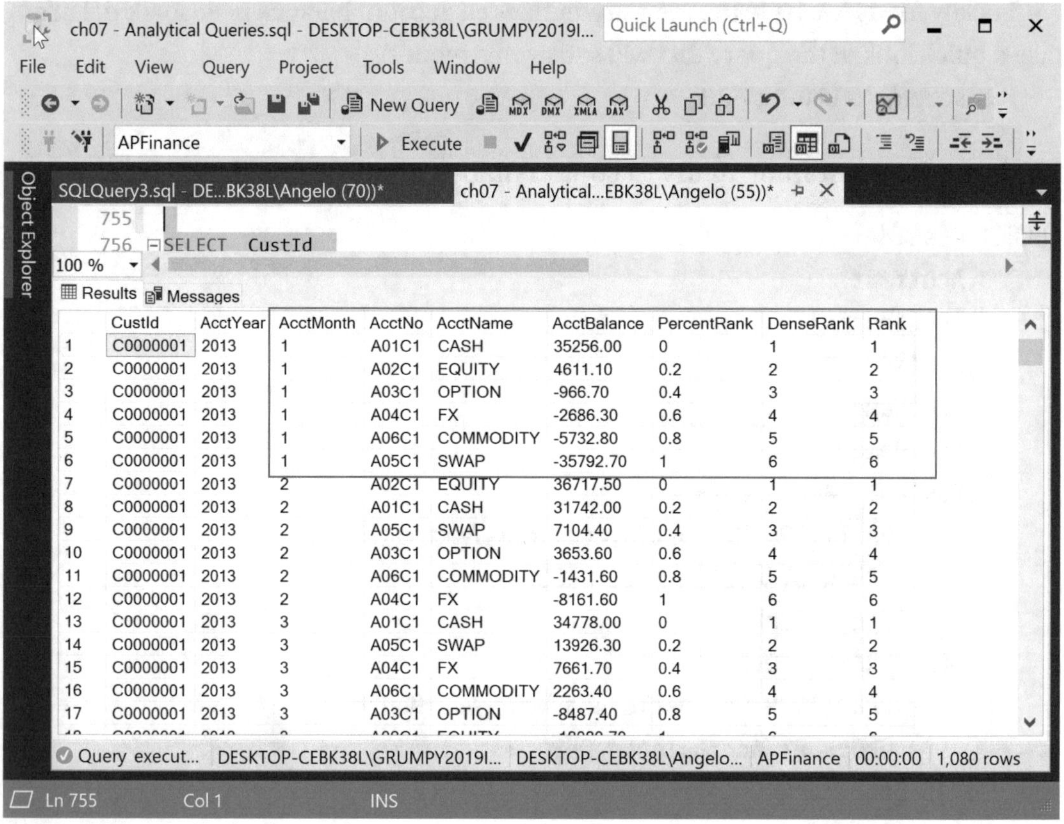

Figure 7-23. *Customer account rank analysis*

We can see that the cash account is top for January 2013 with the equity account at number 2, but not by much. The rest of the accounts show negative balances. What does this mean? Maybe they all cashed out and the profits are in the cash account.

February shows all positive account positions except for commodity and FX (foreign exchange) accounts.

These are interesting patterns that our analyst needs to study and understand for them to make investment suggestions based on the findings or analyze why things went wrong.

Let's turn our attention to some performance tuning and strategies.

Performance Considerations

Figure 7-24 is the first cut of the estimated query plan. The plan is fairly large and has a pattern that repeats itself twice, one for each window function in the query.

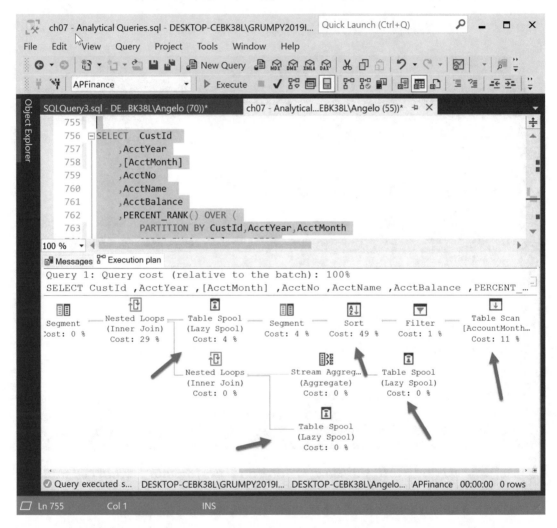

Figure 7-24. *Estimated query plan for the rank analysis report*

Notice the absence of window spool tasks. Remember that we are using a memory-enhanced table in the query, so the table scan is of no concern at 11% cost. There is also a sort task at 49%, which is a bit high. Next, we see three table spools, two at 0% cost and one at 4% cost. These are followed by a nested loops (inner join) task at a cost of 29% that merges the data streams.

Next are segment, sequence project, and compute scalar tasks, and the pattern repeats itself for the second window function. Do run this plan and see what these tasks do. Also get more explanation in the StmtText column in the PROFILE statistics.

Let's look at IO and TIME statistics next. Please refer to Figure 7-25.

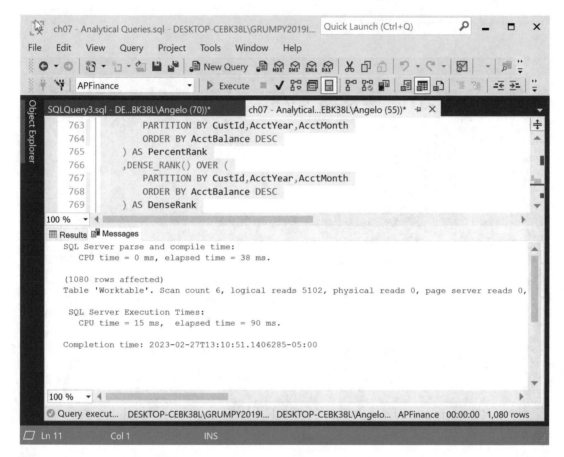

Figure 7-25. *IO and TIME statistics for rank analysis (memory-enhanced table)*

As this is a memory-enhanced table, there are no statistics for the AccountMonthlyBalancesMem table, only for the work table.

Based on the plan and statistics we looked at, we can take advantage of this strategy for all our examples. Let's move on to the next set of functions.

PERCENTILE_CONT() and PERCENTILE_DISC()

What do these two functions do? Well, recall that they work on continuous and discrete sets of data.

Recall what continuous data is. It is data that changes over specific periods of time, like sales amounts of time periods over time, for example, total sales in a 12-month period for a specific product. Continuous data is data measured over a period of time, like boiler temperatures in equipment over a 24-hour period.

Summing up each boiler temperature to arrive at a total makes no sense. Summing up the boiler temperatures and then dividing them by the time period to get an average does.

What is discrete data anyway?

Discrete data values are numbers that have values that you can count over time, like investment account balances over days, months, and years. You can add these up to get a grand total and average and identify trends in account balances over time.

In conclusion, the PERCENTILE_CONT() function works on a continuous set of data based on a required percentile so as to return a value within the data set that satisfies the percentile (you need to supply a percentile as a parameter, like .25, .50, .75, etc.).

With this function the value is interpolated. It usually does not exist, so it is introduced. Or by coincidence it could use one of the values in the data set if the numbers line up correctly.

The PERCENTILE_DISC() function works on a discrete set of data based on a required percentile so as to return a value within the data set that satisfies the percentile. With this function the value exists in the data set; it is not interpolated like in the PERCENTILE_ CONT() function (both functions return a FLOAT(53) data type value).

PERCENTILE_CONT()

Let's start off with a simple example before we dive into some analysis of our financial data. Our analyst submits the following business requirements for this next report:

Create a report that shows option (a type of derivative instrument) account balances by customer, year, month, account number and name, and continuous percentiles for 25%, 50%, 75%, and 90% percentiles. Initially we will show only our friend customer C0000001, but the report will be used later on for more than one customer. The query needs to be flexible so we can meet all these requirements. Lastly, the query needs to run in less than one second (our first performance requirement)!

Please refer to Listing 7-14 for the solution.

Listing 7-14. Customer Continuous Percentiles Query

```
SELECT CustId
      ,AcctYear
      ,AcctMonth
      ,AcctNo
      ,AcctName
      ,AcctBalance
      ,PERCENTILE_CONT(.25)
            WITHIN GROUP (ORDER BY AcctBalance)
            OVER (
            PARTITION BY CustId,AcctYear
            ) AS [PercentCont-25%]
      ,PERCENTILE_CONT(.50)
            WITHIN GROUP (ORDER BY AcctBalance)
            OVER (
            PARTITION BY CustId,AcctYear
            ) AS [PercentCont-50%]
      ,PERCENTILE_CONT(.75)
            WITHIN GROUP (ORDER BY AcctBalance)
            OVER (
            PARTITION BY CustId,AcctYear
            ) AS [PercentCont-75%]
      ,PERCENTILE_CONT(.90)
            WITHIN GROUP (ORDER BY AcctBalance)
            OVER (
            PARTITION BY CustId,AcctYear
      ) AS [PercentCont-90%]
FROM [FinancialReports].[AccountMonthlyBalancesMem]
WHERE Acctname = 'OPTION'
AND CustId = 'C0000001'
GO
```

Notice that this query looks slightly different from our usual queries that include an OVER() clause. The function is included four times in the SELECT clause, but the values for each desired percentile are passed as parameters. Notice also we are using the memory-enhanced table, so this should be fast!

Prior to the appearance of the OVER() clause, we need to include a WITHIN GROUP (ORDER BY ...) clause. Notice that the ORDER BY clause is not included in the OVER() clause. All we see is the PARTITION BY clause that sets up a partition by customer identifier and account year in each of the four cases. Looks like the partition will be sorted by the AcctBalance column for processing in each case.

A simple WHERE clause follows that filters the results for the option account and for customer C0000001. This query can be repurposed to show results for more than one customer and more than one account, so you might need to modify the OVER() clause.

Note   As usual we start off with a small data set to make sure our query works correctly, after which we can increase the scope.

Please refer to the partial results in Figure 7-26.

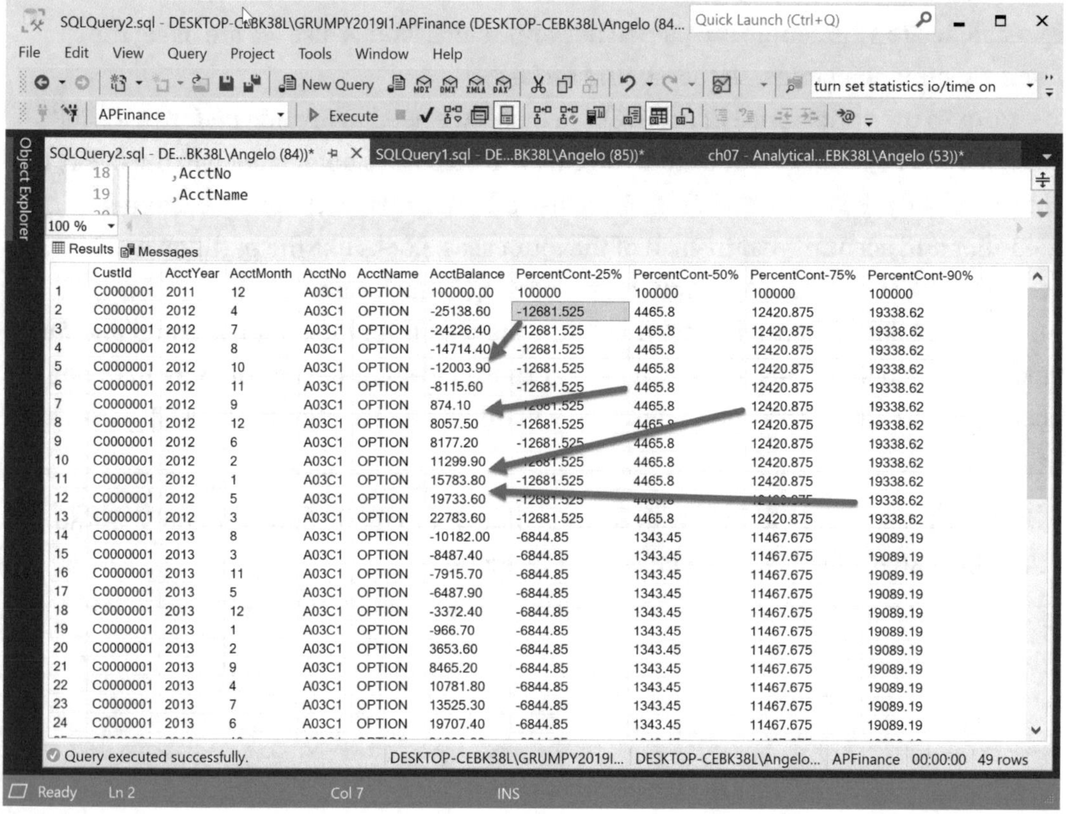

	CustId	AcctYear	AcctMonth	AcctNo	AcctName	AcctBalance	PercentCont-25%	PercentCont-50%	PercentCont-75%	PercentCont-90%
1	C0000001	2011	12	A03C1	OPTION	100000.00	100000	100000	100000	100000
2	C0000001	2012	4	A03C1	OPTION	-25138.60	-12681.525	4465.8	12420.875	19338.62
3	C0000001	2012	7	A03C1	OPTION	-24226.40	-12681.525	4465.8	12420.875	19338.62
4	C0000001	2012	8	A03C1	OPTION	-14714.40	-12681.525	4465.8	12420.875	19338.62
5	C0000001	2012	10	A03C1	OPTION	-12003.90	-12681.525	4465.8	12420.875	19338.62
6	C0000001	2012	11	A03C1	OPTION	-8115.60	-12681.525	4465.8	12420.875	19338.62
7	C0000001	2012	9	A03C1	OPTION	874.10	-12681.525	4465.8	12420.875	19338.62
8	C0000001	2012	12	A03C1	OPTION	8057.50	-12681.525	4465.8	12420.875	19338.62
9	C0000001	2012	6	A03C1	OPTION	8177.20	-12681.525	4465.8	12420.875	19338.62
10	C0000001	2012	2	A03C1	OPTION	11299.90	-12681.525	4465.8	12420.875	19338.62
11	C0000001	2012	1	A03C1	OPTION	15783.80	-12681.525	4465.8	12420.875	19338.62
12	C0000001	2012	5	A03C1	OPTION	19733.60	-12681.525	4465.8	12420.875	19338.62
13	C0000001	2012	3	A03C1	OPTION	22783.60	-12681.525	4465.8	12420.875	19338.62
14	C0000001	2013	8	A03C1	OPTION	-10182.00	-6844.85	1343.45	11467.675	19089.19
15	C0000001	2013	3	A03C1	OPTION	-8487.40	-6844.85	1343.45	11467.675	19089.19
16	C0000001	2013	11	A03C1	OPTION	-7915.70	-6844.85	1343.45	11467.675	19089.19
17	C0000001	2013	5	A03C1	OPTION	-6487.90	-6844.85	1343.45	11467.675	19089.19
18	C0000001	2013	12	A03C1	OPTION	-3372.40	-6844.85	1343.45	11467.675	19089.19
19	C0000001	2013	1	A03C1	OPTION	-966.70	-6844.85	1343.45	11467.675	19089.19
20	C0000001	2013	2	A03C1	OPTION	3653.60	-6844.85	1343.45	11467.675	19089.19
21	C0000001	2013	9	A03C1	OPTION	8465.20	-6844.85	1343.45	11467.675	19089.19
22	C0000001	2013	4	A03C1	OPTION	10781.80	-6844.85	1343.45	11467.675	19089.19
23	C0000001	2013	7	A03C1	OPTION	13525.30	-6844.85	1343.45	11467.675	19089.19
24	C0000001	2013	6	A03C1	OPTION	19707.40	-6844.85	1343.45	11467.675	19089.19

Figure 7-26. *Percentile continuous for various percentiles report*

What stands out in this report? The percentile values are interpolated. Notice that the arrows point to the two closest values, for example, 19338.62 falls between 15783.80 and 19733.60. The value 19338.62 does not exist in the actual column values.

Remember that this is the behavior of the function.

This is not the rule though. There can be data patterns that will result in the percentile continuous values matching the column values. Most of the time this does not happen. Look at the results again for the value $100,000.00. The column value and percentile values are equal. That's because there is only one row in this partition. So this is a scenario that is an exception to the way that the function behaves.

In my opinion, this report does give us a good idea of the spread of values.

Performance Considerations

Let's generate an estimated query plan in the usual manner. We will examine the right portion only as the execution pattern repeats itself four times, once for each `PERCENTILE_CONT()` function call in the SELECT clause (for each percentage parameter). There is actually a fifth repeating pattern right after the initial index seek to set up the data. These include a sort step and a few of the usual table spool tasks.

Please refer to Figure 7-27.

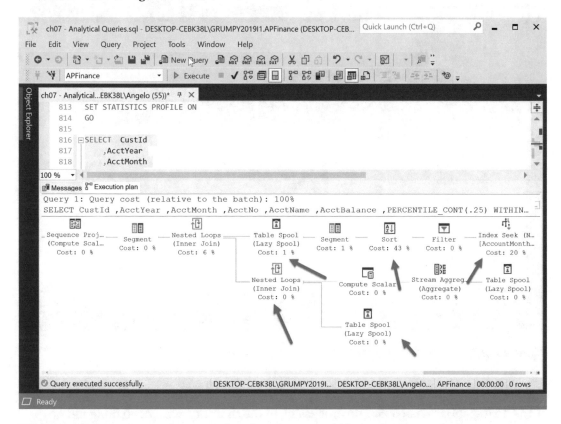

Figure 7-27. *Estimated query plan for the percentile continuous report*

An existing index satisfies the query requirement, so we see an index seek with a cost of 20%. There is the filter for the WHERE clause and the expensive sort at 43%. In the lower two branches, we see two table spools that generate data flows for the nested loops join, all with a cost of 0%.

A second nested loops join merges the top and bottom flows at a cost of 6%. As I said earlier, the remaining plan sections are similar and repeat themselves except there is no index seek and no sort task.

Next, let's look at the TIME and IO statistics. Please refer to Figure 7-28.

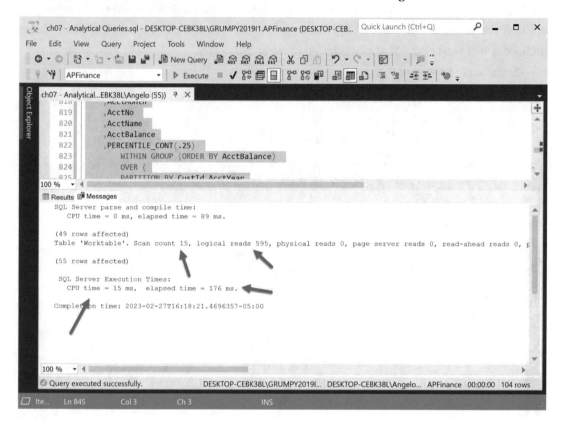

Figure 7-28. *IO and TIME statistics*

As can be seen, we only have statistics for the work table that include a scan count of 15 and logical reads of 595. The processing of the memory-enhanced table did not produce any statistics. Lastly, the SQL Server elapsed execution time is 176 ms.

Please refer to Figure 7-29 for partial `STATISTICS PROFILE` statistics.

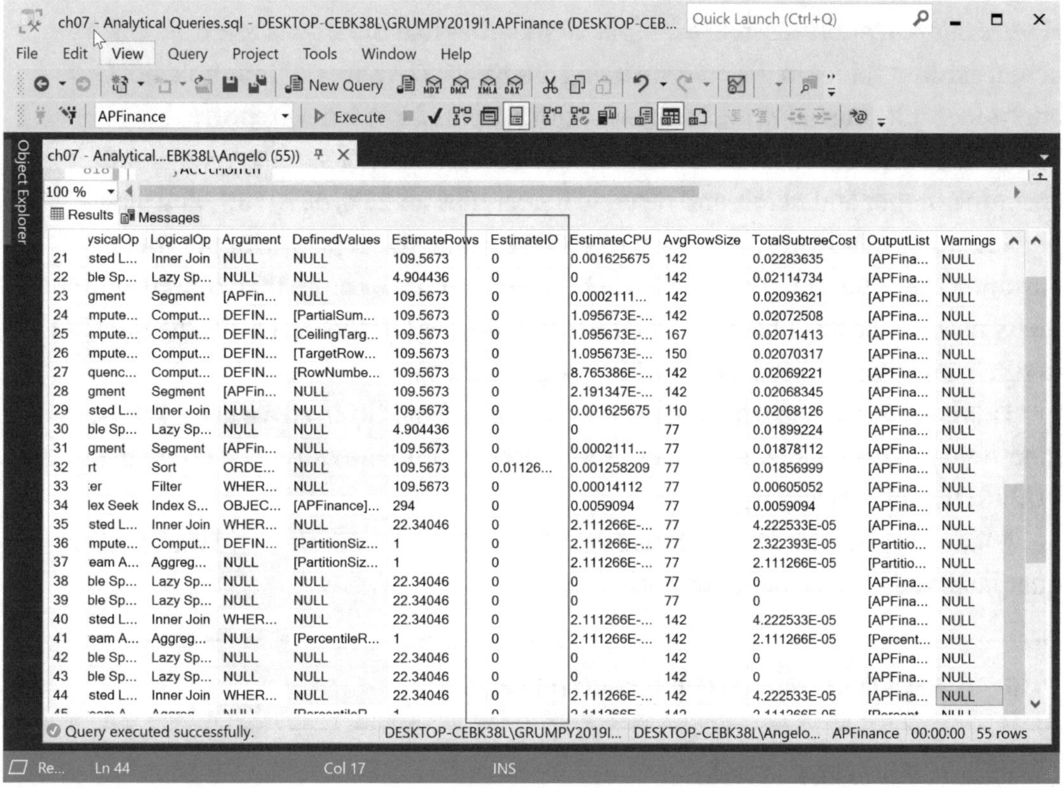

	ysicalOp	LogicalOp	Argument	DefinedValues	EstimateRows	EstimateIO	EstimateCPU	AvgRowSize	TotalSubtreeCost	OutputList	Warnings
21	sted L...	Inner Join	NULL	NULL	109.5673	0	0.001625675	142	0.02283635	[APFina...	NULL
22	ble Sp...	Lazy Sp...	NULL	NULL	4.904436	0	0	142	0.02114734	[APFina...	NULL
23	gment	Segment	[APFin...	NULL	109.5673	0	0.0002111...	142	0.02093621	[APFina...	NULL
24	mpute...	Comput...	DEFIN...	[PartialSum...	109.5673	0	1.095673E-...	142	0.02072508	[APFina...	NULL
25	mpute...	Comput...	DEFIN...	[CeilingTarg...	109.5673	0	1.095673E-...	167	0.02071413	[APFina...	NULL
26	mpute...	Comput...	DEFIN...	[TargetRow...	109.5673	0	1.095673E-...	150	0.02070317	[APFina...	NULL
27	quenc...	Comput...	DEFIN...	[RowNumbe...	109.5673	0	8.765386E-...	142	0.02069221	[APFina...	NULL
28	gment	Segment	[APFin...	NULL	109.5673	0	2.191347E-...	142	0.02068345	[APFina...	NULL
29	sted L...	Inner Join	NULL	NULL	109.5673	0	0.001625675	110	0.02068126	[APFina...	NULL
30	ble Sp...	Lazy Sp...	NULL	NULL	4.904436	0	0	77	0.01899224	[APFina...	NULL
31	gment	Segment	[APFin...	NULL	109.5673	0	0.0002111...	77	0.01878112	[APFina...	NULL
32	rt	Sort	ORDE...	NULL	109.5673	0.01126...	0.001258209	77	0.01856999	[APFina...	NULL
33	:er	Filter	WHER...	NULL	109.5673	0	0.00014112	77	0.00605052	[APFina...	NULL
34	lex Seek	Index S...	OBJEC...	[APFinance]...	294	0	0.0059094	77	0.0059094	[APFina...	NULL
35	sted L...	Inner Join	WHER...	NULL	22.34046	0	2.111266E-...	77	4.222533E-05	[APFina...	NULL
36	mpute...	Comput...	DEFIN...	[PartitionSiz...	1	0	2.111266E-...	77	2.322393E-05	[Partitio...	NULL
37	eam A...	Aggreg...	NULL	[PartitionSiz...	1	0	2.111266E-...	77	2.111266E-05	[Partitio...	NULL
38	ble Sp...	Lazy Sp...	NULL	NULL	22.34046	0	0	77	0	[APFina...	NULL
39	ble Sp...	Lazy Sp...	NULL	NULL	22.34046	0	0	77	0	[APFina...	NULL
40	sted L...	Inner Join	WHER...	NULL	22.34046	0	2.111266E-...	142	4.222533E-05	[APFina...	NULL
41	eam A...	Aggreg...	NULL	[PercentileR...	1	0	2.111266E-...	142	2.111266E-05	[Percent...	NULL
42	ble Sp...	Lazy Sp...	NULL	NULL	22.34046	0	0	142	0	[APFina...	NULL
43	ble Sp...	Lazy Sp...	NULL	NULL	22.34046	0	0	142	0	[APFina...	NULL
44	sted L...	Inner Join	WHER...	NULL	22.34046	0	2.111266E-...	142	4.222533E-05	[APFina...	NULL

***Figure 7-29.** Profile statistics for the percentile continuous query*

I do not know if you noticed it, but with memory-enhanced tables, we see 0 or very low `EstimateIO` statistics but higher `EstimateCPU` statistics. This is a partial output as there are 55 steps in total, but all the IO is zero except for one step! The `EstimateCPU` and `TotalSubtreeCost` vary per step, the highest for the `WHERE` clause and for the nested loops joins.

As far as performance analysis plans are concerned, I think this one is very interesting and illustrates the power of memory-enhanced tables.

Question Do the higher `EstimateCPU` statistics imply you will need powerful CPUs when taking advantage of memory-enhanced tables? Something to consider.

PERCENTILE_DISC Function

Here is our percentile discrete example. Remember that the values returned will be actual existing values in the column being analyzed. They are not interpolated. Our analyst submits the following business requirements for this next report:

Create a report that shows option account balances by customer, year, month, account number and name, and discrete percentiles for 25%, 50%, 75%, and 90% percentiles by modifying the prior query. Once again, this report is for our friend customer C0000001, but the report will be used later on for more than one customer. The query needs to be flexible so we can meet all these requirements. Lastly, the query needs to run in less than one second, just like the prior query.

This is great. We can copy the prior example and modify it, so we use a CTE vs. the memory-enhanced table. This gives us a chance to compare estimated query plans and STATISTICS PROFILE statistics.

A little copy and paste action, use the CTE from prior examples, and we have our query that satisfies the requirement.

Tip   Always code with an eye toward reusability.

Please refer to Listing 7-15.

Listing 7-15. Percentile Discrete, Using the CTE Scheme

```
WITH MonthlyAccountBalances (
AcctYear,AcctMonth,CustId, PrtfNo, AcctNo, AcctName, AcctBalance
)
AS
(
SELECT YEAR(PostDate)  AS AcctYear
      ,MONTH(PostDate) AS AcctMonth
      ,CustId
      ,PrtfNo
      ,AcctNo
      ,AcctName
      ,SUM(AcctBalance)
FROM Financial.Account
```

```
GROUP BY YEAR(PostDate)
      ,MONTH(PostDate)
      ,CustId
      ,PrtfNo
      ,AcctNo
      ,AcctName
)

SELECT CustId
      ,AcctYear
      ,AcctMonth
      ,AcctNo
      ,AcctName
      ,AcctBalance
      ,PERCENTILE_DISC(.25)
            WITHIN GROUP (ORDER BY AcctBalance)
            OVER (
            PARTITION BY CustId,AcctYear
            ) AS [PercentDisc-25%]
      ,PERCENTILE_DISC(.50)
            WITHIN GROUP (ORDER BY AcctBalance)
            OVER (
            PARTITION BY CustId,AcctYear
            ) AS [PercentDisc-50%]
      ,PERCENTILE_DISC(.75)
            WITHIN GROUP (ORDER BY AcctBalance)
            OVER (
            PARTITION BY CustId,AcctYear
            ) AS [PercentDisc-75%]
      ,PERCENTILE_DISC(.90)
            WITHIN GROUP (ORDER BY AcctBalance)
            OVER (
            PARTITION BY CustId,AcctYear
      ) AS [PercentDisc-90%]
FROM FinancialReports.AccountMonthlyBalances
WHERE Acctname = 'OPTION'
```

```
AND CustId = 'C0000001'
ORDER BY CustId
      ,AcctYear
GO
```

As can be seen, we want to generate the discrete percentiles of 25%, 50%, 75%, and also 90% percentiles for account balance spreads for our client and for all years. Once again notice the WITHIN GROUP clause that is specific to this function. The only other function that uses this is the PERCENTILE_CONT() function we just studied (not sure why Microsoft implemented these functions this way, but they must have had a good reason).

Lastly, I included an ORDER BY clause at the end of the query, so output is neatly sorted by customer and year. This is not redundant as the ORDER BY clause in the WITHIN GROUP clause sorts by the account balances. Let's check out the results and keep in mind that these values are not interpolated – they actually exist!

Please refer to the partial results in Figure 7-30.

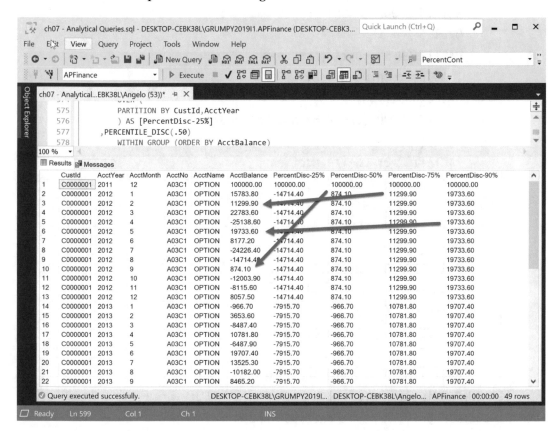

Figure 7-30. *Percentile discrete analysis for account balances*

464

As can be seen the results are **not interpolated**. Each of the percentile values can be traced back to an actual account balance value. For example, looking at the last arrow, the value 19733.60 can be found in the Account Balance column. Notice the difference between this behavior and the behavior of the percentile continuous calculation in the prior example.

Let's see some performance analysis against this query.

Performance Considerations

As usual, let's run an estimated query plan to establish a performance baseline. The plan is a long one, so it will be broken up into several screenshots. The following is the first part starting at the right-hand side. We will take inventory of each section of the plan and then see if we can make some observations and conclusions. Starting at the right-hand sign, we see the following steps in Figure 7-31.

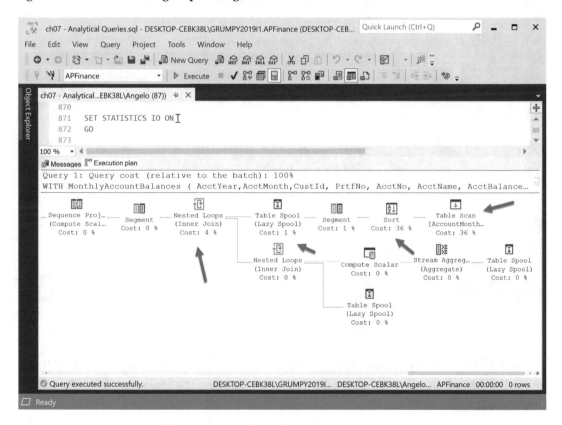

Figure 7-31. *Estimated execution plan, part A*

This looks familiar. Notice that there are no index recommendations although all we see are one table scan against the AccountMonthlyBalances CTE, one sort at a cost of 36%, and two table spools and a nested loops join albeit with a zero cost. The last table spool displays 1% cost (recall table spools temporarily store data in TEMPDB for retrieval by other tasks).

Next, we see another nested loops join operation at a cost of 4%. Let's look at the next section of the plan and see if this pattern repeats itself.

Please refer to Figure 7-32.

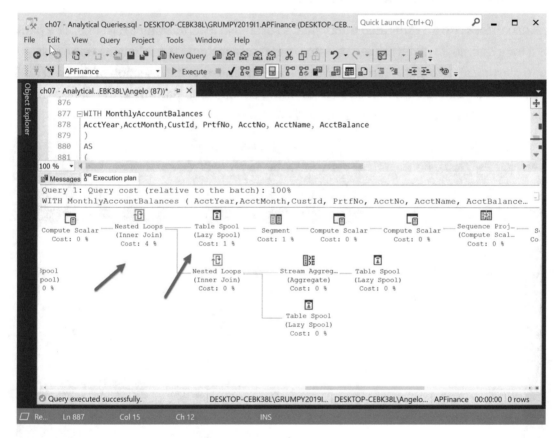

Figure 7-32. *Estimated execution plan, part B*

Looks like we pretty much have the same pattern as the first part of the plan we just discussed except there are no sort tasks or table scans. Can anyone guess what the other sections will look like? Let's see.

Please refer to Figure 7-33.

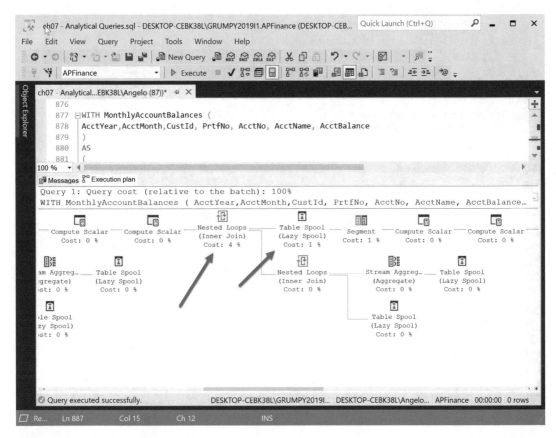

Figure 7-33. *Estimated execution plan, part C*

Yep, same pattern: table spool tasks, some compute scalar tasks, and a nested loops join operation at 4%. If you have not guessed by now, the pattern will repeat itself four times, one for each PERCENTILE_DISC() function call. My initial assumption is that these patterns can only be improved with fast memory and disk or memory-enhanced tables. Let's see the next section.

Please refer to Figure 7-34.

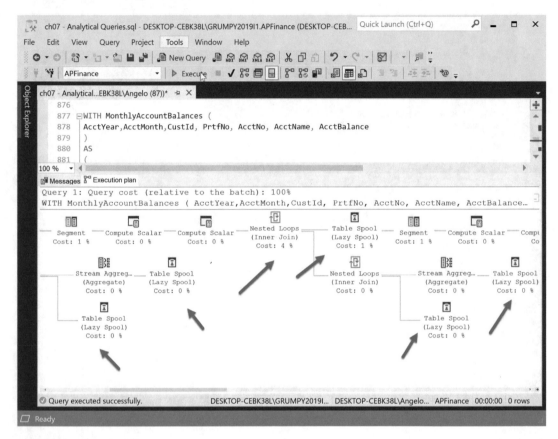

Figure 7-34. *Estimated execution plan, part D*

Same pattern here also: the table spools and nested loops joins are the predominant tasks. I checked the PROFILE STATISTICS, and almost all the table spools have zero IO, so looks like spooling is performed in memory. Also, if you look at the hints for each task, you will see that the cluster of table spools in each section has the same node ID, so it looks like there is reuse of the spooled data in the temporary tables. This seems very efficient.

Let's look at the last part of the plan. Please refer to Figure 7-35.

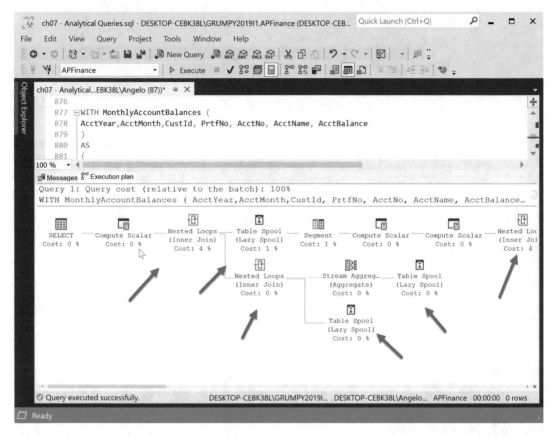

Figure 7-35. *Estimated execution plan, part E*

Same pattern again, with one final table spool, this time at a cost of 1%, and one final nested loops join to merge results (4% cost). The table spool shows a cost of 1% although the STATISTICS PROFILE statistics show 0 IO.

Initial conclusions are that the only way to improve this plan is by

1) Getting more memory

2) Placing TEMPDB on a solid-state drive

3) Loading the Account table into memory via a memory-enhanced table

All good suggestions, but is it worth it?

For one last item, let's generate a live query plan so we can see the flow of data between all the tasks for the last, left-hand side of the plan. Please refer to Figure 7-36.

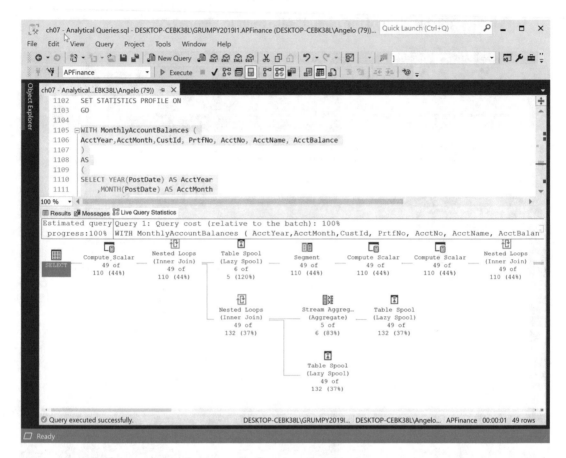

Figure 7-36. *Live query plan*

Here we see the number of rows in the various branches with the final count of 49 rows, which is what we saw in the results of the query pane.

Tip   Always compare estimated query plans against live query plans to make sure values and operators are not too far off.

Recall scheme 3? Let's see how it works. We will create several memory-enhanced tables, one for each year of account balances, and see what that buys us in terms of performance improvement, albeit at the cost of increased complexity.

Multi-memory-enhanced-table Strategy

We will create five memory-enhanced tables, one for each of the years from 2011 to 2015. We will load them, create a VIEW object that ties all five together via UNION ALL operators, and modify the prior query so it goes against the memory-enhanced tables via the new VIEW. Let's start by creating and loading the tables.

Note   Once memory-enhanced tables are created and loaded, you do not need to reload them if you reboot your laptop. They are loaded automatically. You only need to add new rows. Make sure you do not accidentally add duplicate rows!

Listing 7-16 is the code for the first one. The code for the remaining tables is the same except each year part of the table name changes to reflect the remaining years, 2012, 2013, 2014, and 2015.

Listing 7-16. Create Five Memory-Enhanced Tables

```
CREATE TABLE [FinancialReports].[AccountMonthlyBalancesMem2011]
(
        [AcctYear]      [int] NULL,
        [AcctMonth]     [int] NULL,
        [CustId]        [varchar](8) NOT NULL,
        [PrtfNo]        [varchar](8) NOT NULL,
        [AcctNo]        [varchar](64) NOT NULL,
        [AcctName]      [varchar](24) NOT NULL,
        [AcctBalance]   [decimal](10, 2) NOT NULL,

        INDEX [ieMonthlyAcctBalanceMemory2011] NONCLUSTERED
        (
        [CustId],
        [AcctYear],
        [AcctMonth],
        [PrtfNo],
        [AcctNo]
        )
)WITH (MEMORY_OPTIMIZED = ON,DURABILITY = SCHEMA_ONLY)
GO
```

This table is typical for all five tables; all that changes is the name of course. We need to declare an index within the CREATE TABLE command and set the memory-optimized parameters to ON:

```
WITH ( MEMORY_OPTIMIZED = ON , DURABILITY = SCHEMA_ONLY )
```

In the script for this chapter, you will find the DDL command for all five CREATE TABLE commands. Next, we need to load the tables.

Listing 7-17 is the TSQL DML command to insert rows into the first memory table for account balances in 2011. The remaining insert statements are identical except that we change the name of the table to reflect the year we are loading and the WHERE clause predicate to reflect the remaining years' worth of data we need to load.

Listing 7-17. Load the Five Memory-Enhanced Tables

```
INSERT INTO [FinancialReports].[AccountMonthlyBalancesMem2011]
SELECT YEAR(PostDate) AS AcctYear
      ,MONTH(PostDate) AS AcctMonth
      ,CustId
      ,PrtfNo
      ,AcctNo
      ,AcctName
      ,CONVERT(DECIMAL(10,2),SUM(AcctBalance)) AS AcctBalance
FROM Financial.Account
WHERE YEAR(PostDate) = 2011
GROUP BY YEAR(PostDate)
      ,MONTH(PostDate)
      ,CustId
      ,PrtfNo
      ,AcctNo
      ,AcctName
GO
```

This is a standard INSERT/SELECT command with a GROUP BY clause as we are summing up the account balances by year, month, customer, and portfolio. Our last step is to create the TSQL VIEW object.

Yes, these incur costs as far as performance, but since this is historical data, you only need to load them once per year. As new years roll by, you load them once. Now we need to tie all these tables together into one VIEW object.

Listing 7-18 is the DDL command to create a VIEW that connects all the tables via five queries inter-connected with UNION ALL commands.

Listing 7-18. Create the View Based on the Five Memory-Enhanced Tables

```
CREATE VIEW [FinancialReports].[AccountMonthlyBalancesMemView]
AS
SELECT [AcctYear]
      ,[AcctMonth]
      ,[CustId]
      ,[PrtfNo]
      ,[AcctNo]
      ,[AcctName]
      ,[AcctBalance]
FROM [FinancialReports].[AccountMonthlyBalancesMem2011]
UNION ALL
SELECT [AcctYear]
      ,[AcctMonth]
      ,[CustId]
      ,[PrtfNo]
      ,[AcctNo]
      ,[AcctName]
      ,[AcctBalance]
FROM [FinancialReports].[AccountMonthlyBalancesMem2012]
UNION ALL
SELECT [AcctYear]
      ,[AcctMonth]
      ,[CustId]
      ,[PrtfNo]
      ,[AcctNo]
```

```
     ,[AcctName]
     ,[AcctBalance]
FROM [FinancialReports].[AccountMonthlyBalancesMem2013]
UNION ALL
SELECT [AcctYear]
     ,[AcctMonth]
     ,[CustId]
     ,[PrtfNo]
     ,[AcctNo]
     ,[AcctName]
     ,[AcctBalance]
FROM [FinancialReports].[AccountMonthlyBalancesMem2014]
UNION ALL
SELECT [AcctYear]
     ,[AcctMonth]
     ,[CustId]
     ,[PrtfNo]
     ,[AcctNo]
     ,[AcctName]
     ,[AcctBalance]
FROM [FinancialReports].[AccountMonthlyBalancesMem2015]
GO
```

Large script, but it is just a series of four UNION ALL commands that tie the rows from all five tables together via identical queries except for the table names.

Tip   UNION ALL is preferred vs. UNION, which removes duplicates and costs more in terms of performance. Your data should not have unexpected duplicates or triplicates or more!

Last but not least, Listing 7-19 is our query that includes the PERCENTILE_DISC() function that now references the VIEW instead of one large table.

Listing 7-19. Account Balance Report from a Memory Table–Based View

```
SELECT CustId
      ,AcctYear
      ,AcctMonth
      ,AcctNo
      ,AcctName
      ,AcctBalance
      ,PERCENTILE_DISC(.25)
            WITHIN GROUP (ORDER BY AcctBalance)
            OVER (
            PARTITION BY CustId,AcctYear
            ) AS [PercentDisc-25%]
      ,PERCENTILE_DISC(.50)
            WITHIN GROUP (ORDER BY AcctBalance)
            OVER (
            PARTITION BY CustId,AcctYear
            ) AS [PercentDisc-50%]
      ,PERCENTILE_DISC(.75)
            WITHIN GROUP (ORDER BY AcctBalance)
            OVER (
            PARTITION BY CustId,AcctYear
            ) AS [PercentDisc-75%]
      ,PERCENTILE_DISC(.90)
            WITHIN GROUP (ORDER BY AcctBalance)
            OVER (
            PARTITION BY CustId,AcctYear
      ) AS [PercentDisc-90%]
FROM [FinancialReports].[AccountMonthlyBalancesMemView]
WHERE Acctname = 'OPTION'
AND CustId = 'C0000001'
ORDER BY CustId
      ,AcctYear
GO
```

No need to explain the code as we discussed it earlier. All that changes is that we are referencing a VIEW.

The results are the same as the CTE query, so in the interest of space, I will not display another screenshot here.

Performance Considerations

We have a very interesting, estimated execution plan, so make sure you read this next section.

Please refer to Figure 7-37.

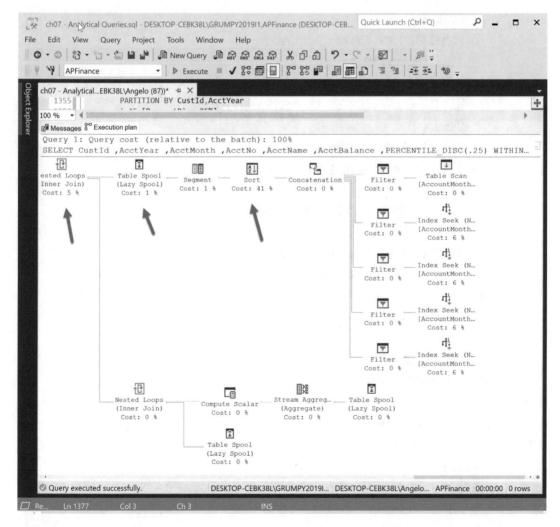

Figure 7-37. *Estimated execution plan for the multi-memory-enhanced-table strategy – part A*

Yes, it is interesting and more complex. We see index seeks against the memory-enhanced tables, but as they are in memory, this should improve performance. We see our old friends, the table spool tasks at 0% cost, the sort at 41% cost, the 1% cost table spool, and the first nested loops inner join. Recall we learned that the temporary tables created by the table spool operator can be reused (*shared* might be a better term) so we do not have a whole bunch of temporary tables taking up space on TEMPDB.

So this gives us one strategy to improve performance: if you cannot eliminate the table spools, then have your dba place TEMPDB on a solid-state drive – this should help.

Let's look at the second part of the plan.

Please refer to Figure 7-38.

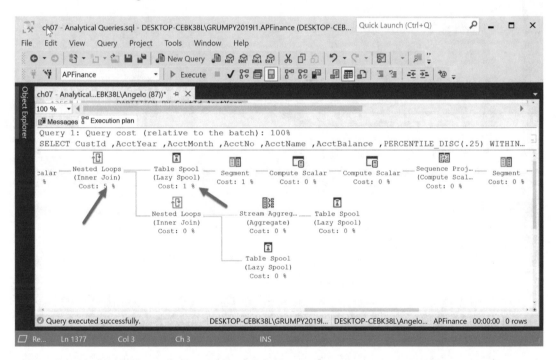

Figure 7-38. *Estimated execution plan for the multi-memory-enhanced-table strategy – part B*

Here we go again, same pattern: table spool tasks at 0%, one at 1%, and the nested loops join at 5%. The remaining sections of the plan are the same, just as was the case with the CTE approach, so it looks like all we did was make the query more complex to load and introduce some logic to create and load memory-enhanced tables.

This strategy would be ideal for a server with multiple CPUs so that the query engine can execute the logic that retrieves data from each table in parallel. More hardware solutions, but more money to implement them!

Let's get the complete performance picture. What about IO and TIME statistics? Please refer to Figure 7-39.

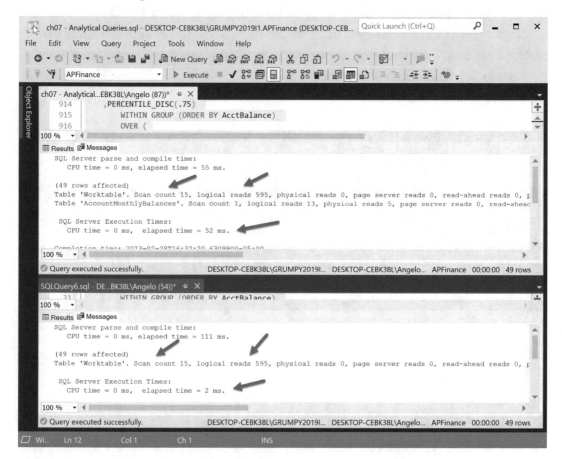

Figure 7-39. *Multiple-memory-enhanced-table scheme IO and TIME statistics*

Now we are getting somewhere. The work table statistics did not change, but the account balances table statistics are not displayed in the memory table query statistics. That's because everything is performed in memory, so it looks like a significant improvement as we eliminated a whole set of operations. Lastly, the SQL Server execution times went from 52 ms to 2 ms, so another significant improvement. SQL Server parse and compile times doubled with the memory-enhanced table approach. You can't win them all.

So this scheme looks promising.

For one last item, let's compare the IO and TIME statistics across the three schemes, CTE approach, report table approach, and memory-enhanced table approach.

Please refer to Figure 7-40.

Figure 7-40.   *Comparing IO and TIME statistics*

A little hard to see, but let's compare some vital statistics. Reading from left to right, we see the IO and TIME statistics for the CTE/QUERY approach, the report table approach, and finally the memory-enhanced table approach.

Parse and compile times went from 88 ms to 74 ms to 0 ms – big improvement.

Work table IO values were the same for all three except for the memory-enhanced tables where the IO went up, from 595 to 610.

Lastly, the SQL Server execution times went from 179 ms to 59 ms to 1 ms!

Looks conclusive to me, but these tests need to be performed several times, using DBCC to clear cache and noting the performance statistics also.

Finally, let's quickly look at STATISTICS PROFILE statistics results.

Please refer to Figure 7-41.

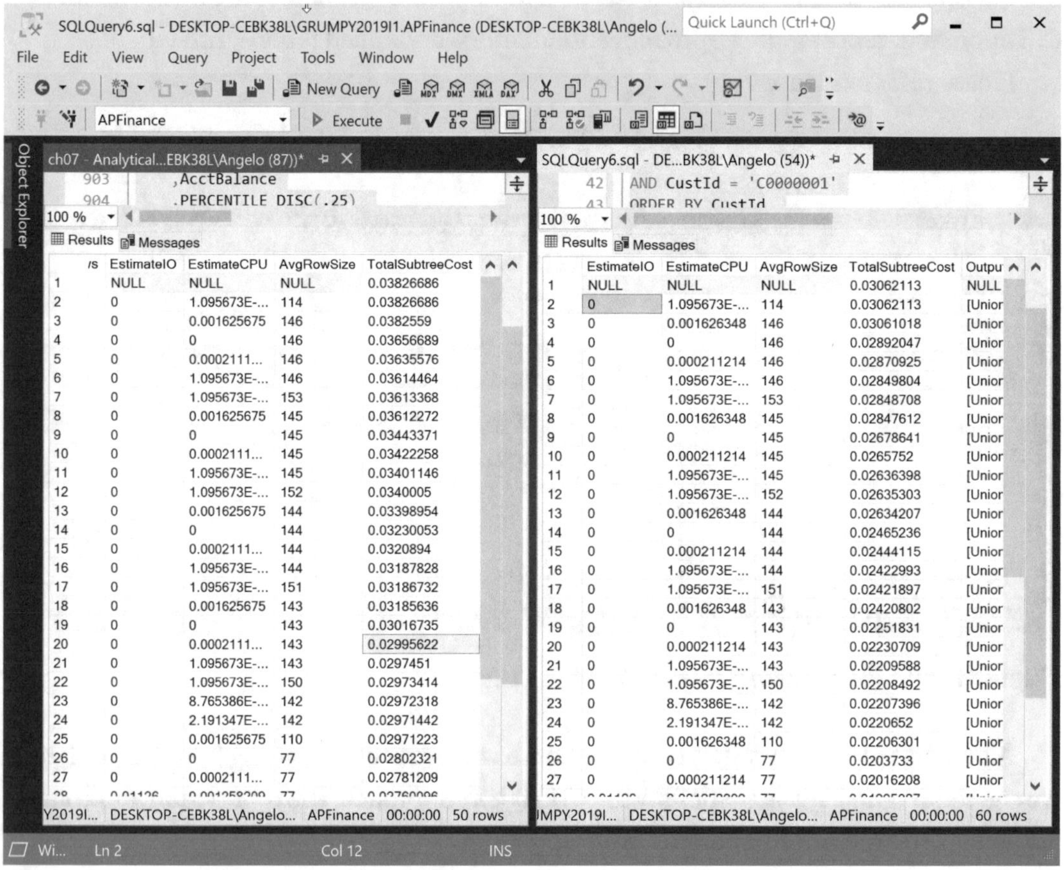

Figure 7-41. *Comparing profile statistics between the two schemes*

Does not look like much of a change except that the CTE approach generated 50 steps, while the memory-enhanced table approach generated 60 steps.

Verdict: The memory-enhanced table approach seems to work as far as performance improvement but with added complexity to implement and maintain as tables need loading and new tables need to be created and loaded. As stated earlier, if one has a solid-state drive where they can implement TEMPDB, then performance should significantly improve even more. Additionally, very fast memory, and lots of it, is also recommended.

In conclusion, we need to keep things in perspective. This analysis was performed by running queries and estimated query plans on a personal laptop. Certainly we are limited in what we can do to increase performance, but I think we have seen what strategies are available to us and what statistics to look at to see which one worked better. As stated earlier, when you perform this work on a company development server, you have more options available to you. I leave you with a question:

Was it worth the effort of creating and loading all the memory-enhanced tables and creating the view? Also creating the CTE/query and report table approaches? I think so. You got to not only see some strategies you can consider, but you also learned how the operators in the query plans do and how temporary tables are used (and shared) to spool data.

Wrapping it up here are the strategies we discussed and one we did not:

1) CTE/query approach

2) Preloaded report table approach

3) Single memory-enhanced table approach

4) Several-memory-enhanced-table approach (created on separate disks)

5) **Partitioned tables (not discussed)**

The last one, partitioned tables, are a way of creating a table that is spread across multiple fast disks by criteria, like each disk stores one year's worth of data. This works very well for environments that have SAN (Storage Area Network) disks and servers with multiple CPUs.

Note   In a company production environment, make sure you perform analysis related to which tables you place in memory. These should be tables used by 80% of the query/report/script catalog so they can be shared. Memory is limited and comes at a high price tag, so perform the analysis and design to correctly place tables in either memory or physical storage whether it be disk or SAN.

Summary

We covered a lot of ground. We did our usual exercise of presenting queries and results for the analytical functions, but we discussed new techniques for analyzing performance and possible solutions.

We revisited IO, TIME, and STATISTIC PROFILE statistics and saw how the details behind them tell us a lot about how the query is processed and what parts of the query are linked to the tasks in the estimated query plan.

Lastly, we took one look at real-time performance plans to see how long each task took to execute its logic but also to examine the flow of rows between each task.

We looked at the usual strategy of replacing the CTE block of a query with a preloaded table but also implemented single memory-enhanced tables and then multiple memory-enhanced tables.

Lastly, we touched upon how hardware such as SAN disk arrays and memory also plays a part in improving performance. We also hinted at using partitioned tables to spread data across multiple drives to increase performance, like dedicating a drive for each year in a particular set of data.

CHAPTER 8

Plant Use Case: Aggregate Functions

This is the first of three chapters that will use a power plant business use case scenario. A picture will give us an idea of the equipment we need to monitor with our window aggregate functions. Please refer to Figure 8-1.

Power Plant

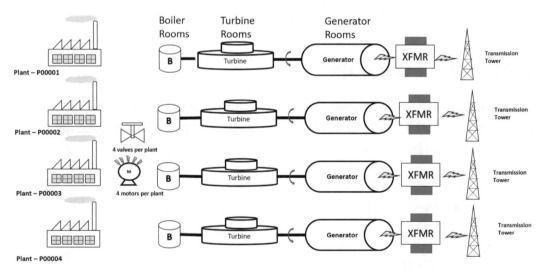

Figure 8-1. *Conceptual diagram of a power plant*

We will store information for four power plants in our database. Basically, some fuel such as oil or coal is used to heat a furnace, which in turn boils water until it turns to steam. The steam spins a turbine, which spins a generator, which produces electricity, which is finally transferred to a transformer that steps up the voltage so it can be distributed to the regions each plant supports.

A. Bobak, *SQL Server Analytical Toolkit*, https://doi.org/10.1007/978-1-4842-8667-8_8

Each power plant will have four valves, four motors, one boiler, one turbine, and one generator, which we need to monitor for failures. Failure statistics will be stored in a small data warehouse implemented as a snowflake schema. Simple, but it will give us a lot of data to analyze via the aggregate window functions.

How will we do this?

In this chapter we will take a slightly different approach. We will eliminate the use of a CTE in almost all the examples. As I was creating the code for this chapter, I did use a CTE, which was getting repetitive, so I took the code and used it for a script that populates a report table. Plus, from prior chapters we learned that the preloaded report table approach yields greater performance when dealing with historical data. This approach will be used by most examples.

Also, we will concentrate on report output and graphing the results with minimal discussions on performance tuning until the end of the chapter. At the end, we will create a complex query that uses most of the aggregate functions and see what the report looks like for the example.

We will try a couple of approaches and see which one works better. You might be surprised by the conclusion!

Aggregate Functions

Here is our list of the aggregate functions under this category. We have seen them in prior chapters, but I list them just in case you jumped directly to this chapter before reading the other chapters:

COUNT() and SUM()

MAX() and MIN()

AVG()

GROUPING()

STRING_AGG()

STDEV() and STDEVP()

VAR() and VARP()

Some of the functions we will cover together as they are naturally associated, like the MIN() and MAX() functions, STDEV() and STDEVP(), and VAR() and VARP(). We will use them with the OVER() clause, and also without, under certain circumstances.

Since this is the first chapter dealing with the power plant database called APPlant, we will take a look at a conceptual model of the database and also a few simple data dictionaries that describe the entities, attributes, and relationships.

Recall that a conceptual data model is the first data model that is created with cooperation of business analysts so as to define the business scope of the database. The conceptual model consists of entities, attributes, and relationships. Following the conceptual model, a logical normalized model is created and finally a physical database model. This would be for transactional databases. For data warehouses and data marts, we would go from the conceptual right to the physical STAR or SNOWFLAKE schema design.

In our case, we will go from the conceptual data model to a SNOWFLAKE schema data warehouse model, which is highly denormalized. Our database will contain three fact tables, several dimension tables, and two report tables. The fact and dimension tables all include surrogate keys to establish the STAR relationships, while the two report tables do not. They are denormalized tables based on the STAR tables and are used to generate reports. A few category and type tables exist, which will turn the model from a STAR schema into a SNOWFLAKE schema, for example, the relationship between the Equipment dimension and the EquipmentType dimension.

Lastly, we will see toward the end of the chapter that there is a mistake in the design. An important link or association table is missing. We will address this, so stay tuned.

Data Model

The following is our simple conceptual model that will give you a basic understanding of how all the tables tie together once we start creating our window aggregate queries. I will present three models, one for each of the principal fact tables. Let's start with the equipment failure conceptual model.

Please refer to Figure 8-2.

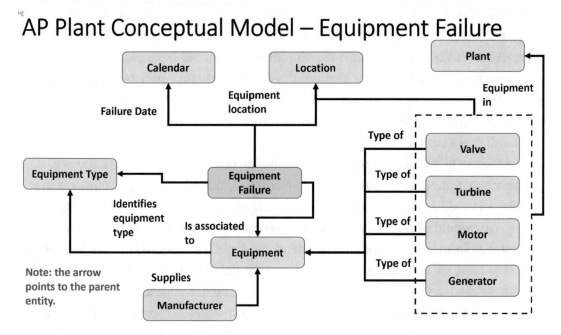

Figure 8-2. *Plant conceptual data model – equipment failure subject area*

In the center we see the Equipment Failure entity, which will form the basis of the Equipment Failure fact table. This entity as the name implies stores the number of failures of each piece of equipment together with the date of the failure.

The Equipment entity is composed of attributes that record the equipment ID, serial number, the type of equipment, and the manufacturer identifier (in the form of a surrogate key). This entity links to the Manufacturer entity and to one of the four entities that make up the equipment we are going to track.

Valves, turbines, motors, and generators are types of equipment, so we see the type-subtype relationship between these entities and the Equipment entity.

Note   When we implement this in the physical database, it will modify our model from a STAR schema to a SNOWFLAKE schema.

On the upper right-hand side of the model, we see the Plant entity, which contains basic information like the plant name and identifier.

The Equipment Failure entity has an association with the Location entity, so we know the location of the failed piece of equipment. The Valve, Turbine, Motor, and Generator entities also have an association with the Location entity.

Second to last is the Calendar entity, which provides dates for the equipment failures.

Lastly, there are many discussions and debates along the line of which model is better, the STAR or SNOWFLAKE model. I tend to go for the SNOWFLAKE model for smaller data volumes as it is more expressive and allows you to understand the relationships between the objects being stored. Also, SNOWFLAKE models allow for easier and more accurate updates in case a category or type codes need to be updated, especially when you need to implement slowly changing dimensions.

Closing comment: The STAR schema tends to perform better for large volumes of data if implemented correctly. The SNOWFLAKE schema is more expressive and easier to maintain. Choose the model that best fits the business requirements and implementation.

Let's look at the data model for the second entity that will be used to create a fact table called EquipmentStatusHistory. This fact table will store historical data that shows when a piece of equipment was online or offline together with the reasons for each status.

Please refer to Figure 8-3 for the conceptual model.

AP Plant Conceptual Model – Equipment Status History

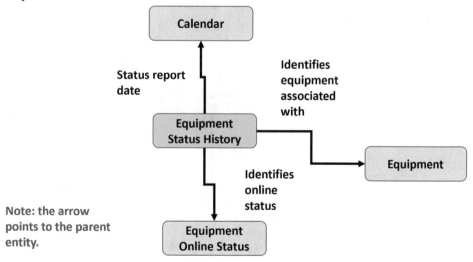

Figure 8-3. *Plant conceptual data model – equipment status history subject area*

This is a much simpler model. It has only one fact entity and the three dimension entities. The Equipment entity, as was the case in the prior model, is used to identify the piece of equipment whose status history is being tracked. The Calendar entity provides the date for the status assignment, and the last entity, the Equipment Online Status entity, has attributes that identify one of three codes that identify the status of the piece of equipment, which are identified in Table 8-1.

Table 8-1. *Equipment Online Status Codes*

Online Status Code	Online Status Description
0001	Online – normal operation
0002	Offline – maintenance
0003	Offline – fault

Only three status codes but enough data for us to perform some interesting analysis. Our last model is the plant expense subject area. Please refer to Figure 8-4.

AP Plant Conceptual Model – Plant Expense

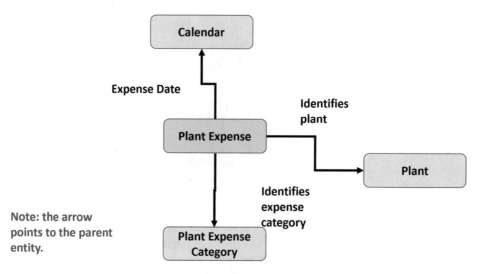

Figure 8-4. *Plant conceptual data model – plant expense subject area*

Another simple subject area model with only one "fact" entity and three "dimension" entities. I called out these names in quotes because the terms really only apply to physical tables, but for our purposes we understand that the entities are used to give us an understanding of the business model that will be used to build the data warehouse. Later, when the database is implemented, they will become either fact tables or dimension tables.

Back to the subject area model. Here we wish to record plant expenses over time, so of course there is an association between the Plant Expense entity and the Calendar entity to record the expense date. The association between plant expenses and plant is obvious. Lastly, the Plant Expense Category entity is used to identify the type of expense using one of the following codes shown in Table 8-2.

Table 8-2. *Plant Expense Codes*

Expense Code	Expense Description
01	Salary
02	Repairs
03	Machinery
04	Building maintenance

Having created three simple models, let's create some simple data dictionaries that provide descriptions of the entities, their attributes, and the relationships between the entities (these are called business rules).

They are a bit long and, to be truthful, boring to read. Just take a look at them and use them for reference as needed. At least you know where to find them.

Data Dictionaries

Entity Data Dictionary

Our first data dictionary is used to describe the purpose of each business entity. These are the business objects that define the purpose of the database.

Please refer to Table 8-3.

Table 8-3. *Entity Data Dictionary*

Entity Name	Description
Calendar	Basic calendar for date reporting.
EquipFailureManufacturer	Associates failed equipment with the manufacturer.
Equipment	The equipment in the plant, in our scenario either a valve, motor, or turbine.
EquipmentFailure	Fact table containing historical equipment failure rates.
EquipmentFailureStatistics	Denormalized report table containing production keys and failure history.
EquipmentStatusHistory	Fact table containing history of equipment online or offline status. How often was equipment online and offline?
EquipmentType	Identifies the type of equipment.
EquipOnlineStatusCode	Identifies status codes of equipment operation: 0001 – online normal operation, 0002 – offline for maintenance, 0003 – offline due to failure.
Generator	Dimension table storing attributes of a generator, like voltage.
Location	Dimension table storing attributes of a location.
Manufacturer	The manufacturer for a particular piece of equipment.
Motor	Dimension table storing attributes of a motor such as current and voltage.
Plant	The power plant that contains the equipment we are monitoring and measuring.
PlantBudget	Budget allocated to a plant.
PlantExpense	Expenses incurred by a power plant.
PlantExpenseCategory	Identifies categories of expenses: 01 – employee salaries; 02 – equipment and plant repairs; 03 – capital expenses associated with machinery, like new motors; 04 – plant building maintenance.
TempSensorLog	Report table containing temperature history of equipment. Also identifies the hour where a temperature was not recorded.
Turbine	Dimension table storing attributes of a turbine such as rpm.
Valve	Dimension table storing attributes of a valve.

Next is the entity attribute data dictionary, a long list, but it is worth becoming familiar with the attributes that will be used to create the table columns. You do not need to read this table one row at a time; just skim through it. You can refer back to it, if you do not know what a column in the tables you are using in a query means. Consider it a reference table. I could have put it in an appendix, but I think it is more useful here. Saves jumping back and forth between the chapter and the appendix.

Entity Attribute Data Dictionary

This data dictionary should also contain a column to at least identify the primary and foreign keys and other keys (PK, primary key; FK, foreign key; IE, inversion entry key; AK, alternate key), but in the interest of space, I omitted the column. I did highlight the proposed surrogate keys by using italics and bold font. These are unique keys that can be used as primary keys when the database is implemented.

This basic information should be enough to guide you when you create the JOIN clauses in your queries.

The descriptions are brief but should provide a basic understanding of the business attributes that will become the physical columns in our data warehouse tables. Please refer to Table 8-4.

Table 8-4. *Entity Attribute Data Dictionary*

Entity	Attribute	Description
Calendar	CalendarDate	The individual date of the year.
Calendar	CalendarDay	The day part of the date (1–28, 1–29, 1–30, or 1–31 depending on the month).
Calendar	***CalendarKey***	The unique surrogate key.
Calendar	CalendarMonth	The month part of the date (1–12).
Calendar	CalendarQuarter	The quarter part of the date (1–4).
Calendar	CalendarYear	The year part of a date, that is, 2016.
EquipFailureManufacturer	CalendarDate	The failure calendar date.
EquipFailureManufacturer	***Equipment***	The name of the failed equipment.

(continued)

Table 8-4. (*continued*)

Entity	Attribute	Description
EquipFailureManufacturer	Location	The location of the failed equipment.
EquipFailureManufacturer	***Manufacturer***	The name of the manufacturer of the failed equipment.
EquipFailureManufacturer	NormalTemp	Normal operating temperature of the equipment.
EquipFailureManufacturer	OverUnderTemp	Not normal operating temperature of the equipment.
EquipFailureManufacturer	Plant	The name of the plant.
EquipFailureManufacturer	TempAlert	Times an alert occurred.
Equipment	EquipAbbrev	Abbreviation for the equipment.
Equipment	EquipmentId	Identifier for the piece of equipment.
Equipment	***EquipmentKey***	Unique surrogate key for the piece of equipment.
Equipment	EquipmentTypeKey	Foreign surrogate key to the Equipment Type outrigger table.
Equipment	ManufacturerKey	Foreign surrogate key to the equipment Manufacturer table.
Equipment	SerialNo	The serial number of the piece of equipment.
EquipmentFailure	***CalendarKey***	Foreign surrogate key to the Calendar entity.
EquipmentFailure	***EquipmentKey***	Foreign surrogate key to the Equipment entity.
EquipmentFailure	***EquipmentTypeKey***	Foreign surrogate key to the Equipment Type entity.
EquipmentFailure	Failure	Records the number of failures for a particular date.

(*continued*)

Table 8-4. (*continued*)

Entity	Attribute	Description
EquipmentFailure	*LocationKey*	Foreign surrogate key to the Location entity.
EquipmentFailureStatistics	CalendarMonth	The month part of the failure date.
EquipmentFailureStatistics	CalendarYear	The year part of the failure year.
EquipmentFailureStatistics	EquipmentId	The equipment unique identifier number.
EquipmentFailureStatistics	FailureEvents	The number of failure events.
EquipmentFailureStatistics	LocationName	The name of the location in the plant where the failure occurred.
EquipmentFailureStatistics	MonthName	The name of the month for the failure date.
EquipmentFailureStatistics	PlantName	The name of the plant where the failure occurred.
EquipmentFailureStatistics	QuarterName	The quarter part of the date.
EquipmentFailureStatistics	SumEquipmentFailure	The total number of failures on a particular date.
EquipmentStatusHistory	*CalendarKey*	Foreign surrogate key to the Calendar entity.
EquipmentStatusHistory	*EquipmentKey*	Foreign surrogate key to the Equipment entity.
EquipmentStatusHistory	EquipOnlineStatus	Online status code.
EquipmentStatusHistory	*EquipOnlineStatusKey*	Foreign surrogate key to the Equipment Online Status entity.
EquipmentType	EquipmentDescription	The description of the piece of equipment.
EquipmentType	EquipmentType	The type of equipment, that is, valve, motor, etc.
EquipmentType	*EquipmentTypeKey*	Surrogate key of the Equipment Type entity.

(continued)

493

Table 8-4. (*continued*)

Entity	Attribute	Description
EquipOnlineStatusCode	EquipOnlineStatusCode	Online status code for the piece of equipment.
EquipOnlineStatusCode	EquipOnlineStatusDesc	Online status description associated with the status code.
EquipOnlineStatusCode	***EquipOnlineStatusKey***	Surrogate key to the Equipment entity.
Generator	Amp	The amperage value for the generator.
Generator	EquipmentId	The unique equipment identifier for the generator.
Generator	GeneratorId	The unique identifier for the generator associated with the generic equipment ID.
Generator	***GeneratorKey***	Surrogate key for the Generator entity.
Generator	GeneratorName	The name of the generator.
Generator	Kva	The kilovolt amperage rating for the generator.
Generator	LocationId	The location identifier for the generator.
Generator	PlantId	The plant identifier for the generator.
Generator	Temperature	The operational temperature rating for the generator.
Generator	Voltage	The voltage of the generator.
Location	LocationId	The location identifier for the location.
Location	***LocationKey***	Surrogate key for the Location entity.
Location	LocationName	The name of the location.
Location	PlantId	The plant identifier for the location. A plant has many locations.
Manufacturer	ManufacturerId	The unique identifier for the equipment manufacturer.

(*continued*)

Table 8-4. (*continued*)

Entity	Attribute	Description
Manufacturer	***ManufacturerKey***	Surrogate key for the Manufacturer entity.
Manufacturer	ManufacturerName	The name of the manufacturer.
Motor	EquipmentId	Unique identifier for a piece of equipment.
Motor	LocationId	Unique identifier for the location of the motor.
Motor	MotorId	Unique identifier for the motor.
Motor	***MotorKey***	Surrogate key for the Motor entity.
Motor	MotorName	The name of the motor.
Motor	PlantId	The plant identifier for the motor.
Motor	Rpm	The revolutions per minute rating of the motor.
Motor	Voltage	The voltage rating of the motor.
Plant	PlantDescription	A description of the plant.
Plant	PlantId	A unique identifier for the plant.
Plant	***PlantKey***	Surrogate key for the Plant entity.
Plant	PlantName	The name of the plant.
PlantBudget	Budget	The budget assigned for the plant.
PlantBudget	CalendarYear	The year part of the date.
PlantBudget	***PlantBudgetKey***	Surrogate key for the Plant Budget entity.
PlantBudget	PlantId	Unique identifier for the plant.
PlantExpense	***CalendarDateKey***	The calendar date the expense was incurred.
PlantExpense	ExpenseAmount	The amount of the expense.
PlantExpense	***PlantExpenseCategoryKey***	Foreign surrogate key to the Plant Expense Category entity.

(*continued*)

Table 8-4. (*continued*)

Entity	Attribute	Description
PlantExpense	***PlantKey***	Foreign surrogate key to the Plant entity.
PlantExpenseCategory	ExpenseCategoryCode	The category code for the expense.
PlantExpenseCategory	ExpenseCategoryDesc	The description associated with the expense category code.
PlantExpenseCategory	***PlantExpenseCategoryKey***	Surrogate key for the Plant Expense Category entity.
TempSensorLog	***BoilerId***	Unique identifier for the boiler.
TempSensorLog	IslandGapGroup	The name for the island or gap in a series of measurement values.
TempSensorLog	ReadingHour	The hour the reading took place.
TempSensorLog	SensorId	A unique identifier for the sensor.
TempSensorLog	Temperature	The temperature logged by the sensor.
Turbine	Amps	The amperage rating of the turbine.
Turbine	EquipmentId	A unique identifier assigned to the piece of equipment.
Turbine	LocationId	A unique identifier assigned to the location.
Turbine	PlantId	A unique identifier assigned to the plant.
Turbine	Rpm	The revolutions per minute rating for the turbine.
Turbine	***TurbineId***	A unique identifier assigned to the turbine.
Turbine	TurbineKey	The surrogate key for the turbine.
Turbine	TurbineName	The name of the turbine.
Turbine	Voltage	The voltage rating of the turbine.
Valve	EquipmentId	A unique identifier assigned to the piece of equipment.

(*continued*)

Table 8-4. (*continued*)

Entity	Attribute	Description
Valve	LocationId	A unique identifier assigned to the location.
Valve	PlantId	A unique identifier assigned to the plant.
Valve	SteamPsi	The pounds per square inch rating of the valve.
Valve	SteamTemp	The steam temperature of the steam passing through a valve measured by a sensor.
Valve	ValveId	The unique identifier of the valve.
Valve	***ValveKey***	The surrogate key used for the Valve entity row.
Valve	ValveName	The name of the valve.

Yes, this was a long one, but it provides minimal information that lets you understand what each attribute is used for.

Note   There are two items to mention. An outrigger table (see Kimball's *The Data Warehouse Toolkit: The Definitive Guide to Dimensional Modeling*) is a table linked to a dimension table, for example, if you have a product table, then a product subcategory and product category are considered outriggers or the snowflake tables in a SNOWFLAKE schema. Second, technically, since these are business-type data dictionaries, you would not see any surrogate keys mentioned, but I took some liberties. Mainly, I generated the list of columns by querying system tables to generate the list of columns so I could put them in a document table.

One more item: There will be several report tables that are denormalized and do not have surrogate keys. They are used to stage data to increase performance for our aggregate window queries. They are

- Reports.EquipFailureManufacturer

- Reports.EquipmentDailyStatusHistoryByHour

- Reports.EquipmentFailureStatistics

- Reports.EquipmentFailureStatisticsMem

- Reports.EquipmentRollingMonthlyHourTotals

- Reports.EquipmentStatusHistoryByHour

- Reports.TempSensorLog

Watch out for these. As can be seen they will be assigned to the Reports schema.

Lastly, three data dictionaries follow to describe the relationships between each of the fact tables and the dimension tables in the three subject areas. Let's start off with the equipment failure fact table model.

Entity Relationship Data Dictionary: Equipment Failure Subject Area

In Table 8-5 are the business rules that link the facts and dimensions for the equipment failure subject area.

Table 8-5. *Entity Relationship Data Dictionary – Equipment Failure*

Parent Entity	Relationship	Child Entity
Calendar	Identifies failure date in	Equipment Failure
Location	Identifies equipment location in	Equipment Failure
Equipment	Is associated with	Equipment Failure
Equipment Type	Identifies equipment type in	Equipment Failure
Valve	Can be/Is a type of	Equipment
Turbine	Can be/Is a type of	Equipment
Motor	Can be/Is a type of	Equipment
Generator	Can be/Is a type of	Equipment

(continued)

Table 8-5. (*continued*)

Parent Entity	Relationship	Child Entity
Plant	Is location for	Valve
Plant	Is location for	Turbine
Plant	Is location for	Motor
Plant	Is location for	Generator
Location	Identifies location of a	Valve
Location	Identifies location of a	Turbine
Location	Identifies location of a	Motor
Location	Identifies location of a	Generator
Manufacturer	Is supplier of	Equipment
Equipment Type	Identifies type of	Equipment

These rules that link the entity objects together are usually derived through sessions between the data modeler or data architecture modeler and various business stakeholders. They are key to deriving the primary to foreign keys that will be assigned in the physical tables when the database is created.

Usually there will be two rules implemented as verb phrases: one verb phrase from the parent entity to the child entity and one verb phrase from the child entity to the parent entity, for example:

- A valve is in a plant.

- A plant is the location of a valve.

This is also true for master/detail structures like the following:

- A piece of equipment can be a generator.

- A generator is an example of a piece of equipment.

I did not go into a rigorous implementation of the data dictionaries but aimed to give you just enough description of the business rules between business objects to understand the relationships that link them together.

Entity Relationship Data Dictionary: Equipment Status History

Next, we examine the business rules that link the facts and dimensions for the equipment status history subject area.

Please refer to Table 8-6.

Table 8-6. *Entity Relationship Data Dictionary – Equipment Status History*

Parent Entity	Relationship	Child Entity
Calendar	Identifies status date for the status history in	Equipment Status History
Equipment	Identifies equipment in the status history in	Equipment Status History
Equipment Online Status	Identifies equipment online status code (on, maintenance, or offline) in	Equipment Status History

These simple relationships provide a minimal explanation of the business rules that link two business objects. These are key to understand as you will use them when you create the JOIN clauses in your queries. These of course are in the conceptual model but apply to our physical data warehouse also and will be used to implement the physical primary surrogate and foreign keys.

Note   I did not mention the cardinalities between the entities, for example, zero, one, or more to zero, one, or more rows. For example, equipment online status codes appear zero, one, or more times in the Equipment Status History entity. This was done to save some space. You can figure out the cardinalities by writing some queries between tables.

Entity Relationship Data Dictionary: Plant Expense

Next let's examine the relationships between the Plant Expense fact table and the dimension tables in the plant expense subject area.

Please refer to Table 8-7.

Table 8-7. *Entity Relationship Data Dictionary – Plant Expense*

Parent Entity	Relationship	Child Entity
Calendar	Identifies expense date in	Plant Expense
Plant	Expenses identified in	Plant Expense
Plant Expense Category	Identifies expense category for a plant in	Plant Expense

The information contained in the conceptual data models and data dictionaries is very important as it will assist you in understanding not only what is in the database tables but what the columns mean for when you create your scripts and queries, particularly when you need to join tables together.

Note   More extensive entity/relationship models would list the attributes in each entity. Again, this would take a lot of space, so I did not include them. You should try to use a tool like PowerPoint to create these models as this is another valuable skill and tool to have under your belt.

A fourth data dictionary that shows the primary and foreign keys in each entity should be included, but you can identify them from the physical database tables. They are the surrogate keys used to implement the relationships, for example, `EquipmentId` will link any fact table to the `Equipment` dimension. You can identify these keys by looking at the table columns.

Lastly, these are models of the simplest data warehouse design using a SNOWFLAKE schema approach. The intent is to give you enough tables to use in queries, not only the ones in the book but also on your own when you practice. Learning how to design star and snowflake schemas would take an entire book!

COUNT() Function

Let's start with the COUNT() aggregate function. We will start off by examining the TSQL query that would have been used in the CTE part of the query. We will use it to insert rows into a report table that will replace the CTE in order to simulate a report table that is loaded once a day or week as we are interested in queries that analyze historical data.

The query we will use is called a STAR query because it joins several dimension tables to a fact table. We are assuming we can use it to load the report table that will be used for several queries.

This is another reason the conceptual models are useful; they show us how to code the JOIN clauses for our queries.

Pop Quiz   Historically, queries that included fact tables and linked dimension tables were called STAR queries. Is a query that references dimensions and outrigger dimensions a SNOWFLAKE query?

Please refer to Listing 8-1a.

Listing 8-1a. The Repurposed CTE Component

```
SELECT
      C.CalendarYear
      ,C.QuarterName
      ,C.MonthName
      ,C.CalendarMonth
      ,P.PlantName
      ,L.LocationName
      ,E.EquipmentId
      ,COUNT(Failure) AS FailureEvents
      ,SUM(Failure) AS SumEquipmentFailure
FROM DimTable.CalendarView C
JOIN FactTable.EquipmentFailure EF
      ON C.CalendarKey = EF.CalendarKey
JOIN DimTable.Equipment E
      ON EF.EquipmentKey = E.EquipmentKey
JOIN DimTable.Location L
      ON L.LocationKey = EF.LocationKey
JOIN DimTable.Plant P
      ON L.PlantId = P.PlantId
GROUP BY
      C.CalendarYear
```

```
        ,C.QuarterName
        ,C.MonthName
        ,C.CalendarMonth
        ,P.PlantName
        ,L.LocationName
        ,E.EquipmentId
ORDER BY
        C.CalendarYear
        ,C.QuarterName
        ,C.CalendarMonth
        ,P.PlantName
        ,L.LocationName
GO
```

This query is used to create an INSERT command to populate a table called Reports.
EquipmentFailureStatistics.

I placed it in the Reports schema because it's not really a fact table but a report
staging table.

The SELECT clause includes two aggregate functions, the COUNT() function that is
used to pre-count the failure events and the SUM() function to add up the failures. These
are plain vanilla aggregate function usage; no OVER() clause is used.

A GROUP BY clause is used to group data by year, month, plant name, location name,
and equipment ID.

Lastly, notice the JOIN clauses; there are four of these, so five dimension tables are
used in the FROM clause. The single fact table is the EquipmentFailure table, which has
only 5,906 rows, so not a lot of rows to process.

Let's see the results in Figure 8-5.

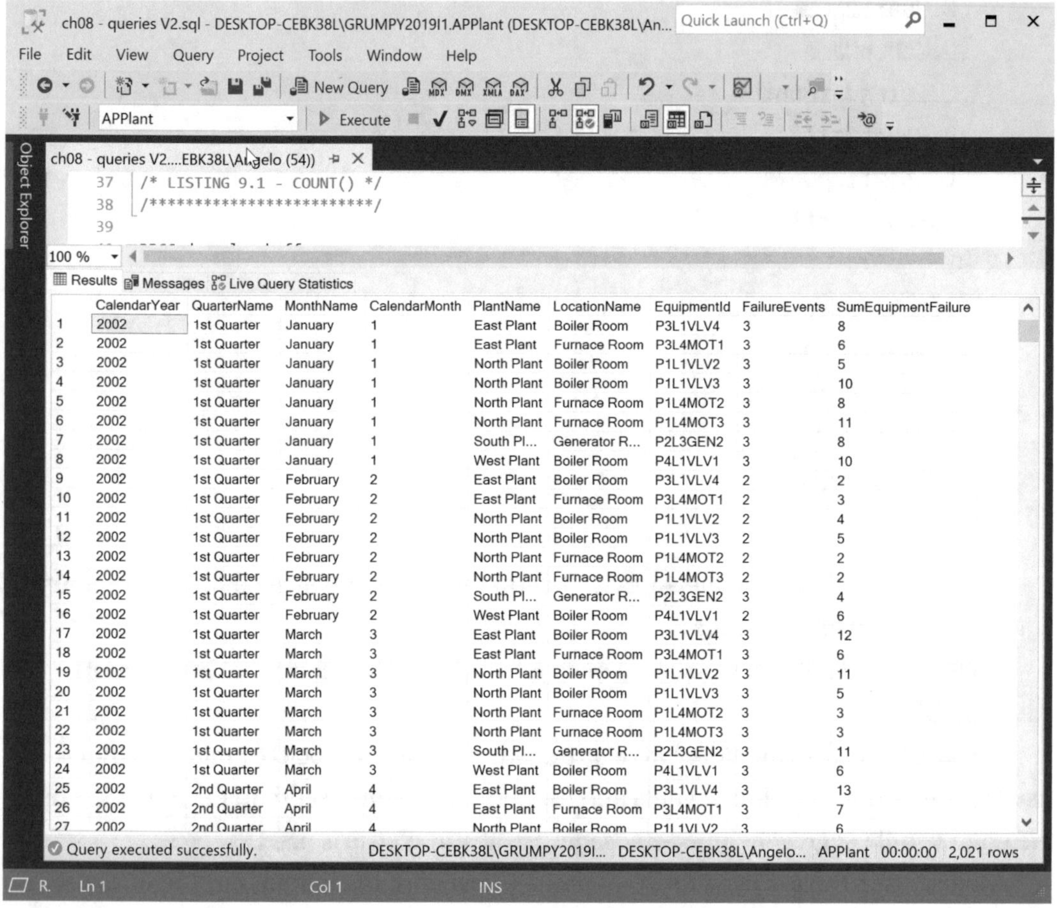

Figure 8-5. *Loading the equipment failure statistics report table*

Not many rows were generated, only 2,021 rows. This did generate an interesting live execution plan though. Fortunately, in a real production environment, it would be loaded once a day, at off hours, so as not to affect business users running reports against the data. Let's create an estimated query plan in the usual manner.

Please refer to Figure 8-6 for the baseline estimated query plan.

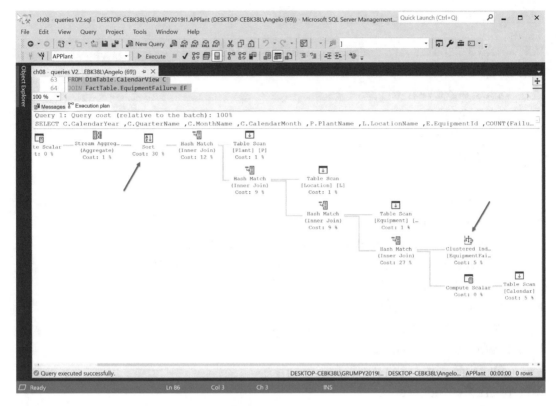

Figure 8-6. *Live query plan for a STAR query*

Lots of table scans, but as the tables involved have small numbers of rows, we do not need to worry about them. We do see a clustered index scan, which is to be expected because we want to load all the rows into the report table.

Hash match joins are being used at each stage where the data flows need to be merged, and finally a sort at a cost of 30% is shown. This sort was incurred by the final ORDER BY clause in the query. This needed to be included because it is required to sort the date parts (year, quarter, month, etc.) correctly. I tried the query without it, and February dates appeared before January dates, so might as well take the performance hit at this time. Makes the resulting report easier to understand.

We will use this query in an INSERT statement to load our report table. Please refer to Listing 8-1b.

Listing 8-1b. Insert into the Report Table

```
INSERT INTO Reports.EquipmentFailureStatistics
SELECT
      C.CalendarYear
      ,C.QuarterName
      ,C.MonthName
      ,C.CalendarMonth
      ,P.PlantName
      ,L.LocationName
      ,E.EquipmentId
      ,CONVERT(INTEGER,COUNT(Failure)) AS CountFailureEvents
      ,CONVERT(INTEGER,SUM(Failure)) AS SumEquipmentFailure
FROM DimTable.CalendarView C
JOIN FactTable.EquipmentFailure EF
      ON C.CalendarKey = EF.CalendarKey
JOIN DimTable.Equipment E
      ON EF.EquipmentKey = E.EquipmentKey
JOIN DimTable.Location L
      ON L.LocationKey = EF.LocationKey
JOIN DimTable.Plant P
      ON L.PlantId = P.PlantId
GROUP BY
      C.CalendarYear
      ,C.QuarterName
      ,C.MonthName
      ,C.CalendarMonth
      ,P.PlantName
      ,L.LocationName
      ,E.EquipmentId
ORDER BY
      C.CalendarYear
      ,C.QuarterName
```

```
      ,C.CalendarMonth
      ,P.PlantName
      ,L.LocationName
GO
```

This query generated 2,021 rows.

Let's try our first aggregate function query against this report table. The following specifications were submitted by our analyst:

We need some information concerning the East Plant, specifically the boiler room. The report should list the total count of failure events for the plant location by month. Each event can have one or more failures for the specific piece of equipment. For example, if a failure event occurred on January 5, 2002, and January 7, 2002, these would count as two events.

If on January 5, 2002, there occurred three failures for a particular valve, the sum of the failures would be 3; this counts as one event. If on January 7, 2002, there occurred another four failures for the same valve, then the total sum of failures for that valve would be 7 (3 + 4) on that date. This counts as another event. So now we have reported on two events with a total sum of seven failures.

We also need to report the rolling monthly totals of the count of events and of the sum of failures as the months roll by. The report needs to be grouped by year, quarter, month, plant, and location.

Again, please note the distinction between the count of events vs. the sum of failures. As I said, an event can have one, two, or more failures for the piece of equipment associated with that event on any specific date.

Lastly, more than one piece of equipment can fail on any date, so we include those counts (as monthly counts and rolling monthly counts) also.

We are not distinguishing by equipment ID in the report.

Let's see the code that will satisfy this report.

Please refer to Listing 8-2.

Listing 8-2. The Report Query, Rolling Failures by Quarter

```
SELECT
      PlantName
      ,LocationName
      ,CalendarYear
      ,QuarterName
```

```
        ,MonthName
        ,CountFailureEvents
        ,SUM(CountFailureEvents) OVER (
                PARTITION BY PlantName,LocationName,CalendarYear,QuarterName
                ORDER BY CalendarMonth
        ) AS SumCountFailureEvents
        ,SumEquipmentFailure
        ,SUM(SumEquipmentFailure) OVER (
                PARTITION BY PlantName,LocationName,CalendarYear,QuarterName
                ORDER BY CalendarMonth
        ) AS SumEquipFailureFailureEvents
        ,COUNT(CountFailureEvents) AS CountFailureEvents
        ,COUNT(SumEquipmentFailure) OVER (
                PARTITION BY PlantName,LocationName,CalendarYear,QuarterName
                ORDER BY CalendarMonth
        ) AS CountEquipFailureFailureEvents
FROM Reports.EquipmentFailureStatistics
WHERE PlantName = 'East Plant'
AND LocationName = 'Boiler Room'
GROUP BY
        PlantName
        ,LocationName
        ,CalendarYear
        ,QuarterName
        ,MonthName
        ,CalendarMonth
        ,CountFailureEvents
        ,SumEquipmentFailure
GO
```

The SUM() aggregate function and the COUNT() function appear twice in the query. All four cases use an OVER() clause with a PARTITION BY clause and an ORDER BY clause. The partition is created by plant name, location, year, and quarter in each case. The partition is ordered by the calendar month. The columns CountFailureEvents and SumEquipmentFailure are included so you can see how the rolling totals and sums are derived. Let's see the results.

Please refer to Figure 8-7.

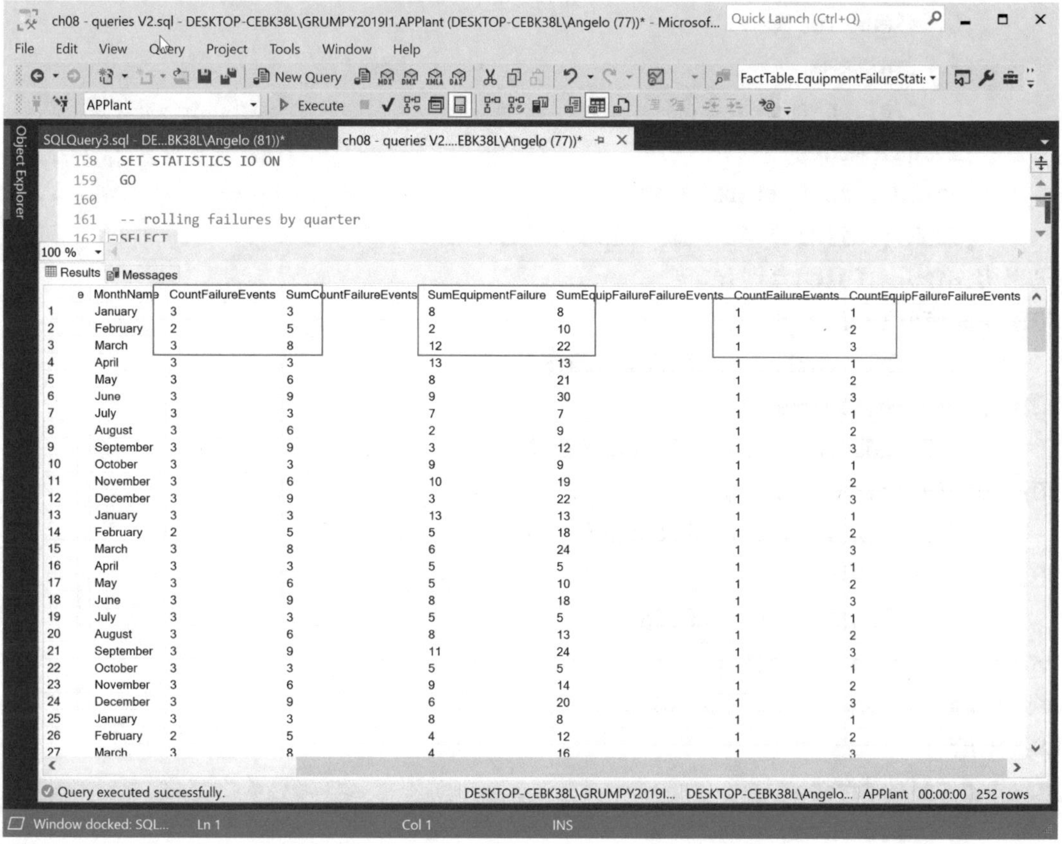

Figure 8-7. *Report query results*

Checking out the results in the boxes clearly shows the rolling values add up correctly. The COUNT() function results repeat themselves for each quarter. The query is a bit confusing, as we are trying to differentiate the count of events vs. sum of failures, so I chose to make the names more descriptive.

In cases like this, it is good to validate the results by looking at the individual raw counts and sums. Listing 8-3 is a query that will accomplish this.

Listing 8-3. Validate Counts and Sums

```
SELECT
    C.CalendarYear
    ,C.QuarterName
```

```
      ,C.MonthName
      ,C.CalendarMonth
      ,C.CalendarDate
      ,P.PlantName
      ,L.LocationName
      ,E.EquipmentId
      ,Failure AS FailureEvent
      ,COUNT(Failure) AS SumEquipmentFailure
FROM DimTable.CalendarView C
JOIN FactTable.EquipmentFailure EF
      ON C.CalendarKey = EF.CalendarKey
JOIN DimTable.Equipment E
      ON EF.EquipmentKey = E.EquipmentKey
JOIN DimTable.Location L
      ON L.LocationKey = EF.LocationKey
JOIN DimTable.Plant P
      ON L.PlantId = P.PlantId
WHERE PlantName = 'East Plant'
AND LocationName = 'Boiler Room'
GROUP BY
      C.CalendarYear
      ,C.QuarterName
      ,C.MonthName
      ,C.CalendarMonth
      ,C.CalendarDate
      ,P.PlantName
      ,L.LocationName
      ,E.EquipmentId
      ,Failure
ORDER BY
      C.CalendarYear
      ,C.QuarterName
      ,C.CalendarMonth
```

```
        ,C.CalendarDate
        ,P.PlantName
        ,L.LocationName
GO
```

I modified the query used to insert rows into the Reports.EquipmentFailure Statistics table and added the calendar date and also equipment to get it to the lowest granularity. Figure 8-8 is a partial comparison of both query results side by side that shows how the counts and sums added up.

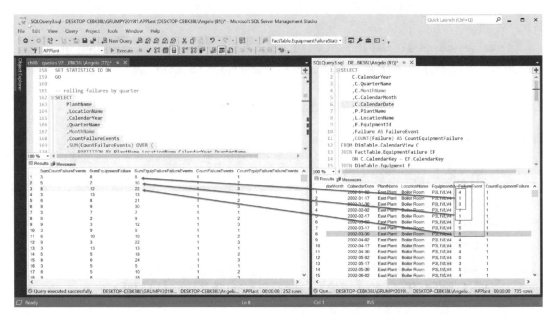

Figure 8-8. *Validation report for counts and sum of failures*

As can be seen, the total number of failures for January is 8 (4 + 1 + 3), and this lines up to the original report value under the SumEquipmentFailure report column. Same can be said for February and March if you perform the additions.

The number of events corresponds to the dates, one event per day, as we did not lower the granularity to the hour in the day. So the count will always be one per day, and the sum by quarter will be the sum of the days.

Hopefully this example shows the semantic difference between counting and summing up values and events.

AVG( ) Function

Next, we generate some averages for the East Plant, for equipment in the furnace room, with the AVG( ) function. Here are our business requirements submitted by our analyst:

A report is needed that shows rolling averages by month, quarter, and year. The report should show the results for the furnace room in the East Plant. Results should be grouped by plant, location, year, quarter, and month. The averages should be for monthly equipment failure rates.

We are not interested in the failure rate for individual equipment in this report. Lastly, the report should refer to the years 2002 and 2003 (our lowest level of granularity is by month).

Looks like we need to use the AVG( ) function three times and define different partitions for each. Listing 8-4 is the query that will satisfy this request.

Listing 8-4. Plant Failures by Rolling Month, Quarter, and Year

```
SELECT PlantName
      ,LocationName
      ,CalendarYear
      ,QuarterName
      ,MonthName
      ,CalendarMonth
      ,SumEquipmentFailure
      ,AVG(CONVERT(DECIMAL(10,2),[SumEquipmentFailure])) OVER (
            PARTITION BY CalendarYear
            ORDER BY CalendarYear,CalendarMonth
      ) AS RollingAvgMon
      ,AVG(CONVERT(DECIMAL(10,2),[SumEquipmentFailure])) OVER (
            PARTITION BY CalendarYear,QuarterName
            ORDER BY CalendarYear
      ) AS RollingAvgQtr
      ,AVG(CONVERT(DECIMAL(10,2),[SumEquipmentFailure])) OVER (
            PARTITION BY CalendarYear
            ORDER BY CalendarYear
      ) AS RollingAvgYear
FROM Reports.EquipmentFailureStatistics
WHERE CalendarYear IN(2002,2003)
```

512

```
AND Plantname = 'East Plant'
AND LocationName = 'Furnace Room'
GO
```

Yes, each of the three OVER() clauses has three different PARTITION and ORDER BY clauses.

For rolling monthly failures, the partition is defined for the calendar year and ordered by the year and month.

For rolling quarterly failures, the partition is defined for the calendar year and quarter and ordered by the year and quarter.

Lastly, for rolling yearly failures, the partition is defined for the calendar year and ordered by the calendar year (notice the month column is not required). This gives totally different results than the first partition definition.

Try playing around with the combinations to see the results. Make sure to copy and paste the results in an Excel spreadsheet and validate them. Are the results what you expected?

Please refer to the partial results in Figure 8-9.

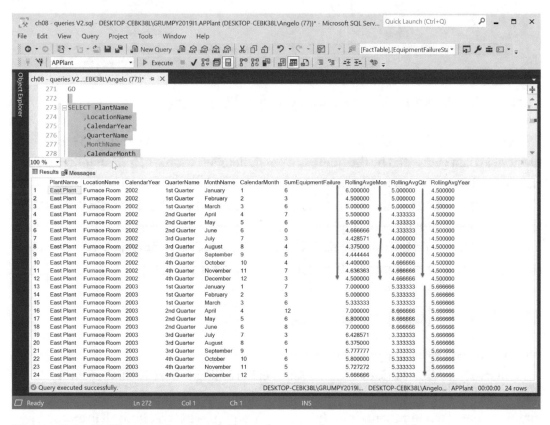

Figure 8-9. *Rolling equipment failures by month, quarter, and year*

Figure 8-10 is a graph created with Microsoft Excel showing the three rolling averages and the failure rates. Seems like the trend is going down; the fewer failures, the better!

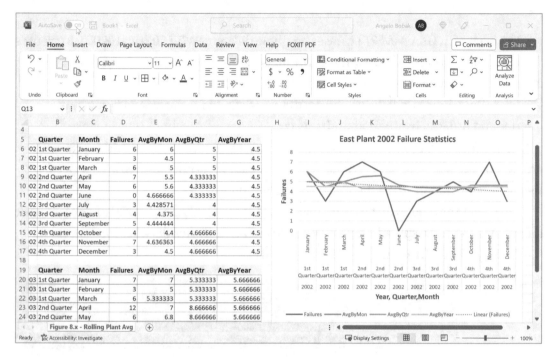

Figure 8-10. *Monthly, quarterly, and yearly rolling averages*

A visual is a good starting point to give you an overall view of the results. The numerical data generated by the query is a good, detailed source of information that you can use to dive deeper into the analysis.

Let's continue our analysis and build up our failure reports by looking at the smallest and largest failures at the monthly level to start building a profile of failure events. This data by the way could be used in a series of SSRS (SQL Server Reporting Services) web reports or Power BI dashboards. More tools to consider learning!

MIN() and MAX() Functions

Why do we want to build a profile of failures?

Obviously, we want to know what equipment is failing, where it is failing, and how often. We want to determine if a piece of equipment is failing due to improper maintenance or environmental conditions, if the equipment is faulty, or if the wrong equipment was placed in a specific area or application. Or maybe all these!

514

For example, is the equipment rated to operate in the temperatures of a specific plant location?

We also want to know if a piece of equipment failed, like a motor or valve, and if it was fixed and after a few weeks or months it failed again. We want to know why. Plant maintenance procedures and staff also come into play. Did the equipment fail multiple times in a day? Were there multiple attempts to fix it?

We want to get the best equipment; we want to aim for correct equipment installation and also the least number possible of failures. We also want a solid management program with knowledgeable technicians that can fix issues quickly. We want to answer these questions:

- How often did failures occur?

- How many attempts were made to fix the issue?

- Is the issue recurring?

- Is a single vendor providing faulty equipment?

- What is the root cause of the failures?

This is why a good set of reports and graphs that give us a profile of failure statistics and events will help us achieve these goals and answers to the questions related to equipment failure. The aim is to identify and then reduce or even prevent entirely equipment failure.

Our next report will deal with maximum and minimum failures for equipment for a specific time period. Here are the business requirements our analyst supplied for this profile report:

Please provide a report that shows the rolling monthly minimum and maximum failure rates occurring in the furnace room for the East Plant. Let's keep the report for this location as we want to see the high-level values first and then drill down to the equipment level.

Please refer to Listing 8-5 for maximum and minimum failure statistics.

Listing 8-5. Rolling Monthly Maximum and Minimum Plant Failures

```
SELECT PlantName
      ,LocationName
      ,CalendarYear
      ,QuarterName
```

```
        ,MonthName
        ,SumEquipmentFailure
        ,MIN(SumEquipmentFailure) OVER (
                PARTITION BY PlantName,LocationName,CalendarYear,QuarterName
                ORDER BY CalendarMonth
        ) AS RollingMin
        ,MAX(SumEquipmentFailure) OVER (
                PARTITION BY PlantName,LocationName,CalendarYear,QuarterName
                ORDER BY CalendarMonth
        ) AS RollingMax
FROM Reports.EquipmentFailureStatistics
WHERE CalendarYear IN(2002,2003)
AND Plantname = 'East Plant'
AND LocationName = 'Boiler Room'
GO
```

This is a simple query that uses the same OVER() clause for both the MIN() and the MAX() function. The partition is set up by plant name, location name, year, and quarter. The ORDER BY clause is by year and month. The WHERE clause filters include the years 2002 and 2003 and of course the furnace room for the East Plant. Let's see the results in Figure 8-11.

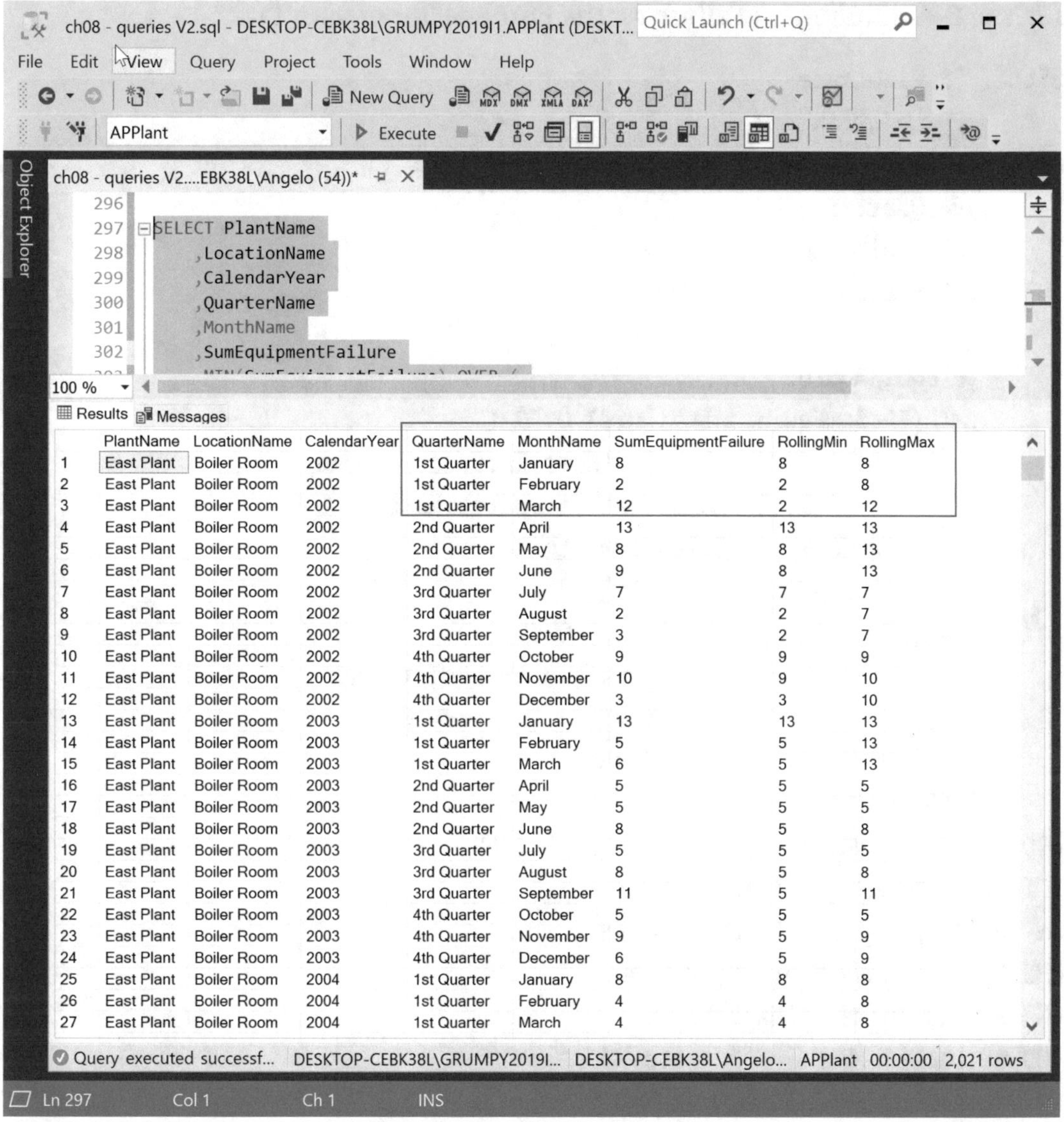

Figure 8-11. *Rolling monthly equipment failures for the East Plant*

We can see that the monthly rolling calculation is working, but we can see we have a lot of failures at this location. Is it one piece of equipment or more than one?

Let's run the following query in Listing 8-6 to narrow things down.

Listing 8-6. Equipment Failure for the Furnace Room, East Plant

```
SELECT EF.PlantName
      ,EF.LocationName
      ,EF.CalendarYear
      ,EF.QuarterName
      ,EF.MonthName
      ,EF.SumEquipmentFailure
      ,EF.EquipmentId
      ,E.EquipAbbrev
      ,MIN(EF.SumEquipmentFailure) OVER (
            PARTITION BY EF.PlantName,EF.LocationName,
                  EF.CalendarYear,EF.QuarterName
            ORDER BY EF.CalendarMonth
      ) AS RollingMin
      ,MAX(EF.SumEquipmentFailure) OVER (
            PARTITION BY EF.PlantName,EF.LocationName,
                  EF.CalendarYear,EF.QuarterName
            ORDER BY EF.CalendarMonth
      ) AS RollingMax
FROM Reports.EquipmentFailureStatistics EF
JOIN [DimTable].[Equipment] E
ON EF.EquipmentId = E.EquipmentId
WHERE EF.CalendarYear IN(2002,2003)
AND EF.Plantname = 'East Plant'
AND EF.LocationName = 'Furnace Room'
GO
```

The MIN() and MAX() functions are used with OVER() clauses that define partitions by plant name, location name, calendar year, and quarter. Partitions are ordered by calendar month.

A WHERE clause is used to include data only for two years, for the East Plant and the furnace room.

Please refer to Figure 8-12 for query results.

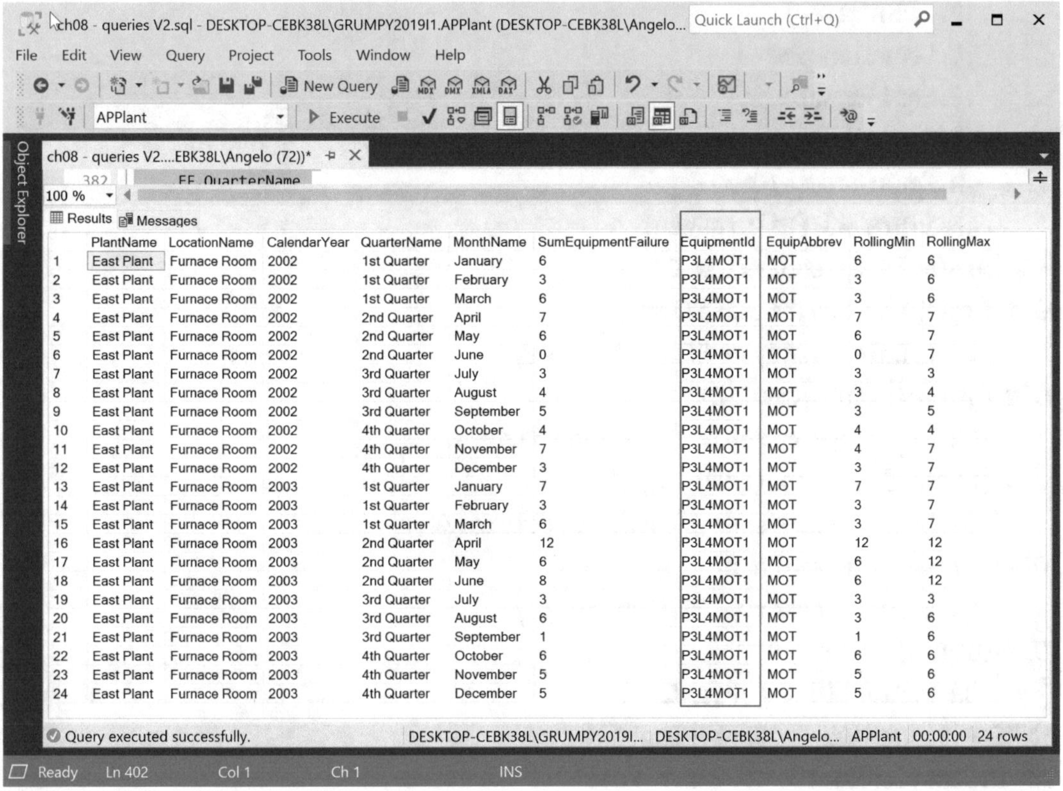

	PlantName	LocationName	CalendarYear	QuarterName	MonthName	SumEquipmentFailure	EquipmentId	EquipAbbrev	RollingMin	RollingMax
1	East Plant	Furnace Room	2002	1st Quarter	January	6	P3L4MOT1	MOT	6	6
2	East Plant	Furnace Room	2002	1st Quarter	February	3	P3L4MOT1	MOT	3	6
3	East Plant	Furnace Room	2002	1st Quarter	March	6	P3L4MOT1	MOT	3	6
4	East Plant	Furnace Room	2002	2nd Quarter	April	7	P3L4MOT1	MOT	7	7
5	East Plant	Furnace Room	2002	2nd Quarter	May	6	P3L4MOT1	MOT	6	7
6	East Plant	Furnace Room	2002	2nd Quarter	June	0	P3L4MOT1	MOT	0	7
7	East Plant	Furnace Room	2002	3rd Quarter	July	3	P3L4MOT1	MOT	3	3
8	East Plant	Furnace Room	2002	3rd Quarter	August	4	P3L4MOT1	MOT	3	4
9	East Plant	Furnace Room	2002	3rd Quarter	September	5	P3L4MOT1	MOT	3	5
10	East Plant	Furnace Room	2002	4th Quarter	October	4	P3L4MOT1	MOT	4	4
11	East Plant	Furnace Room	2002	4th Quarter	November	7	P3L4MOT1	MOT	4	7
12	East Plant	Furnace Room	2002	4th Quarter	December	3	P3L4MOT1	MOT	3	7
13	East Plant	Furnace Room	2003	1st Quarter	January	7	P3L4MOT1	MOT	7	7
14	East Plant	Furnace Room	2003	1st Quarter	February	3	P3L4MOT1	MOT	3	7
15	East Plant	Furnace Room	2003	1st Quarter	March	6	P3L4MOT1	MOT	3	7
16	East Plant	Furnace Room	2003	2nd Quarter	April	12	P3L4MOT1	MOT	12	12
17	East Plant	Furnace Room	2003	2nd Quarter	May	6	P3L4MOT1	MOT	6	12
18	East Plant	Furnace Room	2003	2nd Quarter	June	8	P3L4MOT1	MOT	6	12
19	East Plant	Furnace Room	2003	3rd Quarter	July	3	P3L4MOT1	MOT	3	3
20	East Plant	Furnace Room	2003	3rd Quarter	August	6	P3L4MOT1	MOT	3	6
21	East Plant	Furnace Room	2003	3rd Quarter	September	1	P3L4MOT1	MOT	1	6
22	East Plant	Furnace Room	2003	4th Quarter	October	6	P3L4MOT1	MOT	6	6
23	East Plant	Furnace Room	2003	4th Quarter	November	5	P3L4MOT1	MOT	5	6
24	East Plant	Furnace Room	2003	4th Quarter	December	5	P3L4MOT1	MOT	5	6

Figure 8-12. *Equipment failure listing*

Looks like equipment P3L4MOT1 is the failing equipment. This is a motor and is a significant problem. As can be seen by the name, it is located in plant 3 and location 4, and the name of the motor is MOT1. These are unacceptable numbers of failures each month. We need to see what days these failures occur on.

Let's create another query, this time at the equipment and date level of granularity. Please refer to Listing 8-7.

Listing 8-7. Equipment Failure by Date

```
SELECT
      C.CalendarYear
      ,C.QuarterName
      ,C.CalendarDate
      ,C.MonthName
      ,C.CalendarMonth
```

```
        ,P.PlantName
        ,L.LocationName
        ,E.EquipmentId
        ,E.EquipAbbrev
        ,M.ManufacturerName
        ,Failure AS FailureEvent
FROM DimTable.CalendarView C
JOIN FactTable.EquipmentFailure EF
        ON C.CalendarKey = EF.CalendarKey
JOIN DimTable.Equipment E
        ON EF.EquipmentKey = E.EquipmentKey
JOIN [DimTable].[Manufacturer] M
        ON E.ManufacturerKey = M.ManufacturerKey
JOIN DimTable.Location L
        ON L.LocationKey = EF.LocationKey
JOIN DimTable.Plant P
        ON L.PlantId = P.PlantId
WHERE P.PlantName = 'East Plant'
AND L.LocationName = 'Furnace Room'
AND C.CalendarYear = '2002'
AND C.MonthName = 'April'
ORDER BY
        C.CalendarYear
        ,C.QuarterName
        ,C.CalendarMonth
        ,C.CalendarDate
        ,P.PlantName
        ,L.LocationName
GO
```

This query was modified from the original query we used to load the failure report table. The date and equipment ID fields were added to arrive at the lowest level of granularity. We could have made things even more interesting by recording the hours for each 24-hour-day period.

In a real application, the analyst would examine all months at this level, but let's limit our investigation to the month of April for the year 2002 so we can make sure our logic works.

Executing this query yields the following report in Figure 8-13.

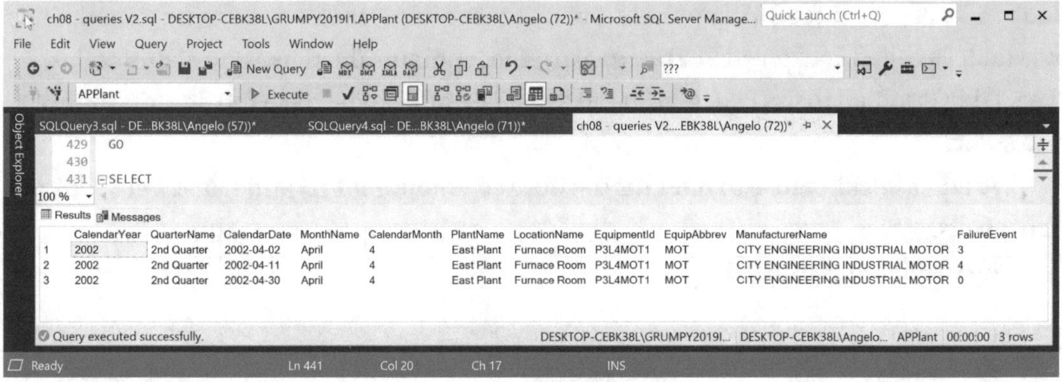

Figure 8-13. *P3L4MOT1 equipment failure by date*

We now see that the motor equipment identifier P3L4MOT1 failed on April 2 three times and April 11 four times. This motor was manufactured by the CITY ENGINEERING INDUSTRIAL MOTOR company. Now we are getting somewhere. We now need to determine why the motor failed on these dates. If you have not guessed already, this means another query. Please refer to Listing 8-8.

Listing 8-8. Check Equipment Status

```
SELECT CV.CalendarDate
    ,E.EquipmentId
    ,E.EquipAbbrev
    ,EOLSC.EquipOnlineStatusCode
    ,EOLSC.EquipOnlineStatusDesc
FROM APPlant.FactTable.EquipmentStatusHistory ESH
JOIN DimTable.Equipment E
    ON ESH.EquipmentKey = E.EquipmentKey
JOIN DimTable.EquipOnlineStatusCode EOLSC
    ON ESH.EquipOnlineStatusKey = EOLSC.EquipOnlineStatusKey
JOIN DimTable.CalendarView CV
    ON ESH.CalendarKey = CV.CalendarKey
```

```
WHERE E.EquipmentId = 'P3L4MOT1'
AND CV.CalendarDate IN('2002-04-02','2002-04-11','2002-04-30')
GO
```

This query checks for the status codes assigned on each of the three dates we need to look at. There are three `JOIN` operators, which means four tables need to be joined. The key table is called `EquipmentStatusHistory` that contains the dates and status codes. This table is joined to the `EquipOnlineStatusCode` table to retrieve the descriptions of the codes.

We add a `WHERE` clause to filter for the desired equipment ID and dates. Let's look at the results.

Please refer to Figure 8-14.

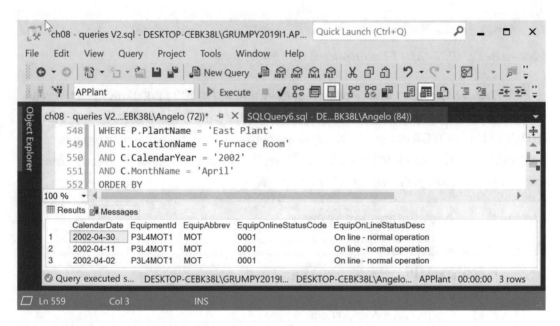

Figure 8-14. *Equipment failure status by day*

Wow, look at that! All "online – normal operation" status in the results.

This poses a problem. Looks like the motor was online for those three days. Fortunately, we also have a status history table by hour for those days (a lower level of granularity). Maybe we had some other events on one or more of the hours on those dates.

Let's execute the following query in Listing 8-9 for each of the three days.

Listing 8-9. Equipment Failure by Hour

```
SELECT ESBH.StatusDate
     ,ESBH.EquipmentId
     ,E.EquipAbbrev
     ,M.MotorName
     ,M.Rpm
     ,M.Voltage
     ,ESBH.StatusHour
     ,ESBH.EquipOnlineStatusCode
     ,EOLSC.EquipOnlineStatusDesc
FROM Reports.EquipmentStatusHistoryByHour ESBH
JOIN DimTable.Equipment E
ON ESBH.EquipmentId = E.EquipmentId
JOIN DimTable.Motor M
ON E.EquipmentId = M.EquipmentId
JOIN DimTable.EquipOnlineStatusCode EOLSC
ON ESBH.EquipOnlineStatusCode = EOLSC.EquipOnlineStatusCode
WHERE ESBH.EquipmentId = 'P3L4MOT1'
AND StatusDate = '2002-04-02'
GO
```

This query references the report table called `Reports.EquipmentStatusHistory ByHour` that contains equipment status by hour over 24-hour periods. Lots of joins between various tables to get some detailed information related to equipment and equipment status codes and descriptions.

Nothing fancy, just plain old TSQL coding with joins and filters. No aggregate window functions and OVER() clauses.

Executing this query yields the results for the first date we are examining.

Please refer to Figure 8-15.

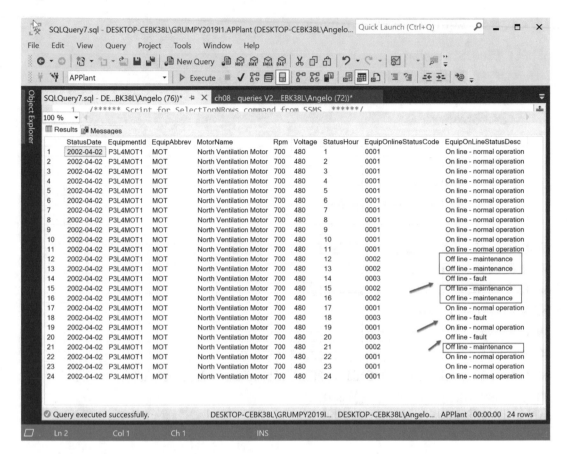

Figure 8-15. *Equipment failure for April 2, 2002*

Here are the 24-hour results for April 2, 2002. Yes, indeed, we do have three faults. That matches our results from the prior query. Notice also that the faults started to occur when there was a maintenance period for hours 12 and 13. The crew tried to get the motor online and it failed.

Some more maintenance occurred at hours 15 and 16, and another fault occurred at hour 18 after one hour of normal operation. Another fault occurred in hour 20 followed by one hour of maintenance at hour 21. After that the motor operated normally, so it looks like they got the issue finally fixed after three maintenance sessions (it would be interesting if we had data to show the names of the maintenance crew that were assigned to these failure events), until April 11, that is. Let's see what happened on that date.

Executing this query after we modify the WHERE clause, so it filters for April 11, 2002, yields the following results in Figure 8-16.

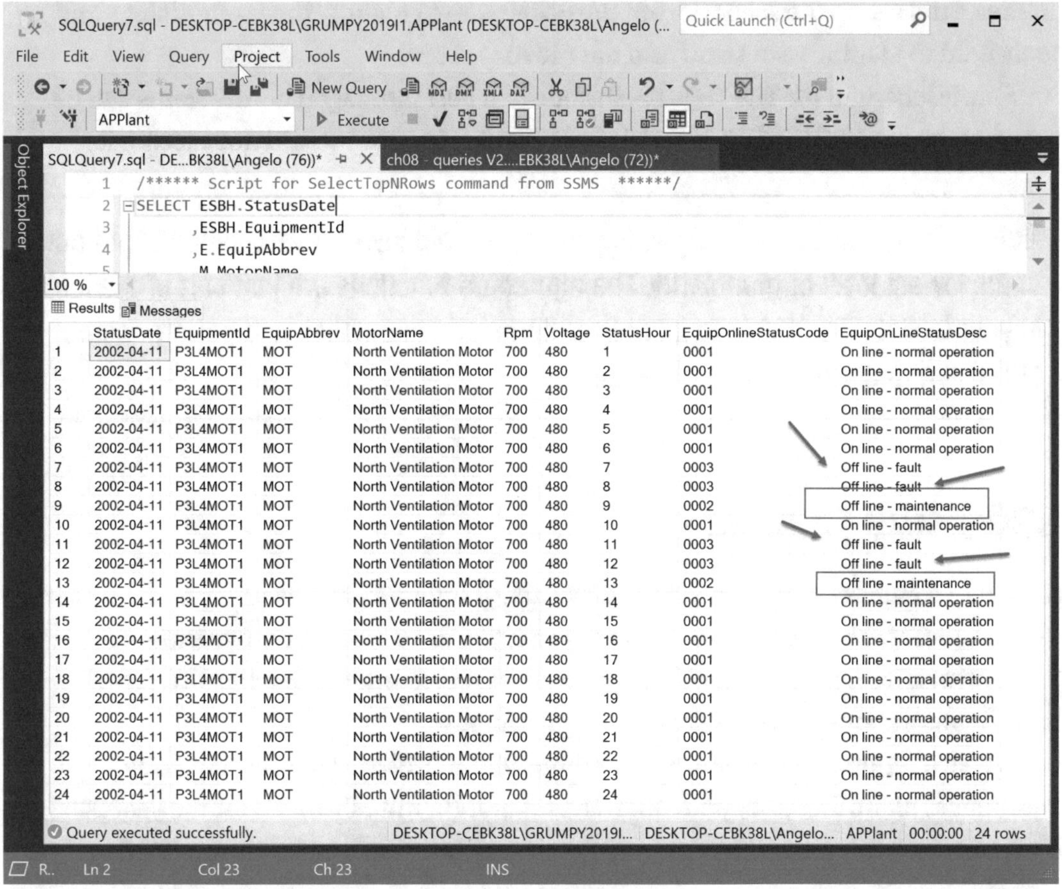

	StatusDate	EquipmentId	EquipAbbrev	MotorName	Rpm	Voltage	StatusHour	EquipOnlineStatusCode	EquipOnLineStatusDesc
1	2002-04-11	P3L4MOT1	MOT	North Ventilation Motor	700	480	1	0001	On line - normal operation
2	2002-04-11	P3L4MOT1	MOT	North Ventilation Motor	700	480	2	0001	On line - normal operation
3	2002-04-11	P3L4MOT1	MOT	North Ventilation Motor	700	480	3	0001	On line - normal operation
4	2002-04-11	P3L4MOT1	MOT	North Ventilation Motor	700	480	4	0001	On line - normal operation
5	2002-04-11	P3L4MOT1	MOT	North Ventilation Motor	700	480	5	0001	On line - normal operation
6	2002-04-11	P3L4MOT1	MOT	North Ventilation Motor	700	480	6	0001	On line - normal operation
7	2002-04-11	P3L4MOT1	MOT	North Ventilation Motor	700	480	7	0003	Off line - fault
8	2002-04-11	P3L4MOT1	MOT	North Ventilation Motor	700	480	8	0003	Off line - fault
9	2002-04-11	P3L4MOT1	MOT	North Ventilation Motor	700	480	9	0002	Off line - maintenance
10	2002-04-11	P3L4MOT1	MOT	North Ventilation Motor	700	480	10	0001	On line - normal operation
11	2002-04-11	P3L4MOT1	MOT	North Ventilation Motor	700	480	11	0003	Off line - fault
12	2002-04-11	P3L4MOT1	MOT	North Ventilation Motor	700	480	12	0003	Off line - fault
13	2002-04-11	P3L4MOT1	MOT	North Ventilation Motor	700	480	13	0002	Off line - maintenance
14	2002-04-11	P3L4MOT1	MOT	North Ventilation Motor	700	480	14	0001	On line - normal operation
15	2002-04-11	P3L4MOT1	MOT	North Ventilation Motor	700	480	15	0001	On line - normal operation
16	2002-04-11	P3L4MOT1	MOT	North Ventilation Motor	700	480	16	0001	On line - normal operation
17	2002-04-11	P3L4MOT1	MOT	North Ventilation Motor	700	480	17	0001	On line - normal operation
18	2002-04-11	P3L4MOT1	MOT	North Ventilation Motor	700	480	18	0001	On line - normal operation
19	2002-04-11	P3L4MOT1	MOT	North Ventilation Motor	700	480	19	0001	On line - normal operation
20	2002-04-11	P3L4MOT1	MOT	North Ventilation Motor	700	480	20	0001	On line - normal operation
21	2002-04-11	P3L4MOT1	MOT	North Ventilation Motor	700	480	21	0001	On line - normal operation
22	2002-04-11	P3L4MOT1	MOT	North Ventilation Motor	700	480	22	0001	On line - normal operation
23	2002-04-11	P3L4MOT1	MOT	North Ventilation Motor	700	480	23	0001	On line - normal operation
24	2002-04-11	P3L4MOT1	MOT	North Ventilation Motor	700	480	24	0001	On line - normal operation

***Figure 8-16.** Equipment failure for April 11, 2002*

This time we see the four faults. The first two occurred in hours 7 and 8. One hour of maintenance to attempt to fix it in hour 9 followed, and then there was one hour of normal operation in hour 10. Two more faults occurred in hours 11 and 12, and there was another hour of maintenance in hour 13 where it looks like they finally got the issue fixed.

So it looks like a meeting with the vendor is in order so as to find out why the maintenance did not go smoothly on these dates (also with the maintenance crew).

Why did we go through all these queries?

I was trying to illustrate how using the aggregate functions with the `OVER()` clauses can help find interesting patterns and anomalies that could suggest problems. In this case we identified a lot of failures on certain dates and used lower-level-granularity queries to drill down into the failures until we discovered that a vendor's maintenance crew was the issue.

Yes, I did set up the scenario, but the goals were to identify the methodology and tools used to identify issues and find out the root cause!

One last comment: The TSQL batch code to insert the staged hourly status can be found at the beginning of the script for this chapter in case you did not execute it.

Note   Sometimes your analysis tools are plain old lower-level queries to get down to the lowest level of granularity. The aggregate functions are just part of the tools in your toolkit. The trick is to identify patterns and anomalies and perform some drill-down to discover the root cause.

GROUPING() Function

Next, let's create some rollups via the GROUPING() function. Not as pretty and easy to read as Excel pivot tables, but it is available to you if you are a glutton for punishment!

As usual, we need to start off with a business requirements specification. Our friendly analyst supplies the following requirements:

I need a report showing summary rollups for failures at all levels, that is, by equipment, plant, location, year, quarter, and month. Initially this is for the year 2002 only, but the query should be written in such a manner as to be easily modified to include all years or just one product. The report needs to sort so that all NULL values float to the top, so the rollups are easy to read; otherwise, they are all over the report, and figuring out what the rollups are becomes impossible.

Additionally, create a query that will deliver the low-level data and copy and paste the results of the query in a Microsoft Excel spreadsheet with a pivot table so the results can be validated.

Wow, two specifications this time: one for the report and one for a spreadsheet to test and validate the query report. Listing 8-10 is the code we delivered to the analyst.

Listing 8-10. Equipment Failure Summary Rollups

```
SELECT
      PlantName
      ,LocationName
      ,CalendarYear
```

```
    ,QuarterName
    ,MonthName
    ,CalendarMonth
    ,EquipmentId
    ,SumEquipmentFailure
    ,SUM(SumEquipmentFailure) AS FailedEquipmentRollup
    ,GROUPING(SumEquipmentFailure) AS GroupingLevels
FROM [Reports].[EquipmentFailureStatistics]
WHERE CalendarYear = 2002
GROUP BY
    PlantName,LocationName,CalendarYear,QuarterName,MonthName,
    CalendarMonth,
    EquipmentId,SumEquipmentFailure WITH ROLLUP
ORDER BY PlantName,LocationName,CalendarYear,QuarterName,
    CalendarMonth,EquipmentId ASC,
            (
            CASE
                    WHEN SumEquipmentFailure IS NULL THEN 0
            END
          ) DESC,
    GROUPING(SumEquipmentFailure) DESC
GO
```

The SUM() aggregate function is used together with the GROUPING() function in the SELECT clause. The GROUP BY clause declares WITH ROLLUP at the end so we can generate subtotals at all the required levels. Lastly, the ORDER BY clause has a CASE block that tests the value of the SumEquipmentFailure column. If it is NULL, it replaces it with 0 and sorts the resulting values generated by the CASE block in descending order. Pretty nice trick if I say so myself.

The results will tell us if the trick worked. Please refer to the partial results in Figure 8-17.

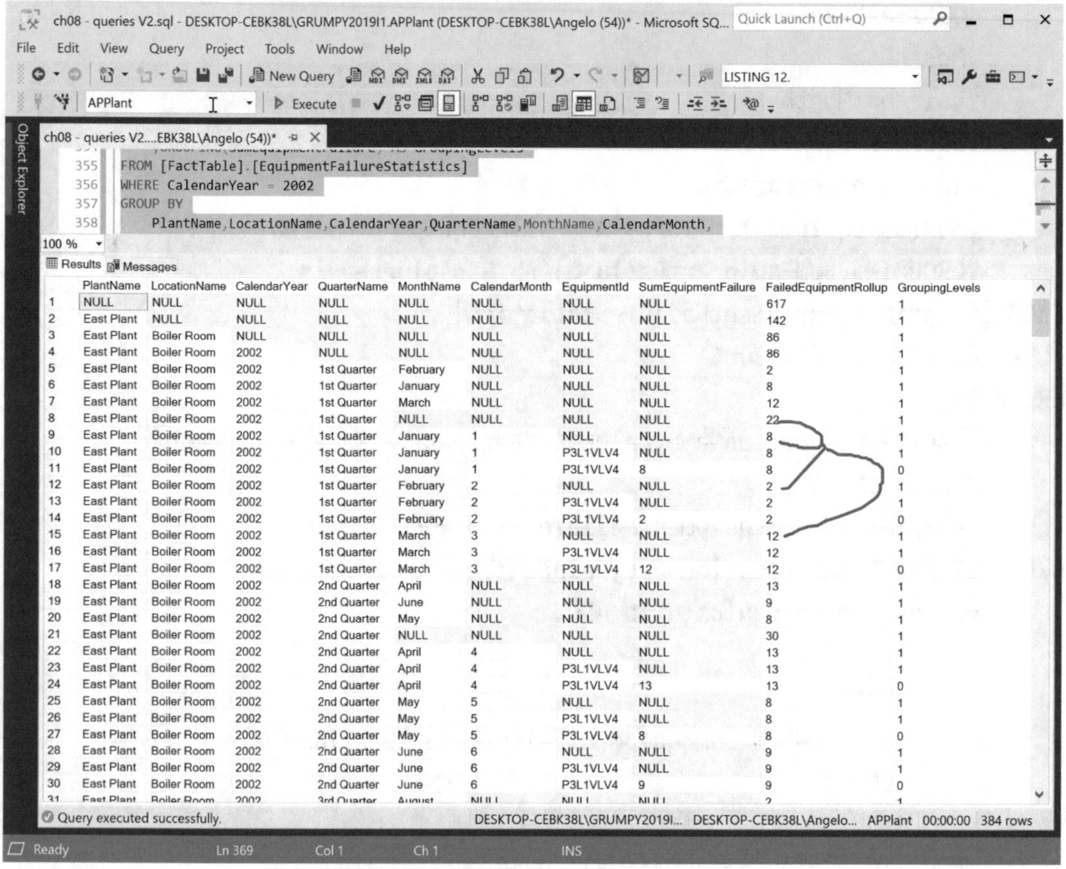

Figure 8-17. *Failure totals with rollups*

Does it roll up? If you look at the squiggly lines, we can see the following result: 22 = 8 + 2+ 12! So it does work. But this is hard to read and understand. Imagine if we had more than one equipment to report on. We can get lost quickly.

I took the query we just discussed and removed the SUM() and GROUPING() functions; plus I removed the GROUP BY clause and the CASE block in the ORDER BY clause in order to have a query that would generate the raw data.

I copied and pasted the results of the query into a Microsoft Excel spreadsheet and then created a pivot table. This can be seen in Figure 8-18.

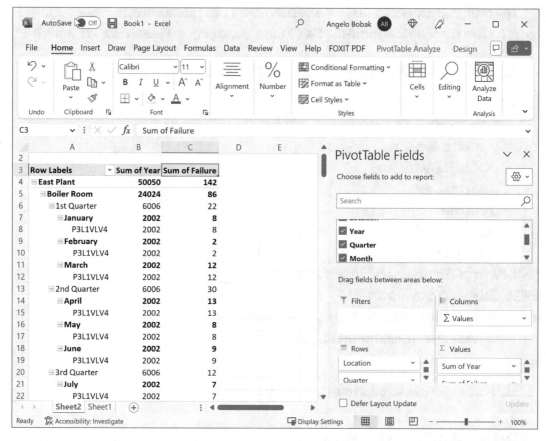

Figure 8-18. *Equipment failure summary pivot table report*

This is a lot better. You can drag and drop the columns you need to generate the rollups and can collapse or expand the levels. I think this option works a lot better than the GROUPING() function, but at least you have two tools you can use to generate rollups, TSQL and Excel.

Note   For very large volumes of data, you can load them into SSAS cubes and perform multidimensional data by connecting to the cubes with Excel or third-party web tools that connect to multidimensional cubes created with SSAS (SQL Server Analysis Services).

The code that generated the low-level results for this spreadsheet is in the TSQL script for this chapter after the GROUPING() function example. Please refer to Listing 8-11.

Listing 8-11. Pivot Table Query for Microsoft Excel

```
SELECT
      PlantName
      ,LocationName
      ,CalendarYear
      ,QuarterName
      ,MonthName
      ,EquipmentId
      ,SumEquipmentFailure
FROM Reports.EquipmentFailureStatistics
WHERE CalendarYear = 2002
ORDER BY PlantName,LocationName,CalendarYear,QuarterName,
      CalendarMonth,EquipmentId ASC
GO
```

Next, we aggregate some strings to generate plant information streams.

STRING_AGG() Function

The following example will generate a set of data for one day's worth of temperature results for one piece of equipment. A table variable is used to store the hour values for a day, and the STRING_AGG() function is used in a query that retrieves the required columns. The results are assembled in a temporary table for processing. You will use this technique if you need to send data streams from the database to an external application, like a website or another database platform. The streams use commas and colons as delimiters so the data can be parsed at the other end.

Tip   You can use the logic developed in our scripts and place them in functions so that you can repeatedly use them with different parameters. This really would be a powerful set of tools.

Back to our example. Please refer to Listing 8-12 for the first part of the script.

Listing 8-12. Assemble a Temperature Data Stream

```
DROP TABLE IF EXISTS #GeneratorTemperature
GO

DECLARE @Hour TABLE (
      SampleHour SMALLINT
);

INSERT INTO @Hour VALUES
(1),(2),(3),(4),(5),(6),(7),(8),(9),(10),(11),(12),
(13),(14),(15),(16),(17),(18),(19),(20),(21),(22),(23),(24);

SELECT G.PlantId
      ,G.LocationId
      ,G.EquipmentId
      ,G.GeneratorId
      ,G.GeneratorName
      ,C.CalendarDate
      ,H.SampleHour
      ,UPPER (
            CONVERT(INT,CRYPT_GEN_RANDOM(1)
                  ) * 5) AS Temperature
      ,G.Voltage
  INTO #GeneratorTemperature
  FROM DimTable.Generator G
  CROSS JOIN DimTable.Calendar C
  CROSS JOIN @Hour H
  ORDER BY G.PlantId
      ,G.LocationId
      ,G.EquipmentId
      ,G.GeneratorId
      ,G.GeneratorName
      ,C.CalendarDate
      ,H.SampleHour;
```

The small table variable is loaded with the values 1–24, which represent the hours of a day. A temporary table is created on the fly by retrieving the necessary columns together with a randomly generated temperature value. This is done with the CRYPT_GEN_RANDOM() function, which I think works better in a query than the RAND() function.

Two CROSS JOINS are performed against the Generator, Calendar, and @Hour tables in order to create a cross product of all the values for our test data set. I included an ORDER BY clause so we can sort the data. This part alone generated 736,320 rows of data and took 15 seconds to run. We will not perform any estimated query plan against this, as it is only for demonstration purposes.

The second part of the script is shown in Listing 8-13.

Listing 8-13. Create the Message Stream

```
SELECT STRING_AGG('<msg start->PlantId:'+ PlantId
        + ',Location:' + LocationId
        + ',Equipment:' + EquipmentId
        + ',GeneratorId:' + GeneratorId
        + ',Generator Name:' + GeneratorName
        + ',Voltage:' + CONVERT(VARCHAR,Voltage)
        + ',Temperature:' + CONVERT(VARCHAR,Temperature)
        + ',Sample Date:' + CONVERT(VARCHAR,CalendarDate)
        + ',Sample Hour:' + CONVERT(VARCHAR,SampleHour) + '>','!')
FROM #GeneratorTemperature
WHERE CalendarDate = '2005-11-30'
AND PlantId = 'PP000001'
GROUP BY PlantId
    ,LocationId
    ,EquipmentId
    ,GeneratorId
    ,GeneratorName
    ,CalendarDate
    ,SampleHour
GO

DROP TABLE IF EXISTS #GeneratorTemperature
GO
```

Running the query a second time took only one second. A lot of the data and the query plan must have been in cache.

As can be seen, the query has a WHERE clause to filter out information for only one day and one plant ID.

The STRING_AGG() function is passed the strings you want to concatenate separated by commas. It is a good idea to format the query by listing the text strings in vertical fashion so you can see where to place the delimiters.

In Figure 8-19 are the results.

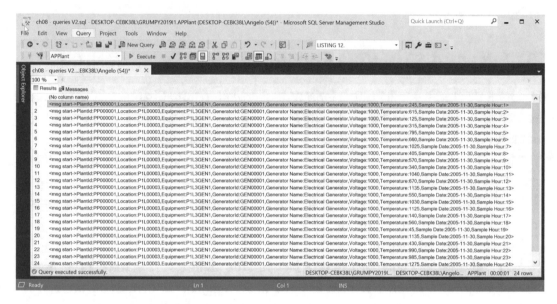

Figure 8-19. *Equipment statistics data stream*

Let's pull out the first row so you can see how the row was formatted:

```
<msg start->
PlantId:PP000001,
Location:P1L00003,
Equipment:P1L3GEN1,
GeneratorId:GEN00001,
Generator Name:Electrical Generator,
Voltage:1000,
Temperature:130,
Sample Date:2005-11-30,
Sample Hour:1
>
```

As I said, this has uses for transmitting data streams between platforms. See if you can modify the script so it loads a second table and then try to write a query that tokenizes each of the elements in the row stream so you can load them into a new table, like a table in a different database or even a different instance of SQL Server.

This technique can also be used to populate websites so analysts and other business stakeholders can view results via their web browsers. For this you need some Java programming skills!

This skill will one day come in handy! So add it to your knowledge base. Now it's statistics time.

STDEV() and STDEVP() Functions

As a reminder, recall our definition for standard deviation from Chapter 2 (skip this if you are comfortable with how this works and go right to the business specification for the example query). If you need a refresher or this is new to you, continue reading:

The STDEV() function calculates the statistical standard deviation of a set of values when the entire population is not known or available.

The STDEVP() function calculates the statistical standard deviation when the entire population of the data set is known.

But what does this mean?

What this means is how close or how far apart the numbers are in the data set from the arithmetic **MEAN** (which is another word for average). In our context the average or **MEAN** is the sum of all data values in question divided by the number of data values. There are other types of **MEAN** like geometric, harmonic, and weighted **MEAN**, which I will briefly cover in Appendix B.

Recall how the standard deviation is calculated.

Here is the algorithm to calculate the standard deviation, which you can implement on your own with TSQL if you are adventurous:

- Calculate the average (**MEAN**) of all values in question and store it in a variable. You can use the AVG() function.

- Loop around the value set. For each value in the data set

  - Subtract the calculated average.

  - Square the result.

- Take the sum of all the preceding results and divide the value by the number of data elements minus 1. (This step calculates the variance; to calculate for the entire population, do not subtract 1.)

- Now take the square root of the variance, and that's your standard deviation.

Or you could just use the STDEV() function! There is a STDEVP() function as mentioned earlier. As stated earlier, the only difference is that STDEV() works on part of the population of the data. This is when the entire data set is not known. The STDEVP() function works on the entire data set, and this is what is meant by the entire population of data.

Once again here is the syntax:

```
SELECT STDEV( ALL | DISTINCT ) AS [Column Alias]
FROM [Table Name]
GO
```

You need to supply a column name that contains the data you want to use. You can optionally supply the ALL or DISTINCT keyword if you wish. If you leave these out, just like the other functions, the behavior will default to ALL. Using the DISTINCT keyword will ignore duplicate values in the column (make sure this is what you want to do).

Here come the business requirements for this next report supplied to us by our business analyst:

A quarterly rolling standard deviation report for equipment failures is required. Right now, we are interested in equipment located in the boiler room of the east power plant. This is a summary report, so individual equipment failures are not required, just totals so the information in the EquipmentFailureStatistics table can be used.

Recall that the EquipmentFailureStatistics table takes the place of the original CTE query approach we were using. This is preloaded nightly. This approach proved to increase performance as we do not have to generate totals each time the query is run.

Listing 8-14 is the TSQL script that satisfies the business requirements from the analyst.

Listing 8-14. Quarterly Rolling Equipment Failure Standard Deviations

```
SELECT PlantName     AS Plant
      ,LocationName AS Location
      ,CalendarYear AS [Year]
      ,QuarterName  AS [Quarter]
      ,MonthName            AS [Month]
      ,SumEquipmentFailure AS EquipFailure
      ,STDEV(SumEquipmentFailure) OVER (
            PARTITION BY PlantName,LocationName,CalendarYear,QuarterName
            ORDER BY PlantName,LocationName,CalendarYear
      ) AS StDevQtr
      ,STDEVP(SumEquipmentFailure) OVER (
            PARTITION BY PlantName,LocationName,CalendarYear,QuarterName
            ORDER BY PlantName,LocationName,CalendarYear
      ) AS StDevpQtr
FROM Reports.EquipmentFailureStatistics
WHERE CalendarYear IN(2002,2003)
AND PlantName = 'East Plant'
AND LocationName = 'Boiler Room'
ORDER BY PlantName,LocationName,CalendarYear,CalendarMonth
GO
```

This is an easy requirement to fulfill. The STDEV() function uses an OVER() clause that defines a partition by plant name, location name, year, and quarter. The partition is sorted by plant name, location name, and calendar year.

The STDEVP() function was also included, and it uses the same OVER() clause as the STDEV() function.

If you are getting used to these concepts, you might be asking why include the plant name and location name if we have a WHERE clause that filters for one plant and one location only.

That's true. We do not need the plant name and location name in the PARTITION BY clause. I modified the query to exclude them and got the same results. I also got the same estimated execution plan as shown in the Figure 8-20.

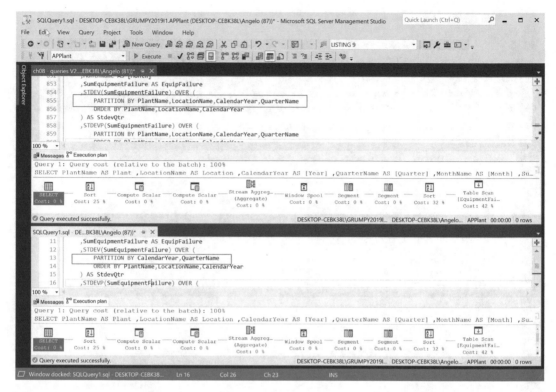

Figure 8-20. *Comparing execution plans*

As can be seen, there's the same estimated execution plan for both queries. The `IO` and `TIME` statistics were the same for both, so no harm done. Good coding practice dictates to use the least amount of code to get the desired and correct results. This query can be used for more than one plant and more than one location, so if you have the full `PARTITION BY` clause, you can use it as is. Otherwise, comment it out and use the smaller one if all you will ever use this query on is one plant and location.

I modified the query so all plants and locations were used, and the change to the estimated query plan was minor; a window spool step was added at a cost of 1%. So keep all these items in consideration when creating your scripts and queries.

Back to the query, please refer to the partial results in Figure 8-21.

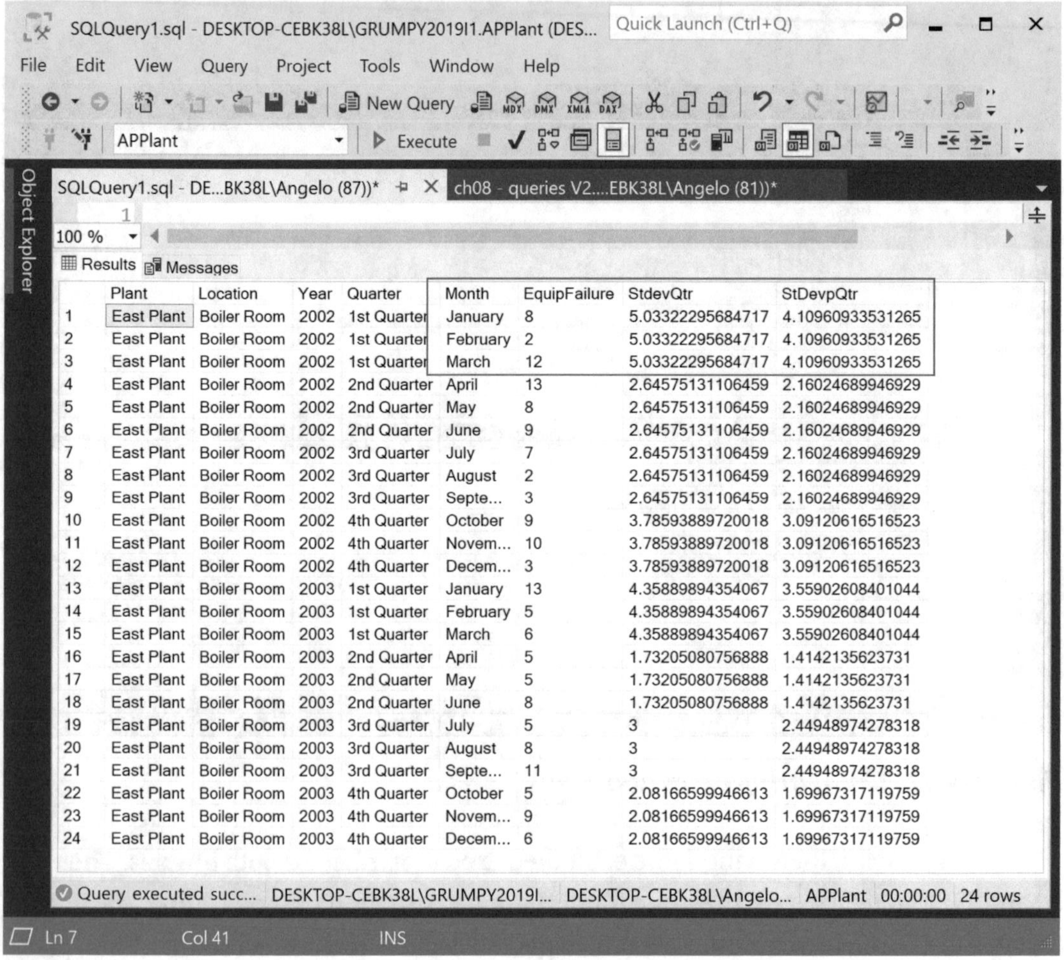

	Plant	Location	Year	Quarter	Month	EquipFailure	StdevQtr	StDevpQtr
1	East Plant	Boiler Room	2002	1st Quarter	January	8	5.03322295684717	4.10960933531265
2	East Plant	Boiler Room	2002	1st Quarter	February	2	5.03322295684717	4.10960933531265
3	East Plant	Boiler Room	2002	1st Quarter	March	12	5.03322295684717	4.10960933531265
4	East Plant	Boiler Room	2002	2nd Quarter	April	13	2.64575131106459	2.16024689946929
5	East Plant	Boiler Room	2002	2nd Quarter	May	8	2.64575131106459	2.16024689946929
6	East Plant	Boiler Room	2002	2nd Quarter	June	9	2.64575131106459	2.16024689946929
7	East Plant	Boiler Room	2002	3rd Quarter	July	7	2.64575131106459	2.16024689946929
8	East Plant	Boiler Room	2002	3rd Quarter	August	2	2.64575131106459	2.16024689946929
9	East Plant	Boiler Room	2002	3rd Quarter	Septe...	3	2.64575131106459	2.16024689946929
10	East Plant	Boiler Room	2002	4th Quarter	October	9	3.78593889720018	3.09120616516523
11	East Plant	Boiler Room	2002	4th Quarter	Novem...	10	3.78593889720018	3.09120616516523
12	East Plant	Boiler Room	2002	4th Quarter	Decem...	3	3.78593889720018	3.09120616516523
13	East Plant	Boiler Room	2003	1st Quarter	January	13	4.35889894354067	3.55902608401044
14	East Plant	Boiler Room	2003	1st Quarter	February	5	4.35889894354067	3.55902608401044
15	East Plant	Boiler Room	2003	1st Quarter	March	6	4.35889894354067	3.55902608401044
16	East Plant	Boiler Room	2003	2nd Quarter	April	5	1.73205080756888	1.4142135623731
17	East Plant	Boiler Room	2003	2nd Quarter	May	5	1.73205080756888	1.4142135623731
18	East Plant	Boiler Room	2003	2nd Quarter	June	8	1.73205080756888	1.4142135623731
19	East Plant	Boiler Room	2003	3rd Quarter	July	5	3	2.44948974278318
20	East Plant	Boiler Room	2003	3rd Quarter	August	8	3	2.44948974278318
21	East Plant	Boiler Room	2003	3rd Quarter	Septe...	11	3	2.44948974278318
22	East Plant	Boiler Room	2003	4th Quarter	October	5	2.08166599946613	1.69967317119759
23	East Plant	Boiler Room	2003	4th Quarter	Novem...	9	2.08166599946613	1.69967317119759
24	East Plant	Boiler Room	2003	4th Quarter	Decem...	6	2.08166599946613	1.69967317119759

***Figure 8-21.** Rolling quarterly failures by plant, location, and year*

Notice the slight difference between `STDEV()` and `STDEVP()` results.

Pop Quiz Why are values generated by the `STDEVP()` function always smaller?

We will use the standard deviation results to generate a bell curve. The bell curve is generated by values called normal distributions, which are derived from the standard deviation, mean, and each individual value in the data set.

Let's see what graphs this data generates in a Microsoft Excel spreadsheet. Please refer to Figure 8-22.

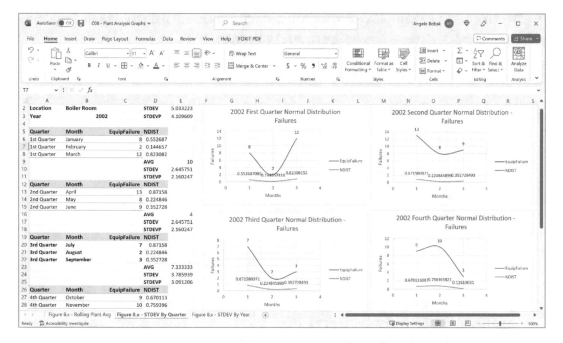

Figure 8-22. *Standard deviation by quarter for boiler room equipment failure*

This time I used the multi-graph approach, that is, one bell curve graph for each year-quarter combination. A bit wonky, as we have inverted bell curves and one regular (sort of) bell curve.

Note   We will discuss the steps to take in the Microsoft Excel spreadsheet to generate normal distributions, mean values, standard deviations, and bell curves in Appendix B.

What can be seen is a graph for each quarter for 2002. I had to generate the average for the quarters as the normal distribution function (NORM.DIST) in Excel. This function uses the value, the MEAN (average), and the standard deviation generated by the SQL Server query.

Try to generate these graphs by yourself, but I will include the spreadsheet in the folder that contains the code for this chapter. Reminder, this is a good skill to have besides your TSQL skills.

Let's try this analysis by year. Please refer to Listing 8-15.

Here is one final example where we will generate the normal distributions from the yearly standard deviations for two years, 2002 and 2003.

Listing 8-15. Yearly Standard Deviation

```
SELECT PlantName AS Plant
      ,LocationName AS Location
      ,CalendarYear AS [Year]
      ,QuarterName  AS [Quarter]
      ,MonthName    AS Month
      ,SumEquipmentFailure As TotalFailures
      ,STDEV(SumEquipmentFailure) OVER (
            PARTITION BY PlantName,LocationName,CalendarYear
            ORDER BY CalendarYear
      ) AS RollingStdev
      ,STDEVP(SumEquipmentFailure) OVER (
            PARTITION BY PlantName,LocationName,CalendarYear
            ORDER BY CalendarYear
      ) AS RollingStdevp
FROM Reports.EquipmentFailureStatistics
WHERE CalendarYear IN(2002,2003)
AND PlantName = 'East Plant'
AND LocationName = 'Boiler Room'
ORDER BY PlantName,LocationName,CalendarYear,CalendarMonth
GO
```

Notice the square brackets in some of the column aliases in the SELECT clause. The aliases used are reserved words in SQL Server, so using the brackets eliminates the colored warnings. No harm is done if you use the keywords, but this is one way around looking at all those colors.

One minor change: The PARTITION BY clause was modified so it does not include the quarter name, and the ORDER BY clause was modified so it only includes the year. The same argument as before applies; since we are filtering for one plant and location only, we really do not need these in the PARTITION BY clause.

Assignment   Once you are comfortable with how the query works, modify it for more than one year and location. This is a good way to develop and test queries that can be used in Microsoft SSRS (SQL Server Reporting Services) reports.

Here are the results we will copy and paste into a Microsoft Excel spreadsheet. Please refer to Figure 8-23.

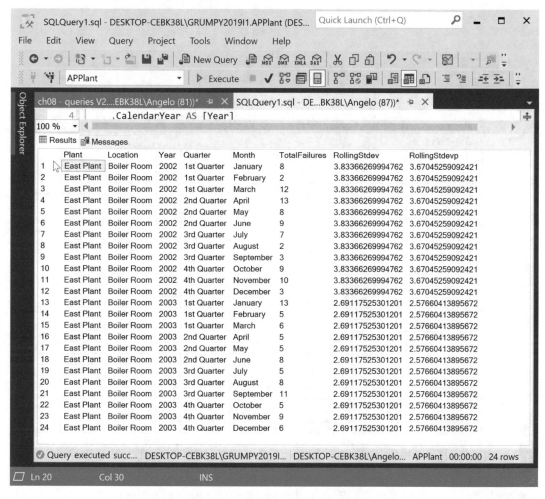

Figure 8-23. *Yearly failure standard deviation for the East Plant, boiler room*

Okay, nice numbers, but as we always say, a picture is worth a thousand words, or in our case a thousand numbers. Each year has its own standard deviation value for all months. In other words, each month has the same standard deviation value.

A little copy and paste action plus some creative formatting into a Microsoft Excel spreadsheet yields the following report in Figure 8-24 that includes the normal distribution graphs for both years.

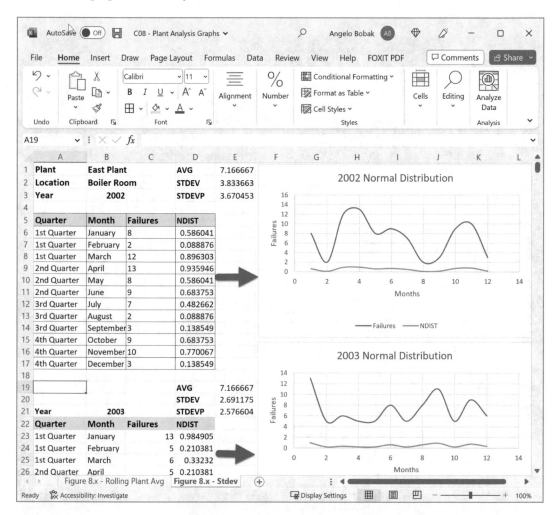

Figure 8-24. *Normal distribution bell curve for equipment failures*

Better but still a wonky bell curve. Each is a set of bell curves if we want to get picky. In other words, instead of the entire year, you can generate bell curves for each quarter in the calendar year.

More statistical analysis, this time with the variance functions, is on deck.

Pop Quiz Answer   Why are values generated by the STDEVP() function always smaller? Because they have more data to work with.

VAR() and VARP() Functions

Here are the definitions from Chapter 2 for these functions. Again, skip them if you are comfortable with how they work and go to the business specifications for the example query.

Read them if you read Chapter 1 and then jumped directly here because your interest is engineering or power plant management or some other related discipline.

The VAR() and VARP() functions are used to generate the variance for a set of values in a data sample. The same discussion related to the data sets used in the STDEV()and STDEVP() examples applies to the VAR() and VARP() functions relative to the population of the data.

The VAR() function works on a partial set of the data when the entire data set is not known or available, while the VARP() function works on the whole data set population when it is known.

So, if your data set consists of ten rows, the function names ending with P will include all ten rows when processing the values; the ones not ending in the letter P will look at N – 1 or 10 – 1 = 9 rows.

Okay, now that we got that out of the way, what does variance mean anyway?

For us developers and non-statisticians, here's a simple explanation:

Informally, if you take all the differences of each value from the mean (average) and square the results and then add all the results and finally divide them by the number of data samples, you get the variance.

Reminder   If you take the square root of the variance, you get the standard deviation, so there is some sort of relationship going on!

For our engineering scenario, let's say we had some data that forecasted equipment online day targets for one of our pieces of plant equipment (like a motor) during any particular year. When equipment status history is finally generated for the motor, we want to see how close or far the online days were to or from the forecast. It answers the following questions:

- Did we miss the online event targets?

- Did we meet our online event targets?

- Did we exceed our online event targets (like no downtime)?

This statistical formula is also discussed in Appendix B by the way.

The next examples will show how the VAR() and VARP() functions can be used with three different PARTITION BY clauses.

Example 1: Rolling Variance

Here come the requirements for our first variance report:

Create a rolling variance report for plant location equipment failures. Set up a report by plant, location, quarter, and month. Rolling variance should include any previous row and the current row in the generation of the variance value as results change from month to month.

A simple report, we use one of the queries from the standard deviation examples, modify it a bit, and add a ROWS clause in the PARTITION BY clause.

Please refer to Listing 8-16 for the new query.

Listing 8-16. Rolling Variance by Plant, Location, Year, and Month

```
SELECT PlantName
    ,LocationName
    ,CalendarYear
    ,QuarterName
    ,MonthName
    ,CalendarMonth
    ,SumEquipmentFailure
    ,VAR(SumEquipmentFailure) OVER (
        PARTITION BY PlantName,LocationName,CalendarYear
        ORDER BY CalendarMonth
```

```
            ROWS BETWEEN UNBOUNDED PRECEDING AND CURRENT ROW
      ) AS RollingVar
      ,VARP(SumEquipmentFailure) OVER (
            PARTITION BY PlantName,LocationName,CalendarYear
            ORDER BY CalendarMonth
            ROWS BETWEEN UNBOUNDED PRECEDING AND CURRENT ROW
      ) AS RollingVarp
FROM [Reports].[EquipmentFailureStatistics]
WHERE CalendarYear IN(2002,2003)
AND PlantName = 'East Plant'
AND LocationName = 'Boiler Room'
ORDER BY PlantName,LocationName,CalendarYear,CalendarMonth
GO
```

As can be seen we added the ROWS clause and kept the WHERE clause to limit the results for the boiler room in the East Plant. This query can be modified to include all plants and locations if we want to pop the results into a Microsoft Excel pivot table.

Please refer to the partial results in Figure 8-25.

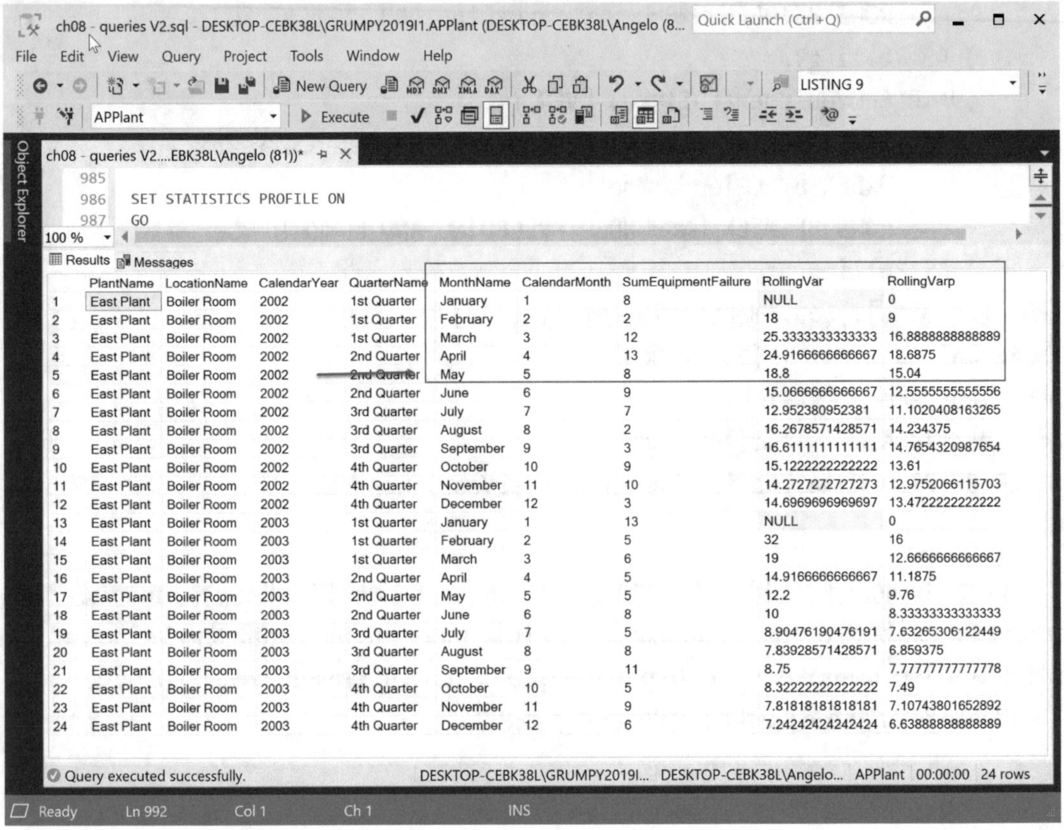

	PlantName	LocationName	CalendarYear	QuarterName	MonthName	CalendarMonth	SumEquipmentFailure	RollingVar	RollingVarp
1	East Plant	Boiler Room	2002	1st Quarter	January	1	8	NULL	0
2	East Plant	Boiler Room	2002	1st Quarter	February	2	2	18	9
3	East Plant	Boiler Room	2002	1st Quarter	March	3	12	25.3333333333333	16.8888888888889
4	East Plant	Boiler Room	2002	2nd Quarter	April	4	13	24.9166666666667	18.6875
5	East Plant	Boiler Room	2002	2nd Quarter	May	5	8	18.8	15.04
6	East Plant	Boiler Room	2002	2nd Quarter	June	6	9	15.0666666666667	12.5555555555556
7	East Plant	Boiler Room	2002	3rd Quarter	July	7	7	12.952380952381	11.1020408163265
8	East Plant	Boiler Room	2002	3rd Quarter	August	8	2	16.2678571428571	14.234375
9	East Plant	Boiler Room	2002	3rd Quarter	September	9	3	16.6111111111111	14.7654320987654
10	East Plant	Boiler Room	2002	4th Quarter	October	10	9	15.1222222222222	13.61
11	East Plant	Boiler Room	2002	4th Quarter	November	11	10	14.2727272727273	12.9752066115703
12	East Plant	Boiler Room	2002	4th Quarter	December	12	3	14.6969696969697	13.4722222222222
13	East Plant	Boiler Room	2003	1st Quarter	January	1	13	NULL	0
14	East Plant	Boiler Room	2003	1st Quarter	February	2	5	32	16
15	East Plant	Boiler Room	2003	1st Quarter	March	3	6	19	12.6666666666667
16	East Plant	Boiler Room	2003	2nd Quarter	April	4	5	14.9166666666667	11.1875
17	East Plant	Boiler Room	2003	2nd Quarter	May	5	5	12.2	9.76
18	East Plant	Boiler Room	2003	2nd Quarter	June	6	8	10	8.33333333333333
19	East Plant	Boiler Room	2003	3rd Quarter	July	7	5	8.90476190476191	7.63265306122449
20	East Plant	Boiler Room	2003	3rd Quarter	August	8	8	7.83928571428571	6.859375
21	East Plant	Boiler Room	2003	3rd Quarter	September	9	11	8.75	7.77777777777778
22	East Plant	Boiler Room	2003	4th Quarter	October	10	5	8.32222222222222	7.49
23	East Plant	Boiler Room	2003	4th Quarter	November	11	9	7.81818181818181	7.10743801652892
24	East Plant	Boiler Room	2003	4th Quarter	December	12	6	7.24242424242424	6.63888888888889

Figure 8-25. *Rolling variance for plant location failures*

The box defines a partition for five rows that include values for January, February, March, April, and May. The current row for processing is row 5 for this frame. The first VAR() value is NULL, and the remaining values are not NULL until a new year is encountered. I copied the 24 rows into Microsoft Excel and was able to duplicate the results with Excel functions. If you do this, remember that VAR($G$1,G5) is not the same as VAR($G$1:G5). You need to use the semicolon in order to define an ever-increasing range that includes all prior rows:

```
VAR($G$1:G5)
VAR($G$1:G6)
VAR($G$1:G7)
```

And so on. The comma version will only include two rows each time. Wrong! I forgot this and it drove me nuts. Could not figure out why the SQL code generated different values than Microsoft Excel. Easy trap to fall into.

By the way, try this code snippet in the preceding query, and you will see that if we take the square root of the variance, we do indeed get the standard deviation:

```
      ,SQRT(
            VAR(SumEquipmentFailure) OVER (
            PARTITION BY PlantName,LocationName,CalendarYear
            ORDER BY CalendarMonth
            ROWS BETWEEN UNBOUNDED PRECEDING AND CURRENT ROW
            )
      ) AS MyRollingStdev
      ,STDEV(SumEquipmentFailure) OVER (
            PARTITION BY PlantName,LocationName,CalendarYear
            ORDER BY CalendarMonth
            ROWS BETWEEN UNBOUNDED PRECEDING AND CURRENT ROW
      ) AS RollingStdev
Next example.
```

Example 2: Variance by Quarter

Here come the requirements for our second variance report:

Create a variance report by quarter for plant location equipment failures. Set up a report by plant, location, quarter, and month. The variance should include any previous row and any following row within the partition.

We just copy and paste the prior query and make some minor adjustments.

Please refer to Listing 8-17.

Listing 8-17. Variance Report by Quarter for Plant and Location

```
SELECT PlantName           AS Plant
      ,LocationName        AS Location
      ,CalendarYear        AS Year
      ,QuarterName         AS Quarter
      ,MonthName           AS Month
      ,SumEquipmentFailure AS EquipFailure
      ,VAR(SumEquipmentFailure) OVER (
            PARTITION BY PlantName,LocationName,CalendarYear,QuarterName
            ORDER BY PlantName,LocationName,CalendarYear
            ROWS BETWEEN UNBOUNDED PRECEDING AND UNBOUNDED FOLLOWING
      ) AS VarByQtr
      ,VARP(SumEquipmentFailure) OVER (
            PARTITION BY PlantName,LocationName,CalendarYear,QuarterName
            ORDER BY PlantName,LocationName,CalendarYear
            ROWS BETWEEN UNBOUNDED PRECEDING AND UNBOUNDED FOLLOWING
      ) AS VarpByQtr
FROM Reports.EquipmentFailureStatistics
WHERE CalendarYear IN(2002,2003)
AND PlantName = 'East Plant'
AND LocationName = 'Boiler Room'
ORDER BY PlantName,LocationName,CalendarYear,CalendarMonth
GO
```

Notice the PARTITION BY clause. The partition is defined by plant, location, year, and quarter. The partition is sorted by plant name, location name, and year for processing.

Again, there is the WHERE clause that limits the result set to two years and to one plant and one location within the plant. I did this to keep the result set small, but once you practice, you can remove these. Right now, as the query stands, we do not need the plant name and location name in the `PARTITION BY` and `ORDER BY` clauses as we previously discussed because of the `WHERE` clause.

Please refer to Figure 8-26 for the results.

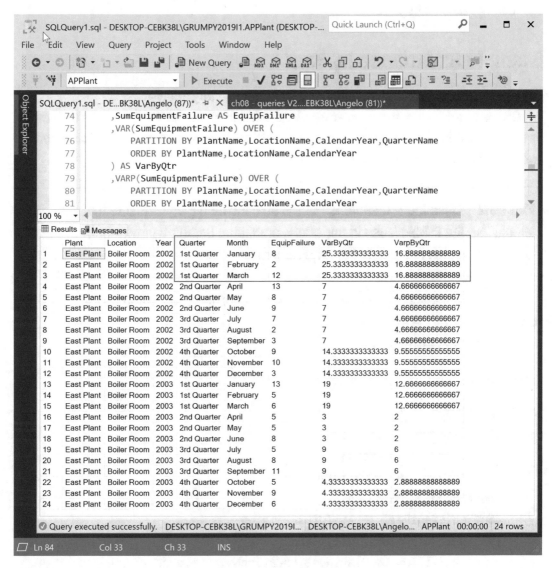

Figure 8-26. *Variance for plant location failures by quarter*

As can be seen, each quarter gets its own variance result. This was also validated in Microsoft Excel.

Last but not least, let's build up our variance statistical profile report set by calculating variance by year.

Example 3: Variance by Year

Here's a short requirements specification from our friendly analyst:

Create a variance report by year for plant location equipment failures. Set up a report by plant, location, quarter, and month. The variance should include any previous row and any following row within the partition.

Pop Quiz Does the last sentence of the specification give us a clue that we should use the ROWS clause in our query, or will the default behavior of the OVER() clause we implement satisfy the requirement?

Once again we just copy and paste the prior query and make some minor adjustments to the OVER() clause.

Please refer to Listing 8-18.

Listing 8-18. Rolling Variance by Year

```
SELECT PlantName    AS Plant
      ,LocationName AS Location
      ,CalendarYear AS [Year]
      ,QuarterName  AS [Quarter]
      ,MonthName    AS Month
      ,SumEquipmentFailure As TotalFailures
      ,VAR(SumEquipmentFailure) OVER (
            PARTITION BY PlantName,LocationName,CalendarYear
            ORDER BY CalendarYear
      ) AS VarByYear
      ,VARP(SumEquipmentFailure) OVER (
            PARTITION BY PlantName,LocationName,CalendarYear
            ORDER BY CalendarYear
```

```
    ) AS VarpByYear
FROM Reports.EquipmentFailureStatistics
WHERE CalendarYear IN(2002,2003)
AND PlantName = 'East Plant'
AND LocationName = 'Boiler Room'
ORDER BY PlantName,LocationName,CalendarYear,CalendarMonth
GO
```

This time we omit the ROWS clause and set up a partition by plant name, location, and year. The partition is ordered by the year. With this query we have a nice time profile of the variance of failures, namely, the rolling monthly variances, the variances by quarter, and finally the variances by year. Let's look at the results in Figure 8-27.

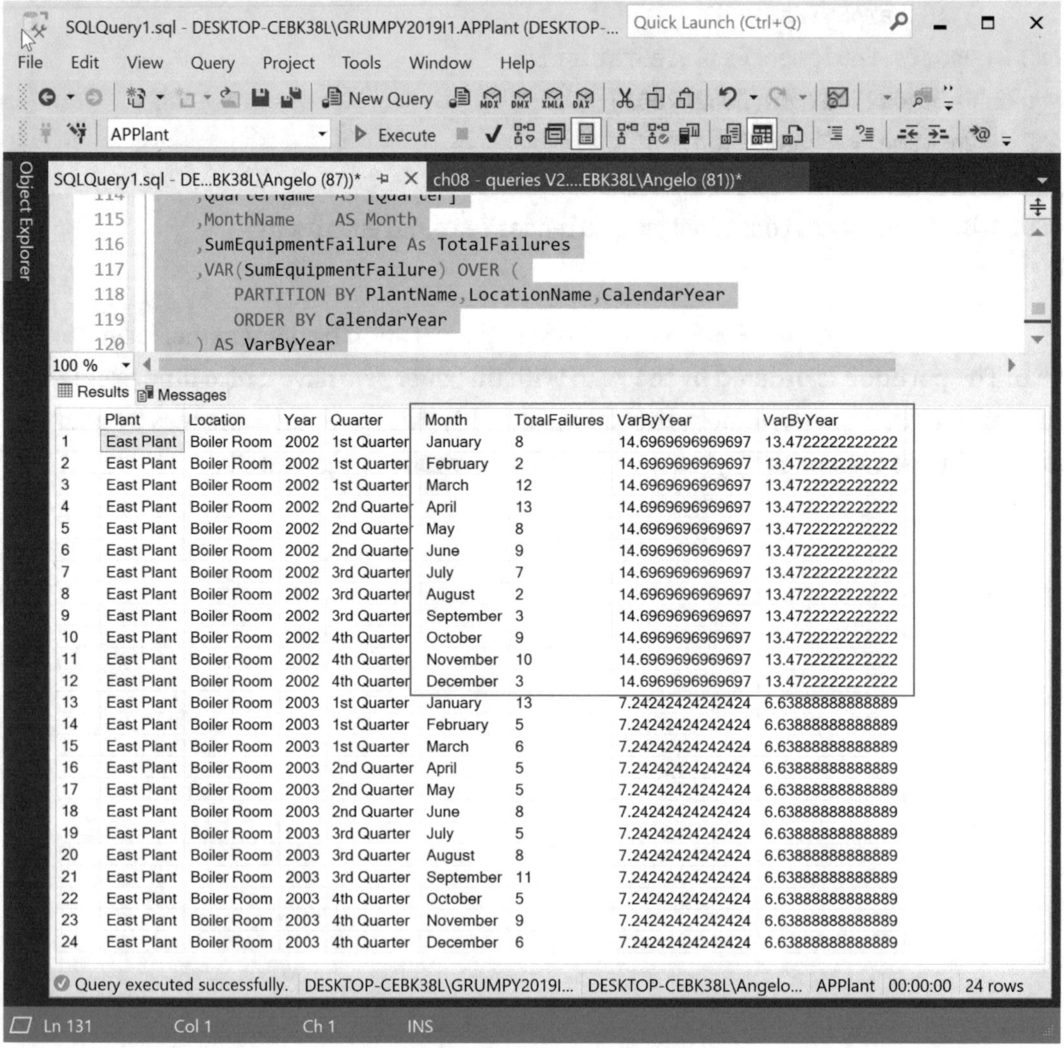

Figure 8-27. *Plant location equipment failure variance by year*

Each year has a single variance value for all months. Notice that the VARP() function generated a smaller variance in each year. That's because it took into account the entire population of the data set.

Pop Quiz Answer   Does the last sentence of the specification give us a clue that we should use the ROWS clause in our query, or will the default behavior of the OVER() clause we implement satisfy the requirement? No, looks like the default behavior satisfies the requirement. When you include a PARTITION BY clause and an ORDER BY clause, you include all prior rows and following rows relative to the current row being processed.

Performance Considerations

Let's close out this chapter with a little performance tuning. We will use the statistical aggregate functions and the average function in one query to see the complexity of the estimated query plan that we will generate.

Recall we are going to use a report table that replaced the CTE, so performance should be acceptable even without indexes. This table has only 2,021 rows, so it is very small and could very well be implemented as a memory-enhanced table.

Before we try the memory-optimized table approach, let's proceed and see if indexes help or hinder our performance goals. We will even try a clustered index.

The requirement is to create a query that can be used to provide a statistical profile of plant failures. The AVG(), STDEV() and STDEVP(), and VAR() and VARP() functions will be used. We want rolling values by plant, location, calendar year, and month.

Let's examine the query.

Please refer to Listing 8-19.

Listing 8-19. Plant Failure Statistical Profile

```
SELECT PlantName
      ,LocationName
      ,CalendarYear
      ,QuarterName
      ,MonthName
      ,CalendarMonth
      ,SumEquipmentFailure
      ,AVG(CONVERT(DECIMAL(10,2),SumEquipmentFailure)) OVER (
            PARTITION BY PlantName,LocationName,CalendarYear
```

553

```
            ORDER BY CalendarMonth
    ) AS RollingAvg
    ,STDEV(SumEquipmentFailure) OVER (
            PARTITION BY PlantName,LocationName,CalendarYear
            ORDER BY CalendarMonth
    ) AS RollingStdev
    ,STDEVP(SumEquipmentFailure) OVER (
            PARTITION BY PlantName,LocationName,CalendarYear
            ORDER BY CalendarMonth
    ) AS RollingStdevp
    ,VAR(SumEquipmentFailure) OVER (
            PARTITION BY PlantName,LocationName,CalendarYear
            ORDER BY CalendarMonth
    ) AS RollingVar
    ,VARP(SumEquipmentFailure) OVER (
            PARTITION BY PlantName,LocationName,CalendarYear
            ORDER BY CalendarMonth
    ) AS RollingVarp
FROM [Reports].[EquipmentFailureStatistics]
GO
```

The partitions are set up the same way for all the OVER() clauses. The partitions
are all using an ORDER BY clause that sorts the partition by the CalendarMonth column.
The ORDER BY clause is used to define the order in which the window functions will be
applied to the partition. Let's check the results before looking at the query plan.

Please refer to Figure 8-28.

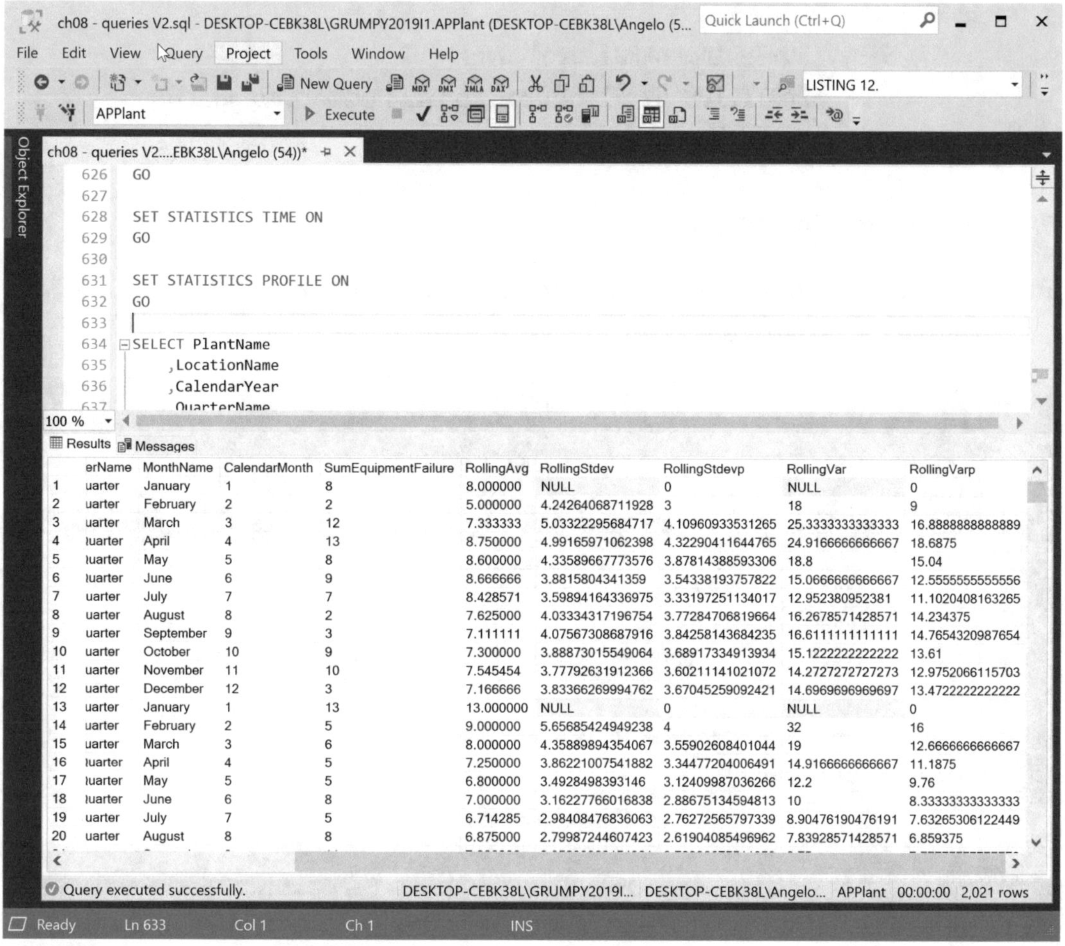

Figure 8-28. *Equipment failure profile*

Works as expected, rolling value results across the rows. Notice the NULL results when we are processing the first row of each partition. Also notice the zero values when the functions are processing the entire data set or, in other words, the entire data set is known.

You can override the NULL results by using a CASE block to display 0 instead of NULL if you want to. This is shown in the following code snippet:

```
CASE WHEN STDEV(SumEquipmentFailure) OVER (
        PARTITION BY PlantName,LocationName,CalendarYear
        ORDER BY CalendarMonth
    ) IS NULL THEN 0
```

555

```
ELSE
    STDEV(SumEquipmentFailure) OVER (
        PARTITION BY PlantName,LocationName,CalendarYear
        ORDER BY CalendarMonth
    )
END AS RollingStdev
```

A little busy but it does the trick. Here comes the estimated query plan. Please refer to Figure 8-29.

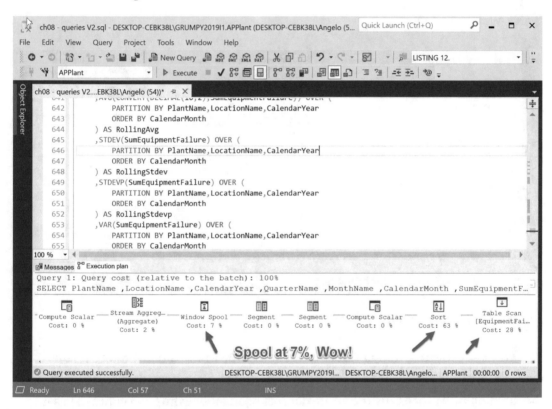

Figure 8-29. *Baseline estimated query plan*

We see a table scan, an expensive sort at 63% cost, and a hefty window spool task at 7% cost.

I created this index to see if the window spool cost would go down:

```
CREATE INDEX iePlantequipmentFailureProfile
ON [Reports].[EquipmentFailureStatistics](
```

```
    PlantName
    ,LocationName
    ,CalendarYear
    ,CalendarMonth
    )
GO
```

The rule of thumb is to create an index for the columns referenced in the PARTITION BY clause and the ORDER BY clause to increase performance.

Guess what? The estimated query plan remained the same, so I will not show it. The lesson here is that when the table is small, like a couple of thousand rows, creating indexes will not help. They will just take up space, so might as well not create them. They will slow down performance when loading the table, modifying rows in the table, and deleting rows in the table.

Next, I created a clustered index to see if the window spool cost would go down:

```
CREATE CLUSTERED INDEX ieClustPlantequipmentFailureProfile
ON Reports.EquipmentFailureStatistics(
    PlantName
    ,LocationName
    ,CalendarYear
    ,CalendarMonth
    )
GO
```

Figure 8-30 is the second estimated query plan.

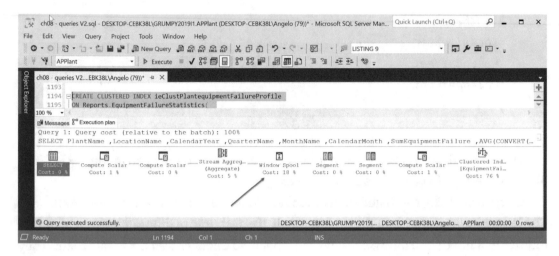

Figure 8-30. *Estimated query plan using a clustered index*

This is worse, or is it? The window spool went up to 18%, but the sort step disappeared. A clustered index scan was introduced at a cost of 76%.

How about the IO and TIME statistics? Please refer to Figure 8-31.

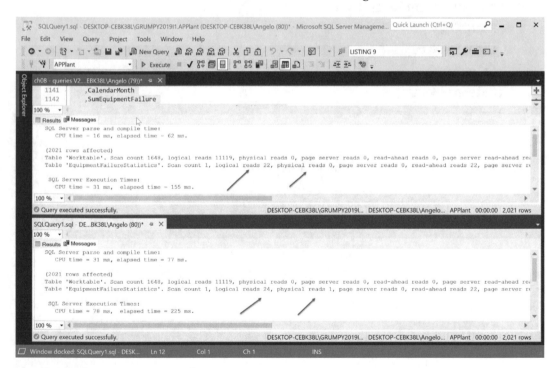

Figure 8-31. *Comparing IO and TIME statistics*

Wow, no improvement here either. The statistics on the work table remained the same, while the logical reads and physical reads went up a bit with the clustered index. The SQL Server execution times went up in both the CPU time and elapsed time. What does this teach us?

If it works, don't fix it. Do not create clustered indexes or other types of indexes if they are not used.

Memory-Optimized Table Approach

Well, our first two attempts to improve performance did not yield impressive results. So now we try the memory-optimized table approach. Recall the main steps:

- Create a file group for the memory-optimized table.

- Create one or more files and assign them to the file group.

- Create the table with the directives to make it a memory-optimized table.

- Load the memory-optimized table.

- Reference the table in your queries.

I suggest you keep a template script for the first four steps so you can quickly create the objects you need for a new memory-optimized table.

Tip   Create a folder with templates you might need to create scripts, database objects, and not to mention system table queries to retrieve metadata about database objects, like listing the columns in a table. Include templates for creating stored procedures, functions, and triggers.

Let's look at the code.

Create a File and File Group

Let's start with creating the file group we need. Please refer to Listing 8-20.

Listing 8-20. Step 1: Create the File Group for the Memory-Optimized Table

```
/*********************************************/
/* CREATE FILE GROUP FOR MEMORY OPTIMIZED TABLE */
/*********************************************/

ALTER DATABASE [APPlant]
ADD FILEGROUP AP_PLANT_MEM_FG
CONTAINS MEMORY_OPTIMIZED_DATA
GO
```

The only difference between this ALTER command and other ALTER commands used to create memory-optimized tables is that we add the directive CONTAINS MEMORY_OPTIMIZED_DATA.

The next step is to add one file to the file group. Please refer to Listing 8-21.

Listing 8-21. Step 2: Add a File to the File Group for the Memory-Optimized Table

```
/**************************************************/
/* ADD FILE TO FILE GROUP FOR MEMORY OPTIMIZED TABLE */
/**************************************************/

ALTER DATABASE [APPlant]
ADD FILE
(
    NAME = AP_PLANT_MEME_DATA,
    FILENAME = 'D:\APRESS_DATABASES\AP_PLANT\MEM_DATA\AP_PLANT_MEM_
DATA.NDF'
)
TO FILEGROUP AP_PLANT_MEM_FG
GO
```

This is a standard ALTER DATABASE command for adding a file, but notice that there are no size parameters. This is a characteristic of files associated with memory-optimized tables.

Two points:

Question: Why are we creating physical files if the table is in memory.

Answer: We need physical space in case we run out of memory and need somewhere to park the data when the server is shut off.

Make sure the FILENAME identifier points to a drive on your own laptop or workstation.

Question: If we need a physical file stored on disk, what should we use?

Answer: If management can afford the price tag or if you can, use a solid-state drive for these types of files.

Next step.

Create the Memory-Optimized Table

Now that we have created the file group and one file, we need to create the actual memory-optimized table. Please refer to Listing 8-22.

Listing 8-22. Step 3: Create the Memory-Optimized Table

```
CREATE TABLE Reports.EquipmentFailureStatisticsMem(
        CalendarYear            SMALLINT        NOT NULL,
        QuarterName             VARCHAR(11)     NOT NULL,
        MonthName               VARCHAR(9)      NOT NULL,
        CalendarMonth           SMALLINT        NOT NULL,
        PlantName               VARCHAR(64)     NOT NULL,
        LocationName            VARCHAR(128)    NOT NULL,
        EquipmentId             VARCHAR(8)      NOT NULL,
        CountFailureEvents      INT             NOT NULL,
        SumEquipmentFailure     INT             NOT NULL
INDEX [ieEquipFailStatMem]      NONCLUSTERED
(
        PlantName       ASC,
        LocationName    ASC,
        CalendarYear    ASC,
        CalendarMonth   ASC,
        EquipmentId     ASC
)
)WITH (MEMORY_OPTIMIZED = ON , DURABILITY = SCHEMA_ONLY )
GO
```

The command starts off like any other normal CREATE TABLE command, but we need to specify a non-clustered index and add the directive to make this a memory-optimized table.

Load the Memory-Optimized Table

Now that we have created the file group, one file attached to the file group, and the memory-optimized table itself, we need to load the actual memory-optimized table.

Please refer to Listing 8-23.

Listing 8-23. Step 4: Load the Memory-Optimized Table

```
/************************************/
/* LOAD THE MEMORY OPTIMIZED TABLE */
/************************************/

TRUNCATE TABLE Reports.EquipmentFailureStatisticsMem
GO

INSERT INTO Reports.EquipmentFailureStatisticsMem
SELECT * FROM Reports.EquipmentFailureStatistics
GO
```

Standard INSERT command from a query that retrieves rows from the original table to the new memory-optimized table. Use this coding style just for development; remember SELECT * is frowned upon.

Lastly, I will not show the code for the query as it is the same query we discussed in the beginning of the section; all we did was change the name of the table in the FROM clause to

```
Reports.EquipmentFailureStatisticsMem
```

Estimated Query Plan

We generate an estimated query plan for both queries in horizontal query plane windows in SSMS. This makes it easy to compare both plans.

Please refer to Figure 8-32.

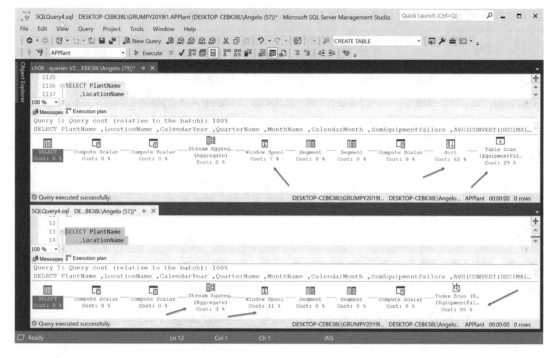

Figure 8-32. *Comparing estimated query plans*

The table scan in the first plan becomes an index scan in the plan for the memory-optimized table. The sort step disappeared, but look at the window spool task. It went up from 7% to 31%. The stream aggregate task went up by 1%.

Let's look at the cost statistics for both plans before we pass judgement. Please refer to Table 8-8.

Table 8-8. *Comparing Estimated Query Plan Step Costs*

Non-memory		Memory		
Step	**Cost**	**Step**	**Cost**	**Comment**
Table scan	29%	Index scan	85%	Replaced table scan with index scan
Sort	62%	N/A	N/A	Eliminated
Compute scalar	0%	Compute scalar	0%	No change
Segment	0%	Segment	0%	No change
Segment	0%	Segment	0%	No change
Window spool	**7%**	**Window spool**	**11%**	**Went up**
Stream aggregate	**2%**	**Stream aggregate**	**3%**	**Went up**
Compute scalar	0%	Compute scalar	0%	No change
Compute scalar	0%	Compute scalar	0%	No change
SELECT	0%	SELECT	0%	No change

Not much improvement for a lot of work. As a matter of fact, it looks like we went a little backward in terms of performance improvement. The window spool cost went up by 3%, and the stream aggregate step went up 1% in terms of cost. The sort step was eliminated though, but we have an index scan.

Let's look at the IO and TIME statistics for both queries. Please refer to Figure 8-33.

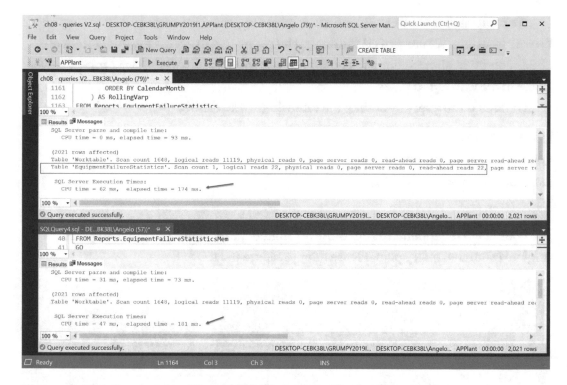

Figure 8-33. *Comparing IO and TIME statistics for both strategies*

The table statistics disappeared in the memory-optimized query, so this is a positive step since the table is in memory. The work table statistics are identical for both queries. SQL Server execution time went up a bit in the memory-optimized table, so more bad news.

Tip   This type of analysis should be performed multiple times, each time using DBCC to clear memory cache to make sure the results are consistent.

The verdict is in: all this work and not a lot of improvement, so this scenario was not a candidate for memory-optimized tables either.

One last item to discuss: Our original query returned all the rows, which is usually not a good idea. No one can possibly analyze thousands of rows in a result set. The ideal strategy is to write queries that use WHERE clause filters to reduce the results and give the indexes a good workout.

We went back to the analyst, and we received approval to narrow the data scope by filtering the data. We convinced the analyst by promising to access the query from an SSRS report that allowed them to choose the year, plant, and location via drop-down list boxes in the web report page.

I modified the query by adding the following WHERE clause:

```
WHERE CalendarYear = 2003
AND PlantName = 'East Plant'
AND LocationName = 'Boiler Room'
GO
```

Look at what happened as far as the estimated query plans; the top plan is the original estimated query plan. Please refer to Figure 8-34.

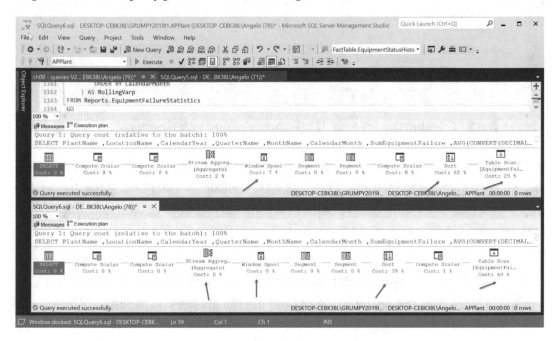

Figure 8-34. *Comparing query plans where a WHERE clause is introduced*

Here is our analysis:

- The table scan went up from 29% to 60%.

- The sort step went down from 62% to 39%.

- The window spool step went down from 7% to 0%.

- The stream aggregate went down from 2% to 0%.

So it looks like a simple query change like adding a `WHERE` clause delivered more performance improvement than all the indexes and memory-optimized tables!

In conclusion, when a business analyst supplies you with a business requirement, it does not hurt to ask questions and negotiate. Maybe you can suggest improvements like filters and so on that will make the query results easier to use and improve performance. Make sure you ask how the report is used and by whom.

Do this before you embark on a lengthy coding and testing journey such as creating indexes on small tables and, worse yet, creating memory-optimized tables. After all, memory is a limited resource. You will run out eventually if you clog it up with a memory-optimized table!

Last but not least are the `IO` and `TIME` statistics. You will like these; the original query statistics are at the top of the screenshot.

Please refer to Figure 8-35.

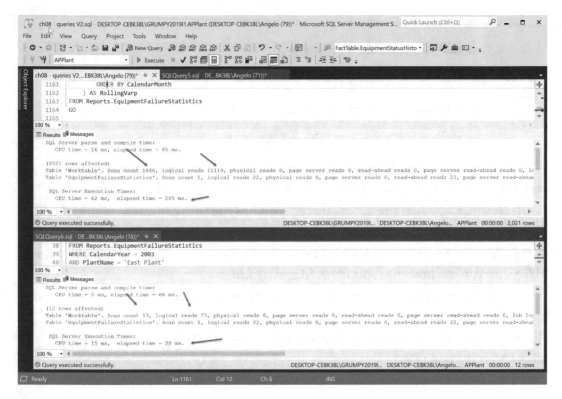

Figure 8-35. *Comparing IO and TIME statistics*

Looks like these statistics improved also. Let's take a look.

Please refer to Table 8-9.

Table 8-9. *IO and Time Statistics Side by Side*

	Original	With WHERE
Statistics	IO	IO
Work Table		
Scan count	1648	13
Logical reads	11119	73
Read-ahead reads	0	0

(continued)

Table 8-9. (*continued*)

	Original	With WHERE
EquipFailureTbl		
Scan count	1	1
Logical reads	22	22
Read-ahead reads	22	22
	TIME	**TIME**
Parse CPU	16 ms	0 ms
Parse elapsed	85 ms	66 ms
Exec CPU	62 ms	15 ms
Exec elapsed	285 ms	28 ms

Very interesting results. The IO statistics for the EquipmentFailureStatistics report table remained the same in both queries. But there is a considerable difference when we look at the work table statistics for both queries:

- Scan count went from 1,648 to 13 in the WHERE clause version of the query.

- Logical reads went down from 11,119 to 73 in the WHERE clause query.

- Read-ahead reads stayed at 0 for both versions of the query.

Lastly, let's look at CPU and elapsed times for both queries:

Parse CPU time went from 16 ms to 0 ms in the WHERE clause query.

Parse elapsed time went from 85 ms to 66 ms in the WHERE clause query.

Execution CPU time went from 62 ms to 15 ms in the WHERE clause query.

Execution elapsed time went from 285 ms to 28 ms in the WHERE clause query.

Looks like adding the WHERE clause significantly improved performance. This makes sense as we are working with smaller result sets.

Modifying the query is another technique we need to consider during our performance analysis sessions. Is that table scan without a WHERE clause really needed?

Seven-Million-Row Query: Performance Tuning

Let's look at one more query. We have been writing queries all along on small- to medium-size tables. This allowed us to learn and interpret estimated query plans and the various statistics. Let's try a table with seven-million-plus rows to see what the query plan looks like! The code to create and load this table is available in the folder for this chapter, and the file is called

"ch08 – load large equipment status history.sql"

Now while creating this query, I found a design flaw in the APPlant database. I did not do it on purpose, but I used it to make lemonade out of lemons.

What do I mean?

I found an opportunity to discuss how sometimes you are given requirements for a query that cannot be satisfied unless you find roundabout ways to join all the tables you need. As you pursue the roundabout trials and errors, you realize that the database design is flawed. Something is missing. You get the query to work, but it performs badly in terms of performance, and you have to check you did not introduce data anomalies like duplicates.

Specifically, I found that my design had no way to directly link the Equipment table to the Location table. The way the database is designed, one needs to link the Equipment table to either the Valve, Motor, Turbine, or Generator table, so you can link these to the Location table. Stay tuned for the fix.

Let's look at workarounds first.

Here is the first cut of the query with the workaround. The query is based on a CTE, so I split the discussion into two parts.

Please refer to Listing 8-24 for the CTE block.

Listing 8-24. Seven-Million-Row Window Aggregate Query

```
WITH EquipStatusCTE (
StatusYear,StatusMonth,EquipmentId,EquipAbbrev,
    EquipOnlineStatusCode,EquipOnLineStatusDesc,StatusCount
)
```

```
AS (
SELECT YEAR(ESBH.CalendarDate) AS StatusYear
      ,MONTH(ESBH.CalendarDate) AS StatusMonth
      ,ESBH.EquipmentId
      ,E.EquipAbbrev
      ,ESBH.EquipOnLineStatusCode
      ,EOLSC.EquipOnLineStatusDesc
      ,COUNT(*) AS StatusCount
FROM Reports.EquipmentDailyStatusHistoryByHour ESBH
JOIN DimTable.Equipment E
      ON ESBH.EquipmentId = E.EquipmentId
JOIN DimTable.EquipOnlineStatusCode EOLSC
      ON ESBH.EquipOnlineStatusCode = EOLSC.EquipOnlineStatusCode
GROUP BY YEAR(ESBH.CalendarDate)
      ,MONTH(ESBH.CalendarDate)
      ,ESBH.EquipmentId
      ,E.EquipAbbrev
      ,ESBH.EquipOnLineStatusCode
      ,EOLSC.EquipOnLineStatusDesc
)
```

As can be seen we are joining two dimension tables to a denormalized report table called `Reports.EquipmentDailyStatusHistoryByHour`. The number of the status codes that identify whether a piece of equipment is either online or offline due to maintenance or faults is what we want to count. Some standard reporting columns are included like the year, month, equipment ID, abbreviation (type of equipment), status code, and description.

Since we are using the `COUNT()` function without an `OVER()` clause, we need to include a `GROUP BY` clause. This will add to the processing time.

Here comes the query part. This is where the workaround to retrieve equipment locations is solved by a series of `UNION` query blocks nested in a query that makes up an inline table.

Please refer to Listing 8-25.

Listing 8-25. Equipment Status Code Totals Rollup Query

```
SELECT ES.StatusYear AS ReportYear
       ,ES.StatusMonth AS Reportmonth
       ,EP.LocationId
       ,ES.EquipmentId
       ,StatusCount
       ,SUM(StatusCount) OVER(
           PARTITION BY ES.EquipmentId,ES.StatusYear
           ORDER BY ES.EquipmentId,ES.StatusYear,ES.StatusMonth
           ) AS SumStatusEvent
       ,ES.EquipOnlineStatusCode
       ,ES.EquipOnLineStatusDesc
       ,EP.PlantId
       ,L.LocationId
       ,P.PlantName
       ,P.PlantDescription
       ,ES.EquipAbbrev
       ,EP.UnitId
       ,EP.UnitName
       ,E.SerialNo
FROM EquipStatusCTE ES
INNER JOIN DimTable.Equipment E
ON ES.EquipmentId = E.EquipmentId
INNER JOIN (
      SELECT PlantId,LocationId,EquipmentId,MotorId AS UnitId,
            MotorName AS UnitName
      FROM DimTable.Motor
UNION
      SELECT PlantId,LocationId,EquipmentId,ValveId AS UnitId,
            'Valve - ' + ValveName AS UnitName
      FROM DimTable.Valve
UNION
      SELECT PlantId,LocationId,EquipmentId,TurbineId AS UnitId,
            TurbineName AS UnitName
      FROM DimTable.Turbine
```

```
UNION
     SELECT PlantId,LocationId,EquipmentId,
            GeneratorId AS UnitId,GeneratorName AS UnitName
     FROM DimTable.Generator
) EP
     ON ES.EquipmentId = EP.EquipmentId
INNER JOIN DimTable.Plant P
     ON EP.PlantId = P.PlantId
INNER JOIN DimTable.Location L
     ON EP.PlantId = L.PlantId
AND EP.LocationId = L.LocationId
WHERE ES.StatusYear = 2002
AND ES.EquipOnlineStatusCode = '0001'
GO
```

Yes, it's messy!

First of all, we use the SUM() aggregate function with an OVER() clause to set up the rolling calculations. The partition is defined by equipment ID and status. The ORDER BY clause defines the logical order for processing by the equipment ID, status year, and status month columns (let's just say the ORDER BY clause sorts the rows in the order that is required for processing to achieve the desired results).

Now for the workaround. In the FROM clause, I defined an inline table by retrieving the columns required to link a piece of equipment to the correct plant and location by a series of three UNION blocks. The queries retrieve the required columns from the Motor, Valve, Turbine, and Generator tables.

The results are given an alias of EP, which is used in the FROM clause to join the other tables to this inline table via the EquipmentId column. Not pretty but it works. But how fast?

This is the design flaw. There should be an association table or dimension table to store this data, so it does not have to be assembled each time this query is run. We will fix this shortly.

Let's look at the results of this query. Please refer to Figure 8-36 for the partial results.

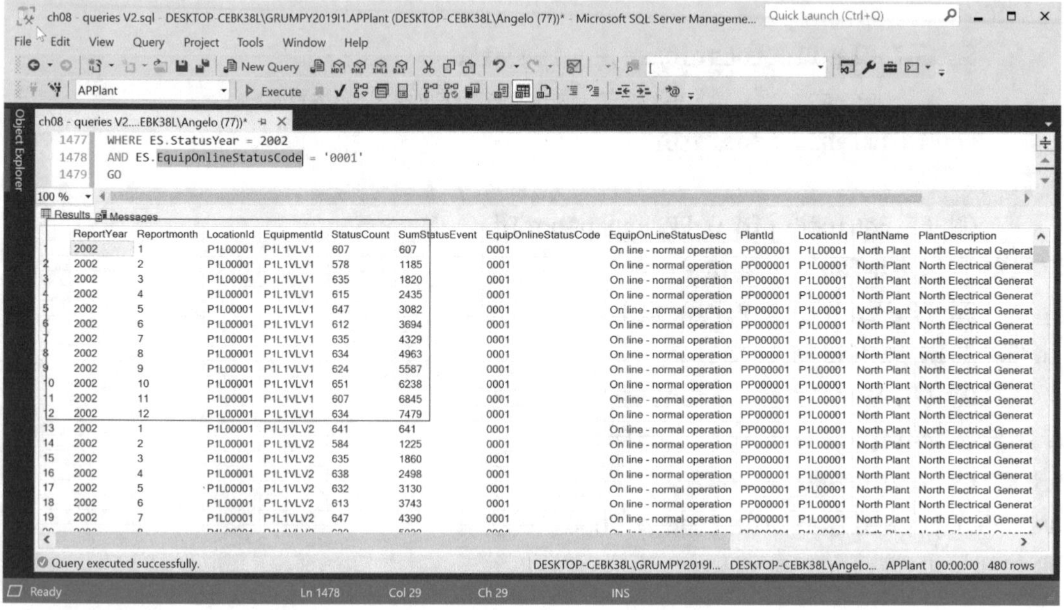

Figure 8-36. *Rolling equipment status summaries by month*

Looks like the rolling sums are working. A quick check shows the rolling summaries work: 607 + 578 = 1185 for the first two months' flow. When the equipment ID changes, the failures are reset for January, and the rolling process begins again.

The important columns are up front, so you can see the year and month information, the location, and the equipment ID, followed by the counts and rolling summaries, and the detail columns come next.

This query took two seconds to run. Not bad considering we are going against a seven-million-row-plus table on a laptop.

Let's look at the baseline query plan; it will be a large one. Please refer to the following three figures so we can understand what is going on in terms of the estimated execution plan and flow.

Please refer to Figure 8-37.

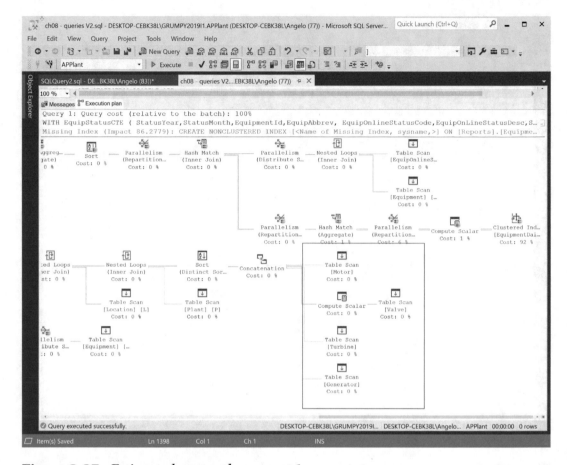

Figure 8-37. *Estimated query plan – part 1*

Notice the clustered index scan is being used. This comes at a cost of 92%, so most of the work is performed here. Notice the table scans for the Valve, Motor, Generator, and Turbine tables. These are all 0% cost. These are small tables so there is no performance issue. There are a lot of table scans and also sorts, but it seems that all these costs are 0. Let's look at the next section of the plan.

Please refer to Figure 8-38.

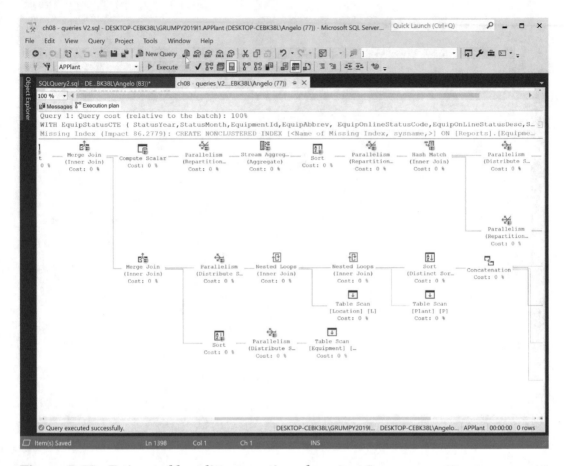

Figure 8-38. *Estimated baseline execution plan, part 2*

By the way, I call it a baseline execution plan because it is the first plan we generate and analyze so we can come up with some enhancements to increase performance. Then we generate another estimated execution plan to see if our enhancements worked.

By now we pretty much know what the tasks are, and they are all 0% cost although the flow looks very complex with multiple branches with sorts, nested loops joins, parallelism, and other tasks.

There are different tasks to merge the flow like hash match joins, concatenation, and merge joins.

Let's look at the last section of the plan. Please refer to Figure 8-39.

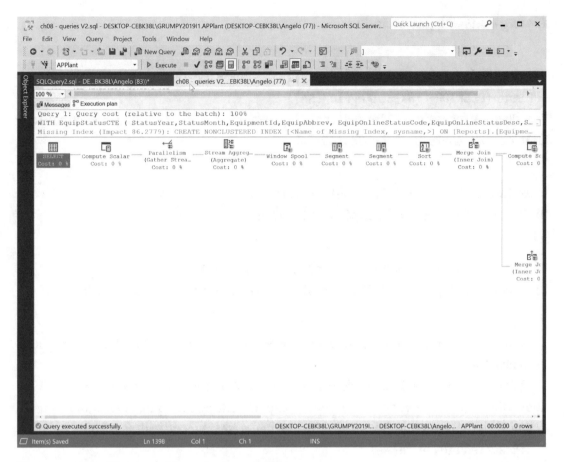

Figure 8-39. *Estimated query plan, part 3*

The last section of the plan is a single stream and branch. Dare we say that the leftmost section is the trunk and as it flows to the right we encounter the branches? Just like an inverted tree structure.

Back to reality, I almost forgot that the estimated execution plan is suggesting an index. Although we did use the clustered index, it did not do the job. Let's see what the index looks like in Listing 8-26.

Listing 8-26. Suggested Index

```
/*
Missing Index Details from SQLQuery7.sql - DESKTOP-CEBK38L\GRUMPY2019I1.
APPlant (DESKTOP-CEBK38L\Angelo (54))
The Query Processor estimates that implementing the following index could
improve the query cost by 88.1301%.
*/

DROP INDEX IF EXISTS ieEquipOnlineStatusCode
ON Reports.EquipmentDailyStatusHistoryByHour
GO

CREATE NONCLUSTERED INDEX ieEquipOnlineStatusCode
ON Reports.EquipmentDailyStatusHistoryByHour(EquipOnlineStatusCode)
GO
```

If we create this index, it claims execution will be improved by 86.27%. Sounds good. Let's give it a try.

We copy and paste the suggested index template, give it a name, and execute it. (I added the DROP INDEX code in case you are experimenting with query plans and need to create, drop, and recreate the index.) Now we generate a second estimated query plan.

Please refer to Figure 8-40.

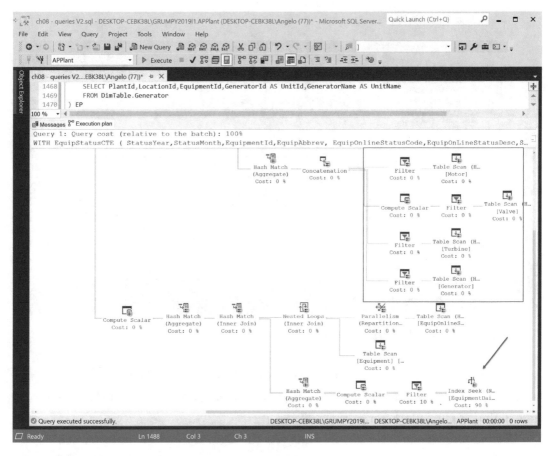

Figure 8-40. *Updated estimated query plan*

Notice the clustered index scan has been replaced by an index seek task on the new index. A seek is better than a scan, remember? We still have those four table scans on the Valve, Motor, Turbine, and Generator tables though (by the way, the query took two seconds to run).

An interesting set of IO and TIME statistics are generated by this query. Please refer to Figure 8-41.

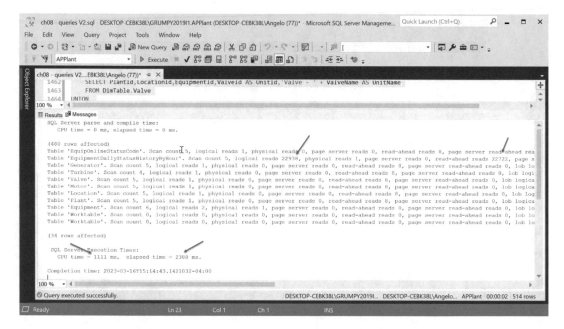

Figure 8-41. *IO and TIME statistics*

I will not go through all the results, but notice the logical reads on the physical table coming in at 22,958 and read-ahead reads coming in at 22,722. Lastly, CPU execution time is 1111 ms and elapsed time is 2388 ms.

This is a large table, and the only thing you could do to increase performance is to create the table as a partitioned table where the years are spread out across multiple physical disks or LUNS on a SAN. This is about the only way to manage performance when you are dealing with millions of rows or more of data! This strategy works well when data can be split by years, one disk per year.

Next, we correct the design flaw by creating and loading an association table that links equipment to plant and location. We can consider this as a dimension table as it can be used directly on its own with fact tables or it can be used to link two other dimensions together. Please refer to Listing 8-27.

Listing 8-27. Create and Load the Association Dimension

```
DROP TABLE IF EXISTS [DimTable].[PlantEquipLocation]
GO

CREATE TABLE [DimTable].[PlantEquipLocation](
    [EquipmentKey] [int] NOT NULL,
```

```
        [PlantId]      [varchar](8) NOT NULL,
        [LocationId]   [varchar](8) NOT NULL,
        [EquipmentId]  [varchar](8) NOT NULL,
        [UnitId]       [varchar](8) NOT NULL,
        [UnitName]     [varchar](136) NOT NULL
) ON [AP_PLANT_FG]
GO

TRUNCATE TABLE DIMTable.PlantEquipLocation
GO

INSERT INTO DIMTable.PlantEquipLocation
SELECT MotorKey AS EquipmentKey,PlantId,LocationId,EquipmentId,
        MotorId AS UnitId,MotorName AS UnitName
FROM DimTable.Motor
UNION
        SELECT ValveKey AS EquipmentKey,PlantId,LocationId,EquipmentId,
            ValveId AS UnitId,'Valve - ' + ValveName AS UnitName
FROM DimTable.Valve
UNION
        SELECT TurbineKey AS EquipmentKey,PlantId,LocationId,EquipmentId,
            TurbineId AS UnitId,TurbineName AS UnitName
FROM DimTable.Turbine
UNION
SELECT GeneratorKey AS EquipmentKey,PlantId,LocationId,EquipmentId,
            GeneratorId AS UnitId,GeneratorName AS UnitName
FROM DimTable.Generator
GO
```

The script includes the DDL command to create the table and the INSERT/SELECT statement required to load the table. The SELECT statement is simply the four queries connected by the UNION operator to load the necessary columns required to link the equipment ID to the plant and location.

So opportunities to solve design problems or enhance a database design surface when writing queries to satisfy business analyst or other user requirements. Always look for better, more efficient, and simpler ways to achieve a goal.

Complex, hard-to-read code may impress your managers and friends, but simple code that does the job will make your users happy. After all, that's what it is all about. Okay, I will get off my soapbox.

As a homework assignment, modify the query we just discussed and replace the UNION code block with the name of this table in the FROM clause. Rerun the estimated query plan and IO and TIME statistics to see what improvements, if any, in performance are realized with this design enhancement.

Also, pull out the CTE code and use it to build a report table. Run the query with the aggregate function against the report table. Any performance improvements?

What do the estimated query plan and TIME/IO statistics show?

I included the code for this homework assignment in the script for this chapter. Try it out on your own first and look at the code if you get stuck. The estimated query plan for this revision is drastically different and shorter.

Summary

Hopefully you found this chapter interesting. We put the aggregate functions through their paces in a plant engineering data warehouse scenario. Along the way we performed some detailed and interesting performance analysis and not only found opportunities to improve the queries so as to achieve better performance but also discovered a design flaw in the database.

No, I did not do this on purpose. I actually made a design mistake, but it was a good opportunity to demonstrate that these things occur in a real development environment and that it provides opportunities to not only fix but improve a design.

Lastly, we created some other design artifacts in the form of data dictionaries, which helped us understand what is in the database so that when we create queries, we understand the semantics of the data and relationships between tables.

On to Chapter 9!

Plant Use Case: Ranking Functions

In this chapter, we use the ranking window functions to perform some analysis on the power plant database. There are only four functions, but we will check out performance as usual and also create a few Microsoft Excel pivot tables and graphs to take our analysis a step further.

Our query performance analysis will utilize most of the tools we used in our prior chapters and will also show an example that leads us down the wrong path, which we will correct.

Sometimes, when you are coding without proper design considerations, you might pick an approach that you feel makes sense only to discover that there was a simpler way of solving the coding requirement if only you would have thought it out.

Sometimes the easiest solutions are the best!

Ranking Functions

Here are the four functions. Hopefully this is not the first time you have seen them as we have covered them in prior chapters. If you need a refresher on this topic, check out Appendix A where each function is described in detail:

- RANK()
- DENSE_RANK()
- NTILE()
- ROW_NUMBER()

© Angelo Bobak 2023
A. Bobak, *SQL Server Analytical Toolkit*, https://doi.org/10.1007/978-1-4842-8667-8_9

The first two functions are similar in that they are used to rank results, the difference being that they treat duplicate values and subsequent rankings differently. Take this into consideration when you are trying to solve a problem that requires ranking results.

RANK() Function

Recall that the RANK() function returns the number of rows before the current row plus the current row based on the value being ranked. Let's see what the business analysts' specifications are:

A report is needed to rank the number of equipment failures by plant, location, year, quarter, and month. They need to be in descending order for the year 2008. The report query needs to be easily modified so it can include more years in the results.

Simple query, so we will start by taking the CTE approach.

Please refer to Listing 9-1.

Listing 9-1. Ranking Sums of Equipment Failures

```
WITH FailedEquipmentCount
(
      CalendarYear, QuarterName,[MonthName], CalendarMonth, PlantName,
      LocationName, SumEquipFailures
)
AS (
      SELECT
              C.CalendarYear
              ,C.QuarterName
              ,C.[MonthName]
              ,C.CalendarMonth
              ,P.PlantName
              ,L.LocationName
              ,SUM(EF.Failure) AS SumEquipFailures
      FROM DimTable.CalendarView C
      JOIN FactTable.EquipmentFailure EF
      ON C.CalendarKey = EF.CalendarKey
      JOIN DimTable.Equipment E
      ON EF.EquipmentKey = E.EquipmentKey
```

```
        JOIN DimTable.Location L
        ON L.LocationKey = EF.LocationKey
        JOIN DimTable.Plant P
        ON L.PlantId = P.PlantId
        GROUP BY
                C.CalendarYear
                ,C.QuarterName
                ,C.[MonthName]
                ,C.CalendarMonth
                ,P.PlantName
                ,L.LocationName
)
SELECT
        PlantName
        ,LocationName
        ,CalendarYear
        ,QuarterName
        ,[MonthName]
        ,SumEquipFailures
        ,RANK() OVER (
                PARTITION BY PlantName,LocationName
                ORDER BY SumEquipFailures DESC
        ) AS FailureRank
FROM FailedEquipmentCount
WHERE CalendarYear = 2008
GO
```

The CTE logic is a little complex. A STAR query approach is used, and five dimension tables need to be linked via four JOIN predicates. Notice that the surrogate keys are used and that we join to the CalendarView VIEW instead of the Calendar dimension so we can pull out some names for the date object.

Note   Performing joins on surrogate keys is faster than performing joins on production keys that have an alphanumeric data type. Integer joins are faster.

The SUM() aggregate function is used to total up the failures, and we include the mandatory GROUP BY clause.

Looking at the OVER() clause, a partition is set up by plant name and location name. The year is not included in the partition definition as our WHERE clause filters out results for 2008 only. To include more years in the results, don't forget to add the year in the PARTITION BY clause.

Lastly, notice the ORDER BY clause sorts results in DESC (descending) order. That is, it specifies to process results in the partition by the SumEquipFailures column in descending order.

Let's see what the results look like.

Please refer to the partial results in Figure 9-1.

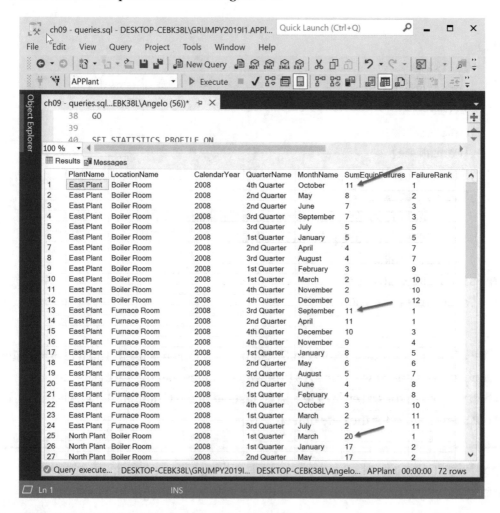

Figure 9-1. *Failure rank by year and location*

Looking at the boiler room results, it seems October was the top-ranking month as far as failures go, so this needs to be investigated. Our analyst needs to see which piece or pieces of equipment suffered the most failures.

For the furnace room, looks like we have a tie. September and April were the worst months for equipment failure in the furnace room.

Let's chart the results for a visual that will help us see the rankings for 2008.

Please refer to Figure 9-2.

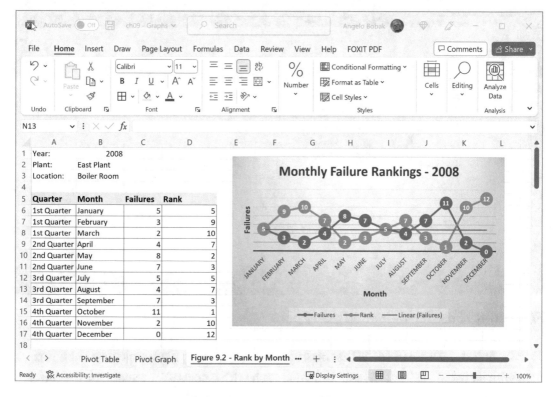

Figure 9-2. *Monthly failures for 2008*

This seems to be a nice graph as it clearly shows the rankings by month and also the failures. Our analyst will be happy.

How about performance? Let's see how this query does by performing our usual analysis with the tools available in our toolbox.

Performance Considerations

We generate our baseline estimated query performance plan in the usual manner. It is a big one.

Please refer to Figure 9-3.

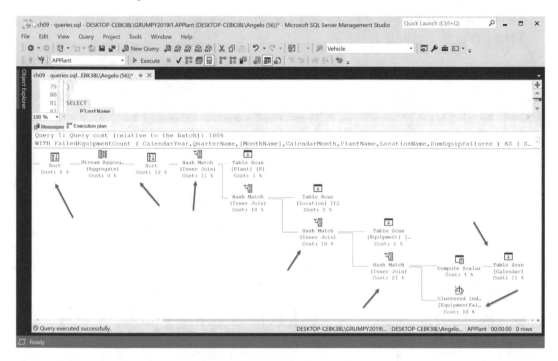

Figure 9-3. *Baseline estimated query plan*

The first thing we can see is that no index was suggested. We see a clustered index scan with a cost of 10% plus five table scans, each with a cost of 1% except for the table scan on the Calendar table, which chimes in at a cost of 11%.

Seems like all the streams of rows are joined via a series of hash match joins at costs ranging from 21% to 10%. A sort task follows with a cost of 12% followed by a stream aggregate task at 0% cost and then a sort step at 8%. The rest of the tasks in the plan are at 0% cost.

Since this is a query that works on historical data, we can conclude that we can replace the CTE portion with the report table strategy we have been using in several of our examples.

Let's take a peek at the IO and TIME statistics.

Please refer to Figure 9-4.

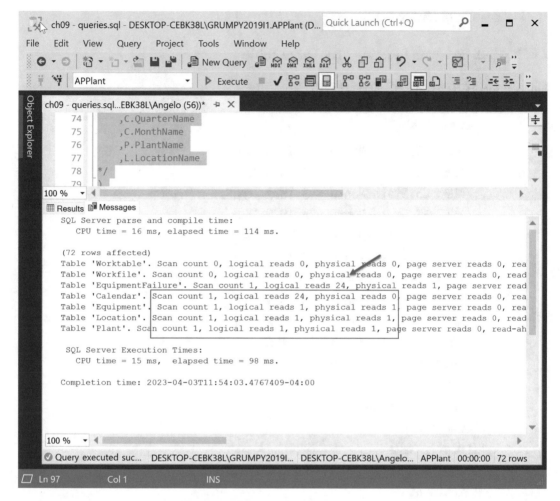

Figure 9-4. *IO and TIME statistics*

The scan counts on physical tables have a value of 1, while the logical reads on the EquipmentFailure fact table and the Calendar dimension have values of 24. The remaining dimension tables used in the STAR query have logical read values of 1, and all fact and dimension tables except for the Calendar dimension have physical reads of 1.

Elapsed time for the SQL execution time is 98 ms.

Let's look at a query that accesses a table with over 30,000 rows and see what type of performance we get. This query is similar to the prior query, but we add more columns in the SELECT clause to produce a more detailed report.

We will filter out rows for the year 2002 and equipment status code "0001" only to limit the size of the report.

Please refer to Listing 9-2.

Listing 9-2. Perform Ranking on a Large Table

```
SELECT ES.StatusYear AS ReportYear
      ,ES.StatusMonth AS Reportmonth
       ,CASE
            WHEN ES.StatusMonth = 1 THEN 'Jan'
            WHEN ES.StatusMonth = 2 THEN 'Feb'
            WHEN ES.StatusMonth = 3 THEN 'Mar'
            WHEN ES.StatusMonth = 4 THEN 'Apr'
            WHEN ES.StatusMonth = 5 THEN 'May'
            WHEN ES.StatusMonth = 6 THEN 'Jun'
            WHEN ES.StatusMonth = 7 THEN 'Jul'
            WHEN ES.StatusMonth = 8 THEN 'Aug'
            WHEN ES.StatusMonth = 9 THEN 'Sep'
            WHEN ES.StatusMonth = 10 THEN 'Oct'
            WHEN ES.StatusMonth = 11 THEN 'Nov'
            WHEN ES.StatusMonth = 12 THEN 'Dec'
            END AS MonthName
        ,EP.LocationId
        ,L.LocationName
        ,ES.EquipmentId
        ,StatusCount
        ,RANK() OVER(
            PARTITION BY ES.EquipmentId --,ES.StatusYear
            ORDER BY ES.EquipmentId ASC,ES.StatusYear ASC,StatusCount DESC
            ) AS NoFailureRank
        ,ES.EquipOnlineStatusCode
        ,ES.EquipOnLineStatusDesc
        ,EP.PlantId
        ,P.PlantName
        ,P.PlantDescription
        ,ES.EquipAbbrev
```

```
        ,EP.UnitId
        ,EP.UnitName
        ,E.SerialNo
FROM Reports.EquipmentRollingMonthlyHourTotals ES -- 30,000 rows plus
INNER JOIN DimTable.Equipment E
ON ES.EquipmentId = E.EquipmentId
INNER JOIN [DimTable].[PlantEquipLocation] EP
ON ES.EquipmentId = EP.EquipmentId
INNER JOIN DimTable.Plant P
ON EP.PlantId = P.PlantId
INNER JOIN DimTable.Location L
ON EP.PlantId = L.PlantId
AND EP.LocationId = L.LocationId
WHERE ES.StatusYear = 2002
AND ES.EquipOnlineStatusCode = '0001'
GO
```

We see a CASE block has been included in order to generate the month names, and we are joining the large report table to four dimension tables. To limit the size of the report, we only want to see rankings for equipment that has an online status code of "0001." This stands for "online – normal operation."

We also want to see data for the calendar year 2002, so notice that the year column was commented out in the OVER() clause PARTITION BY clause. If you want more years, just comment it out and modify the WHERE clause filter.

Let's see the results.

Please refer to Figure 9-5.

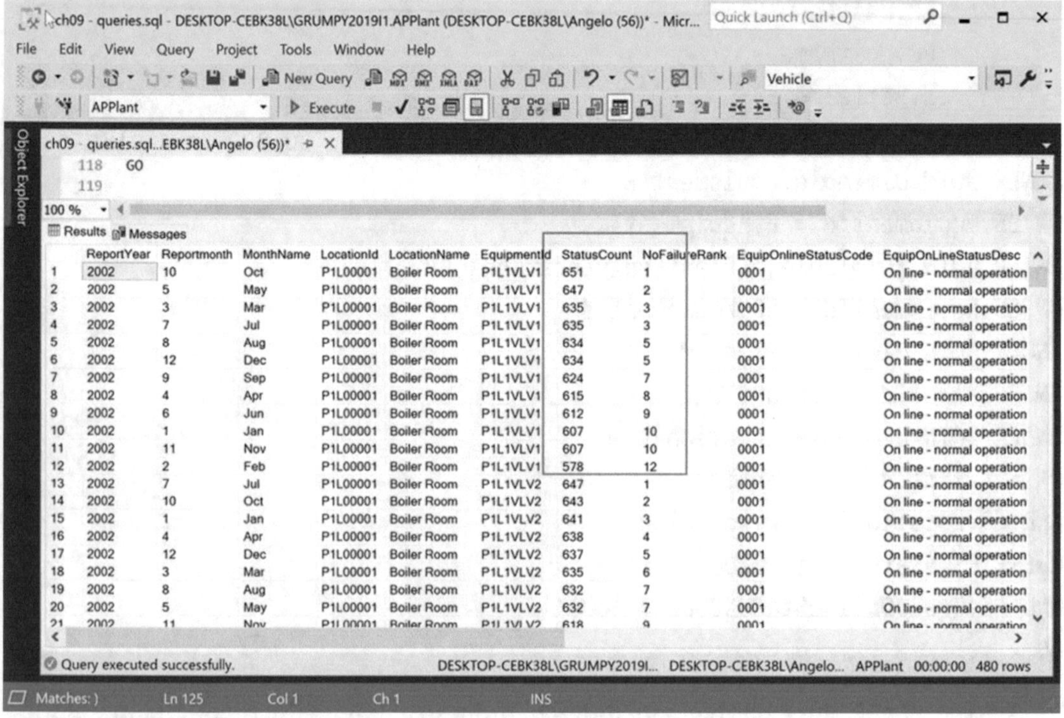

	ReportYear	Reportmonth	MonthName	LocationId	LocationName	EquipmentId	StatusCount	NoFailureRank	EquipOnlineStatusCode	EquipOnLineStatusDesc
1	2002	10	Oct	P1L00001	Boiler Room	P1L1VLV1	651	1	0001	On line - normal operation
2	2002	5	May	P1L00001	Boiler Room	P1L1VLV1	647	2	0001	On line - normal operation
3	2002	3	Mar	P1L00001	Boiler Room	P1L1VLV1	635	3	0001	On line - normal operation
4	2002	7	Jul	P1L00001	Boiler Room	P1L1VLV1	635	3	0001	On line - normal operation
5	2002	8	Aug	P1L00001	Boiler Room	P1L1VLV1	634	5	0001	On line - normal operation
6	2002	12	Dec	P1L00001	Boiler Room	P1L1VLV1	634	5	0001	On line - normal operation
7	2002	9	Sep	P1L00001	Boiler Room	P1L1VLV1	624	7	0001	On line - normal operation
8	2002	4	Apr	P1L00001	Boiler Room	P1L1VLV1	615	8	0001	On line - normal operation
9	2002	6	Jun	P1L00001	Boiler Room	P1L1VLV1	612	9	0001	On line - normal operation
10	2002	1	Jan	P1L00001	Boiler Room	P1L1VLV1	607	10	0001	On line - normal operation
11	2002	11	Nov	P1L00001	Boiler Room	P1L1VLV1	607	10	0001	On line - normal operation
12	2002	2	Feb	P1L00001	Boiler Room	P1L1VLV1	578	12	0001	On line - normal operation
13	2002	7	Jul	P1L00001	Boiler Room	P1L1VLV2	647	1	0001	On line - normal operation
14	2002	10	Oct	P1L00001	Boiler Room	P1L1VLV2	643	2	0001	On line - normal operation
15	2002	1	Jan	P1L00001	Boiler Room	P1L1VLV2	641	3	0001	On line - normal operation
16	2002	4	Apr	P1L00001	Boiler Room	P1L1VLV2	638	4	0001	On line - normal operation
17	2002	12	Dec	P1L00001	Boiler Room	P1L1VLV2	637	5	0001	On line - normal operation
18	2002	3	Mar	P1L00001	Boiler Room	P1L1VLV2	635	6	0001	On line - normal operation
19	2002	8	Aug	P1L00001	Boiler Room	P1L1VLV2	632	7	0001	On line - normal operation
20	2002	5	May	P1L00001	Boiler Room	P1L1VLV2	632	7	0001	On line - normal operation
21	2002	11	Nov	P1L00001	Boiler Room	P1L1VLV2	618	9	0001	On line - normal operation

***Figure 9-5.** Ranking failures for 2002*

As can be seen, October 2002 was the best month for the equipment being online for the boiler room in plant PL00001, which is the North Plant. The equipment in question is valve P1L1VLV1. This is a large report, so it is an ideal candidate for a pivot table so you can narrow down the scope of equipment being analyzed and change it by using filters!

Please refer to Figure 9-6.

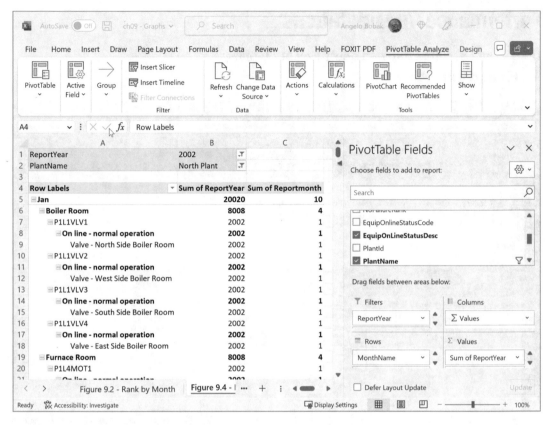

Figure 9-6. *Detailed plant online status pivot table*

This is a lot more flexible and powerful as we can slice and dice the results and also limit certain parameters like plant, location, and equipment.

Note   If your company does not use SSAS (SQL Server Analysis Services) to create multidimensional cubes, Microsoft Excel pivot tables are a good substitute.

Try this approach on your own and modify the query so that it returns data for all years, plants, and locations. All equipment also! Now we see that our toolkit includes not only the window functions but also Microsoft Excel so we can produce graphs and pivot tables.

Let's see how this query performs.

Performance Considerations

Here is our baseline estimated query execution plan.

Please refer to Figure 9-7.

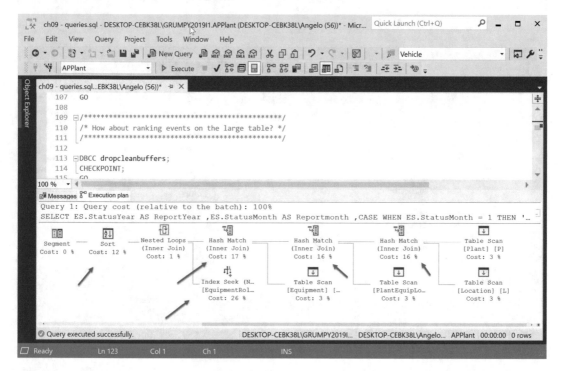

Figure 9-7. *Large table estimated query plan*

We see an index seek on the large report table at a cost of 26%, but look at all those hash match joins to merge rows from the dimension tables and all the table scans on the dimension tables. One nested loops task at 1% follows and then a sort with a cost of 12%.

Pop Quiz   When we say cost of a task, what does it really mean? Percentage of time the task took to execute? Answer to follow later.

Let's check out the IO and TIME statistics for this query.

Please refer to Figure 9-8.

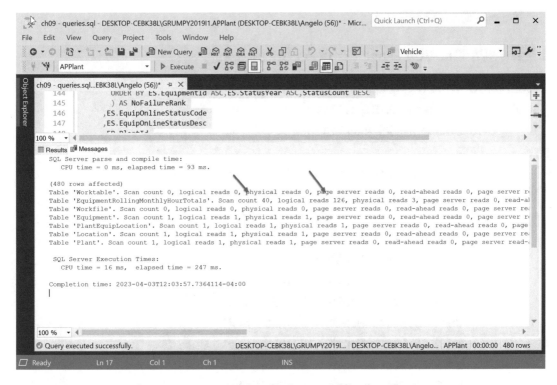

Figure 9-8. *IO and TIME statistics for the large table query*

The report table has high values for scan count (40) and logical reads (126), with physical reads of 3. The dimension tables all have scan counts, logical reads, and physical reads of 1. The work file all have 0 values for these parameters.

As stated earlier, this is a historical report, so clearly, to improve performance, we want to stage most of the data with a one-time load of all historical data. As each month rolls by, we only need to perform a monthly load to catch up on the data. We could do weekly loads also if the users need data to be updated weekly or even daily.

Pop Quiz Answer   *When we say cost of a task, what does it really mean? Percentage of time the task took to execute?* No, this is a blended value as to how much CPU, memory, and IO are used plus some other parameters.

Let's modify our strategy so we extract the code to load all the data onto yet another report table.

Please refer to Listing 9-3.

Listing 9-3. One-Time Load of the Report Table

```
INSERT INTO Reports.EquipmentMonthlyOnLineStatus
SELECT ES.StatusYear AS ReportYear
     ,ES.StatusMonth AS Reportmonth
      ,CASE
            WHEN ES.StatusMonth = 1 THEN 'Jan'
            WHEN ES.StatusMonth = 2 THEN 'Feb'
            WHEN ES.StatusMonth = 3 THEN 'Mar'
            WHEN ES.StatusMonth = 4 THEN 'Apr'
            WHEN ES.StatusMonth = 5 THEN 'May'
            WHEN ES.StatusMonth = 6 THEN 'Jun'
            WHEN ES.StatusMonth = 7 THEN 'Jul'
            WHEN ES.StatusMonth = 8 THEN 'Aug'
            WHEN ES.StatusMonth = 9 THEN 'Sep'
            WHEN ES.StatusMonth = 10 THEN 'Oct'
            WHEN ES.StatusMonth = 11 THEN 'Nov'
            WHEN ES.StatusMonth = 12 THEN 'Dec'
            END AS MonthName
      ,EP.LocationId
      ,L.LocationName
      ,ES.EquipmentId
      ,StatusCount
      ,ES.EquipOnlineStatusCode
      ,ES.EquipOnLineStatusDesc
      ,EP.PlantId
      ,P.PlantName
      ,P.PlantDescription
      ,ES.EquipAbbrev
      ,EP.UnitId
      ,EP.UnitName
      ,E.SerialNo
-- Created table with:
-- INTO  Reports.EquipmentMonthlyOnLineStatus
```

```
FROM Reports.EquipmentRollingMonthlyHourTotals ES -- 30,240 rows plus
INNER JOIN DimTable.Equipment E
ON ES.EquipmentId = E.EquipmentId
INNER JOIN [DimTable].[PlantEquipLocation] EP
ON ES.EquipmentId = EP.EquipmentId
INNER JOIN DimTable.Plant P
ON EP.PlantId = P.PlantId
INNER JOIN DimTable.Location L
ON EP.PlantId = L.PlantId
AND EP.LocationId = L.LocationId
ORDER BY ES.StatusYear
      ,ES.StatusMonth
        ,EP.LocationId
        ,ES.EquipmentId
        ,ES.EquipOnlineStatusCode
        ,EP.PlantId
GO
```

There's that big CASE block. Also, notice all the JOIN predicates in the FROM clause. You can cheat and create the table by first using the INTO <table name> right before the FROM clause. Comment it out after the table is created and use the usual INSERT/SELECT to load it. Always check the data types of the columns to make sure they are consistent with the data types of the columns that were retrieved from the tables used in the query.

This strategy allows us to use the ORDER BY clause to sort the data in a manner that will help the queries we will execute against it.

After this table is created and loaded, we would load it once a month, so a script that adds the month filter in the where clause would need to be created. This can be placed in a SQL Agent scheduled job so the table can be loaded incrementally, a month, week, or even day at a time.

Reminder   SQL Agent is the job execution process. You can use this to schedule execution of queries, scripts, external programs, and SSIS ETL packages.

For this strategy create a small table that logs the last date the table was loaded. This way, the query script can use the last date as a filter condition and request rows that are one day more than the last date stored. Remember to add the new load date so it works correctly the next time you run the query.

Let's look at the new and improved query.

Please refer to Listing 9-4.

Listing 9-4. The New and Improved Query

```
SELECT ReportYear
      ,Reportmonth
        ,MonthName
        ,LocationId
        ,LocationName
        ,EquipmentId
        ,StatusCount
        ,RANK() OVER(
            PARTITION BY EquipmentId,ReportYear
            ORDER BY LocationId,EquipmentId ASC,ReportYear ASC,
                StatusCount DESC
            ) AS NoFailureRank
        ,EquipOnlineStatusCode
        ,EquipOnLineStatusDesc
        ,PlantId
        ,PlantName
        ,PlantDescription
        ,EquipAbbrev
        ,UnitId
        ,UnitName
        ,SerialNo
FROM Reports.EquipmentMonthlyOnLineStatus -- 30,240 rows plus
WHERE ReportYear = 2002
AND EquipOnlineStatusCode = '0001'
AND EquipmentId = 'P1L1VLV1'
ORDER BY ReportYear
      ,Reportmonth
```

598

```
        ,LocationId
        ,EquipmentId
        ,EquipOnlineStatusCode
        ,PlantId
GO
```

No surprises. It is the same query with the RANK() function, but this time it uses our new and improved table. We can expect to get improved performance. Or can we?

We will narrow down the results to one piece of equipment.

Please refer to the partial results in Figure 9-9.

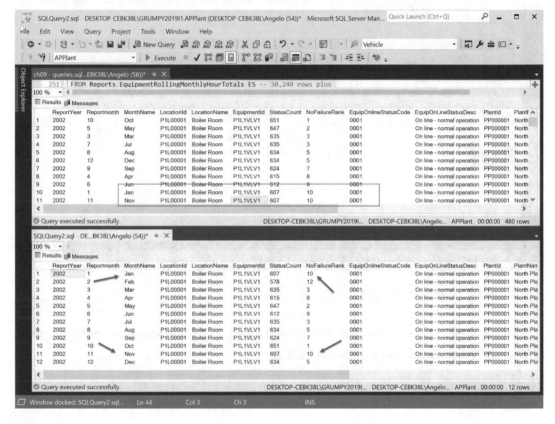

Figure 9-9. *Comparing query reports*

Keep in mind that I added a filter to report on only one piece of equipment. As you can see, the results match, but the first report is sorted by rank, whereas the second report is sorted by month. It all depends on how the analyst wants the results sorted.

Performance Considerations

You know the drill. Let's create a baseline estimated query plan so we can see what performance conditions exist.

Please refer to Figure 9-10.

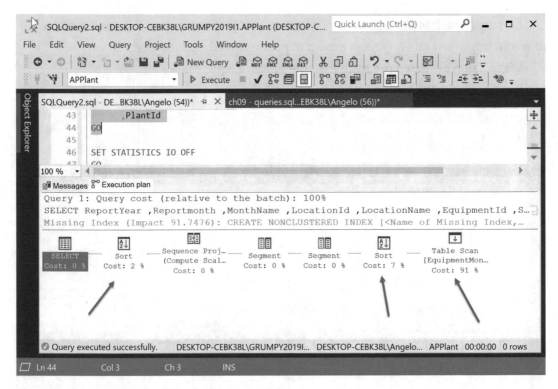

Figure 9-10. *Estimated baseline query plan*

A table scan plus two sorts and a suggested index. The table scan chimes in at a cost of 91%, so that is something we should look at!

Let's create the index. Couldn't hurt!

Please refer to Listing 9-5.

Listing 9-5. Recommended Index

```
/*
Missing Index Details from SQLQuery2.sql - DESKTOP-CEBK38L\GRUMPY2019I1.
APPlant (DESKTOP-CEBK38L\Angelo (54))
The Query Processor estimates that implementing the following index could
improve the query cost by 91.7476%.
*/
CREATE NONCLUSTERED INDEX ieEquipmentMonthlyOnLineStatus
ON Reports.EquipmentMonthlyOnLineStatus (ReportYear,EquipmentId,EquipOnline
StatusCode)
INCLUDE (
     Reportmonth,MonthName,LocationId,LocationName,StatusCount,
     EquipOnLineStatusDesc,PlantId,PlantName,PlantDescription,
     EquipAbbrev,UnitId,UnitName,SerialNo
     )
GO
```

The estimated query plan analyzer states the performance will improve by 91.7476%, so let's create it and run another estimated query plan.

Please refer to Figure 9-11.

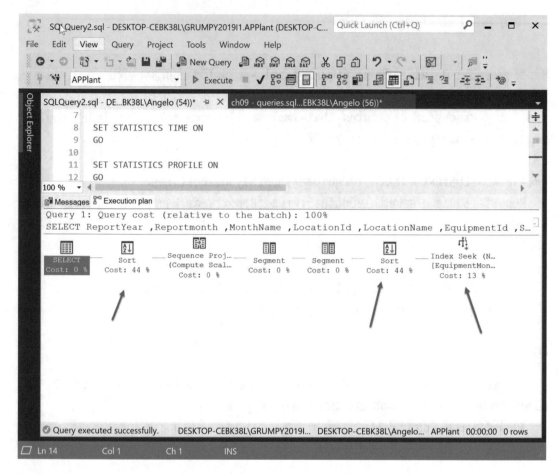

Figure 9-11. *Updated execution plan with the index*

So the index was used (cost of 13%). This is good news, but we have two sorts each with a cost of 44%. Let's look at the IO and TIME statistics.

Please refer to Figure 9-12.

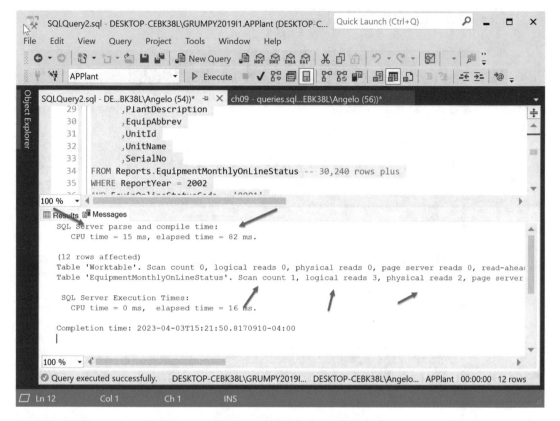

Figure 9-12. *Improved query IO and TIME statistics*

Not too bad: a scan count of 1, logical reads of 3, and physical reads of 2 on the new report table. All the statistics on the work table are zero, and the SQL Server execution elapsed time is 16 ms.

Looking back at the query, we see we used an ORDER BY clause. Is it necessary? I took the query, commented out the ORDER BY clause, and compared the estimated execution plans side by side, and this is what I got.

Please refer to Figure 9-13.

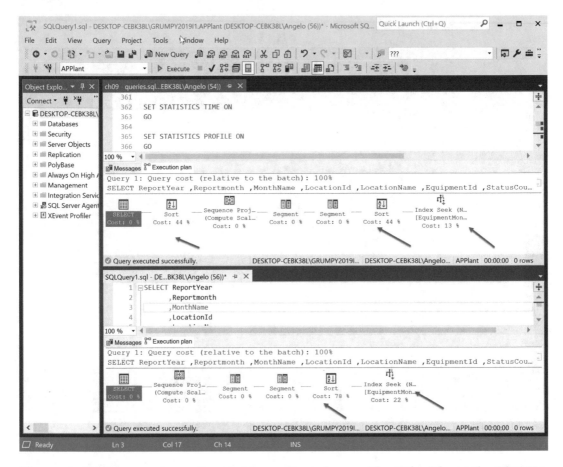

Figure 9-13. *Comparing estimated execution plans with and without the ORDER BY clause*

A sort step was eliminated, but the index seek went up to 22% and the remaining sort step went up to 78%. Is this better? Let's see what the TIME and IO statistics tell us.

Please refer to Figure 9-14.

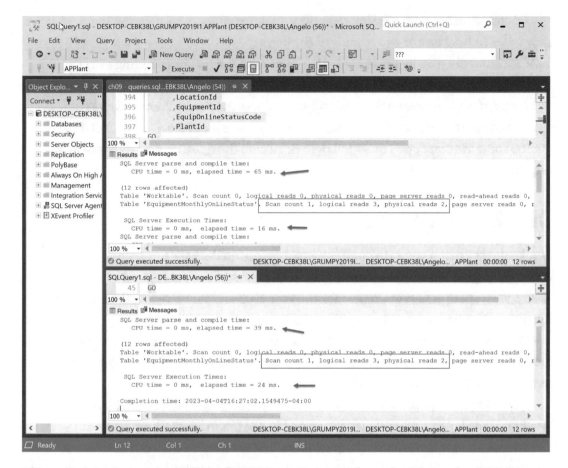

Figure 9-14. *Comparing IO and TIME statistics with and without the ORDER BY clause*

Some improvement in the parse and compile times, and execution times went up in the query without the ORDER BY clause. This test needs to be run two or three times, each time running the DBCC command to clear the cache in each session to make some sort of decision. You can try it out yourself, but what it comes down to, if this was a real production scenario, is how the analyst wants the report sorted.

As we saw in other chapters, sometimes the easiest solution is the best.

What do I mean by this?

Instead of creating a few indexes and doing extensive testing, just split up the query and pop the base data in a report table (if the data is historical in nature) and then modify the report query to access data from the new table.

Of course, see what the estimated query analyzer suggests as far as an index with the new table. At least one or two rounds of comparing estimated query plans plus IO and TIME statistics are usually warranted. Also, check out the results of setting STATISTICS PROFILE to ON if you need to dig deeper.

Lastly, if we understand the analyst's or business user's requirements, we can craft a high-performance solution. Make sure you ask a lot of questions. Even if you get the specs via email, pop an email right back with the questions so you have all the requirements and expectations on hand.

DENSE_RANK() Function

Our business analyst wants to compare rankings, using the RANK() function and the DENSE_RANK() function. Also, the month needs to be spelled out instead of just a number. Lastly, we want to generate a report for the year 2002 and for valve P1L1VLV 1 in plant 1, location 1. Seems like it has been giving the engineers a bit of trouble.

The query, once tested, needs to be expanded to include all years, plants, and locations so we can pop the results onto a Microsoft Excel pivot table (which you will do as a homework exercise, surprise!).

Decoding the equipment ID tells us the valve is in plant 1, the North Plant, and location 1, the boiler room. We are interested in all status descriptions.

We want to use the prior new report table, and we want to spell out the quarter name. Sadly our little Calendar table does not have the quarters spelled out, so we need to compensate with an inline table, which we will find out is the wrong direction to take.

Homework   Using the DDL code for this chapter, see if you can modify the Calendar dimension so it uses more verbose names for quarters and months instead of just numbers.

Sometimes, when designing a query, one makes assumptions as to the logic just to later find out there is a simpler solution! We will see that this is the case shortly!

Please refer to Listing 9-6.

Listing 9-6. RANK() vs. DENSE_RANK() Report

```
SELECT
        EMOLS.PlantName
        ,EMOLS.LocationName
        ,EMOLS.ReportYear
        ,MSC.CalendarQuarter AS ReportQuarter
        ,EMOLS.[MonthName] AS ReportMonth
        ,EMOLS.StatusCount
        ,EMOLS.EquipOnlineStatusCode
        ,EMOLS.EquipOnLineStatusDesc
        -- skips next value in sequence in case of ties
        ,RANK()  OVER (
        PARTITION BY EMOLS.PlantName,EMOLS.LocationName --,EMOLS.ReportYear,
                EMOLS.EquipOnlineStatusCode
        ORDER BY EMOLS.StatusCount DESC
        ) AS FailureRank

        -- preserves sequence even with ties
        ,DENSE_RANK()  OVER (
        PARTITION BY EMOLS.PlantName,EMOLS.LocationName --,EMOLS.ReportYear,
                EMOLS.EquipOnlineStatusCode
        ORDER BY EMOLS.StatusCount DESC
        ) AS FailureDenseRank

FROM Reports.EquipmentMonthlyOnLineStatus EMOLS
INNER JOIN (
        SELECT DISTINCT [CalendarYear]
                ,[CalendarQuarter]
                ,CASE
                        WHEN CalendarMonth = 1 THEN 'Jan'
                        WHEN CalendarMonth = 2 THEN 'Feb'
                        WHEN CalendarMonth = 3 THEN 'Mar'
                        WHEN CalendarMonth = 4 THEN 'Apr'
                        WHEN CalendarMonth = 5 THEN 'May'
```

```
                    WHEN CalendarMonth = 6 THEN 'Jun'
                    WHEN CalendarMonth = 7 THEN 'Jul'
                    WHEN CalendarMonth = 8 THEN 'Aug'
                    WHEN CalendarMonth = 9 THEN 'Sep'
                    WHEN CalendarMonth = 10 THEN 'Oct'
                    WHEN CalendarMonth = 11 THEN 'Nov'
                    WHEN CalendarMonth = 12 THEN 'Dec'
            END AS CalendarMonthName
        FROM [DimTable].[Calendar]
        ) AS MSC
        ON (
        EMOLS.ReportYear = MSC.CalendarYear
        AND EMOLS.MonthName = MSC.CalendarMonthName
        )
WHERE EMOLS.ReportYear = 2002
AND EMOLS.EquipmentId = 'P1L1VLV1'
GO
```

First things first, notice I commented out the year column in both PARTITION BY clauses. Remember, this report will be expanded to include all years, so it can be uncommented when you try it out and load the data into Microsoft Excel for the pivot table.

Next, we see that we are trying to satisfy the requirement by joining to the Calendar dimension and generating the quarter names via an inline table, the first thing that came to mind but the wrong design although it looks slick if you like complex code that overcomplicates the solution!

Finally, for both the RANK() and DENSE_RANK() functions, both the PARTITION BY and ORDER BY clauses are the same. The partition is created by plant name, location name, report year, and equipment online status code. The partition is ordered by the status count value in descending order.

Pop Quiz   Why is the inline table a bad solution?

Let's see the results in Figure 9-15.

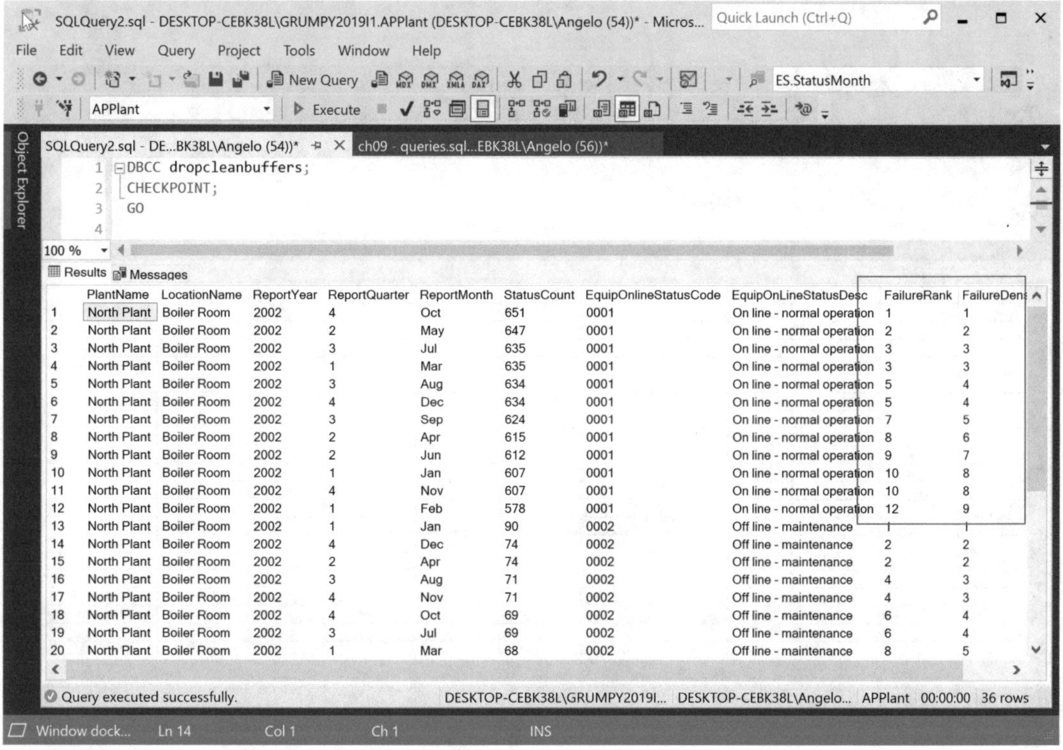

Figure 9-15. *Results using the inline table*

Results look correct. October 2002 was the best month in terms of the times equipment was online. February was the worst month. Notice the familiar pattern between the RANK() and DENSE_RANK() functions when duplicates occur. Let's see the performance landscape for this query.

Let's look at a simple graph to see the trends in the three categories of equipment online/offline status. Please refer to Figure 9-16.

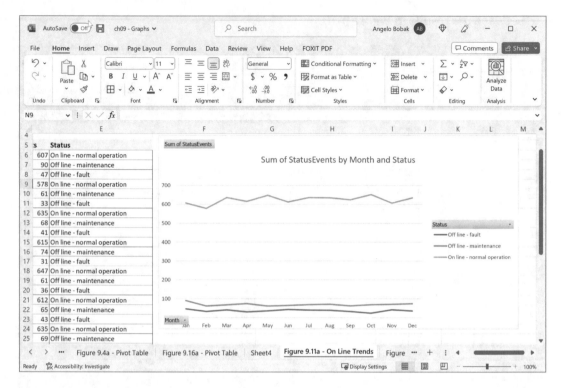

Figure 9-16. *Online trends*

The good news is that this plant seems to have a good record of online equipment. Offline for maintenance seems a bit high but not too bad, and offline status for faults is around an average of 50.

Tip   I had to do a bit of formatting after I copied and pasted the results from the SQL Server SSMS query plane to Excel, so when you write your queries, think ahead as to how you will use the results, so you minimize the formatting and renaming of columns in the spreadsheet.

The first time I ran this query, it took five seconds. Running it a second time, it ran in under one second, so this is a good example where the data was cached and you need to use DBCC to clear cache each time you run IO and TIME statistics to evaluate performance. If you do not run DBCC, you will get misleading statistics!

Performance Considerations

Time for our performance analysis session for this query. As usual, let's first create a baseline estimated execution plan.

Please refer to Figure 9-17.

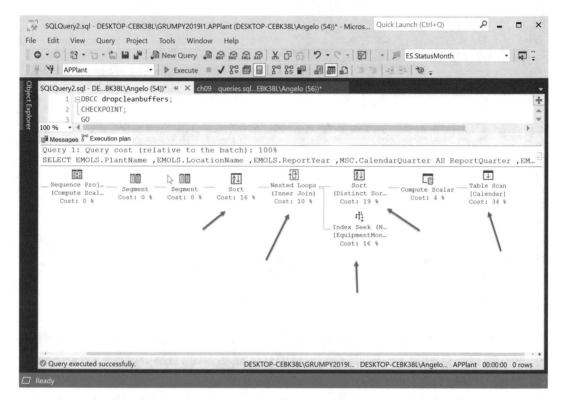

Figure 9-17. *RANK() vs. DENSE_RANK() estimated execution plan*

We have an index seek on the report table with a cost of 16%, but the JOIN clause to the Calendar dimension is causing problems.

We see a table scan with a cost of 34% and a sort task with a cost of 19%.

Both streams are merged with a nested loops (inner join) task at a cost of 10% followed by a sort task with a cost of 16%. That inline table was not a good solution from a performance perspective. This is where I went wrong!

Let's see what the IO and TIME statistics tell us.

Please refer to Figure 9-18.

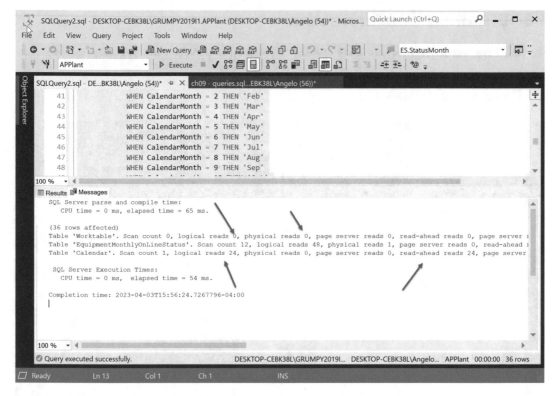

Figure 9-18. *RANK() vs. DENSE_RANK() IO and TIME statistics*

What we can see is that for the work table, all values are 0 so we will not worry about them.

For our report table, the `EquipmentMonthlyOnLineStatus` table, we see a scan count of 12; logical reads chime in at 48 and physical reads at 1.

Lastly, the Calendar dimension table incurs 1 scan count, 24 logical reads, and 24 read-ahead reads. This is one performance issue we might want to address.

Why not use the `CASE` block in the `SELECT` clause of the query instead of the inline table? The `JOIN` to the Calendar dimension makes no sense. Bad coding! Let's see a better solution.

Please refer to Listing 9-7.

Listing 9-7. A Better Solution

```
SELECT
     EMOLS.PlantName
     ,EMOLS.LocationName
     ,EMOLS.ReportYear
```

```
          ,CASE
                WHEN EMOLS.[MonthName] = 'Jan' THEN 'Qtr 1'
                WHEN EMOLS.[MonthName] = 'Feb' THEN 'Qtr 1'
                WHEN EMOLS.[MonthName] = 'Mar' THEN 'Qtr 1'
                WHEN EMOLS.[MonthName] = 'Apr' THEN 'Qtr 2'
                WHEN EMOLS.[MonthName] = 'May' THEN 'Qtr 2'
                WHEN EMOLS.[MonthName] = 'Jun' THEN 'Qtr 2'
                WHEN EMOLS.[MonthName] = 'Jul' THEN 'Qtr 3'
                WHEN EMOLS.[MonthName] = 'Aug' THEN 'Qtr 3'
                WHEN EMOLS.[MonthName] = 'Sep' THEN 'Qtr 3'
                WHEN EMOLS.[MonthName] = 'Oct' THEN 'Qtr 4'
                WHEN EMOLS.[MonthName] = 'Nov' THEN 'Qtr 4'
                WHEN EMOLS.[MonthName] = 'Dec' THEN 'Qtr 4'
          END AS CalendarMonthName
          ,EMOLS.[MonthName] AS ReportMonth
          ,EMOLS.StatusCount
          ,EMOLS.EquipOnlineStatusCode
          ,EMOLS.EquipOnLineStatusDesc
          -- skips next value in sequence in case of ties
          ,RANK()  OVER (
          PARTITION BY EMOLS.PlantName,EMOLS.LocationName,EMOLS.ReportYear,
                EMOLS.EquipOnlineStatusCode
          ORDER BY StatusCount DESC
          ) AS FailureRank

          -- preserves sequence even with ties
          ,DENSE_RANK()  OVER (
          PARTITION BY EMOLS.PlantName,EMOLS.LocationName,EMOLS.ReportYear,
                EMOLS.EquipOnlineStatusCode
          ORDER BY EMOLS.StatusCount DESC
          ) AS FailureDenseRank
FROM Reports.EquipmentMonthlyOnLineStatus EMOLS
WHERE EMOLS.ReportYear = 2002
AND EMOLS.EquipmentId = 'P1L1VLV1'
GO
```

We modified the query by taking out the inline table, which eliminated the JOIN operation. The CASE block was added so we can determine the quarter based on the month we are in. Let's run a new estimated query plan and compare it with the old one generated by the original query.

Please refer to Figure 9-19.

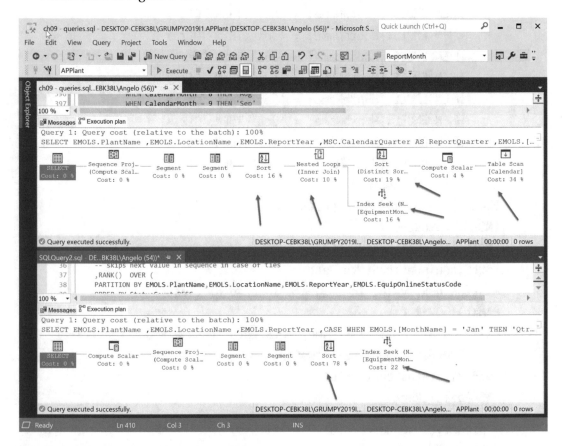

Figure 9-19. *Comparing query plans*

The new scheme shows significant improvement and simplification of the query plan. The Calendar dimension branch of the execution tree is gone. Our index seek, though, went up from 16% to 22%, and the first sort step went way up from 19% to 78%. The last sort step was eliminated. We need to compare IO and TIME statistics between both strategies.

Please refer to Figure 9-20.

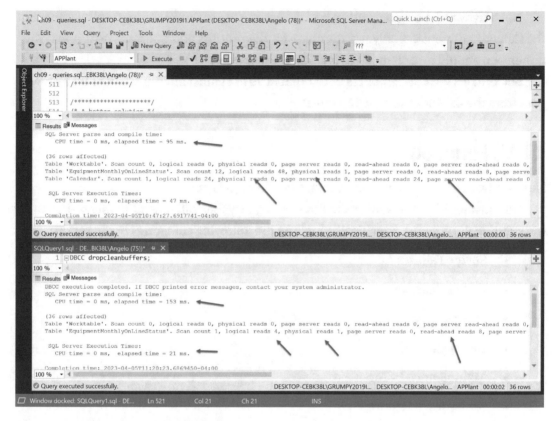

Figure 9-20. *Comparing IO and TIME statistics between both strategies*

Let's compare the statistics:

SQL Server parse and compile times went from 95 ms (old query) to 153 ms (worse).

SQL Server execution times went from 47 ms (old query) down to 21 ms (good improvement).

On the report table, the scan count went from 12 (old query) to 1 (good improvement).

On the report table, the logical reads went from 48 (old query) to 4 (good improvement).

On the report table, the read-ahead reads went from 24 (old query) to 8 (good improvement).

So it looks like the new strategy works better for us, in terms of performance, than the old query with the inline table.

Not only was the original code a bad design but we were able to bypass the issue that the Calendar dimension only has numbers for the months and quarters, no names. This dimension should be enhanced to include these names.

Looks like a CASE statement is better than a JOIN against an inline table that retrieves data from the Calendar dimension table!

NTILE Function

Next, we want to create some online classification buckets.

Our business analyst wants us to develop a query that will rate plants according to the status of equipment. Specifically, include the following categories in the report:

- Severe Failures

- Critical Failures

- Moderate Failures

- Investigate Failures

- Maintenance Failures

- No Issues to Report

That is, place the failure statistics into buckets that represent the preceding categories. The report should include the plant name, location name, calendar year, quarter, month, and total sum of equipment failures by month.

The following is the first attempt, which uses a two chained CTE approach. I split up the code into three sections as it is fairly long. Let's start off with the first CTE.

Note   Recall that by chained CTEs I mean that a lower CTE refers to the CTE above it and so on (CTE n − 1 refers to CTE n).

Please refer to Listing 9-8a.

Listing 9-8a. First CTE – Summing Logic

```
WITH FailedEquipmentCount
(
CalendarYear,QuarterName,[MonthName],CalendarMonth,PlantName,LocationName,
SumEquipFailures
)
AS (
SELECT
      C.CalendarYear
      ,C.QuarterName
      ,C.[MonthName]
      ,C.CalendarMonth
      ,P.PlantName
      ,L.LocationName
      ,SUM(EF.Failure) AS SumEquipFailures
FROM DimTable.CalendarView C
JOIN FactTable.EquipmentFailure EF
ON C.CalendarKey = EF.CalendarKey
JOIN DimTable.Equipment E
ON EF.EquipmentKey = E.EquipmentKey
JOIN DimTable.Location L
ON L.LocationKey = EF.LocationKey
JOIN DimTable.Plant P
ON L.PlantId = P.PlantId
GROUP BY
      C.CalendarYear
      ,C.QuarterName
      ,C.[MonthName]
      ,C.CalendarMonth
      ,P.PlantName
      ,L.LocationName
),
```

The first CTE retrieves the desired dimension columns and uses the SUM() function to add up the failures by month. Notice the inclusion of the mandatory GROUP BY clause. Also note that we need to JOIN five tables, one fact table and four dimension tables.

Reminder   Remember, if you need to link up N tables with JOIN clauses, you need N − 1 joins. So if you need to link five tables, you need four JOIN clauses in your query.

Next, we create another CTE that contains the NTILE() function and refer it to the first CTE we discussed earlier.

Please refer to Listing 9-8b.

Listing 9-8b. NTILE() Bucket CTE

```
FailureBucket (
PlantName,LocationName,CalendarYear,QuarterName,[MonthName],CalendarMonth,
SumEquipFailures,MonthBucket)
AS (
SELECT
      PlantName,LocationName,CalendarYear,QuarterName,[MonthName],
            CalendarMonth,SumEquipFailures,
      NTILE(5) OVER (
      PARTITION BY PlantName,LocationName
      ORDER BY SumEquipFailures
      ) AS MonthBucket
FROM FailedEquipmentCount
WHERE CalendarYear = 2008
)
```

We need to create five buckets, so the NTILE() function is used, and we specify a partition based on the plant name and location name columns. The partition is sorted by the values in the SumEquipFailures column.

Last but not least is the query to generate the report.

Please refer to Listing 9-8c.

Listing 9-8c. Categorize the Failure Buckets

```
SELECT PlantName,LocationName,CalendarYear,QuarterName,[MonthName],
SumEquipFailures,
CASE
      WHEN MonthBucket = 5 AND SumEquipFailures <> 0 THEN 'Severe Failures'
      WHEN MonthBucket = 4 AND SumEquipFailures <> 0 THEN 'Critical
      Failures'
      WHEN MonthBucket = 3 AND SumEquipFailures <> 0 THEN 'Moderate
      Failures'
      WHEN MonthBucket = 2 AND SumEquipFailures <> 0 THEN 'Investigate
      Failures'
      WHEN MonthBucket = 1 AND SumEquipFailures <> 0 THEN 'Maintenance
      Failures'
      WHEN MonthBucket = 1 AND SumEquipFailures = 0
            THEN 'No issues to report'
      ELSE 'No Alerts'
END AS AlertMessage
FROM FailureBucket
GO
```

To the rescue is our old friend, the CASE block. We simply print out the categories that are based on the tile number. A simple solution but a little bit complex because we needed to use two chained CTE blocks. Let's look at the results before we analyze performance.

Please refer to the partial results in Figure 9-21.

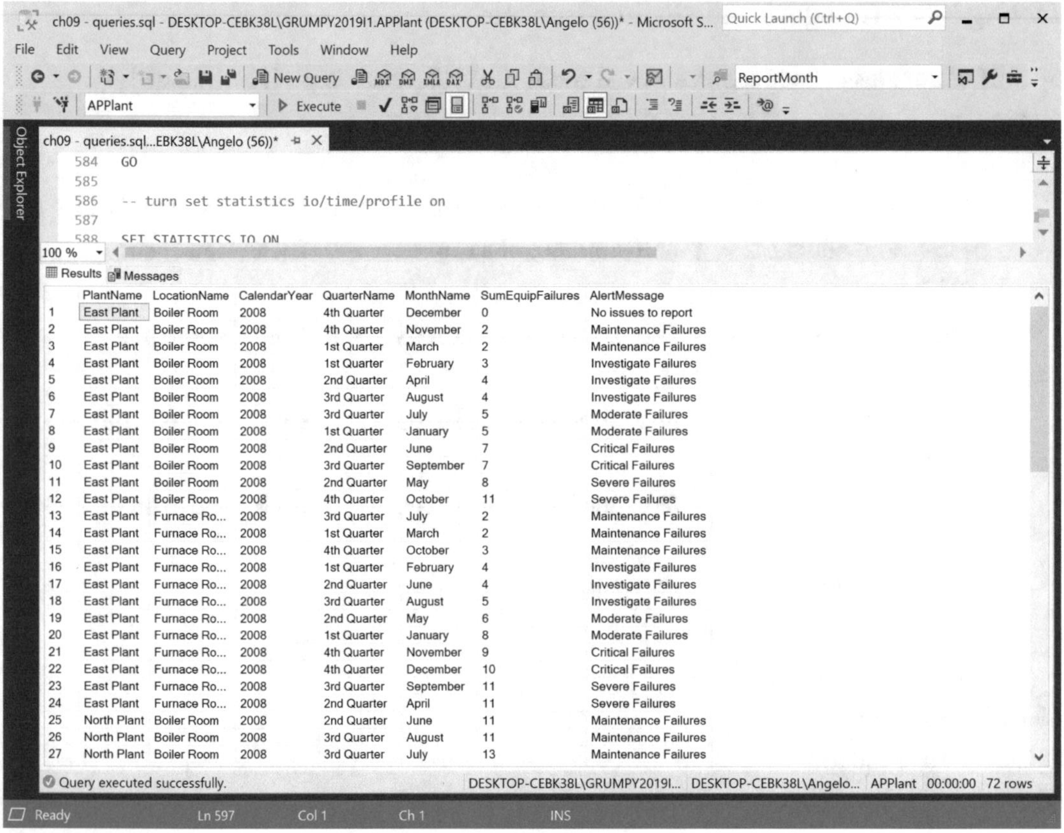

Figure 9-21. *Equipment failure bucket report*

The report looks good and shows the failures by buckets and plant, location, and calendar year. If we take out the WHERE clause and run the report for all years, we generate 1,517 rows. These could be copied and pasted into a Microsoft Excel pivot table so we could do some fancy slicing and dicing to really understand the results.

And that is exactly what we are going to do. Please refer to Figure 9-22.

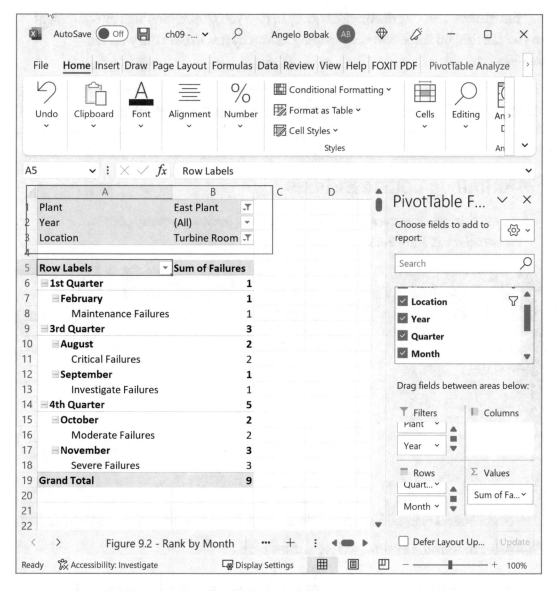

Figure 9-22. *Failure category bucket pivot table*

This is a very useful technique and tool to add to your toolkit as it will allow your business users and analysts the capability of viewing results similar to a multidimensional cube. The filter area highlighted in the figure lets you select the plant, year, and location so you can drill down into the data.

You can even modify the original query to give you failure numbers by equipment so you can drill up and down the grouping hierarchy just like we did with the GROUPING clause!

Homework   Try creating a pivot graph to complete this pivot table.

Performance Considerations

So how did this query perform? Here is our baseline estimated query plan.

Please refer to Figure 9-23.

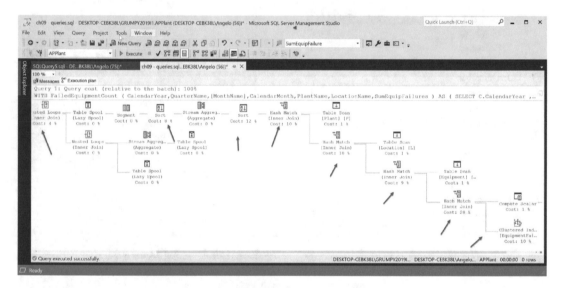

Figure 9-23. *Baseline estimated execution plan*

Lots of branches and table scans because of the joins to the dimension tables. The clustered index on the fact table helps and the cost is only 10%. There are four hash match joins (20%, 9%, 10%, and 10%) to merge the rightmost branches, a sort at 12%, another sort at 8%, and a final nested loops join at a cost of 4%.

Table spools are at 0%. Clearly our first objective is to simplify the query so the estimated query plan becomes less complex.

Now we use one of our existing report tables we created earlier. Listing 9-9 is the modified query.

Listing 9-9. Improved Query

```
WITH FailedEquipmentCount
AS (
SELECT
      PlantName,LocationName,CalendarYear,QuarterName,[MonthName]
      ,CalendarMonth,SumEquipmentFailure,
      NTILE(5) OVER (
      PARTITION BY PlantName,LocationName
      ORDER BY SumEquipmentFailure
      ) AS MonthBucket
FROM Reports.EquipmentFailureStatistics
WHERE CalendarYear = 2008
)

SELECT
      PlantName, LocationName, CalendarYear, QuarterName, [MonthName],
      SumEquipmentFailure,
CASE
            WHEN MonthBucket = 5 AND SumEquipmentFailure <> 0
            THEN 'Severe Failures'
            WHEN MonthBucket = 4 AND SumEquipmentFailure <> 0
            THEN 'Critical Failures'
            WHEN MonthBucket = 3 AND SumEquipmentFailure <> 0
            THEN 'Moderate Failures'
            WHEN MonthBucket = 2 AND SumEquipmentFailure <> 0
            THEN 'Investigate Failures'
            WHEN MonthBucket = 1 AND SumEquipmentFailure <> 0
            THEN 'Maintenance Failures'
            WHEN MonthBucket = 1 AND SumEquipmentFailure = 0
            THEN 'No issues to report'
      ELSE 'No Alerts'
END AS AlertMessage
FROM FailedEquipmentCount
GO
```

Using one of the existing report tables (`Reports.EquipmentFailureStatistics`) eliminates the first CTE. Let's see the revised estimated execution plan.

Please refer to Figure 9-24.

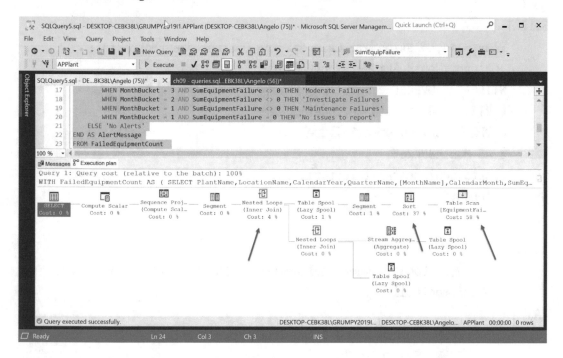

Figure 9-24. *Revised estimated query plan*

We can see the query plan was simplified. The clustered index seek was eliminated because we no longer use the table. What appears now is a table scan at a cost of 58% on the report table. Is this better? This is better as the report table is all set up with the data we need and it has only about 2,000 rows. If we really wanted to increase performance, we could make this a memory-optimized table. Everything is faster when you pop it in memory, right?

We also see a sort task with a cost of 37%, a table spool at 1%, and two nested loops joins with costs of 0% and 4% as we navigate right to left on the plan.

I guess this is better because there are fewer branches and steps. Let's look at the `IO` and `TIME` statistics.

Please refer to Figure 9-25.

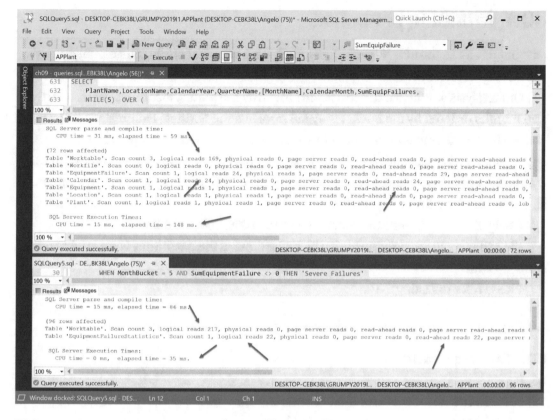

Figure 9-25. *Comparing IO and TIME statistics*

The new table in the improved, simpler query had 22 logical reads vs. 24 logical reads with the original EquipmentFailure table. Modest improvement but we will take it. Also, execution went down from 148 ms to 35 ms.

Conclusion: Think before you code. Also prototype.

Keep track of the report tables you create so you can reuse them for different queries! Sometimes you might come up with two report tables that are almost the same but differ because they need one or two more columns. Merge the table so you only have one table that can be used for multiple queries.

Lastly, always compare results when you improve a query. Make sure that the new improved query results are exactly the same as the old query. Improvements are great, but if the results do not match, they are meaningless.

Oh yes, the report tables solve a lot of problems, but they take time to load. Too many report tables, even if they are loaded at off-hours, can become a problem as the load window gets tighter and tighter.

ROW_NUMBER() Function

Last but not least, the ROW_NUMBER() function. Let's see what our business analysts' specifications look like:

Our analysts would like to see a report that lists the plants, locations, years, quarters, and months ordered by the bucket they fall into, like the "Critical Failures" bucket. Of course, the number of equipment failures for each month needs to appear in the report.

As each bucket has one or more months to report on, the analysts would like to see the number of the reporting event, not the month number but the sequence number or slot number, if you like, of the bucket.

Lastly, the report needs to cover three years: 2008, 2009, and 2010.

We will use a little trick for this in the definitions of the partitions. Here is what we came up with.

Please refer to Listing 9-10.

Listing 9-10. Failure Event Buckets with Bucket Slot Number

```
WITH FailedEquipmentCount
AS (
SELECT PlantName, LocationName, CalendarYear, QuarterName, [MonthName],
      CalendarMonth, SumEquipmentFailure,
      NTILE(5)  OVER (
      PARTITION BY PlantName,LocationName
      ORDER BY SumEquipmentFailure
      ) AS MonthBucket
FROM [Reports].[EquipmentFailureStatistics]
)

SELECT PlantName,LocationName,CalendarYear,QuarterName,[MonthName]
      ,CASE
            WHEN MonthBucket = 5 AND SumEquipmentFailure <> 0
            THEN 'Severe Failures'
            WHEN MonthBucket = 4 AND SumEquipmentFailure <> 0
            THEN 'Critical Failures'
            WHEN MonthBucket = 3 AND SumEquipmentFailure <> 0
            THEN 'Moderate Failures'
            WHEN MonthBucket = 2 AND SumEquipmentFailure <> 0
```

```
            THEN 'Investigate Failures'
            WHEN MonthBucket = 1  AND SumEquipmentFailure <> 0
            THEN 'Maintenance Failures'
            WHEN MonthBucket = 1  AND SumEquipmentFailure = 0
            THEN 'No issues to report'
      ELSE 'No Alerts'
      END AS StatusBucket
      ,ROW_NUMBER()  OVER (
      PARTITION BY (
      CASE
            WHEN MonthBucket = 5 AND SumEquipmentFailure <> 0
            THEN 'Severe Failures'
            WHEN MonthBucket = 4 AND SumEquipmentFailure <> 0
            THEN 'Critical Failures'
            WHEN MonthBucket = 3 AND SumEquipmentFailure <> 0
            THEN 'Moderate Failures'
            WHEN MonthBucket = 2 AND SumEquipmentFailure <> 0
            THEN 'Investigate Failures'
            WHEN MonthBucket = 1  AND SumEquipmentFailure <> 0
            THEN 'Maintenance Failures'
            WHEN MonthBucket = 1  AND SumEquipmentFailure = 0
            THEN 'No issues to report'
      ELSE 'No Alerts'
      END
      )
      ORDER BY SumEquipmentFailure
      ) AS BucketEventNumber
      ,SumEquipmentFailure AS EquipmentFailures
FROM FailedEquipmentCount
WHERE CalendarYear IN (2008,2009,2010)
GO
```

The trick is that a CASE block was used instead of the column name in the PARTITION BY clause. This illustrates one of the many ways you can specify partitions. A neat little trick to place in your back pocket.

Please refer to the partial results in Figure 9-26.

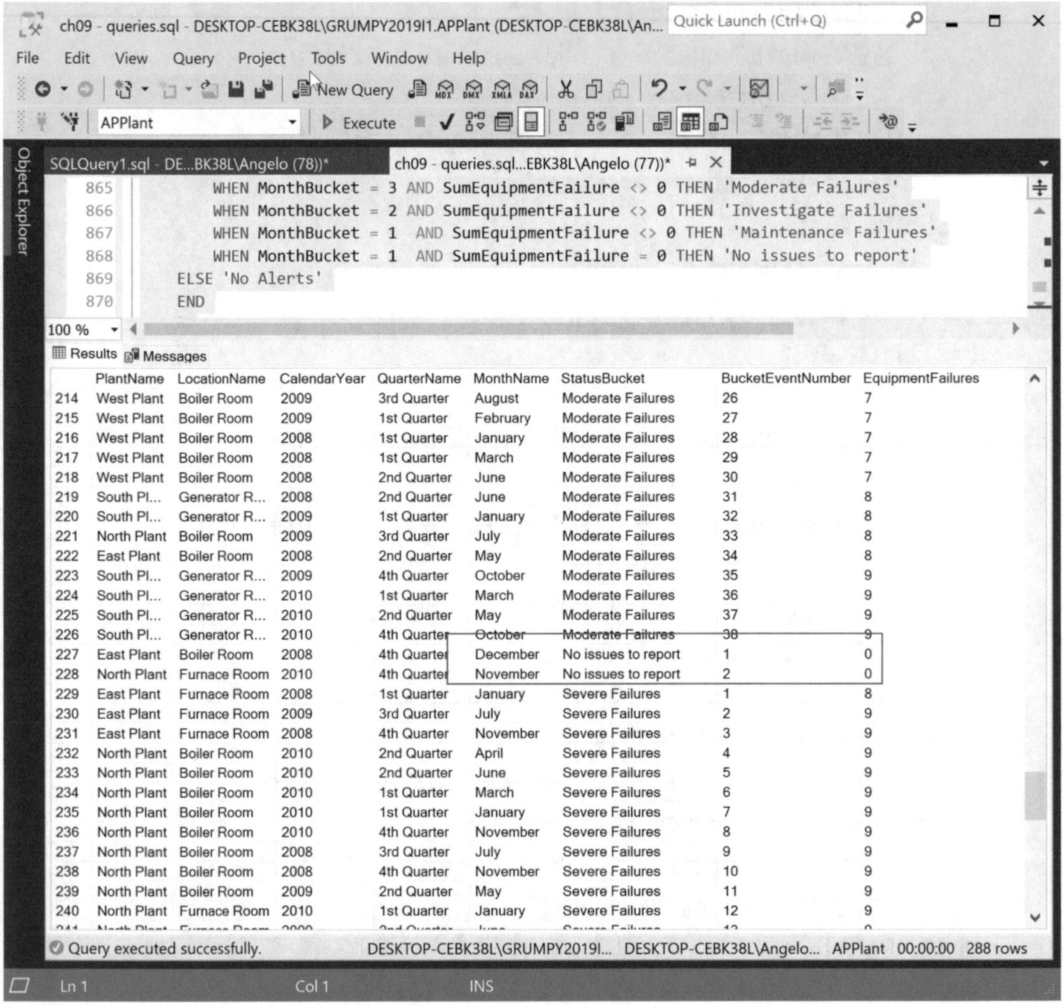

Figure 9-26. *Failure event bucket report with bucket slot number*

Okay, it works. Probably not a very valuable report, but I wanted to illustrate a variation of specifying the PARTITION BY clause.

On the other hand, the results can be used to create a nice pivot table. Let's copy and paste the results into a Microsoft Excel spreadsheet and create the pivot table.

Please refer to the partial results in Figure 9-27.

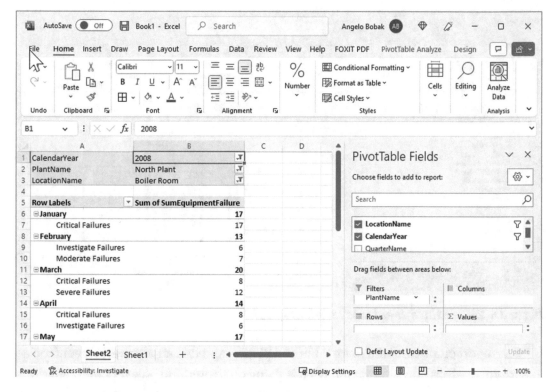

Figure 9-27. *Failure event bucket pivot table*

By dragging and dropping column names into the Filters, Columns, Rows, or Values list box, you can perform some powerful slicing and dicing of the data and traverse up and down the hierarchies of totals. The filters are very handy as you can focus on parts of the data at a time. Here we are looking at results for the year 2008, the North Plant, and the boiler room location.

You can also create a pivot graph with the same data.

Please refer to the partial results in Figure 9-28.

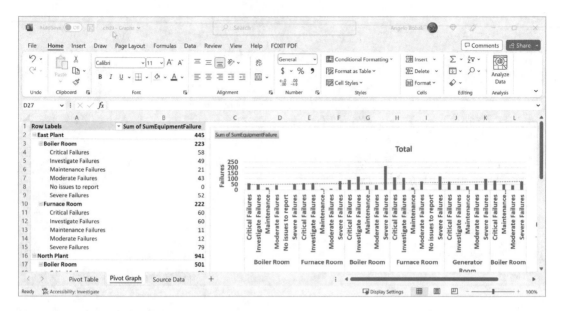

Figure 9-28. *Failure event bucket pivot graph*

The report output and the pivot table and pivot graph deliver a powerful set of tools to perform analysis and understand what is going on in our little power plant scenario. I definitely recommend that you run these queries, copy the results to Microsoft Excel, and generate the graphs and pivot tables. This is a powerful set of tools that will enhance your career, help you create powerful reports, and also help you solve performance issues.

Performance Considerations

This query was slightly different from the other queries in that we used two CASE blocks and used the IN( ) clause to specify some filters for the years we wanted to include in the results. Let's create our baseline estimated query plan to see if there is anything out of the ordinary.

Please refer to Figure 9-29.

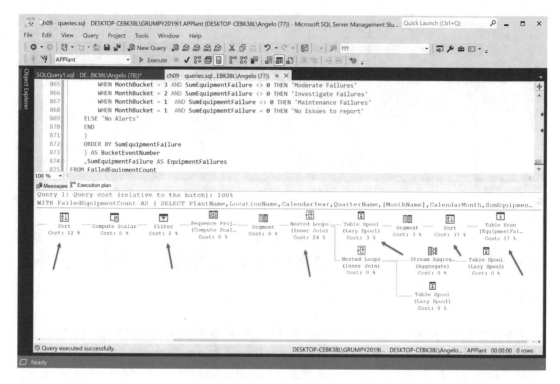

Figure 9-29. *Baseline estimated query plan*

Starting from the right side to the left side in the plan, we see one table scan at a cost of 17%. Since this is a small table, about 2,000 rows, we should not be concerned. Next, we see three table spools, two with costs of 0% and one with a cost of 3%. A couple of nested loops joins merge the rows from the branches. The first one is at a cost of 0%, but the second one is fairly high, chiming in at 24%. The rest of the flow includes a filter task (for the years) and a sort step at 12%.

Since we are not dealing with large volumes of data, I think we can let things stand. Let's look at the IO and TIME statistics followed by the STATISTICS PROFILE so we have a complete picture of the performance steps.

Please refer to Figure 9-30.

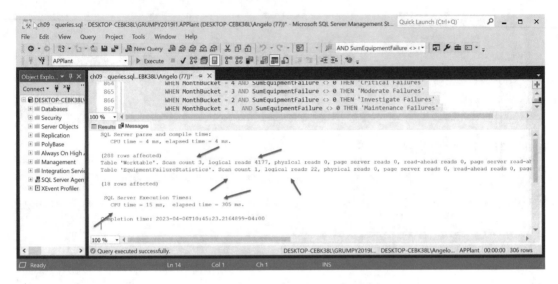

Figure 9-30. *IO and TIME statistics*

Looking at these statistics, we see the work table is getting a lot of action, a scan count of 3 and logical reads at 4,177. The report table has a scan count of 1 and 22 logical reads. Lastly, the SQL Server execution elapsed time is 305 ms, so under 1 s to return 306 rows.

Next are the `STATISTICS PROFILE` results.

Please refer to Figure 9-31.

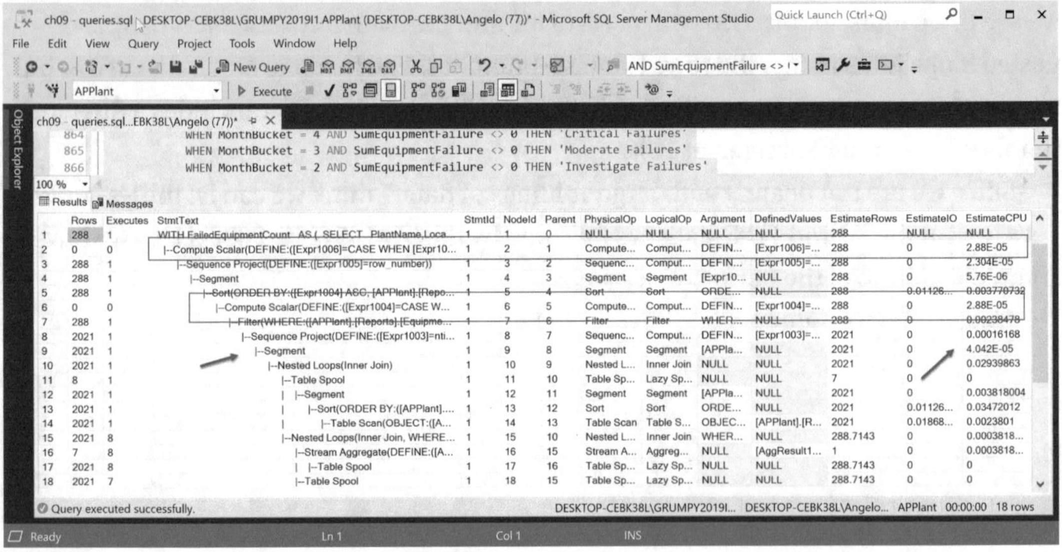

Figure 9-31. *Profile statistics report*

Pretty low values except for two `EstimateCPU` values. The first two are for the `CASE` blocks at 2.88E-05. Then we have a value of 4.042E-05 for a segment task, which includes the `GROUP BY` logic processing.

In conclusion, this was a query that had a few interesting steps, but the row count that needed to be processed was low, so there is no need for performance tuning.

Summary

This wraps up the chapter. We covered our usual application of the ranking functions and did our performance analysis using the techniques and tools from prior chapters. We also were able to leverage the data in some reports to create some interesting graphs and also pivot tables. I hope you found the power plant scenario interesting and will try out the code examples.

One final word: If you are or aspire to become a data architect that oversees multiple developers, you need to keep track of the proliferation of report tables. You do not want duplicates and missed opportunities for code and table sharing!

Good peer review sessions will ensure the team is aligned and up to date as to what is available to use in each of the individual development efforts. You do not need to reinvent the wheel and keep adding more and more report tables and indexes.

This applies to memory-enhanced tables; there are just so many you can create before you run out of memory!

Plant Use Case: Analytical Functions

Analytical Functions

We are going to add a couple of new tools to our toolkit in this chapter. We will take a quick look at Report Builder and also Analysis Services' multidimensional cubes.

Report Builder allows you to connect to your databases and create sophisticated web reports via SQL Server Reporting Services. This way, your business analysts can go right to a website and examine the reports they need.

Analysis Services is a powerful platform that is a part of the SQL Server BI stack and allows you to build multidimensional cubes based on data stored in a data warehouse, like the APPlant data warehouse we have been working with.

Let's continue with our usual analysis by applying the window functions belonging to this category against our data warehouse before we look at Report Builder and SSAS cubes.

As a reminder, the following are the eight functions for the analytical function category sorted by name. Recall that they can be used in queries together with the OVER() clause to generate valuable reports for our business analysts.

They should be familiar to you, if you already read Chapter 3 and/or 7. If not, take the time to read what they do:

- CUME_DIST(): **Does not** support the window frame specifications

- FIRST_VALUE() and LAST_VALUE(): **Do** support the window frame specifications

- LAG() and LEAD(): **Do not** support the window frame specifications

© Angelo Bobak 2023
A. Bobak, *SQL Server Analytical Toolkit*, https://doi.org/10.1007/978-1-4842-8667-8_10

- PERCENT_RANK(): **Does not** support the window frame specifications

- PERCENTILE_CONT(): **Does not** support the window frame specifications

- PERCENTILE_DISC(): **Does not** support the window frame specifications

The same approach as the prior chapter will be taken. The FIRST_VALUE() and LAST_VALUE() functions will be discussed together.

I describe what the function does, review the specifications supplied by our business analyst, and then present the code and the results that hopefully satisfy the business requirements. Feel free to bypass the descriptions if you already read them in Chapters 3 and 7 (by now you know how the functions work, but the application is to a new business scenario).

Next, our usual performance analysis is performed so we can see what improvements, if any, are required. We now have several performance-enhancing tricks we can try out in case our original queries are too slow:

- Add indexes.

- Add denormalized report tables.

- Create the supporting table as a memory-enhanced table.

- Modify the query.

As we saw in the prior chapters, improvements can usually be achieved by adding indexes or even creating a denormalized report table so that when the query using the analytical window functions is executed, it goes against preloaded data avoiding joins, calculations, and complex logic. When all else fails, try making the supporting table in a query a memory-enhanced table.

One last comment: In this group of analytical functions, you can only specify the ROWS and RANGE window frame specifications for the FIRST_VALUE() and LAST_VALUE() functions. The other functions do not support the window frame specifications. There will be a pop quiz later in the chapter, so don't forget.

CUME_DIST() Function

Recall from Chapter 7 that this function calculates the relative position of a value within a data set like a table, partition, or table variable loaded with test data. It calculates the probability that a random value, like a failure rate, is less than or equal to a particular failure rate for a piece of equipment. It is cumulative because it takes all values into account when calculating the distinct value for a specific value.

If you are a bit unsure of this, go back to Chapter 7 or Appendix A and review how this function works and how it is applied.

Here are the business specifications supplied by our business analyst friend:

Our analyst wants us to calculate the cumulative distribution of the sum of failures across the East Plant, specifically for the boiler room for the year 2002. The report needs to be sorted in ascending order of the calculated cumulative distribution values.

Additionally, the plant name, location, year, calendar quarter, and month name need to be included in the report.

Lastly, use the results to prepare a line chart in Microsoft Excel that shows the failure trends by month but also includes the corresponding cumulative distribution values.

Note   Graphing the cumulative distribution values makes no sense as they will go linearly up or down, but showing the failures in let's say a bar or line chart does.

This is a fairly simple query to write, so let's take a look. Please refer to Listing 10-1.

Listing 10-1. Cumulative Distribution Report for Failures

```
WITH FailedEquipmentCount
(
        CalendarYear, QuarterName, MonthName, CalendarMonth, PlantName,
        LocationName, SumEquipFailures
)
AS (
SELECT
        C.CalendarYear
        ,C.QuarterName
        ,C.MonthName
        ,C.CalendarMonth
```

```
      ,P.PlantName
      ,L.LocationName
      ,SUM(EF.Failure) AS SumEquipFailures
FROM DimTable.CalendarView C
JOIN FactTable.EquipmentFailure EF
ON C.CalendarKey = EF.CalendarKey
JOIN DimTable.Equipment E
ON EF.EquipmentKey = E.EquipmentKey
JOIN DimTable.Location L
ON L.LocationKey = EF.LocationKey
JOIN DimTable.Plant P
ON L.PlantId = P.PlantId
GROUP BY
      C.CalendarYear
      ,C.QuarterName
      ,C.MonthName
      ,C.CalendarMonth
      ,P.PlantName
      ,L.LocationName
)
SELECT
      PlantName, LocationName, CalendarYear, QuarterName,
      MonthName, CalendarMonth
      ,SumEquipFailures
      ,FORMAT(CUME_DIST() OVER (
       ORDER BY SumEquipFailures
      ),'P') AS CumeDist
FROM FailedEquipmentCount
WHERE CalendarYear = 2002
AND LocationName = 'Boiler Room'
AND PlantName = 'East Plant'
GO
```

Notice that I used the FORMAT() function so the results are printed as percentages instead of values between 0 and 1 (makes more sense to me anyway). The OVER() clause just contains an ORDER BY clause that arranges the equipment failure totals in ascending order.

Pop Quiz   Do you remember what the default PARTITION BY behavior is for an OVER() clause that only has an ORDER BY clause? Does this function allow window specifications in the OVER() clause? Answer at the end of the section.

If you want to include more power plants, locations, and years, modify the OVER() clause as follows:

```
,FORMAT(CUME_DIST() OVER (
        PARTITION BY PlantName,LocationName,CalendarYear
        ORDER BY CalendarMonth,SumEquipFailures
),'P') AS CumeDist
```

Make sure you remove the WHERE clause to get all results.

In Figure 10-1 are the partial results for both queries.

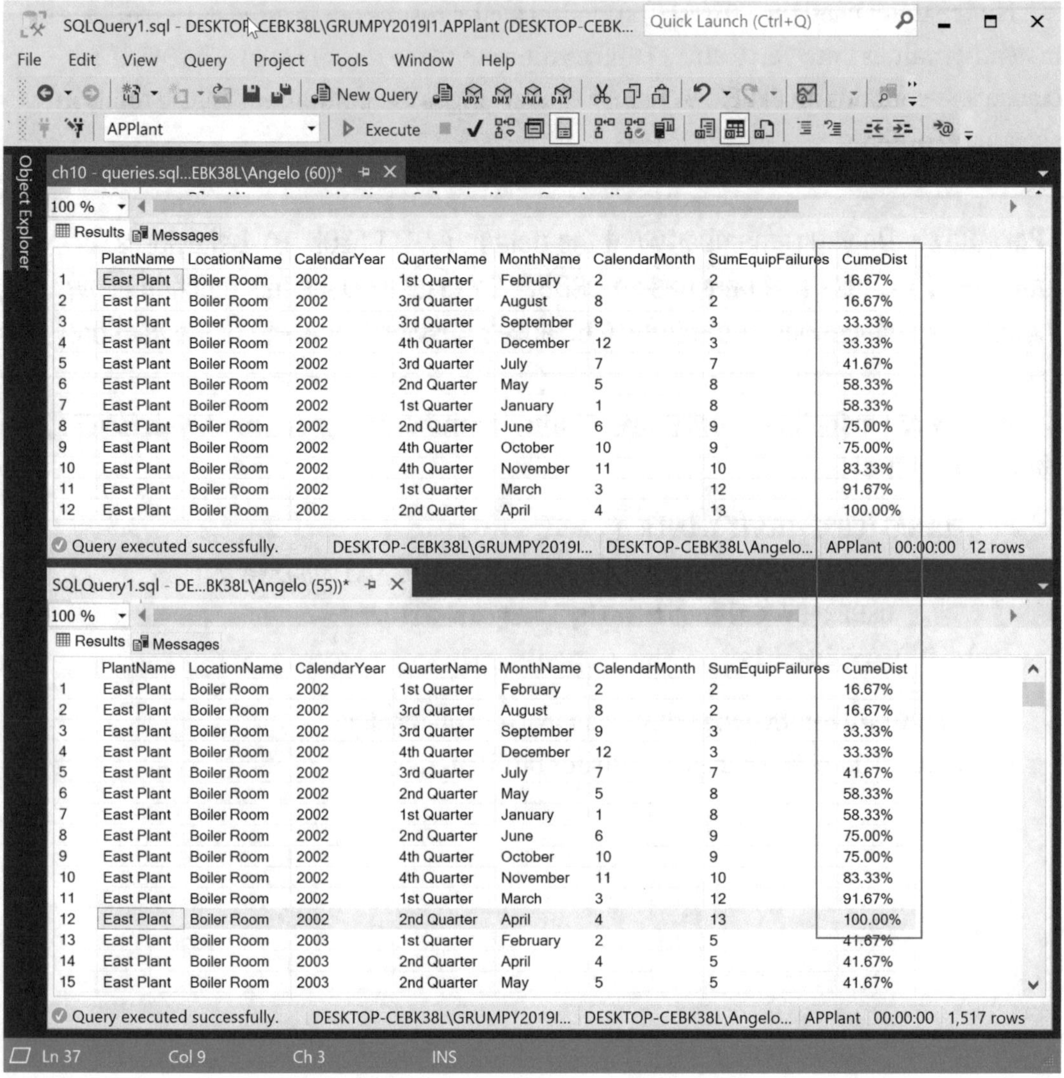

	PlantName	LocationName	CalendarYear	QuarterName	MonthName	CalendarMonth	SumEquipFailures	CumeDist
1	East Plant	Boiler Room	2002	1st Quarter	February	2	2	16.67%
2	East Plant	Boiler Room	2002	3rd Quarter	August	8	2	16.67%
3	East Plant	Boiler Room	2002	3rd Quarter	September	9	3	33.33%
4	East Plant	Boiler Room	2002	4th Quarter	December	12	3	33.33%
5	East Plant	Boiler Room	2002	3rd Quarter	July	7	7	41.67%
6	East Plant	Boiler Room	2002	2nd Quarter	May	5	8	58.33%
7	East Plant	Boiler Room	2002	1st Quarter	January	1	8	58.33%
8	East Plant	Boiler Room	2002	2nd Quarter	June	6	9	75.00%
9	East Plant	Boiler Room	2002	4th Quarter	October	10	9	75.00%
10	East Plant	Boiler Room	2002	4th Quarter	November	11	10	83.33%
11	East Plant	Boiler Room	2002	1st Quarter	March	3	12	91.67%
12	East Plant	Boiler Room	2002	2nd Quarter	April	4	13	100.00%

	PlantName	LocationName	CalendarYear	QuarterName	MonthName	CalendarMonth	SumEquipFailures	CumeDist
1	East Plant	Boiler Room	2002	1st Quarter	February	2	2	16.67%
2	East Plant	Boiler Room	2002	3rd Quarter	August	8	2	16.67%
3	East Plant	Boiler Room	2002	3rd Quarter	September	9	3	33.33%
4	East Plant	Boiler Room	2002	4th Quarter	December	12	3	33.33%
5	East Plant	Boiler Room	2002	3rd Quarter	July	7	7	41.67%
6	East Plant	Boiler Room	2002	2nd Quarter	May	5	8	58.33%
7	East Plant	Boiler Room	2002	1st Quarter	January	1	8	58.33%
8	East Plant	Boiler Room	2002	2nd Quarter	June	6	9	75.00%
9	East Plant	Boiler Room	2002	4th Quarter	October	10	9	75.00%
10	East Plant	Boiler Room	2002	4th Quarter	November	11	10	83.33%
11	East Plant	Boiler Room	2002	1st Quarter	March	3	12	91.67%
12	East Plant	Boiler Room	2002	2nd Quarter	April	4	13	100.00%
13	East Plant	Boiler Room	2003	1st Quarter	February	2	5	41.67%
14	East Plant	Boiler Room	2003	2nd Quarter	April	4	5	41.67%
15	East Plant	Boiler Room	2003	2nd Quarter	May	5	5	41.67%

Figure 10-1. *Cumulative distribution reports for failures*

I highlighted in a red box the first year results for both queries so you can see that the modified partition works correctly.

So how would you interpret these results?

Looking at the value for March 2002, the East Plant boiler room, we would interpret the value as meaning 91.67% of the values in the partition will be less than or equal to 13 failures. It is as simple as that!

Let's create the Microsoft Excel spreadsheet requested by our analyst next. Please refer to Figure 10-2.

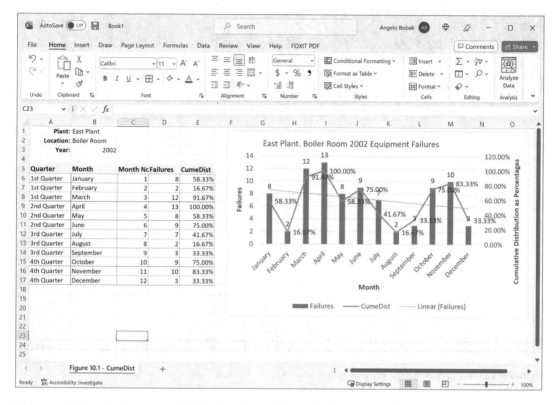

Figure 10-2. *Failure trends for the East Plant boiler room for 2002*

After a little bit of formatting, we have a nice report with a table that shows the failure rates for the 12 months in 2002. The graph shows that failures are trending down as the year progresses with April having the highest number of failures. The cumulative distribution values are printed for each month next to the failure numbers, so this comes in handy when interpreting the graph.

To generate this report, I needed to sort the data table by month number. If you do not want to see these numbers, just the month names, you could hide the column.

Just to give you a taste of a web reporting tool called Report Builder (that Microsoft makes available for free), Figure 10-3 is the data we just generated with the query (minus the WHERE clause so we can filter locations in the report) in a simple Report Builder report using a filter so we see the South Plant.

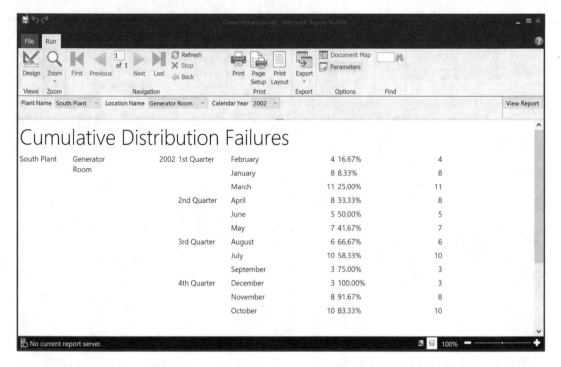

Figure 10-3. *Report Builder report*

This is a preview of the report that was created by the tool. I will go into more detail in the next chapter on how to actually create the report and publish it to a SSRS (SQL Server Reporting Services) website. For now, just be aware this is another important tool to add to your toolkit.

Here is what the report looks like after it was published to a Microsoft Reporting Services website on my laptop. Notice we filter the location for North Plant. Please refer to Figure 10-4.

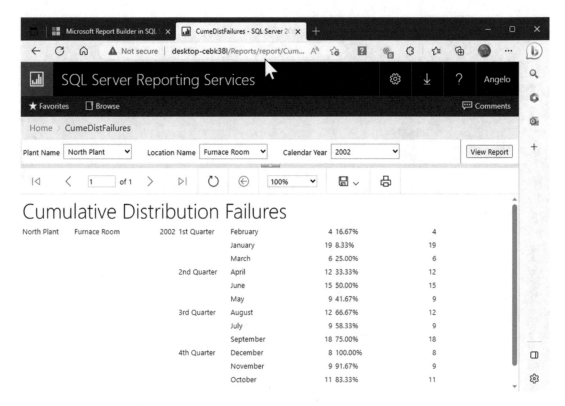

Figure 10-4. Published report on Reporting Services

Reporting Services affectionately known as SSRS is a complete web server platform that allows you to create and publish sophisticated reports and dashboards to a website. This is a powerful facility that allows business users and analysts quick access to the reports they ask for.

The good news is that you can install it on your laptop or workstation so you can prototype your reports!

Back to the sample report, notice that it has three drop-down list boxes so you can view the report for any plant, location, and year. More on this later!

Performance Considerations

Let's create our usual baseline estimated query plan to see how well this query runs (or not).

This is split into three parts as it is a large, estimated query plan. Please refer to Figure 10-5a for the first part.

643

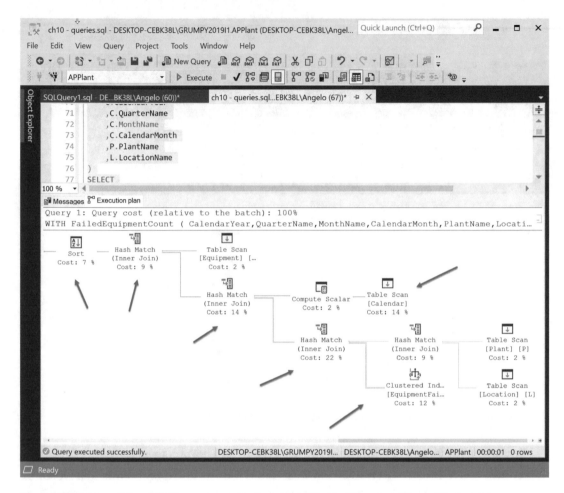

Figure 10-5a. *Plant failure cumulative distribution query plan, first part*

This looks like a very busy plan. Reading from right to left and starting at the bottom, we see two table scans for the Plant and Location tables, each with a cost of 2%. These could be candidates for memory-enhanced tables or maybe some denormalization by creating a Plantlocation table that combines the two tables.

Next is a clustered index scan against the equipment failure report table followed by two hash match tasks with costs of 22% and 14%.

Moving up to the middle branch, we see a table scan against the Calendar table with a cost of 14% and a small compute scalar task at 2% flowing into the hash match tasks that combine these two streams. The last task in this portion of the plan is a sort task with a cost of 7%.

So we see a lot of activity and opportunities for performance improvement. Maybe take the time to download the script and examine the query plan as we discuss it. Make sure you place the mouse pointer over each task to see the pop-up panel that gives you details related to what each task does.

Let's move to the middle part of the estimated query plan. Please refer to Figure 10-5b.

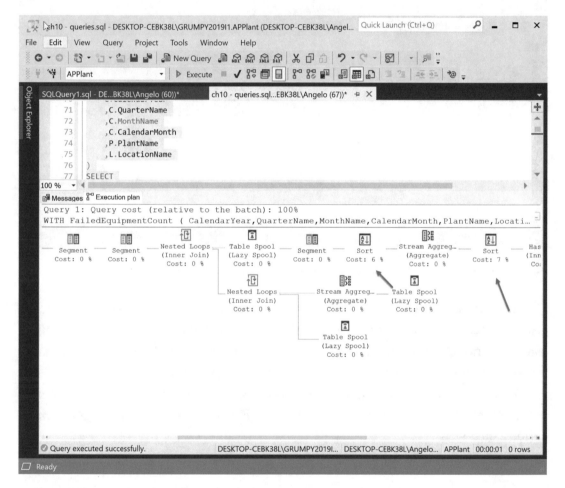

Figure 10-5b. *Middle section of the estimated query plan*

There's the sort task at 7% cost as a frame of reference with the prior section we just discussed.

Lots of tasks but all at zero cost except for a second sort step, which costs 6%. Notice the thickness of the lines as the data flows move along right to left. This indicates how much data is passing through. The thinner the line, the smaller the number of rows.

Typically, the lines will get thinner as you go, but when branches merge, they could get thicker as more rows pass through.

Last but not least is the remaining part of the estimated execution plan. Please refer to Figure 10-5c.

Figure 10-5c. *Remaining section of the estimated query plan*

Notice the sort task with the 6% cost as a frame of reference from the prior section we just examined. This portion of the plan is also busy, but all the costs are 0%, so we will leave well enough alone. Just notice the window spool task at 0%. Usually, 0% cost means the window spool is performed in memory, so that's a good thing.

This task is required so the data being processed can be referenced multiple times.

Let's look at the IO and TIME statistics next to get a better sense of what is going on performance-wise.

Please refer to Figure 10-6.

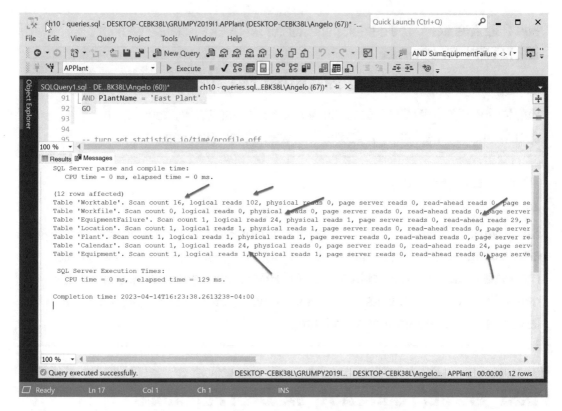

Figure 10-6. *Cumulative distribution failures – IO and TIME statistics*

First things first, SQL Server execution elapsed time is 129 ms, under 1 s, so we are okay here. We do see a scan count of 16 and 102 logical reads for the work table. The work file has zero values, so we will not worry about this set of statistics.

The `EquipmentFailure` table has 24 logical reads, 1 physical read- and 29 read ahead reads, so clearly something is going on here.

All the dimension tables except for the Calendar table have low values. The `Calendar` dimension table has 24 logical reads and 24 read-ahead reads.

So what is our conclusion? I would recommend you try to load the `Calendar` table into memory by recreating it as a memory-enhanced table and also replace the `CTE` with a report table as the results are historical, so we want to minimize the impact of generating this base data.

Also try to create a denormalized table that combines plant and location information and create the table as a memory-enhanced table.

Here is the answer to the pop quiz:

Pop Quiz Answer   The default behavior for an OVER() clause without a PARTITION BY clause is

RANGE BETWEEN UNBOUNDED PRECEDING AND CURRENT ROW

But, for the CUME_DIST() function, you cannot insert a window frame specification in the OVER() clause. This was a trick question.

Here is some homework:

Homework   Attempt the preceding recommendation on your own. Make sure you create memory-enhanced tables with different names, like CalendarMEM. Also, run some extensive query plan and IO/TIME statistics analysis between both methods to see if the results match. Remember to drop the memory-enhanced tables when you are finished with them.

FIRST_VALUE() and LAST_VALUE() Functions

Here are our next window functions. Remember these?

Given a set of sorted values, the FIRST_VALUE() function will return the first value in the partition. So, depending on whether you sort it in ascending order or descending order, the value will change! This is the value of the first position of the data partition, not the smallest value in the data set.

Given a set of sorted values, the LAST_VALUE() function will return the last value in the partition. Depending on how you sort the partition, the value will also be different as discussed earlier.

Both functions return whatever data type was used as an argument by the way; it could be an integer, it could be a string, etc. If this is a little unclear, create your own table variable with ten or more rows and write some queries against it so you can see how these window functions behave. Practicing a little will make the usage and results of the window functions clear.

Let's dive right into the example. Here is our business specification:

Our analyst wants to see the first and last values for each group of failures by year, location, and month. This has to include all calendar years and plants, so the result will be fairly large. The month names need to be spelled out, and so do the quarter names.

Looking at this specification, we realize we need to use the `Calendar VIEW`, which has the `CASE` blocks for quarter and month names – this could be expensive. Let's check out the query.

Please refer to Listing 10-2.

Listing 10-2. First and Last Values for Failures by Plant, Location, and Year

```
WITH FailedEquipmentCount
(
      CalendarYear, QuarterName, MonthName, CalendarMonth, PlantName,
      LocationName, SumEquipFailures
)
AS (
SELECT
      C.CalendarYear
      ,C.QuarterName
      ,C.MonthName
      ,C.CalendarMonth
      ,P.PlantName
      ,L.LocationName
      ,SUM(EF.Failure) AS SumEquipFailures
FROM DimTable.CalendarView C
JOIN FactTable.EquipmentFailure EF
ON C.CalendarKey = EF.CalendarKey
JOIN DimTable.Equipment E
ON EF.EquipmentKey = E.EquipmentKey
JOIN DimTable.Location L
ON L.LocationKey = EF.LocationKey
JOIN DimTable.Plant P
ON L.PlantId = P.PlantId
GROUP BY
      C.CalendarYear
      ,C.QuarterName
```

```
        ,C.MonthName
        ,C.CalendarMonth
        ,P.PlantName
        ,L.LocationName
)

SELECT PlantName, LocationName, CalendarYear, QuarterName,
MonthName,   CalendarMonth, SumEquipFailures
      ,FIRST_VALUE(SumEquipFailures) OVER (
            PARTITION BY PlantName,LocationName,CalendarYear,QuarterName
            ORDER BY CalendarMonth
      ) AS FirstValue
      ,LAST_VALUE(SumEquipFailures) OVER (
            PARTITION BY PlantName,LocationName,CalendarYear,QuarterName
            ORDER BY CalendarMonth
      ) AS LastValue
FROM FailedEquipmentCount
GO
```

We are already starting to see a trend taking shape here. Our first query and this query seem to be using the same CTE block, so we might definitely want to consider using a common report table based on the CTE instead of reinventing the CTE each time we need it. This is a simple but effective performance strategy and also simplifies the coding required for all our solutions.

As far as the logic for the OVER() clause, we see that a partition is set up by plant, location, year, and quarter for both the FIRST_VALUE() and LAST_VALUE() functions. The same is true for the ORDER BY clause used in both functions. The rows in the partition are sorted by month.

Let's see the results generated by this query. Please refer to the partial results in Figure 10-7.

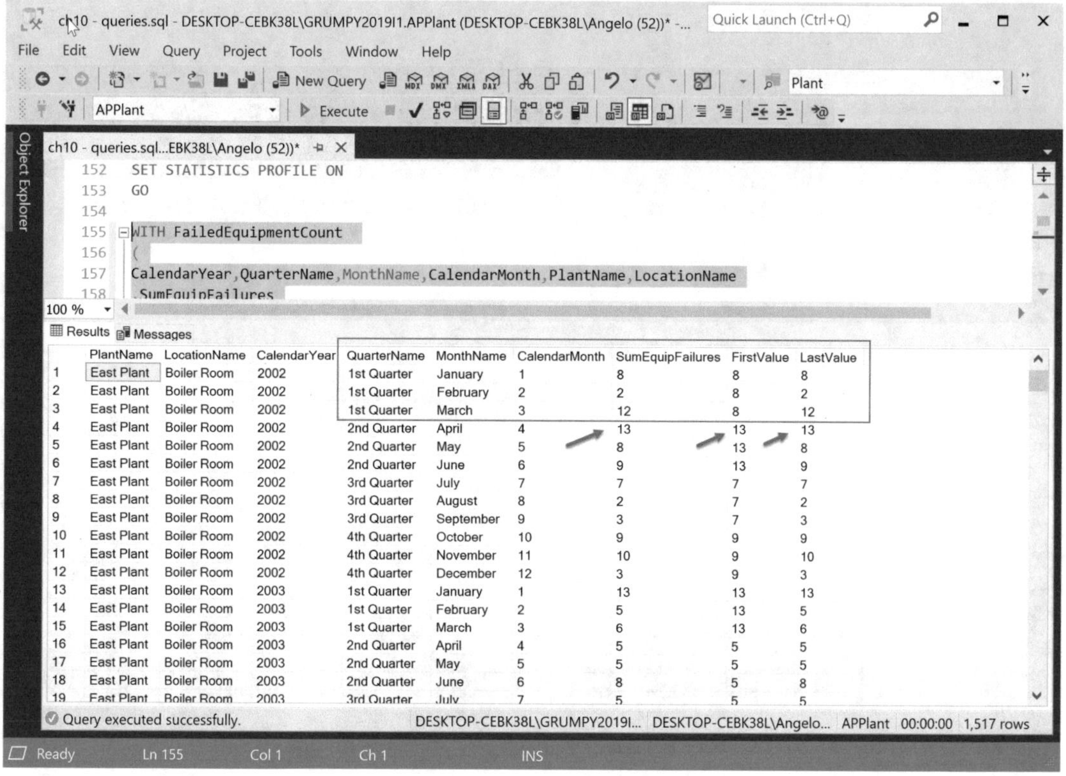

Figure 10-7. *First and last failure values by quarter*

As can be seen, everything lines up nicely. Notice that for the first row for each quarter partition, the first and last values are the same. As we move to the second month in the quarter, the values change, and the same is true for the third month. Going into the next quarter, the first and last values are reset and are the same. This report gives us a nice indication of the failure spreads, from smallest to largest on a calendar quarter basis.

Performance Considerations

Let's create our baseline estimated query plan to see what the damage is.

Please refer to Figure 10-8.

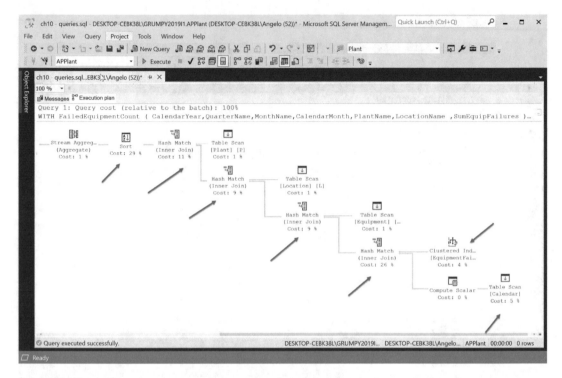

Figure 10-8. *Baseline estimated query plan for first and last values*

Looking at the high-value costs, we see a clustered index scan at 4% (not bad, but a seek would have been better), a whole bunch of hash match joins with costs ranging from 26% to 9%, and three table scans at 1%. Finally, at the top left, we see an expensive sort chiming in at 29% and not seen in the plan a window spool at 2% (appears on the left side of the plan).

So as not to waste time, I tried a test whereby I denormalized by combining the Plant and Location tables and substituting the new table in the query, but the results were actually a little worse, so let's forget about that strategy. You can try it on your own though if you want.

I think all we could do to improve performance is to replace the CTE block with a report table as previously discussed.

LAG() Function

We are lagging behind, so let's look at the next function, the LAG() function (bad pun!).

Recall what the LAG() function does:

This function is used to retrieve the previous value of a column relative to the current row column value (usually within a time dimension). For example, given this month's failure value for plant equipment, like a motor, pull last month's failure value for the same equipment ID or specify some offset, like three months ago or even a year or more depending on how much historical data has been collected and what your business analyst wants to see.

Use this logic to calculate the difference. Did failures go up or down? Critical measurement to monitor to spot failing equipment! Maybe the manufacturer needs a stern talking too?

Here are the specifications for the query our analyst wants:

A query is needed that reports failure rates one month behind so the failure can be compared to the current month's failure rate. The report needs to specify the year, quarter, and month calendar values together with the plant name and location. Each month's value should be a total of all failures for the month. So the report is historical in nature.

Please refer to Listing 10-3.

Listing 10-3. Last Month's Equipment Failures

```
WITH FailedEquipmentCount
(
      CalendarYear, QuarterName, MonthName, CalendarMonth, PlantName,
      LocationName, SumEquipFailures
)
AS (
SELECT
      C.CalendarYear
      ,C.QuarterName
      ,C.MonthName
      ,C.CalendarMonth
      ,P.PlantName
      ,L.LocationName
      ,SUM(EF.Failure) AS SumEquipFailures
FROM DimTable.CalendarView C
JOIN FactTable.EquipmentFailure EF
ON C.CalendarKey = EF.CalendarKey
```

```
JOIN DimTable.Equipment E
ON EF.EquipmentKey = E.EquipmentKey
JOIN DimTable.Location L
ON L.LocationKey = EF.LocationKey
JOIN DimTable.Plant P
ON L.PlantId = P.PlantId
GROUP BY
        C.CalendarYear
        ,C.QuarterName
        ,C.MonthName
        ,C.CalendarMonth
        ,P.PlantName
        ,L.LocationName
)
SELECT
        PlantName, LocationName, CalendarYear, QuarterName, MonthName,
        CalendarMonth, SumEquipFailures,
        LAG(SumEquipFailures,1,0) OVER (
                PARTITION BY PlantName, LocationName, CalendarYear, QuarterName
                ORDER BY CalendarMonth
        ) AS LastFailureSum
FROM FailedEquipmentCount
WHERE CalendarYear > 2008
GO
```

The CTE query has a five-way table JOIN plus the SUM() function and the obligatory GROUP BY clause. You know this could cause performance problems.

The OVER() clause has a PARTITION BY clause and an ORDER BY clause. This means the default behavior for the frame is

```
RANGE BETWEEN UNBOUNDED PRECEDING AND CURRENT ROW
```

Or does it?

Turns out the LAG() and LEAD() functions also cannot have a window frame. Easy trap to fall into! So when you use this or the LEAD() function, forget about window frame definitions. It makes sense if you think about how the functions work. They either look forward or backward and that's it!

Back to the query. The partition is set up by plant name, location name, year, and quarter. The partition is sorted by the month (numeric value).

Lastly, notice the parameters used with the LAG() function; the failure value is included, 1 indicates go back one month (second parameter), and 0 (third parameter) means replace any NULL values with 0. Try changing the value of the number of months to skip to get a feel for how the function works.

Please refer to the partial results in Figure 10-9.

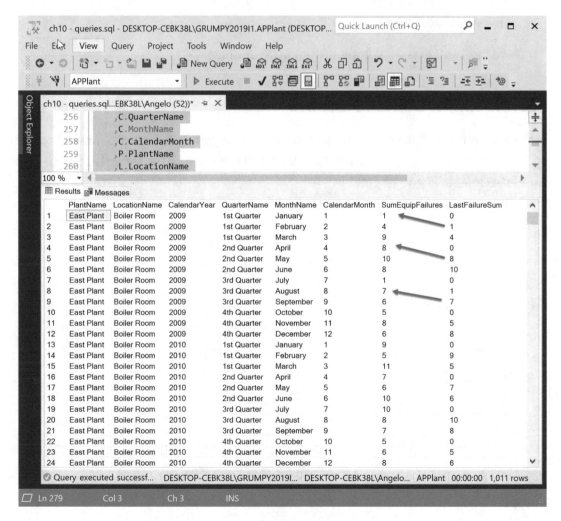

Figure 10-9. *Monthly lag report for equipment failures*

Notice the zero values. These mark the beginning of the partitions, so the prior rows (months) are ignored. Everything lines up though.

If we add a calculation that subtracts the lag value from the current month's failure value, we could see the change in the failure rates. Try out this code snippet:

```
,SumEquipFailures -
     LAG(SumEquipFailures,1,0) OVER
     (
     PARTITION BY PlantName,LocationName,CalendarYear,QuarterName
     ORDER BY CalendarMonth
     ) AS FailureChange
```

Good indicator to see if failures are going up or down.

Performance Considerations

Let's create our baseline estimated query performance plan in the usual manner. I split up the estimated execution plan into two parts so we can see all the steps this time.

Please refer to Figure 10-10a.

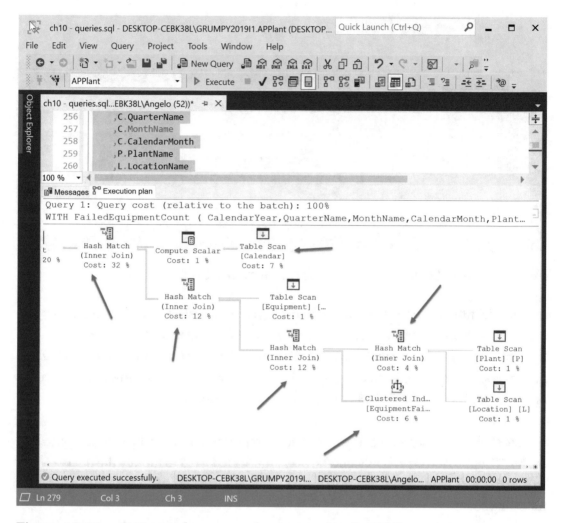

Figure 10-10a. *Estimated query performance plan for LAG – part 1*

Starting at the top right-hand side, we see the familiar table scan on the Calendar dimension with a cost of 7%. We see a clustered index scan (a seek is better) with a cost of 6%, but we are retrieving all the rows in the CTE, so it cannot be helped.

Next, we see four hash match join operations with costs of 4%, 12%,12%, and 32%. Lots of processing here! Let's check out the left section of the plan.

Please refer to Figure 10-10b.

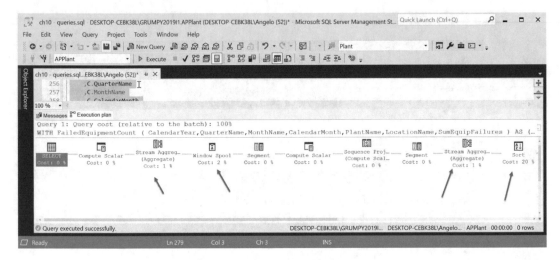

Figure 10-10b. *Estimated query performance plan for LAG – part 2*

Starting from the right side, we see a sort step at 20% and a stream aggregate step at 1%. Let's ignore the 0% tasks, and there is our window spool at 2% cost followed by a stream aggregate task with a cost of 1%. Remember that the window spool at 2% indicates spooling is performed in physical storage. The only performance improvement enhancement is to see if TEMPDB can be placed on a solid-state drive.

So a lot of steps but are they all necessary?

Recall that the specifications stated that the report is a monthly historical report, so the CTE approach is probably not the best approach. I tried to modify the query so that the SUM( ) function appears in the base query and not the CTE query with the following code:

```
SUM(EquipFailures) AS SumEquipFailures,
LAG(SUM(EquipFailures),1,0) OVER (
        PARTITION BY PlantName,LocationName,CalendarYear,QuarterName
        ORDER BY CalendarMonth
) AS LastFailureSum
```

Although it shows it is possible to insert an aggregate function inside the LAG( ) function, it did not result in any improvements as far as the estimated query plan was concerned. I also removed the GROUP BY clause and included it in the base query. This code is available in the script in case you want to try it out.

But let's look at the IO and TIME statistics before we come to any conclusions.

Please refer to Figure 10-11.

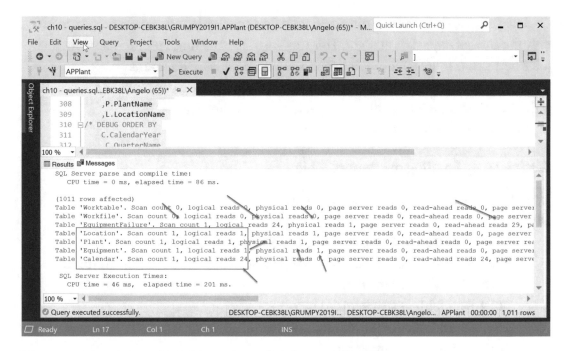

Figure 10-11. *IO and TIME statistics for the LAG() report*

We are seeing a trend here. Imagine you are the manager that is responsible for all the developers that are creating these reports and you keep seeing the same values over and over because of course the same CTE is being used for the analysis. The same CTE block sums up failures, but different analytical window functions are used in the various queries.

The EquipmentFailure report table has a scan count of 1, logical reads of 24, physical reads of 1, and read-ahead reads of 29. Notice the boxed section showing scan count and logical read values of 1 and our old friend the Calendar table showing 24 logical reads.

If this is not enough evidence that we need to put the CTE in a report table!

Furthermore, let's revise our requirement scenario so it reflects that multiple users, up to 10–20, will run this report after it is published to a SSRS website. The report will need filters so the users can pick the year.

Right now, the report returns 1,011 rows for a single year. Most likely the users will want to filter out not only the year but the plant, location, and year. Thinking ahead and implementing the CTE as a report table that is loaded once a month makes sense. An added wrinkle might be that users want the historical data, but they want daily loads for the current month, so the loading strategy needs to be revised:

Load all prior months one time, and then for each day in the current month, load only the current day's data at night. This way we offload all those performance-intensive joins between the fact table and the four dimension tables at off hours.

Having 20 users execute a CTE-based query several times a day makes no sense. Here is the INSERT command we will use to perform a one-time load.

Please refer to Listing 10-4.

Listing 10-4. Loading the Report Table

```
INSERT INTO Reports.PlantSumEquipFailures
SELECT
     C.CalendarYear, C.QuarterName, C.MonthName, C.CalendarMonth,
     P.PlantName,L.LocationName, SUM(EF.Failure) AS SumEquipFailures
     FROM DimTable.CalendarView C
     JOIN FactTable.EquipmentFailure EF
     ON C.CalendarKey = EF.CalendarKey
     JOIN DimTable.Equipment E
     ON EF.EquipmentKey = E.EquipmentKey
     JOIN DimTable.Location L
     ON L.LocationKey = EF.LocationKey
     JOIN DimTable.Plant P
     ON L.PlantId = P.PlantId

-- Here is the daily load
-- use the WHERE clause below to get the current day's values.
-- WHERE C.CalendarDate = CONVERT(DATE,GETDATE())

-- Here is the one time get all prior months load
-- use the WHERE clause below to get the all values for the last month.
-- WHERE MONTH(C.CalendarDate) = DATEDIFF(mm,MONTH(CONVERT(DATE,GE
TDATE())))

GROUP BY
     C.CalendarYear
     ,C.QuarterName
     ,C.MonthName
```

```
        ,C.CalendarMonth
        ,P.PlantName
        ,L.LocationName
ORDER BY
        C.CalendarYear
        ,C.QuarterName
        ,C.MonthName
        ,P.PlantName
        ,L.LocationName
GO
```

The INSERT statement is straightforward. It is based on the CTE so I will not discuss it again, but notice the commented-out WHERE clauses.

The first one can be used with the INSERT query to get the current day's failures. The second WHERE clause can be used for the one-time load to retrieve failures for all the prior months. So now the off-hour query should be fast as it only retrieves the current day's failures.

Listing 10-5 is the revised query that can be easily used by an SSRS report that has filters for not only the year but the plant and location.

Listing 10-5. Querying the Report Table

```
SELECT
        PlantName, LocationName, CalendarYear, QuarterName,
        MonthName, CalendarMonth, SumEquipFailures,
        LAG(SumEquipFailures,1,0) OVER (
                PARTITION BY PlantName,LocationName,CalendarYear,QuarterName
                ORDER BY CalendarMonth
        ) AS LastFailureSum
FROM Reports.PlantSumEquipFailures
WHERE CalendarYear > 2008
GO
```

Very simple. Let's see the new revised plan.

Please refer to Figure 10-12a.

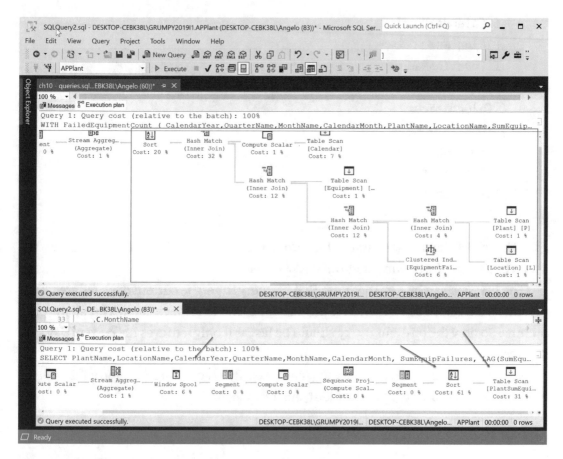

Figure 10-12a. *Original vs. new query plan – part 1*

The new estimated execution plan has fewer steps and no hash match joins although the sort step and window spool step went up considerably as can be seen in the last section of the plan. Maybe this is better or not!

Please refer to Figure 10-12b.

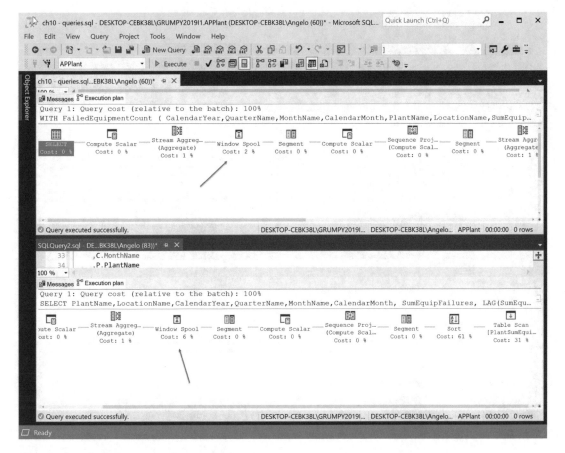

Figure 10-12b. *Original vs. new query plan – part 2*

The sort step increased to 41% and the window spool increased from 2% to 6%. What do the IO and TIME statistics look like?

Please refer to Figure 10-12c.

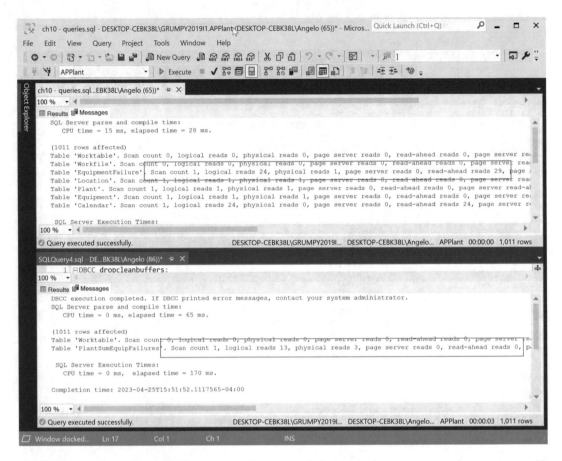

Figure 10-12c. *IO and TIME statistics comparison*

A picture is worth a thousand words. Less statistics and only one work table in the new query vs. a work table and work file in the old query.

Logical reads went from 24 in the old query to 13 in the new query. Physical reads went up though, from 1 to 3, but read-ahead reads went down from 29 to 0.

Now looking at all these statistics gets a little tedious, but you need to learn and practice this discipline as it will be put to good use when you are creating production-level queries in your work environment.

Pop Quiz   How do you turn on IO and TIME statistics?

LEAD( ) Function

The counterpart of the LAG( ) function is the LEAD( ) function. This function is used to retrieve the next value relative to the current row column. For example, given this month's sales amounts for a product, pull next month's sales amounts for the same product or specify some offset, like three months in the future or even a year in the future (historical data of course; one cannot see into the future from the present, and neither can your database if the data does not exist).

So this function can only be used in a historical context up to the current date. No future values exist yet!

Here comes the business analyst requirements:

Our analyst wants to see the same type of report as the prior report, but this time they want to see the change from the current month (in time) to the next month that follows. Luckily for us we now have the report table and do not need to use a CTE-based query.

Please refer to Listing 10-6.

Listing 10-6. Equipment Failure LEAD Query

```
SELECT
     PlantName,LocationName,CalendarYear,QuarterName,MonthName
     ,CalendarMonth,SumEquipFailures
     ,LEAD(SumEquipFailures,1,0) OVER (
          PARTITION BY PlantName,LocationName,CalendarYear,QuarterName
          ORDER BY CalendarMonth
     ) AS NextFailureSum
FROM [Reports].[PlantSumEquipFailures]
WHERE CalendarYear > 2008
GO
```

All we did was copy and paste the prior query, and instead of using the LAG( ) function, we used the LEAD( ) function. Let's check out the results in Figure 10-13.

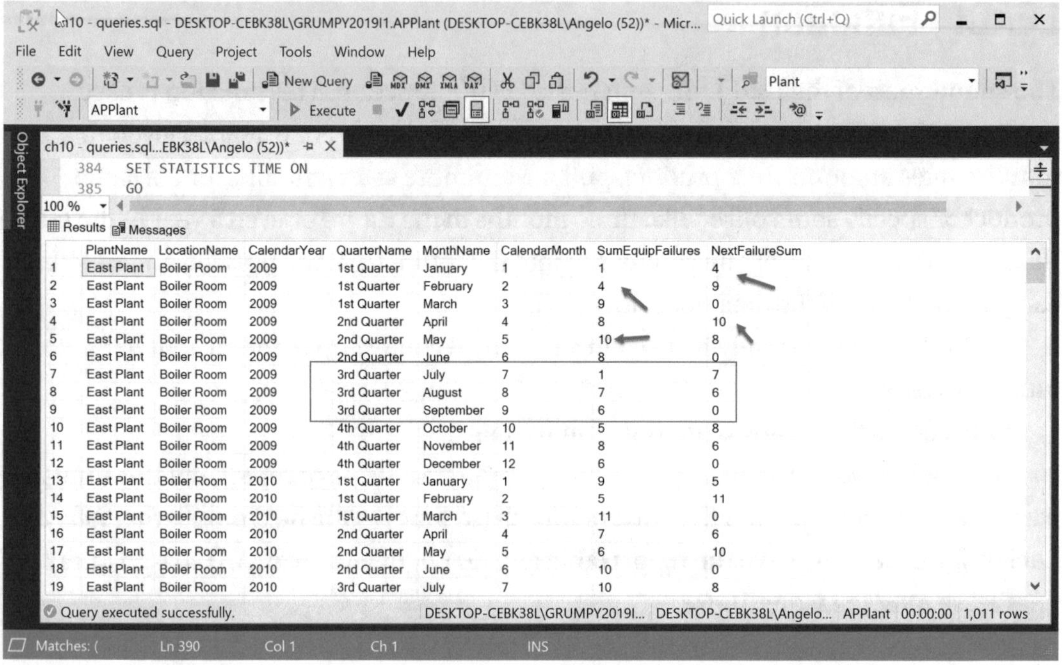

Figure 10-13. *Current month's and next month's failures*

As can be seen, this function also works well. Notice how the values for the next month are reset when a new quarter is processed, just how we specified it in the `PARTITION BY` clause.

Homework   Using this function and the prior function, add logic to calculate the difference between the prior month's failure value and the current month's failure value. Also do the same for the `LEAD()` function. That is, calculate the difference between the current value and next month's value. Create one large query and do some performance analysis. The code for this homework is included in the script for this chapter in case you get stuck.

Performance analysis time!

Performance Considerations

Let's create a baseline estimated query plan in the usual manner. Keep in mind that you cannot specify RANGE or ROW window frames for the LEAD() or LAG() function. So either you look ahead relative to the current month or look back from the current month.

Back to the query. Both a PARTITION BY and an ORDER BY clause are included in the OVER() clause.

I think we can deduce what tasks will have high percentage values. Please refer to Figure 10-14.

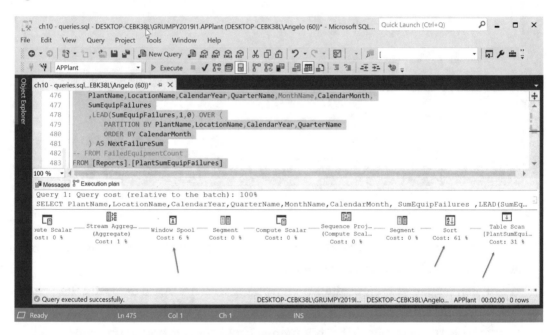

Figure 10-14. *Baseline estimated query plan for the LEAD function query*

If you guessed that we would see expensive sort and window spool tasks, you are correct. Notice the sort task chimes in at 61% and the window spool task chimes in at 6%. The table we are working with is too small for an index, so we will wrap things up by looking at the IO and TIME statistics.

Please refer to Figure 10-15.

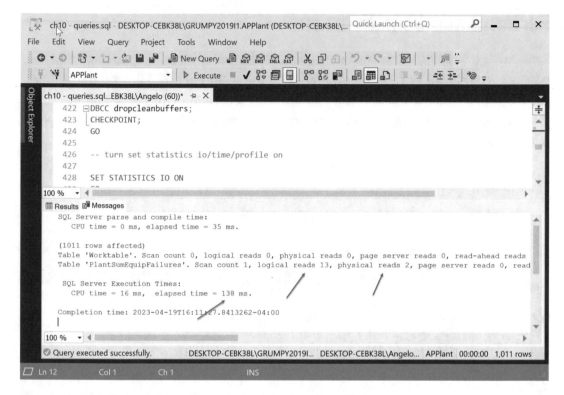

Figure 10-15. *IO and TIME statistics for the LEAD() function*

These are not too bad: all zero values for the work table and 1 scan count, 13 logical reads, and 2 physical reads for the report table.

I think that we can now agree that when working with historical data with the window functions, we should always consider some sort of denormalized report table as the base table replacing any CTE. This approach will deliver good performance, especially when multiple users execute the same query via a report or directly with SSMS.

Pop Quiz Answer   How do you turn on IO and TIME statistics?

Execute the following commands:

```
SET STATISTICS IO ON

GO

SET STATISTICS TIME ON
```

```
GO
```

To turn them off, just use the OFF keyword instead of ON.

PERCENT_RANK() Function

Next on deck is the PERCENT_RANK() function. What does it do? We covered this in Chapter 3, but in case you need a reminder or are not reading the chapters in sequential order, here is a review.

Using a data set like results from a query, partition, or table variable, this function calculates the relative rank of each individual value relative to the entire data set (as a percentage). You need to multiply the results (the value returned is a float) by 100.00 or use the FORMAT() function to convert the results to a percentage format (this function returns a FLOAT(53) data type by the way).

Also, it works a lot like the CUME_DIST() function; at least that's what the Microsoft documentation says.

You will also see that it works a lot like the RANK() function (in theory anyway). The RANK() function returns the position of the value, while PERCENT_RANK() returns a percentage value. We will use this function together with the CUME_DIST() function just discussed so we can compare the results and also see what effect it will have on the query plan we will generate shortly.

Our friendly business analyst supplies the following specifications:

A report is needed that shows percentage rank and cumulative distribution by plant, location, year, quarter, and month for all plants and all years. Values should be calculated based on groupings of year, plant name, and location.

Based on these simple specifications, in Listing 10-7 is the TSQL query.

Listing 10-7. Percent Rank vs. Cumulative Distribution

```
SELECT
     PlantName,LocationName,CalendarYear,QuarterName,
         MonthName,CalendarMonth
     ,SumEquipFailures
```

```
        ,FORMAT(PERCENT_RANK() OVER (
            PARTITION BY CalendarYear,PlantName,LocationName
            ORDER BY CalendarYear,PlantName,LocationName,
            SumEquipFailures
        ),'P') AS PercentRank

        ,FORMAT(CUME_DIST() OVER (
            PARTITION BY CalendarYear,PlantName,LocationName
            ORDER BY CalendarYear,PlantName,LocationName,
            SumEquipFailures
        ),'P') AS CumeDist
FROM [Reports].[PlantSumEquipFailures]
GO
```

Notice that there is no CTE used this time. We are now using the precalculated report table for the sum of failures in order to simplify the query and increase performance.

The partition is being set up by the year, plant name, and location name for both functions. The ORDER BY clause is declared using the year, plant name, and location name and the equipment failure sum values. Let's see how the returned values of the window functions compare.

Please refer to the partial results in Figure 10-16.

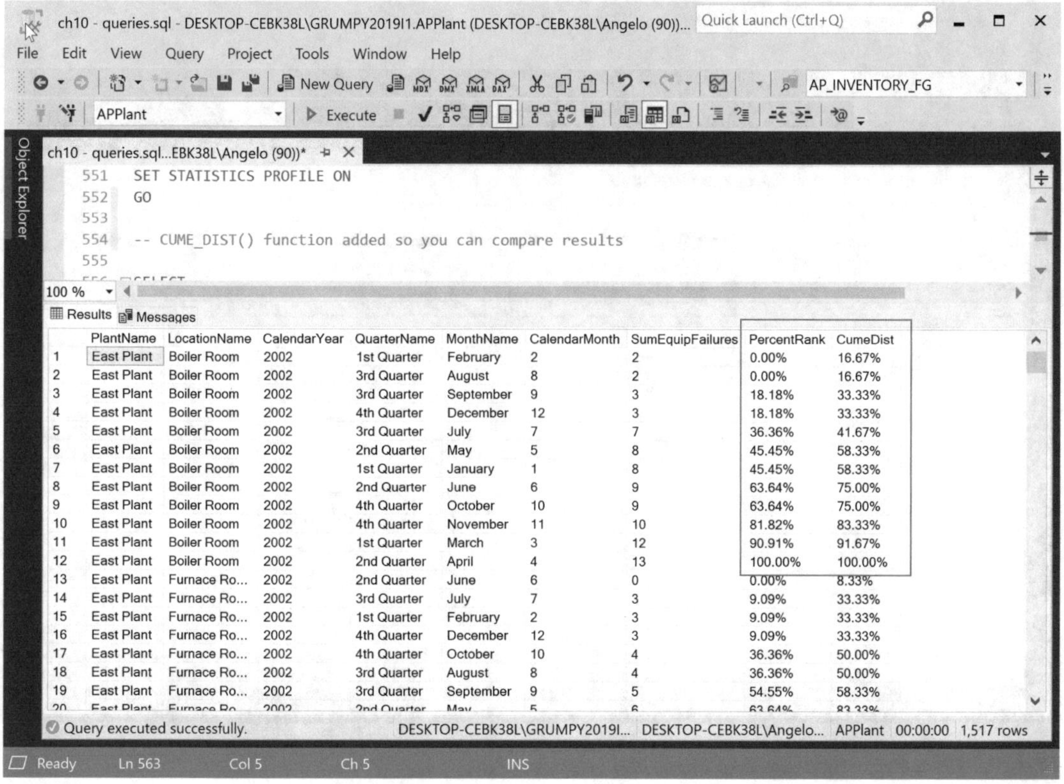

Figure 10-16. *Percent rank and cumulative distribution for equipment failures*

The results for both functions are similar but not by much. The partitions are set up correctly, and we see the results for the month by plant, location, and year. Looks like we had the most failures in April and no failures at all for February and August for 2002 for the boiler room.

Let's see the estimated performance plan for this query.

Performance Considerations

Let's create our baseline estimated execution plan so we can perform some analysis. We need to break up the plan into two parts again as it is rather large.

Please refer to Figure 10-17a.

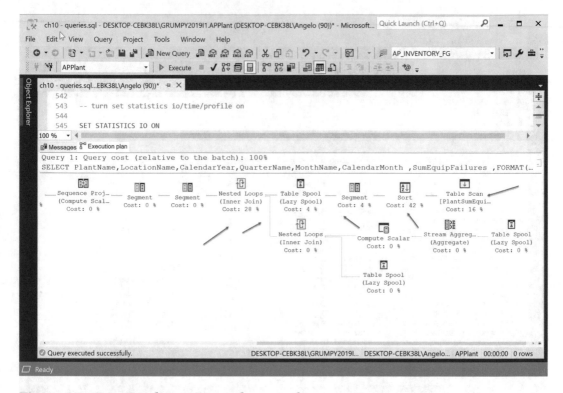

Figure 10-17a. *Baseline estimated query plan, part 1*

First, I need to say the report table has only 1,517 rows so it is not big, but it will be interesting to see how the estimated performance plan breaks it up into tasks. Second, the query ran in less than one second, so why bother? Practice makes perfect, that's why!

We can see a lot of activity. The table scan comes in at 16% followed by a sort (42% cost), segment (4% cost), and table spool (4% cost) before it is merged with the bottom branch with a nested loops join at 28%. Seems that in our performance analysis travels, the sorts and joins always seem to have the higher costs. We are using small tables, but if the tables had large volumes of rows, we would need to consider indexes on the dimension surrogate keys.

Let's see the second half of the plan.

Please refer to Figure 10-17b.

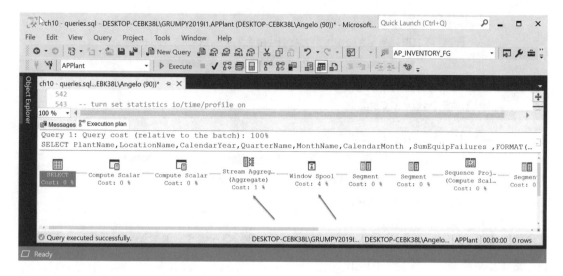

Figure 10-17b. *Baseline estimated query plan, part 2*

Moving from right to left, we see a window spool at 4% cost and a stream aggregate at 1% cost.

We are returning a lot of rows, so we see a lot of window spool action. Remember that window spool tasks with higher-than-zero percentages mean the spooling is performed on a physical disk. So this table could be a candidate for a memory-enhanced table.

I tried to add a clustered index based on the columns in the PARTITION BY clause and the ORDER BY clause, but the improvement was minimal, so I will not even bother to show the results. I do include the code for the index in the script for this chapter.

Let's turn our attention to the fact that this query has 1,517 rows. What happens when you add a WHERE clause to filter out only one year? Will our analyst allow a modification, so we can filter out rows based on the year, plant, and location? Let's modify the query and try it out.

Please refer to listing 10-8.

Listing 10-8. Adding a WHERE clause

```
SELECT
     PlantName,LocationName,CalendarYear,QuarterName,MonthName,
     CalendarMonth
     ,SumEquipFailures
```

673

```
       ,FORMAT(PERCENT_RANK() OVER (
             PARTITION BY CalendarYear,PlantName,LocationName
             ORDER BY CalendarYear,PlantName,LocationName,
             SumEquipFailures
       ),'P') AS PercentRank

       ,FORMAT(CUME_DIST() OVER (
             PARTITION BY CalendarYear,PlantName,LocationName
             ORDER BY CalendarYear,PlantName,LocationName,
             SumEquipFailures
       ),'P') AS CumeDist
FROM [Reports].[PlantSumEquipFailures]
WHERE PlantName = 'East Plant'
AND LocationName = 'Boiler Room'
AND CalendarYear = 2002
GO
```

As you can see, the query is hard-coded, so it filters the results for plant "East Plant," the boiler room, and the year 2002. The analyst wants all the years, plants, and locations, but let's see the performance plan after this modification.

Please refer to Figure 10-18.

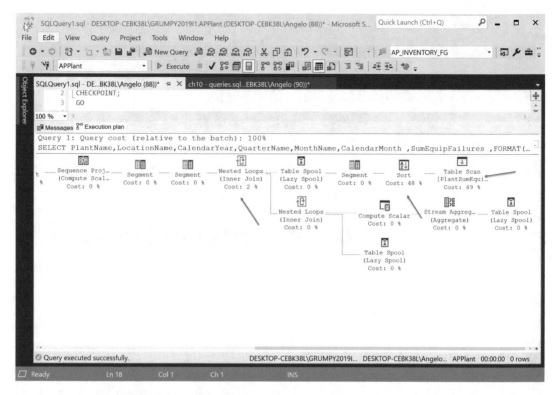

Figure 10-18. *Estimated performance plan for the WHERE clause*

Not much improvement in the table scan and sort. (I tried to also add a clustered index, but things did not get better.) The nested loops join went down to 2%, so this is an improvement. The window spool that is not shown went down to 0%, so at least some improvement there.

Here are the IO/TIME statistics between the query without the WHERE clause and the new and improved query.

Please refer to Figure 10-19.

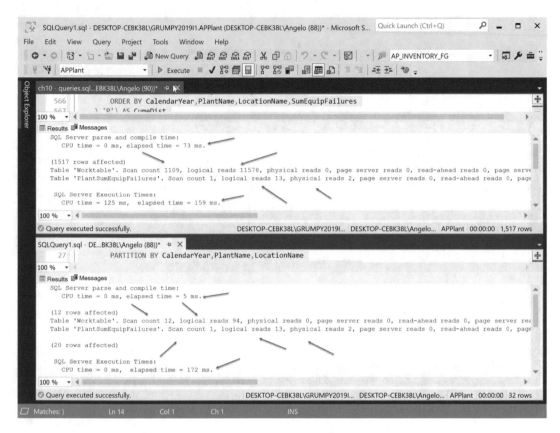

Figure 10-19. *Original and revised IO and TIME statistics*

Clearly the logical reads went down dramatically. The elapsed SQL Server execution time went up by 11 seconds though. So what does all this tell us? Do we really need all the years?

Let's ask our analyst to make sure. Porting this query with the WHERE clause to a web report created with Report Builder that includes filters for the plant, location, and year will deliver a faster-performing report. We can also give the user the option to include or not include all years, plants, and locations. This way, you can give your analyst what they want and give them options as to how much data will be generated in the report.

Pop Quiz   Which analytical window functions do not support window frame specifications (ROWS and RANGE)? Answer in Appendix A or the beginning of this chapter. Maybe the question is which two functions support this specification?

It's time for the percentile continuous and discrete functions.

PERCENTILE_CONT Function

We will discuss the PERCENTILE_CONT() function first, and what does this function do?

Well, if you read Chapter 7, you will recall that it work on a continuous set of data.

Recall what continuous data is. Continuous data is data measured over a period of time, like boiler temperatures in equipment over a 24-hour period or equipment failures over a period of time.

Example: Summing up each boiler temperature to arrive at a total makes no sense. Summing up the boiler temperatures and then dividing them by the time period to get an average does. **So we can say continuous data can be measured by using formulas like average and so on.**

In conclusion, the PERCENTILE_CONT() function works on a continuous set of data based on a required percentile so as to return a value within the data set that satisfies the percentile (you need to supply a percentile as a parameter, like .25, .50, .75, etc.).

With this function the value is interpolated. It usually does not exist, so it is interpolated. Or by coincidence it could use one of the values in the data set if the numbers line up correctly.

Here come the business requirements for this query:

We are interested in the daily temperature profile for "Boiler 2" in plant "PP000002" for the year 2004. The report needs to show the percentile continuous values for 25%, 50%, and 75%.

Simple specification, but this will be interesting as the table has over 2.5 million rows. Please refer to Listing 10-9.

Listing 10-9. Percentile Continuous Analysis for Monthly Failures

```
SELECT PlantId
      ,LocationId
      ,LocationName
      ,BoilerName
      ,CalendarDate
      ,Hour
      ,BoilerTemperature
      ,PERCENTILE_CONT(.25) WITHIN GROUP (ORDER BY BoilerTemperature)
```

```
      OVER (
      PARTITION BY CalendarDate
      ) AS [PercentCont .25]
      ,PERCENTILE_CONT(.5) WITHIN GROUP (ORDER BY BoilerTemperature)
      OVER (
      PARTITION BY CalendarDate
      ) AS [PercentCont .5]
      ,PERCENTILE_CONT(.75) WITHIN GROUP (ORDER BY BoilerTemperature)
      OVER (
      PARTITION BY CalendarDate
      ) AS [PercentCont .75]
FROM EquipStatistics.BoilerTemperatureHistory
WHERE PlantId = 'PP000002'
AND YEAR(CalendarDate) = 2004
AND BoilerName = 'Boiler 2'
GO
```

The report is simple enough. The PERCENTILE_CONT() function is used three times, once for each desired percentile. The data is continuous as we are profiling it over time, specifically days and hours of the day. Let's see the results.

Please refer to the partial results in Figure 10-20.

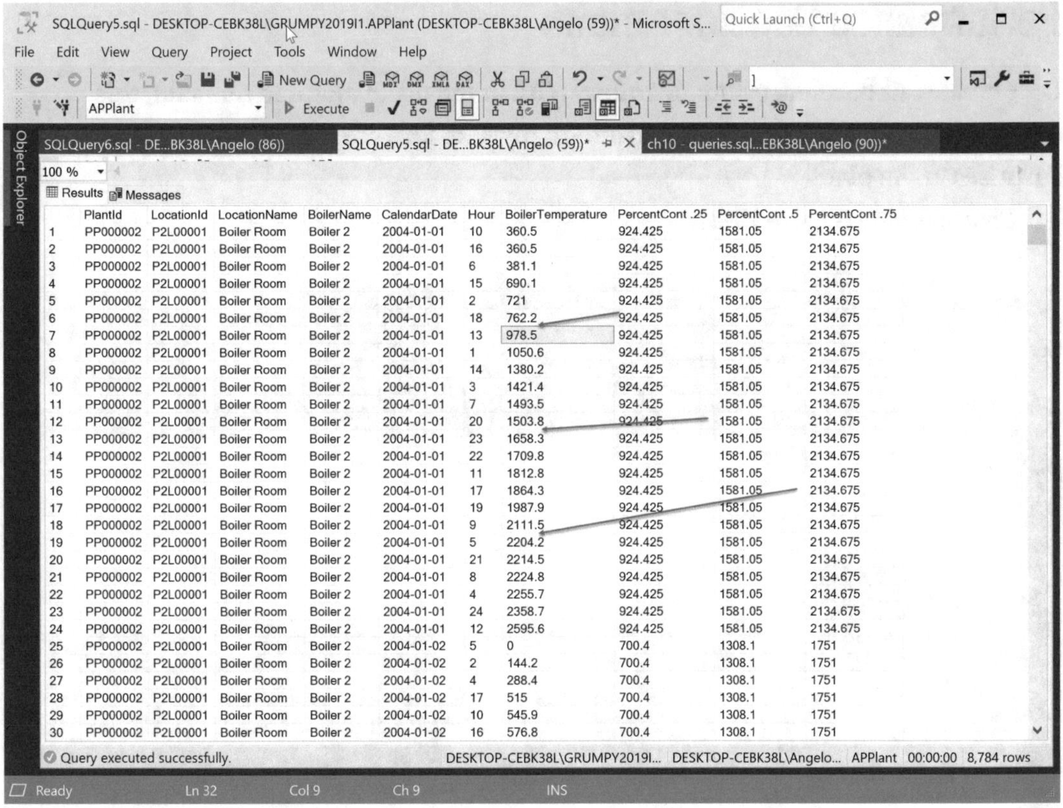

	PlantId	LocationId	LocationName	BoilerName	CalendarDate	Hour	BoilerTemperature	PercentCont .25	PercentCont .5	PercentCont .75
1	PP000002	P2L00001	Boiler Room	Boiler 2	2004-01-01	10	360.5	924.425	1581.05	2134.675
2	PP000002	P2L00001	Boiler Room	Boiler 2	2004-01-01	16	360.5	924.425	1581.05	2134.675
3	PP000002	P2L00001	Boiler Room	Boiler 2	2004-01-01	6	381.1	924.425	1581.05	2134.675
4	PP000002	P2L00001	Boiler Room	Boiler 2	2004-01-01	15	690.1	924.425	1581.05	2134.675
5	PP000002	P2L00001	Boiler Room	Boiler 2	2004-01-01	2	721	924.425	1581.05	2134.675
6	PP000002	P2L00001	Boiler Room	Boiler 2	2004-01-01	18	762.2	924.425	1581.05	2134.675
7	PP000002	P2L00001	Boiler Room	Boiler 2	2004-01-01	13	978.5	924.425	1581.05	2134.675
8	PP000002	P2L00001	Boiler Room	Boiler 2	2004-01-01	1	1050.6	924.425	1581.05	2134.675
9	PP000002	P2L00001	Boiler Room	Boiler 2	2004-01-01	14	1380.2	924.425	1581.05	2134.675
10	PP000002	P2L00001	Boiler Room	Boiler 2	2004-01-01	3	1421.4	924.425	1581.05	2134.675
11	PP000002	P2L00001	Boiler Room	Boiler 2	2004-01-01	7	1493.5	924.425	1581.05	2134.675
12	PP000002	P2L00001	Boiler Room	Boiler 2	2004-01-01	20	1503.8	924.425	1581.05	2134.675
13	PP000002	P2L00001	Boiler Room	Boiler 2	2004-01-01	23	1658.3	924.425	1581.05	2134.675
14	PP000002	P2L00001	Boiler Room	Boiler 2	2004-01-01	22	1709.8	924.425	1581.05	2134.675
15	PP000002	P2L00001	Boiler Room	Boiler 2	2004-01-01	11	1812.8	924.425	1581.05	2134.675
16	PP000002	P2L00001	Boiler Room	Boiler 2	2004-01-01	17	1864.3	924.425	1581.05	2134.675
17	PP000002	P2L00001	Boiler Room	Boiler 2	2004-01-01	19	1987.9	924.425	1581.05	2134.675
18	PP000002	P2L00001	Boiler Room	Boiler 2	2004-01-01	9	2111.5	924.425	1581.05	2134.675
19	PP000002	P2L00001	Boiler Room	Boiler 2	2004-01-01	5	2204.2	924.425	1581.05	2134.675
20	PP000002	P2L00001	Boiler Room	Boiler 2	2004-01-01	21	2214.5	924.425	1581.05	2134.675
21	PP000002	P2L00001	Boiler Room	Boiler 2	2004-01-01	8	2224.8	924.425	1581.05	2134.675
22	PP000002	P2L00001	Boiler Room	Boiler 2	2004-01-01	4	2255.7	924.425	1581.05	2134.675
23	PP000002	P2L00001	Boiler Room	Boiler 2	2004-01-01	24	2358.7	924.425	1581.05	2134.675
24	PP000002	P2L00001	Boiler Room	Boiler 2	2004-01-01	12	2595.6	924.425	1581.05	2134.675
25	PP000002	P2L00001	Boiler Room	Boiler 2	2004-01-02	5	0	700.4	1308.1	1751
26	PP000002	P2L00001	Boiler Room	Boiler 2	2004-01-02	2	144.2	700.4	1308.1	1751
27	PP000002	P2L00001	Boiler Room	Boiler 2	2004-01-02	4	288.4	700.4	1308.1	1751
28	PP000002	P2L00001	Boiler Room	Boiler 2	2004-01-02	17	515	700.4	1308.1	1751
29	PP000002	P2L00001	Boiler Room	Boiler 2	2004-01-02	10	545.9	700.4	1308.1	1751
30	PP000002	P2L00001	Boiler Room	Boiler 2	2004-01-02	16	576.8	700.4	1308.1	1751

***Figure 10-20.** Percentile continuous boiler temperature*

The query ran very fast. It ran under one second, and 8,784 rows were returned. Notice how each of the percentiles is interpolated. The arrows show the slot where they should appear if they existed. For example, for the 25th percentile, the value 924.425 falls between the values 762.2 and 978.5. This means 25% of the values in the partition are less than or equal to this nonexistent value.

Just follow the arrows for the other examples. Lastly, as a reminder, this query does not allow you to specify the ROWS or RANGE window frame specifications. No use trying it out as you will get a syntax error message:

```
Msg 10752, Level 15, State 1, Line 52
The function 'PERCENTILE_CONT' may not have a window frame.
```

Okay, how does this query perform on 2.5 million rows?

Performance Considerations

Let's create our baseline estimated query plan in the usual manner. As I said, this query is hitting a table with over 2.5 million rows, so let's see if there are any significant surprises in the plan!

Please refer to Figure 10-21a.

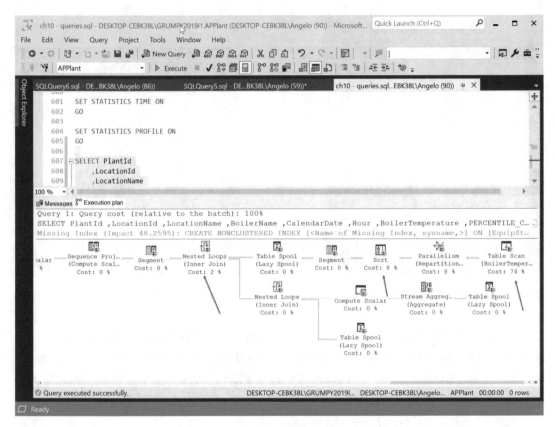

Figure 10-21a. *Baseline estimated query plan*

Right away we can see that an index is missing, so we will create it in a little bit. We have a table scan on the 2.5-million-row table at a cost of 74%, so that's very expensive. It is followed by a parallelism repartition task at 8% and a sort at 8%.

The lower-level branch tasks all cost 0%, so we will not discuss them. We do see a nested loops join task chiming in at 2% that joins these two branches (data flow branches). We need to do something about that expensive table scan.

Next is one of three sections that are repeated for each call of the PERCENTILE_CONT()
function. As they are all the same, I will briefly just discuss the first one.

Please refer to Figure 10-21b.

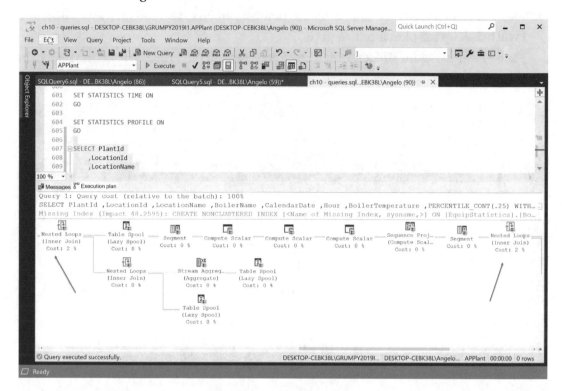

Figure 10-21b. *Plan section for the first PERCENTILE_CONT function call*

The first nested loops join at the right is the one we saw on the left side of the section
we just discussed. It is here as a frame of reference. All our other tasks are 0% cost, and
we see another nested loops join on the left hand side with a cost of 2%. Let's create the
suggested index right away so we can see if there is any improvement.

Please refer to Listing 10-10.

Listing 10-10. Suggested Index for the Large Table Query

```
CREATE NONCLUSTERED INDEX [iePlantBoilerLocationDateHour]
ON [EquipStatistics].[BoilerTemperatureHistory] ([PlantId],[BoilerName])
INCLUDE
([LocationId],[LocationName],[CalendarDate],[Hour],[BoilerTemperature])
GO
```

The estimated query plan generator states that creating this index will improve cost by 48.25%. Let's see.

The INDEX CREATE command took 11 seconds to run! That's because we have 2.5 million rows in this table. Let's see what the new estimated execution plan looks like.

I will break up the plan into two parts.

Please refer to Figure 10-22a, which compares the old plan (on top) and the new plan (on the bottom).

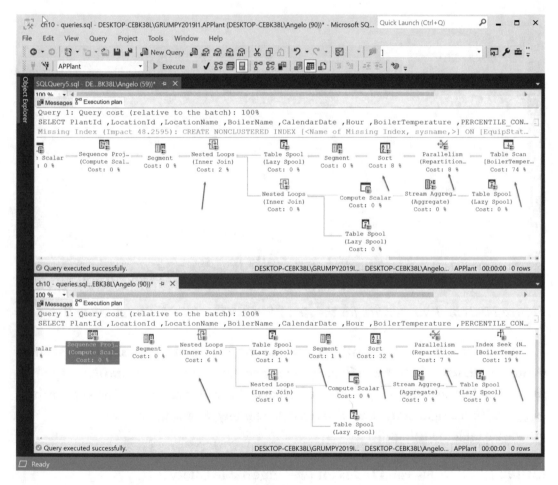

Figure 10-22a.   *Old vs. new estimated index plan*

Based on the preceding screenshot, let's list the old and new values and see how we did. Starting from right to left

- The old estimated query performance plan showed a table scan (74% cost) vs. the new plan's index seek (19%) – **major improvement**.

- The old parallelism task went down from 8% to 7% – **meh**.

- The sort went up from 8% to 32. Not good, but we see this happens when we introduce an index – **major disappointment**.

- The segment tasks went up from 0% to 1% – **no worries**.

- The nested loops joins went from 2% to 6%, but we knew this would happen because of the index. We saw this pattern before – **disappointing**.

Let's look at the next plan section for the window function. Please refer to Figure 10-22b.

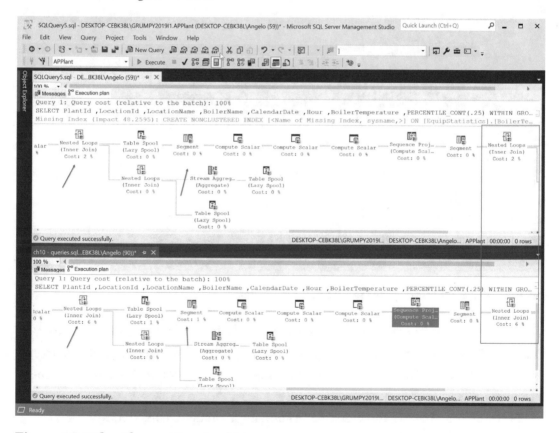

Figure 10-22b. *Plan section 2, old vs. new*

I put a box around the nested loops join tasks as a point of reference for the first section that we just discussed. The only two tasks that changed are the segment and far-right nested loops join. Both went up. The segment task went up from 0% to 1%, and the nested loops join went up from 2% to 6%. We will have to run the IO and TIME statistics for both to see what they tell us as no improvement was gained here.

Please refer to Figure 10-23.

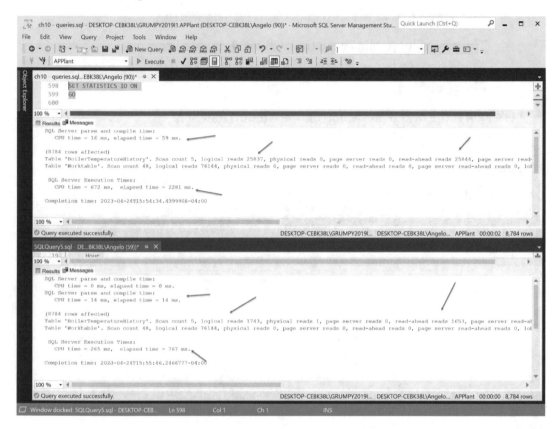

Figure 10-23. *IO and Time statistics, old vs. new*

Here is where we see some real improvements. All the critical statistics went down. First, SQL Server execution time went down from 2281 ms to 767 ms. Logical reads went down from 25,837 to 1,743. This is really big! Finally, read-ahead reads went down from 25,844 to 1,653.

Based on our analysis, the index was a good improvement, considering we are working with the 2.5-million-row table.

Let's check out the cousin of this function, the PERCENTILE_DISC() function.

PERCENTILE_DISC( ) Function

Here is our percentile discrete example. The PERCENTILE_DISC() function works on a discrete set of data based on a required percentile so as to return a value within the data set that satisfies the percentile.

Remember discrete data?

Discrete data values are numbers that have values that you can count over time. Also recall with this function the value exists in the data set; it is not interpolated like in the PERCENTILE_CONT() function.

Our analyst submits the following business requirements for this next report:

This time we need to calculate the percentile discrete values for the sum of failures by month. That is, each month we add up the failures and then generate the 25th, 50th, and 75th discrete percentiles. Results need to be by plant name, location in the plant, year, quarter, and month.

Results need to be for the year 2008. We will revert back to our reliable CTE-based query this time.

Please refer to Listing 10-11.

Listing 10-11. Percentile Discrete Analysis for Monthly Equipment Failures

```
WITH FailedEquipmentCount
(
      CalendarYear, QuarterName, [MonthName], CalendarMonth, PlantName,
      LocationName, SumEquipFailures
)
AS (
SELECT
      C.CalendarYear
      ,C.QuarterName
      ,C.[MonthName]
      ,C.CalendarMonth
      ,P.PlantName
      ,L.LocationName
      ,SUM(EF.Failure) AS SumEquipFailures
FROM DimTable.CalendarView C
JOIN FactTable.EquipmentFailure EF
```

```
ON C.CalendarKey = EF.CalendarKey
JOIN DimTable.Equipment E
ON EF.EquipmentKey = E.EquipmentKey
JOIN DimTable.Location L
ON L.LocationKey = EF.LocationKey
JOIN DimTable.Plant P
ON L.PlantId = P.PlantId
GROUP BY
      C.CalendarYear
      ,C.QuarterName
      ,C.[MonthName]
      ,C.CalendarMonth
      ,P.PlantName
      ,L.LocationName
)
SELECT
      PlantName,LocationName,CalendarYear,QuarterName,[MonthName]
      ,CalendarMonth,SumEquipFailures

      -- actual value form list
      ,PERCENTILE_DISC(.25) WITHIN GROUP (ORDER BY SumEquipFailures)
            OVER (
                  PARTITION BY PlantName,LocationName,CalendarYear
      ) AS [PercentDisc .25]

      ,PERCENTILE_DISC(.5) WITHIN GROUP (ORDER BY SumEquipFailures)
            OVER (
                  PARTITION BY PlantName,LocationName,CalendarYear
      ) AS [PercentDisc .5]

      ,PERCENTILE_DISC(.75) WITHIN GROUP (ORDER BY SumEquipFailures)
            OVER (
                  PARTITION BY PlantName,LocationName,CalendarYear
      ) AS [PercentDisc .75]
FROM FailedEquipmentCount
WHERE CalendarYear = 2008
GO
```

Notice once again we are using the fact table and also dimension tables, so we see a number of JOIN clauses. The partition is set up by plant, location, and year for each of the percentile calculations. The ORDER BY clause references the SumEquipFailures column in each of the three calls to the PERCENTILE_DISC() function (remember, the returned values should exist this time).

Please refer to the partial results in Figure 10-24.

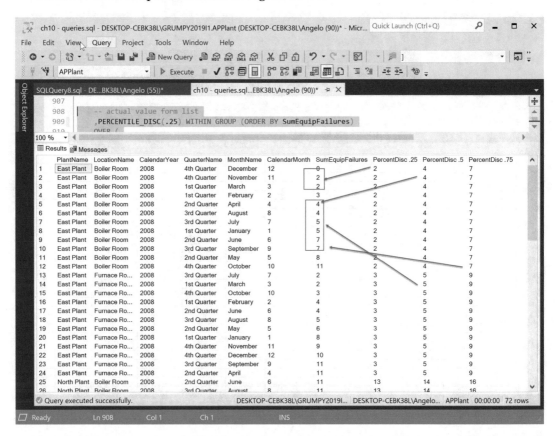

Figure 10-24. *Percentile discrete analysis for plant equipment failures*

Yes, the values returned exist, and we see we have some ties, perfectly normal as it is possible to have the same numbers of total monthly failures for equipment. It could be different equipment, but they can be the same or different.

Performance Considerations

Let's see how different or the same the discrete percentile function works vs. the continuous percentile function in terms of performance. Let's generate our usual baseline estimated query plan and break it up into parts as it will be a large one.

Please refer to Figure 10-25a.

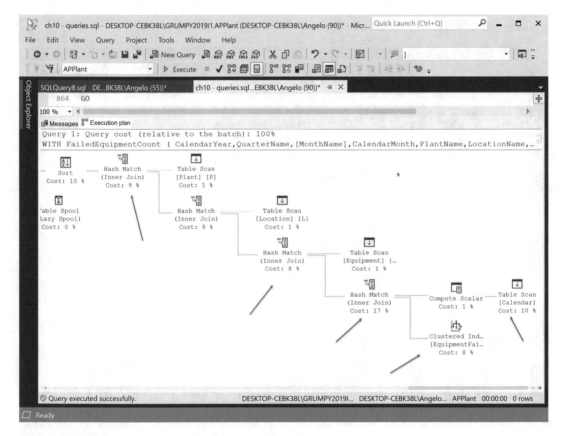

Figure 10-25a. *Estimated baseline query plan, part 1*

The table scan task with the 10% cost looks familiar, and so do the 1% table scans of the dimension tables. One thing I failed to mention is that we are using a VIEW for the date information called CalendarView.

This is something to examine. We are using this VIEW as it contains the text names of months, quarters, etc. Basically, a VIEW translates to executing a query, in this case the underlying query that the VIEW is based on. This is a bit costly to use, so we would like to replace this VIEW with a more detailed Calendar dimension that contains all the text values. We would also want to create an index or two on the new Calendar table (this will be a homework assignment at the end of the section, so think about it).

The lower right-hand clustered index scan has a cost of 8%. Working up the lower branch, we see the hash match joins with respective costs of 17%, 8%, and 8% with a final hash match join with a cost of 9%.

Lastly, we see a sort task to the left at 10% plus a table spool (lazy spool) task at a 0% cost. I think that the final conclusion for performance planning for this type of pattern is to either place the smaller dimensions with 100 or fewer rows in memory-optimized tables or else on very-high-speed solid-state drives. At least, TEMPDB should always be on a solid-state drive.

The existing dimension tables are too small to benefit from indexes, and as can be seen, none were recommended by the query plan tool. Let's look at the next section of the plan as we navigate to the left-hand side.

Please refer to Figure 10-25b.

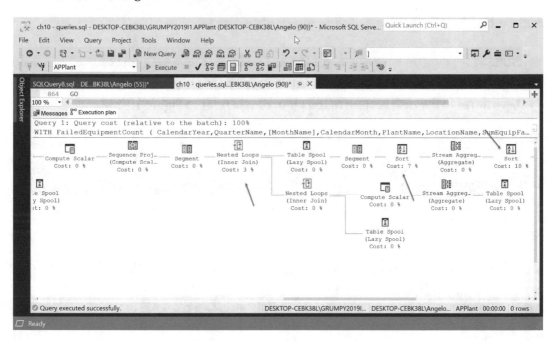

Figure 10-25b. *Estimated baseline query plan, part 2*

There's the sort task from the last screenshot as a point of reference. I will just call out the non-zero-cost tasks. A sort appears (7% cost), followed by a nested loops join task to merge the flow of the branches (3%). Let's move to the next section.

Please refer to Figure 10-25c.

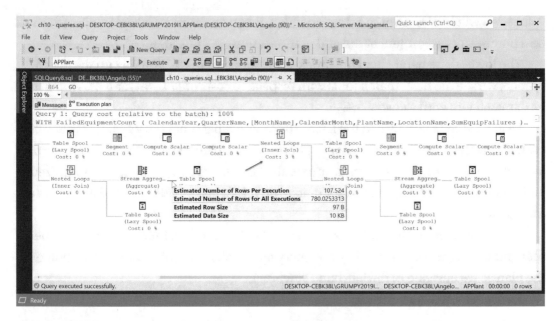

Figure 10-25c. *Estimated baseline query plan, part 3*

Nothing is going on here except for the nested loops join with a cost of 3%. Notice the thickness of the lines indicating how many rows are passing through.

Let's move to the next section of the plan. Please refer to Figure 10-25d.

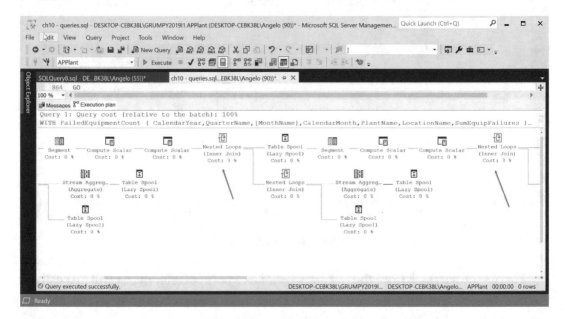

Figure 10-25d. *Estimated baseline query plan, part 4*

We can see the patterns start to repeat themselves. Notice that as branches merge, the number of rows after the nested loops joins goes up. Makes sense, right?

I am not showing the last section as it includes the nested loops join and all other tasks are 0% cost.

We can now expect that the estimated query plan patterns when we use a function multiple times will usually be the same. This assumption may or may not help you with any performance improvement. The good news is that if you try out a fix that improves one section, it most likely improves the others.

Last but not least, let's generate the IO and TIME statistics.

Please refer to Figure 10-26.

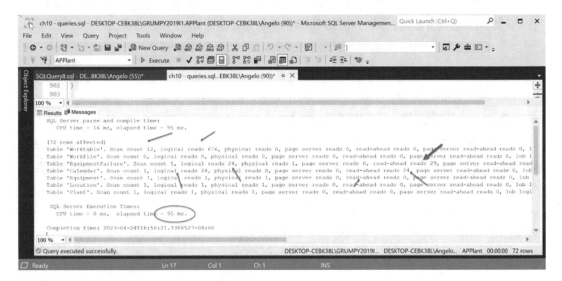

Figure 10-26. *IO and time statistics*

As can be seen, all the dimensions involved in the query have low values except for the Calendar dimension, which has logical reads at 24, physical reads at 1, and read-ahead reads at 24. All other values are either 0 or 1 except for the work table, which has scan count at 12 and logical reads at 676.

Lastly, the EquipmentFailure table has scan count of 1, logical reads of 24, physical read of 1, and read-ahead reads of 29.

SQL Server elapsed execution time is 85 ms. Not too bad.

Note   This window function does not support ROWS or RANGE window frame specifications in the OVER() clause, so this will not be an option to try to increase performance.

Here comes a tip and some homework. First the tip:

Tip   Each time you generate a plan, do place the mouse over some of the lines so you can see the estimated number of rows flowing between tasks. This can help identify bottlenecks sometimes. If too many rows are going through, maybe you need to redefine the WHERE clause filters or add some new predicates.

And here is the homework I promised:

Homework   Using the SELECT query that makes up the VIEW, create a new Calendar dimension called CalendarEnhanced. Use this new table in the queries we just discussed and run some estimated query plans. If indexes are suggested, create them and see what improvements in performance are achieved. The solution is in the script for this chapter.

Let's try our report table approach.

Our Usual Report Table Solution

We are going to create a report table called EquipFailPctCountDisc so we can preload it, pay the penalties for all the JOIN clauses once, and assume the table is loaded off hours once a day.

The following is the INSERT statement to load the table.

Please refer to Listing 10-12.

Listing 10-12. Loading the Report Table

```
INSERT INTO Reports.EquipFailPctContDisc
SELECT
      C.CalendarYear
      ,C.QuarterName
      ,C.[MonthName]
      ,C.CalendarMonth
      ,P.PlantName
      ,L.LocationName
      ,SUM(EF.Failure) AS SumEquipFailures
FROM DimTable.CalendarView C
JOIN FactTable.EquipmentFailure EF
ON C.CalendarKey = EF.CalendarKey
JOIN DimTable.Equipment E
ON EF.EquipmentKey = E.EquipmentKey
JOIN DimTable.Location L
ON L.LocationKey = EF.LocationKey
JOIN DimTable.Plant P
ON L.PlantId = P.PlantId
GROUP BY
      C.CalendarYear
      ,C.QuarterName
      ,C.[MonthName]
      ,C.CalendarMonth
      ,P.PlantName
      ,L.LocationName
ORDER BY
      C.CalendarYear
      ,C.QuarterName
      ,C.MonthName
      ,P.PlantName
      ,L.LocationName
GO
```

Now our query performance should improve. We use the query in the CTE block in an INSERT command to load a new report table. We will use this for the percentile query.

Listing 10-13 is the modified query that accesses the new table.

Listing 10-13. Modified Percentile Query

```
SELECT
        PlantName,LocationName,CalendarYear,QuarterName
            ,[MonthName],CalendarMonth,SumEquipFailures

        -- actual value from list
        ,PERCENTILE_DISC(.25) WITHIN GROUP (ORDER BY SumEquipFailures)
            OVER (
            PARTITION BY PlantName,LocationName,CalendarYear
            ) AS [PercentDisc .25]

        ,PERCENTILE_DISC(.5) WITHIN GROUP (ORDER BY SumEquipFailures)
            OVER (
            PARTITION BY PlantName,LocationName,CalendarYear
            ) AS [PercentDisc .5]

        ,PERCENTILE_DISC(.75) WITHIN GROUP (ORDER BY SumEquipFailures)
        OVER (
            PARTITION BY PlantName,LocationName,CalendarYear
            ) AS [PercentDisc .75]
FROM Reports.EquipFailPctContDisc
WHERE CalendarYear = 2008
GO
```

I will not walk through it again as we discussed it earlier, so let us see what the estimated performance plan is for this revised strategy and query. One comment I would like to make is related to my naming standards. I tried to opt for descriptive column names in the query, so it is clear what the semantic value is. I am sure you can come up with better names.

Let's compare the right-hand side of the old and new estimated query plans.

Please refer to Figure 10-27.

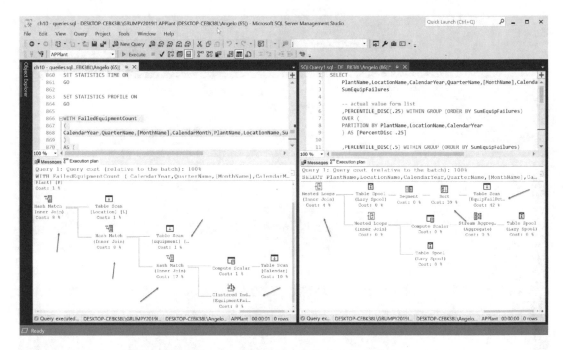

Figure 10-27. *Comparing the first section of the old and new plans*

I am only showing the rightmost-hand side of the estimated query plans as this is where all the JOINS were. As can be seen in the new plan, the JOINS are gone, but we see a table scan at 42%.

This replaces the clustered index seek on the fact table. We also see a new sort at 39% and a nested loops join task at 4% cost in the new estimated query plan.

Of course, we do not have a clustered index created on the new table yet, but it only contains 1,517 rows, which is what the CTE returned in the old query. Also, the estimated query plan tool did not suggest an index, so we will leave it as is.

All that is left to look at are the TIME and IO statistics. This will tell us if the report table comes to our rescue again.

Please refer to Figure 10-28.

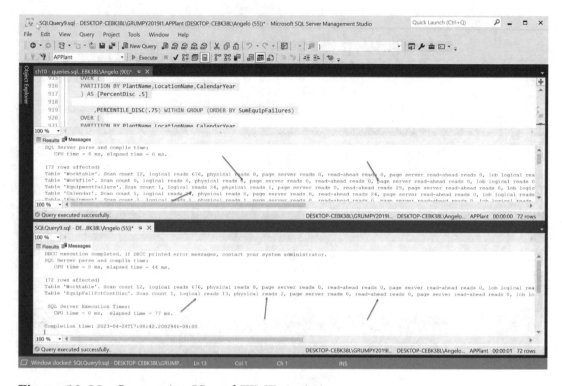

Figure 10-28. *Comparing IO and TIME statistics*

We can see that there are a lot less statistics because all the table joins were eliminated for the dimension tables as they are now in the INSERT statement that would run once a day at off hours.

The work table scan count and logical reads remained the same. The read-ahead reads went down from 29 to 0. Lastly, the new report table shows logical reads went down from 24 to 13 but physical reads went up by 1. Lastly though, the SQL Server execution elapsed time went down from 98 ms to 30 ms.

So, it looks like this approach worked again.

Reminder   When running these types of tests that compare two approaches, remember to always run DBCC to clear buffers in each session so you get accurate results. Run the queries and compare three or four times to get an average to make sure there were no anomalies.

As you might be thinking, we have a few report tables now, so when using this approach, see if you can merge similar tables so as to reduce the number of report tables. Things could get complicated very fast with the proliferation of the tables. We want our physical database architecture to be as simple and compact as possible.

We are going to change gears and look at a new approach for our analysis. We are now going to look at another tool you can add to your toolkit for analyzing data. SQL Server Analysis Services (SSAS) is a powerful platform for creating multidimensional cubes based on millions of rows of data!

SQL Server Analysis Services

We are now going to introduce multidimensional structures called cubes so we can perform slicing and dicing of data and also drill up and down the various hierarchies defined by our dimensions.

Slicing and dicing refers to the ability to change perspectives by introducing and removing dimension columns as we view the cube, just like we did with pivot tables.

Navigating hierarchies refers to the ability to view data by different levels of aggregations, like by year, quarter, month, week, and day and then in reverse.

As I stated earlier, these are similar to pivot tables except they can process millions of rows, which would challenge an Excel spreadsheet. The additional good news is that the Excel spreadsheets can connect to cubes and create pivot tables based on them (the number of rows processed might be an issue as there are limits).

Lastly, if you know how to use Power BI, you can connect Power BI to SSAS multidimensional cubes in order to create some impressive scorecards, dashboards, and reports, so your toolbox is really evolving!

Here is a little taste! Please refer to Figure 10-29.

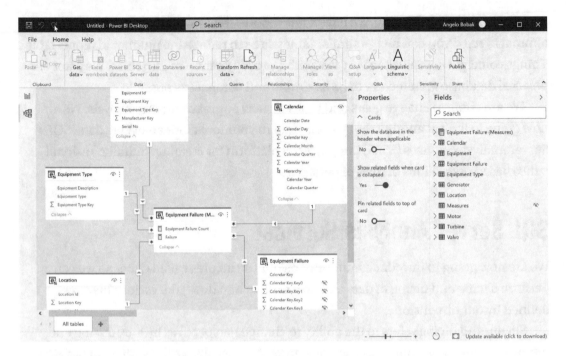

Figure 10-29. *Power BI showing the APPlant STAR schema*

In the last three chapters of the book, when we apply our window functions to an inventory database business model, we will take a detailed look at Power BI. For now, let's see the main steps involved in creating a multidimensional cube with SSAS.

There are several steps involved:

- Create a data source.

- Create a data view.

- Define dimensions.

- Define the cube.

- Define the dimension usage (per dimension).

- Build and deploy the cube.

- Browse the cube.

- Create a pivot table with the cube.

Creating a data source is basically identifying the SSAS server and supplying credentials to enable users and developers to connect to the cube (or create one).

The data source view is basically a STAR or SNOWFLAKE schema of the tables that will be used to create and populate the view, so this is a necessary component. There are two models that you can create with SSAS; one is the tabular model and the other is the multidimensional model. You define the type when you first install SSAS on your laptop, workstation, or server. We will concentrate on the multidimensional model.

Once you have the source and data view defined, you need to identify the tables you will use as dimensions and the fact table you will use to populate the measures in the cube.

Once these tasks are completed you need to link the dimensions to the cube; this is called defining the dimension usage.

Once all these tasks are completed, you need to build the cube and deploy the cube to the SSAS server instance installed on your environment (you can have both tabular and multidimensional instances installed if you have enough memory, CPU, and disk space). Now you can browse the cube or connect to it with Microsoft Excel or Power BI.

Please refer to Figure 10-30.

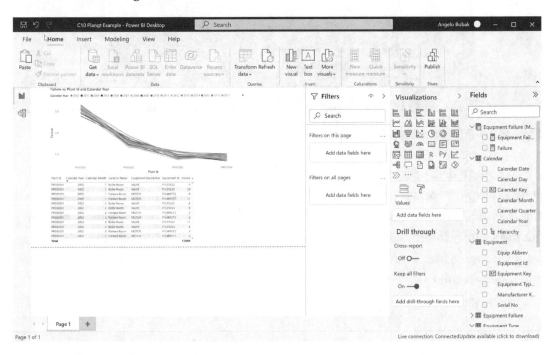

Figure 10-30. *A simple Power BI dashboard*

This is a simple dashboard created from the cube we are about to build. I wanted to show this now so you can get inspired to learn how to use SSAS! The dashboard is a combination of a report plus some sort of visual graph or chart. Notice all the visuals that are available on the right side of the tool. More on this in the last three chapters!

Let's start off by learning how to create a data source.

Please refer to Figure 10-31.

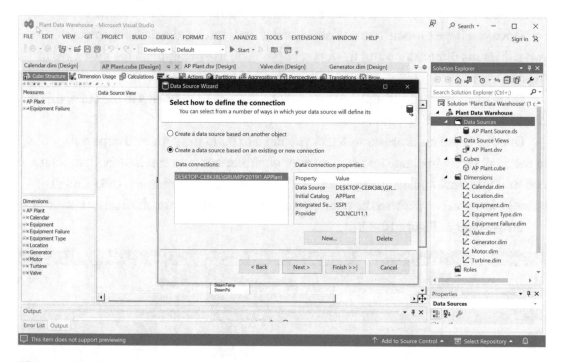

Figure 10-31. *Create a data source*

Creating a data source is very simple. Right-click the Data Sources folder in the Solution Explorer and select "New data source." This panel is displayed. Click the "Create a data source based on an existing or new connection," click the "Next" button, and fill out the security information called "Impersonation information." In other words, you want the tool to impersonate a system user ID and password that has the rights to create and modify a cube.

Please refer to Figure 10-32.

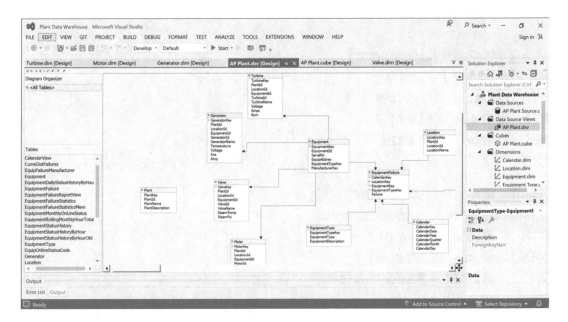

Figure 10-32. *Define the data view*

Next, right-click the Data Source Views folder and select "New data source view." This is where you retrieve all the tables you will need to define dimensions and fact tables. Mostly, you will link the dimension tables to the fact tables via the surrogate keys that were included when we created the database.

Notice that the Valve, Turbine, Generator, and Motor dimensions are not yet connected to the Plant dimension. I made a mistake in the creation of the database in that I did not give these first four tables surrogate keys. I could use the production key, PlantId, to connect them, but that would be a bit dodgy although it would work. I had to go back and add surrogate keys to the tables and populate them with the surrogate key assigned to the Plant table.

I used the following code in Listing 10-14.

Listing 10-14. Adding Surrogate Keys to the Equipment Tables

```
ALTER TABLE [DimTable].[Generator]
ADD PlantKey INTEGER NULL
GO

ALTER TABLE [DimTable].[Generator]
ADD LocationKey INTEGER NULL
```

701

```
GO

UPDATE [DimTable].[Generator]
SET PlantKey = P.PlantKey
FROM [DimTable].[Generator] G
JOIN [DimTable].[Plant] P
ON G.PlantId = P.PlantId
GO

UPDATE [DimTable].[Generator]
SET LocationKey = P.LocationKey
FROM [DimTable].[Generator] G
JOIN [DimTable].[Location] P
ON G.LocationId = P.LocationId
GO

SELECT * FROM [DimTable].[Generator]
GO
```

In the interest of space, I am only showing you the code for the Generator table as the code for the remaining three tables is the same except the names of the tables and columns changed.

I actually made this design mistake when I created the sample database. This is what happens when you rush, but there is usually a solution whereby you can fix your mistakes. The ALTER command lets us add the columns we need, and the UPDATE command lets us initialize the surrogate keys. Now we can complete our cube.

Please refer to Figure 10-33 for the changes and simple query used to verify that all went well.

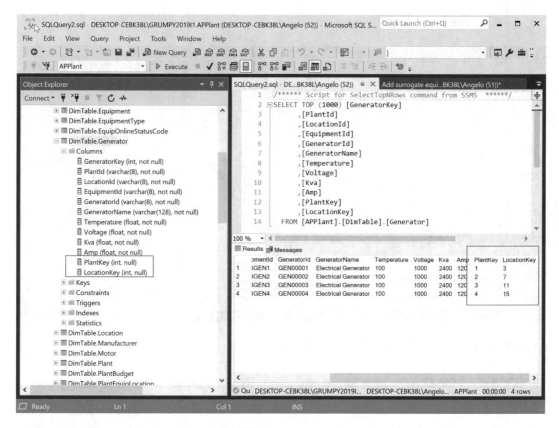

Figure 10-33. *Verifying the table changes*

Our plant and location surrogate keys are initialized. This will make for a more correct SNOWFLAKE schema design and also increase performance when these tables are used in a query (joining numbers vs. alphanumeric values is always faster).

Back to the steps for creating a cube. Please refer to Figure 10-34.

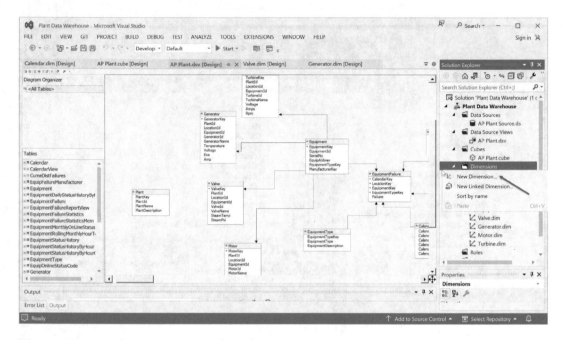

Figure 10-34. *Creating a new dimension*

Right-click the Dimensions folder and select New Dimension. The following panel in Figure 10-35 is presented.

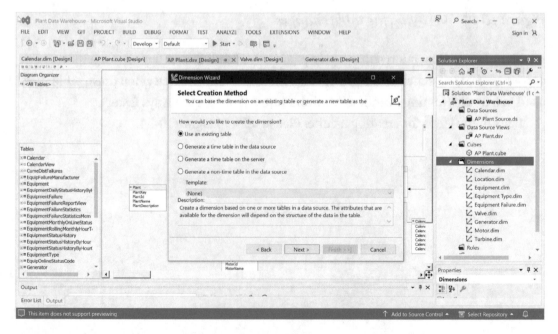

Figure 10-35. *Defining the dimension with the Dimension Wizard*

The panel presented is called the Dimension Wizard, and you simply fill out or select the requested information and click the Next button until you are finished. In this case we are using an existing table to create the dimension tables.

Once we are finished, we need to perform some final steps like identifying hierarchies in the dimensions. Please refer to Figure 10-36.

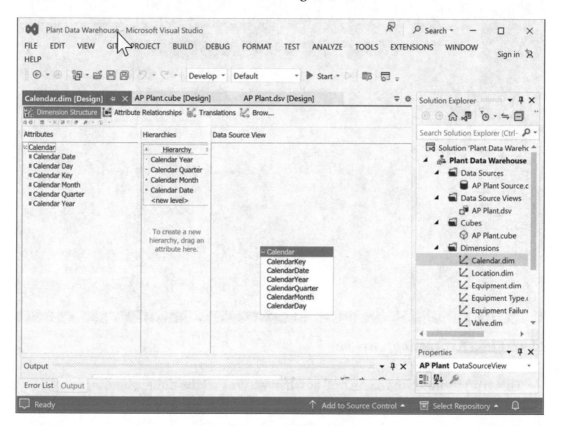

Figure 10-36. *Define dimensions and hierarchies*

This is a simple task, but you need to fully understand your data model. Simply drag the necessary attributes from the Attributes panel and drop them in the Hierarchy section in the correct order. In this case it is pretty simple: the Calendar hierarchy is defined as Calendar Year ➤ Calendar Quarter ➤ Calendar Month ➤ Calendar Date. Simple.

Next, we need to create a cube by identifying the fact table that contains the measures (stuff we can count or measure, like equipment failures) and link it to the dimensions.

If you have not already guessed, you create a new cube by right-clicking with your mouse the Cubes folder and selecting New Cube. You are presented with the panel shown in the Figure 10-37.

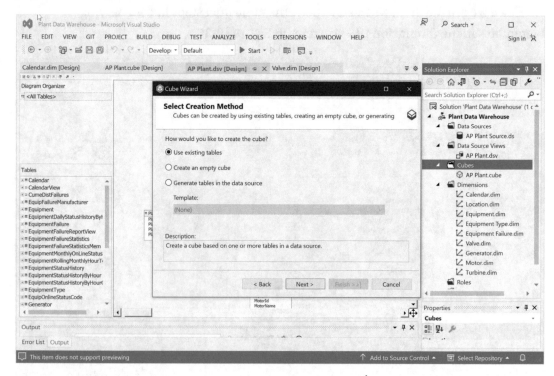

Figure 10-37. *Creating the cube*

This is pretty much the same drill as when we created the dimensions for our multidimensional model. A wizard is presented, and you just follow the steps by clicking the appropriate buttons or filling out a few text boxes.

When you are finished, you are presented with the following model in Figure 10-38.

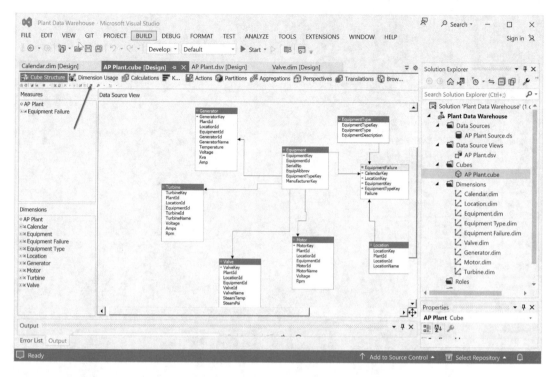

Figure 10-38. *Our dimensions and fact table*

Looks pretty good. We have one more task to complete before we build and deploy our cube to the SSAS server instance. We need to link the dimensions to the cube by using the surrogate keys in the original tables. This is called dimension usage definitions, and all we need to do is to navigate to the Dimension Usage tab indicated by the arrow in the figure. Another design panel is presented that allows us to complete these tasks.

Please refer to Figure 10-39.

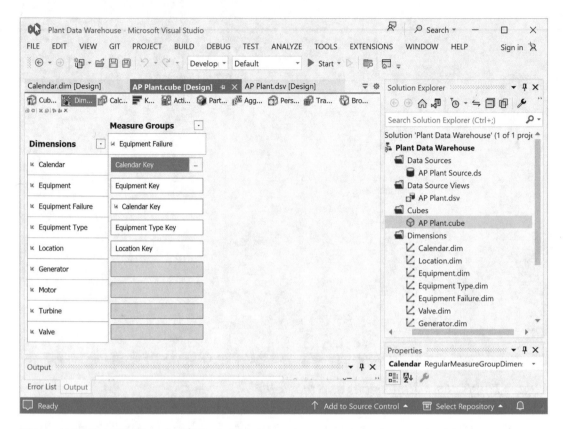

Figure 10-39. *Define the dimension usage*

At the intersection of each dimension and the measure group (the fact table) is a drop-down list box that lets you identify the key that will link the two types of tables. As can be seen, we are using all the surrogate keys. For example, we link the Calendar dimension to the Equipment Failure measure group via the Calendar Key.

This is a simple task, but it illustrates the use of good naming standards. Always use the same names for columns that appear in multiple tables, so it is obvious what they are and what they are used for, like linking a dimension to a fact table. Apples to apples, right?

We are finished. If we completed each task correctly, all we need to do is build and deploy the cube.

Please refer to Figure 10-40.

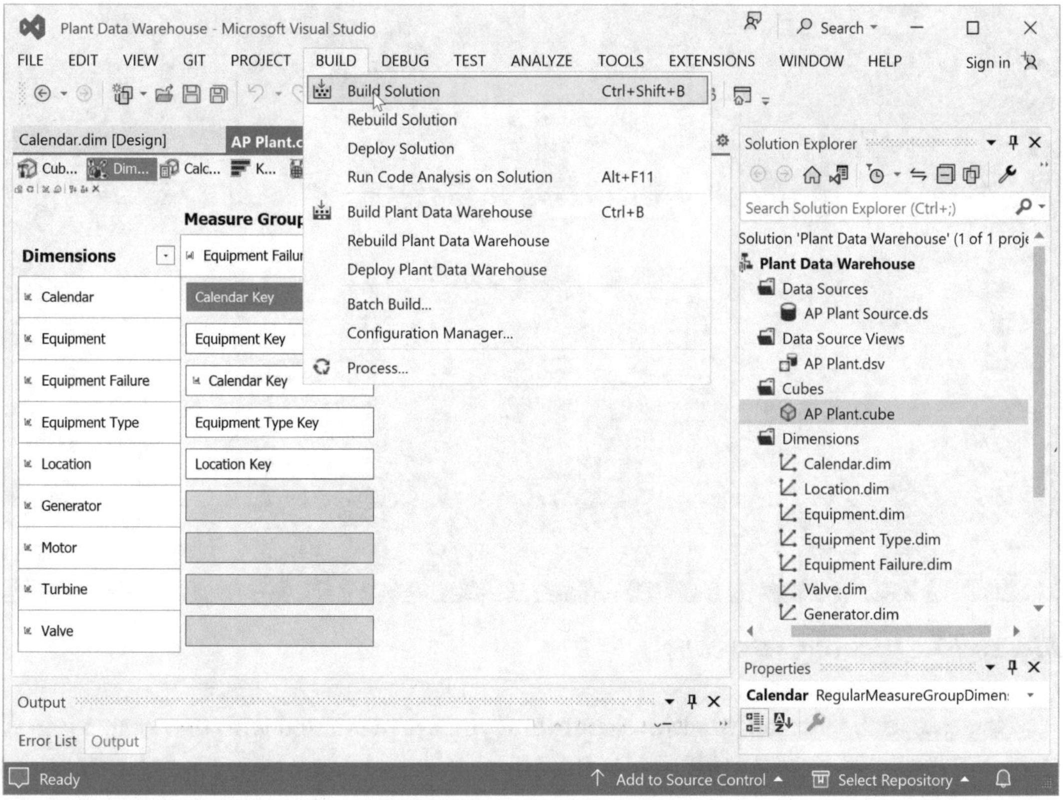

Figure 10-40. *Build and deploy the solution*

Underneath the BUILD menu tab, we see the selections for building and deploying the cube. First, you need to build the cube and next deploy the cube. You will see a lot of messages flash across the panel, and hopefully you will not see any red messages that indicate something went wrong.

If all went well, you are ready to browse the cube.

Please refer to Figure 10-41.

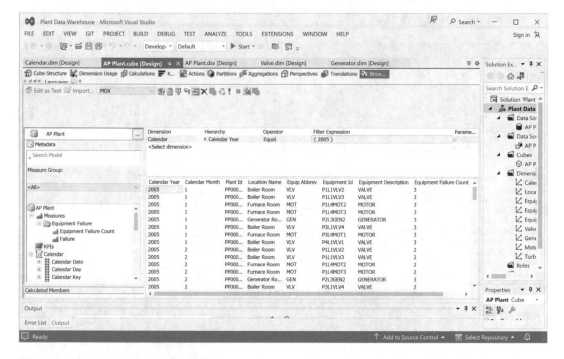

Figure 10-41. *Browse the cube*

Navigate to the last tab, the Browse tab, and you are presented with this design panel. Drag and drop the desired attributes in the dimension and measure group. Also drag and drop the hierarchy attributes as needed. You may need to refresh the panel as you go, but this is the final result.

We can even create queries in a language called MDX. Click the Design button in the menu bar and view the MDX query that was generated from the report we just created. Copy and paste it to an MDX query panel and SSMS query panel and execute it (MDX stands for Multidimensional Expressions).

Here are the results. Please refer to Figure 10-42.

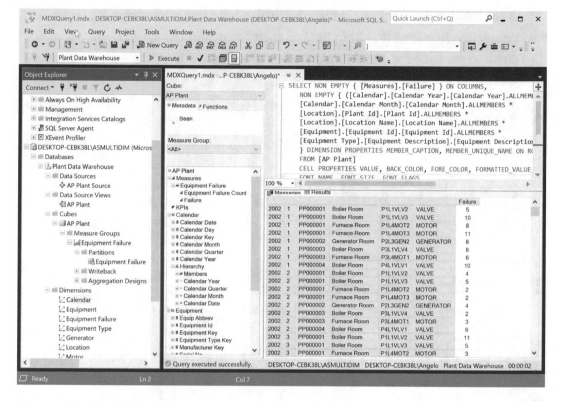

***Figure 10-42.** MDX query generated by Visual Studio*

One word of caution: MDX is a complicated and hard-to-learn language, but it could be a valuable tool in your SQL Server toolkit. Many good resources exist on MDX, so search the Web or Apress books on this topic on Amazon!

In conclusion, the browser is a nice tool as far as testing for us developers, but it is not a tool for our users. You really do not want to install a copy of Visual Studio, the development environment that contains all the tools we just used, on a business analyst's desk.

Fortunately, we can use our old friend, the Microsoft Excel spreadsheet application, to connect to the cube, pull in some data, and perform slicing and dicing just like when we created pivot tables and graphs. Let's take a look.

Note   Analysis Services has its own version of a SQL-like language called MDX that allows you to query cubes in SSMS just like when you write queries with TSQL – another language you might want to add to your toolkit.

All you need to do is define a connection to the cube server by navigating to the Data tab and selecting a connection from the first menu selection called Get Data.

Click it to generate the panel shown in the following.

Please refer to Figure 10-43.

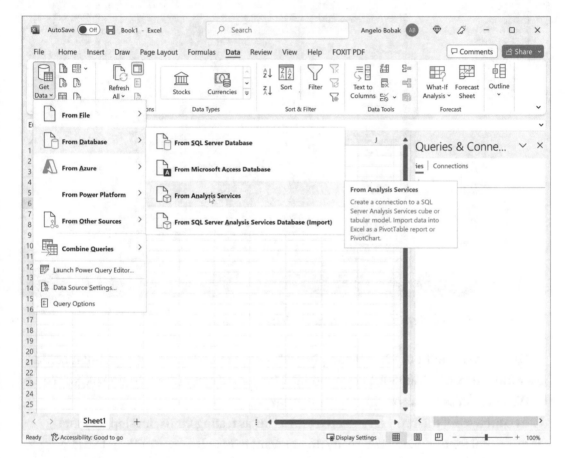

Figure 10-43. *Connecting Microsoft Excel to Analysis Services*

From the drop-down menu that appears, just make the necessary selections by clicking "From Database" and then "From Analysis Services." A panel will appear. Just enter the server's name and the security credentials (make sure you use the impersonation credentials you used when creating the cube), and your connection is defined.

Two more panels appear that will let you select the cube and also save the connection information in a file.

You are finally presented with a small panel that will let you specify where in Excel you want to place the imported data and also where the pivot table will be created.

You can now drag and drop the desired columns and values to create the pivot tables as shown in Figure 10-44.

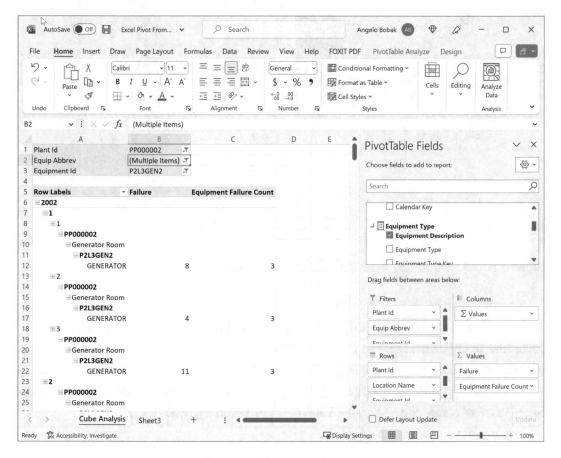

Figure 10-44. *Microsoft Excel pivot table based on the cube*

Looks familiar, doesn't it? You can add filters and change the combinations as required so as to drill up and down the hierarchies you defined.

Lastly, just to whet your appetite again, I will show you a Power BI report and dashboard created by connecting Power BI to the cube we just created.

Please refer to Figure 10-45.

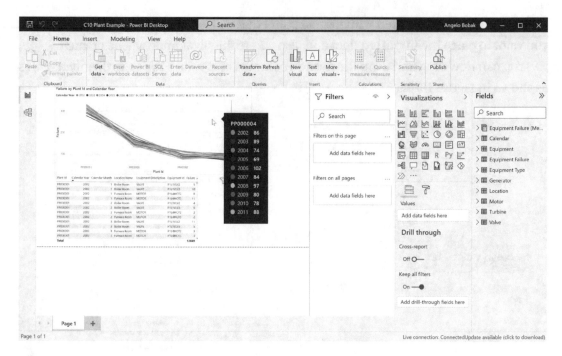

Figure 10-45. *Power BI dashboard for power plant failures*

Learning and creating Power BI dashboards, scorecards, and reports would take an entire book. But as you can see, this is a very powerful tool to add to your toolkit. I will cover how to create a simple scorecard or two in the last three chapters so that you have some basic skills on how to use `Power BI`.

The Apress publisher has a good book on this topic called *Beginning Microsoft Power BI: A Practical Guide to Self-Service Analytics*, third edition, by Dan Clark.

You might want to check this out.

Summary

We covered a lot of ground in this chapter. I hope you enjoyed it and found it useful. We performed our usual study of the analytical functions that come with SQL Server, but this time we applied them to our power plant scenario in order to see the type of analysis one would perform to identify failure rates for critical equipment that is part of a power plant.

We also conducted our performance analysis on the queries we wrote and got more practice with the performance analysis tools we learned through the book so far.

Lastly, we took things a bit further by introducing two new tools for your toolkit, namely, `Analysis Services` and `Report Builder`.

We will expand on these tools in the last three chapters when we perform some analysis on an inventory control case study. I will introduce a last tool called SQL Server Integration Services, which is the premier ETL tool that is a part of the Microsoft SQL Server BI stack that shows you how to create process flows that load databases and data warehouses.

See you in Chapter 11 where we take another look at aggregate window functions to track inventory.

Inventory Use Case: Aggregate Functions

This will be an interesting chapter. Not only will we be taking the window aggregate functions through their paces in an inventory management scenario but we will also learn how to use Microsoft's premier ETL tool called SSIS to create a process that loads an inventory data warehouse from the inventory database that is used in the queries we will create.

We will create queries that go against the APInventory database and also the APInventoryWarehouse data warehouse so we can compare results and performance. The usual performance analysis discussions will be included, and also you will have a few suggested homework assignments.

Lastly, the scripts for creating and loading both databases are included in the folder for this chapter on the Apress website together with the query scripts. Let's check out what the databases look like with some simple business conceptual models.

The Inventory Database

Let's begin this chapter by describing the database we will use in our examples. The following is our high-level business conceptual model that identifies the business entities for the inventory database.

Please refer to Figure 11-1.

© Angelo Bobak 2023
A. Bobak, *SQL Server Analytical Toolkit*, https://doi.org/10.1007/978-1-4842-8667-8_11

AP Inventory Conceptual Model

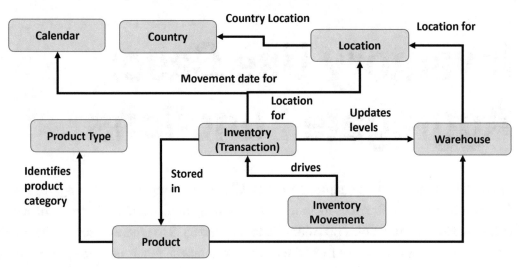

Note: the arrow points to the parent entity.

Figure 11-1. *The inventory database conceptual model*

The main business object is the Inventory table, which stores information related to products moving in and out of inventory. The Inventory Movement business object identifies the quantity on hand of any product over any date. The Product object stores typical information about the products being sold, and the Product Type object identifies the type of product. Lastly, we have the usual Country, Location, and Calendar business objects.

The preceding model identifies the relationships between objects with simple verb phrases. In Table 11-1 are the high-level descriptions of each entity.

Table 11-1. *The Inventory Database Tables*

Entity	Description
Calendar	Contains calendar dates used in transaction events and inventory movement.
Country	Identifies ISO two- and three-character country codes and country names.
Location	Identifies location of inventory and warehouses. A location has one or more inventories. Inventories have one or more warehouses.
Product	Identifies products available in inventory.
Product Type	Identifies the type of a product, like HO Electric Locomotives.
Inventory	Identifies high-level units available in the warehouses associated with this inventory.
Inventory Movement History	Identifies the date of inventory in and out of the warehouses together with the associated product.
Inventory Sales Report	Sales report related to inventory movements.
Inventory Transaction	Identifies the daily in and out movement of products from the warehouses.
Warehouse	Detailed information related to the products in inventory, the current levels, and the reorder amounts.

This simple data dictionary identifies the business entities in the conceptual model together with the descriptions of each table. I will not include a detailed attribute data dictionary in the interest of space as the column names in the database tables clearly identify what the business data being stored is.

The Inventory Data Warehouse

Next, we look at a second database, which is a data warehouse implemented as a SNOWFLAKE schema. This data warehouse is loaded from the APInventory database. This chapter will include a section on how to use SQL Server Integration Services (SSIS) to load this simple data warehouse.

Figure 11-2 is the simplified SNOWFLAKE schema for our example data warehouse.

AP Inventory Data Warehouse Conceptual Model

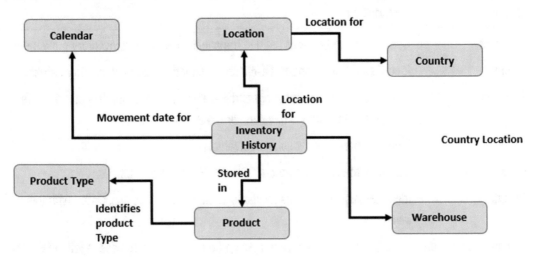

Note: the arrow points to the parent entity.

Figure 11-2. *The inventory data warehouse conceptual model*

The central business object in this model is the Inventory History object, which will be the basis for the fact table in the data warehouse. Historical data related to inventory movement in and out of the APInventory database will be stored in this object.

Surrounding the Inventory History business object are the various entities that will be used to create the dimensions in the data warehouse. These are basically the same as those we discussed in the APInventory database, and the verb phrases clearly discuss the business relationships between the historical Inventory History object and the future dimension tables. For example, a product type identifies the type of a product, like a steam locomotive type.

The following are the physical table names, the assigned schema, and primary key information related to the data warehouse model.

Please refer to Table 11-2.

Table 11-2. *The Inventory Data Warehouse Tables*

Table	Type	Primary Key
Calendar	Dimension	CalendarKey
Country	Dimension	CountryKey
Inventory	Dimension	InvKey
Location	Dimension	LocKey
Product	Dimension	ProdKey
Product Type	Dimension	ProdTypeKey
Warehouse	Dimension	WhKey
Inventory History	Fact	CalendarKey, LocKey, WhKey, ProdKey, ProdTypeKey

Technically, the Warehouse table should include the primary surrogate keys for Location and Inventory, but I took a bit of a design shortcut to simplify the model and load scripts.

Next, I will introduce a new tool for your toolkit. This tool is called SSIS (SQL Server Integration Services), which is Microsoft's premier ETL tool. We will use this tool to load our data warehouse from the transactional inventory database.

All code to create the two databases is available in the chapter folder on the Apress website for this book.

Loading the Inventory Data Warehouse

SSIS, which stands for SQL Server Integration Services, is a graphical tool that allows one to create process flows called control flows and data flows for any data integration process. The control flows identify the steps and order that tasks need to be executed in order to extract, transform, and load data, while the data flows allow the data architect to specify the flow of data and the low-level transformation tasks. The flow could be sequential or have many branches depending on how the data needs to be distributed.

SSIS creates applications called packages. Within each package are a series of low-level tasks that define and execute the low-level steps to process the data. We can say, then, that SSIS is composed of control and data flows, packages, and tasks. An SSIS project can have one or more packages, which can call each other to define very complex data integration and transformation processes.

A toolbox is available with all the tasks one might need, and also a Solution Explorer panel allows one to define the data sources and connections one needs. Let's explore the main facilities available with this tool.

Please refer to Figure 11-3.

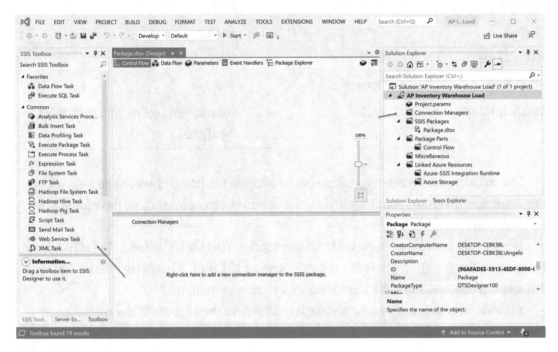

Figure 11-3. *Creating the project*

This is not meant to be an exhaustive tutorial on SSIS. The intent is to give you just enough familiarity with the tool so you can create simple load processes and explore the tool on your own. Many excellent books and courses are available to take you to the next skill level.

The GUI is composed of four main panels:

- SSIS Toolbox: Contains all low-level tasks that you need in your flows, like FTP tasks.

- Design area: This contains the GUI used to create control flows and data flows and identify parameters used in ETL packages and also event handlers and a package explorer.

- Solution Explorer: This is where you identify project parameters, connection managers, SSIS packages you will need, package parts and folders for miscellaneous components, and linked Azure resources.

- Connection Managers: This identifies connections to databases, other servers, and even the SSAS (SQL Server Analysis Services) server.

Right-clicking the folder in the Solution Explorer will present menus to add new components like connections.

Let's start creating our package (our project will have only one package). The first task is to create a connection manager.

Please refer to Figure 11-4.

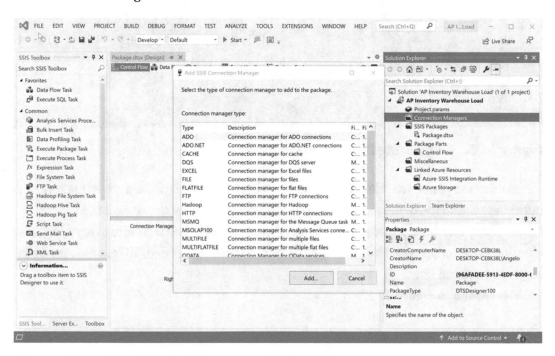

Figure 11-4. *Adding the connection manager*

Right-click the Connection Managers folder, and you are presented with a pop-up dialog box that allows you to view and select the connection manager you require. For our purposes we need the first one called ADO (Active X Data Objects), so we select that one and click the Add button. Notice the list box, many data sources like Hadoop and Excel.

Next, we need to define security information. Please refer to Figure 11-5.

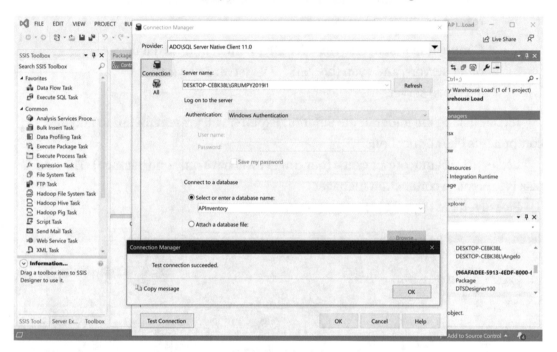

Figure 11-5. *Configuring the connection manager*

In this next panel, we select the server we wish to connect to, the authentication information, and the database we wish to connect to. In our case we want the APInventory database. We also want to test the connection to make sure everything was identified correctly. As can be seen, the test connection succeeded. We are now ready to create our control flow.

Please refer to Figure 11-6.

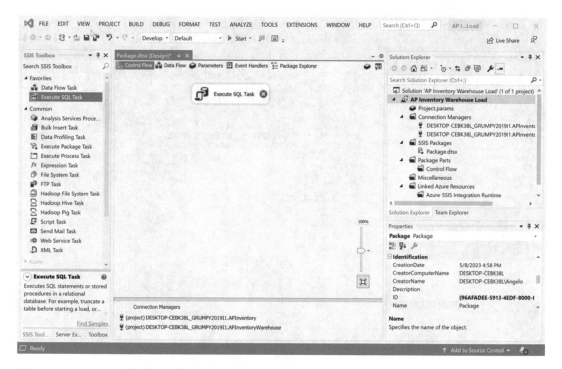

Figure 11-6. *Adding a SQL task*

The simple package we will create will be a series of tasks connected by control flow connections (we do this in the Control Flow tab). Each of these tasks will execute a simple TSQL INSERT command that pulls data from a table in the APInventory database and loads it into the corresponding table in the APInventoryWarehouse data warehouse.

I will walk through the steps required to complete the first step. All other steps are identical except that the TSQL INSERT command changes to reflect the table pairs we are using.

We begin by dragging and dropping Execute SQL Task from the SSIS Toolbox to the Design area.

Next, we need to establish the connection to a database in the new task we just added.

Please refer to Figure 11-7.

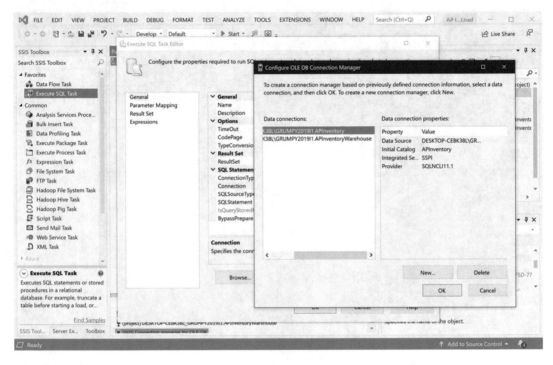

Figure 11-7.   Adding the connection in the SQL task

Double-click the task, and a pop-up dialog box called the Execute SQL Task Editor appears. Where it says "SQL Statement," click the small down arrow button to the right of the Connection text box and select "New connection."

The Connection Manager panel appears, and you need to fill out the SQL server name we will use and the authentication (user ID and password) method. Lastly select the database to connect to, and don't forget to test the connection. If you already created a connection, you can simply select the existing connection instead of selecting "New connection," and you're good to go.

Click the OK button and we proceed to the next step.

Please refer to Figure 11-8.

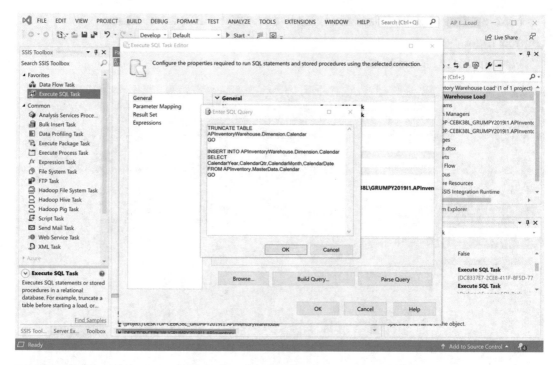

Figure 11-8. *Adding the TSQL INSERT query to the SQL task*

Now we need to add the following small TSQL script to the task.
Please refer to Listing 11-1.

Listing 11-1. Load the Calendar Dimension

```
TRUNCATE TABLE APInventoryWarehouse.Dimension.Calendar
GO

INSERT INTO APInventoryWarehouse.Dimension.Calendar
SELECT CalendarYear,CalendarQtr,CalendarMonth,CalendarDate
FROM APInventory.MasterData.Calendar
GO
```

As you can see, it has two commands, one to truncate the Calendar table and
the other the INSERT/SELECT combo that pulls data from the Calendar table in the
APInventory transaction database and loads it to the Calendar dimension table in the
APInventoryWarehouse data warehouse.

It is recommended to use the fully qualified table name, that is,
<database>.<schema>.<table name>, so you do not run into issues!

Next, we drag and drop the remaining Execute SQL Tasks we need and connect them.

Please refer to Figure 11-9.

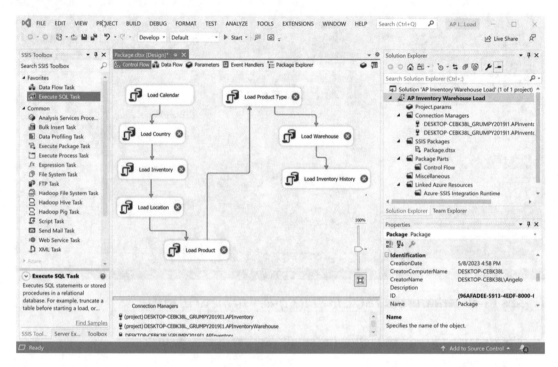

Figure 11-9. *Add other Execute SQL Tasks*

Notice all the small red circles with the X labels indicating an error. That's because we have not added the connection and query information yet. You do not want to see these when you run the package as they indicate an error.

Each time you drop a new Execute SQL Task into the Design pane, you will see a small down-pointing green arrow. This is the arrow that you click and drag and attach to the next task in the control flow sequence so as to establish the path. If it is the last task in the sequence, you do not have to attach the green arrow to anything; you can just ignore it.

You can also drag and drop another Execute SQL Task and call it something like Log Error. Inside you would place some TSQL logic that traps SQL Server error codes and messages and inserts them into a log table. You would click the original SQL task so that the little green arrow appears, and you would attach it to the log error task.

Lastly, you would right-click the new green arrow and click "Failure" in the pop-up menu. This will turn the arrow color to red. With this flow, if an error occurs, the flow is redirected to this task, the error details are loaded into a log table, and you have the choice of continuing to the next tasks or stopping execution entirely.

The following is the completed package after I loaded all the tasks and established the connections and inserted the queries.

Please refer to Figure 11-10.

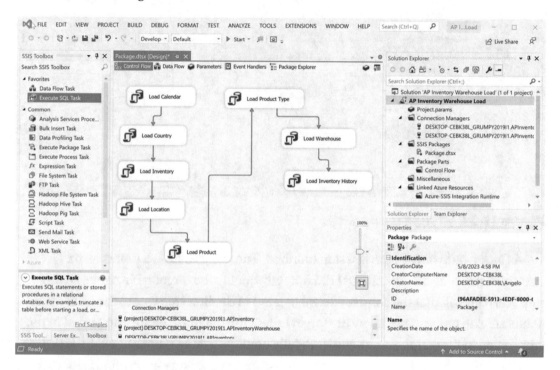

Figure 11-10. *Completing the package*

This was fairly easy, "a walk in the park" as they say! All that is left to do is run the package by clicking the green arrow labeled "Start" in the middle of the menu bar, right below the ANALYZE selection.

The package runs and we should see all green arrows next to each task.

Please refer to Figure 11-11.

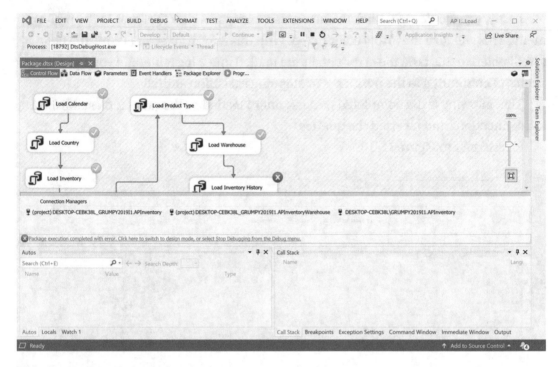

Figure 11-11. *Testing the package*

As luck would have it, the last step bombed. Turns out there is a syntax error. I examined the code and found out I did not fully qualify the target table name, so the step failed. Recall that the connection we created in the first step was to the transaction database. The tasks do not know the target unless the database name is included in the table name in the script. So remember to fully qualify all table names in all the tasks!

Correcting the syntax error, we rerun the script to make sure everything is loaded correctly. This is not necessary as you can execute each task individually by right-clicking the task and selecting "Execute Task."

Please refer to Figure 11-12.

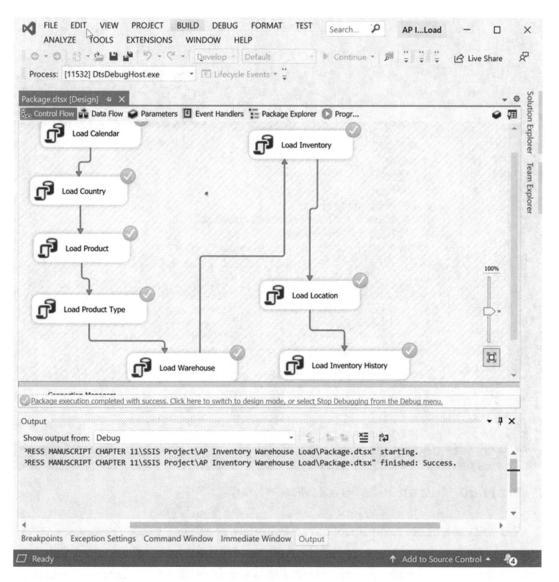

Figure 11-12. *The second test run – success*

Perfect. Everything ran without problems, and our data warehouse should be loaded. Let's see what is inside the fact table just to make sure.

Please refer to Figure 11-13.

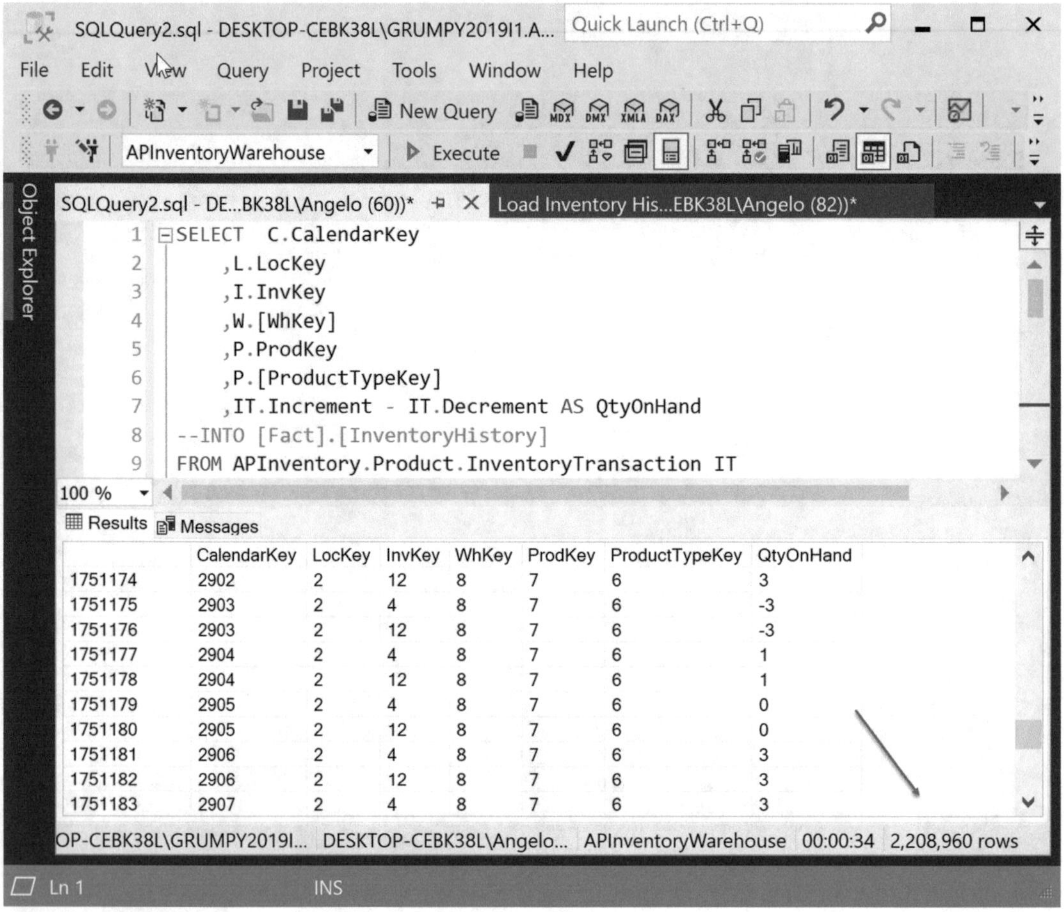

Figure 11-13. *Fact table dump – 2,208,960 rows*

We execute a query to dump all the rows, and we see that 2,208,960 rows were inserted. Not bad. We also see that all the columns are the numerical values for the surrogate keys, and the last QtyOnHand column shows how many products are available. All these surrogate keys mean we will have to join the fact table to the dimension tables to see what all this data really is!

Surrogate tables are used because joining tables over numerical columns is faster than joining tables with alphanumeric or text columns.

I created a TSQL VIEW that we will use later on that shows actual product values like product names and identifiers. By querying this VIEW object, we can see that the joins worked. The TSQL VIEW aggregates on a monthly level, so all we get is 24,192 rows.

Let's see what the results are when we use the VIEW in a simple SELECT query.

Please refer to Figure 11-14.

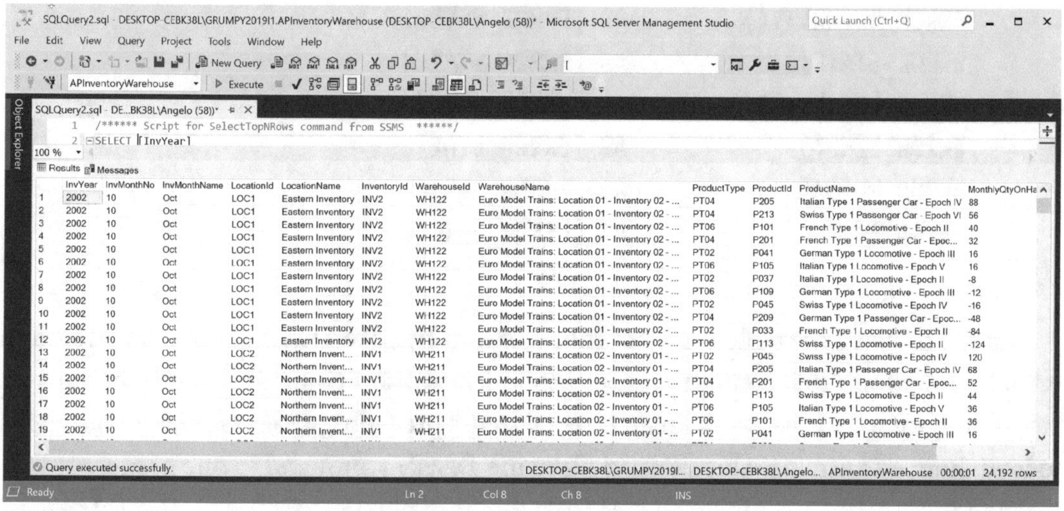

Figure 11-14. *Monthly inventory totals on hand*

As can be seen, all the descriptive information is available, and we can use this TSQL VIEW in many report queries and scenarios.

One last comment on SSIS: Earlier I mentioned you can add error trapping. Listing 11-2 is a simple table definition and a simple INSERT command to trap SQL Server system-level error codes and messages.

Listing 11-2. Error Trapping Kit

```
CREATE TABLE [dbo].[ErrorLog](
      [ErrorNo]          [int] NULL,
      [ErrorSeverity]    [int] NULL,
      [ErrorState]       [int] NULL,
      [ErrorProc]        [nvarchar](128) NULL,
      [ErrorLine]        [int] NULL,
      [ErrorMsg]         [nvarchar](4000) NULL,
      [ErrorDate]        [datetime]
)
GO

INSERT INTO [dbo].[ErrorLog]
SELECT
      ERROR_NUMBER()          AS [ERROR_NO]
```

```
        ,ERROR_SEVERITY()        AS [ERROR_SEVERITY]
        ,ERROR_STATE()           AS [ERROR_STATE]
        ,ERROR_PROCEDURE()       AS [ERROR_PROC]
        ,ERROR_LINE()            AS [ERROR_LINE]
        ,ERROR_MESSAGE()         AS [ERROR_MSG]
        ,GETDATE()               AS [ERROR_DATE];
GO
```

As can be seen, the table can store the error number, severity, and other usual system error information when something goes wrong. These error codes and messages are stored in system parameters that can be used in the preceding INSERT/SELECT command to load them into our log table. We can place the INSERT command in one or more Execute SQL Tasks and redirect all flow from the normal tasks to the error task (in case of errors) as shown in the Figure 11-15.

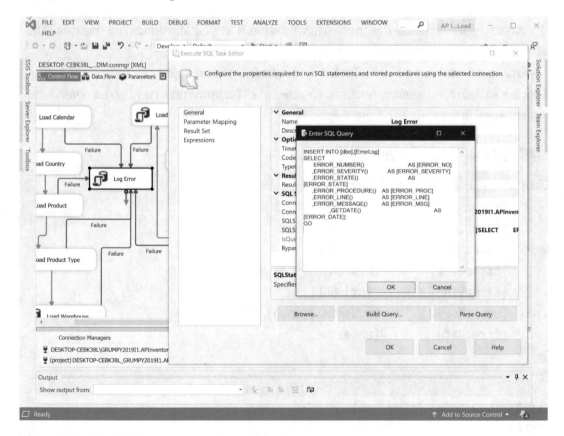

Figure 11-15. *SSIS error trapping added*

This way, if something bombs out, we have all the error codes, error messages, the line number of the buggy code, and also the date of the error.

We can now move to our aggregate function analysis, but I think you have enough SSIS knowledge under your belt to create simple packages with tasks to perform some simple ETL processing!

Aggregate Functions

We will take a brief look at the aggregate window functions one more time. They should be familiar to you as we covered them in all the prior three business scenarios. This time we apply them to our inventory control scenario so we can see how a business analyst would monitor inventory levels and trap events like low or no product inventory so that corrective measures can be taken – or maybe high inventory levels, which indicate a product is not selling!

The following are the window aggregate functions we will discuss in this chapter:

- COUNT(): Window frame specifications support

- SUM(): Window frame specifications support

- MAX(): Window frame specifications support

- MIN(): Window frame specifications support

- AVG(): Window frame specifications support

- STDEV(): Window frame specifications support

- VAR(): Window frame specifications support

As you can see, they all support the ROWS and RANGE window frame specifications.

Our first analysis combines the first five functions in one query as we covered these and are familiar with how they work. In other words, we will not dedicate a single query to each.

COUNT( ), SUM( ), MAX( ), MIN( ), and AVG( ) Functions

We saw these before and the names imply what they do, so I will not include a description of each function at this time. Let's check out our first set of business specifications supplied by our business analyst:

Our business analyst wants us to create an inventory movement profile report, that is, a report that shows the count, minimum, and maximum of quantities plus the totals and averages for product inventory levels less than 50 units. Additionally, the information has to be for all years, for location "LOC1," for inventory "INV1", and for product "P033".

All results need to be rolling results by month with the exception of the SUM() results, which only need to take into account the current month total plus the carry-over total from last month.

Please refer to Listing 11-3 for our solution.

Listing 11-3. Inventory Level Profile Report

```
WITH InventoryMovement (
        LocId, InvId, WhId, ProdId, AsOfYear, AsOfMonth, AsOfDate, InvOut,
        InvIn, Change, QtyOnHand
)
AS(
SELECT LocId
    ,InvId
    ,WhId
    ,ProdId
    ,AsOfYear
    ,AsOfMonth
    ,AsOfDate
    ,InvOut
    ,InvIn
    ,(InvIn - InvOut) AS Change
    ,(InvIn - InvOut) + LAG(QtyOnHand,1,0) OVER (
            PARTITION BY LocId,InvId,WhId,ProdId
            ORDER BY [AsOfDate]
        )AS QtyOnHand -- 05/31/2022
FROM APInventory.Product.WarehouseMonthly
)

SELECT LocId
    ,InvId
    ,WhId
```

```
        ,ProdId
        ,AsOfYear
        ,AsOfMonth
        ,AsOfDate
        ,InvOut
        ,InvIn
        ,Change
        ,QtyOnHand
        ,COUNT(*) OVER (
                PARTITION BY AsOfYear,LocId,InvId,WhId,ProdId
                ORDER BY LocId,InvId,WhId,AsOfMonth
                ROWS BETWEEN UNBOUNDED PRECEDING AND CURRENT ROW
          ) AS WarehouseLT50
        ,MIN(QtyOnHand) OVER (
                PARTITION BY AsOfYear,LocId,InvId,WhId,ProdId
                ORDER BY LocId,InvId,WhId,AsOfMonth
                ROWS BETWEEN UNBOUNDED PRECEDING AND CURRENT ROW
           ) AS MinMonthlyQty
        ,MAX(QtyOnHand) OVER (
                PARTITION BY AsOfYear,LocId,InvId,WhId,ProdId
                ORDER BY LocId,InvId,WhId,AsOfMonth
                ROWS BETWEEN UNBOUNDED PRECEDING AND CURRENT ROW
           ) AS MaxMonthlyQty
        ,SUM(QtyOnHand) OVER (
                PARTITION BY AsOfYear,LocId,InvId,WhId,ProdId
                ORDER BY LocId,InvId,WhId,AsOfMonth
                --ROWS BETWEEN UNBOUNDED PRECEDING AND CURRENT ROW
                ROWS BETWEEN 1 PRECEDING AND CURRENT ROW
            ) AS Rolling2MonthlyQty
        ,AVG(QtyOnHand) OVER (
                PARTITION BY AsOfYear,LocId,InvId,WhId,ProdId
                ORDER BY LocId,InvId,WhId,AsOfMonth
                ROWS BETWEEN UNBOUNDED PRECEDING AND CURRENT ROW
            ) AS AvgMonthlyQty
FROM InventoryMovement
```

737

```
WHERE QtyOnHand <= 50
AND LocId = 'LOC1'
AND InvId = 'INV1'
AND ProdId ='P033'
ORDER BY LocId
    ,InvId
    ,WhId
    ,ProdId
    ,AsOfYear
    ,AsOfMonth
GO
```

As can be seen, we are setting up partitions by year as we are not filtering the results for one year only.

I did include a ROWS window frame clause for all aggregate functions to generate the rolling month effect:

ROWS BETWEEN UNBOUNDED PRECEDING AND CURRENT ROW

For the SUM() aggregate function, I used the following window frame clause so the monthly sum carries over the prior monthly total:

ROWS BETWEEN 1 PRECEDING AND CURRENT ROW

Recall this means that the aggregate function will work on the current row and all previous rows to produce the "rolling" effect for the COUNT(), MAX(), MIN(), and AVG() functions. As mentioned earlier the SUM() function "rolls" for two months.

The ORDER BY clause orders the partition by inventory ID and month. If we want to include more than one year, we will have to include the year here and in the PARTITION BY clause.

Now for the WHERE clause. As we are filtering results for one inventory location, we do not really need to call out the inventory ID. Try this query out by running it as is, generating an estimated query plan, and then take out the inventory ID from the OVER() clause and generate a second estimated query plan. See if they change or if they are the same.

Lastly, we see the WHERE clause that can easily be changed if we want to include all locations, inventories, and warehouses in the result. A SQL Server Reporting Services (SSRS) report can be created that provides filters for these dimensions so the business analyst can pick and choose what they want.

Please refer to the partial results in Figure 11-16.

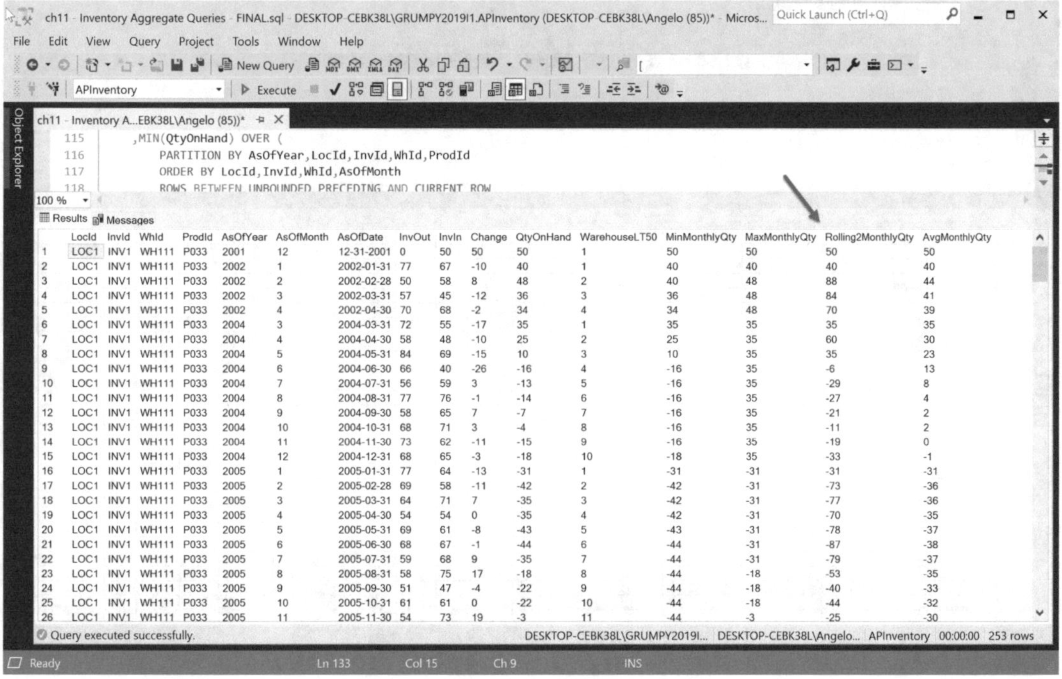

	LocId	InvId	WhId	ProdId	AsOfYear	AsOfMonth	AsOfDate	InvOut	InvIn	Change	QtyOnHand	WarehouseLT50	MinMonthlyQty	MaxMonthlyQty	Rolling2MonthlyQty	AvgMonthlyQty
1	LOC1	INV1	WH111	P033	2001	12	12-31-2001	0	50	50	50	1	50	50	50	50
2	LOC1	INV1	WH111	P033	2002	1	2002-01-31	77	67	-10	40	1	40	40	40	40
3	LOC1	INV1	WH111	P033	2002	2	2002-02-28	50	58	8	48	2	40	48	88	44
4	LOC1	INV1	WH111	P033	2002	3	2002-03-31	57	45	-12	36	3	36	48	84	41
5	LOC1	INV1	WH111	P033	2002	4	2002-04-30	70	68	-2	34	4	34	48	70	39
6	LOC1	INV1	WH111	P033	2004	3	2004-03-31	72	55	-17	35	1	35	35	35	35
7	LOC1	INV1	WH111	P033	2004	4	2004-04-30	58	48	-10	25	2	25	35	60	30
8	LOC1	INV1	WH111	P033	2004	5	2004-05-31	84	69	-15	10	3	10	35	35	23
9	LOC1	INV1	WH111	P033	2004	6	2004-06-30	66	40	-26	-16	4	-16	35	-6	13
10	LOC1	INV1	WH111	P033	2004	7	2004-07-31	56	59	3	-13	5	-16	35	-29	8
11	LOC1	INV1	WH111	P033	2004	8	2004-08-31	77	76	-1	-14	6	-16	35	-27	4
12	LOC1	INV1	WH111	P033	2004	9	2004-09-30	58	65	7	-7	7	-16	35	-21	2
13	LOC1	INV1	WH111	P033	2004	10	2004-10-31	68	71	3	-4	8	-16	35	-11	2
14	LOC1	INV1	WH111	P033	2004	11	2004-11-30	73	62	-11	-15	9	-16	35	-19	0
15	LOC1	INV1	WH111	P033	2004	12	2004-12-31	68	65	-3	-18	10	-18	35	-33	-1
16	LOC1	INV1	WH111	P033	2005	1	2005-01-31	77	64	-13	-31	1	-31	-31	-31	-31
17	LOC1	INV1	WH111	P033	2005	2	2005-02-28	69	58	-11	-42	2	-42	-31	-73	-36
18	LOC1	INV1	WH111	P033	2005	3	2005-03-31	64	71	7	-35	3	-42	-31	-77	-36
19	LOC1	INV1	WH111	P033	2005	4	2005-04-30	54	54	0	-35	4	-42	-31	-70	-35
20	LOC1	INV1	WH111	P033	2005	5	2005-05-31	69	61	-8	-43	5	-43	-31	-78	-37
21	LOC1	INV1	WH111	P033	2005	6	2005-06-30	68	67	-1	-44	6	-44	-31	-87	-38
22	LOC1	INV1	WH111	P033	2005	7	2005-07-31	59	68	9	-35	7	-44	-31	-79	-37
23	LOC1	INV1	WH111	P033	2005	8	2005-08-31	58	75	17	-18	8	-44	-18	-53	-35
24	LOC1	INV1	WH111	P033	2005	9	2005-09-30	51	47	-4	-22	9	-44	-18	-40	-33
25	LOC1	INV1	WH111	P033	2005	10	2005-10-31	61	61	0	-22	10	-44	-18	-44	-32
26	LOC1	INV1	WH111	P033	2005	11	2005-11-30	54	73	19	-3	11	-44	-3	-25	-30

Figure 11-16. *Inventory level profile report*

Looks good. The rolling month values all add up. Since it is a monthly rolling process, warehouse items less than or equal to 50 on hand increase by one as the months proceed as we are only looking at one product. The sum of quantities on hand adds up incrementally for the current month and prior month, so that works.

Check out a few rows and do the math as a simple form of validation. You can even copy and paste the results in a Microsoft Excel spreadsheet and perform the validation there. The max and min levels also look correct.

Keep in mind that we started off with positive inventory levels, so the sum of increments and decrements does not necessarily reflect the quantity on hand. In other words, quantity on hand is what is left after you add the original levels to the number of items coming into inventory and subtract the number of items going out.

Performance Considerations

Let's run our usual baseline estimated query plan to see how this query behaves performance-wise. Please refer to Figure 11-17.

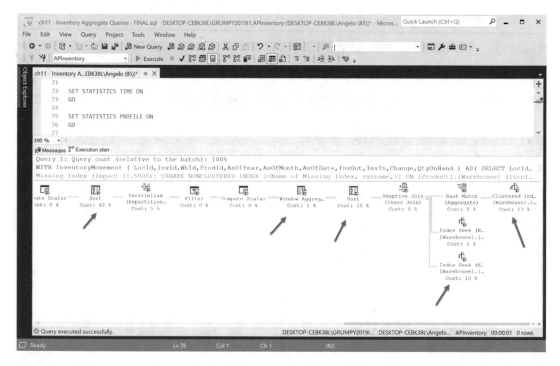

Figure 11-17. *Inventory movement profile baseline estimated query plan*

We will not spend a lot of time on this, but it looks like our index coverage is not quite there. We see a clustered index seek on the Warehouse table at a cost of 13% and another two index seeks on the Warehouse table with costs of 10% and 1%. This is because we are using a VIEW called WarehouseMonthly, which is based on the Warehouse table.

We also see we still have to create the following index:

```
CREATE NONCLUSTERED INDEX ieInvOutInvInAsOfYearAsOfMonth
ON [Product].[Warehouse] ([LocId],[InvId],[ProdId])
INCLUDE ([InvOut],[InvIn],[AsOfYear],[AsOfMonth])
GO
```

Lastly, we see two sort tasks with costs of 15% and 42%. All other task costs are low. The following are the IO and TIME statistics for this query.

Please refer to Figure 11-18.

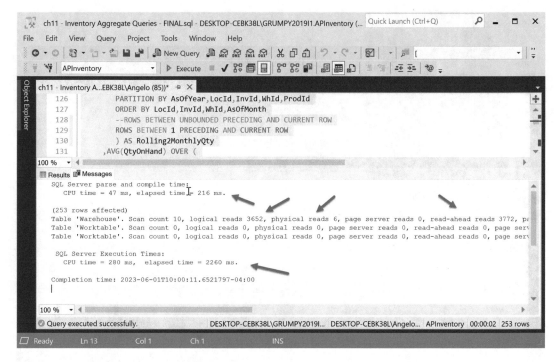

Figure 11-18. *IO and TIME statistics*

The work table values are all zero so we will ignore these. For the Warehouse table though, we see a scan count of 10, logical reads of 3,652, physical reads of 6, and read-ahead reads of 3,772. The query execution time came in at 2260 ms. This is a busy query indeed.

Let's create the recommended index and see what improvements, if any, are gained.

Please refer to Figure 11-18a for the revised estimated query plan.

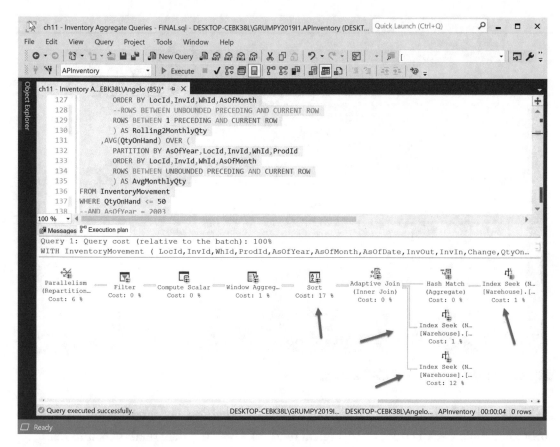

Figure 11-18a. *Revised estimated query plan*

This looks a little better. The first index seek task went down; the other tasks are more all less the same, so some modest improvement with the new index. Let's check out the IO and TIME statistics.

Please refer to Figure 11-18b for the new IO and TIME statistics.

Figure 11-18b. *Revised estimated query plan IO and TIME statistics*

Things look a lot better here. The Warehouse table counts went down considerably, and so did the execution time. So it seems the recommended index paid off.

Let's take a quick look at the AVG() function again.

AVG() Function

As the name implies, this function is used to calculate averages, but you knew that. Let's check out our business specifications supplied by our friendly business analyst:

This time our analyst wants to see rolling averages for the inventory movement values, specifically inventory in and inventory out. The rolling averages are by year instead of month.

Listing 11-4 is the code for the query. Not shown is the CTE because it is the same as the CTE in the prior query.

Listing 11-4. Rolling Yearly Averages for Inventory Movement In and Out

```
SELECT MovementDate
      ,InvId
      ,LocId
      ,WhId
      ,ProdId
      ,Increment
      ,Decrement
      ,AVG(CONVERT(DECIMAL(10,5),Increment)) OVER (
            ORDER BY MovementDate
            ROWS BETWEEN UNBOUNDED PRECEDING AND UNBOUNDED FOLLOWING
            ) AS MonthlyAvgIncr
      ,AVG(CONVERT(DECIMAL(10,5),Decrement)) OVER (
            ORDER BY MovementDate
            ROWS BETWEEN UNBOUNDED PRECEDING AND UNBOUNDED FOLLOWING
            ) AS MonthlyAvgDecr
  FROM APInventory.Product.InventoryTransaction
  WHERE ProdId = 'P033'
  AND YEAR(MovementDate) = 2010
  AND MONTH(MovementDate) = 5
  AND WhId = 'WH112'
  GO
```

A ROWS window frame specification is used in order to define the range for processing. In this query we define the range over the year. For each current row, we (or should we say SQL Server) consider all preceding rows and all succeeding rows within the partition.

Note we are filtering for the year 2010 as per the business analyst requirements, but once we test this, it can be modified for more years, months, and all locations, inventories, warehouses, and products. This would make it a good candidate for an SSRS report created with Report Builder.

Please refer to the partial results in Figure 11-19 together with a nice graph.

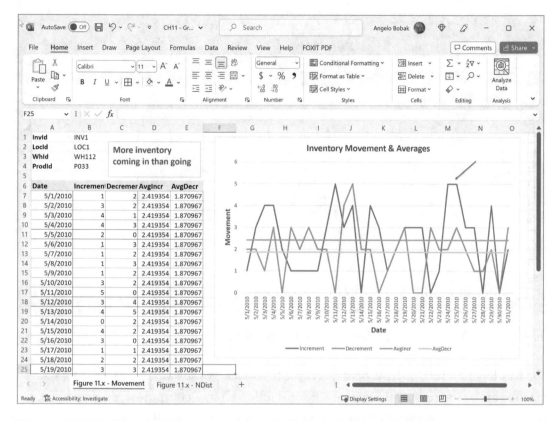

Figure 11-19. *Inventory movement analysis line chart and report*

We can see a lot of swings, both up and down, in the inventory movement. This can be interpreted as there are a lot of sales. The business analyst needs to determine though that the inventory out movement does not result in many instances of backorders. This would mean a customer needs to wait for the item they ordered, which could lead to unhappy customers, which in turn could lead to lost customers.

Performance Considerations

Next, we need to analyze performance as usual, so let's generate our baseline estimated query plan to see if this query needs some work or some indexes to support it for optimal performance.

Please refer to Figure 11-20.

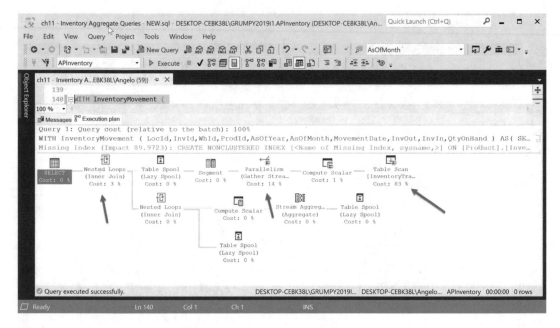

Figure 11-20. *Baseline estimated execution plan*

Not too complex, but this plan has three branches. We see an expensive table scan starting at the top right coming in at 83% cost followed by a compute scalar (cost 1%), a parallelism task to gather streams of data (14% cost), and a nested loops join at 3%.

And of course, an index is suggested that impacts the cost by 89.97%. Let's create the index and see if this helps.

Please refer to Listing 11-5.

Listing 11-5. Suggested Index

```
CREATE NONCLUSTERED INDEX ieWhIdProdIdMovementDate
ON APInventory.Product.InventoryTransaction (WhId,ProdId)
INCLUDE (Increment,Decrement,MovementDate,InvId,LocId)
GO
```

Notice only two columns, the WhId and ProdId columns, but we need to include the other columns such as the Increment, Decrement, MovementDate, InvId, and LocId columns in the INCLUDE clause.

After creating the index and rerunning the estimated query plan, we generate the revised estimated query plan. Please refer to Figure 11-21.

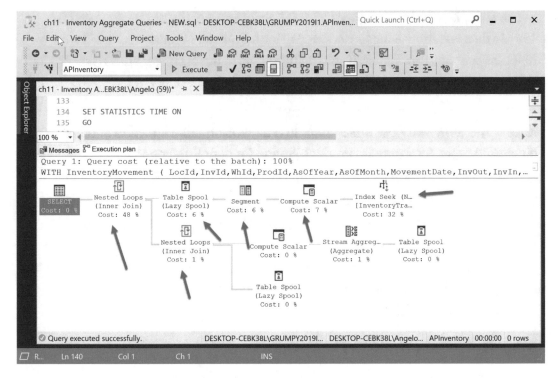

Figure 11-21. *The revised estimated query plan*

We can see the improvement. The table scan and parallelism task went away, and now an index seek appears with a cost of 32%. We see a table spool and of course the nested loops join with a cost of 48%. Once again, the estimated query analyzer came up with a good suggestion.

Homework   Try creating the index as a clustered index to see if there are significant improvements. Create the index and check out the estimated query plan and the IO and TIME statistics.

STAR query time (or should I say SNOWFLAKE query time)?

Data Warehouse Query

Let's see if querying a SNOWFLAKE data warehouse database will yield interesting performance characteristics compared with the query plans and statistics of the transactional database.

747

The query is called a STAR query. This is because it uses the dimensions and fact table of a data warehouse that is designed using a STAR data model. Yes, it will work on a SNOWFLAKE schema also. I guess if you include some of the outrigger tables (like PRODUCT TYPE) connected to a dimension, you can call it a SNOWFLAKE query.

Note   The term *outrigger table* refers to a dimension table linked to another dimension table, for example, Product Type to Product Subtype to Product. If you are unfamiliar with data warehouse models, check the Internet for articles that describe the different combinations of models and approaches used to design a data warehouse. For example, see the approaches Kimball takes vs. Inmon.

We will take the specifications of the query we just discussed and modify them a bit, so we report on the sums and averages of the inventory levels.

Please refer to Listing 11-6.

Listing 11-6.   Example SNOWFLAKE Query

```
SELECT C.CalendarYear
        ,C.CalendarMonth
        ,C.CalendarDate AS AsOfDate
        ,L.LocId
        ,L.LocName
        ,I.InvId
        ,W.WhId
        ,P.ProdId
        ,P.ProdName
        ,QtyOnHand
        ,SUM(IH.QtyOnHand) OVER (
                ORDER BY C.CalendarMonth
                ROWS BETWEEN 1 PRECEDING AND CURRENT ROW
            ) AS RollingSum
        ,AVG(SUM(CONVERT(DECIMAL(10,2),IH.QtyOnHand))) OVER (
                ORDER BY C.CalendarMonth
                ROWS BETWEEN 1 PRECEDING AND CURRENT ROW
            ) AS RollingAvgOnHand
FROM ApInventoryWarehouse.Fact.InventoryHistory IH
```

```
JOIN Dimension.Calendar C
     ON IH.CalendarKey = C.CalendarKey
JOIN Dimension.Location L
     ON IH.LocKey = L.LocKey
JOIN Dimension.Warehouse W
     ON IH.WhKey = W.WhKey
JOIN Dimension.Inventory I
     ON IH.InvKey = I.InvKey
JOIN Dimension.Product P
     ON IH.ProdKey = P.ProdKey
JOIN Dimension.ProductType PT
     ON P.ProductTypeKey = PT.ProductTypeKey
WHERE C.CalendarYear = 2010

-- Was: AND C.[CalendarDate] = AsOfDate - last day of the month
-- changed to below as Warehouse table now contains all days of the month
-- prior version contained only the last day of the month

AND C.[CalendarDate] = EOMONTH(CONVERT(VARCHAR,YEAR(C.
[CalendarDate])),(MONTH(C.[CalendarDate]) - 1))

AND L.LocId = 'LOC1'
AND I.InvId = 'INV1'
AND W.WhId = 'WH111'
AND P.ProdName LIKE 'French Type 1 Locomotive %'
AND P.ProdId = 'P101'
GROUP BY C.CalendarYear
     ,C.CalendarMonth
     ,C.CalendarDate
     ,L.LocId
     ,L.LocName
     ,I.InvId
     ,W.WhId
     ,P.ProdId
     ,P.ProdName
     ,QtyOnHand
GO
```

Right away we notice we need a GROUP BY clause. But look at all of those JOIN predicates!

Pop Quiz   Why are we calling this a SNOWFLAKE query?

Both the SUM() and AVG() functions utilize an OVER() clause that contains an ORDER BY clause but no PARTITION BY clause. A window frame clause is included:

ROWS BETWEEN 1 PRECEDING AND CURRENT ROW

This only considers the current row and prior row being processed in the partition.

Do you remember what the default window frame behavior is if we do not include a window frame specification?

Here is a refresher from the table we discussed in Chapter 1 that identifies the default behavior of the OVER() clause under different combinations of ORDER BY and PARTITION BY clauses. Please refer to Table 11-3.

Table 11-3. *Default ROWS and RANGE Behavior*

ORDER BY	PARTITION BY	Default Frame
No	No	ROWS BETWEEN UNBOUNDED PRECEDING AND UNBOUNDED FOLLOWING
No	Yes	ROWS BETWEEN UNBOUNDED PRECEDING AND UNBOUNDED FOLLOWING
Yes	No	RANGE BETWEEN UNBOUNDED PRECEDING AND CURRENT ROW
Yes	Yes	RANGE BETWEEN UNBOUNDED PRECEDING AND CURRENT ROW

A little review never hurts. Saves flipping back to Chapter 1 anyway.

Let's see what the estimated query plan looks like.

Performance Considerations

We can expect an interesting, estimated query plan as we have all those joins.

Let's generate our plan in the usual manner by using the selection under the Query menu option.

I need to split it into two sections as it is a large one.

Please refer to Figure 11-22a.

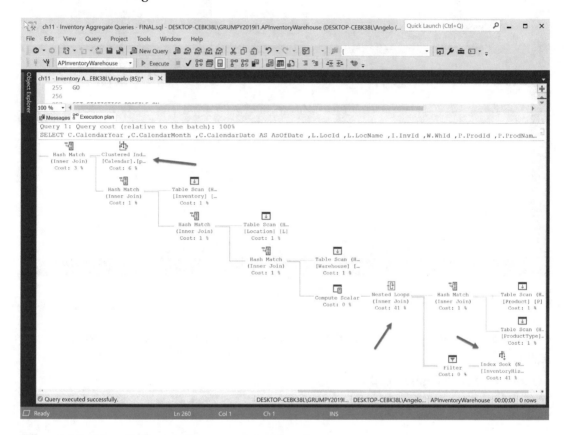

Figure 11-22a. *Estimated query plan part A*

Here is the first section of the plan. These types of queries will have a series of table scans on the dimensions if the row count is small, so an index would not be used, even if you created one!

We do have an index seek on the fact table, which is a good thing (seek better than a scan). Looks like we had a usable index lying around that satisfied the OVER( ) clause columns and also the WHERE clause predicate columns, so an index was not suggested.

All the streams are joined by a hash match join (3% cost). We do have a nested loops join task with a cost of 41% that joins the streams from the Product, ProductType, and InventoryHistory tables. Let's look at the second part of the query plan.

Please refer to Figure 11-22b.

751

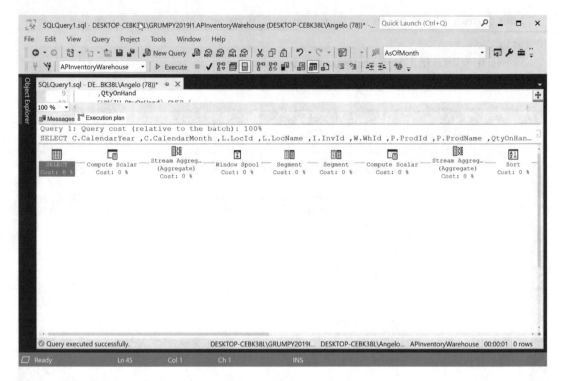

Figure 11-22b. *Estimated query plan part B*

Nothing to worry about here. All zero costs, so let's check out the TIME and IO statistics.

Please refer to Figure 11-23.

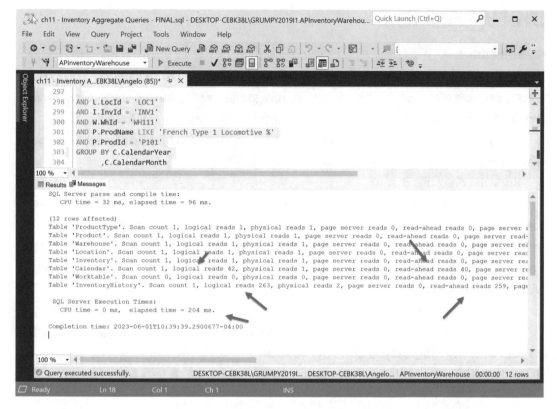

Figure 11-23. *IO and TIME statistics*

Let's just discuss the higher-value statistics. Our Calendar dimension generated around 42 logical reads and 40 read-ahead reads, so this is an area to focus on for performance improvement. We can place it on a high-speed or solid-state disk or a memory-enhanced table.

The work table has all 0 values so we will not worry about this.

Lastly, the fact table InventoryHistory has a scan count of 1 and 263 logical reads, 2 physical reads, and 259 read-ahead reads. This is to be expected as the table has 736,320 rows, so it is rather large (at least for a laptop).

Tip and Homework   Since the data is historical in nature, we should put the results in a report table like we did in prior examples. This way we take the performance hit only once when the table is loaded at off hours. Try it on your own and perform the necessary analysis.

STDEV( ) Function

Statistics time!

As a reminder, recall our definition for standard deviation from Chapter 2 (skip this if you are comfortable with how this works and go right to the business specification for the example query). If you need a refresher or this is new to you, continue reading:

The STDEV( ) function calculates the statistical standard deviation of a set of values when the entire population **is not known** or available.

But what does this mean?

What this means is how close or how far apart the numbers are in the data set from the arithmetic **MEAN** (which is another word for average). In our context the average or **MEAN** is the sum of all data values in question divided by the number of data values. There are other types of **MEAN** like geometric, harmonic, and weighted **MEAN**, which I will briefly cover in Appendix B.

Recall how the standard deviation is calculated.

Here is the algorithm to calculate the standard deviation, which you can implement on your own with TSQL if you are adventurous:

- Calculate the average (**MEAN**) of all values in question and store it in a variable. You can use the AVG( ) function.

- Loop around the value set. For each value in the data set

  - Subtract the calculated average.

  - Square the result.

- Take the sum of all the preceding results and divide the value by the number of data elements minus 1. (This step calculates the variance; to calculate for the entire population, do not subtract 1.)

- Now take the square root of the variance, and that's your standard deviation.

Or you could just use the STDEV( ) function! There is a STDEVP( ) function. The only difference is that STDEV( ) works on part of the population of the data. This is when the entire data set is not known. The STDEVP( ) function works on the entire data set, and this is what is meant by the entire population of data.

Once again here is the syntax:

```
SELECT STDEV( ALL | DISTINCT ) AS [Column Alias]
FROM [Table Name]
GO
```

You need to supply a column name that contains the data you want to use. You can optionally supply the ALL or DISTINCT keyword if you wish. If you leave these out, just like the other functions, the behavior will default to ALL. Using the DISTINCT keyword will ignore duplicate values in the column (make sure this is what you want to do).

Here come the business requirements for this next report supplied to us by our business analyst:

A report is needed that displays the average and standard deviation for inventory quantity on hand. Results need to be for the year "2003", location "LOC1", inventory "INV1", and product "PO33". The report data needs to be used to create a normal distribution graph in a Microsoft Excel spreadsheet.

Here is the code we came up with to deliver the specification requirements.

Please refer to Listing 11-7.

Listing 11-7. Average and Standard Deviation Report for Inventory Quantities

```
WITH YearlyWarehouseReport (
     AsOfYear,AsOfQuarter,AsOfMonth,AsOfDate,LocId,InvId,WhId,
     ProdId,AvgQtyOnHand
     )
AS
(
SELECT AsOfYear
     ,DATEPART(qq,AsOfDate) AS AsOfQuarter
     ,AsOfMonth
     ,AsOfDate
     ,LocId
     ,InvId
     ,WhId
     ,ProdId
     ,AVG(QtyOnHand) AS AvgQtyOnHand
FROM Product.Warehouse
GROUP BY AsOfYear
     ,DATEPART(qq,AsOfDate)
```

```
        ,AsOfMonth
        ,AsOfDate
        ,LocId
        ,InvId
        ,WhId
        ,ProdId
)
SELECT AsOfYear
        ,AsOfQuarter
        ,AsOfMonth
        ,AsOfDate
        ,LocId
        ,InvId
        ,WhId
        ,ProdId
        ,AvgQtyOnHand AS MonthlyAvgQtyOnHand
        ,AVG(AvgQtyOnHand) OVER (
                PARTITION BY AsOfYear,WhId
                ORDER BY AsOfYear
                ) AS YearlyAvg
        ,STDEV(AvgQtyOnHand) OVER (
                PARTITION BY AsOfYear,WhId
                ORDER BY AsOfYear
                ) AS YearlyStdev
FROM YearlyWarehouseReport
WHERE AsOfYear = 2003
  AND LocId = 'LOC1'
  AND InvId = 'INV1'
  AND ProdId ='PO33'
GO
```

The OVER() clause partition is by year and warehouse ID although the year is not needed as we are filtering results for 2003 only. The business analyst wants to use this code to include more years, so we include this column so the query can be modified. Additionally, we want the average and standard distribution values to be for the entire year as we will use the data in a Microsoft Excel spreadsheet to generate a normal distribution curve. Note that we will get results for two warehouses. Each inventory has two warehouses. Each location has one inventory.

Please refer to the partial results in Figure 11-24.

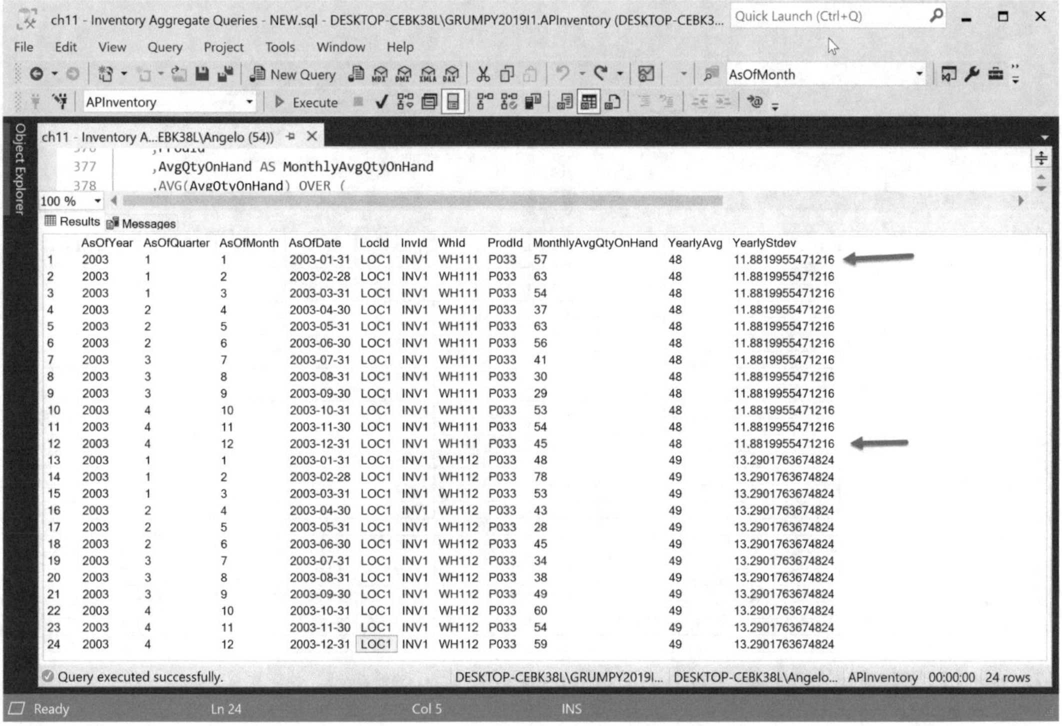

	AsOfYear	AsOfQuarter	AsOfMonth	AsOfDate	LocId	InvId	WhId	ProdId	MonthlyAvgQtyOnHand	YearlyAvg	YearlyStdev
1	2003	1	1	2003-01-31	LOC1	INV1	WH111	P033	57	48	11.8819955471216
2	2003	1	2	2003-02-28	LOC1	INV1	WH111	P033	63	48	11.8819955471216
3	2003	1	3	2003-03-31	LOC1	INV1	WH111	P033	54	48	11.8819955471216
4	2003	2	4	2003-04-30	LOC1	INV1	WH111	P033	37	48	11.8819955471216
5	2003	2	5	2003-05-31	LOC1	INV1	WH111	P033	63	48	11.8819955471216
6	2003	2	6	2003-06-30	LOC1	INV1	WH111	P033	56	48	11.8819955471216
7	2003	3	7	2003-07-31	LOC1	INV1	WH111	P033	41	48	11.8819955471216
8	2003	3	8	2003-08-31	LOC1	INV1	WH111	P033	30	48	11.8819955471216
9	2003	3	9	2003-09-30	LOC1	INV1	WH111	P033	29	48	11.8819955471216
10	2003	4	10	2003-10-31	LOC1	INV1	WH111	P033	53	48	11.8819955471216
11	2003	4	11	2003-11-30	LOC1	INV1	WH111	P033	54	48	11.8819955471216
12	2003	4	12	2003-12-31	LOC1	INV1	WH111	P033	45	48	11.8819955471216
13	2003	1	1	2003-01-31	LOC1	INV1	WH112	P033	48	49	13.2901763674824
14	2003	1	2	2003-02-28	LOC1	INV1	WH112	P033	78	49	13.2901763674824
15	2003	1	3	2003-03-31	LOC1	INV1	WH112	P033	53	49	13.2901763674824
16	2003	2	4	2003-04-30	LOC1	INV1	WH112	P033	43	49	13.2901763674824
17	2003	2	5	2003-05-31	LOC1	INV1	WH112	P033	28	49	13.2901763674824
18	2003	2	6	2003-06-30	LOC1	INV1	WH112	P033	45	49	13.2901763674824
19	2003	3	7	2003-07-31	LOC1	INV1	WH112	P033	34	49	13.2901763674824
20	2003	3	8	2003-08-31	LOC1	INV1	WH112	P033	38	49	13.2901763674824
21	2003	3	9	2003-09-30	LOC1	INV1	WH112	P033	49	49	13.2901763674824
22	2003	4	10	2003-10-31	LOC1	INV1	WH112	P033	60	49	13.2901763674824
23	2003	4	11	2003-11-30	LOC1	INV1	WH112	P033	54	49	13.2901763674824
24	2003	4	12	2003-12-31	LOC1	INV1	WH112	P033	59	49	13.2901763674824

***Figure 11-24.** Average and standard deviation report for inventory quantities*

Results are for a single year but for two warehouses. Notice how the average and standard deviation values are the same for all months for each inventory. Let's copy these results and pop them into a Microsoft Excel spreadsheet so we can create a graph showing a nice bell curve.

Please refer to the partial results in Figure 11-25.

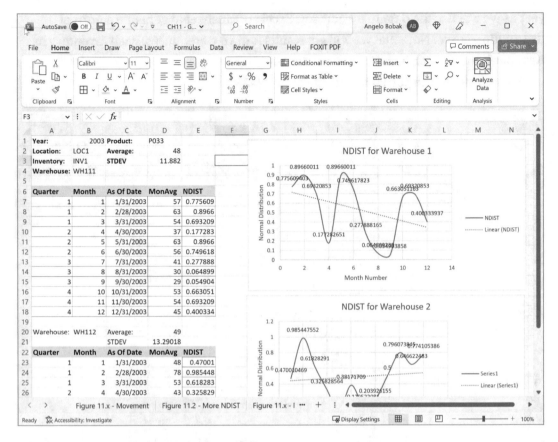

Figure 11-25. *Bell curves for monthly average quantities*

Alright, it looks like a number of bell curves, not one. We can see that the averages have a wide range of differences as we proceed month to month with the trend going down for warehouse WH111 and going up for warehouse WH112. It will be interesting to see what the variance results look like, but let's perform our usual baseline estimated performance analysis first.

Performance Considerations

This time we will look at the baseline plan only, and I leave it up to you to try any enhancements like replacing the CTE portion of the query with a reporting table like we did in prior chapters.

Here is our baseline estimated query performance plan in Figure 11-26.

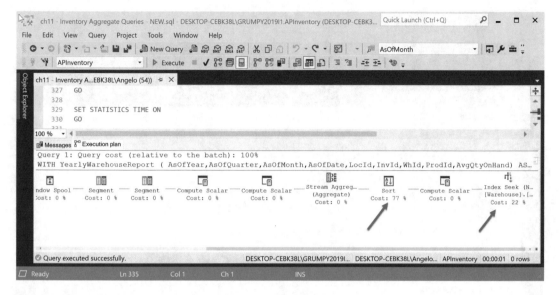

Figure 11-26. *Baseline estimated execution plan*

Looks like a straightforward linear plan with most of the activity on the Warehouse table. We have an index seek task with a cost of 22% and a pretty expensive sort at 77%. All the other tasks chime in at 0% cost. Let's look at the IO and TIME performance statistics.

Please refer to Figure 11-27.

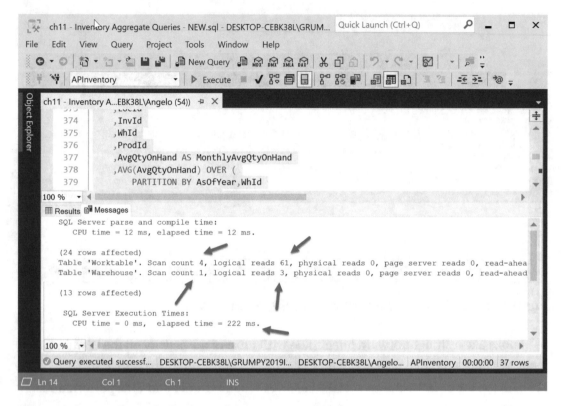

Figure 11-27. IO and TIME performance statistics

The query took 222 ms to run, under 1 s. Most of the activity is with the work table. We can see a scan count value of 4 and 61 logical reads. The Warehouse table has a scan count of 1 plus logical reads of 3. All other values are zero.

Looks like all we can do to improve queries of this type that have high sort values is to ensure that TEMPDB is placed on a high-speed drive or better yet a solid-state drive. Sometimes all you can do is add more hardware!

Data Warehouse Query

Let's see if querying a SNOWFLAKE database will yield interesting performance characteristics compared with the query plans and statistics of the transactional database. I think you know by now that the query will use a number of JOIN clauses, and this will make for an interesting estimated query plan.

The business specifications are mostly the same except there is no WHERE clause. We calculate the standard deviation and average of the sum of quantities.

Please refer to Listing 11-8.

Listing 11-8. Another Example SNOWFLAKE Query

```
SELECT C.CalendarYear
      ,C.CalendarMonth
      ,C.CalendarDate
      ,L.LocId
      ,L.LocName
      ,I.InvId
      ,W.WhId
      ,P.ProdId
      ,P.ProdName
      ,SUM(QtyOnHand) As QtyOnHandSum
      ,AVG(CONVERT(DECIMAL(10,2),IH.QtyOnHand)) OVER (
              PARTITION BY C.CalendarYear
              ORDER BY C.CalendarMonth
        ) AS RollingAvg
      ,STDEV(SUM(IH.QtyOnHand)) OVER (
              PARTITION BY C.CalendarYear
              ORDER BY C.CalendarMonth
        ) AS RollingStdev
FROM Fact.InventoryHistory IH
JOIN Dimension.Calendar C
     ON IH.CalendarKey = C.CalendarKey
JOIN Dimension.Location L
     ON IH.LocKey = L.LocKey
JOIN Dimension.Warehouse W
     ON IH.WhKey = W.WhKey
JOIN Dimension.Inventory I
     ON IH.InvKey = I.InvKey
JOIN Dimension.Product P
     ON IH.ProdKey = P.ProdKey
JOIN Dimension.ProductType PT
     ON P.ProductTypeKey = PT.ProductTypeKey
WHERE L.LocId = 'LOC1'
```

```
AND I.InvId = 'INV1'
AND W.WhId = 'WH112'
AND P.ProdName LIKE 'French Type 1 Locomotive%'
AND P.ProdId = 'P101'
AND C.CalendarYear = 2002
GROUP BY C.CalendarYear
        ,C.CalendarMonth
        ,C.CalendarDate
        ,L.LocId
        ,L.LocName
        ,I.InvId
        ,W.WhId
        ,P.ProdId
        ,P.ProdName
        ,QtyOnHand
GO
```

As can be seen, a LIKE predicate is used, which will cost us performance-wise. We see all the JOIN clauses as we need data from six dimension tables. The specifications for the STDEV() function and OVER() clause call for a partition defined by year and an ORDER BY clause sorted by calendar month. We also need a mandatory GROUP BY clause and pass the window functions the SUM(QtyOnHand) to generate the results. We also had to convert the quantity on hand to calculate the average to a DECIMAL data type as it is so small we get fractional parts.

Let's check out the results.

Please refer to Figure 11-28.

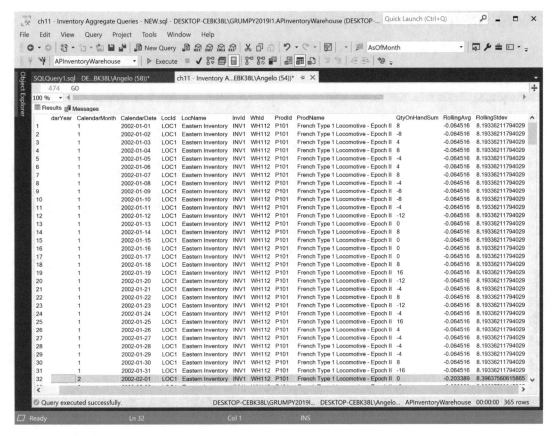

Figure 11-28. *Rolling monthly standard deviations*

The results look good. As the standard deviation and average were calculated over the year, this will allow us to generate some interesting bell curves if we copy and paste the results into a Microsoft Excel spreadsheet. Consider this a homework assignment and try it on your own.

Homework   Try modifying the SUM( ) function that calculates two month's rolling totals.

I know you have been waiting to see the query plan, so here it is.

Performance Considerations

Let's run our baseline execution plan in the usual manner. Our last plan has a linear configuration. I think you know this plan will have a multi-branch configuration with many joins.

Please refer to Figure 11-29.

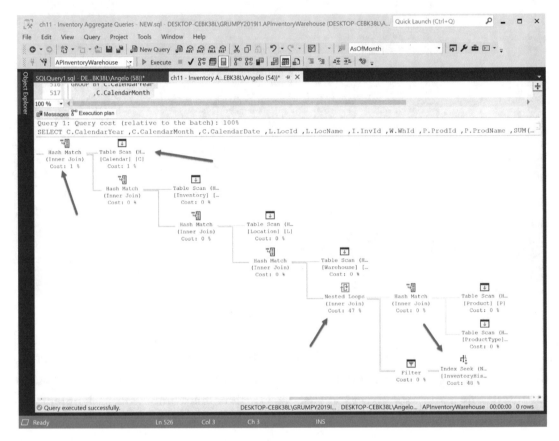

Figure 11-29. *Baseline estimated execution plan*

Sure enough, we have a lopsided tree configuration. All the way at the bottom right-hand side, we see an index seek with a cost of 48%, a nested loops join task with a cost of 47%, and, last but not least at the top, a table scan on the Calendar table with a cost of 1% and a hash match join with a cost of 1%.

Not a bad plan considering the size and complexity of the query and tables. Let's look at the IO and TIME statistics for this complex query.

Please refer to Figure 11-30.

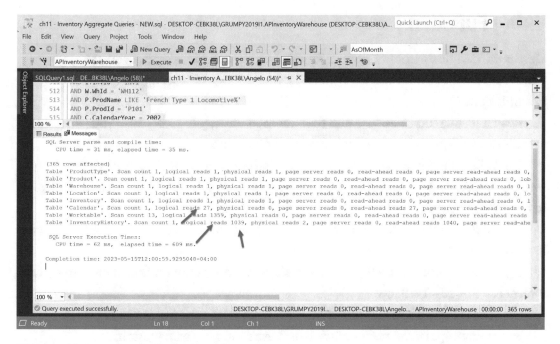

Figure 11-30. *Estimated IO and TIME statistics*

I highlighted the higher values only. Our friend the Calendar table has the usual 27 logical reads and 27 read-ahead reads, while our work table has 13 scans and 1,359 reads.

Looking at the InventoryHistory table, we see only 1 scan and 1,039 logical reads, 2 physical reads, and 1,040 read-ahead reads. Recall this table has 2,945,280 rows, so the results are pretty good performance-wise.

Based on these statistics and also past statistics, since our Calendar table is small, I would place this in a memory-enhanced table.

Homework   Replace the LIKE predicate with an "=" predicate using the full correct name of the product. Create a new estimated query plan together with IO and TIME statistics to see if there are any improvements. Don't forget to run the DBCC command.

More statistics coming right up!

VAR( ) Function

Here is the definition from Chapter 2 for this function. Again, skip it if you are comfortable with how it works and go to the business specifications for the example query.

Read it if you read Chapter 1 and then jumped directly here because your interest is inventory management or some other related discipline.

The VAR() function is used to generate the variance for a set of values in a data sample. The same discussion related to the data set used in the STDEV() example applies to the VAR() function relative to population of the data.

The VAR() function works on a partial set of the data when the entire data set is not known or available.

So, if your data set consists of ten rows, these functions not ending in the letter P will look at N – 1 or 10 – 1 = 9 rows.

Okay, now that we got that out of the way, what does variance mean anyway?

For us developers and non-statisticians, here's a simple explanation:

Informally, if you take all the differences of each value from the mean (average) and square the results and then add all the results and finally divide them by the number of data samples, you get the variance.

Reminder   If you take the square root of the variance, you get the standard deviation, so there is some sort of relationship going on!

It answers the following questions:

- Did we miss our targets?

- Did we meet our targets?

- Did we exceed our targets?

Here are the business requirements for this next query:

The business analyst wants to use the prior query we developed for the standard deviation and adopt it so we can calculate the variance, the standard deviation, and the average. Also, they want to see the standard deviation calculated in two ways, the first by the STDEV() function and the second method to be used as validation, by taking the square root of the variance.

Here is the code we came up with in Listing 11-9.

Listing 11-9. Average, Variance, and Standard Deviation Report

```
WITH YearlyWarehouseReport (
      AsOfYear,AsOfQuarter,AsOfMonth,AsOfDate,LocId,InvId
      ,WhId,ProdId,AvgQtyOnHand
      )
AS
(
SELECT AsOfYear
      ,DATEPART(qq,AsOfDate) AS AsOfQuarter
      ,AsOfMonth
      ,AsOfDate
      ,LocId
      ,InvId
      ,WhId
      ,ProdId
      ,AVG(QtyOnHand) AS AvgQtyOnHand
FROM Product.Warehouse
GROUP BY AsOfYear
      ,DATEPART(qq,AsOfDate)
      ,AsOfMonth
      ,AsOfDate
      ,LocId
      ,InvId
      ,WhId
      ,ProdId
)
SELECT AsOfYear
      ,AsOfQuarter
      ,AsOfMonth
      ,AsOfDate
      ,LocId
      ,InvId
      ,WhId
```

```
      ,ProdId
      ,AvgQtyOnHand AS MonthlyAvgQtyOnHand
      ,AVG(AvgQtyOnHand) OVER (
            PARTITION BY AsOfYear,WhId
            ORDER BY AsOfYear
            ) AS YearlyAvg
      ,VAR(AvgQtyOnHand) OVER (
            PARTITION BY AsOfYear,WhId
            ORDER BY AsOfYear
            ) AS YearlyVar
      ,STDEV(AvgQtyOnHand) OVER (
            PARTITION BY AsOfYear,WhId
            ORDER BY AsOfYear
            ) AS YearlySTDEV
      /* just to prove that the square root of the variance */
      /* is the standard deviation */
      ,SQRT(
                  VAR(AvgQtyOnHand) OVER (
                  PARTITION BY AsOfYear,WhId
                  ORDER BY AsOfYear
                  )
            ) AS MyYearlyStdev
FROM YearlyWarehouseReport
WHERE AsOfYear = 2003
  AND LocId = 'LOC1'
  AND InvId = 'INV1'
  AND ProdId ='P033'
GO
```

We are back to our usual CTE-based query. For all three functions, we set up an
OVER() clause that includes a PARTITION BY and an ORDER BY clause. The partitions are
set up by year and warehouse ID; the ORDER BY clause is set up by year.

Once again, we start off with a small data set and use a WHERE clause, so we only get
results for the year "2003", location "LOC1", inventory "INV1", and product "P033" (these
can be removed once the data is validated).

The query can then be used for a Reporting Services (SSRS) report that includes filters so the analyst can slice and dice the report any way they want.

Please refer to the partial results in Figure 11-31.

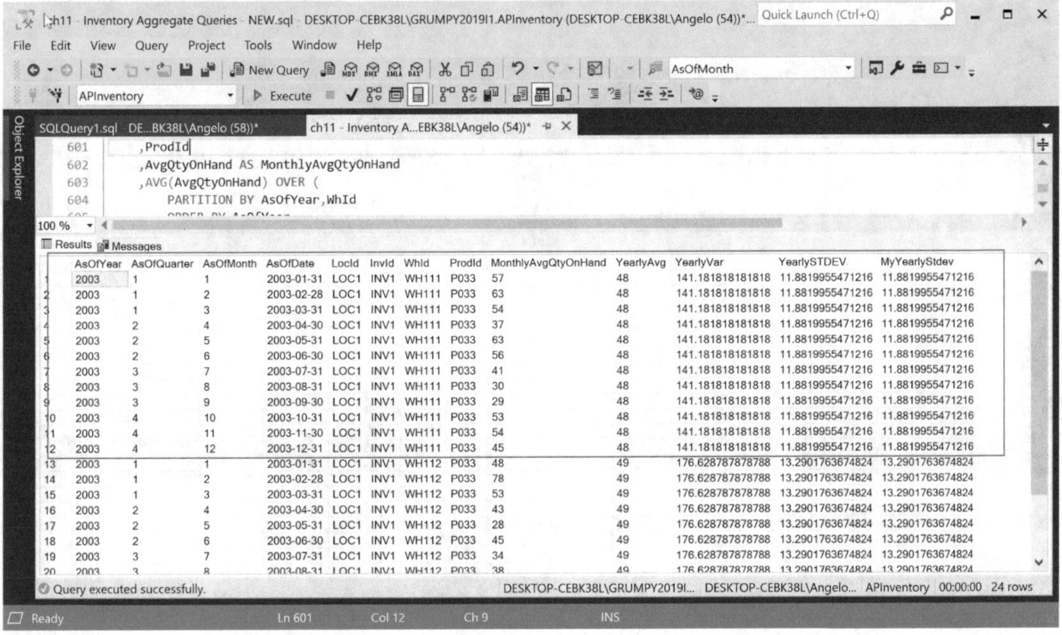

Figure 11-31. *Average, variance, and standard deviation report*

Looking at the report, we see the partitions worked as expected and that the two standard deviation calculations matched, so we not only validated the results but proved that if we take the square root of the variance, we get the standard deviation.

Performance Considerations

Next, we examine the estimated baseline execution plan for this query.

Please refer to Figure 11-32.

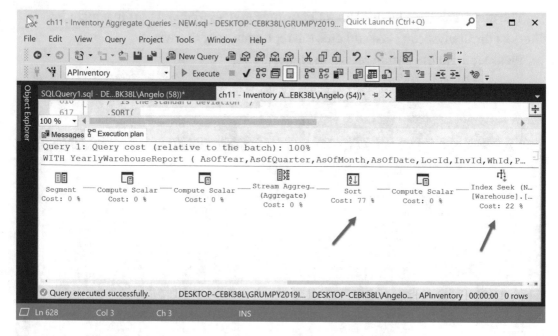

Figure 11-32. *Baseline estimated query plan*

We have a covering index, so we see an index seek with a cost of 22%. And as expected the usual expensive sort chimes in at 77%. An index was not recommended by the estimated query execution plan, so this query is ready for production.

As was the case with prior queries, a solid-state drive would mitigate the expensive sorts. I encourage you to try creating your own indexes to see if they can get the sort down, but 99% of the time, the estimated query plan tool gets it right, so to increase performance you need to consider physical improvements like faster disk or memory-enhanced tables.

Last but not least, we might as well check the TIME and IO statistics.

Please refer to Figure 11-33.

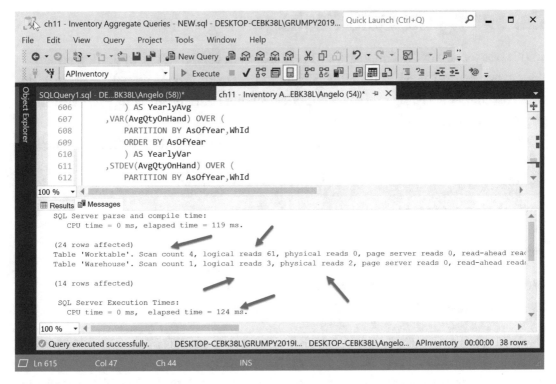

Figure 11-33. IO and TIME statistics

Our SQL Server execution time is 124 ms. The work table has the highest statistics with 4 scans and 61 logical reads. This was due to the high sort task.

The Warehouse table has a scan count value of 1, 3 logical reads, and 2 physical reads. So this looks like a well-performing query.

Let's now turn our attention to another tool, SSIS, which is the premier ETL tool that is a part of the Microsoft SQL Server BI stack. SSIS stands for SQL Server Integration Services.

Note Throughout the book we have taken the CTE portion of a query and sometimes created a report table. Consider that in real production environments, this would be a good solution because the report table could be accessed by many different types of queries or SSRS reports. Each of these queries and reports could simultaneously be executed by multiple users. So, if the data is historical and can be at least one day old, eliminate the CTE and go with a heavily indexed report table.

Enhancing the SSIS Package

In this last section of the chapter, we will enhance our SSIS package by adding tasks to create indexes and a task to load the SSAS multidimensional cube we will be using in the next two chapters. Listing 11-10 is the TSQL script that contains the code to create the required indexes.

Each CREATE statement will go in a dedicated Execute SQL Task in the SSIS package.

Listing 11-10. TSQL Script for SSIS Execute TSQL Tasks

```
/******************/
/* CLUSTERED INDEX */
/******************/

DROP INDEX IF EXISTS pkInventoryHistory
ON APInventoryWarehouse.Fact.InventoryHistory
GO

CREATE UNIQUE CLUSTERED INDEX pkInventoryHistory
ON APInventoryWarehouse.Fact.InventoryHistory (LocKey,InvKey,WhKey,ProdKey,
CalendarKey)
GO

/*********************/
/* NON CLUSTERED INDEX */
/*********************/

DROP INDEX IF EXISTS ieProdProdTypeQty
ON APInventoryWarehouse.Fact.InventoryHistory
GO

CREATE NONCLUSTERED INDEX ieProdProdTypeQty
ON APInventoryWarehouse.Fact.InventoryHistory (ProdKey)
INCLUDE (ProductTypeKey,QtyOnHand)
GO

/***************/
/* Calendar Key */
/***************/
```

```
DROP INDEX IF EXISTS ieInvHistCalendar
ON APInventoryWarehouse. Fact.InventoryHistory
GO

CREATE NONCLUSTERED INDEX ieInvHistCalendar
ON APInventoryWarehouse. Fact.InventoryHistory (CalendarKey)
GO

/***************/
/* Location Key */
/***************/

DROP INDEX IF EXISTS ieInvHistLocKey
ON APInventoryWarehouse.Fact.InventoryHistory
GO

CREATE NONCLUSTERED INDEX ieInvHistLocKey
ON APInventoryWarehouse.Fact.InventoryHistory (LocKey)
GO

/****************/
/* Inventory Key */
/****************/

DROP INDEX IF EXISTS ieInvHistInvKey
ON APInventoryWarehouse.Fact.InventoryHistory
GO

CREATE NONCLUSTERED INDEX ieInvHistInvKey
ON APInventoryWarehouse.Fact.InventoryHistory (InvKey)
GO

/****************/
/* Warehouse Key */
/****************/

DROP INDEX IF EXISTS ieInvHistWHKey
ON APInventoryWarehouse.Fact.InventoryHistory
GO
```

```
CREATE NONCLUSTERED INDEX ieInvHistWHKey
ON APInventoryWarehouse.Fact.InventoryHistory (WHKey)
GO

/**************/
/* Product Key */
/**************/

DROP INDEX IF EXISTS ieInvHistProdKey
ON APInventoryWarehouse.Fact.InventoryHistory
GO

CREATE NONCLUSTERED INDEX ieInvHistProdKey
ON APInventoryWarehouse.Fact.InventoryHistory (ProdKey)
GO

/********************/
/* Product Type Key */
/********************/

DROP INDEX IF EXISTS ieInvHistProdTypeKey
ON APInventoryWarehouse.Fact.InventoryHistory
GO

CREATE NONCLUSTERED INDEX ieInvHistProdTypeKey
ON APInventoryWarehouse.Fact.InventoryHistory (ProductTypeKey)
GO

UPDATE STATISTICS Fact.InventoryHistory
GO
```

As you can see, these are standard commands for creating indexes. We will copy and paste each individual CREATE INDEX command into its own Execute SQL Task. We could include all in one Execute SQL Task, but it is best to break them up, so in case there is an error, it will be easy to isolate and correct.

I included a DROP INDEX command in case the index already exists so it can be rebuilt. Also, the last command is an UPDATE STATISTICS command to make sure the statistics on the InventoryHistory table are up to date after all the indexes have been created. An example of table statistics is how many rows are loaded in the table, so make sure to keep these up to date.

Tip   It is always good policy to update statistics when you create or rebuild indexes; INSERT, UPDATE, or DELETE rows; or TRUNCATE tables.

We will not discuss the tasks for creating each individual index because the steps are identical to the ones we discussed, which were used to insert rows from the transactional inventory database to the inventory warehouse. These are identical except instead of INSERT commands, we are using CREATE INDEX commands just discussed.

For example, refer to Figure 11-34.

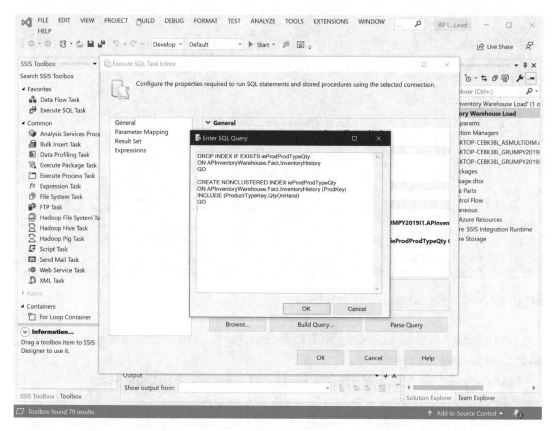

Figure 11-34. *Execute SQL Task for creating a clustered index*

This is typical of all the Execute SQL Tasks in the package we are discussing. Here we are creating a clustered index on the `InventoryHistory` table. Again, notice the table names are fully qualified. This is important if the query you are using refers to objects in different databases. You can only use one connection in Execute SQL Tasks, so qualify every table, index, etc.

All the other Execute SQL Tasks are similar to this except for the different indexes being created.

Figure 11-35 is a screenshot of the semi-completed SSIS package.

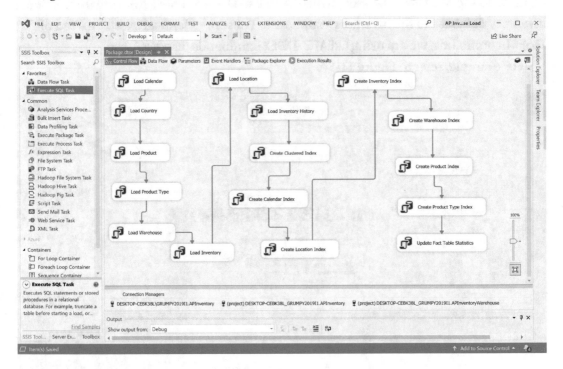

Figure 11-35. *Enhanced SSIS package*

This is what we want to end up with as far as adding all the Execute steps for building all the indexes we discussed.

Lastly, we also want to add a task to build and redeploy an existing SSAS multidimensional cube. Let's discuss how this is accomplished.

We need to add a task for processing the multidimensional cube we built in one of our chapters.

We drag it from the Common task area in the SSIS Toolbox and place it at the end of the control flow all the way on the lower right-hand side of the Design area.

Please refer to Figure 11-36.

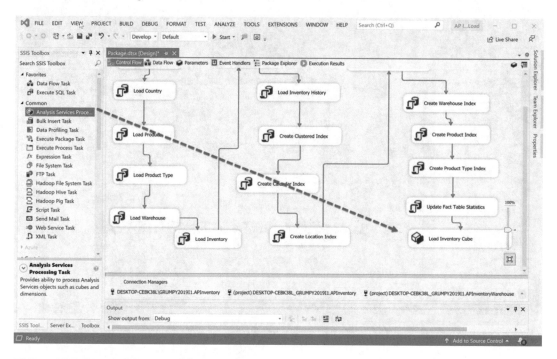

Figure 11-36. *Adding an SSAS task*

Now we need to configure it. The first step is to create and establish a connection to the SSAS server. Double-clicking the task presents the connection dialog box.

Please refer to Figure 11-37.

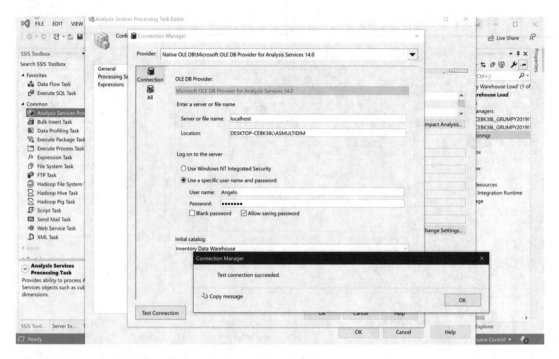

Figure 11-37. *Defining connection to an SSAS server*

A name for the SSAS server is filled out, in this case "localhost," although a better name would have been nicer, like Inventory Cube Server. Next, we add the actual physical name of the SSAS in the Location text box. Lastly, we supply username and password credentials and test the connection. Looks familiar, doesn't it? Same concept as when we defined a connection to SQL Server and a database.

Next, we need to identify the properties and objects we need to process. Click "Processing Settings," and the following panel is presented.

Please refer to Figure 11-38.

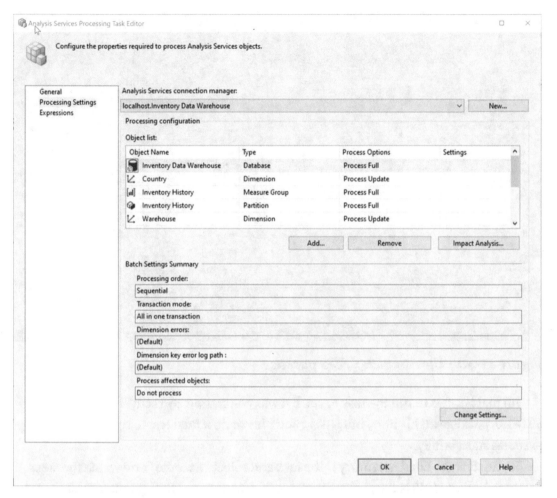

Figure 11-38. *Identifying objects to process*

Add or remove any objects (fact tables and dimensions). Also place your mouse over any process option list box to change the settings.

Click the OK button. This should complete the configuration, and the task is added at the end of the control flow as shown in Figure 11-39.

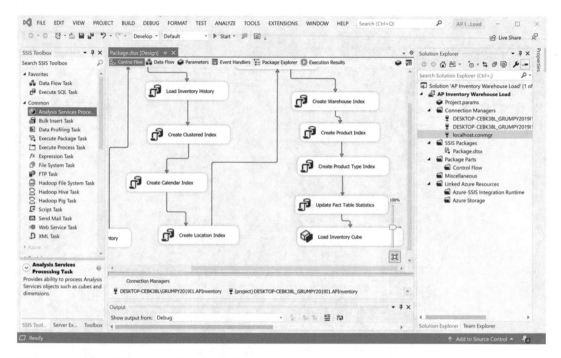

Figure 11-39. *The completed SSIS package*

All that is left is to run the task to see if it was configured correctly. No need to run all of the package at this time, but it is a good idea to do a final test to see that everything executes successfully.

Right-click the Load Inventory Cube task and select "Execute" and watch the steps progress. Figure 11-40 is the process running. You cannot see the spinning circle, but you will see it when you try this on your own.

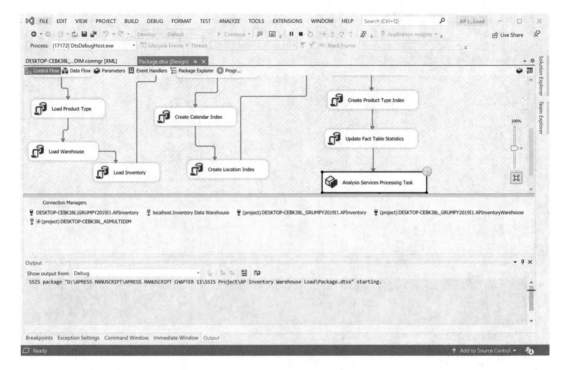

Figure 11-40. *Running the completed SSIS package*

If everything ran without error, you would see a green checkmark on the task indicating success as can be seen in Figure 11-41.

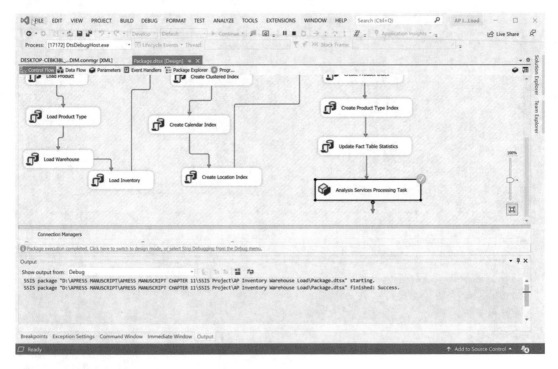

Figure 11-41. *The new SSAS task ran to completion*

This is what you want to see; any red **X** marks indicate something is wrong, so you need to address them. In this simple package, it is either an incorrect connection or security information or a syntax error in the query.

That's it. You can now add SSIS as one of your tools in your SQL Server BI toolkit. Let's wrap up this chapter.

Tip   I encourage you to download the latest community edition of Visual Studio and add the SSIS and data tools/components. At the time of this writing, it should be Visual Studio 2022 Community Edition.

Learn the basics, build something, break it, and fix it. That is how one learns!

Summary

This was our first chapter in a series of three chapters that deal with inventory management. We checked out some simple business data models so we could become familiar with the databases we used in the queries.

We also continued to learn about SSIS, Microsoft's premier ETL tool that is used to create ETL processes to extract, transform, and load data from one source to the other. We used this tool to extract inventory movement data from a transactional database that stored inventory levels for products related to model trains and loaded the transaction data into a data warehouse so we could analyze it.

We went through our usual creation of queries that use aggregate window functions and performed our analysis using the estimated query plan tool and also the various performance statistics settings available with SQL Server.

Finally, we enhanced our SSIS process: called a package to not only load the tables but to also create indexes and process an SSAS multidimensional cube.

See you in the next chapter where we check out the ranking window functions against our inventory databases.

CHAPTER 12

Inventory Use Case: Ranking Functions

Welcome to the third to last chapter of the book. I hope you are enjoying the journey. This time we launch the four ranking window functions against our inventory database and the inventory data warehouse. We will perform the usual analysis of the results and do some performance analysis, but we will also go through the steps to create a web report with Microsoft Report Builder and publish it to a website implemented with SSRS (SQL Server Reporting Services), Microsoft's premiere reporting architecture.

But there is more. We will also go through the steps of creating a simple Power BI report and dashboard that will deliver some pretty neat charts and reports. We will publish these to a personal SSRS server website and a Power BI server website that runs on your laptop or workstation.

Lastly, this chapter will include a large number of figures as I feel that one can learn from diagrams, charts, and screen prints even better than reading text (well, almost). A picture is worth a thousand words! Easy to follow, too, as you are trying these tools out.

As far as the window functions are concerned, we will proceed in the usual manner and define the business specifications, create and run the query, analyze the data, and then generate estimated query plans and IO and TIME statistics in order to analyze query performance.

After all, we are following a cookbook approach to learn the window functions, so we follow the recipe just described. By the way there are nine code listings but 59 figures and one diagram. We have a lot of material to cover.

© Angelo Bobak 2023
A. Bobak, *SQL Server Analytical Toolkit*, https://doi.org/10.1007/978-1-4842-8667-8_12

Ranking Functions

Here are our four functions in this category:

- RANK(): **Does not** support ROWS and RANGE specifications

- DENSE_RANK(): **Does not** support ROWS and RANGE specifications

- NTILE(): **Does not** support ROWS and RANGE specifications

- ROW_NUMBER(): **Does not** support ROWS and RANGE specifications

The first two are used to rank results in partitions, and the third lets you define buckets in the data returned by a query. The NTILE() function works well in categorization schemes.

The last function, ROW_NUMBER(), is our workhorse function used to generate row numbers within the entire data set or partitions within the data set. We will use it for a different purpose though. This function can be used to generate random data with some tricky logic.

None of these functions support the ROW and RANGE frame specifications, and if you understand what these functions do, it makes sense.

We will start off by performing our usual analysis of our ranking functions. That is, we look at the business specifications submitted by our friendly business analyst, create a script to support the requirements, and run the script to check out the results. Next, we generate estimated query plans as we have been doing in order to analyze and, if possible, improve the performance of the queries. Of course, we will also checkout IO, TIME, and PERFORMANCE STATISTICS to further dive deeper into the execution steps and costs.

Let's start by discussing the RANK() function that we will apply our inventory data to.

RANK() Function

Recall that the RANK() function returns the number of rows before the current row plus the current row based on the value being ranked. Let's see what the business analyst's specifications are:

The analyst wants to see some statistical data on the movement out of inventory.

Specifically, the report needs to assign a rank of movements for product "P209" in warehouse "WH111," inventory "INV1," and location "LOC1." The report needs to spell out the calendar quarter and month names as opposed to just displaying numbers.

Initially this report needs to be for the year 2010, but as usual, once the report is bug-free, it needs to be easily modified to support multiple years, products, etc. and even used for SSRS reports or Power BI scorecards.

By the way product "P209" is a "Swiss type 1 passenger car" for any model train buffs out there.

Here is the CTE-based query that will address this requirement.

Please refer to Listing 12-1.

Listing 12-1. Ranking Inventory Movement Out for Swiss Passenger Car Models

```
WITH MonthlyInventoryMovement (
      MovementYear,MovementQuarter,MovQtrName,MovementMonth,MovMonthName
      ,InvId,LocId,WhId,ProdId,MonthlyDecrementMovement
)
AS
(
SELECT YEAR(MovementDate) AS MovementYear
      ,DATEPART(qq,MovementDate) AS MovementQuarter
      ,CASE
            WHEN DATEPART(qq,MovementDate) = 1 THEN '1st Quarter'
            WHEN DATEPART(qq,MovementDate) = 2 THEN '2nd Quarter'
            WHEN DATEPART(qq,MovementDate) = 3 THEN '3rd Quarter'
            WHEN DATEPART(qq,MovementDate) = 4 THEN '4th Quarter'
      END AS MovQtrName
      ,MONTH(MovementDate) AS MovementMonth
      ,CASE
            WHEN MONTH(MovementDate)  = 1 THEN 'Jan'
            WHEN MONTH(MovementDate)  = 2 THEN 'Feb'
            WHEN MONTH(MovementDate)  = 3 THEN 'Mar'
            WHEN MONTH(MovementDate)  = 4 THEN 'Apr'
            WHEN MONTH(MovementDate)  = 5 THEN 'May'
            WHEN MONTH(MovementDate)  = 6 THEN 'June'
            WHEN MONTH(MovementDate)  = 7 THEN 'Jul'
```

```
              WHEN MONTH(MovementDate)   = 8 THEN 'Aug'
              WHEN MONTH(MovementDate)   = 9 THEN 'Sep'
              WHEN MONTH(MovementDate)   = 10 THEN 'Oct'
              WHEN MONTH(MovementDate)   = 11 THEN 'Nov'
              WHEN MONTH(MovementDate)   = 12 THEN 'Dec'
       END AS MovMonthName
      ,InvId,LocId,WhId,ProdId
      ,SUM(Decrement) AS MonthlyDecrementMovement
FROM Product.InventoryTransaction
GROUP BY YEAR(MovementDate)
      ,DATEPART(qq,MovementDate)
      ,MONTH(MovementDate)
      ,InvId
      ,LocId
      ,WhId
      ,ProdId
)
SELECT MovementYear
      ,MovementQuarter
      ,MovQtrName
      ,MovementMonth
      ,MovMonthName
      ,LocId
      ,InvId
      ,WhId
      ,ProdId
      ,MonthlyDecrementMovement
      ,RANK() OVER (
            PARTITION BY MovementYear
            ORDER BY MonthlyDecrementMovement DESC
            ) AS DecRank
FROM MonthlyInventoryMovement
WHERE MovementYear = 2010
AND LocId = 'LOC1'
AND InvId = 'INV1'
```

```
AND WhId = 'WH111'
AND ProdId = 'P209'
GO
```

The first thing we notice is the two large CASE statements in the SELECT clause. As highlighted in the business requirements, the business analyst wants to see the names of the quarters and months spelled out, so we need this logic to decode the numerical values for these data parts. Our Calendar table is fairly simple, so a definite performance improvement would be to add the text versions of quarters and months so we could improve performance by avoiding CASE blocks in the SELECT clause.

Homework   Using two ALTER and one UPDATE commands, add columns to support the text version of calendar quarters and calendar months. Implement this modification in the APInventory.MasterData.Calendar table and the APInventory.Dimension.Calendar dimension table.

Next, let's check out the OVER() clause. We see a PARTITION BY clause that references the MovementYear column and an ORDER BY clause that references the MonthlyDecrementMovement column. The values are sorted in descending order.

Remember that we do not need to refer to the year column in the PARTITION BY clause as we have a WHERE clause that filters results for the year 2010 only. Our analyst will want to see data for all years, so this will come in handy when this request is made. Having it there now does not seem to affect performance. Try it on your own, and run a baseline estimated query plan with the column and then without the column. Don't forget to run the DBCC command for each test.

Once all the years are available (if we take out the filter in the WHERE clause), this query would generate results that would make it an ideal candidate for an SSRS report created with Report Builder. All the columns that were in the WHERE clause would be used for report filters.

Let's check out the results. Please refer to the partial results in Figure 12-1.

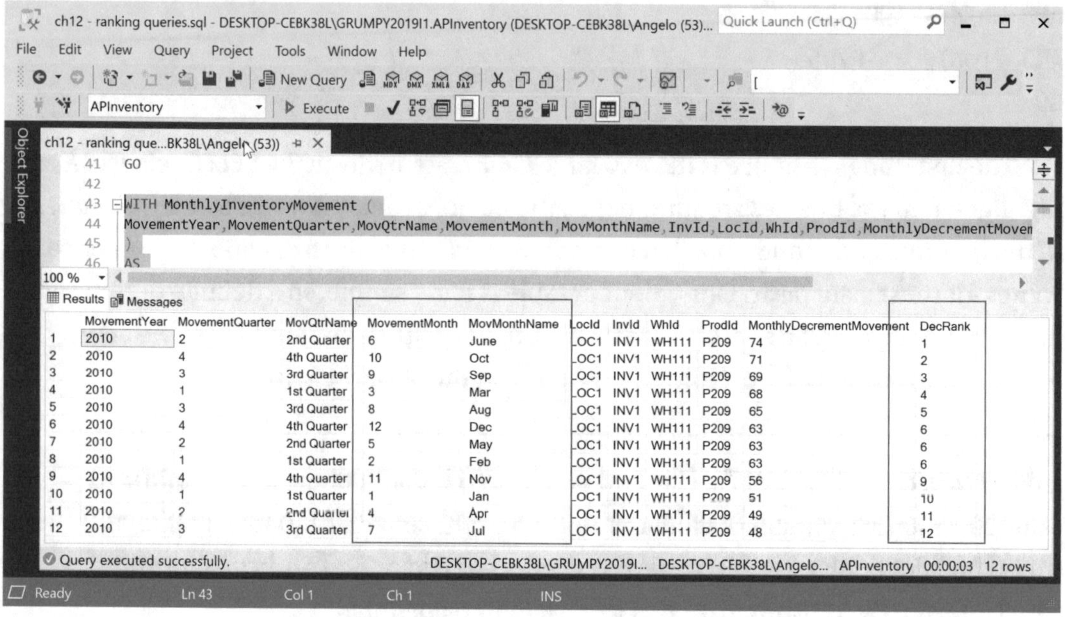

Figure 12-1. Inventory movement out rank report

Looking at the report, we see that June had the highest number of items moving out of inventory, which translates into sales, so this is a good thing. July has the lowest rank for items moving out. This would translate into diminishing sales. As a follow-up, the analyst would see how many items came back into inventory to make sure the inventory manager is not over-ordering items.

Let's copy the results into a Microsoft Excel spreadsheet and create a line chart to see the rank and inventory movement trend. Data reports and graphs are optimal tools to analyze data.

Please refer to Figure 12-2.

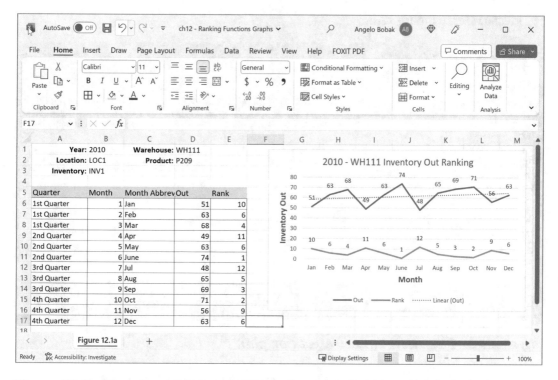

Figure 12-2. *Inventory movement out rank report chart*

Looking at the graph, we see some swings, but the trend line (dashed line) is moving up, so sales are decent. We would not want to see the movement of products going out trending down, which would indicate that the products are staying in inventory. This graph will most likely be in grayscale or black and white in the book, but if you are viewing this as an eBook, the colors are a helpful aid in analyzing the data.

Let's see how this query performs.

Performance Considerations

Let's run our usual baseline execution plan and see if any indexes are needed. We will also look at some IO and TIME statistics to really understand how the query performs. Remember when you try this on your own to also generate a live query plan so you can compare the costs between the estimated plan and the live plan. Any major differences indicate a problem possibly caused by stale table statistics.

Pop Quiz   How do you update table statistics?

Please refer to Figure 12-3.

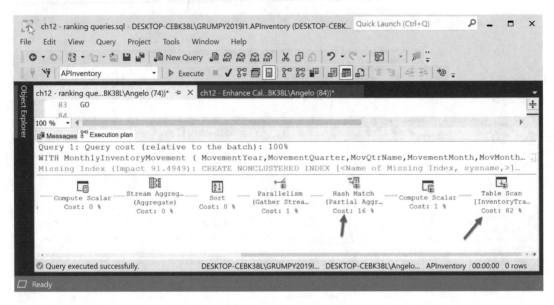

Figure 12-3. *Baseline estimated execution plan*

We see that an index is suggested so we will need to create it. Starting from right to left, we see a table scan with a cost of 82%, a compute scalar with a cost of 1%, and a hash match join task with a cost of 16%. Finally, there's a parallelism (gather stream) task with a cost of 1%, and all other costs are 0%. Clearly some improvement is needed so let's create the recommended index.

Please refer to Listing 12-2.

Listing 12-2. Suggested Index for the Inventory Transaction Table

```
CREATE NONCLUSTERED INDEX ieInvIdLocIdWhIdProdIdDecDate
ON Product.InventoryTransaction (InvId,LocId,WhId,ProdId)
INCLUDE (Decrement,MovementDate)
GO

UPDATE STATISTICS Product.InventoryTransaction
GO
```

If we compare the columns in the suggested index with the columns in the query's WHERE clause and OVER() clause, we see that the columns are covered (this is called a covering index).

The InvId, LocId, WhId, and ProdId columns appear in the WHERE clause, so they need to be included in the index. The index also has an INCLUDE predicate, which calls out the Decrement and MovementDate columns, which cover the columns mentioned in the PARTITION BY and ORDER BY clauses.

So we are covered! Pun intended, sorry.

We can conclude then, based on experience from the prior chapter examples, that with these types of window-based queries, the rule of thumb is to make sure we have an index based on columns in the OVER() clause and any columns appearing in the WHERE clause.

Let's create the index and check out the new estimated query plan.

Please refer to Figure 12-4.

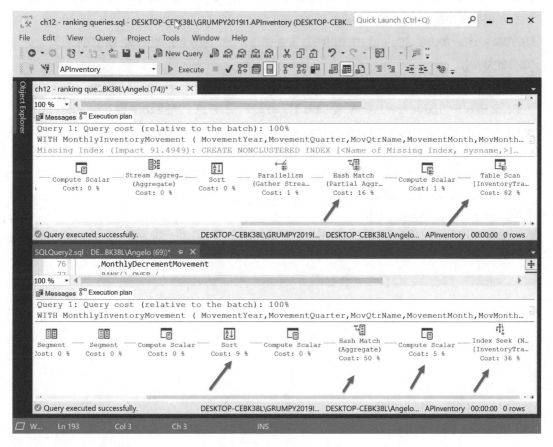

Figure 12-4. *Comparing before and after estimated execution plans*

This screenshot compares the before and after estimated execution plans. Familiar pattern, anyone? An index seek task replaced the table scan, but the cost of both the hash match and sort went up.

Always a trade-off.

Also, the sort step went from 0% to 9%, so this step and the hash match should be the focus of any further performance tuning. I said it before and I will say it again: make sure TEMPDB is implemented on a very fast drive or else a magnetic disk drive.

Side Comment   We are looking at performance tuning from the query and covering index point of view. But performance tuning involves tuning vertical hardware scaling and horizontal hardware scaling. Vertical scaling means we look at our server hardware and see if we need more fast disks, more CPUs, or more memory. Horizontal scaling means adding more servers for parallel processing (expensive).

Back to our analysis. A last suggestion is to get rid of the CASE blocks by enhancing the Calendar table to include the text versions of the quarter and month data parts. If you did the homework I suggested earlier, modify this query so it just uses the new calendar date part columns and see what sort of improvement is realized.

Let's look at the IO and TIME statistics next.

Please refer to Figure 12-5.

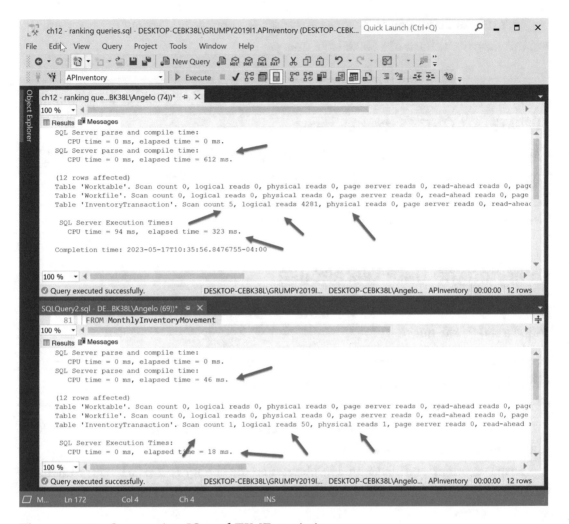

Figure 12-5. *Comparing IO and TIME statistics*

As expected, performance statistics improved. SQL Server parse and compile times went down. SQL Server execution time went down from 612 ms to 18 ms. Good improvement. Focusing on the InventoryTransaction table, scan count went down from 5 to 1, logical reads went down from 4,281 to 50, but physical reads went up from 0 to 1.

Overall major improvements. Based on this analysis, the suggested improvements are to try out the Calendar table changes and to also replace the CTE logic with a preloaded report table. Try this out on your own, perform the analysis, and see if performance improves or not. The added twist is that you will have to modify the CTE query so it joins the Calendar table so you can use the new columns. Will this be better? Will you need a new index on the Calendar table?

Let's turn our attention to the inventory data warehouse and try out the same query but this time in the form of a SNOWFLAKE query.

Pop Quiz Answer   How do you update table statistics? By running the command

```
UPDATE STATISTICS <table name>
```

Make sure to do this often, especially after many INSERT, UPDATE, and DELETE queries. You might even want to add it to a SQL Agent task that runs periodically and updates all the statistics for all the tables in your database.

Querying the Data Warehouse

Let's now query the fact table in our inventory data warehouse with a similar query called a SNOWFLAKE query.

Pop Quiz   Why is this called a SNOWFLAKE query? Answer coming up shortly.

The business requirements for this query are the same as the prior query, so I will not elaborate further on these.

Based on our prior discussion, I implemented the change to the Calendar dimension table, so it includes text columns to support quarter and month names. This should speed things up.

The query is still CTE based, and because it is executed against a SNOWFLAKE schema, it contains many table joins. The listing is fairly long, but I suggest you walk through it to understand all the logic and fine points.

Please refer to Listing 12-3.

Listing 12-3. SNOWFLAKE Query for Ranking Report

```
WITH WarehouseCTE (
     InvYear,InvMonthNo,InvMonthMonthName,LocationId,LocationName
          ,InventoryId,WarehouseId,WarehouseName,ProductId,ProductName
     ,ProductType,ProdTypeName,MonthlyQtyOnHand
)
AS ( -- returns 24,192 rows
```

```
SELECT C.CalendarYear        AS InvYear
       ,C.CalendarMonth      AS InvMonthNo
       ,C.CalendarTxtMonth   AS InvMonthMonthName
       ,L.LocId              AS LocationId
       ,L.LocName            AS LocationName
       ,I.InvId              AS InventoryId
       ,W.WhId               AS WarehouseId
       ,W.WhName             AS WarehouseName
       ,P.ProdId             AS ProductId
       ,P.ProdName           AS ProductName
       ,P.ProdType           AS ProductType
       ,PT.ProdTypeName      AS ProdTypeName
      ,SUM(IH.[QtyOnHand]) AS MonthlyQtyOnHand
FROM [Fact].[InventoryHistory] IH -- WITH (INDEX(ieInvHistCalendar))
JOIN [Dimension].[Location] L
     ON IH.LocKey = L.LocKey
JOIN [Dimension].[Calendar] C
     ON IH.CalendarKey = C.CalendarKey
JOIN [Dimension].[Warehouse] W
     ON IH.WhKey = W.WhKey
JOIN [Dimension].[Inventory] I
     ON IH.[InvKey] = I.InvKey
JOIN [Dimension].[Product] P
     ON IH.ProdKey = P.ProdKey
JOIN [Dimension].[ProductType] PT
     ON P.ProductTypeKey = PT.ProductTypeKey
GROUP BY C.CalendarYear
        ,C.CalendarMonth
        ,C.CalendarTxtMonth
        ,L.LocId
        ,L.LocName
        ,I.InvId
        ,W.WhId
        ,W.WhName
        ,P.ProdId
        ,P.ProdName
```

```
        ,P.ProdType
        ,PT.ProdTypeName
)
SELECT InvYear,InvMonthNo,InvMonthMonthName,LocationId,LocationName
        ,InventoryId,WarehouseId, ProductType,ProductId,ProdTypeName
        ,MonthlyQtyOnHand
        ,RANK() OVER (
            ORDER BY MonthlyQtyOnHand DESC
            ) QtyOnHandRank
FROM WarehouseCTE
WHERE InvYear= 2010
AND LocationId = 'LOC1'
AND InventoryId = 'INV1'
AND WarehouseId = 'WH111'
AND ProductId = 'P041'
GO
```

This query has it all: a CTE, multiple joins in the form of a SNOWFLAKE query, and a WHERE clause.

Notice that I removed the PARTITION BY clause for this version of the query as the WHERE clause filter limits the results to a single year, location, inventory, warehouse, and product. All that is required is to include the ORDER BY clause that refers to the monthly sum of quantity on hand for this inventory location. Let's check out the results.

Please refer to the partial results in Figure 12-6.

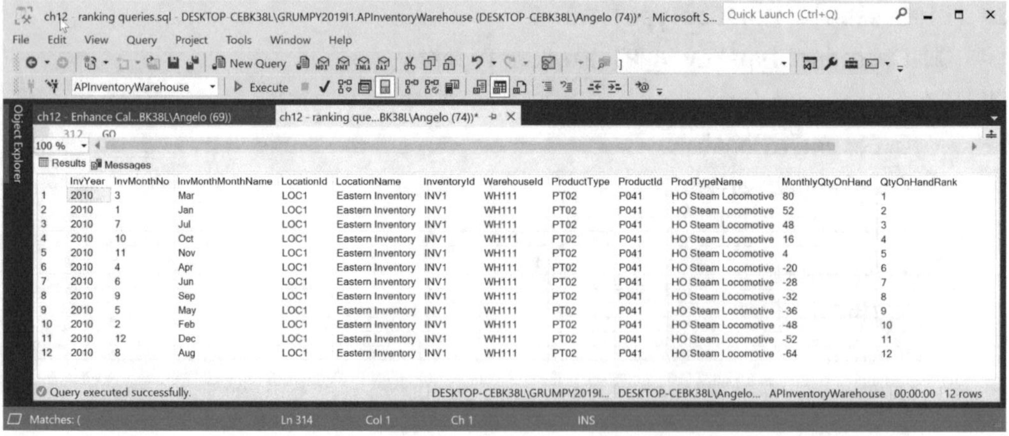

	InvYear	InvMonthNo	InvMonthMonthName	LocationId	LocationName	InventoryId	WarehouseId	ProductType	ProductId	ProdTypeName	MonthlyQtyOnHand	QtyOnHandRank
1	2010	3	Mar	LOC1	Eastern Inventory	INV1	WH111	PT02	P041	HO Steam Locomotive	80	1
2	2010	1	Jan	LOC1	Eastern Inventory	INV1	WH111	PT02	P041	HO Steam Locomotive	52	2
3	2010	7	Jul	LOC1	Eastern Inventory	INV1	WH111	PT02	P041	HO Steam Locomotive	48	3
4	2010	10	Oct	LOC1	Eastern Inventory	INV1	WH111	PT02	P041	HO Steam Locomotive	16	4
5	2010	11	Nov	LOC1	Eastern Inventory	INV1	WH111	PT02	P041	HO Steam Locomotive	4	5
6	2010	4	Apr	LOC1	Eastern Inventory	INV1	WH111	PT02	P041	HO Steam Locomotive	-20	6
7	2010	6	Jun	LOC1	Eastern Inventory	INV1	WH111	PT02	P041	HO Steam Locomotive	-28	7
8	2010	9	Sep	LOC1	Eastern Inventory	INV1	WH111	PT02	P041	HO Steam Locomotive	-32	8
9	2010	5	May	LOC1	Eastern Inventory	INV1	WH111	PT02	P041	HO Steam Locomotive	-36	9
10	2010	2	Feb	LOC1	Eastern Inventory	INV1	WH111	PT02	P041	HO Steam Locomotive	-48	10
11	2010	12	Dec	LOC1	Eastern Inventory	INV1	WH111	PT02	P041	HO Steam Locomotive	-52	11
12	2010	8	Aug	LOC1	Eastern Inventory	INV1	WH111	PT02	P041	HO Steam Locomotive	-64	12

***Figure 12-6.** SNOWFLAKE query results*

Looks like it worked. It returned 12 rows in under one second. This table has 736,320 rows, so I think we can be impressed with the performance so far. Let's check out our baseline estimated query plan for this query.

Please refer to Figure 12-7.

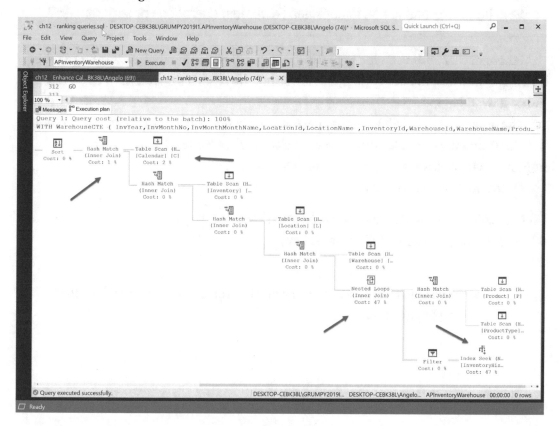

Figure 12-7. *Baseline estimated query plan*

Look at all those joins!

Starting from the lower-right bottom branch, we see an index seek on the fact table, so we are off to a good start. There is a high-cost nested loops join task between the Product dimension and the Product Type outrigger dimension, but these tables are small in terms of row count, so this is not an issue. We could consider denormalizing the Product table and adding the product type columns in order to eliminate this join. Something to consider.

Note   Combining the outrigger table with the main dimension will eliminate joins but on the other hand will make the dimension table larger, which means fewer rows can be fitted in a memory page when the query is processed. So there are always trade-offs one needs to consider. Fewer join operations or larger tables?

I thought that adding an index to the `Calendar` dimension table and adding a table hint might improve the query plan, but it was not the case. If the index is a clustered index, it would help.

The other hash joins all have zero cost so we will not concern ourselves with them.

Our friend the `Calendar` dimension clocks in with a 2% table scan. Can this be improved with a clustered index on the surrogate key?

Check out the following code:

```
CREATE UNIQUE CLUSTERED INDEX ieCalendarKey
ON [Dimension].[Calendar](CalendarKey)
GO
```

Let's create this clustered index and look at a revised estimated execution plan. Please refer to Figure 12-8.

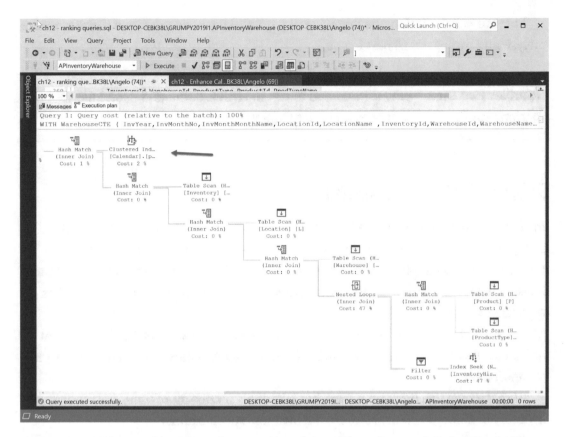

Figure 12-8. *Revised estimated execution plan with a clustered Calendar index*

I am happy to say this worked. A table hint was not needed because the clustered index was used.

We see a clustered index scan on the Calendar table. This dimension has 7,670 rows, so we can come up with a conclusion for a table of this size. That is, create a clustered index on the surrogate key, and most likely it will be used!

Pop Quiz Answer   Why is this called a SNOWFLAKE query?

Answer   Because it joins the dimension and outrigger tables that makes the model look like a SNOWFLAKE. If a database is implemented as a STAR schema, then the query against it would be a STAR query. Makes sense?

For your consideration, the following pop quiz is so you do not forget if performance analysis and tuning are all new to you.

Another Pop Quiz   Why do we need to run the DBCC command each time we perform performance analysis and query tuning?

DENSE_RANK() Function

Recall from Chapter 4 that the DENSE_RANK() function works almost the same as the RANK() function. In case of ties, the dense rank value is the same for the ties, but the next value after the duplicate values is simply assigned the next rank number. So, if the dense rank for the ties is 4, the next rank assigned to the next higher value is 5. Makes sense to me.

Not so for the RANK() function results. If you have five tie values of 6 and the current rank for the ties is 4, then the next rank is 9 (the 5 tie values + the current rank value 4 = 9). Strange indeed.

We have seen this function before, but let's apply it once more time, this time to our inventory business scenario. Here are the business specifications:

Basically, the business analyst likes the prior report we gave them. This time they want to create the same report but add the DENSE_RANK() function in order to see how ties are treated by each function.

This is an easy requirement to fill as we just copy and paste the prior query and add the new ranking window function. TSQL is all about code reusability! Why reinvent the wheel?

Here is what we came up with.

Please refer to Listing 12-4.

Listing 12-4. Comparing Ranking Between RANK() and DENSE_RANK()

```
WITH MonthlyInventoryMovement (
MovementYear,MovementQuarter,MovementMonth,InvId,LocId,WhId
     ,ProdId,MonthlyDecrementMovement
)
AS
(
```

```
SELECT YEAR(MovementDate)         AS MovementYear
      ,DATEPART(qq,MovementDate) AS MovementQuarter
      ,MONTH(MovementDate)        AS MovementMonth
      ,InvId
      ,LocId
      ,WhId
      ,ProdId
      ,SUM(Decrement) AS MonthlyDecrementMovement
FROM Product.InventoryTransaction
GROUP BY YEAR(MovementDate)
      ,DATEPART(qq,MovementDate)
      ,MONTH(MovementDate)
      ,InvId
      ,LocId
      ,WhId
      ,ProdId
)
SELECT MovementYear
      ,MovementQuarter
      ,MovementMonth
      ,LocId
      ,InvId
      ,WhId
      ,ProdId
      ,MonthlyDecrementMovement
      ,RANK() OVER (
            PARTITION BY MovementYear
            ORDER BY MonthlyDecrementMovement DESC
            ) AS DecRank
      ,DENSE_RANK() OVER (
            PARTITION BY MovementYear
            ORDER BY MonthlyDecrementMovement DESC
            ) AS DecDEnseRank
FROM MonthlyInventoryMovement
WHERE MovementYear = 2010
```

```
AND LocId = 'LOC1'
AND InvId = 'INV1'
AND WhId = 'WH111'
AND ProdId = 'P209'
GO
```

Nothing to go over as we have seen the query. The partition is set up for the movement year, and the ORDER BY clause orders the monthly decrement sums in descending order. The WHERE clause can be removed when this query is used to fill up a report table with filters that allow business users the capability of selecting what values to use.

Note   In my opinion, the coding style I used to list each column on a single line makes the query easy to understand. That's just my opinion. I know it takes up more space, especially in a book like this!

Back to our discussion. What we are interested in seeing is how the ties are handled in an inventory business scenario. Let's look at the results.

Please refer to the partial results in Figure 12-9.

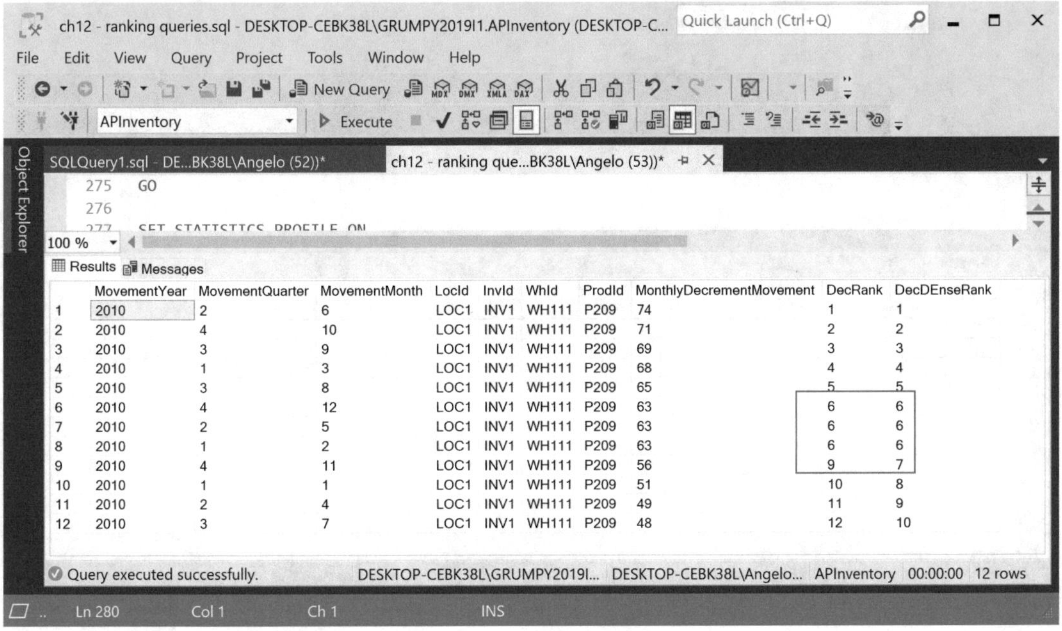

Figure 12-9. *Rank vs. dense rank*

The ties in the decrement values occur for the months of February (2), May (5), and December (12). In my opinion, and I stated this before, the DENSE_RANK() values make more sense as the tie values have, well, the same rank value. So we can state that for the months of February, May, and December, inventory decreased by 63 units for product "P209."

Skipping two rank values in the case of the RANK() function is confusing. Why do we want to use this?

I have not seen a reason to use this when I researched the Internet as to why this function should be used and under what circumstances. If you know, please send me an email!

Let's move on to performance analysis and tuning for this function.

Performance Considerations

Figure 12-10 is our baseline estimated query plan.

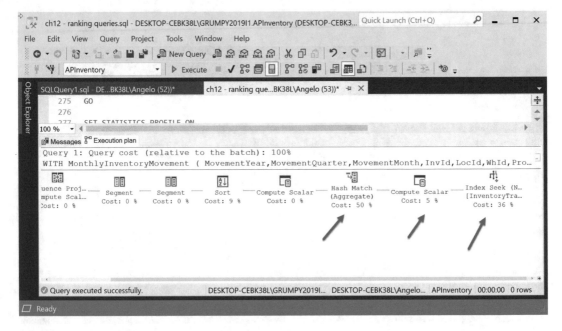

Figure 12-10. *Estimated baseline query plan*

Seems like the indexes we built are paying off. The index seek comes in at a cost of 36%. Next are a compute scalar with a cost of 5% and a hash match join task with a cost of 50%. Lastly, a sort task is at 9%.

I think we need to look at the STATISTICS PROFILE values to see what exactly is going on in these three tasks. Let's list the expression used in these tasks so we know what to look at:

Compute Scalar: Expr 1003, Expr 1004, Expr 1005

Hash Match: Expr 1004, Expr 1005

Sort: Expr 1004, Expr 1005, Expr 1006

So what logic is used in these expressions?
Please refer to Figure 12-11.

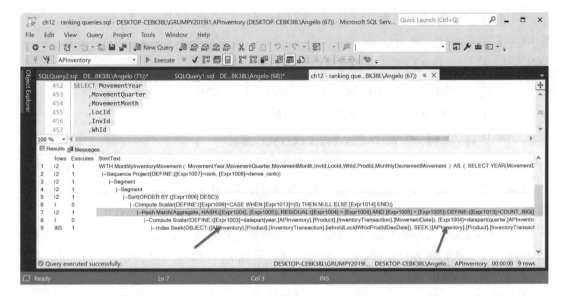

Figure 12-11. *Statistics profile for the RANK() vs. DENSE_RANK() Query*

This set of statistics really gets into the detail of how the query is processed and where the problem spots might be. If we look at the statement text column, we can see what is going on.

Not shown in the figure is Expr1003 but here are the values:

> [**Expr1003**]=datepart(year,[APInventory].[Product].
> [InventoryTransaction].[MovementDate]),

Here are expressions Expr1004 and Expr1005:

> [**Expr1004**]=datepart(quarter,[APInventory].[Product].
> [InventoryTransaction].[MovementDate])

> [**Expr1005**]=datepart(month,[APInventory].[Product].
> [InventoryTransaction].[MovementDate])

This tells us that the DATEPART() function calls are contributing to the high cost for the hash match task. We can resolve this by using the Calendar table to retrieve these values although we will incur the cost of a JOIN on the Calendar table (the clustered index we created will help).

Here is Expr1006 used in the sort task and compute scalar task:

> [**Expr1006**]=CASE WHEN [Expr1013]=(0) THEN NULL ELSE
> [Expr1014] END))

Looks like this is for a CASE code block? Clearly, pulling out those date parts is cranking up costs in these tasks. But wait. We need to take a look at Expr1013 and Expr1014. We are really getting into the weeds here:

> [**Expr1013**]=COUNT_BIG([APInventory].[Product].
> [InventoryTransaction].[Decrement])

> [**Expr1014**]=SUM([APInventory].[Product].
> [InventoryTransaction].[Decrement])

Looks like these last two steps are used to calculate the sum of decrements in the SELECT clause of the CTE and also the count so the rank can be calculated.

So this was a good performance analysis session. We started off with the estimated query plan, and then we moved on to the STATISTICS PROFILE statistics.

But what does all this tell us?

It tells us the performance hits are centered around the extraction of the year, quarter, and month parts and also the SUM() function in the CTE. Clearly the recommended solution is to replace the CTE logic with an INSERT/SELECT statement that loads a reporting table so we can isolate these performance costs once a day at off hours! Additionally, use the enhanced Calendar table so pulling out the date parts is not required each time the report table is loaded.

Homework   Based on this new knowledge, use the query in the CTE to create a report table. Then modify the base query so it refers to the report table vs. the CTE and perform the usual performance tuning analysis. Make sure you JOIN to the Calendar table to get the date parts in the logic that loads the report table. See if it is more efficient than using the DATEPART() function! Don't forget to create the clustered index on the Calendar table.

Here is the answer to the pop quiz.

Another Pop Quiz Answer   Why do we need to run the DBCC command each time we perform performance analysis and query tuning?

Answer   Because we want to clear memory cache that might contain old query plans. If you do not, your query might refer to objects already in memory, and your performance analysis results will be compromised. Always start with a clean slate.

NTILE( ) Function

Remember the definition of this function? The NTILE( ) function allows you to divide a set of rows in a data set into tiles or buckets (I like this term better). If you have a data set of 12 rows and you want to assign four tiles, each tile will have three rows. If you have a data set with 13 rows and want four tiles, the last tile will have more rows than the others. No way to control this.

Anyway, this function comes in handy when you want to create buckets of data to analyze inventory levels so as to categorize and prioritize what product needs to be reordered first.

Note   If you have an odd number of rows, like 15, and you decide to create four tiles, you will get tiles with four rows, four rows, four rows, and three rows. So there is no way to control what goes into a tile; it just divides them up.

Let's check out the business requirements submitted by our friendly business analyst:

Our business analyst wants to set up a classification scheme so that they can be alerted as to what products need to be refilled in inventory on a priority basis. The report needs to print out the following messages together with the year, month, location, inventory, warehouse, and of course products remaining:

- Bucket 1: Order First High Priority Alert

- Bucket 2: Order Second High Priority Alert

- Bucket 3: Order Third High Priority Alert

- Bucket 4: Order Fourth Medium Priority Alert

- Bucket 5: Order Fifth Low Priority Alert

The analyst wants to start off by generating a report for year "2005," location "LOC1," inventory "INV1," and warehouse "WH111." All products need to be included (I did not use the actual names in order to save space, but a user most likely will want to see names, not just codes).

Here is what we came up with to address this requirement.

Please refer to Listing 12-5.

Listing 12-5. Setting Reorder Priority Alerts

```
WITH InventoryReOrderAlert (
 InventoryYear,InventoryMonth,LocId,InvId,WhId,ProdId,ItemsRemaining
)
AS
(
SELECT YEAR(MovementDate)  AS InventoryYear
      ,MONTH(MovementDate) AS InventoryMonth
      ,IT.LocId
      ,IT.InvId
      ,IT.WhId
      ,IT.ProdId
      ,SUM(IT.Increment - IT.Decrement) AS ItemsRemaining
FROM APInventory.Product.InventoryTransaction IT
GROUP BY YEAR(MovementDate)
      ,MONTH(MovementDate)
      ,IT.LocId
      ,IT.InvId
      ,IT.WhId
      ,IT.ProdId
)
SELECT
      InventoryYear
      ,InventoryMonth
      ,LocId
      ,InvId
      ,WhId
      ,ProdId
```

```
    ,ItemsRemaining
    ,CASE
        WHEN NTILE(5) OVER(
            ORDER BY ItemsRemaining ASC
        ) = 1 THEN 'Order First High Priority Alert'
        WHEN NTILE(5) OVER(
            ORDER BY ItemsRemaining ASC
        ) = 2 THEN 'Order Second High Priority Alert'
        WHEN NTILE(5) OVER(
            ORDER BY ItemsRemaining ASC
        ) = 3 THEN 'Order Third High Priority Alert'
        WHEN NTILE(5) OVER(
            ORDER BY ItemsRemaining ASC
        ) = 4 THEN 'Order Fourth, Medium Priority Alert'
        WHEN NTILE(5) OVER(
            ORDER BY ItemsRemaining ASC
        ) = 5 THEN 'Order Fifth Low Priority Alert'
        END AS AlertMessage
FROM InventoryReOrderAlert
WHERE InventoryYear = 2005
  AND Locid = 'LOC1'
  AND InvId = 'INV1'
  AND WhId = 'WH111'
ORDER BY InventoryMonth
GO
```

We use a CTE approach again, and based on our prior analysis, we know what to expect when we perform our tuning analysis on this query. As in prior queries, we are using a WHERE clause to filter the results so that we can analyze and debug (if necessary) on a small set of data.

Our focus is on the CASE block in the SELECT statement:

```
    ,CASE
        WHEN NTILE(5) OVER(
        ORDER BY ItemsRemaining ASC
        ) = 1 THEN 'Order First High Priority Alert'
```

```
WHEN NTILE(5) OVER(
ORDER BY ItemsRemaining ASC
) = 2 THEN 'Order Second High Priority Alert'
WHEN NTILE(5) OVER(
ORDER BY ItemsRemaining ASC
) = 3 THEN 'Order Third High Priority Alert'
WHEN NTILE(5) OVER(
ORDER BY ItemsRemaining ASC
) = 4 THEN 'Order Fourth, Medium Priority Alert'
WHEN NTILE(5) OVER(
ORDER BY ItemsRemaining ASC
) = 5 THEN 'Order Fifth Low Priority Alert'
END AS AlertMessage
```

Wow, the CASE block calls the NTILE() function five times. This should be interesting as far as the estimated query performance plan is concerned.

Let's make sure the query works and check out the results. Then we can look at performance analysis and tuning.

Please refer to Figure 12-12.

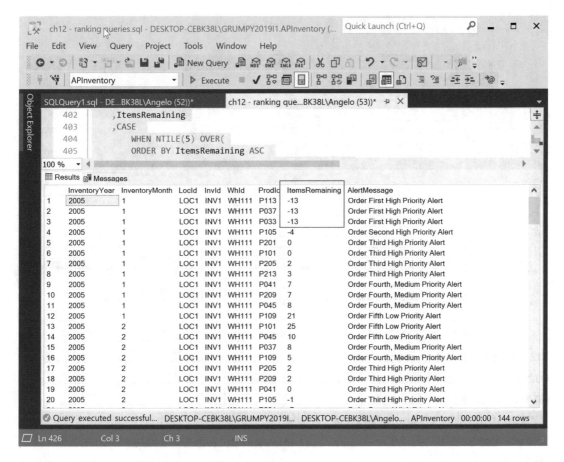

Figure 12-12.   Assigning reorder priorities

Looks like it works. We see three products, P113, P037, and P033, that have –13 items remaining. That means that 13 orders are back-ordered because of no stock in inventory. So the logic works, and our warehouse manager is pulling out their hair! Not a good thing.

Let's see how this query behaves.

Performance Considerations

Let's begin by generating our baseline execution plan.

Please refer to Figure 12-13.

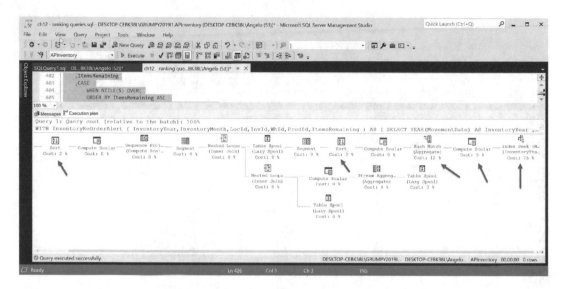

Figure 12-13. *Baseline estimated execution plan for the NTILE report*

No indexes were identified as missing, so we are off to a good start. Starting at the upper right-hand side in the plan, we see an index seek at 75% cost followed by a compute scalar at 9% and a hash match at 12%. Next is a sort at 2% cost (I am ignoring 0% cost tasks for our discussion).

Our lower-branch costs are all 0%. Amazingly enough the nested loops join task that merges both branches is also 0%. All the way at the left-hand side we see a sort task at 2%.

Note   Merge joins are better than nested loops joins. With a nested loops join task, for each row in the outer table, all rows in the inner table need to be scanned. Merge joins line them up so you do not have to scan all rows to search for matches! That's why you need indexes to help the seeks and scans.

Conclusion: This is a well-performing query so let's leave it alone. But curiosity got the better of me. Let's pop the CTE logic into an INSERT/SELECT query so we can load a report table. Listing 12-6 is the CREATE DDL statement for the table followed by the code to load it.

Listing 12-6. Report Table Replacing the CTE Logic

```
USE APInventory
GO

DROP SCHEMA IF EXISTS Reports
GO

CREATE SCHEMA Reports
GO

DROP TABLE IF EXISTS Reports.InventoryReOrderAlert
GO

CREATE TABLE Reports.InventoryReOrderAlert(
     InventoryYear int  NULL,
     InventoryMonth int NULL,
     LocId varchar(4)   NOT NULL,
     InvId varchar(4)   NOT NULL,
     WhId varchar(5)    NOT NULL,
     ProdId varchar(4)  NOT NULL,
     ItemsRemaining     int NULL
) ON AP_INVENTORY_FG
GO
INSERT INTO Reports.InventoryReOrderAlert
SELECT YEAR(MovementDate) AS InventoryYear
       ,MONTH(MovementDate) AS InventoryMonth
       ,IT.LocId
       ,IT.InvId
       ,IT.WhId
       ,IT.ProdId
       ,SUM(IT.Increment - IT.Decrement) AS ItemsRemaining
FROM APInventory.Product.InventoryTransaction IT
GROUP BY YEAR(MovementDate)
     ,MONTH(MovementDate)
     ,IT.LocId
     ,IT.InvId
     ,IT.WhId
```

```
        ,IT.ProdId
ORDER BY YEAR(MovementDate)
        ,MONTH(MovementDate)
        ,IT.LocId
        ,IT.InvId
        ,IT.WhId
        ,IT.ProdId
GO
```

I used the same name for the table as the name for the CTE. Notice I assigned it to the Reports schema. The code to load this table is just the CTE query that is now a part of an INSERT/SELECT command. You have seen these before.

The base query does not change as I used the name of the CTE as the name of the new table, so let's dive into the revised estimated query plan.

Please refer to Figure 12-14.

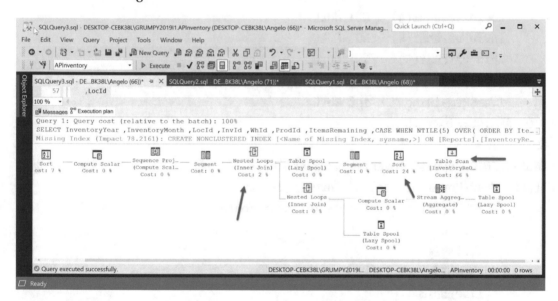

Figure 12-14. *Baseline estimated query plan showing an index is needed*

As luck would have it, we need to create an index to support this strategy as we see a table scan with a cost of 66% and an expensive sort step at 24%. Let's build the suggested index shown in Listing 12-7.

Listing 12-7. Create DDL for the Suggested Index

```
CREATE NONCLUSTERED INDEX
ieYearLocIdInvIdWhIdInventoryMonthProdIdItemsRemaining
ON Reports.InventoryReOrderAlert (InventoryYear,LocId,InvId,WhId)
INCLUDE (InventoryMonth,ProdId,ItemsRemaining)
GO
```

Long index name, I know, but it is descriptive.

SQL Server states in the suggested index message that if we build this index the query cost will improve by 78.2161%. Not bad.

Let's create it and regenerate the estimated query plan.

Please refer to Figure 12-15.

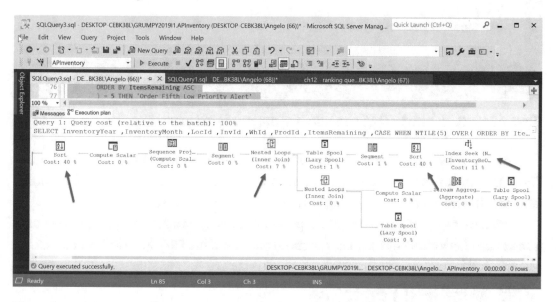

Figure 12-15. *Revised estimated execution plan*

The table scan was replaced by the index seek, but of course as we have seen many times before, the sorts went up and so did the nested loops join tasks. Might as well look at the IO and TIME statistics.

Please refer to Figure 12-16.

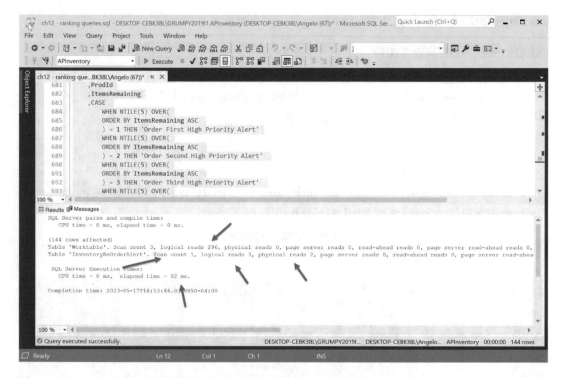

Figure 12-16. *IO and TIME statistics for the modified query*

These look like good statistics. First of all, the query ran under 1 s at 82 ms. Very fast. The work table has a value of 3 for the scan count and 296 logical reads. The new report table also has a scan count of 1, 3 logical reads, and 2 physical reads.

I think we can state again that our report table strategy is a good way to go when data is historical. Let's check out our last function, our friend the ROW_NUMBER() function.

ROW_NUMBER() Function

Last but not least is the ROW_NUMBER() function. As a reminder, I include the description. If you read the other chapters in sequence and you understand how this function works, you do not need to read the next paragraph. Skip it if you like.

This function, if you recall, does just what the name implies. Given a set of rows in the data result set, it will assign a sequential number to each row in the data set (or partition).

Depending on how you set up the OVER() clause and the PARTITION BY and ORDER BY clauses, it will assign row numbers to each partition in the result set. That is, each time you move to the next partition, the row number starts at 1 and goes on sequentially until the last row in the partition.

This time we will use the ROW_NUMBER() function to help us generate some random quantities so we can use them for testing purposes. In prior chapters we used this function to help us with the islands and gaps data phenomena!

Lastly, the calendar quarters and months need to be spelled instead of just numerical values, so a CASE block will be needed.

The business requirements for this next script are as follows:

This requirement is a bit of a change of pace. The analyst does not want actual results but wants us to generate some random inventory movements so that they can create some prototype Microsoft Excel graphs and charts. They are happy with data for the year "2002," product "P101," location "LOC1," inventory "INV1," and warehouse "WH112."

This is a great specification as it allows us to use the ROW_NUMBER() function to generate random data instead of just generating row numbers for a result set or partition. Valuable mechanism, but we have seen examples of this before.

Anyway, here is what we came up with to fulfill this requirement. Please refer to Listing 12-8.

Listing 12-8. Generating Random Values Using ROW_NUMBER()

```
WITH InventoryMovement (
AsOfYear,AsOfQuarter,AsOfMonth,AsOfDate,LocId,InvId,WhId,ProdId,Inv
Out,InvIn
)
AS
(
SELECT AsOfYear
    ,CASE
        WHEN DATEPART(qq,AsOfDate)  = 1 THEN 'Qtr 1'
        WHEN DATEPART(qq,AsOfDate)  = 2 THEN 'Qtr 2'
        WHEN DATEPART(qq,AsOfDate)  = 3 THEN 'Qtr 3'
        WHEN DATEPART(qq,AsOfDate)  = 4 THEN 'Qtr 4'
      END AS AsOfQuarter
    ,CASE
        WHEN MONTH(AsOfDate)  = 1 THEN 'Jan'
        WHEN MONTH(AsOfDate)  = 2 THEN 'Feb'
        WHEN MONTH(AsOfDate)  = 3 THEN 'Mar'
        WHEN MONTH(AsOfDate)  = 4 THEN 'Apr'
```

```
            WHEN MONTH(AsOfDate)   = 5 THEN 'May'
            WHEN MONTH(AsOfDate)   = 6 THEN 'June'
            WHEN MONTH(AsOfDate)   = 7 THEN 'Jul'
            WHEN MONTH(AsOfDate)   = 8 THEN 'Aug'
            WHEN MONTH(AsOfDate)   = 9 THEN 'Sep'
            WHEN MONTH(AsOfDate)   = 10 THEN 'Oct'
            WHEN MONTH(AsOfDate)   = 11 THEN 'Nov'
            WHEN MONTH(AsOfDate)   = 12 THEN 'Dec'
     END AS AsOfMonth
    ,AsOfDate
    ,LocId
    ,InvId
    ,WhId
    ,ProdId
    ,ROUND(CEILING(RAND(ROW_NUMBER() OVER (
    PARTITION BY AsOfYear,DATEPART(qq,AsOfDate)
    ORDER BY AsOfDate
    ))
    * 85 *
    RAND(ROW_NUMBER() OVER (
    PARTITION BY AsOfYear
    ORDER BY AsOfDate
    ) * 1900
    )),1) AS InvOut
    ,ROUND(CEILING(RAND(ROW_NUMBER() OVER (
    PARTITION BY AsOfYear,DATEPART(qq,AsOfDate)
    ORDER BY AsOfDate
    ))
    * 100 *
    RAND(ROW_NUMBER() OVER (
    PARTITION BY AsOfYear
    ORDER BY AsOfDate
    ) * 100000
    )),1) AS InvIn
FROM Product.Warehouse
```

```
WHERE AsOfYear = 2002
AND ProdId = 'P101'
AND Locid = 'LOC1'
AND InvId = 'INV1'
AND WhId = 'WH112'
GROUP BY AsOfYear
      ,DATEPART(qq,AsOfDate)
      ,AsOfMonth
      ,AsOfDate
      ,LocId
      ,InvId
      ,WhId
      ,ProdId
)
SELECT AsOfYear
      ,AsOfQuarter
      ,AsOfMonth
      ,AsOfDate
      ,LocId
      ,InvId
      ,WhId
      ,ProdId
      ,InvOut
      ,InvIn
      ,InvIn - InvOut AS QtyOnHand
FROM InventoryMovement
ORDER BY AsOfYear
      ,AsOfQuarter
      ,AsOfMonth
      ,AsOfDate
      ,LocId
      ,InvId
      ,WhId
      ,ProdId
GO
```

I will not go over the code as we have seen this structure before, but we can focus on the code that generates the inventory out random value. (InvOut):

```
,ROUND(CEILING(
        RAND(ROW_NUMBER() OVER (
            PARTITION BY AsOfYear,DATEPART(qq,AsOfDate)
            ORDER BY AsOfDate
            )
        )
    * 85 *
        RAND(ROW_NUMBER() OVER (
            PARTITION BY AsOfYear
            ORDER BY AsOfDate
            ) * 1900
        )
    ),1
) AS InvOut
```

Working our way from the inside out, the ROW_NUMBER() function uses an OVER() clause that is partitioned by the AsOfYear column and the quarter derived from AsOfDate using the DATEPART() function. The ORDER BY clause orders the partition result set by AsOfDate. The value generated is used as a seed for the RAND() function, and the value generated is multiplied by 85.

Next, we multiply the value just generated by the same logic except the PARTITION BY in the second OVER() clause is defined by the AsOfYear only. This value is used again by the RAND() function as the seed value, and the result is multiplied by 1900.

The last new value is adjusted by using the CEILING() function to get the next highest value, and then we wrap it up by using the ROUND() function to round the final value so no fractional part is included.

This logic is also used to generate a random value for the InvIn column.

Please refer to the partial results in Figure 12-17.

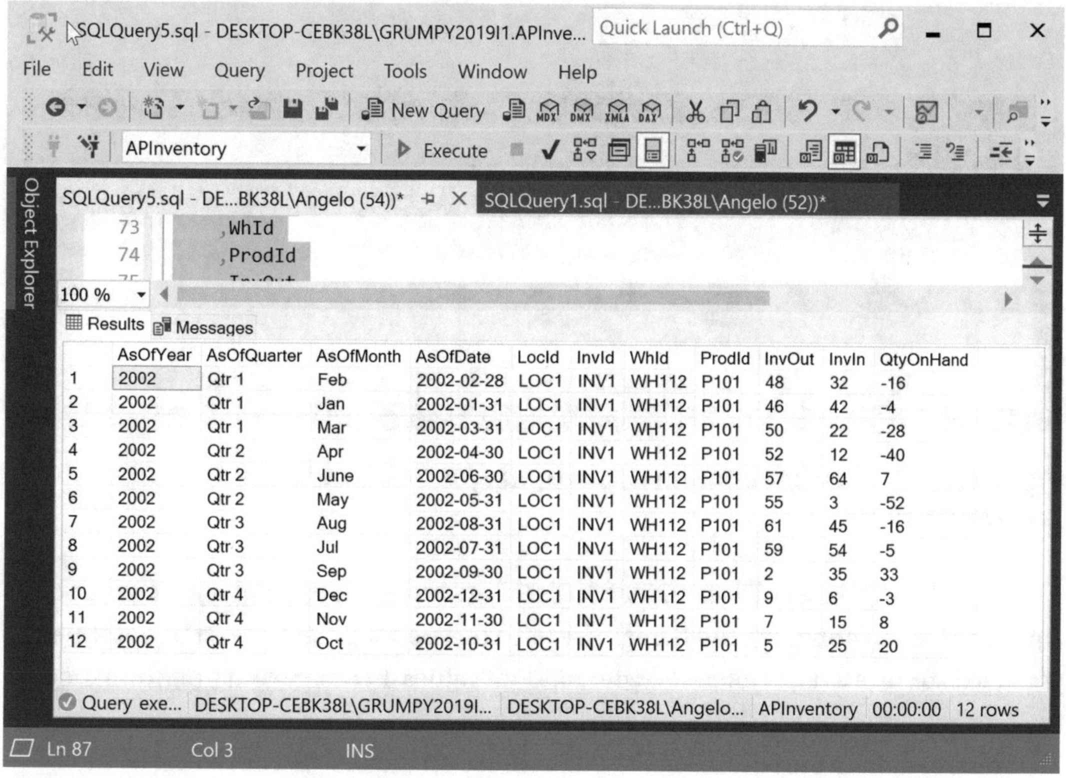

	AsOfYear	AsOfQuarter	AsOfMonth	AsOfDate	LocId	InvId	WhId	ProdId	InvOut	InvIn	QtyOnHand
1	2002	Qtr 1	Feb	2002-02-28	LOC1	INV1	WH112	P101	48	32	-16
2	2002	Qtr 1	Jan	2002-01-31	LOC1	INV1	WH112	P101	46	42	-4
3	2002	Qtr 1	Mar	2002-03-31	LOC1	INV1	WH112	P101	50	22	-28
4	2002	Qtr 2	Apr	2002-04-30	LOC1	INV1	WH112	P101	52	12	-40
5	2002	Qtr 2	June	2002-06-30	LOC1	INV1	WH112	P101	57	64	7
6	2002	Qtr 2	May	2002-05-31	LOC1	INV1	WH112	P101	55	3	-52
7	2002	Qtr 3	Aug	2002-08-31	LOC1	INV1	WH112	P101	61	45	-16
8	2002	Qtr 3	Jul	2002-07-31	LOC1	INV1	WH112	P101	59	54	-5
9	2002	Qtr 3	Sep	2002-09-30	LOC1	INV1	WH112	P101	2	35	33
10	2002	Qtr 4	Dec	2002-12-31	LOC1	INV1	WH112	P101	9	6	-3
11	2002	Qtr 4	Nov	2002-11-30	LOC1	INV1	WH112	P101	7	15	8
12	2002	Qtr 4	Oct	2002-10-31	LOC1	INV1	WH112	P101	5	25	20

Figure 12-17. *Generating random values with the ROW_NUMBER function*

As we can see, the random values seem reasonable. At the end of the day, this is just a bunch of logic to manipulate numbers to achieve the desired results. I just want to give an example of the `ROW_NUMBER()` function. that is used for other purposes besides generating row numbers inside a partition.

Performance Considerations

Let's take a final peek at estimated query plans by generating the baseline plan for this query.

Please refer to Figure 12-18.

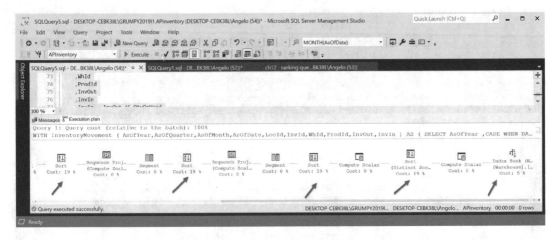

Figure 12-18. *Baseline estimated query plan for the random value generation query*

We will just. take a quick examination of this query as it is one not typically found in a production environment, but it might be used in a development environment when you need some test data generated with random values. In other words, performance is not that important unless the query runs for hours.

Starting right to the left side in the plan, we see an index scan with a cost of 5%, a sort with a cost of 19%, a second sort with a cost of 19%, and yet two more sort tasks with costs of 19%. Not seen is a fifth sort task all the way to the left with the same cost of 19%.

So we can deduce that this type of query will need to sort the interim results many times. As usual I ignored the zero-cost tasks although when you perform your analysis, you should be aware of them, especially tasks that you do not know what they do!

This is just one way of generating random data, and when you study the scripts for loading the databases used in this book, you will see other techniques that use SQL Server functions for generating random data..

Create an SSRS Report

We are now going to create a report using Report Builder and publish it on the SSRS website. This report needs to satisfy the following requirements submitted by our analyst:

This is a detailed report that lists all product inventory on hand for all calendar years and months; all locations, inventories, and warehouses need to be included. The names of the months need to be displayed instead of the numbers (yes, another expensive CASE code block).

This report will be used by multiple users, so a set of filters needs to be included that will allow business users the ability to select which years, months, and other dimensions they want to view – in any combination!

In other words, all the data is needed, and this needs to be a dynamic report to support all the possible combinations of filtering choices.

Performance is key. Our business users want everything. They are very demanding. Users cannot be expected to wait more than 30 seconds (or less) for the report to display results.

We need to base this report on a VIEW based on the following query. Please refer to Listing 12-9.

Listing 12-9. TSQL VIEW to Support Inventory Ranking

```
CREATE VIEW MonthlyInventoryRanks
AS
WITH WarehouseCTE (
        InvYear,InvMonthNo,InvMonthMonthName,LocationId,LocationName
        ,InventoryId,WarehouseId,WarehouseName,ProductId,ProductName
        ,ProductType,ProdTypeName,MonthlyQtyOnHand
)
AS ( -- returns 24,192 rows
SELECT YEAR(C.[CalendarDate])  AS InvYear
        ,MONTH(C.[CalendarDate]) AS InvMonthNo
        ,CASE
                WHEN MONTH(C.[CalendarDate])  = 1 THEN 'Jan'
                WHEN MONTH(C.[CalendarDate])  = 2 THEN 'Feb'
                WHEN MONTH(C.[CalendarDate])  = 3 THEN 'Mar'
                WHEN MONTH(C.[CalendarDate])  = 4 THEN 'Apr'
                WHEN MONTH(C.[CalendarDate])  = 5 THEN 'May'
                WHEN MONTH(C.[CalendarDate])  = 6 THEN 'June'
                WHEN MONTH(C.[CalendarDate])  = 7 THEN 'Jul'
                WHEN MONTH(C.[CalendarDate])  = 8 THEN 'Aug'
                WHEN MONTH(C.[CalendarDate])  = 9 THEN 'Sep'
                WHEN MONTH(C.[CalendarDate])  = 10 THEN 'Oct'
                WHEN MONTH(C.[CalendarDate])  = 11 THEN 'Nov'
                WHEN MONTH(C.[CalendarDate])  = 12 THEN 'Dec'
```

```
        END AS InvMonthMonthName
        ,L.LocId             AS LocationId
        ,L.LocName           AS LocationName
        ,I.InvId             AS InventoryId
        ,W.WhId              AS WarehouseId
        ,W.WhName            AS WarehouseName
        ,P.ProdId               AS ProductId
        ,P.ProdName          AS ProductName
        ,P.ProdType          AS ProductType
        ,PT.ProdTypeName     AS ProdTypeName
        ,SUM(IH.[QtyOnHand]) AS MonthlyQtyOnHand
FROM [Fact].[InventoryHistory] IH -- WITH (INDEX(ieInvHistCalendar))
JOIN [Dimension].[Location] L
     ON IH.LocKey = L.LocKey
JOIN [Dimension].[Calendar] C
     ON IH.CalendarKey = C.CalendarKey
JOIN [Dimension].[Warehouse] W
     ON IH.WhKey = W.WhKey
JOIN [Dimension].[Inventory] I
     ON IH.[InvKey] = I.InvKey
JOIN [Dimension].[Product] P
     ON IH.ProdKey = P.ProdKey
JOIN [Dimension].[ProductType] PT
     ON P.ProductTypeKey = PT.ProductTypeKey
GROUP BY YEAR(C.[CalendarDate])
        ,MONTH(C.[CalendarDate])
        ,L.LocId
        ,L.LocName
        ,I.InvId
        ,W.WhId
        ,W.WhName
        ,P.ProdId
        ,P.ProdName
        ,P.ProdType
        ,PT.ProdTypeName
```

```
)
SELECT InvYear,InvMonthNo,InvMonthMonthName,LocationId,LocationName
     ,InventoryId,WarehouseId,WarehouseName,ProductType,ProductId
     ,ProductName
     ,MonthlyQtyOnHand
     ,RANK() OVER (
          PARTITION BY InvYear,InvMonthNo,InvMonthName,LocationId
                  ,InventoryId,WarehouseId
          ORDER BY MonthlyQtyOnHand DESC
     ) QtyOnHandRank
FROM WarehouseCTE
GO
```

Lots of code, but it is worth the time and effort to review it and understand it, especially if you are new to TSQL coding.

Notice the granularity of the partition specified in the PARTITION BY clause. We specify the year, month (number and year), location, inventory, and warehouse (we are not partitioning down to the product; try it on your own, and also modify the view so it uses the DENSE_RANK() function).

Lastly, an ORDER BY clause is included to sort the processing of the partition values by MonthlyQtyOnHand in descending order.

If performance is slow, we need to recreate the report with SCHEMA binding, which means the VIEW will actually be populated with physical data and exist on disk. This will make any query that references the view fast!

This is another option to use vs. the report table option we discussed many times to replace the CTE. Just modify the VIEW DDL command as per the following code snippet:

```
CREATE OR ALTER VIEW Reports.MonthlyInventoryRanks
-- use below to implement a materialized, schema bound view
WITH SCHEMABINDING
AS
```

Now it's time for a brief tutorial on Report Builder that we will use to build and publish a web-based report to SSRS.

Report Builder Mini Tutorial

Let's take a look at one of the new tools in your BI toolkit called Report Builder. As mentioned earlier in the book, this tool allows you to create reports based on your SQL Server databases and publish them to SSRS, Microsoft's premier web reporting architecture.

This section will take the form of a mini tutorial where we look at the steps required to create and publish the report to SSRS. It will provide you with the basic steps you need to take to create and publish a simple report to SSRS. It is not meant to cover all the features as this would take an entire book and not a section in a chapter. Hopefully, it will inspire you to try the tool and learn how to use it with all its features.

By the way, expect a lot of screenshots!

If you do not have Report Builder, simply search for it with your favorite browser and download it from the Microsoft site. It is free! Once you install it, start it up, and you are greeted with the following screen in Figure 12-19.

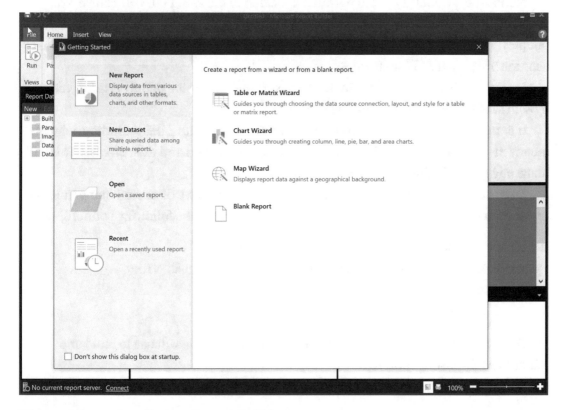

Figure 12-19. Starting up Report Builder

You are greeted with a pop-up panel that lets you choose the type of report or component you want to create. The following are the choices:

- New Report (create)

- New Dataset (create)

- Open (a saved report)

- Recent (open a recent report)

You can also create a report using guided wizards. Clicking New Report displays all the wizards available to guide you through the report creation process.

The choices are

- Table or Matrix Wizard

- Chart Wizard

- Map Wizard

- Blank Report

We will select a new report for our example and use the Table or Matrix Wizard. This will give you access to the report wizards.

The first panel you are presented with allows you to create or use an existing data set. A data set is an object that will contain data from the database you can use in the report.

Tip   If you are new to Report Builder, I recommend you use the wizards as they are easy to use and will give you a sense of what is involved in web-based report creation within a SQL Server architecture.

Please refer to Figure 12-20.

Figure 12-20. *Create a data set*

Here you have a choice of selecting an existing data set or creating a new one. We will click the "Create a dataset" radio button and click the Next button. Another panel is presented.

Please refer to Figure 12-21.

Figure 12-21. *Create a new data source*

Before we create a data set, we need a connection to a data source that will provide the data we need. The data source provides data for the data set.

Click the "New" button. Another panel is presented.

Please refer to Figure 12-22.

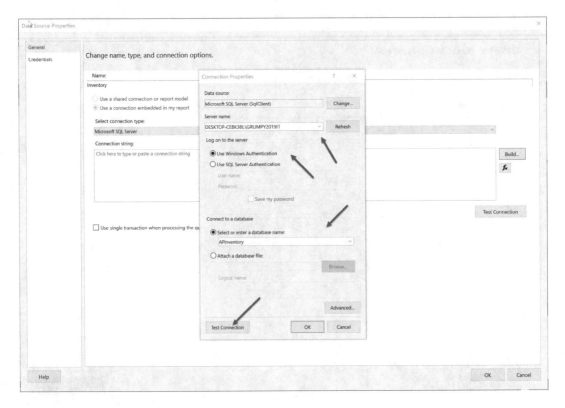

Figure 12-22. *Data source properties*

Provide a name (no spaces please or special characters) for the data set, click the Build button, and a dialog box appears that lets you pick a server from a list and select how to log onto the server. In this case we want to use Windows Authentication instead of SQL Server authentication. Lastly, we select the database we want to connect to and test the connection.

Always test. Remember, there is always time to do things over but never time to do things right!

Click the OK button and the connection creation tasks are complete, and we are back to the original panel.

Please refer to Figure 12-23.

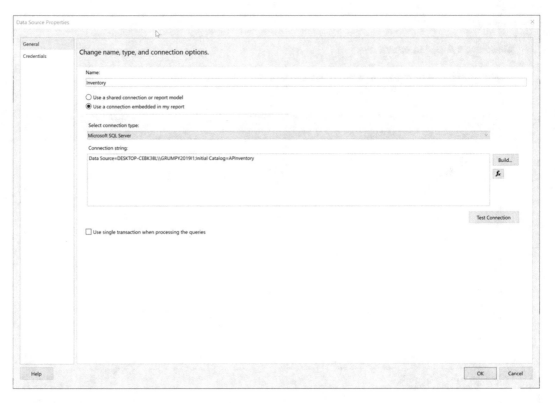

Figure 12-23. *Completed data source properties*

Everything looks good so click the OK button again. You can test the connection here also in case you selected an existing connection instead of creating a new one.

Another panel is presented, which shows the new data source connection.

Please refer to Figure 12-24.

Figure 12-24. *Choose the new connection*

There's the name and you get another chance to test it. This is in case you are creating a report from an existing connection, so you can select it here and test it. Click the "Next" button, and yes, another panel appears for the next step. Maybe a bit confusing as there are different paths to these steps depending on whether you created a new connection or used an existing one.

For the next step, please refer to Figure 12-25.

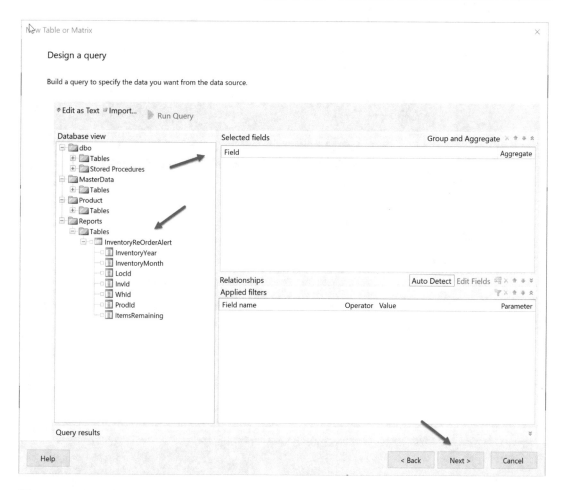

Figure 12-25. *Design a query*

Finally, we get to create the report. Expand the Reports folder and then the Tables folder, and in this case, we only have one table called InventoryReOrderAlert. This table has 24,192 rows, not too big and not too small.

Expand this node to see all the columns. Select all the columns by clicking in the checkbox to the left of each name. The Selected fields panel will display all the columns you selected.

Please refer to Figure 12-26.

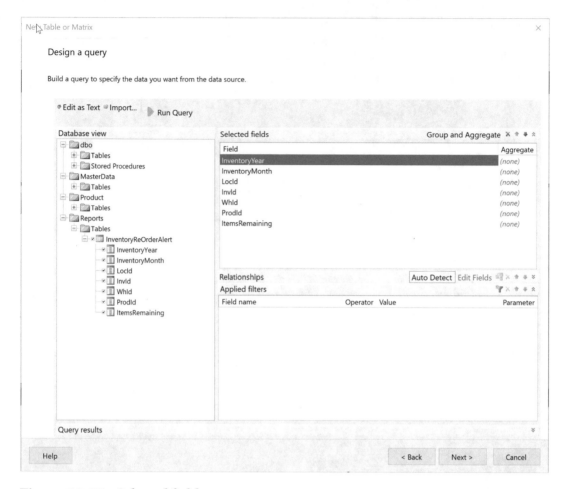

Figure 12-26. Selected fields

We could define filters at this time, but we want to create a drill-down report that uses all the data, so we will not use this part of the panel. Click the Next button to where you will lay out the report.

If you decided to use filters, you would also need to create small data sets, one per column used in the filters (in the form of SELECT DISTINCT <column name> FROM <table name> ORDER BY <column name>).

Tip   Give your data sets clear simple names like ProductId instead of dataset1, dataset2, etc. You know what I mean.

Then you would need to create parameters that link to the data set in order to get the values and finally map the filters in the WHERE clause of the query in the main data set to these parameters.

Next, we need to lay out our report.

Please refer to Figure 12-27.

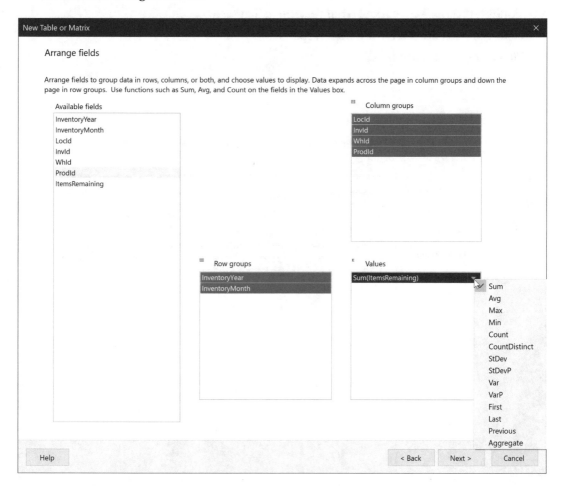

Figure 12-27. *Arrange fields*

This type of report is composed of row groups, column groups, and values. For the row groups, we want to drill down by year and month, so drag the InventoryYear and InventoryMonth columns to this area.

For the column groups, we want to include the LocId, InvId, WhId, and ProdId, so drag these columns into this area.

Lastly, in the Values panel, drag the ItemsRemaining column in this area. If you click the down arrow on the right-hand side, you get a choice of which function you want to apply to this measure (a column you can perform calculations on). Notice some of our old friends like the aggregate and analytical functions.

Lastly, the values appear at intersections of the objects identified in the row groups and column groups. We will see what this looks like next.

Click the Next button for the next step.

Please refer to Figure 12-28.

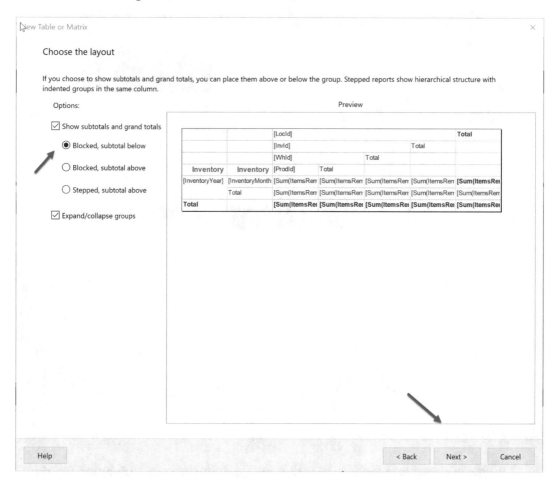

Figure 12-28. *Choose a layout*

Here we get a chance to specify the general layout for the report. If you check "Show subtotals and grand totals," you can specify one of the following layouts:

- Blocked, subtotal below

- Blocked, subtotal above (like the GROUPING function we discussed earlier in the book chapters)

- Stepped, subtotal above

We check Show subtotals and grand totals, and we want a blocked, subtotal below format. You can see it in the preceding figure.

Once you are satisfied with the results, click the Next button for a preview.

Please refer to Figure 12-29.

Figure 12-29. *Preview of the report*

You get to see the layout once more. In case you need to change something, you can click the Back button to go back a step. We are happy with the layout, so click the Finish button.

Now we are ready to test the report. Please refer to Figure 12-30.

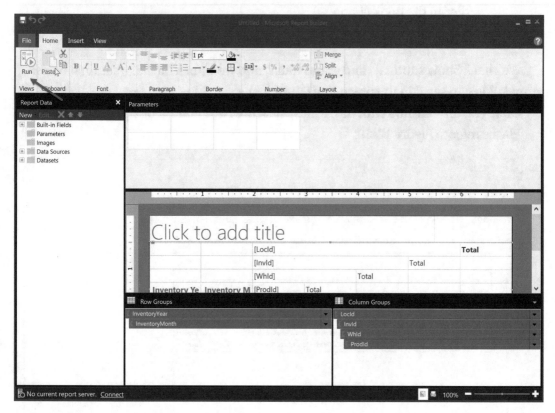

Figure 12-30. *Report layout – run the report*

The report layout is displayed. You could make more modifications like adding parameters to allow business users to further filter the results, like picking one or more products or warehouses. You could add indicators like ones typically found in scorecards or dashboards. I like the little car gauges that show if a value is increasing or decreasing, but I digress.

Now would be a good time to save the report to a folder by the way. I always forget this, and when the parts are published to SSRS, I need to go back and save the report to a folder with a good name so I can then upload it to the SSRS website folder (frustrating).

Click "Save as" and the following panel is presented.

Please refer to Figure 12-31.

Figure 12-31. *Saving the report*

Navigate to the desired folder, give the report a good name, and click the "Save" button.

Next, we take the report for a test drive. Click the "Run" button on the upper left-hand side of the menu bar, and we should see a sample of the report.

Please refer to Figure 12-32.

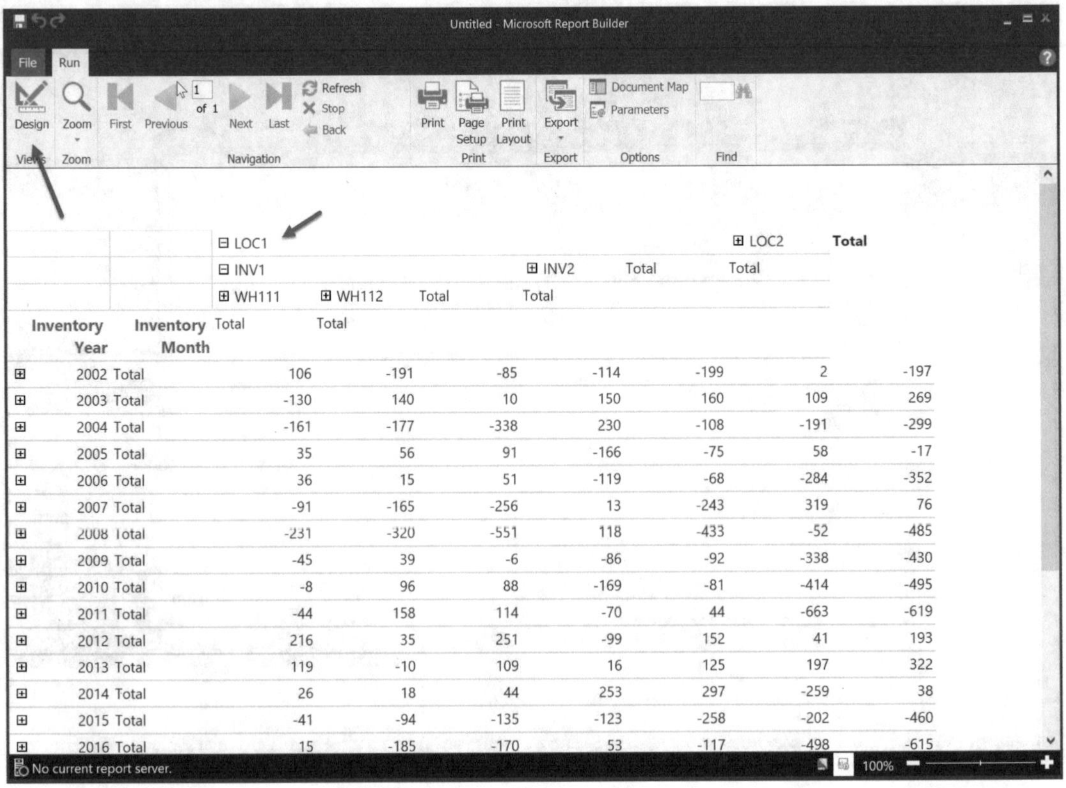

Figure 12-32. Report layout, sample of the report

Everything looks good! Clicking any of the buttons next to the dimensions, like "LOC1," will allow you to drill down into the report. Totals are automatically recalculated to match the level in the hierarchy you are drilling down to.

Please refer to Figure 12-33.

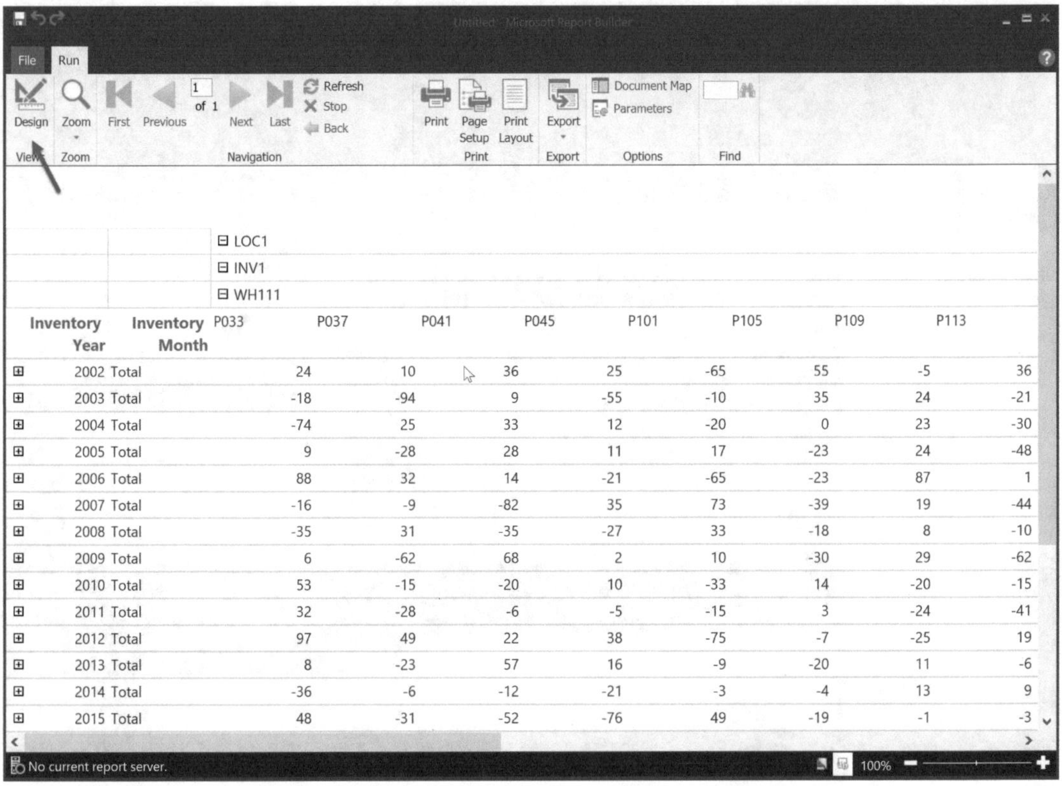

Figure 12-33. *Drill down to products*

Nice drill-down. Talk about data mining. Once you are happy looking at it, you can go back to the Design panel. Click the Design button and you are back to the Design panel.

Now we need to connect to the local SSRS server so we can publish the report.

Note   Download SSRS from the Microsoft site. The installer is called SQLServerReportingServices.exe! Run it and then you need to configure it with the SSRS configuration manager. We will see how to do this in Chapter 14 when we wrap up the book.

Please refer to Figure 12-34.

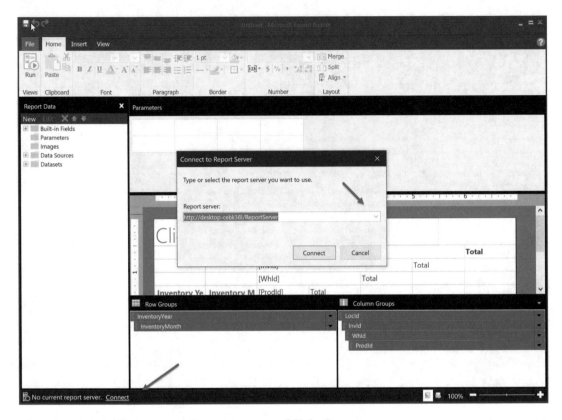

Figure 12-34. *Connect to the server to publish the report*

Now we are ready to publish the report. On the lower left-hand side is the Connect button that we will use to connect to our web server. Click the "Connect" button and supply the name of the SSRS server installed on your laptop, desktop, or actual development server that your company gave you access to.

Pick a server from the drop-down list and press the "Connect" button. Now we are ready to publish the report parts and report.

Please refer to Figure 12-35.

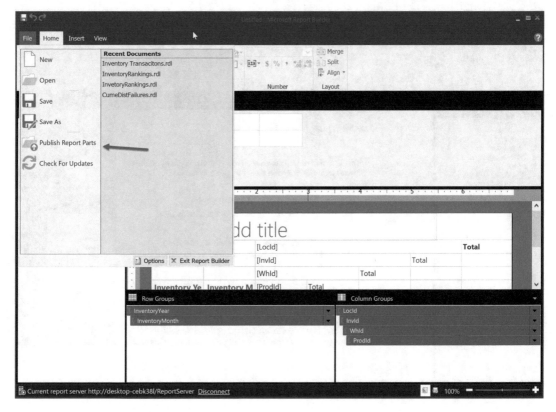

Figure 12-35. *Publish the report*

Switch to the first tab, the "File" tab, and click "Publish Report Parts." Another panel appears that lets you select how you want to publish the report.

Please refer to Figure 12-36.

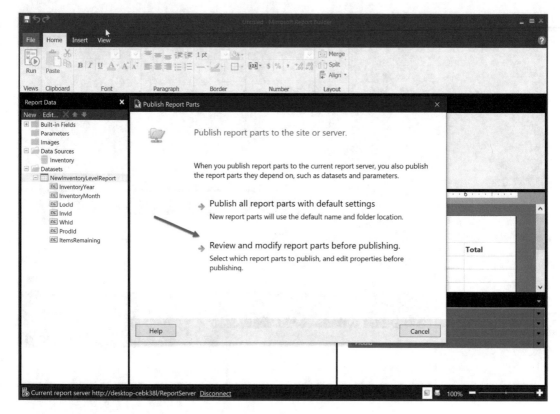

Figure 12-36. *Review parts before publishing, part 1*

We want to review the report parts as we need to rename them to make sure there are no duplicate names used by other reports. These names are basically the Tablix object and the data sets. If you do not change these, when you create more than one report, you will keep on getting publishing errors. Just use good new names each time and you will avoid headaches.

Click the last selection "Review and modify report parts before publishing."

Please refer to Figure 12-37.

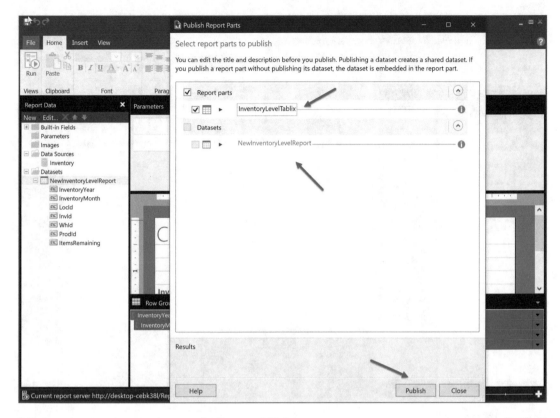

Figure 12-37. *Review parts before publishing, part 2*

I changed the default Tablix1 name to `InventoryLevelTablix`. I am sure this is a new unique name. The data set name is grayed out so you cannot change it. I made sure I gave it a good name during the definition step we discussed earlier.

This looks pretty good, so cross your fingers and press the Publish button. If you received a "Published successfully" message, you're ready to view the report.

Please refer to Figure 12-38.

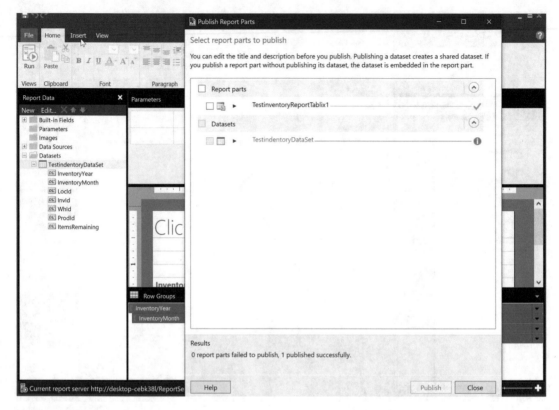

Figure 12-38. *Success*

We did receive a "Published successfully" message. Let's go to our SSRS website and check things out.

Please refer to Figure 12-39.

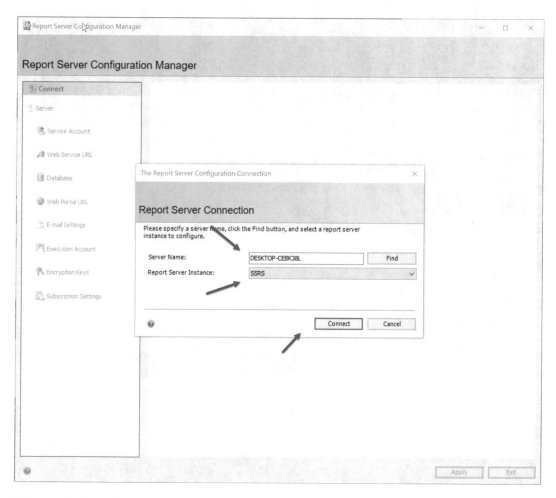

Figure 12-39. *View the report*

I always forget the report server URL.

If you go to the Report Server Configuration Manager tool, you can connect to the server and then view the URL for the web report portal as shown in Figure 12-40.

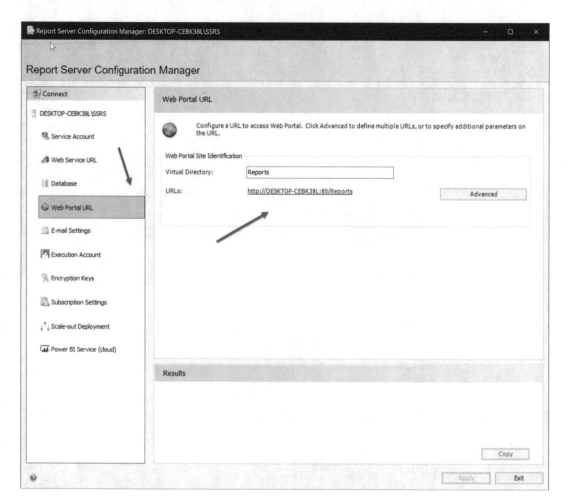

Figure 12-40. *Access the web portal*

Just click the URL to navigate to the web page. You will see the following page in your browser.

Please refer to Figure 12-41.

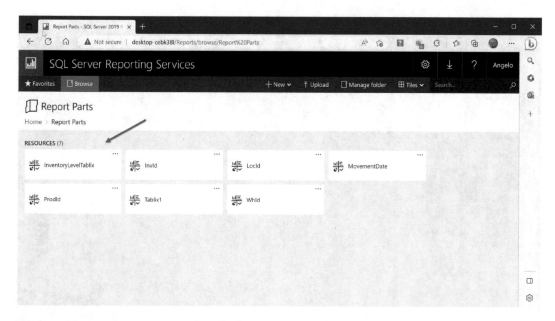

Figure 12-41. *Report parts loaded*

We see the report parts are all there for this and other reports. If we navigate to the Browse tab, we will also see the reports. In our case we need to upload them. Click the "Upload" button in the menu bar.

But before we do this, go back to the Report Builder designer, and save your report if you forgot to do it or changed parts. If you save the report like we did earlier, we are good to go. If not, check out the following screenshot.

Please refer to Figure 12-42.

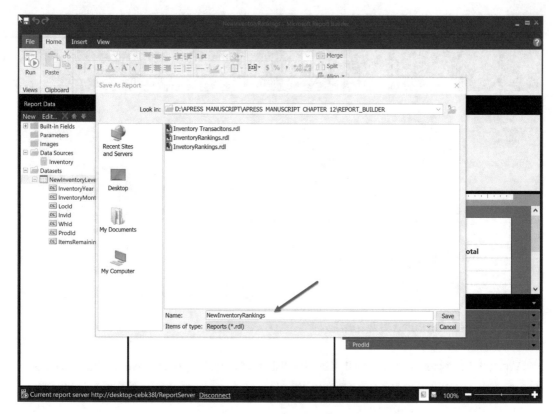

Figure 12-42. *Save the Report Builder report before uploading*

Click the File tab and click Save as. Pick a good unique name in the dialog box and click Save. Make sure you place it in a dedicated folder so that you can remember the name!

I actually forgot to do this when I generated the images in this section, so I had to perform this action here, so might as well review it again. Now you will never forget.

Now we can upload the report.

Please refer to Figure 12-43.

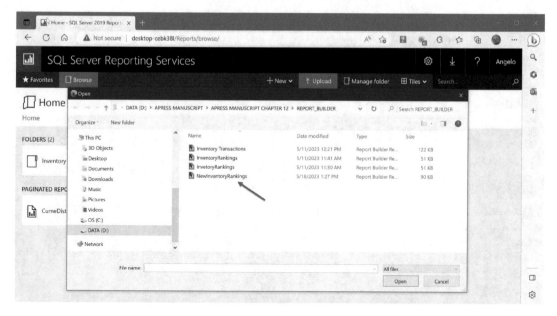

Figure 12-43. *Upload the report*

Click the new file you saved and click the "Open" button. You will get a message that tells you it is loading followed by a loading complete message. The report should appear in the browser as per the following screenshot.

Please refer to Figure 12-44.

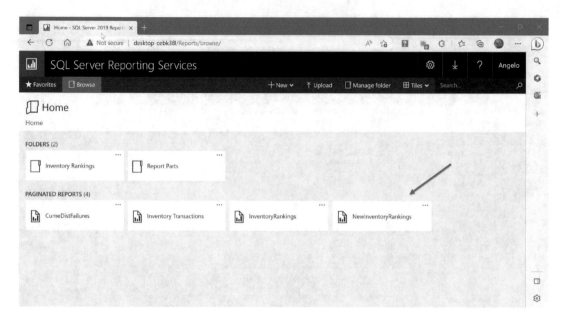

Figure 12-44. *Report added, part 1*

Success, that worked. Let's click the report and take a look.

Please refer to Figure 12-45.

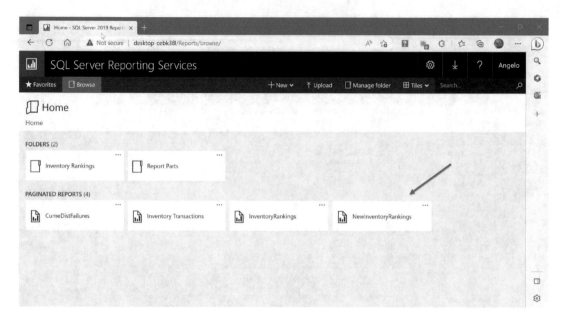

Figure 12-45. *Report added, part 2*

Looks good. Let's click the Inventory Year node for 2002 to expose the results for the months. This is called drilling down by the way. Click the nodes again to drill up.

Please refer to Figure 12-46.

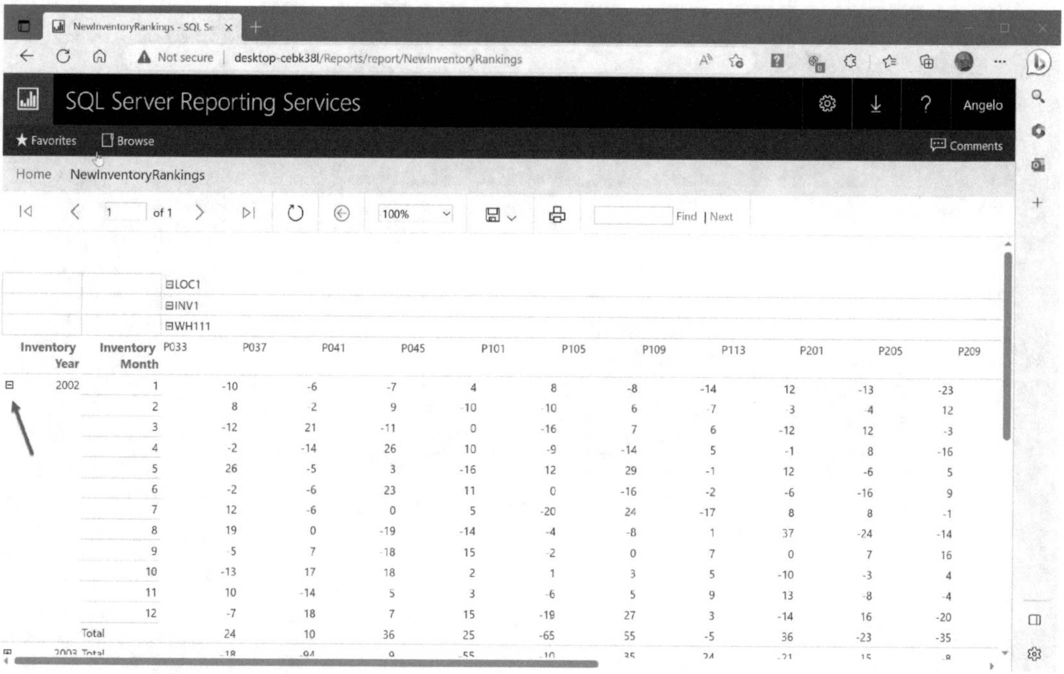

		LOC1										
		INV1										
		WH111										
Inventory Year	Inventory Month	P033	P037	P041	P045	P101	P105	P109	P113	P201	P205	P209
2002	1	-10	-6	-7	4	8	-8	-14	12	-13	-23
	2	8	2	9	-10	-10	6	-7	3	-4	12
	3	-12	21	-11	0	-16	7	6	-12	12	-3
	4	-2	-14	26	10	-9	-14	5	-1	8	-16
	5	26	-5	3	-16	12	29	-1	12	-6	5
	6	-2	-6	23	11	0	-16	-2	-6	-16	9
	7	12	-6	0	5	-20	24	-17	8	8	-1
	8	19	0	-19	-14	-4	-8	1	37	-24	-14
	9	5	7	-18	15	-2	0	7	0	7	16
	10	-13	17	18	2	1	3	5	-10	-3	4
	11	10	-14	5	3	-6	5	9	13	-8	-4
	12	-7	18	7	15	-19	27	3	-14	16	-20
	Total	24	10	36	25	-65	55	-5	36	-23	-35
2003 Total		-18	-94	0	-55	-10	35	24	-21	15	8

Figure 12-46. *Drill down by month*

This worked also. We now have totals by months for all the products. Use the scroll bars to view all years or products. Try this one on your own. We only touched the surface as far as the capabilities of this tool. Charts like line, bar, or pie charts can be included, and you can also design very impressive scorecards and dashboards and publish them to the SSRS site.

Let's take a look at our other tool in our BI toolkit called Power BI.

Create a Power BI Report

In this last section, we will go over a brief tutorial that will show you how to create and publish a report and chart that contains inventory information. The tools we will use are the Power BI free development environment and the free Power BI server for development purposes. These tools can be obtained from the respective Microsoft website and are easy to install. Chapter 14 will show you how to download and install these.

Let's begin with a guided tour of the facilities available to create sophisticated reports, scorecards, and dashboards.

Please refer to Figure 12-47.

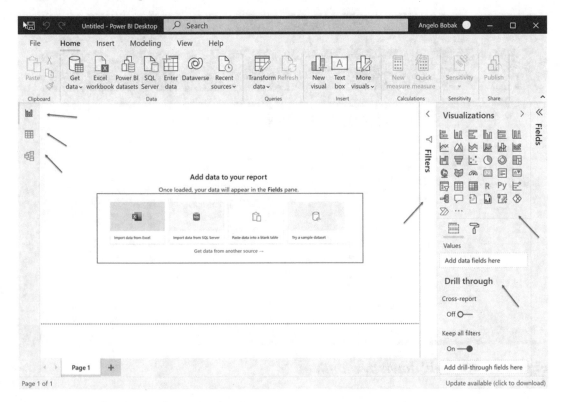

Figure 12-47. *Economy guided tour*

Upon starting Power BI from your desktop, you are presented with this screen. Highlighted are the main tool components we will use. Starting at the left, we see a panel that contains a tool to create a report, view data and views, and create or modify a model based on the data source you will connect to. This model looks much like a STAR or SNOWFLAKE schema like the ones we generated when we created an SSAS multidimensional cube.

The middle panel will be your design panel where you create your report, dashboard, or scorecard. Your project can actually have any combination of these or all of these as you can create multiple pages, each with a unique or related report or visualization.

On the right side, you see a **Visualization** panel that contains icons for all the different charts and graphs you can create. Lastly, the filter section allows you to identify columns in the data that you will use to drill up and down into the report or graph data. As you drill down, the chart values will change to correspond to the different levels and aggregations related to the data aggregation hierarchy.

Okay, how do we build this?

The first step is to define a data source for your graphs and reports. Sounds familiar, like when we create SSAS cubes or SSIS ETL packages or Report Builder reports?

Please refer to Figure 12-48.

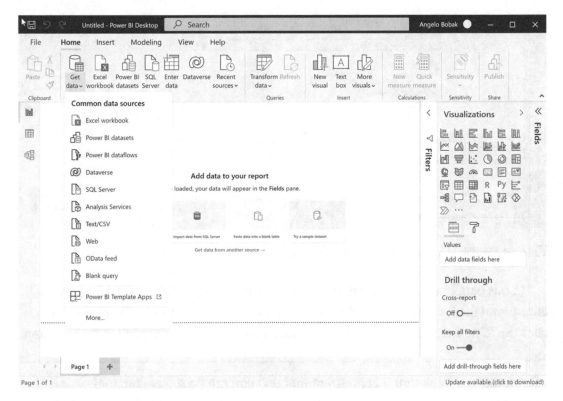

Figure 12-48. *Common available data sources*

Clicking the "**Get data**" selection in the menu bar, on the left-hand side, will produce a list of common data sources we can connect to. In our case we are interested in Analysis Services, so let's click it so we can define a connection and database (in our case an SSAS instance and cube).

Please refer to Figure 12-49.

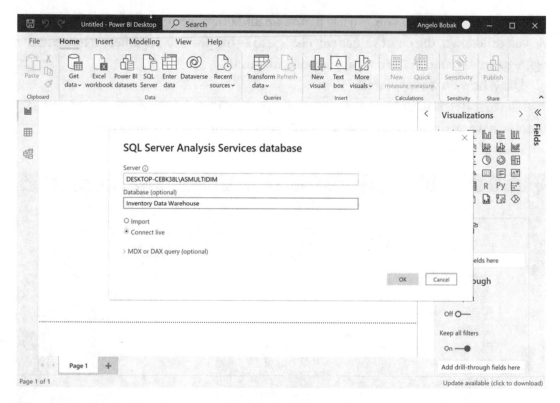

Figure 12-49. *Define the connection and identify the database*

In this panel we need to identify the server, which is an instance of SSAS on my laptop, and the Inventory Data Warehouse cube we created earlier in Chapter 11. We click the "Connect live" radio button and click the OK button to proceed to the next step.

Comment   I think that it is amazing that you can create a development environment on your laptop or desktop that includes the database server, SSAS server, SSIS server, SSRS server and website, and PowerBuilder server and website. There is no limit to what you can learn if you aspire to be a SQL Server data architect.

Next, we check out the data model that will be the foundation of our report and visualization.

Please refer to Figure 12-50.

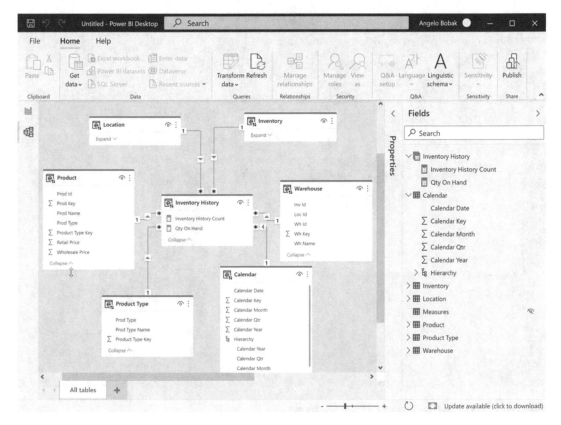

Figure 12-50. *The multidimensional model*

Having connected to the SSAS instance and the desired cube, we can click the data model icon on the left-hand side of the screen to view the multidimensional model of the cube, which looks like a simple STAR schema. Recall that this looks like the model we created in Visual Studio when we created our multidimensional cube on SSAS.

The model looks good so there is nothing for us to do here. We can start creating our report.

Please refer to Figure 12-51.

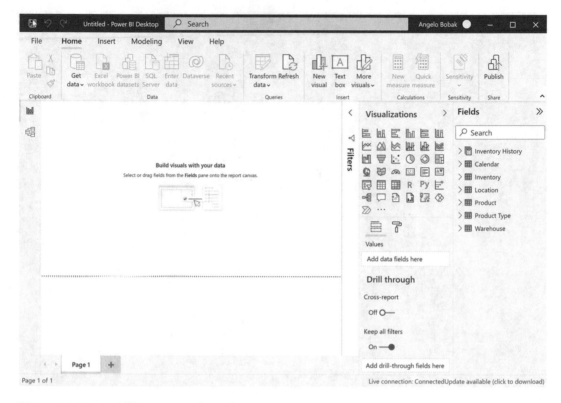

Figure 12-51. *All set up and ready to go*

Yes, we are ready to go!

We need to start adding data columns to our design area so we can start creating our report and graph. Notice that the tab is called Page 1. You can create multiple reports and graphs over multiple tabs in your project. These can be published to the Power BI server. In our last chapter, I will include a brief discussion on how to download and install the developer edition of the Power BI server.

Tip   It has been stated by experts in the design of data visualizations that the human mind can only digest around ten chunks of information. Spread out your visualizations in multiple pages so users can easily jump from one to the other. Don't try to cram everything on one page. I have seen this too many times in real life, and the results are visualizations that take hours to load and then leave users scratching their heads. The colors might be nice, but that's about it as far as usefulness.

Back to our report design. Let's start adding columns to create the basic report. Please refer to Figure 12-52.

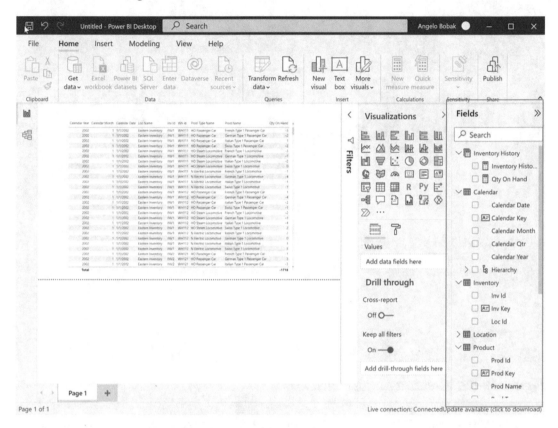

Figure 12-52. *Creating a simple data report*

Once connected, all the dimensions and fact tables together with the columns are available on the right-hand side in the Fields panel. Our goal at this design stage is to create a simple data dump report that has all the key dimension keys (not the surrogate keys) and the quantity-on-hand counts down to the product identifier level together with the location, inventory, and warehouse information. At this stage we do not want users looking at surrogate keys. They are meaningless to them.

Now we are ready to create a chart based on one of the many icons available in the **Visualization** section of Power BI. Let's take a look at an example and explain how it was created.

Please refer to Figure 12-53.

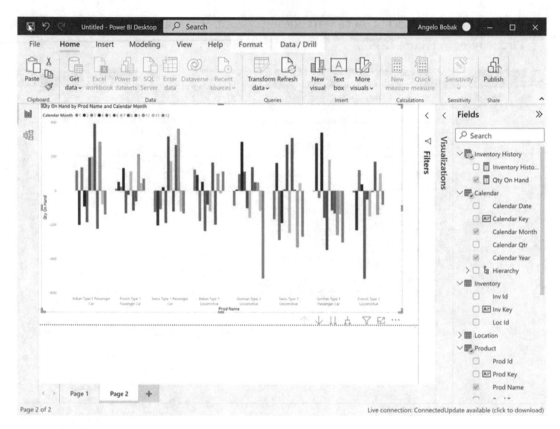

Figure 12-53. *Create a visual chart*

This step is really simple. On Page 1, copy the data report by right-clicking the report and selecting Copy from the drop-down menu. Open a second page and drop the data report in the design area. Next, pick a visualization by clicking it and dragging it over the data report. Your data report is now transformed into the visual chart you selected.

That's all that there is to it, at least to create a basic report and chart. This tool is very powerful and allows you to create filters, drill-downs, and much more. Learning this tool would require an entire book, but with these basic instructions, you can get started. Tons of great online courses are available also that you can pay for, but knowledge comes at a price.

Next, we see how we can drill down into the report.

Please refer to Figure 12-54.

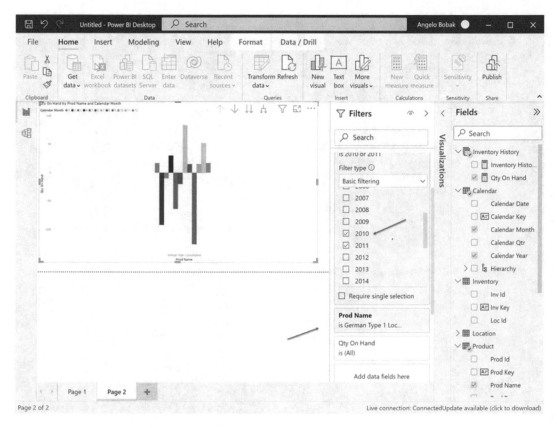

Figure 12-54. *Defining filters*

I did mention drill-down capabilities and features. By default, if you look at the basic filtering panel, you can click or unclick data fields that you want to see. Here we filter for the years "2010" and "2011" and for a single product "German Type 1 Locomotive." Just click the checkboxes, and the visual changes to represent the smaller data scope (this is called slicing and dicing).

Let's look at one more visual you can create, this time with a pie chart.

Please refer to Figure 12-55.

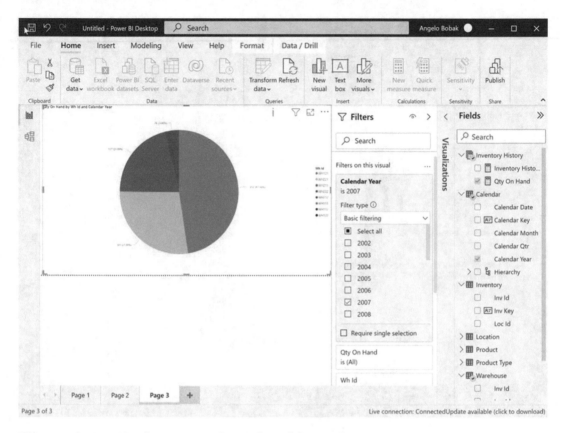

Figure 12-55. *Pie chart, quantity on hand for 2007*

Simply dragging a pie chart icon from the visual section onto a copy of a data report on a new page transforms the data into the pie chart seen in the preceding figure. Your report now has three pages consisting of one old-fashioned row and column report and two graphic charts. All this was accomplished in a few steps.

All that is left is to publish the report to a Power BI server.

Please refer to Figure 12-56.

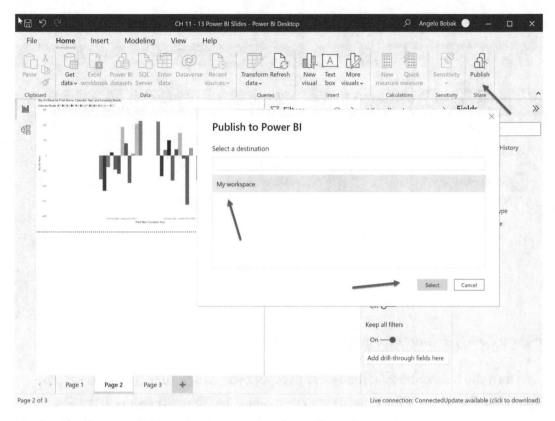

Figure 12-56. *Publishing your report and graphs*

Publishing the report is simple and is similar to what we did earlier to publish a report to SSRS. Simply click the "Publish" button located on the right side of the menu bar, and a pop-up panel appears labeled "Publish to Power BI."

Reminder   Don't forget to save the Power BI report you just created to an easy-to-remember folder with a name that reflects the report content! You know the drill. Go to the "File" selection, select "Save as," etc.

You will see the available destination. In this case there is only one called "My workspace" in my development environment.

Click it to select it, and then click the "Select" button. You will see a message stating that it is being published. Once finished, bring up your browser and enter the URL for your Power BI portal.

Please refer to Figure 12-57.

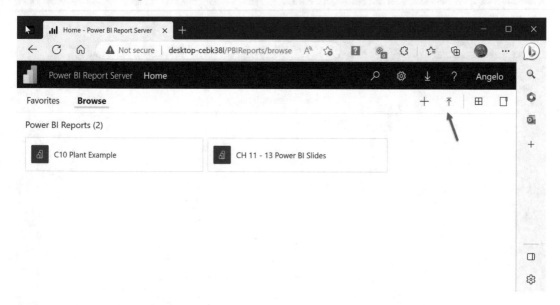

Figure 12-57. *Report and graphs on the server*

Once you have published the Power BI report, you will see it in the "Browse" tab. Uploading the report is the same as when you uploaded the SSRS report to the SSRS portal. Click the upload arrow that appears on the top right of the menu bar and navigate to the folder that has your new Power BI report. Click it and you should see the report as shown in the following.

Please refer to Figure 12-58.

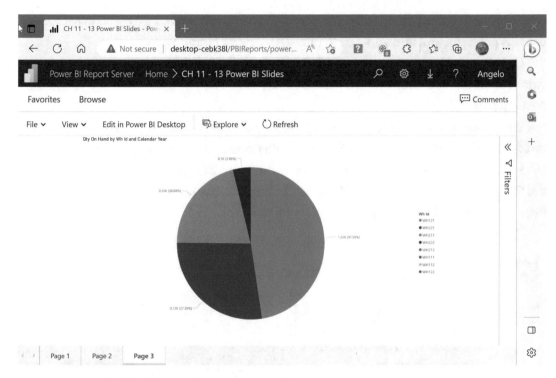

Figure 12-58. *Uploading your report and graphs*

There it is. Our pie chart is looking good, not too shabby.

This is probably not too impressive in black and white, but if you are reading this book in electronic format, you will see the colors.

There you have it. We covered some basic steps, created a simple three-page report or dashboard if you prefer, and published it to a personal Power BI server on your laptop, desktop, or development environment.

In our last chapter, Chapter 14, I will include a section on how to download and install SSRS and Power BI and how to configure each. Mostly, you will just keep the default settings. All you need to fill out is to pick a SQL Server instance where a dedicated database for SSRS and Power BI can be created.

Summary

This was our third to last chapter. I hope you found it interesting. We applied the ranking window functions to our inventory database and inventory data warehouse to put these functions through their paces. We performed the usual performance analysis. Additionally, we created an SSRS report using Report Builder and published it on an SSRS website.

Lastly, we created a simple Power BI report and dashboard so we could learn how to use this powerful tool and added it to our SQL Server BI toolbox.

Our next and last chapter applies the analytical window functions to our inventory data. This should be interesting, so take a break, grab a cup of coffee, and see you in the next chapter.

One more comment: Your SQL Server BI toolkit now consists of the aggregate, ranking, and analytical functions; Visual Studio Community Edition and the SSAS and SSRS components needed to create reports, cubes, and ETL projects; and Power BI and the SSIS, SSAS, SSRS, and Power BI servers. I almost forgot Microsoft Excel that you can use to create reports with graphs and pivot tables.

Please refer to Figure 12-59.

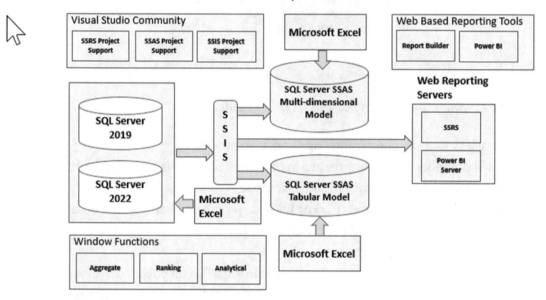

Figure 12-59. *Your SQL Server BI toolkit*

What can I say, a very impressive array of development tools for your BI project and career skills portfolio. These tools and skills are a must for anyone to be a SQL Server data architect or analyst or developer.

I will tell you how to obtain, install, and configure all these components in our last chapter, Chapter 14.

CHAPTER 13

Inventory Use Case: Analytical Functions

Congratulations! You made it to the last chapter dealing with window functions. We do have one more chapter, Chapter 14, which is a summary chapter that includes a review of the tools we used and also where to obtain them.

Analytical Functions

The following are the window analytical functions that should be familiar to you by now unless you jumped directly to this chapter from Chapter 1 because your career focus is inventory management:

- CUME_DIST(): Does not support window frames
- FIRST_VALUE(): **Supports window frames**
- LAST_VALUE(): **Supports window frames**
- LAG(): Does not support window frames
- LEAD(): Does not support window frames
- PERCENT_RANK(): Does not support window frames
- PERCENTILE_CONT(): Does not support window frames
- PERCENTILE_DISC(): Does not support window frames

As we can see, only two functions support window frame specifications. The rest do not support window frames. When you understand how these functions work based on the descriptions in this chapter, it makes sense.

© Angelo Bobak 2023
A. Bobak, *SQL Server Analytical Toolkit*, https://doi.org/10.1007/978-1-4842-8667-8_13

I think these functions are important in inventory analysis, particularly the LAG() and LEAD() functions that can be used to really get into historical inventory movement data.

Lastly, we will perform our usual performance analysis with baseline estimated query plans and IO and TIME statistics. We will also do something new; we will collect all IO and TIME statistics across the analytical functions and graph them to check for similarities and also functions that consume a lot of resources.

Tip   I recommend you perform this last activity with the other category of functions we discussed in the book in order to really understand how they affect your environment. This will give you some indication as to when you can use plain old CTE blocks or rely on some denormalization with report tables or even memory-enhanced tables.

CUME_DIST() Function

Recall from Chapters 7 and 10 that this function calculates the relative position of a value within a data set like a table, partition, or table variable loaded with test data. It calculates the probability that a random value, like a failure rate, is less than or equal to a particular failure rate for a piece of equipment. It is cumulative because it takes all values into account when calculating the distinct value for a specific value.

If you are a bit unsure of this, go back to Chapter 7 or Appendix A and review how this function works and how it is applied.

Let's get started. Here are the business specifications supplied by our business analyst friend:

A warehouse report needs to be created showing the cumulative distribution values for the year "2002," product "P101," location "LOC1," inventory "INV1," and warehouse "WH112." (I did not use the names to save space. If you are curious, check out the master data tables for these.)

In other words, for each month of the year, what is the percentage of average values for each month that are less than or equal to the current month? If the current month is "August" and the average value of the quantity on hand is "46," what percentage of the other values are less than or equal to this value?

Detailed calendar information such as year, quarter, and month needs to appear in the report. The calendar quarters and months need to be spelled out, and the numerical month values need to be included so they can be used to sort results in a Microsoft Excel spreadsheet that we will use to generate a graph for the results.

These requirements have a lot of detail. Hopefully by now you are getting used to receiving written requirements and translating them into some TSQL code. This is typical in a work environment. Sometimes you will get requirements that are very fuzzy. Sometimes you will be left a message on your phone giving you verbal requirements. These are always fun.

Here is what we came up with. Please refer to Listing 13-1.

Listing 13-1. Cumulative Distribution for Monthly Averages

```
WITH YearlyWarehouseReport (
      AsOfYear,AsOfQuarter,AsOfMonth,AsOfDate,LocId
      ,InvId,WhId,ProdId,AvgQtyOnHand
      )
AS
(
SELECT AsOfYear
      ,DATEPART(qq,AsOfDate) AS AsOfQuarter
      ,AsOfMonth
      ,AsOfDate
      ,LocId
      ,InvId
      ,WhId
      ,ProdId
      ,AVG(QtyOnHand) AS AvgQtyOnHand
FROM Product.Warehouse
GROUP BY AsOfYear
      ,DATEPART(qq,AsOfDate)
      ,AsOfMonth
      ,AsOfDate
      ,LocId
      ,InvId
      ,WhId
      ,ProdId
```

```
)
SELECT AsOfYear
     ,AsOfQuarter
          ,CASE
          WHEN DATEPART(qq,AsOfDate) = 1 THEN '1st Quarter'
          WHEN DATEPART(qq,AsOfDate) = 2 THEN '2nd Quarter'
          WHEN DATEPART(qq,AsOfDate) = 3 THEN '3rd Quarter'
          WHEN DATEPART(qq,AsOfDate) = 4 THEN '4th Quarter'
     END AS AsOfQtrName
     ,AsOfMonth
     ,CASE
          WHEN AsOfMonth = 1 THEN 'Jan'
          WHEN AsOfMonth = 2 THEN 'Feb'
          WHEN AsOfMonth = 3 THEN 'Mar'
          WHEN AsOfMonth = 4 THEN 'Apr'
          WHEN AsOfMonth = 5 THEN 'May'
          WHEN AsOfMonth = 6 THEN 'June'
          WHEN AsOfMonth = 7 THEN 'Jul'
          WHEN AsOfMonth = 8 THEN 'Aug'
          WHEN AsOfMonth = 9 THEN 'Sep'
          WHEN AsOfMonth = 10 THEN 'Oct'
          WHEN AsOfMonth = 11 THEN 'Nov'
          WHEN AsOfMonth = 12 THEN 'Dec'
     END AS AvgMonthName
     ,AsOfDate
     ,LocId
     ,InvId
     ,WhId
     ,ProdId
     ,AvgQtyOnHand AS MonthlyAvgQtyOnHand
     ,CUME_DIST() OVER (
          PARTITION BY AsOfYear
          ORDER BY AsOfMonth ASC
     ) AS RollingYearCume
FROM YearlyWarehouseReport
```

```
WHERE AsOfYear = 2002
AND ProdId = 'P101'
AND Locid = 'LOC1'
AND InvId = 'INV1'
AND WhId = 'WH112'
GO
```

In order to satisfy the calendar date part requirements, two CASE blocks are needed to decode the numerical quarter and month values into their respective text values (1 = January and 1 = 1st Quarter and so on).

The OVER() clause includes a PARTITION BY clause that sets up a partition for the year. This is not really required right now as we are filtering for the year "2002" in the WHERE clause, but if we include more years, we want to leave it in. This does not seem to affect performance by the way. Try running estimated query plans with and without the year in the PARTITION BY clause.

Lastly, an ORDER BY clause is included that orders the partition by AsOfMonth column. Let's see the results and then create a graph with Microsoft Excel.

Homework   Modify the query so it retrieves the calendar quarter and month from the new enhanced Calendar table. Also, modify the query so results are calculated for all years. Practice graphing each result set in a Microsoft Excel so as to create a profile of cumulative distribution values across the years. Run our usual estimated query plan and IO and TIME statistics analysis to see if the modified query performs better or worse now that it includes a JOIN to the Calendar table. Lastly, run the plan and IO/TIME statistics to get an average. Make sure you execute DBCC prior to each run.

Please refer to the partial results in Figure 13-1.

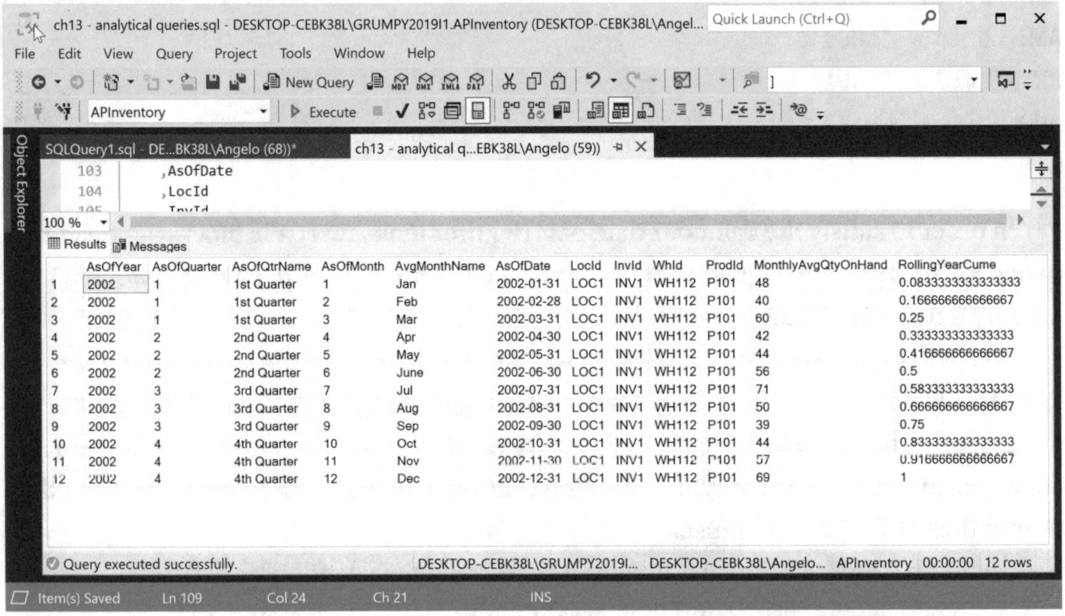

Figure 13-1. *Cumulative distribution analysis for monthly averages*

The RollingYearCume values are sorted in ascending order as the PARTITION BY clause specified that the ORDER BY clause sorted by month, we can clearly see the cumulative distribution will generate a nice upward curving graph. Let's copy this result into a Microsoft Excel spreadsheet.

Please refer to Figure 13-2.

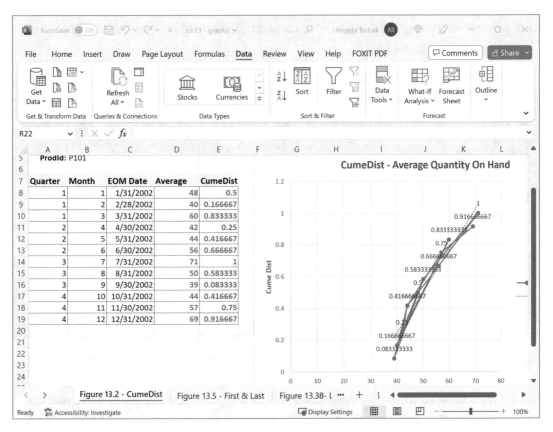

Figure 13-2. *Cumulative distribution, average quantity on hand*

I modified the names a bit and added some formatting to make the report look professional. The results are now sorted by month. We can see an interesting upward-trending graph. If you want to see the actual values less than a specific value and when they occurred, sort the data set by the average values, and you can see the values and the month they occurred in.

Query performance and analysis time.

Performance Considerations

As usual, let's start by generating a baseline estimated query plan in the usual manner.

Please refer to Figure 13-3.

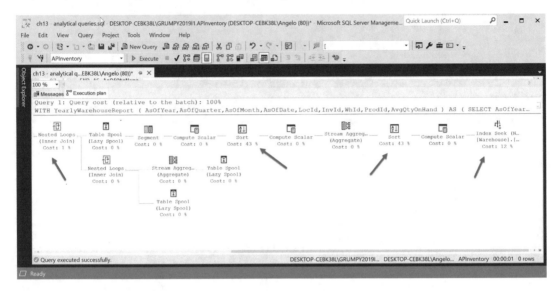

Figure 13-3. *Query plan for cumulative distribution, average quantity on hand*

We will concentrate on the right-hand side of the plan. The left-hand side costs are all zero, so we will not discuss them in order to save some chapter space. Always review these on your own though. Also, make sure to place your mouse on the interesting tasks in order to display the details panel. As a developer or architect, you need to become an expert in query performance tuning!

Back to the plan, no indexes were suggested. So we are off to a good start. We see an index seek with a cost of 12% (remember, an index seek is better than a scan). The next expensive tasks are the two sort tasks chiming in at 43%; the remaining tasks are all zero costs except for the nested loops task on the left side. The lower branch also has zero-cost tasks, and both branches meet at a nested loops (inner join) task.

All in all, it seems like a well-performing query. Let's look at the IO and TIME statistics.

Please refer to Figure 13-4.

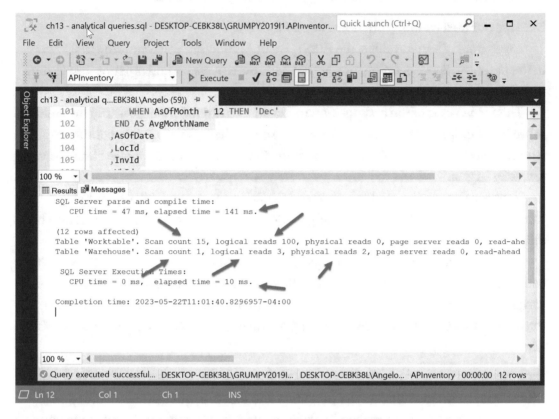

Figure 13-4. *IO and* TIME *statistics for cumulative distribution, average quantity on hand*

The parse and compile time took 141 ms and the elapsed execution time took 10 ms. As can be seen, there is a scan count of 15 scans and 100 logical reads. The Warehouse table has a scan count of 1, a logical read value of 3, and a physical read value of 2 (remember this pattern, 1-3-2; it will pop up again).

Checking the row count of the Warehouse table, we see that it contains 24,288 rows. Not too large but still enough to warrant the physical reads.

If we modify this query so that there is no WHERE clause, we can use it in an SSRS report. We need to consider using our tried-and-true report table approach with historical data whenever there is a benefit (like the CTE returns a couple of hundreds of thousands of rows), that is, preload the table with results from this query so that performance is fast when analysts use the web reports. Might want to add a few indexes to cover the master data columns like location and inventory and the others. This data is loaded once with all available records and, as the days go by, only the new or updated records.

FIRST_VALUE( ) and LAST_VALUE( ) Functions

Here are our next window functions. Remember these?

Given a set of sorted values, the FIRST_VALUE() function will return the first value in the partition. So, depending on whether you sort it in ascending order or descending order, the value will change!

This is the value of the first position in the data partition, not necessarily the smallest value in the data set. This is easy to misunderstand when you are first becoming familiar with this function. One could misinterpret the first value meaning smallest and last value meaning largest.

Next, given a set of sorted values, the LAST_VALUE() function will return the last value in the partition. Again, depending on how you sort the partition, this will also be different as discussed earlier.

Both functions return whatever data type was used as an argument. By the way, it could be an integer, a string, a date, etc. If this is a little unclear, create your own table variable with ten or more rows and write some queries against it so you can see how these window functions behave. Practicing a little will make the usage and results of the window functions clear.

Reminder   When learning new functions like these, always practice with small simple data sets so the function behavior clearly stands out. Also, it will help you understand how the query performs when you run the estimated execution plans and the statistics. Nothing like running a query that takes hours to run and returns millions of rows only to find out the logic you used was wrong!

Let's dive right into the example. Here is our business specification:

A report is needed that shows on a month-by-month basis if the average inventory on hand for our favorite location "LOC1," inventory "INV1," warehouse "WH112," and product "P010" is going up or down. The results have to be on a rolling basis. That is, we want the first and last values for the year as the months go by. The calendar year is 2002, so we will need it in the WHERE clause.

The first value will be the value for January and does not change. As the partition grows, in order to consider the next month, the average for the new month will be the last value. We need to calculate the difference as we roll month by month.

Based on these business requirements, this is what we came up with. We will look at the base portion of the query as the CTE used is the same as the prior query. Please refer to Listing 13-2.

Listing 13-2. Rolling First and Last Value Monthly Averages

```
/* CTE goes here */

SELECT AsOfYear
     ,AsOfQuarter
     ,CASE
          WHEN DATEPART(qq,AsOfDate) = 1 THEN '1st Quarter'
          WHEN DATEPART(qq,AsOfDate) = 2 THEN '2nd Quarter'
          WHEN DATEPART(qq,AsOfDate) = 3 THEN '3rd Quarter'
          WHEN DATEPART(qq,AsOfDate) = 4 THEN '4th Quarter'
     END AS AsOfQtrName
     ,AsOfMonth
     ,CASE
          WHEN AsOfMonth = 1 THEN 'Jan'
          WHEN AsOfMonth = 2 THEN 'Feb'
          WHEN AsOfMonth = 3 THEN 'Mar'
          WHEN AsOfMonth = 4 THEN 'Apr'
          WHEN AsOfMonth = 5 THEN 'May'
          WHEN AsOfMonth = 6 THEN 'June'
          WHEN AsOfMonth = 7 THEN 'Jul'
          WHEN AsOfMonth = 8 THEN 'Aug'
          WHEN AsOfMonth = 9 THEN 'Sep'
          WHEN AsOfMonth = 10 THEN 'Oct'
          WHEN AsOfMonth = 11 THEN 'Nov'
          WHEN AsOfMonth = 12 THEN 'Dec'
      END AS AvgMonthName
     ,AsOfDate
     ,LocId
     ,InvId
     ,WhId
     ,ProdId
```

```
      ,AvgQtyOnHand AS MonthlyAvgQtyOnHand
      ,FIRST_VALUE(AvgQtyOnHand) OVER (
            PARTITION BY AsOfYear
            ORDER BY AsOfDate
            ) AS FirstValue
      ,LAST_VALUE(AvgQtyOnHand) OVER (
            PARTITION BY AsOfYear
            ORDER BY AsOfDate
            ) AS LastValue
      ,LAST_VALUE(AvgQtyOnHand) OVER (
            PARTITION BY AsOfYear
            ORDER BY AsOfDate
            ) -   FIRST_VALUE(AvgQtyOnHand) OVER (
                  PARTITION BY AsOfYear
                  ORDER BY AsOfDate
            ) AS Change
FROM YearlyWarehouseReport
WHERE AsOfYear = 2002
AND ProdId = 'P101'
AND Locid = 'LOC1'
AND InvId = 'INV1'
AND WhId = 'WH112'
GO
```

We used the query used for the cumulative distribution example and replaced the CUME_DIST() function with the FIRST_VALUE() and LAST_VALUE() functions. Very easy modification. As you come up with solutions in your day-to-day work activities, you want to reuse queries and scripts as much as possible in order to avoid reinventing the wheel. These are other toolkit components, scripts, and queries you can reuse and modify to satisfy new business requirements.

Tip   Create a dedicated folder so you can store all your scripts, templates, and queries. This way they are readily available for each new coding assignment. These are part of your toolkit too!

Both OVER() clauses have a PARTITION BY clause and an ORDER BY clause. The partition is set up by year, and the ORDER BY clause is set up by AsOfDate instead of AsOfMonth. The AsOfDate happens to be the last day of the month, so it works out. We should have used the AsOfMonth value instead in order to avoid confusion with the query logic, but it works.

Let's see the results. Please refer to the partial results in Figure 13-5.

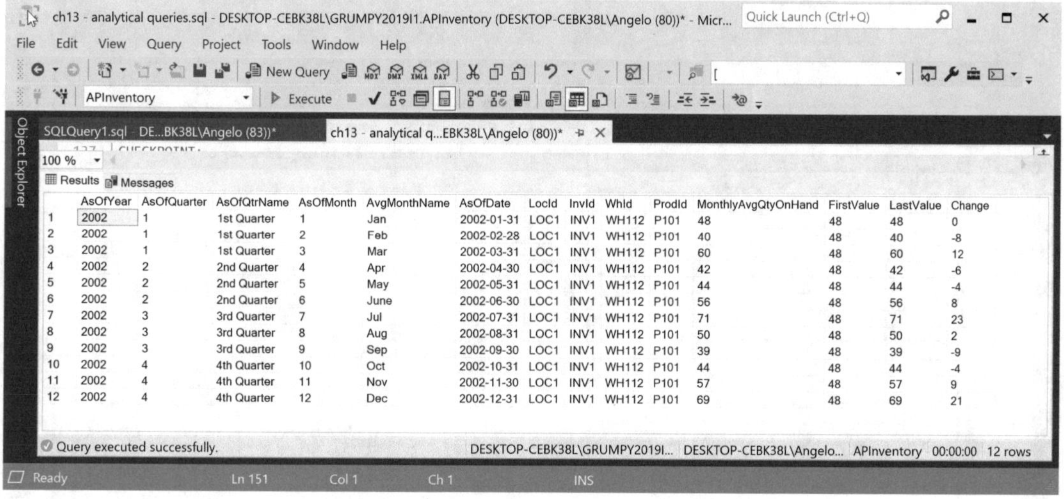

Figure 13-5. *Average monthly quantity, first and last values*

These are good results that an analyst can use to analyze average inventory movement on a month-by-month basis. Granted, this is for only one year, but you know we can remove the filters and use this query to populate a report table that is accessed by an SSRS report that allows users to set up filters that suit the patterns they wish to analyze. Or maybe use it for a pivot table in an Excel spreadsheet or a denormalized table for an SSAS multidimensional cube or even a Power BI scorecard.

That's why performance tuning is important. Returning 12 rows of data for one year is much different than returning thousands of rows for multiple locations, inventories, warehouses, and products (easy to validate too).

Note the change as we go month by month. This data will look impressive in a Microsoft Excel spreadsheet, so let's do our usual copy-and-paste routine.

Please refer to Figure 13-6.

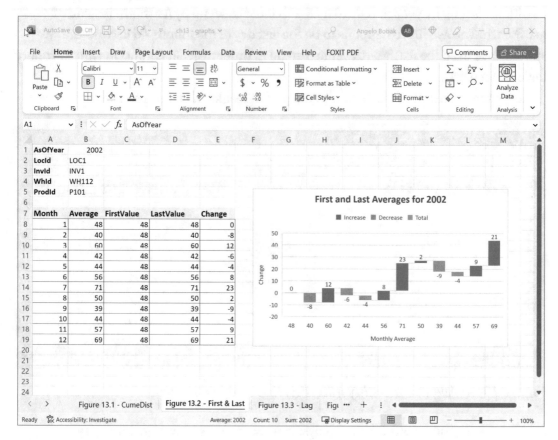

Figure 13-6. *Excel first and last value analysis*

This is another analysis that allows the business analyst to identify if inventory movements are going up or down, which indicates if sales are going up or down. If inventory is going up and not down, then the sales analyst needs to have a meeting with the marketing team to understand why products are not selling and products remain in inventory. This involves a detailed analysis to see the offending stores that have low sales and why (if you are viewing this via an eBook, the colors really stand out)!

Performance Considerations

Here comes our usual baseline estimated execution plan.

Please refer to Figure 13-7.

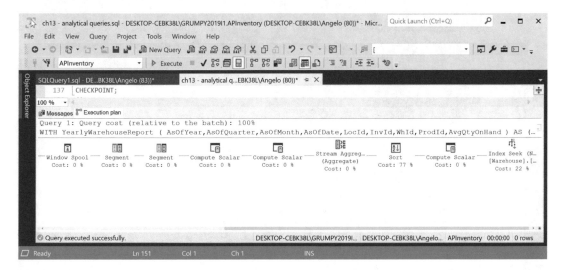

Figure 13-7. *First and last value baseline estimated performance plan*

Seems like a well-performing query. An index was not suggested, and we see by the index seek (22% cost) that an existing index is being used.

This is representative of real production environments, that is, as more and more queries are being created, new index requirements are identified and indexes created, which can then be used by future queries. So, in theory as you create new queries, you do not have to create new indexes each and every time. That's the goal anyway.

Note   Remember that indexes may speed up a query, but they will slow down row insert, update, and delete operations. Indexes need to be updated for these operations, so the more indexes you have, the slower performance you will see.

I almost forgot the 77%-cost sort task is something to look at. The value is high. Setting PROFILE STATISTICS ON, we see that the task is sorting by the columns AsOfDate, an expression, and AsOfMonth.

Turns out the expression is the DATEPART() function used to pull out the calendar quarter. So here is some code that needs to be optimized.

Let's complete the analysis with some IO and TIME statistics. Please refer to Figure 13-8.

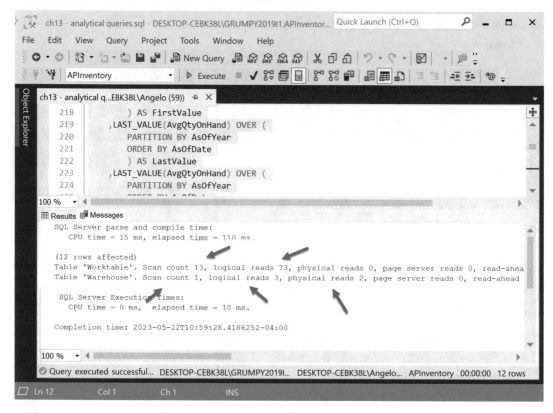

Figure 13-8. *IO and TIME statistics*

The query executes in 10 ms, under a second (parse time is 110 ms). We see the scan count value is at 13 and the logical reads at 73, a bit high.

There's the same pattern for scan counts, logical reads, and physical reads: 1-3-2. Makes sense. We are using the same table.

Possibly joining this query with the enhanced Calendar table we created earlier to pull out the calendar quarter text, instead of generating it by pulling it out of the date, might be a better solution, but the cost of a table JOIN needs to be considered. This strategy would also remove the CASE blocks used to decode the quarter and month numerical values to their corresponding text values.

Tip   Avoid using functions like DATEPART( ) wherever possible, especially in WHERE clause predicates. This will affect performance.

Not one to lag behind, we move to our next function. (Yes, bad pun. I can see the technical reviewer rolling her eyes!)

LAG( ) Function

Recall what the LAG( ) function does:

This function is used to retrieve the previous value of a column relative to the current row column value (usually within a time dimension), in our case inventory movement dates.

This query can help answer questions like

How does the current month compare to the prior month or prior quarter or prior year?

This is a counterpart of the LEAD( ) function, which we will look at next. Here are the specifications for the query our analyst wants:

Our business analyst wants us to modify the prior query so that it shows the current average monthly inventory values but also the prior one-month, prior three-month, and prior 12-month averages (for the same month):

- 1: Go back one month.

- 3: Go back three months (one quarter relative to the current month).

- 12: Go back 12 months (one year relative to the current month).

The same date objects are required, that is, spell out the quarter and month names. And also filter by the same product, location, inventory, and warehouse. The report has to span three years: 2002, 2003, and 2004.

Copying and pasting our last query and making a few modifications delivers the following query. Note the CTE is not copied, as it is the same one as the prior example.

Please refer to Listing 13-3.

Listing 13-3. Current Month Movements vs. Last Month, Quarter, and Year

```
/* CTE goes here */

SELECT AsOfYear
      ,AsOfQuarter
      ,CASE
            WHEN DATEPART(qq,AsOfDate) = 1 THEN '1st Quarter'
```

```
        WHEN DATEPART(qq,AsOfDate) = 2 THEN '2nd Quarter'
        WHEN DATEPART(qq,AsOfDate) = 3 THEN '3rd Quarter'
        WHEN DATEPART(qq,AsOfDate) = 4 THEN '4th Quarter'
END AS AsOfQtrName
,AsOfMonth
,CASE
        WHEN AsOfMonth = 1 THEN 'Jan'
        WHEN AsOfMonth = 2 THEN 'Feb'
        WHEN AsOfMonth = 3 THEN 'Mar'
        WHEN AsOfMonth = 4 THEN 'Apr'
        WHEN AsOfMonth = 5 THEN 'May'
        WHEN AsOfMonth = 6 THEN 'June'
        WHEN AsOfMonth = 7 THEN 'Jul'
        WHEN AsOfMonth = 8 THEN 'Aug'
        WHEN AsOfMonth = 9 THEN 'Sep'
        WHEN AsOfMonth = 10 THEN 'Oct'
        WHEN AsOfMonth = 11 THEN 'Nov'
        WHEN AsOfMonth = 12 THEN 'Dec'
 END AS AvgMonthName
,AsOfDate
,LocId
,InvId
,WhId
,ProdId
,AvgQtyOnHand AS MonthlyAvgQtyOnHand
,LAG(AvgQtyOnHand,1,0) OVER (
        ORDER BY AsOfDate
        ) AS PriorMonthAverage
,LAG(AvgQtyOnHand,3,0) OVER (
        ORDER BY AsOfDate
        ) AS PriorQuarterAverage
,LAG(AvgQtyOnHand,12,0) OVER (
        ORDER BY AsOfDate
        ) AS PriorYearAverage
,AvgQtyOnHand -LAG(AvgQtyOnHand,1,0)
```

```
            OVER (
                    ORDER BY AsOfDate
                    ) AS Change
        ,AvgQtyOnHand -LAG(AvgQtyOnHand,1,0)
            OVER (
                    PARTITION BY AsOfYear
                    ORDER BY AsOfDate
                    ) AS Change
FROM YearlyWarehouseReport - the CTE
WHERE AsOfYear = 2002
AND ProdId = 'P101'
AND Locid = 'LOC1'
AND InvId = 'INV1'
AND WhId = 'WH112'
GO
```

Let's analyze the three LAG() functions.

Firstly, there is no partition set up for the year as we want the results to span all the years in the data set. Each partition has an ORDER BY clause that refers to the AsOfDate column. The only difference is that each LAG() function uses a different parameter to specify the time period we need to go back to:

> 1: Go back one month.

> 3: Go back three months (one quarter relative to the current month).

> 12: Go back 12 months (one year relative to the current month).

The column names are misleading as they do not represent actual quarters or years but represent going back three or 12 months (oh well, I leave it up to you to come up with better names).

The "0" value parameter indicates that we need to display "0" instead of a NULL value when we reach the beginning of a boundary. That is, no prior rows exist to satisfy the span of time required by the LAG() function.

Lastly, for the current/prior month combination, we include a simple formula to calculate the change by subtracting the value returned by the LAG() function from the current month's average value.

Homework   Modify the query so it also generates the percent change between the current month and last month. If you are ambitious, do the same for a quarter and year (extra points!). Hint: You can reuse the code that generated the change value and modify it, so it returns a percentage. Might also want to add logic to display six months back. This would be a powerful report. Check out the estimated performance query plan for this.

Please refer to the partial results in Figure 13-9.

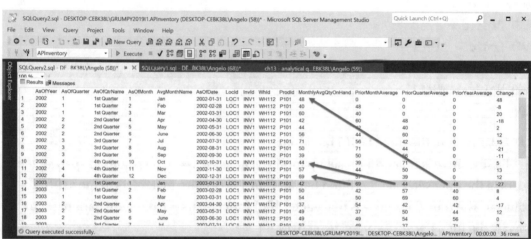

Figure 13-9. *This month vs. last month, quarter, and year averages*

The arrows indicate that the lag effect is working. Notice all the zero values that indicate no prior rows exist to satisfy the earlier time references.

This is a good report as it allows the analyst to look at historical inventory movements and compare them to the current movements. If past movements were indicative of sales, that is, lots of movements out of inventory, what conditions were prevalent vs. this month's lower movements? Graphing these is a good visualization of inventory movements.

Based on my comment, let's generate two types of graphs in a Microsoft Excel spreadsheet with this report.

Please refer to the partial results in Figure 13-10.

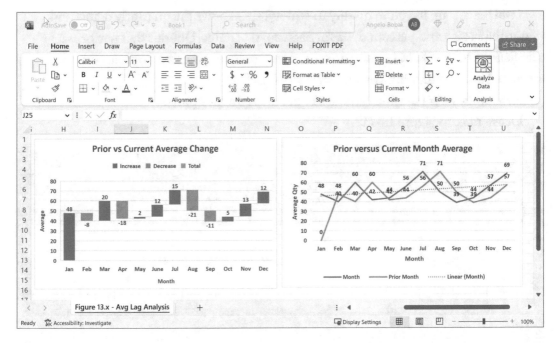

Figure 13-10. *Prior vs. current average analysis (monthly only)*

The first graph shows increase vs. decrease in terms of a bar chart, and the second graph uses a line chart together with the values against the months. We can see that both types of movements are trending up, which indicates healthy sales order volumes. You would not want to see large inventory movements into inventory and little if no inventory moving out.

Tip   As a SQL developer or data integration architect, it is strongly recommended that you learn strong Microsoft Excel skills so you can further deliver quality reports to your users. Comes in handy when you are verifying query results too!

Performance Considerations

This will be a large plan as we are using the LAG() function four times, three times to generate the prior month, quarter, and year values plus once in a calculation to show the differences between this month and last month.

I will leave it up to you to modify the query, so it shows the quarterly deltas and yearly deltas also and percent delta for each span of dates. Queries that perform calculations in the SELECT clause are very performance intensive, so check out the estimated query plans and IO and TIME statistics.

Here comes our baseline estimated query plan in four parts.

Please refer to Figure 13-11.

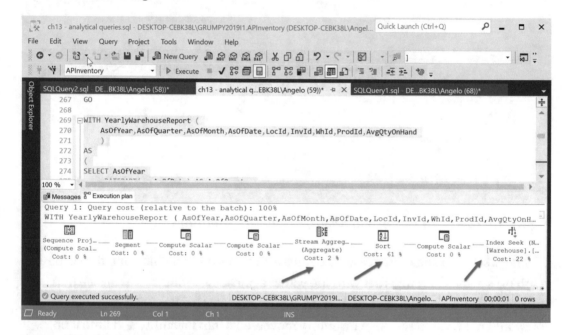

Figure 13-11. *Estimated baseline execution plan, part 1*

Once again, no index is suggested, so we are off to a good start. Sometimes, you will see a plan that uses a table index, but another index is also suggested. We see the index seek task at 22% and the expensive sort task at 61% cost. We saw this pattern earlier, and we know who the culprit is. It's the DATEPART() function, and we know what to do to improve performance. Sometimes performance improvement involves rewriting the query. Or sometimes rewriting the query will make things worse!

Here comes part 2 of the plan. Please refer to Figure 13-12.

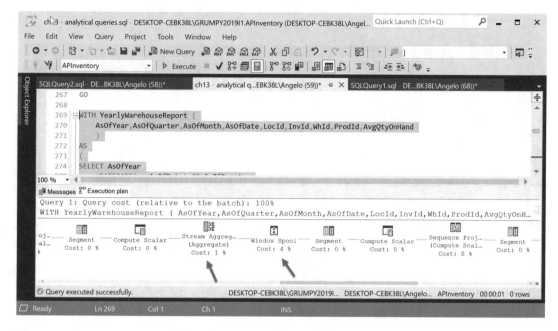

Figure 13-12. *Estimated baseline execution plan, part 2*

A window spool appears at a cost of 4%. This is for the first LAG() function. We can expect to see this pattern for the other LAG() functions in the plan.

Recall that a non-zero value means spooling occurs on physical storage as opposed to memory, so it will be slower. Not much to do here except to hope that your system administrator installs a fast solid-state disk for TEMPDB.

Here comes part 3. Please refer to Figure 13-13.

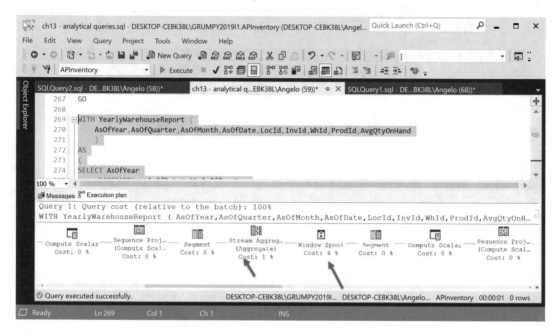

Figure 13-13. *Estimated baseline execution plan, part 3*

I told you. Here is another window spool task with the same 4% cost, followed by the stream aggregate with a cost of 1%. One more plan section to go.

Please refer to Figure 13-14.

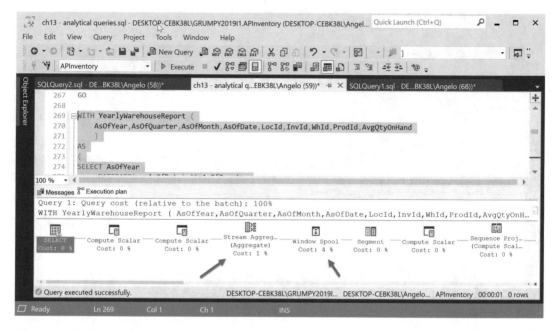

Figure 13-14. *Estimated baseline execution plan, part 4*

Here is the last section with the last window spool task and a stream aggregate task with the same costs, 4% and 1%.

A possible option is to preload the LAG() results in a fast table with keys so the query can perform a JOIN to retrieve the values. Since the data is historical, this might be an option.

Last but not least, to complete the analysis picture, here come the IO and TIME statistics.

Remember, sometimes if we are still stuck and do not have enough information to assess performance issues, we might even need to set the PROFILE STATISTICS on and dig deeper.

Please refer to Figure 13-15.

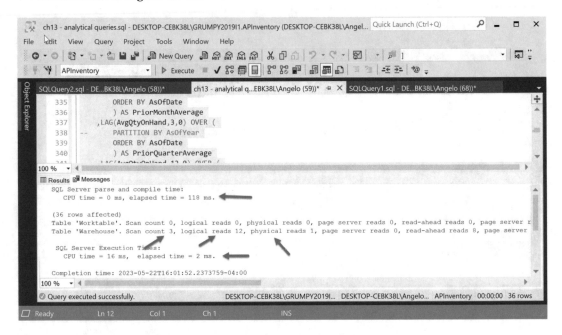

Figure 13-15. *IO and TIME statistics for the LAG() function*

This is interesting. Despite the repeated use of the LAG() function, the IO and TIME statistics are low! The SQL Server execution time came in at 2 ms. Work table values are all zero, and the Warehouse table values reflect a low scan count (3), low logical reads (12), and low physical reads (1). Looks like our pattern changed with this function.

Note   I do not think I mentioned it, but generating all these statistics also takes its toll on performance, so keep this in mind. Do it only on a development environment.

Now that we were done lagging, let's lead (hey, last recipe chapter, so I get to have the bad puns come in)!

LEAD() Function

The counterpart of the LAG() function is the LEAD() function. This function is used to retrieve the next value relative to the current row column (in case of date objects, future dates within a historical set of data).

Here are our business specifications:

Our analyst wants the same exact report as the "looking back in time" report, but this time we need to "look ahead."

We will take a different approach and use the CTE portion of the query to create a VIEW. An option is available to create what is called a schema-bound VIEW that actually stores the results of the query on physical storage.

Here is the DDL command for creating the TSQL VIEW.

Please refer to Listing 13-4.

Listing 13-4. VIEW Replacing the CTE

```
CREATE OR ALTER VIEW [Reports].[MonthlySumInventory]
-- use below to implement a materialized view
-- WITH SCHEMABINDING
AS
WITH WarehouseCTE (
        InvYear,InvQuarterName,InvMonthNo,InvMonthName,LocationId
        ,LocationName,InventoryId,WarehouseId,WarehouseName,ProductId
        ,ProductName,ProductType,ProdTypeName,MonthlySumQtyOnHand
)
AS (
SELECT YEAR(C.[CalendarDate])  AS InvYear
        ,CASE
                WHEN DATEPART(qq,C.[CalendarDate]) = 1 THEN 'Qtr 1'
                WHEN DATEPART(qq,C.[CalendarDate]) = 2 THEN 'Qtr 2'
                WHEN DATEPART(qq,C.[CalendarDate]) = 3 THEN 'Qtr 3'
                WHEN DATEPART(qq,C.[CalendarDate]) = 4 THEN 'Qtr 4'
        END AS AsOfQtrName
        ,MONTH(C.[CalendarDate]) AS InvQuarterName
        ,CASE
                WHEN MONTH(C.[CalendarDate])  = 1 THEN 'Jan'
                WHEN MONTH(C.[CalendarDate])  = 2 THEN 'Feb'
                WHEN MONTH(C.[CalendarDate])  = 3 THEN 'Mar'
                WHEN MONTH(C.[CalendarDate])  = 4 THEN 'Apr'
                WHEN MONTH(C.[CalendarDate])  = 5 THEN 'May'
                WHEN MONTH(C.[CalendarDate])  = 6 THEN 'June'
```

```
                      WHEN MONTH(C.[CalendarDate])   = 7 THEN 'Jul'
                      WHEN MONTH(C.[CalendarDate])   = 8 THEN 'Aug'
                      WHEN MONTH(C.[CalendarDate])   = 9 THEN 'Sep'
                      WHEN MONTH(C.[CalendarDate])   = 10 THEN 'Oct'
                      WHEN MONTH(C.[CalendarDate])   = 11 THEN 'Nov'
                      WHEN MONTH(C.[CalendarDate])   = 12 THEN 'Dec'
             END AS InvMonthMonthName
          ,L.LocId              AS LocationId
          ,L.LocName            AS LocationName
          ,I.InvId              AS InventoryId
          ,W.WhId               AS WarehouseId
          ,W.WhName             AS WarehouseName
          ,P.ProdId             AS ProductId
          ,P.ProdName           AS ProductName
          ,P.ProdType           AS ProductType
          ,PT.ProdTypeName      AS ProdTypeName
        ,SUM(IH.[QtyOnHand]) AS MonthlySumQtyOnHand
FROM [Fact].[InventoryHistory] IH
JOIN [Dimension].[Location] L
      ON IH.LocKey = L.LocKey
JOIN [Dimension].[Calendar] C
      ON IH.CalendarKey = C.CalendarKey
JOIN [Dimension].[Warehouse] W
      ON IH.WhKey = W.WhKey
JOIN [Dimension].[Inventory] I
      ON IH.[InvKey] = I.InvKey
JOIN [Dimension].[Product] P
      ON IH.ProdKey = P.ProdKey
JOIN [Dimension].[ProductType] PT
      ON P.ProductTypeKey = PT.ProductTypeKey
GROUP BY YEAR(C.[CalendarDate])
        ,MONTH(C.[CalendarDate])
        ,DATEPART(qq,C.[CalendarDate])
        ,L.LocId
        ,L.LocName
```

```
        ,I.InvId
        ,W.WhId
        ,W.WhName
        ,P.ProdId
        ,P.ProdName
        ,P.ProdType
        ,PT.ProdTypeName
)
SELECT InvYear,InvQuarterName,InvMonthNo,InvMonthName,LocationId
      ,LocationName,InventoryId,WarehouseId,WarehouseName,ProductType
      ,ProductId,ProductName,MonthlySumQtyOnHand
FROM WarehouseCTE
GO
```

Nothing much to explain as we saw the query before. It demonstrates you can use the CTE to create a TSQL VIEW object. The following code is commented out:

```
-- use below to implement a materialized view
--WITH SCHEMABINDING
```

We will run the example without the schema binding, but as homework you should recreate the VIEW object with schema binding enabled and see what the estimated performance plan looks like. Also check the IO and TIME statistics with the schema-bound VIEW.

The following is the query that references the new VIEW.

Please refer to Listing 13-5.

Listing 13-5. LEAD Example That Uses the Table VIEW

```
SELECT [InvYear]
      ,InvQuarterName
      ,InvMonthNo
      ,[InvMonthName]
      ,[LocationId]
      ,[LocationName]
      ,[InventoryId]
      ,[WarehouseId]
      ,[ProductType]
```

897

```
        ,[ProductId]
        ,MonthlySumQtyOnHand
        ,LEAD(MonthlySumQtyOnHand,1,0) OVER (
                PARTITION BY [LocationId],[InventoryId],[WarehouseId],
                [ProductId]
                ORDER BY InvYear,InvMonthNo,[LocationId],[InventoryId]
                        ,[WarehouseId],[ProductId]
                ) AS NextMonthSum
        ,LEAD(MonthlyAvgQtyOnHand,3,0) OVER (
                PARTITION BY [LocationId],[InventoryId],[WarehouseId],
                [ProductId]
                ORDER BY InvYear,InvMonthNo,[LocationId],[InventoryId]
                        ,[WarehouseId],[ProductId]
                ) AS Skip3MonthSum
        ,LEAD(MonthlyAvgQtyOnHand,12,0) OVER (
                PARTITION BY [LocationId],[InventoryId],[WarehouseId],
                [ProductId]
                ORDER BY InvYear,InvMonthNo,[LocationId],[InventoryId]
                        ,[WarehouseId],[ProductId]
                ) AS Skip12MonthSum
        ,MonthlySumQtyOnHand - LEAD(MonthlySumQtyOnHand,1,0)
                OVER (
                PARTITION BY [LocationId],[InventoryId],[WarehouseId],
                [ProductId]
                ORDER BY InvYear,InvMonthNo,[LocationId],[InventoryId]
                        ,[WarehouseId],[ProductId]
                        ) AS Change
FROM [Reports].[MonthlySumInventory]
GO
```

This is the same exact logic as the LAG() example except we replaced LAG() with LEAD(). Once again reusability of code leads to efficient developer time. I also used better column names for the LEAD() function.

Please refer to the partial results in Figure 13-16.

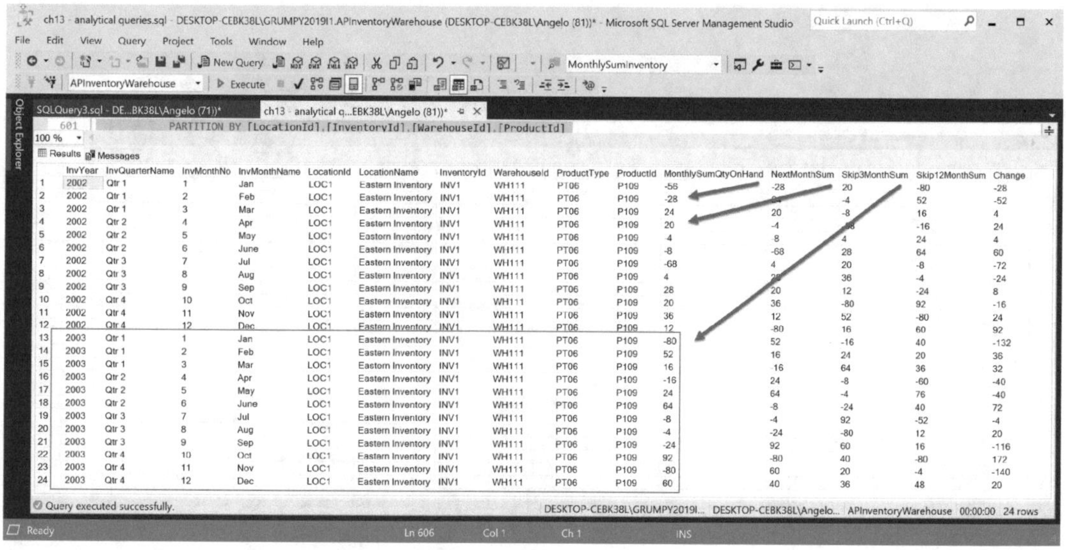

	InvYear	InvQuarterName	InvMonthNo	InvMonthName	LocationId	LocationName	InventoryId	WarehouseId	ProductType	ProductId	MonthlySumQtyOnHand	NextMonthSum	Skip3MonthSum	Skip12MonthSum	Change
1	2002	Qtr 1	1	Jan	LOC1	Eastern Inventory	INV1	WH111	PT06	P109	-56	-28	20	-80	-28
2	2002	Qtr 1	2	Feb	LOC1	Eastern Inventory	INV1	WH111	PT06	P109	-28	24	-4	52	-52
3	2002	Qtr 1	3	Mar	LOC1	Eastern Inventory	INV1	WH111	PT06	P109	24	20	-8	16	4
4	2002	Qtr 2	4	Apr	LOC1	Eastern Inventory	INV1	WH111	PT06	P109	20	-4	-68	-16	24
5	2002	Qtr 2	5	May	LOC1	Eastern Inventory	INV1	WH111	PT06	P109	-4	-8	4	24	4
6	2002	Qtr 2	6	June	LOC1	Eastern Inventory	INV1	WH111	PT06	P109	-8	-68	28	64	60
7	2002	Qtr 3	7	Jul	LOC1	Eastern Inventory	INV1	WH111	PT06	P109	-68	4	20	-8	-72
8	2002	Qtr 3	8	Aug	LOC1	Eastern Inventory	INV1	WH111	PT06	P109	4	28	36	-4	-24
9	2002	Qtr 3	9	Sep	LOC1	Eastern Inventory	INV1	WH111	PT06	P109	28	20	12	-24	8
10	2002	Qtr 4	10	Oct	LOC1	Eastern Inventory	INV1	WH111	PT06	P109	20	36	-80	92	-16
11	2002	Qtr 4	11	Nov	LOC1	Eastern Inventory	INV1	WH111	PT06	P109	36	12	52	-80	24
12	2002	Qtr 4	12	Dec	LOC1	Eastern Inventory	INV1	WH111	PT06	P109	12	-80	16	60	92
13	2003	Qtr 1	1	Jan	LOC1	Eastern Inventory	INV1	WH111	PT06	P109	-80	52	-16	40	-132
14	2003	Qtr 1	2	Feb	LOC1	Eastern Inventory	INV1	WH111	PT06	P109	52	16	24	20	36
15	2003	Qtr 1	3	Mar	LOC1	Eastern Inventory	INV1	WH111	PT06	P109	16	-16	64	36	32
16	2003	Qtr 2	4	Apr	LOC1	Eastern Inventory	INV1	WH111	PT06	P109	-16	24	-8	-60	-40
17	2003	Qtr 2	5	May	LOC1	Eastern Inventory	INV1	WH111	PT06	P109	24	64	-4	76	-40
18	2003	Qtr 2	6	June	LOC1	Eastern Inventory	INV1	WH111	PT06	P109	64	-8	-24	40	72
19	2003	Qtr 3	7	Jul	LOC1	Eastern Inventory	INV1	WH111	PT06	P109	-8	-4	92	-52	-4
20	2003	Qtr 3	8	Aug	LOC1	Eastern Inventory	INV1	WH111	PT06	P109	-4	-24	-80	12	20
21	2003	Qtr 3	9	Sep	LOC1	Eastern Inventory	INV1	WH111	PT06	P109	-24	92	60	16	-116
22	2003	Qtr 4	10	Oct	LOC1	Eastern Inventory	INV1	WH111	PT06	P109	92	-80	40	-80	172
23	2003	Qtr 4	11	Nov	LOC1	Eastern Inventory	INV1	WH111	PT06	P109	-80	60	20	-4	-140
24	2003	Qtr 4	12	Dec	LOC1	Eastern Inventory	INV1	WH111	PT06	P109	60	40	36	48	20

***Figure 13-16.** Current vs. next month, quarter, and year with the LEAD function*

The report works. Arrows point to the correct future-month, three-month-forward, and 12-month-forward values. The boxed area represents the next year, so the logic crosses year boundaries, which is what we want.

This is a good report that allows analysts to evaluate inventory movement over the years. The only issue is that the data is only good for the past and the current calendar date. No future data exists yet!

Let's see what these results look like in a graph.

Please refer to Figure 13-17.

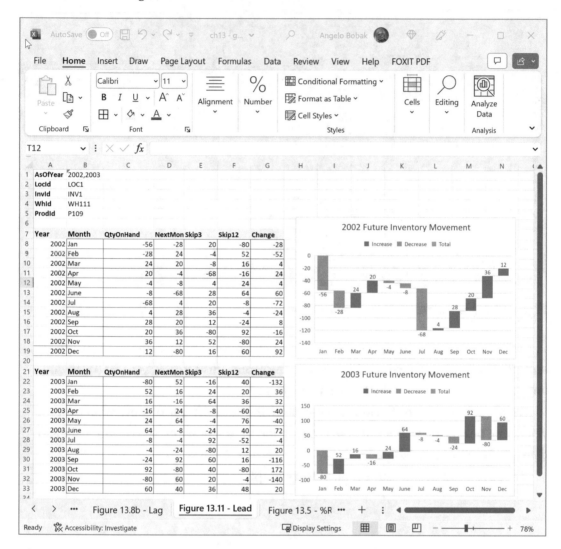

Figure 13-17. *Next month, quarter, and year inventory movements*

This is the same type of graph as the one generated for the LAG( ) function analysis, but it does show future movements relative to a set of years.

Performance Considerations

Here comes our baseline estimated query plan.

Please refer to Figure 13-18.

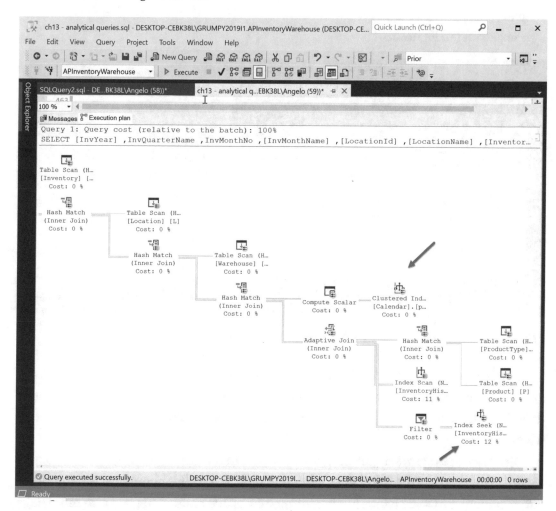

Figure 13-18. *Baseline estimated execution plan – part 1*

We are in data warehouse load, so look at all the different types of join tasks. We have a hash match, an adaptive join, and a series of hash match tasks all at 0% cost. Look at the clustered index scan on the Calendar table (0% cost) and also the index seek task on the InventoryHistory fact table (12% cost).

Tip   Whenever you come across a task in an estimated query plan that you do not know what it does, look it up in the Microsoft documentation. Over time you will understand what all the tasks do.

So far, we are looking good. Let's scroll to the left and examine the next section of the plan.

Please refer to Figure 13-19.

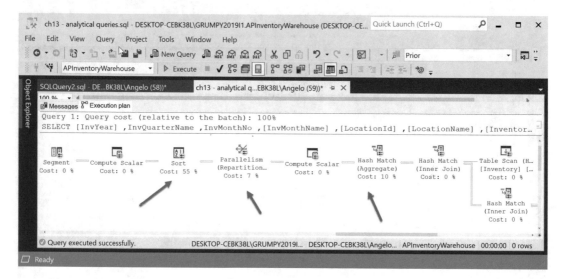

Figure 13-19. *Baseline estimated execution plan – part 2*

Here we get into a bit of trouble with the hash match task (cost 10%), the parallelism task (7%), and a pretty expensive sort task (55%). The sort tasks involve expressions used to pull out the date parts. We saw this in prior examples that this will kick up the costs when sorting.

Again, we see that we might want to add a JOIN to the Calendar dimension so that we do not have to use the DATEPART() function repeatedly. This should get the sort down and eliminate the CASE blocks (but it adds new costs due to the JOIN and also more IO). This change would be performed on the TSQL VIEW we are using.

Let's move to the left again to see more of the estimated execution plan.

Please refer to Figure 13-20.

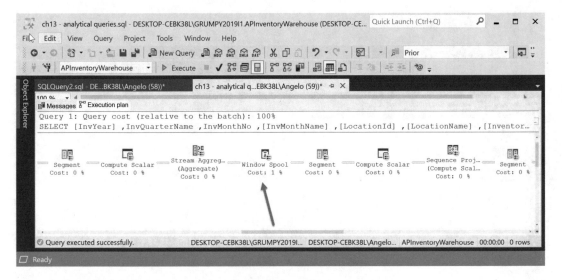

Figure 13-20. *Baseline estimated execution plan – part 3*

Of course, the window spool task had to make an appearance, but it came in with a 1% cost, so not too much to worry about. We are not finished though. Let's scroll to the left again.

Please refer to Figure 13-21.

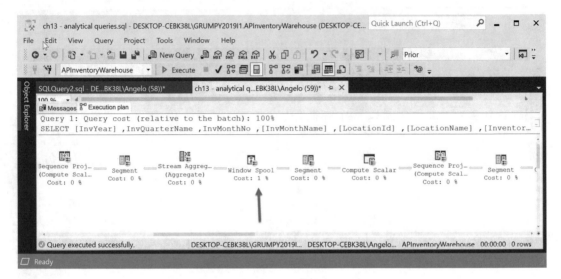

Figure 13-21. *Baseline estimated execution plan – part 4*

Same pattern, but we knew this. There's another window spool task with a cost of 1% for the LEAD() function of course. This spooling is performed on physical disk so data can be repeatedly retrieved. Adding more physical memory could get this down to 0% cost (I think).

Next section. Please refer to Figure 13-22.

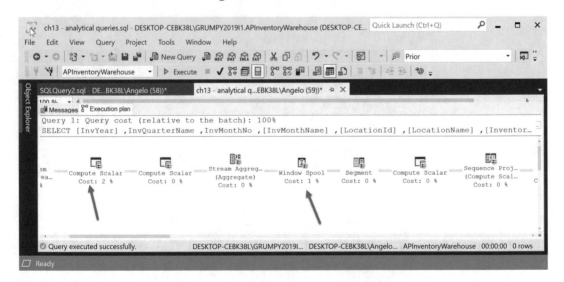

Figure 13-22. *Baseline estimated execution plan – part 5*

The window spool tasks keep coming, once per LEAD() function.

One more time, scroll to the left. Please refer to Figure 13-23.

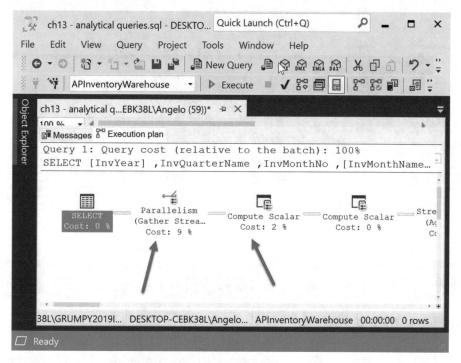

***Figure 13-23.** Baseline estimated execution plan – part 5*

The plan wraps things up with a couple of compute scalar tasks and a final parallelism task with a cost of 9%. This task was for processing the delta calculation to generate the difference between the current inventory movement value and the future value. When not sure, just SET STATISTICS PROFILE ON to see the details behind each task.

Overall, there was a lot of activity, but the query was executed very fast, for two seconds. Remember that the fact table has 2,945,280 rows! Also, the VIEW we used was based on a CTE block with a couple of CASE blocks and two DATEPART() functions.

Homework   Based on our analysis, see if you can restructure the VIEW so it eliminates the CASE blocks and the use of the DATEPART() function.

Let's look at the IO and TIME statistics next for our final step in the analysis process.

Please refer to Figure 13-24.

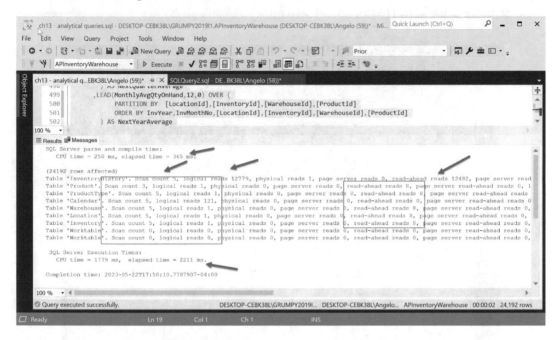

Figure 13-24. *IO and TIME statistics*

Lots of activity this time. The other queries in this chapter have been fairly simple.

I will not go into the details of these statistics but will look at the logical reads and read-ahead reads for the InventoryHistory table. This is to be expected. Also keep in mind that when you run these queries and analyze them on your computing platform, the values will be different based on your hardware.

STAR and SNOWFLAKE queries rely heavily on fact and dimension tables that are indexed. The rule of thumb is to create an index for each surrogate key in the dimension tables and one large key combining all the surrogate keys in the fact table. This assumes the dimension tables contain large volumes of rows. If a table has under 1,000 rows, give or take, the index will not be used and will just take up space.

Note   I purposely kept the queries in this book simple and brief for the most part so you would get comfortable with the usage of the window functions and also with the performance techniques. This particular query that we just covered is typical of what you would find in a real business scenario as it has a lot of moving parts. Also, tables with hundreds of million rows are typical; two million rows are nothing.

PERCENT_RANK( ) Function

Next on deck is the PERCENT_RANK() function.

Recall what it does?

We covered this in Chapters 3, 7, and 10, but in case you need a reminder or are not reading the chapters in sequential order, here is a review (feel free to skip this part and go to the business specifications if you already read this in prior chapters).

Using a data set like results from a query, a partition, or a table variable, this function calculates the relative rank of each individual value relative to the entire data set (as a percentage). You need to multiply the results (the value returned is a float value between 0.0 and 1.0) by 100.00 or use the FORMAT() function to convert the results to a percentage format (this function returns a FLOAT(53) data type by the way).

Recall that it works a lot like the CUME_DIST() and RANK() functions.

The RANK() function returns the position of the value, while PERCENT_RANK() returns a percentage value.

Our friendly business analyst supplies the following specifications:

A report is needed that will show the percentage rank for quantity on hand for product "P101" for the year 2002. In other words, which month had the highest inventory on hand, and which month had the lowest? This is required to further assess how much inventory is not moving. The report needs to refer to location "LOC1," inventory "INV1," and warehouse "WH112." As usual, the quarter and month names need to be spelled out. The results need to be copied to a Microsoft Excel spreadsheet so a Pareto chart can be created to enable visualization for analysis.

Note   A Pareto chart is a chart that includes both bars and lines and was invented by Vilfredo Pareto, an Italian economist.

Here is the query that addresses these requirements. We are back to using the APInventory database instead of the inventory data warehouse database.

Please refer to Listing 13-6.

Listing 13-6. Percent Rank for Product P101 for 2002

```
SELECT AsOfYear
      ,DATEPART(qq,AsOfDate) AS AsOfQuarter
      ,CASE
```

```
            WHEN DATEPART(qq,AsOfDate) = 1 THEN '1st Quarter'
            WHEN DATEPART(qq,AsOfDate) = 2 THEN '2nd Quarter'
            WHEN DATEPART(qq,AsOfDate) = 3 THEN '3rd Quarter'
            WHEN DATEPART(qq,AsOfDate) = 4 THEN '4th Quarter'
        END AS AsOfQtrName
        ,AsOfMonth
        ,CASE
            WHEN AsOfMonth = 1 THEN 'Jan'
            WHEN AsOfMonth = 2 THEN 'Feb'
            WHEN AsOfMonth = 3 THEN 'Mar'
            WHEN AsOfMonth = 4 THEN 'Apr'
            WHEN AsOfMonth = 5 THEN 'May'
            WHEN AsOfMonth = 6 THEN 'June'
            WHEN AsOfMonth = 7 THEN 'Jul'
            WHEN AsOfMonth = 8 THEN 'Aug'
            WHEN AsOfMonth = 9 THEN 'Sep'
            WHEN AsOfMonth = 10 THEN 'Oct'
            WHEN AsOfMonth = 11 THEN 'Nov'
            WHEN AsOfMonth = 12 THEN 'Dec'
         END AS AvgMonthName
        ,EOMONTH(CONVERT(VARCHAR,YEAR(AsOfDate)),(MONTH(AsOfDate) - 1))
        ,LocId
        ,InvId
        ,WhId
        ,ProdId
        ,SUM([InvIn] - [InvOut]) + 50 AS QtyOnHand
        ,FORMAT(PERCENT_RANK() OVER (
            PARTITION BY AsOfYear
            ORDER BY SUM([InvIn] - [InvOut]) + 50
        ),'P') AS PercentRank
FROM Product.Warehouse
WHERE AsOfYear = 2002
AND ProdId = 'P101'
AND Locid = 'LOC1'
AND InvId = 'INV1'
```

```
AND WhId = 'WH112'
GROUP BY AsOfYear,
     DATEPART(qq,AsOfDate),
     AsOfMonth,
     EOMONTH(CONVERT(VARCHAR,YEAR(AsOfDate)),(MONTH(AsOfDate) - 1)),
     Locid,
     InvId,
     WhId,
     ProdId
ORDER BY AsOfYear,
     AsOfMonth,
     Locid,
     InvId,
     WhId,
     ProdId
GO
```

We have seen the WHERE clause and of course the CASE blocks, so we do not need to comment on these except to say we need index coverage for the columns in the OVER() clause and WHERE clause. The CASE block works, well but a join to the enhanced Calendar table should be considered in case there are performance problems.

The function uses an OVER() clause with both a PARTITION BY and an ORDER BY clause. The partition is by year, and the sort that specifies how the function will be applied to the partition is by the quantity on hand value. Notice that a CTE block was not required.

Lastly, the FORMAT() function was used to display the results as a percentage for esthetics (makes the report look pretty).

Please refer to the partial results in Figure 13-25.

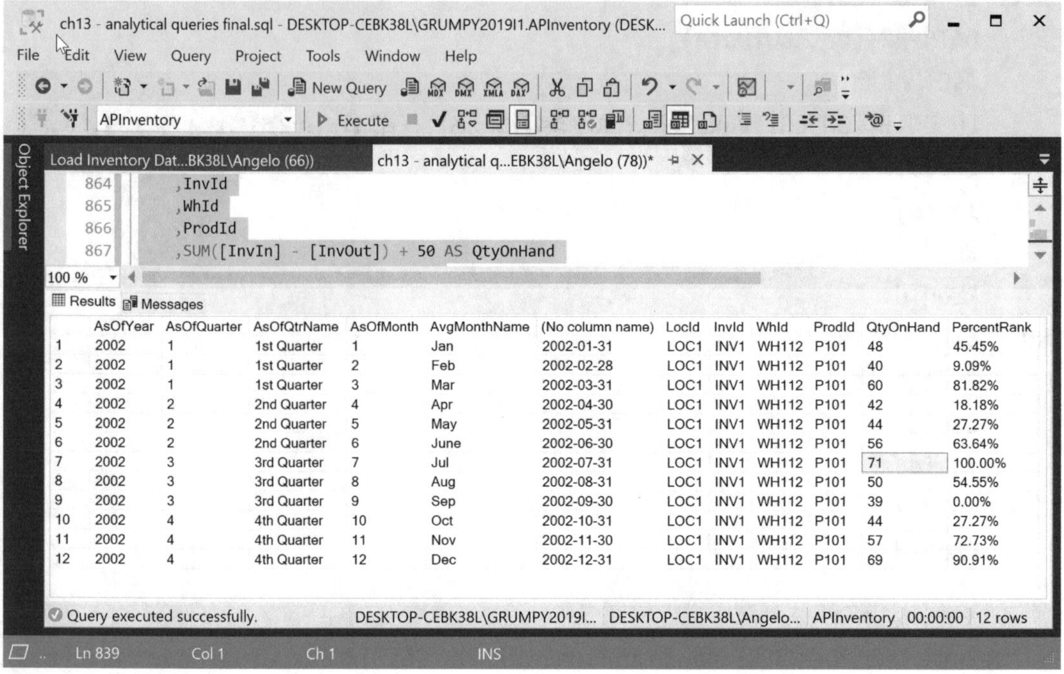

	AsOfYear	AsOfQuarter	AsOfQtrName	AsOfMonth	AvgMonthName	(No column name)	LocId	InvId	WhId	ProdId	QtyOnHand	PercentRank
1	2002	1	1st Quarter	1	Jan	2002-01-31	LOC1	INV1	WH112	P101	48	45.45%
2	2002	1	1st Quarter	2	Feb	2002-02-28	LOC1	INV1	WH112	P101	40	9.09%
3	2002	1	1st Quarter	3	Mar	2002-03-31	LOC1	INV1	WH112	P101	60	81.82%
4	2002	2	2nd Quarter	4	Apr	2002-04-30	LOC1	INV1	WH112	P101	42	18.18%
5	2002	2	2nd Quarter	5	May	2002-05-31	LOC1	INV1	WH112	P101	44	27.27%
6	2002	2	2nd Quarter	6	June	2002-06-30	LOC1	INV1	WH112	P101	56	63.64%
7	2002	3	3rd Quarter	7	Jul	2002-07-31	LOC1	INV1	WH112	P101	71	100.00%
8	2002	3	3rd Quarter	8	Aug	2002-08-31	LOC1	INV1	WH112	P101	50	54.55%
9	2002	3	3rd Quarter	9	Sep	2002-09-30	LOC1	INV1	WH112	P101	39	0.00%
10	2002	4	4th Quarter	10	Oct	2002-10-31	LOC1	INV1	WH112	P101	44	27.27%
11	2002	4	4th Quarter	11	Nov	2002-11-30	LOC1	INV1	WH112	P101	57	72.73%
12	2002	4	4th Quarter	12	Dec	2002-12-31	LOC1	INV1	WH112	P101	69	90.91%

Figure 13-25. *Monthly percent rank for quantity on hand*

The report generates 12 rows, one for each month of 2002. This report can be expanded to include all years so a thorough historical analysis can be made to see how the product sold throughout the years. We can easily see the low and high percentages and all values in between. Looks like July had the highest value for quantity on hand, so that means sales for this product in July were low. It would be interesting to combine this report with the `LAG()` function in order to compare how the product moved in prior years (as luck would have it, we have data from 2002 onward, so no 2001 data except for December).

Let's copy the results into our Microsoft Excel spreadsheet so we can create a Pareto chart.

Please refer to Figure 13-26.

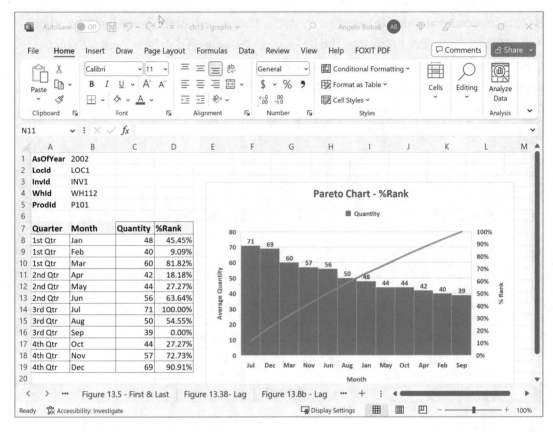

Figure 13-26. *Percentage rank Pareto chart*

I think this is an interesting chart. Notice the bars decrease and the line increases. This shows the high to low inventory on hand amounts together with the percentages. The line is related to the % rank values. Blocks represent the monthly quantities. The data table shows the months are sorted as we expect, but the Pareto chart reorders them, so the bars decrease as we proceed from left to right.

Homework   Modify the query so it generates results for all years and use the LAG() function to display last month's, last quarter's, and last year's results for each of the current months. Perform our usual performance analysis to see how the query performs. Practice your Microsoft Excel skills and generate a graph for each year.

Performance Considerations

Here comes our estimated query plan analysis. The pattern of the branches should be very different as we are now using the inventory transaction database, and we will not see all the joins we saw when we were creating STAR and SNOWFLAKE queries. Also, the Warehouse table has only about 25,000 rows, big difference from the data warehouse fact table with almost one million rows.

Please refer to Figure 13-27.

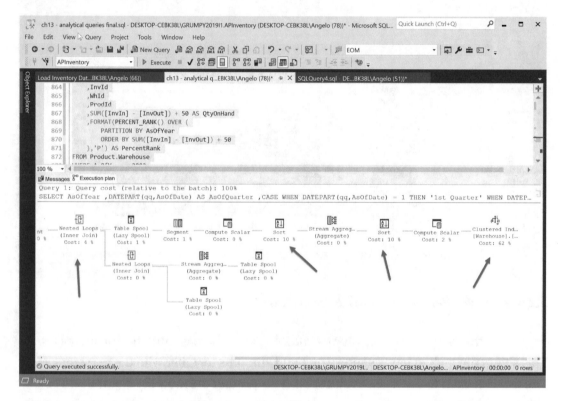

Figure 13-27. *Estimated query execution plan*

As usual, let's concentrate on the high-cost tasks. Right to left we see a clustered index seek with a cost of 62%, so all the indexes we created throughout our examples are paying off. A new index was not suggested, so it looks like we are on the right path.

Here come the IO and TIME statistics. Please refer to Figure 13-28.

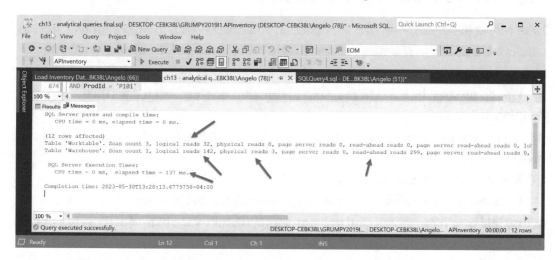

Figure 13-28. *IO and TIME statistics for percent rank calculation*

These are not too bad, but the SQL Server execution time is high, compared with the other times we have been experiencing. It comes in at 137 ms. The parse time is low at 0 ms.

The work table shows 3 scans and 32 logical reads, while the Warehouse table has a scan count value of 1, 142 logical reads, and 3 physical reads.

Our conclusion is that this function takes up more resources than the other window functions we have been using in this chapter, so it needs special attention. Keep in mind we also did some reformatting of the results by using the FORMAT() function. This makes the report look pretty, but do we really need it? Depends on your business analyst!

Homework   Modify the query so that it does not use the FORMAT() function. Also modify the query so it does not use the text date parts, just the numbers. Perform our usual query plan and IO/TIME statistics analysis to see if there is any significant improvement. Don't forget to run DBCC before each run in order to clear cache.

PERCENTILE_CONT( ) Function

Here comes the description of the function. If you read about it in prior chapters and are comfortable with how it works, feel free to skip this discussion. If you need a refresher, please read on!

We will discuss the PERCENTILE_CONT( ) function first, and what does this function do?

Well, if you read Chapters 3, 7, and 10, you will recall that it works on continuous sets of data.

Recall what continuous data is.

Continuous data is data measured over a period of time, like boiler temperatures in equipment over a 24-hour period or equipment failures over a period of time, in our case inventory movement over a period of time.

Summing up each boiler temperature to arrive at a total makes no sense. Summing up the boiler temperatures and then dividing them by the time period to get an average does.

Lastly, summing up the values of inventory in movements makes no sense as 99% of the time out movements occur, so a total monthly sum will not indicate how much inventory there is in stock unless no inventory was removed from stock. Adding up all inventory in movements and then subtracting all the out inventory movements works as it would result in how much inventory is left. So all this can become very tricky.

So we can say continuous data can be measured by using formulas like average and so on.

This works for inventory on hand each day. We can add up all the quantities on hand for each day of the month and then generate the average per month.

In conclusion, the PERCENTILE_CONT( ) function works on a continuous set of data based on a required percentile so as to return a value within the data set that satisfies the percentile (you need to supply a percentile as a parameter, like .25, .50, .75, etc.).

With this function the value is interpolated, meaning it usually does not exist. Or by coincidence it could use one of the values in the data set if the numbers line up correctly. It can happen.

Here come the business requirements for this query:

Our business analyst keeps picking on the same product, location, inventory, and warehouse. This time, they would like us to generate the 25th, 50th, and 75th continuous percentiles across the year 2002 as a starting point. As usual, the quarter and month names need to be spelled out, and the month number needs to be included so when the data is copied to a Microsoft Excel spreadsheet, it can be sorted by month number if necessary.

The quantities on hand need to represent what is left in inventory on the last day of the month. Therefore, the values are calculated by summing up all inventory movements in minus inventory movements out each day of the month, to arrive at the total quantity on the last day of the month.

Here is what we came up with. Please refer to Listing 13-7.

Listing 13-7. Percentile Continuous for Quantity on Hand

```
/*******************************/
/* Generates interpolated value */
/*******************************/
SELECT AsOfYear
      ,DATEPART(qq,AsOfDate) AS AsOfQuarter
      ,CASE
           WHEN DATEPART(qq,AsOfDate) = 1 THEN '1st Quarter'
           WHEN DATEPART(qq,AsOfDate) = 2 THEN '2nd Quarter'
           WHEN DATEPART(qq,AsOfDate) = 3 THEN '3rd Quarter'
           WHEN DATEPART(qq,AsOfDate) = 4 THEN '4th Quarter'
      END AS AsOfQtrName
      ,AsOfMonth
           ,CASE
           WHEN AsOfMonth = 1 THEN 'Jan'
           WHEN AsOfMonth = 2 THEN 'Feb'
           WHEN AsOfMonth = 3 THEN 'Mar'
           WHEN AsOfMonth = 4 THEN 'Apr'
           WHEN AsOfMonth = 5 THEN 'May'
           WHEN AsOfMonth = 6 THEN 'June'
           WHEN AsOfMonth = 7 THEN 'Jul'
           WHEN AsOfMonth = 8 THEN 'Aug'
           WHEN AsOfMonth = 9 THEN 'Sep'
           WHEN AsOfMonth = 10 THEN 'Oct'
           WHEN AsOfMonth = 11 THEN 'Nov'
           WHEN AsOfMonth = 12 THEN 'Dec'
```

```
     END AS AvgMonthName
    ,AsOfDate - last day of the month
    ,LocId
    ,InvId
    ,WhId
    ,ProdId
    ,QtyOnHand
    ,PERCENTILE_CONT(.25) WITHIN GROUP (ORDER BY QtyOnHand)
        OVER (
        PARTITION BY AsOfYear
        ) AS [%Continuous .25]
    ,PERCENTILE_CONT(.5) WITHIN GROUP (ORDER BY QtyOnHand)
        OVER (
        PARTITION BY AsOfYear
        ) AS [%Continuous .5]
    ,PERCENTILE_CONT(.75) WITHIN GROUP (ORDER BY QtyOnHand)
        OVER (
        PARTITION BY AsOfYear
        ) AS [%Continuous .75]
FROM Product.Warehouse
WHERE AsOfYear = 2002
AND ProdId = 'P101'
AND Locid = 'LOC1'
AND InvId = 'INV1'
AND WhId = 'WH112'
GO
```

No CTE this time. Recall that QtyOnHand is the sum of all inventory movements in minus all inventory movements out for the last day of the month. Recall that the format for this function is slightly different. We need to include a WITHIN GROUP clause prior to defining the OVER() clause. Also notice that there is no ORDER BY clause in the OVER() clause as this is specified in the WITHIN GROUP clause. So many clauses!

Lastly, notice that we need to pass the percentile values as a parameter for the window function.

Let's see the results. Please refer to the results in Figure 13-29.

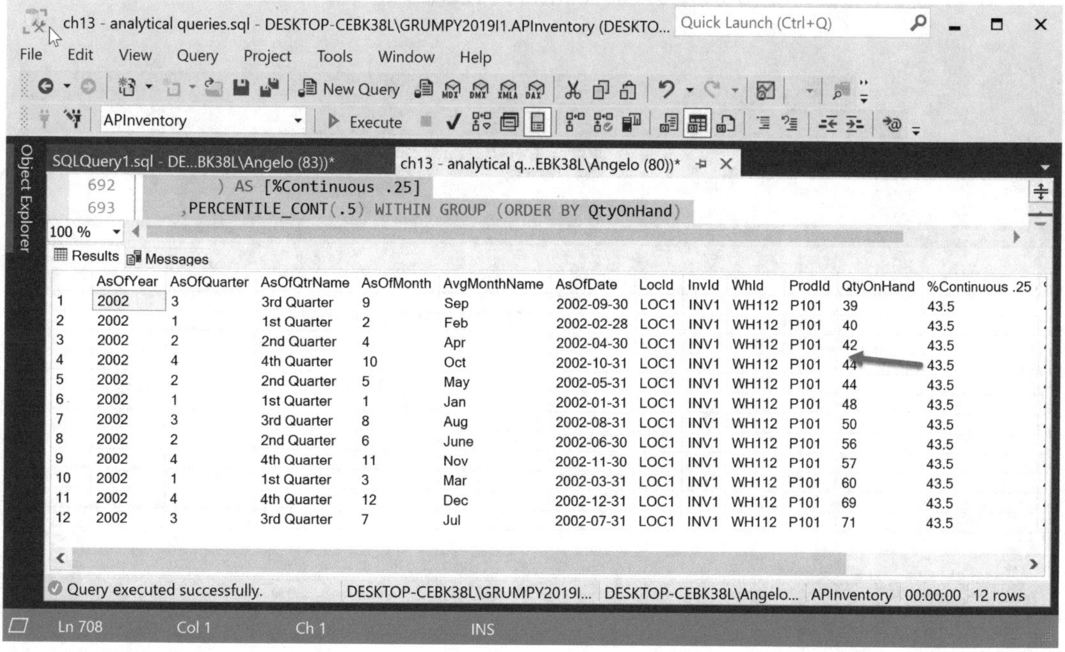

Figure 13-29. *Percentile continuous analysis for monthly quantities on hand*

The report is wide, so we only see the 25th percentile values. Notice that the result is interpolated. It does not exist, but it falls back somewhere between 42 and 44 quantity on hand values. Also 43.5 is a bit incorrect as you cannot have fractional quantities on hand. Might want to use the CEILING() or FLOOR() function to tidy it up. This percentile together with the other percentiles that are not seen gives the analyst a good sense of the spread of the quantities on hand for this year.

Let's use our usual Microsoft Excel spreadsheet and create a visual representation of these results. Please refer to the partial results in Figure 13-30.

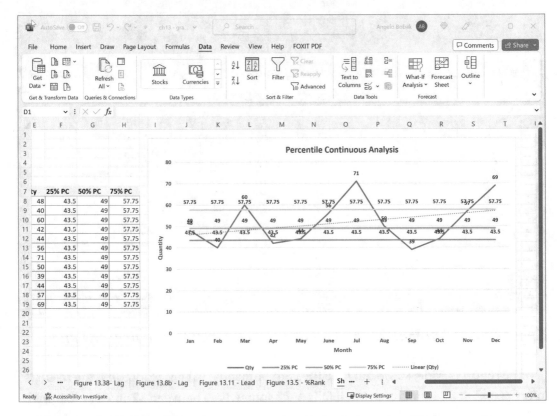

Figure 13-30. *Percentile continuous analysis graph for monthly quantities on hand*

The results are sorted by month on the x axis, and we can see the swings in the inventory quantities. Seems like the months of March, July, and December are when inventory was high.

The months of February, April, May, and September show lower inventory values. Our analyst needs to compare these values vs. sales figures to determine sales performance.

Notice the three horizontal lines that represent the 25th, 50th, and 75th percentiles continuous values. This is a good visualization of the query reports. It packs a lot of valuable information.

Performance Considerations

The following is our estimated query plan. It is divided into five parts (with some overlap for points of reference). We will see that the patterns will repeat themselves for each instance of the PERCENTILE_CONT() function in the SELECT clause. We will discuss the first one and briefly glance at the remaining sections of the estimated execution plan.

Note   If you keep using these functions, you will start to know what to expect as far as query plans and IO and TIME statistics and what to do in order to enhance performance. That's my opinion anyway.

Starting from the right side to the left side, we examine the first section. Please refer to Figure 13-31.

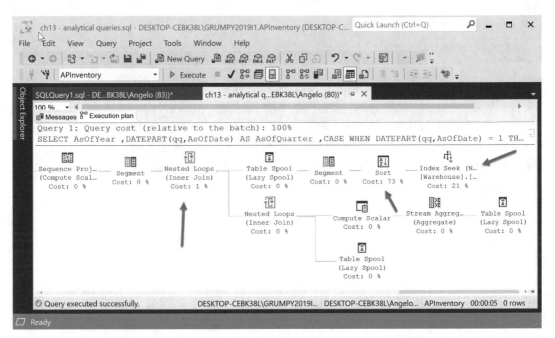

Figure 13-31. *Baseline estimated query plan – part 1*

This is a familiar pattern. We see three branches merging via a nested loops join (cost 1%). The index seek task all the way to the right side of the plan comes in at 21% followed by an expensive sort. The remaining tasks are all at 0% cost. Let's move to the second part of the plan.

Please refer to Figure 13-32 for the next section.

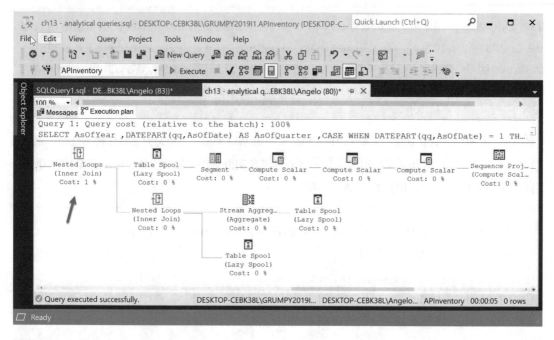

Figure 13-32. *Baseline estimated query plan – part 2*

Once again, three branches merge (the data flows) into one via the nested loops join (1% cost). All the tasks are 0%, which leads us to believe that performance-intensive processing was accomplished in the prior section and that all the data we need is in memory.

Let's see if this is true. Please refer to Figure 13-33.

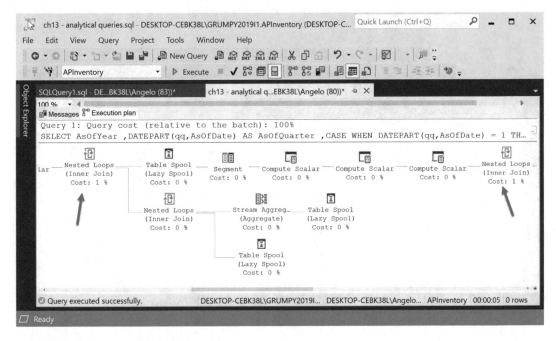

Figure 13-33. Baseline estimated query plan – part 3

Same pattern. Looks like my prior assumption is correct. All the tasks have 0 costs. Remember, these patterns are basically for each execution of the PERCENTILE_CONT() function, but might as well show them all.

Next section. Please refer to Figure 13-34.

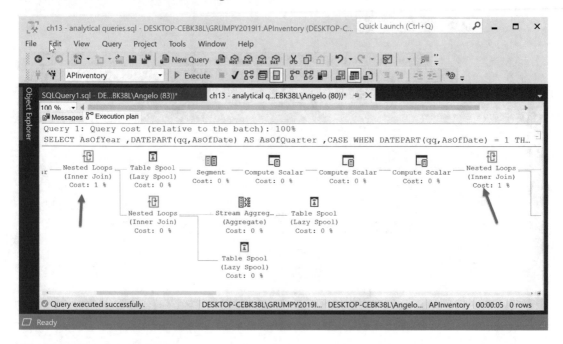

Figure 13-34. *Baseline estimated query plan – part 4*

More of the same. I think we can assume that picking an optimization strategy like replacing the CASE block with an extra JOIN to the Calendar table (so we can retrieve the date parts from the enhanced Calendar table) might help, maybe. We will see!

Subsequent executions of the base query, like from an SSRS report by multiple users, will deliver adequate if not great performance.

Please refer to Figure 13-35.

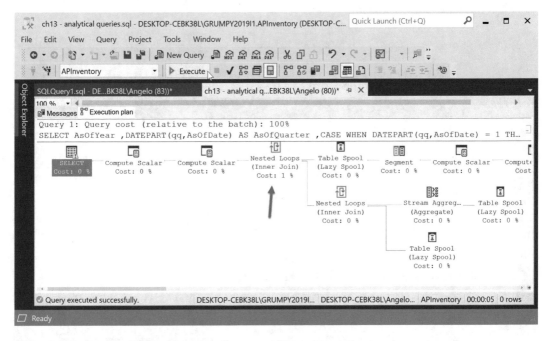

Figure 13-35. *Baseline estimated query plan – part 5*

Last but not least, one final nested loops join task at a cost of 0% (we are not worrying about the 0% costs).

Tip   To get a feel of the data flow volumes between tasks, run a live execution plan. Click the icon in the menu bar below and to the right of the icon that has a small scissor and then run the query.

Conclusion: All the performance tuning needs to be concentrated on the very first right-hand section of the plan we looked at first.

Here are some choices:

- Modify the query so it uses the Calendar table.

- Possibly place the Calendar table in memory. That is, make it a memory-optimized table.

- Some might even go as far as packing the text month and quarter names into the Warehouse table.

Let's take a look at the IO and TIME statistics.

Please refer to Figure 13-36.

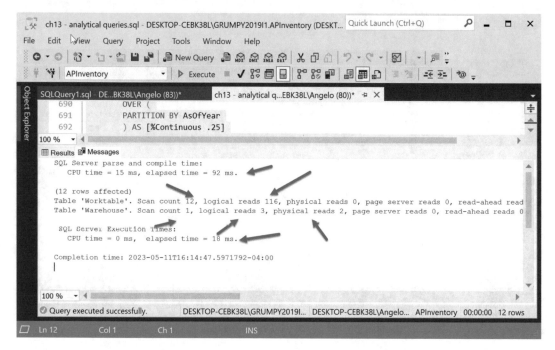

Figure 13-36. *IO and TIME statistics for* `PERCENTILE_CONT`

SQL Server execution time came in at 18 ms. The work table has 12 scans and 116 logical reads. The Warehouse table also has 1 scan, 3 logical reads, and 2 physical reads.

Recall this recommendation for performance tuning: At a minimum, create indexes that cover the columns referenced in the OVER() clause, specifically in the PARTITION BY clause and the ORDER BY clause. Also include columns referenced in the WHERE clause. Your index can then cover these columns.

Note   By the way, calculating PERCENTILE_CONT(.50) generates the median value. Use this in a query with the AVERAGE() function, which generates the MEAN value, and compare them. TSQL does not have a MEDIAN function, but you can create your own.

Tip   Look up the `MERGE` command, which can be used when loading data. It allows you to `INSERT`, `UPDATE`, or `DELETE` new or modified rows. This avoids loading large tables over and over again, which would defeat the purpose of performance tuning. Try adding this to the `INSERT/SELECT` command that loads the report table. Add a few hundred rows to the table you are retrieving rows from and see how the modified `INSERT/SELECT` works when you include the `MERGE` command.

PERCENTILE_DISC() Function

Here is our percentile discrete example. The `PERCENTILE_DISC()` function works on a discrete set of data based on a required percentile so as to return a value within the data set that satisfies the percentile.

Remember discrete data?

Discrete data values are numbers that have values that you can count over time, like investment account deposits over days, months, and years. We can say that in a year one made 1,000 deposits and 700 withdrawals. Adding monthly account balances to get a yearly balance will not work. As the monthly balances go up and down, it would not be a true representation of how much the account earned in a year. What counts is how much money is left in the account on December 31. Adding the starting account balance to all deposits in a year minus all deposits that year makes sense.

You can add these up to get a grand total like total deposits or withdrawals in a year and identify trends in account balances over time (average deposits or average withdrawals).

As another example, summing up the inventory in each warehouse for a location almost makes sense. If a location has two warehouses, you can calculate the sum of total inventory on hand at any given point, total movements in and out for each day, week, month, and so on. Again, total inventory in a year makes no sense; average inventory level values for a year do.

Recall the questions you can ask yourself to differentiate between discrete and continuous data:

- Can I count it (over periods of time)?

- Can I add it up?

- Can I use it in calculations, like averages?

925

Worth repeating:

Continuous data can be measured by using functions like AVG(), SUM(), etc.

Discrete data can be counted by using functions like COUNT().

Back to business. Here are the business specifications for this query supplied by our business analyst:

Our PERCENTILE_CONT() report was such a big hit with the business analyst that they asked us to modify the report so it shows the percentile discrete percentages for the same location, inventory, warehouse, and product. This one is easy, a quick copy-and-paste and changing of the names of the functions. Yes, laziness made us reuse the CASE blocks for the quarter and month data parts.

Lastly, remember that the quantities on hand need to represent what is left in inventory on the last day of the month. Therefore, the values are calculated by summing up all inventory movements in minus inventory movements out each day of the month, to arrive at the total quantity on the last day of the month.

Please refer to Listing 13-8.

Listing 13-8. Percentile Discrete Query for Inventory on Hand

```
SELECT AsOfYear
      ,DATEPART(qq,AsOfDate) AS AsOfQuarter
      ,CASE
            WHEN DATEPART(qq,AsOfDate) = 1 THEN '1st Quarter'
            WHEN DATEPART(qq,AsOfDate) = 2 THEN '2nd Quarter'
            WHEN DATEPART(qq,AsOfDate) = 3 THEN '3rd Quarter'
            WHEN DATEPART(qq,AsOfDate) = 4 THEN '4th Quarter'
      END AS AsOfQtrName
      ,AsOfMonth
          ,CASE
            WHEN AsOfMonth = 1 THEN 'Jan'
            WHEN AsOfMonth = 2 THEN 'Feb'
            WHEN AsOfMonth = 3 THEN 'Mar'
            WHEN AsOfMonth = 4 THEN 'Apr'
            WHEN AsOfMonth = 5 THEN 'May'
            WHEN AsOfMonth = 6 THEN 'June'
            WHEN AsOfMonth = 7 THEN 'Jul'
            WHEN AsOfMonth = 8 THEN 'Aug'
```

```
            WHEN AsOfMonth = 9 THEN 'Sep'
            WHEN AsOfMonth = 10 THEN 'Oct'
            WHEN AsOfMonth = 11 THEN 'Nov'
            WHEN AsOfMonth = 12 THEN 'Dec'
     END AS AvgMonthName
     ,AsOfDate
     ,LocId
     ,InvId
     ,WhId
     ,ProdId
     ,QtyOnHand
     ,PERCENTILE_DISC(.25) WITHIN GROUP (ORDER BY QtyOnHand)
           OVER (
                   PARTITION BY AsOfYear
           ) AS [%Discrete.25]
     ,PERCENTILE_DISC(.5) WITHIN GROUP (ORDER BY QtyOnHand)
           OVER (
                   PARTITION BY AsOfYear
           ) AS [%Discrete .5]
     ,PERCENTILE_DISC(.75) WITHIN GROUP (ORDER BY QtyOnHand)
           OVER (
                   PARTITION BY AsOfYear
           ) AS [%Discrete .75]
FROM Product.Warehouse
WHERE AsOfYear = 2002
AND ProdId = 'P101'
AND Locid = 'LOC1'
AND InvId = 'INV1'
AND WhId = 'WH112'
GO
```

Same logic and structure of the prior example, so let's see the results. Remember the difference between this function and the PERCENTILE_CONT() function results?

Please refer to the partial results in Figure 13-37.

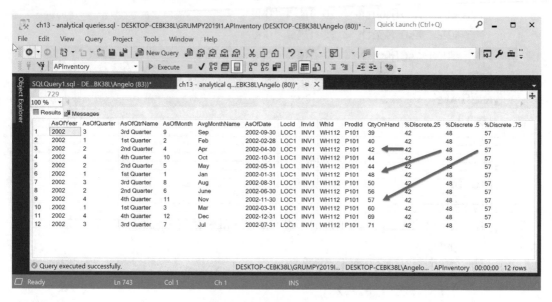

Figure 13-37. *Percentile discrete query results for inventory on hand*

This time the three percentile values point to actual values and not interpolated values. This query checks out so it can be modified for all years, locations, inventories, warehouse, and products. Notice quantities on hand represent the values reported on the last day of the month, not the sum of values each day of the month. This is what makes the data discrete vs. continuous.

Homework   Using this query, modify it so the values returned cover all years, months, locations, inventories, warehouses, and products. Use a WHERE clause filter to first check for the same product and warehouse of the prior query. Once you are sure you modified the PARTITION BY clause correctly, remove the WHERE clause. You will need an ORDER BY query for the entire result set. This query, transformed into a VIEW, would make an ideal candidate for an SSRS report built with the Report Builder tool (or a denormalized report table).

Performance tuning time.

Performance Considerations

We are going to generate our usual estimated query performance plan one more time and look at the results. We will also look at the IO and TIME statistics and then modify the query, so it uses the Calendar table with a JOIN to retrieve the text quarter and month names replacing the CASE blocks. We will then compare the analysis results between both versions of the query.

Pop Quiz   What are the different ways you can generate an estimated query plan using SSMS?

Please refer to Figure 13-38.

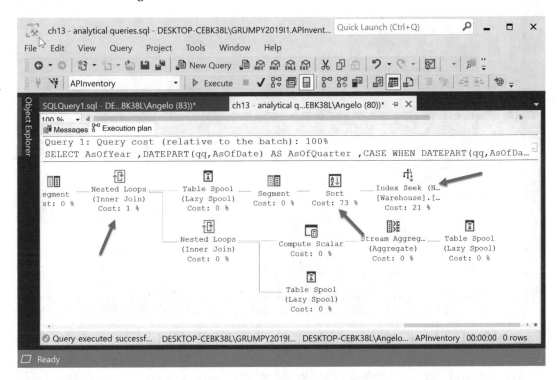

Figure 13-38. *Estimated query plan – right-hand side*

We will only look at the first section of the plan as the remaining sections have the same repeating patterns that we saw when we analyzed the PERCENTILE_CONT query.

The same scenario was generated, an index seek task on the Warehouse table with a cost of 21% and an expensive sort task with a cost of 73%. All other tasks are with 0% cost except the nested loops join task that came in with a 1% cost. We expected this. Let's run our IO and TIME statistics, after which we will modify the query, so it utilizes the Calendar table via a JOIN clause in the query.

Please refer to Figure 13-39.

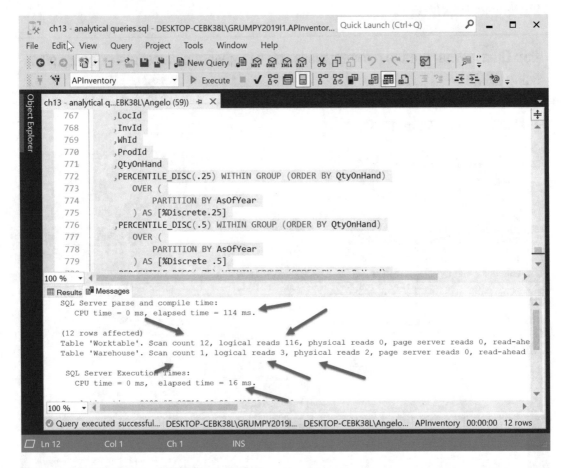

Figure 13-39. *IO and TIME statistics*

Same IO and TIME statistics as PERCENTILE_CONT() except parse and compile time and SQL Server execution times were slightly improved. Notice the physical reads on the Warehouse table (2). Notice the scan count (12) and logical reads (116) on the work table.

Note   When running this type of analysis on your laptop or workstation or even a development server, make sure that there are no background maintenance processes running as these could slow performance and skew results.

Let's modify the query as per Listing 13-9 so it now includes the JOIN to the Calendar table.

Listing 13-9. Enhanced Query, at Least We Hope!

```
CREATE UNIQUE CLUSTERED INDEX pkCalendar
ON MasterData.Calendar (CalendarYear,CalendarDate)
GO
UPDATE STATISTICS MasterData.Calendar
GO

SELECT AsOfYear
      ,C.CalendarQtr
      ,C.CalendarTxtQuarter AS AsOfQtrName
      ,C.CalendarTxtMonth AS AvgMonthName
      ,C.CalendarDate AS AsOfDate
      ,LocId
      ,InvId
      ,WhId
      ,ProdId
      ,QtyOnHand
      ,PERCENTILE_DISC(.25) WITHIN GROUP (ORDER BY QtyOnHand)
          OVER (
                PARTITION BY AsOfYear
          ) AS [%Discrete.25]
      ,PERCENTILE_DISC(.5) WITHIN GROUP (ORDER BY QtyOnHand)
          OVER (
                PARTITION BY AsOfYear
          ) AS [%Discrete .5]
      ,PERCENTILE_DISC(.75) WITHIN GROUP (ORDER BY QtyOnHand)
          OVER (
                PARTITION BY AsOfYear
```

```
        ) AS [%Discrete .75]
FROM MasterData.Calendar C
JOIN Product.Warehouse W
ON W.AsOfDate = C.CalendarDate
WHERE C.CalendarYear = 2002
AND ProdId = 'P101'
AND Locid = 'LOC1'
AND InvId = 'INV1'
AND WhId = 'WH112'
GO
```

The first task that we needed to do was create a unique clustered index on the Calendar table based on the year and calendar date. Next, we updated the table statistics to make sure the query optimizer has the latest and greatest information on the table.

Turning our attention to the modified query, notice the JOIN predicate order; the inner table is the Warehouse table, and the outer table is the Calendar table.

Now let's compare the estimated query plans. Please refer to Figure 13-40.

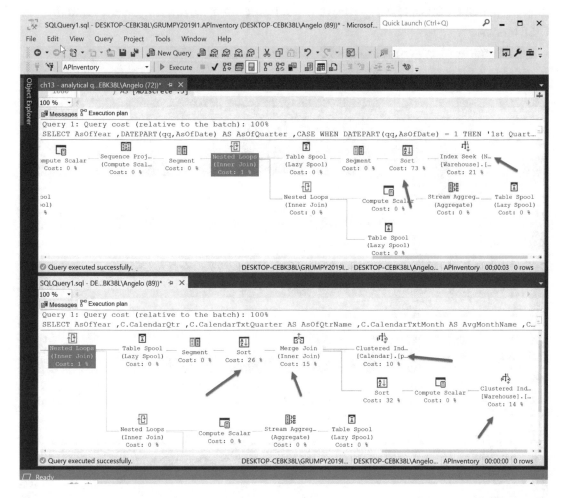

Figure 13-40. *Comparing estimated query plans*

The new plan looks a little better. The clustered index seek cost went down from 21% to 14%, and the sort task went down from 73% to 32%.

So far, so good.

We see the new tasks related to the Calendar table JOIN. We see a clustered index seek with a cost of 10% followed by a merge join task with a cost of 15% and a new sort with a cost of 26%.

Seems like the Warehouse table tasks improved, but we added a few new tasks to process the calendar date text parts we need. Remember we did this to eliminate the CASE blocks.

Is this better? Let's compare the IO and TIME statistics.

Please refer to Figure 13-41.

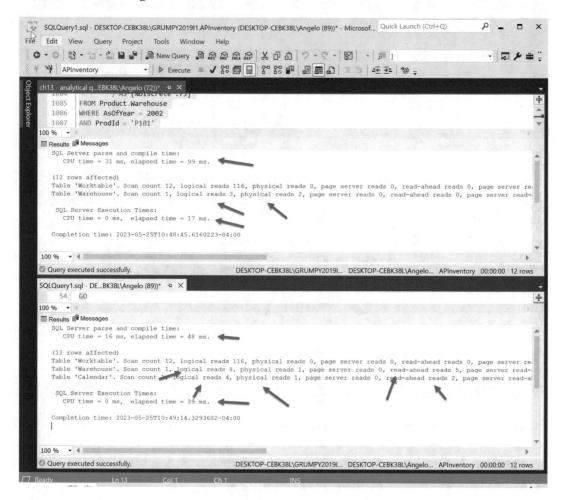

Figure 13-41. *Comparing IO and TIME statistics*

Surprise!

Looks like our scheme did not produce the positive improvements we had hoped for. SQL Server execution time went up from 17 ms to 39 ms, a little more than double. The work table statistics remained the same.

For the Warehouse table, we see a scan count value of 1, logical reads of 8 (this went up), 1 physical read (this went down), and 5 read-ahead reads.

Lastly, we have a new set of statistics for the Calendar table with 1 scan, 4 logical reads, 1 physical read, and 2 read-ahead reads.

Conclusion: Our enhancements did not realize significant improvements. This strategy will work if we place the Calendar table in memory by making it a memory-enhanced table. Otherwise, we should leave this query as is.

Sometimes things do not work out as expected, but it is worth a try in order to learn and see what will work and what will not work.

Let's wrap up our last performance analysis and improvement discussion with an overall look at statistics across all queries.

Pop Quiz Answer   What are the different ways you can generate an estimated query plan using SSMS?

Via the Query selection pop-up menu or via the icons in the top menu bar.

Overall Performance Considerations

We can use Microsoft Excel spreadsheets and graphs to profile IO and TIME statistics to get a sense of how window functions perform related to each other. By collecting all the statistics in a table and graphing them, we can see that the PERCENT_RANK() function ate up a lot of CPU and elapsed execution time compared with the other functions.

We will do this for both the Warehouse table and the work table. Figure 13-42 is the visualization for the statistics of the Warehouse table.

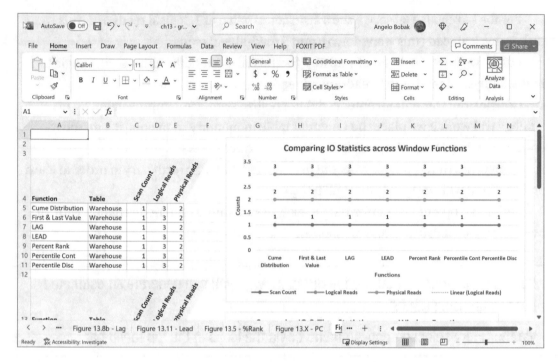

Figure 13-42. *Query portfolio IO statistics for the Warehouse table*

This is interesting. The scan counts, logical read values, and physical read values for all the window functions against the Warehouse table are all the same. Can we say then that all these functions behave consistently the same?

Well, yes, because most of the time we used the same WHERE clause, so this is a factor we can take into consideration.

Let's look at the work table visualization.

Please refer to Figure 13-43.

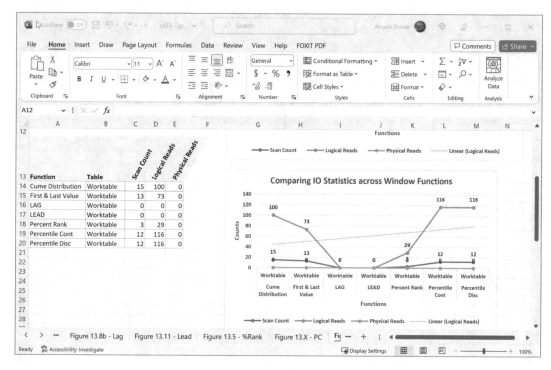

Figure 13-43.   *Query portfolio IO statistics for the work table*

This is a lot more interesting. Looks like the CUME_DIST(), PERCENTILE_CONT(), and PERCENTILE_DISC() functions generate the highest logical reads, while the LAG() and LEAD() functions generate the lowest. The other functions are somewhat in between.

As far as scan counts, we can state the same patterns, and what is interesting is all the physical reads are zero.

Last but not least, let's look at the CPU and elapsed time statistics across all the functions.

Please refer to Figure 13-44.

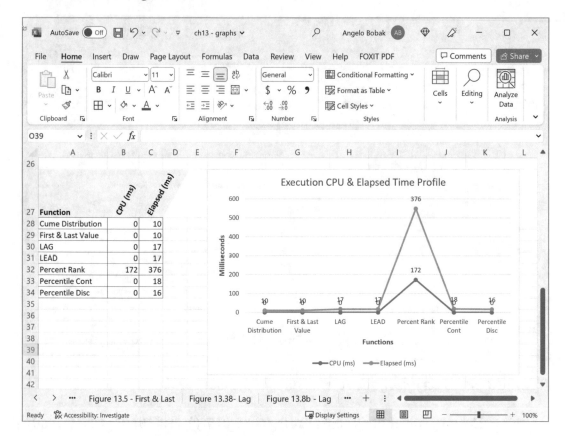

Figure 13-44. *Query portfolio CPU and elapsed time statistics*

Another interesting set of statistics in a visualization. Looks like the PERCENT_RANK() function was using the most CPU resources and took the longest to execute. The rest of the functions were pretty stable.

Conclusion: Generating charts and graphs comparing the resources these window functions use is a good way of isolating the ones that eat up most resources. This is where you would begin performance tuning. In other words, prioritize which queries you need to tune based on this analysis.

Let's wrap up the chapter by looking at a few more SSRS reports to demonstrate the drill-down capabilities. This is more of a show-and-tell discussion, and we will not go through the steps required to create the report as we did this earlier.

Report Builder Examples

In other chapters, we discussed how the simple queries we created and analyzed could be modified to return more data (basically removing the WHERE clause and tweaking the PARTITION BY clause) so they could be utilized as a basis for some sophisticated SSRS web report.

The WHERE clause columns would then be used as filters in Report Builder. This is a powerful concept that allows business analysts great flexibility in what data they wish to view and analyze.

In this section we will take a look at one of the reports I created that references a VIEW we discussed in this chapter. This VIEW is called Reports.MonthlySumInventoryShort. We will use this in a report that will allow us to drill down by years, quarters, and months and also by location, inventory, warehouse, and product and any combination you can think of.

I will not show you how to build it, as we saw in prior chapters, but we will concentrate on the report navigation features.

Please refer to Figure 13-45.

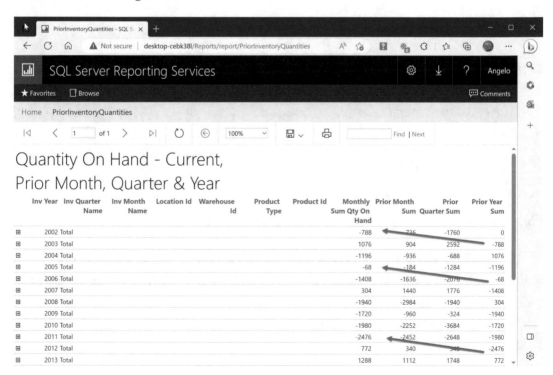

Figure 13-45. *Quantity on hand – prior years' view*

The report was published to my personal SSRS website on my laptop, just like in a real production environment. Amazing!

This first view shows the top-level year data. As can be seen by the arrows, the prior year column points to the correct value in the **Monthly Sum Qty On Hand** column. I wanted to make the field names in the report a bit verbose, so they clearly indicate what they are.

If we click one of the plus sign buttons next to the **Inv Year** column, we drill down to the next level, inventory movements by year and quarter.

Please refer to Figure 13-46.

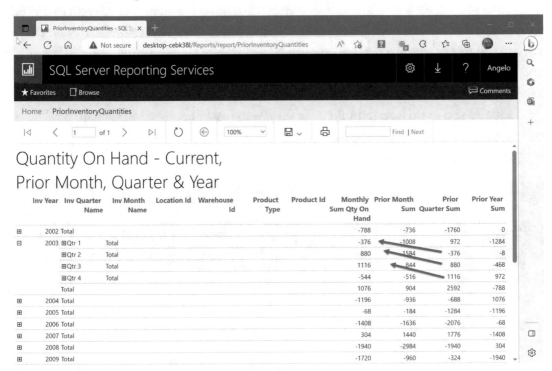

Figure 13-46. *Quantity on hand – prior quarter view*

Success! Drilling down in 2003 displays the four quarters for that year, and once again the values in the Prior Quarter Sum column point to the correct inventory totals in the **Monthly Sum Qty On Hand**.

Let's drill down to the month level.

Please refer to Figure 13-47.

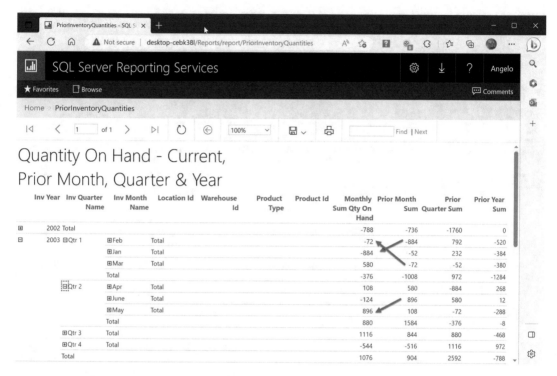

Figure 13-47. *Quantity on hand – prior month view*

Drill-down works well enough, but notice that the months were sorted by name, so the order is clunky. We can always go back, add the numerical month, and republish or just fix the sort order on the column that displays the text values for the month. Yes, for this column you can specify that the values are sorted by the numerical month column, thus overriding the sort on the text values.

Before we do that let's drill down on one of the months to see details at a warehouse level.

Please refer to Figure 13-48.

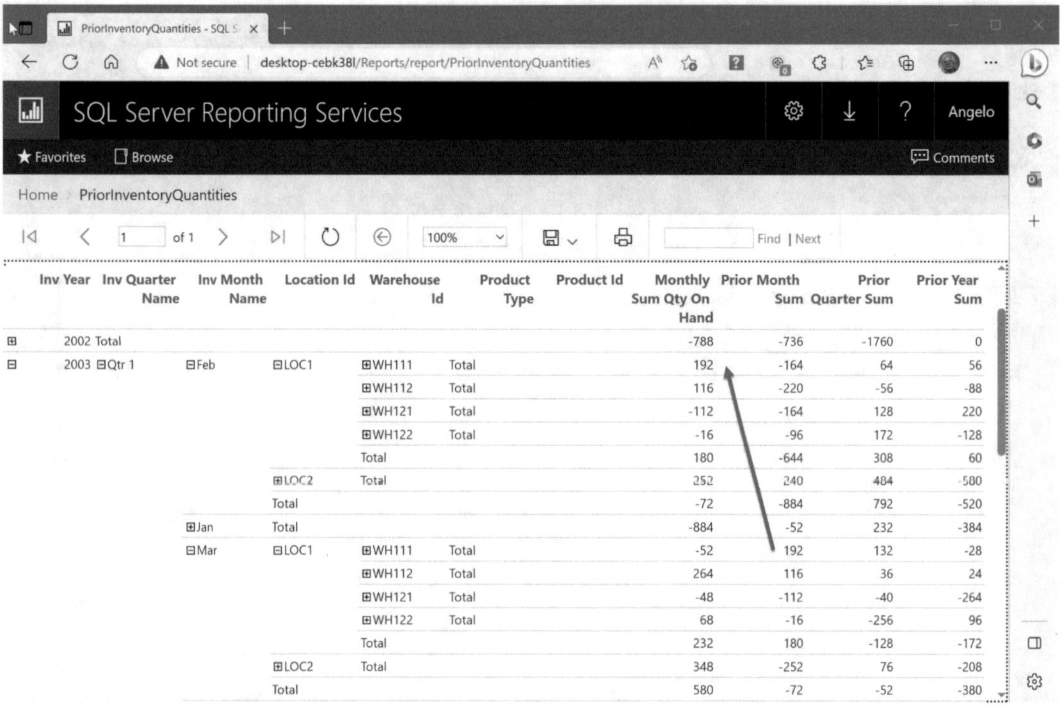

Inv Year	Inv Quarter Name	Inv Month Name	Location Id	Warehouse Id	Product Type	Product Id	Monthly Sum Qty On Hand	Prior Month Sum	Prior Quarter Sum	Prior Year Sum
⊞ 2002	Total						-788	-736	-1760	0
⊟ 2003	⊟Qtr 1	⊟Feb	⊟LOC1	⊞WH111	Total		192	-164	64	56
				⊞WH112	Total		116	-220	-56	-88
				⊞WH121	Total		-112	-164	128	220
				⊞WH122	Total		-16	-96	172	-128
				Total			180	-644	308	60
			⊞LOC2	Total			252	240	484	-580
			Total				-72	-884	792	-520
		⊞Jan	Total				-884	-52	232	-384
		⊟Mar	⊟LOC1	⊞WH111	Total		-52	192	132	-28
				⊞WH112	Total		264	116	36	24
				⊞WH121	Total		-48	-112	-40	-264
				⊞WH122	Total		68	-16	-256	96
				Total			232	180	-128	-172
			⊞LOC2	Total			348	-252	76	-208
			Total				580	-72	-52	-380

Figure 13-48. *Quantity on hand – prior warehouse view*

Looking at March for the first quarter, we can see that the **Prior Month Sum** value for warehouse "WH111" correctly points to the corresponding value in February in the **Monthly Sum Qty On Hand** column.

Let's drill down on one of the warehouses in order to see the results at a product level.

Please refer to Figure 13-49.

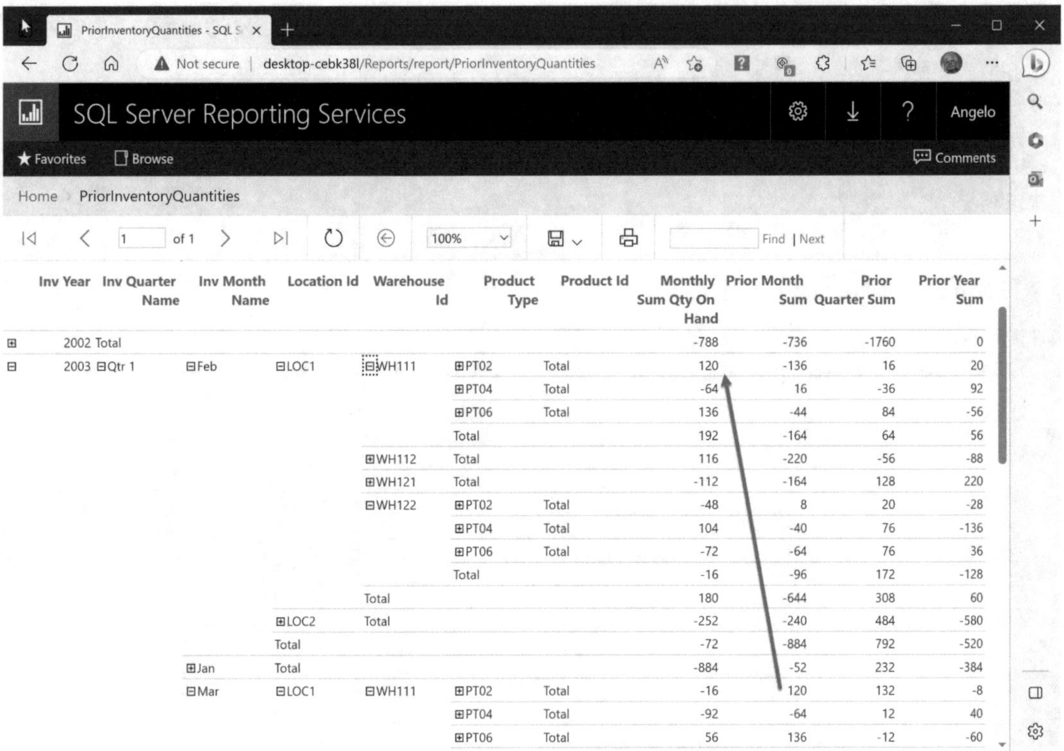

Figure 13-49. *Quantity on hand – prior product view*

This report also shows the correct prior month values but this time at a product level. If it wasn't for the month names being sorted incorrectly, this report would merit top marks!

I fixed the report, and the next screenshot shows the months sorted in correct order.

Please refer to Figure 13-50.

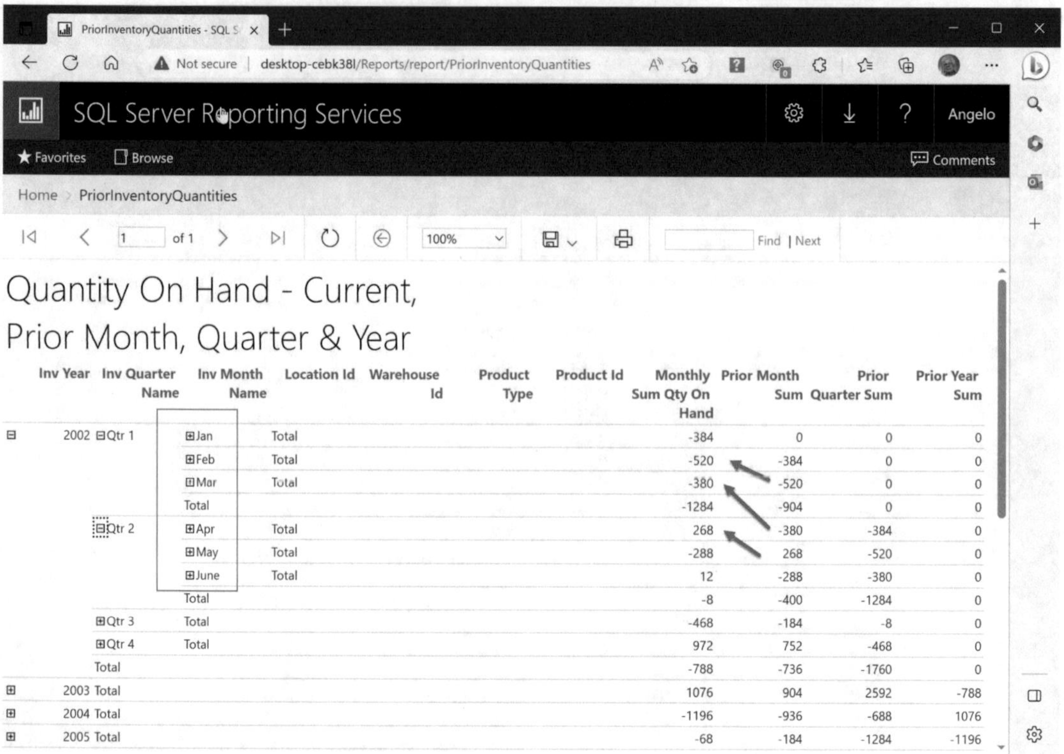

Figure 13-50. *Fixed sort order*

This is a lot better. All that had to be done was click the properties for the month name column and change the sort order, so it sorts by month number as opposed to month name. Visually, you will see the month names in correct order. This is done in Report Builder, or you can even edit the report on the website.

Let's see how this is accomplished.

Please refer to Figure 13-51.

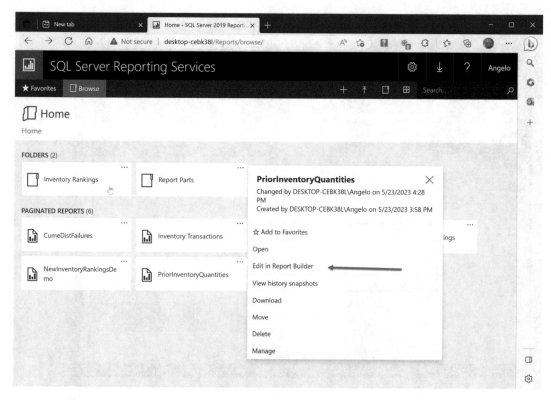

Figure 13-51. *Fixing the month sort order, part 1*

We want to fix the report called **PriorInventoryQuantities**. Click the three dots on the upper right-hand side of the report name to display a menu with several choices. Click "Edit in Report Builder." This will bring up Report Builder if it is already installed on your computer.

Note   If your reports are on an actual server, you need to install Report Builder on the server.

Once it loads in design mode for this report, we simply need to change the sort parameter for this column. This is very easy to do.

Please refer to Figure 13-52.

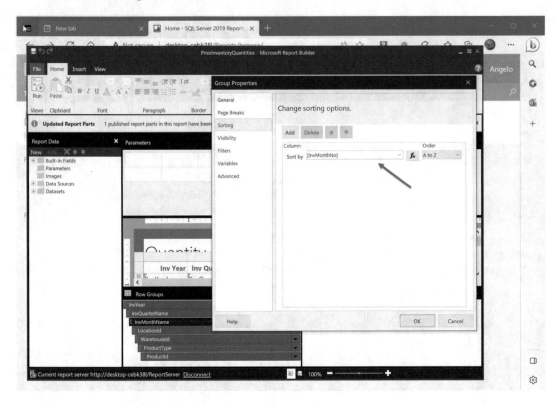

Figure 13-52. *Fixing the month sort order, part 2*

In the Row Groups section at the bottom, find the offending column. Click the down arrow toward the right side and select Group Properties. This displays the dialog box shown in the preceding figure and simply click the Sort by list box and scroll down to the column you want to sort on.

Once you selected it, click the OK button and back out of Report Builder. You will see a yellow Update Report Parts message stating the report was updated on the server. Make sure you save it locally too! You do not want to republish a local report on your laptop and wipe out the changes.

That is all that there is to it. Here comes your homework assignment, after which we will wrap up the chapter.

Homework   Using Report Builder, create any type of report you wish using either the APInventory database or APInventoryWarehouse data warehouse. You do not need to publish it, if you did not install SSRS on your computer or server.

If you have not already downloaded Report Builder, simply perform a web search, and download and install it. This is accomplished with the usual Microsoft installer program, and there are only a few steps involved. The last chapter will show you where to find and download all the tools we used in this book.

Summary

We made it. We are at the end of our journey. In this chapter, we applied the analytical functions one last time to our inventory management scenario. We worked with two databases, an inventory transaction database called APInventory and a data warehouse called APInventoryWarehouse.

We performed our usual script creation based on business analyst specifications and viewed the results. In some cases we used the output of the queries to create interesting Microsoft Excel visualizations.

Lastly, we performed our usual performance analysis and enhancements and learned techniques for analyzing performance statistics in Microsoft Excel spreadsheets.

We wrapped up the chapter with an example demonstrating the drill-down capabilities of SSRS web reports created with Report Builder. We even saw how to correct a mistake in the report after it was published. I wish I could say I did this on purpose, but I actually made the mistake!

As the saying goes, if life gives you lemons, make lemonade.

In the next and last chapter, we wrap things up with a brief summary of what we covered in our adventure. I also show you where to get

- SQL Server Developer (free developer license)

- Visual Studio Community (free developer license)

- SQL Server Data Tools (SSMS) (installed via Visual Studio Installer)

- SSAS server and Visual Studio project support

- SSIS server and Visual Studio project support

- SSMS download installer from the Microsoft site (manage SQL Server and SSAS)

- Report Builder (free developer license)

- Microsoft Excel – you need to pay for this, but it is worth it

SQL Server Data Tools, SSRS, SSAS, and SSIS installers are downloaded from Visual Studio Community Installer. If you do not have a free developer license for SQL Server, make sure you download it and install it first as it will install not only SQL Server instances but also SSAS, SSIS, and SSRS servers.

CHAPTER 14

Summary, Conclusions, and Next Steps

Summary

Congratulations! This is our last chapter. Let's wrap things up by summarizing what we learned and also looking at the tools in our toolkit and where to get them. I might even tell you how to create the databases and load them too! We will also discuss how to set up the folders for storing code and also the physical database files.

I recommend you create the folders, create the database, and load them first so as you read through the chapters, you can follow along and practice.

Our Journey

The goal of this book was to assemble a set of tools that can be used to perform business intelligence analysis, data mining, and data analysis. I affectionately called it the SQL Server toolkit.

I started by identifying three categories of functions available in SQL Server since 2012:

- Aggregate functions

- Ranking functions

- Analytical functions

© Angelo Bobak 2023
A. Bobak, *SQL Server Analytical Toolkit*, https://doi.org/10.1007/978-1-4842-8667-8_14

These, when used with the OVER() clause, are considered window functions and deliver a powerful capability like creating rolling calculations across time spans such as months, days, and years. They even allow us to traverse backward and forward in a set of historical data in order to compare current values (at a point in time) with future or prior values. Last but not least, a few functions are available to perform statistical analysis like standard deviation, variance, and calculating the MEAN and MEDIAN values within a data set.

We applied these functions across four industries so as to get a sense of how one would perform analysis in each:

- Sales use case

- Financial use case

- Power plant equipment monitoring and failure analysis use case

- Inventory movement management use case

Along the way I realized that these tools were only part of the BI data analysis strategy landscape. I introduced multiple tools like SSIS, SSRS, SSAS, Power BI, Report Builder, and Microsoft Excel to help us visualize the results of queries that utilized the window functions.

These tools allowed us to generate, view, and analyze interesting patterns that a business analyst would use to report on performance for their company. Lastly, we took a look at some performance analysis tools such as estimated query plans that allowed us to analyze query performance and come up with strategies to improve performance.

Let's review the tools and wrap things up by showing you how to get them (download, install, and configure)!

But first, a word about the code.

About the Code

The code can be found on the publisher's website for this book. Each business use case has three folders dedicated to it, one per function category. Each chapter folder that is the first chapter in the business scenarios has three subfolders for creating the database and tables (CREATE DATABASE folder), loading the tables (LOAD DATABASE folder), and the queries covered in each chapter (CHAPTER QUERIES folder). The remaining two chapters in the business scenarios have folders for queries only. When you download the code, make sure to create these folders for each chapter on your PC or laptop so you can easily find the TSQL code scripts.

Shown in Figure 14-1 is what they look like.

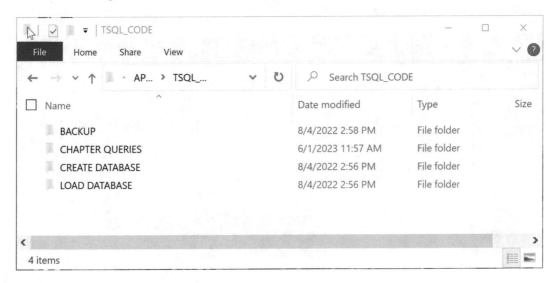

Figure 14-1. *Folder structure for the code*

Now you know where to download the code and where to place it on your device. Creating a BACKUP folder is a good idea so you can save scripts before you modify them just in case you mess things up! Believe me, backup folders have saved me time and again.

About the Database Folders

I set up my databases with a folder for each database, and each database has three subfolders as illustrated in Figure 14-2.

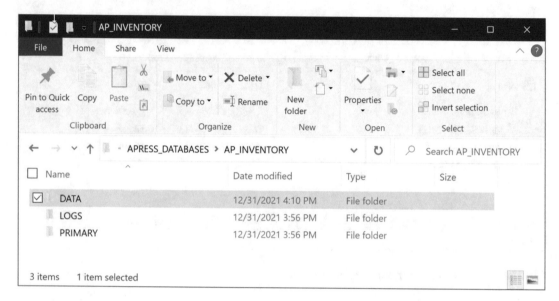

Figure 14-2. *Folder structure for databases*

I created the APRESS_DATABASES folder on my D drive. I recommend you use the same structure I used except, of course, if you have only one drive (drive C:), you can set up the folders there. You need to modify the DDL scripts that create the databases, so they refer to the folder structure you set up on your computer.

If you have a D: drive, save yourself some work and set up the folders as shown. This will save you time. Also, keep the same file group names used in the databases.

Data Used in the Examples

Keep in mind that the data generated for the examples was randomly generated data. I did the best I could to eliminate duplicates and combinations that made no sense. I also took great liberty in chapters that summed up account balances and inventory quantities on hand, so they demonstrated the behavior of the functions.

For example, in the financial database examples

- Assume the monthly value is what is left in the account at the end of the month after all credits and debits. This applies to account- and portfolio-related tables.

- Assume the rolling monthly value adds current month's (after all the debits and credits) value and last month's value so as to represent the total value of the current month.

In some cases, I applied the following ROWS frame specification to process the two-rolling-month scenario:

```
ROWS BETWEEN 1 PRECEDING AND CURRENT ROW
```

For example, in real life an account may have $100 in January and $50 in February. If you think the total is $150, you're wrong if debits occurred in January or February. If no debits occurred, then $150 is the total because it includes the cash in the account carried over from January. So a few factors come into play, like debits and credits to an account.

Here is another scenario. If $50 was debited from the account in February and no credits were made to the account, all that is left over is $50 (current month to date).

The same liberties apply to portfolio balances (in the financial database chapters) and also to the three chapters dealing with inventory movement. Inventory, like the financial scenario, needs to consider inventory moving in and out to realistically report the monthly inventory totals.

The code is liberally commented so all should be clear. The goal is to understand how the window functions work and not simulate a real-world business scenario or realistic data patterns.

If you end up understanding how rolling totals work, you are ahead of the game.

The Toolkit

We previewed this diagram in Chapter 12, but I think it is worth seeing it one more time as it shows you all the tools we discussed and also indirectly shows all the skills you will have once you master these!

All the tools are free except for Microsoft Excel. You need a subscription for this, but it is worth the money. As far as SSAS, SSIS, and SSRS, these are enhancements you can add on once you download and install the latest community edition of Visual Studio. We will discuss how in this last chapter.

Please refer to Figure 14-3.

SQL Server BI Toolkit Development Architecture

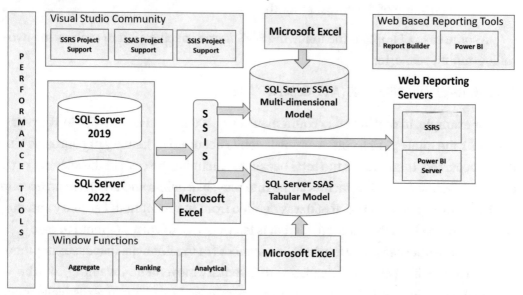

Figure 14-3. *The SQL Server BI toolkit*

As the figure shows, there are a lot of tools, and I think we now have a fundamental knowledge of what they do and how to apply them. Can you believe you learned how to use all of these?

SQL Server

This is the first and most important component tool you need. SQL Server delivers the database engine that allows you to create databases, create and load tables, and of course run all the window functions we just discussed.

Please refer to Figure 14-4.

SQL Server at 10,000 Feet

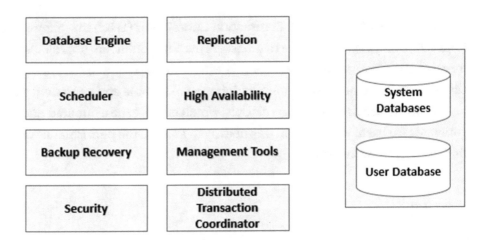

Figure 14-4. *SQL Server at 10,000 feet (give or take a foot or two)*

At a very high level, SQL Server includes the database engine used for storing the database and all related objects plus the query engine, parser, and database file management processes.

Note   Databases are composed of files: files for storing data, files for storing data for the system databases, files for TEMPDB, and files for the transaction logs. All these files can be grouped into buckets called file groups. If you need a refresher on how to create databases and how they are structured, check out my free podcast on creating databases at

www.grumpyolditguy.com

Back to SQL Server. A scheduler is included to schedule queries, scripts, and external programs so they execute at the required time. You can even schedule SSIS packages.

A backup/recovery process is included to back up and restore databases in cases of data corruption or processing errors.

The replication component allows you to replicate databases across servers in a distributed architecture, and the high availability component ensures SQL Server is always up and running.

The tools we learned can take advantage of distributed data, so make sure you understand how this feature works. SSIS can connect all databases in a distributed architecture so as to load them into a centralized database called an ODS (Operational Data Store).

There is also a Distributed Transaction Coordinator (DTC) that allows you to create transactions that span multiple databases. This component ensures all distributed transactions across multiple servers succeed (commit) or are rolled back (rollback) in case of issues so that databases are maintained in a stable and consistent state.

Lastly, we see the management tools, which include the estimated query plan generator we used. We relied on this heavily to analyze the performance of our queries that used window functions.

SSMS

Once we downloaded and installed SQL Server, we needed a tool to help us manage the database and related objects like tables, indexes, views, and query scripts. SQL Server Management Studio is the tool of choice. We used it throughout our journey, so you should be very familiar with it by now. We only touched the surface of its many features and capabilities.

Please refer to Figure 14-5.

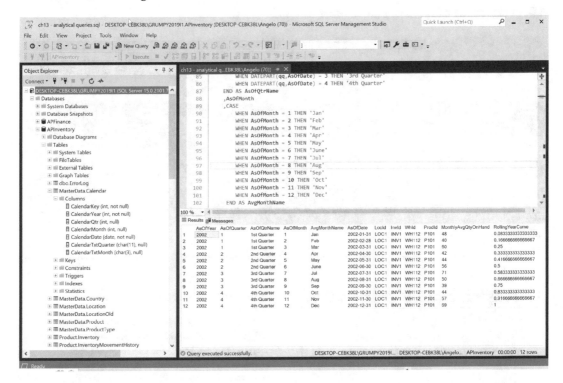

Figure 14-5. *SQL Server Management Studio*

Here we see the familiar Object Explorer, query pane, and results pane windows. Not seen is the Solution Explorer, but you can easily modify the way SSMS looks.

You will need this tool after you install SQL Server 2019 or 2022. This tool will not only let you run the scripts that come from each chapter but also the scripts that create the database used in each chapter and also load the tables. SSMS also has a host of management tools that you will rely on heavily. This is priority number 2! Make sure you download the installer from the Microsoft download site and install this right after you set up SQL Server.

If you would like a brief tutorial on how to use SSMS, check out my free podcast at www.grumpyolditguy.com. Please refer to Figure 14-6.

Figure 14-6. *Grumpy Old IT Guy video training podcasts*

By the way, all podcasts are free, with no registration, but comments are always welcome. If you have ideas on how to improve the content, please let me know.

The best way to use SSMS though is by experimenting with it. Poke around and try the features. In no time, you will rely on this tool for all your programming tasks.

Shortly I will show you where to download and how to install SSMS.

The Window Functions

Here are the functions we reviewed in each business case scenario.

Please refer to Table 14-1.

Table 14-1. *Functions by Category*

Aggregate	Ranking	Analytical
COUNT( )	RANK( )	CUME_DIST( )
COUNT_BIG( )	DENSE_RANK( )	FIRST_VALUE( )
SUM( )	NTILE( )	LAST_VALUE( )
MAX( )	ROW_NUMBER( )	LAG( )
MIN( )		LEAD( )
GROUPING( )		PERCENT_RANK( )
STRING_AGG( )		PERCENTILE_CONT( )
STDEV( )		PERCENTILE_DISC( )
STDEVP( )		
VAR( )		
VARP( )		

These functions, used with OVER( ) clauses, gave us the ability to create powerful queries, like creating moving rolling totals and averages. As you saw, you can also perform some very interesting statistical analysis.

You will rely on these queries to solve complex business problems. You can count on these being discussed in an interview, so make sure you know them inside out! Key is the OVER( ) clause. Make sure you understand what a PARTITION BY clause does, what an ORDER BY clause does, and also what the window frame specifications (ROWS and RANGE) do. Also understand the default behavior as far as window frame specifications when you have OVER( ) clauses that use both a PARTITION BY and an ORDER BY clause or only one of them or even empty OVER( ) clauses. Make sure you have a solid understanding of what a partition within a data set is. If this is a bit fuzzy, make sure you go back to Chapter 1, which explains all.

If you need to review these functions, pick a chapter, and check them out. Make sure you try out the scripts, modify them, break them, and enhance them. That's how one learns.

Lastly, check out the Microsoft documentation on these functions to get really into the details of how they work. Remember, these are your tools, so make sure you know what they do and don't do and how to use them!

Note SQL Server 2022 comes with an alternate syntax for the OVER( ) clause. I briefly covered this in the first few chapters of the book. There are also new window functions such as

APPROX_PERCENTILE_CONT( )

APPROX_PERCENTILE_DISC( )

Definitely worth the time to check out the new features available in Microsoft SQL Server 2022 by referring to the Microsoft SQL Server 2022 website.

The Visual Studio Community License

Visual Studio Community is the free license available to developers for creating all types of applications in the Microsoft development architecture, from C# applications to SSIS packages. This tool is a companion to SSMS, and you will be using it to create SSAS cubes and SSIS ETL packages.

Please refer to Figure 14-7.

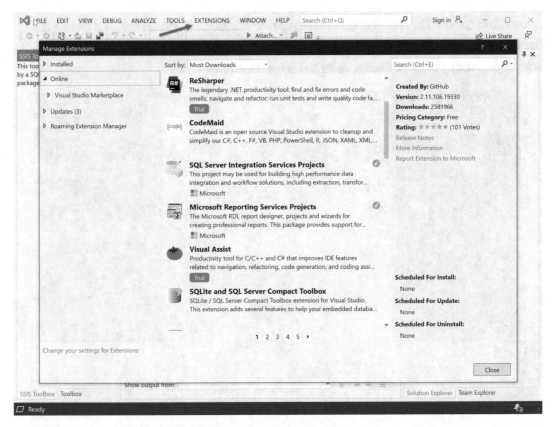

Figure 14-7. *Visual Studio, picking project templates*

I will show you how to download this tool shortly, but let's focus on the
EXTENSIONS selection in the menu bar. This is what you will use to download and
install SSIS, SSAS, and SSRS project support within Visual Studio Community. Just click
EXTENSIONS and then scroll through the list to pick and choose what you want and
need. The list is extensive!

Warning: This will take time to download and install. Again, this will install project
templates so you can develop applications with these tools like SSIS and SSAS projects.
You can create C#, Python, and other types of applications also.

You install the SSAS, SSIS, and SSRS servers when you install SQL Server. As
you navigate through the screens, you will be given a chance to pick and choose the
components you need and configure them by specifying what security credentials to use
and where to place critical folders and files.

SSAS Projects

Remember these? This is what we used to create multidimensional cubes that allow users to slice and dice data and to also navigate up and down through the various levels and aggregations of data. Please refer to Figure 14-8.

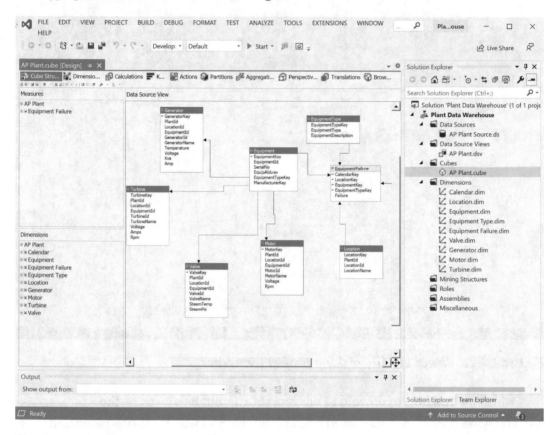

Figure 14-8. *SSAS OLAP model*

Here is the cube structure panel that shows you the structure of the SSAS cube we created in the chapters dealing with the power plant use case. Cubes created with Visual Studio SSAS project templates are created and loaded on the SSAS server instance and can be accessed by Microsoft Excel spreadsheets and Power BI dashboards. You can even create SQL-like queries using a query language called MDX. This is another language you might want to learn.

SSIS Projects

Visual Studio is used to create solutions with one or more projects that contain packages to load databases, create cubes, and any other type of data integration processes. We can say SSIS is what moves the data and SQL Server is what provides the data.

Solutions created with SSIS can be deployed to SQL Server so they can be run on a schedule defined by you, the DBA, or the system administrator, with the SQL Server scheduling tool called SQL Agent.

Please refer to Figure 14-9.

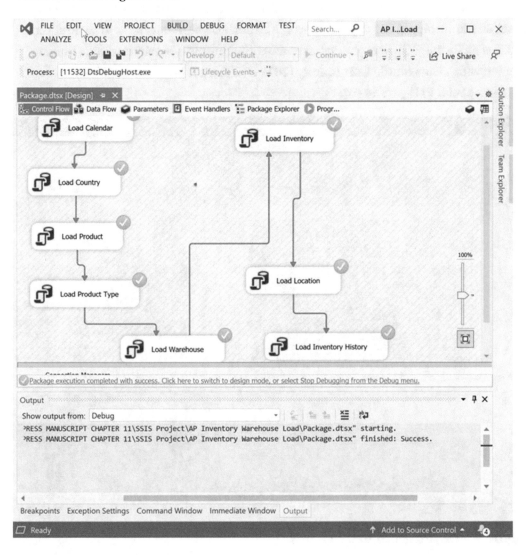

Figure 14-9. *Typical SSIS package*

This shows the successful run of a package we created in the inventory use case chapters. This package retrieves data from the APInventory database and loads it to the APInventoryWarehouse data warehouse. This is a very important tool that you will need to install, learn, and master on your journey to becoming an expert SQL Server developer or data integration architect.

Power BI Web Scorecards, Dashboards, and Reports

Power BI is the tool we used to create some simple reports and scorecards. We only touched the surface of the capabilities of this tool. Power BI connects to various data sources including SQL Server databases and SSAS multidimensional cubes that are created with Visual Studio Community (2019 or 2022; get both – it can't hurt).

Please refer to Figure 14-10.

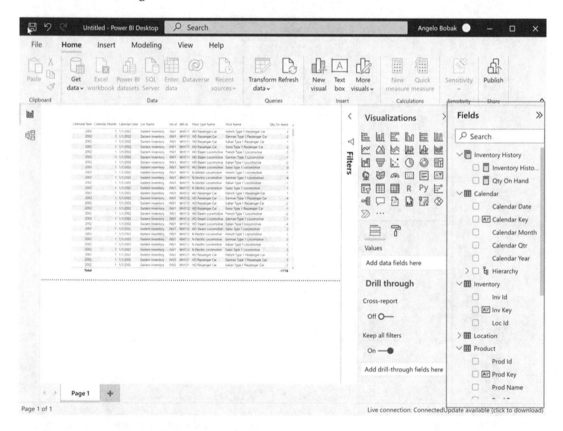

Figure 14-10. *Power BI design panel*

This screenshot shows you the design panel where you can define data sources, retrieve data, and convert it to a number of sophisticated visualizations like pie charts, line charts, and bar charts. This tool can be obtained free of cost for evaluation. We will see how to obtain it shortly.

Microsoft Excel Spreadsheets

Next is our old friend, the Microsoft Excel spreadsheet. We used this to pull in data from our various business use case databases in order to create some interesting reports and visualizations. We also created some powerful pivot tables that allowed us to slice and dice through the data and navigate the various hierarchies defined by the data models for each database.

Please refer to Figure 14-11.

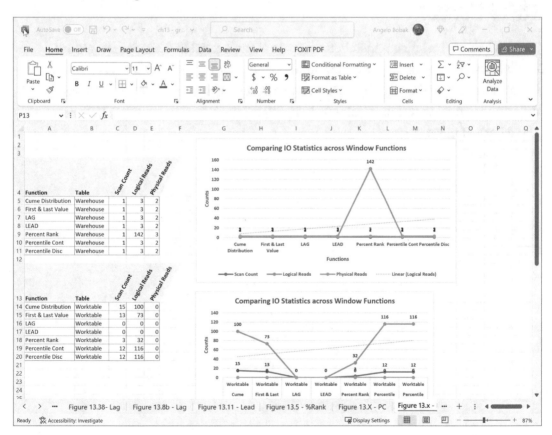

Figure 14-11. *Microsoft Excel visualizations*

This is just an example of one of the visualizations we created and discussed in our inventory chapters. This tool is the first-line tool you would use (or your business analyst) to prepare reports and graphs from the data in the database.

If you are not versed in spreadsheet creation, download and install the tool. It comes at a reasonable price, as it is not free, but well worth the cost so you can add it to your skill set.

Note   Excel also has its own MDX-like programming language called DAX. DAX stands for Data Analysis Expressions. You can use DAX with SSAS, Power BI, and Excel pivot tables.

SSAS Server

The SSAS server instance is installed as a part of the overall SQL Server Developer installation. You can manage and manipulate the cubes created with Visual Studio Community through SSMS. Two types of servers can be installed, either a multidimensional server or tabular (relational) server. When installing SQL Server, install both of these.

Please refer to Figure 14-12.

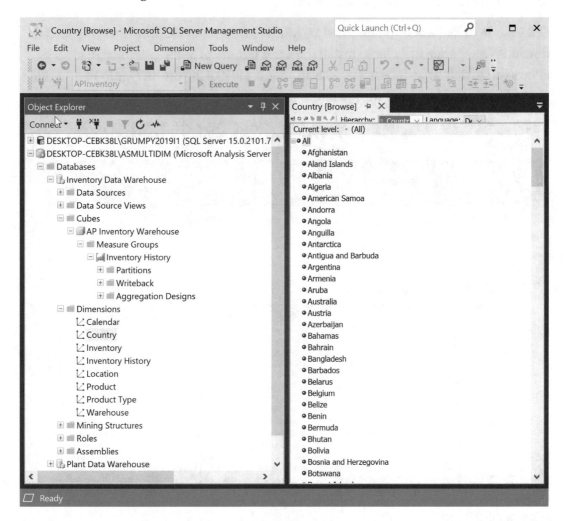

Figure 14-12. *SSAS instance viewed from SSMS*

Here we see the AP Inventory Warehouse cube and the various folders for data sources, data source views, cube measure groups, and dimensions. The right-hand-side panel shows the country names available in the Country dimension. You can use the query pane to create MDX queries.

SSRS Server and Website

SSRS was installed when you installed SQL Server Developer Edition. Now you need to download a separate installer to install the latest edition of SSRS. Once installed SSRS allows you to deliver web-based reports to your users like the one in Figure 14-13, which is one of the sample reports we created.

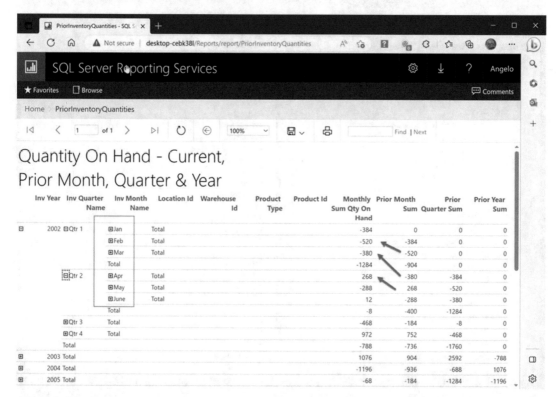

Figure 14-13. *SSRS report*

Once you download and install the SSRS project template components via the **EXTENSION** menu selection in Visual Studio Community, you need to download, install, and configure the SSRS server before you publish a report. You can download the SSRS installer from the Microsoft download site:

www.microsoft.com/en-us/download/details.aspx?id=100122

Run the executable and follow the guided install steps. At the end you can launch the configuration tool. This tool will also be added to your Windows desktop so you can go back and fine-tune it any time.

Please refer to Figure 14-14.

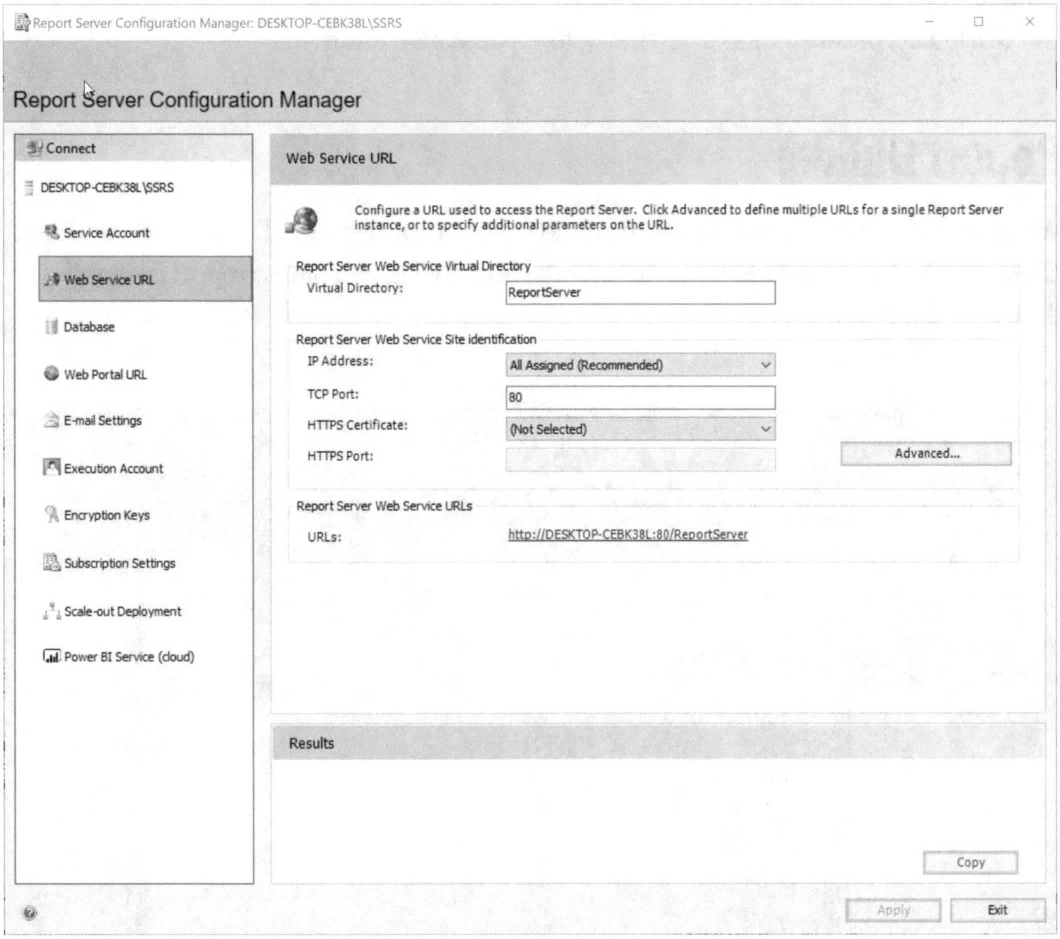

Figure 14-14. *Configuring SSRS*

Configuring SSRS is straightforward. For the most part, you accept all the default parameters. As can be seen in the screenshot, the left-hand side contains a panel with various selections. We see the Web Server URL panel selected, and it shows you the default parameters.

The only one where you need to do some work is the Database selection where you define the database that will store report server and report configuration data and report data. Mostly, you need to just supply some names, click "OK," and watch it run as it builds the databases.

Check out some websites that discuss installing and configuring SSRS in case you are a bit nervous about setting it up. It is easy.

Obviously, perform this setup first before you create reports.

Report Builder

We used this tool in a few chapters to create some simple reports. The process was easy. Choosing the wizard guides you, and before you know it, you have a report that pulls data from your database.

Please refer to Figure 14-15.

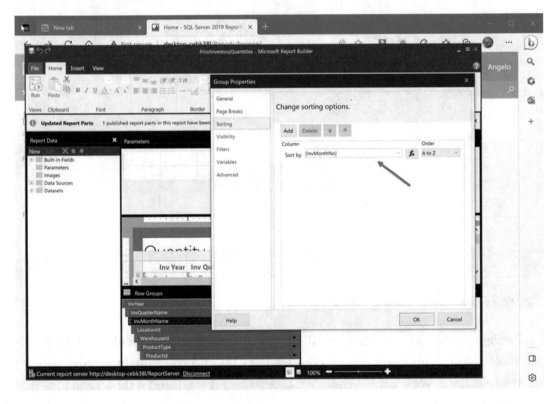

Figure 14-15.   *Report Builder design panel*

We only touched the surface of this report design. You can create reports with graphs and scorecards that contain visualization like gauges that display critical pieces of information and also reports that can drill down to any number of levels of detail. This tool is also free, so there are no excuses in downloading it, installing it, and learning how to use it.

Start with the wizard first, and before you know it, you will be creating reports from scratch with filters, graphs, and drill-downs.

Performance Analysis Tools

I am sure this was your favorite tool to perform performance analysis and enhancements. We used this tool to create estimated execution plans but also real-time execution plans that showed us the steps the SQL Server query engine took to execute a query and also how much data was passed between tasks.

Estimated Query Plans

Please refer to Figure 14-16.

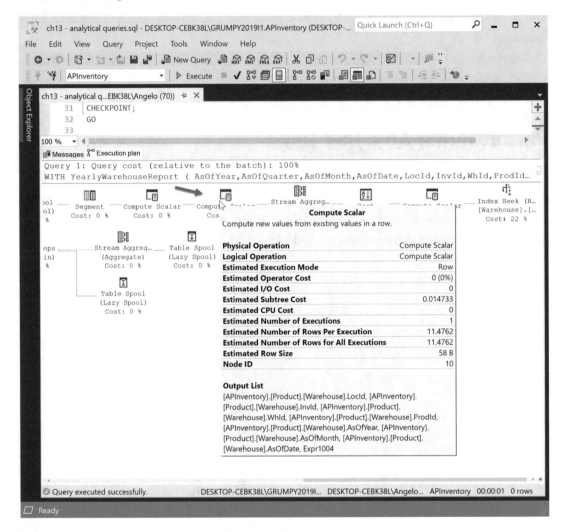

Figure 14-16.   *Typical estimated query plan*

The visual is great, and by placing your mouse over any task, you can see the details involved. This tool coupled with IO, TIME, and PROFILE statistics was powerful that allowed us to evaluate performance and come up with some tuning strategies.

Its cousin, live query plans, takes performance analysis a step further.

Live Query Plans

Please refer to Figure 14-17.

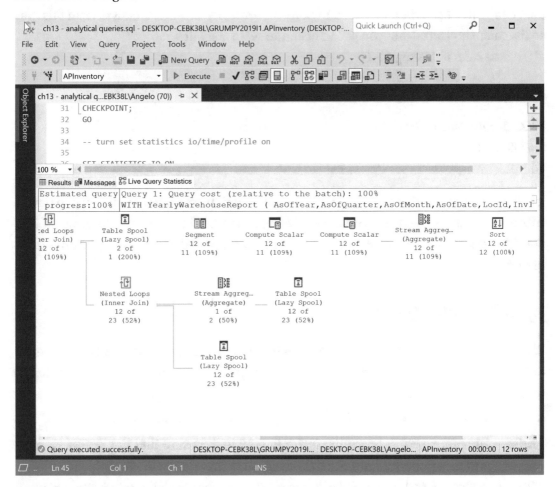

Figure 14-17.   *Live query plan*

This is similar to the estimated query plan tool, but we get to see the actual number of rows that is passed around between each task. If the data volume is large enough and if the query takes minutes to run, you can see the values change dynamically.

Make sure that the estimated costs displayed in both plans are close; otherwise, some serious issues might exist and investigation is required.

DBCC

We used this simple but important tool to clear memory cache that queries use to store data and plans. Each time you need to perform some performance analysis, run this first so the cache is cleared and you get an accurate set of statistics.

Please refer to Listing 14-1.

Listing 14-1. Running DBCC

```
DBCC dropcleanbuffers;
CHECKPOINT;
GO
```

If you do not run this tool, data and plan information are stored in memory so that subsequent runs of the query you are analyzing will return incorrect results as far as IO and TIME statistics because the query is taking advantage of data and query information in memory.

IO and TIME Statistics

We relied on this set of statistics many times in our performance tuning sessions. They tell us how many operations, like disk scan and physical reads, occur and also how long a query takes to parse and run.

Note   The parse step is when the query engine makes sure that the syntax is correct and also knows where the data location is that needs to be retrieved. Objects called parse trees are also created. This is what the estimated query plan tool displays when we look at the plans.

Please refer to Listing 14-2.

Listing 14-2. Setting IO and TIME Statistics On

```
SET STATISTICS IO ON
GO
SET STATISTICS TIME ON
GO
```

Run these to set the statistics on after you run DBCC and then execute the query. Change ON to OFF when you are finished and want to stop displaying statistics.

Please refer to Figure 14-18.

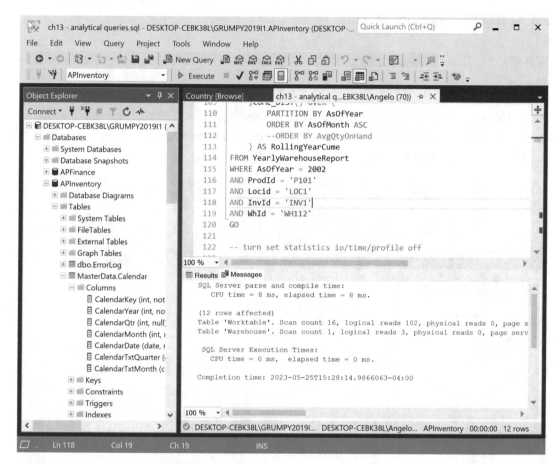

Figure 14-18. *IO and TIME statistics*

I know you remember these. Wouldn't it be nice if all queries ran in 0 ms?

STATISTICS PROFILE

Sometimes you really need to get into the weeds to understand what is going on in the tasks of a query plan. This set of statistics gives you similar information as IO and TIME statistics, but it also shows you the code sections that are generating the statistics.

Setting these statistics on is simple.

Please refer to Listing 14-3.

Listing 14-3. Setting STATISTICS PROFILE ON

```
SET STATISTICS PROFILE ON
GO
SET STATISTICS PROFILE OFF
GO
```

Set these to ON after you set the other statistics or even run them on their own. Make sure you turn them all OFF when finished in a session as these also eat up resources.

Please refer to Figure 14-19.

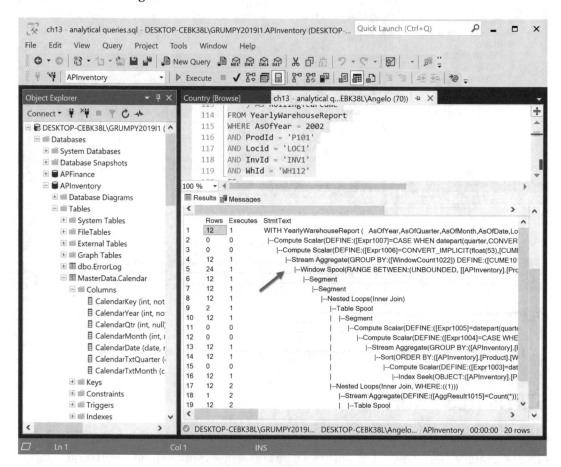

Figure 14-19. *STATISTICS PROFILE report*

Recall how detailed this information is. The following is the statement text that shows what goes on in a window spool task:

```
--Window Spool(
          RANGE BETWEEN:(UNBOUNDED, [[APInventory].[Product].[Warehouse].
          [AsOfMonth]]
      )
)
```

Okay, this completes our review. So how do we get all these nice tools?

Where to Get the Tools

You can find most of these tools by performing a web search that will point you to the correct Microsoft web page. Obtain these free tools from Microsoft only and read the licensing requirements!

SQL Server Developer

As of this writing, here is where you can obtain the developer license for the latest edition of SQL Server, like SQL Server 2022.

Please refer to Figure 14-20.

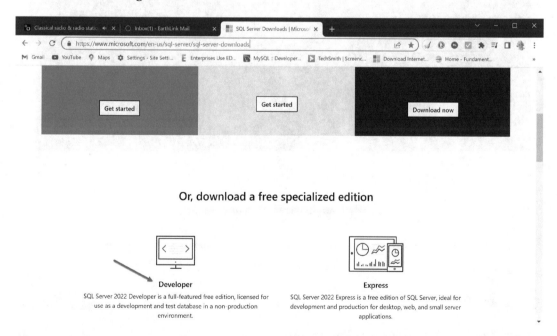

Figure 14-20. *Download SQL Server Developer Edition*

Make sure you download the developer license.

As of this writing, the URL is

`www.microsoft.com/en-us/sql-server/sql-server-downloads`

SQL Server 2019 is also available, with a SQL Server Lite edition called SQL Server Express.

I would recommend you download SQL Server 2022 as it has many new features and also new window functions like `APPROX_PERCENTILE_CONT()` and `APPROX_PERCENTILE_DISC()`.

Visual Studio Community

You need to download and install this first so you can install data tools that include SSMS and also SSAS, SSRS, and SSIS project template support.

Please refer to Figure 14-21.

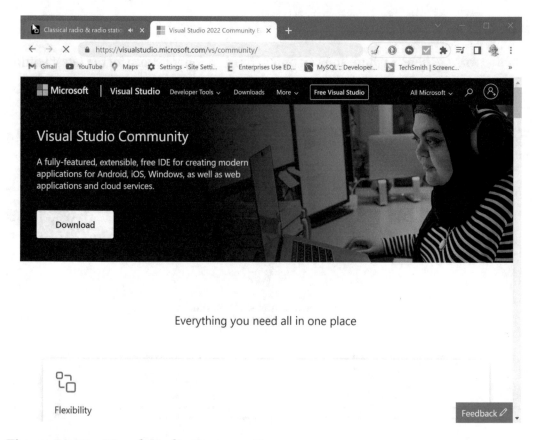

Figure 14-21. *Visual Studio Community*

As of this writing, the URL is

`https://visualstudio.microsoft.com/vs/community/`

Download this and install it. It will take some time. Get the process going and make sure you have a cup of coffee available. Your patience will be rewarded with a powerful development environment.

SQL Server Data Tools

This is available as a workload that you can download with Visual Studio Installer.

The installer is available on your Windows menu when you install the Visual Studio Community license.

Please refer to Figure 14-22.

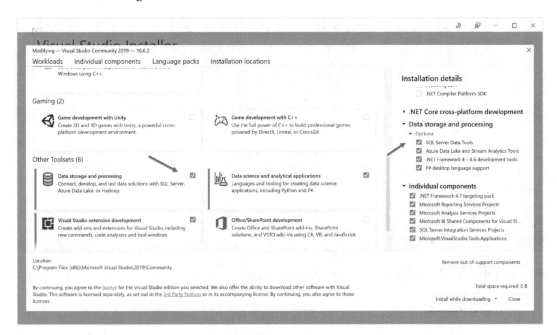

Figure 14-22. *Install the Data storage and processing workload*

Run the Visual Studio Installer tool and pick the features you want.

SQL Server SSAS Project Support

You might want to include this feature also so you can create SQL Server projects with Visual Studio Community Edition.

Please refer to Figure 14-23.

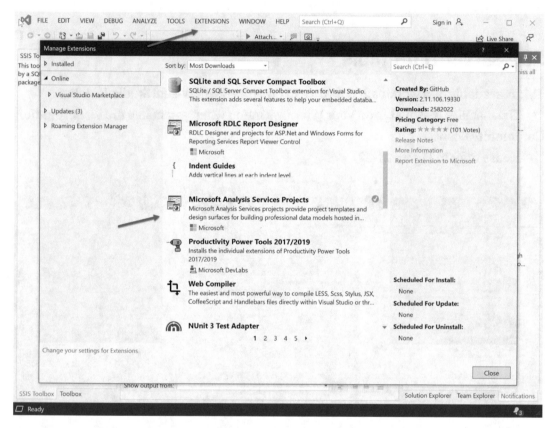

Figure 14-23.   *Download SSAS project support*

Be aware there is a lot of "stuff" to pick, so you might have to scroll quite a bit to find what you need. Be prepared to set aside a couple of hours to download and install all the project templates we discussed in this book.

SQL Server SSIS Project Support

Scroll down to get the SSIS project template.

Please refer to Figure 14-24.

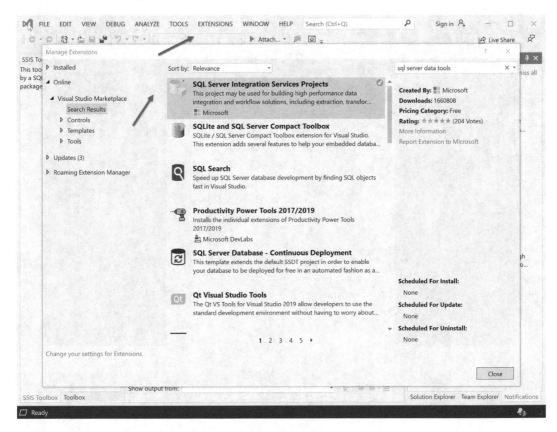

Figure 14-24. *Getting SSIS using Visual Studio*

Just select it and scroll to the next item, which is …

SQL Server SSRS Project Support

Just like the rest, select it and click to download.

Please refer to Figure 14-25.

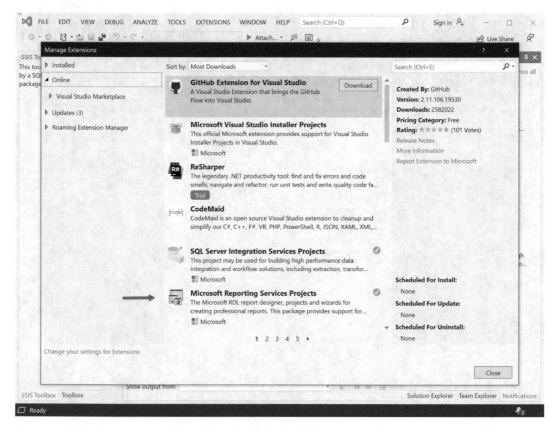

Figure 14-25. *Getting SSRS using Visual Studio*

That wasn't too hard, was it? Next, we need our report development tool.

Report Builder

Perform a search with your favorite web browser and you will find this web page.

Please refer to Figure 14-26.

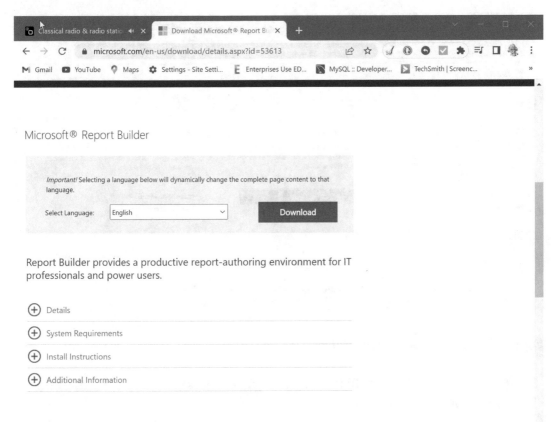

Figure 14-26. *How to get Report Builder*

Perform a search with your favorite browser or else check out the following URL. As of this writing, the URL is

`www.microsoft.com/en-us/download/details.aspx?id=53613`

Just click Download for the installer and then follow the steps. Mostly these are default selections, so accept them. Just click the Next button on each screen.

Three more items to download and install. Next is Power BI Desktop.

Power BI Desktop

We are getting to the finish line.

Please refer to Figure 14-27 for getting Power BI Desktop.

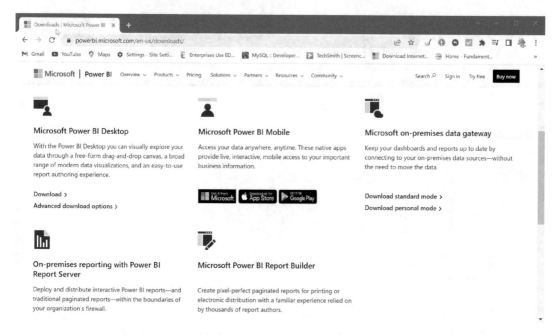

Figure 14-27. *Getting Power BI Desktop*

We are interested in Power BI Desktop and on-premises reporting with Power BI. As of this writing, the URL is

```
https://powerbi.microsoft.com/en-us/downloads/
```

As usual, follow the link, download the installer, run it, and follow the easy instructions.

This site seems to have both Power BI Desktop and the Power BI server by the way.

Power BI Server

Might as well have a personal Power BI server running on your personal learning or development environment, so get this component also.

Please refer to Figure 14-28.

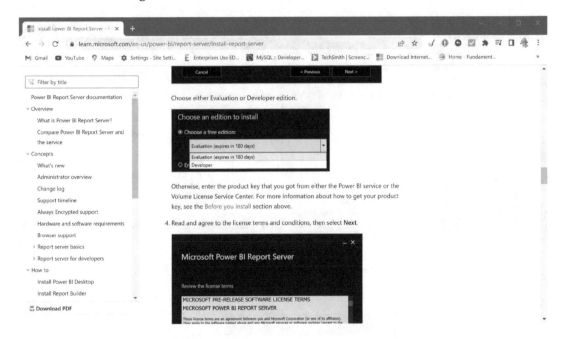

Figure 14-28. *Getting the Power BI server*

Make sure you download the free developer edition and always read the license requirements.

As of this writing, the URL is

```
https://learn.microsoft.com/en-us/power-bi/report-server/install-
report-server
```

Once you install it, the configuration steps look a lot like SSRS.

Last but not least is Microsoft Excel, but if you do not have it already, you need to pay!

Microsoft Excel

This is not free but worth the price! Go to the following website or perform a search with your favorite browser. Purchase only from Microsoft to be safe.

Please refer to Figure 14-29.

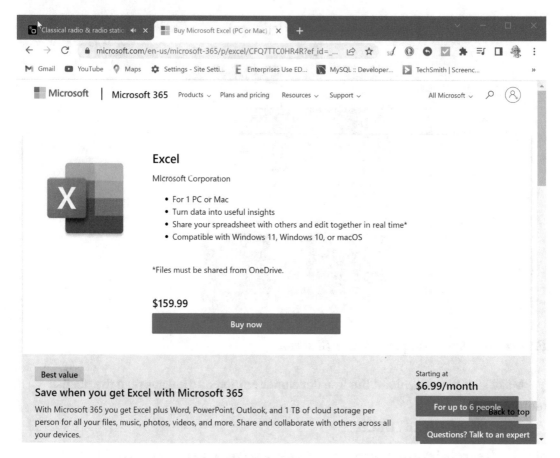

Figure 14-29. *Microsoft Excel subscription*

Check out which option works for you pricing- and feature-wise.

As of this writing, the URL is

`www.microsoft.com/en-us/download/details.aspx?id=53613/`

That's it. Now you know where to get all the tools. If you started off like me when I was learning all these tools, there was a lot of trial and error and poking about, but that is how you learn. Just make sure it is on your own personal environment and not on a production server!

SSMS

I almost forgot SSMS. Simply navigate to the following link to download the installer.

Please refer to Figure 14-30.

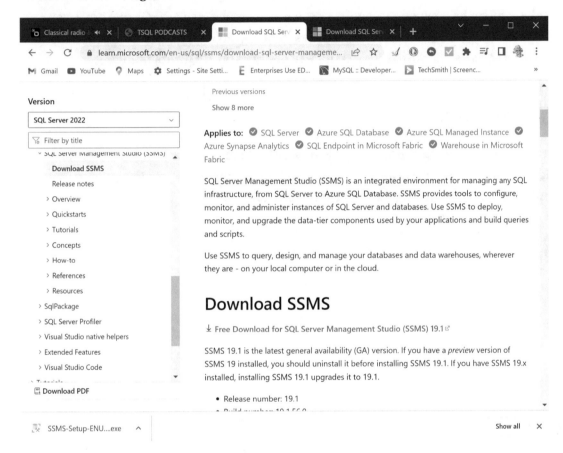

Figure 14-30. *Download SSMS*

As of this writing, the URL is

```
https://learn.microsoft.com/en-us/sql/ssms/download-sql-server-management-
studio-ssms?view=sql-server-ver16
```

Once the installer downloads, run the executable and follow the guided steps to install it on your development computer.

Next Steps

Practice with each of the scripts. Take your time. Download all the tools and, if necessary, research online for tutorials. Most are free, but there are plenty of sites that will offer robust video courses for a fair subscription price. If you are really new to TSQL and need some instruction or a refresher on writing queries, creating tables, etc., please check out my website at `www.grumpyolditguy.com` for free podcasts on TSQL programming.

Thank You!

Thank you for buying and reading this book. Also thank you for your patience. Please feel free to contact me via abobak@grumpyolditguy.com with questions, comments, and suggestions for improving this book in the next edition.

Till the next time …

APPENDIX A

Function Syntax, Descriptions

Here is a brief reference on the description and syntax of the functions we used. Included are the ROWS and RANGE window frame specifications together with the return types in case you need a refresher on how to use them.

I referred to the SQL Server documentation online to prepare the following reference.

The Window Frame Specifications

The following are the ROWS and RANGE window specifications that we used on some of the functions. As the syntax is introduced for each function, I will identify if these are supported or not in the OVER() clause.

In Table A-1 are the ROWS specifications.

© Angelo Bobak 2023
A. Bobak, *SQL Server Analytical Toolkit*, https://doi.org/10.1007/978-1-4842-8667-8

Table A-1. *ROWS Window Frame Specifications*

ROWS UNBOUNDED PRECEDING

ROWS BETWEEN UNBOUNDED PRECEDING AND
CURRENT ROW

ROWS BETWEEN <row number> PRECEDING AND
<row number> FOLLOWING

ROWS BETWEEN <row number> PRECEDING AND
CURRENT ROW

ROWS BETWEEN CURRENT ROW AND UNBOUNDED
FOLLOWING

ROWS BETWEEN CURRENT ROW AND <row number>
FOLLOWING

In Table A-2 are the RANGE specifications.

Table A-2. *RANGE Window Frame Specifications*

RANGE UNBOUNDED PRECEDING

RANGE BETWEEN UNBOUNDED PRECEDING AND CURRENT ROW

RANGE BETWEEN CURRENT ROW AND UNBOUNDED FOLLOWING

I suggest you copy and paste these in a query pane in SSMS (comment them out) and practice writing TSQL queries against a small set of rows, like in a table variable, that uses each one in order to understand the behavior.

ROWS and RANGE Default Behavior

This section was discussed in Chapter 1, but it is worth a review if you are new to these window functions.

By now we understand the syntax of the OVER() clause, and we know that it could include the following clauses:

- PARTITION BY clause

- ORDER BY clause

- ROWS or RANGE clause (see the preceding tables)

- None of these (empty)

The three clauses are optional for most of the window functions. You can leave them out or include one or more as required. There are two scenarios to consider when applying (or not) these clauses.

In our discussions, please keep in mind that most of the clauses work on columns, expressions, and also scalar queries. For example, PARTITION BY <expression> or ORDER BY<expression> implies that the expression could be a column, a list of columns, or even a scalar query that returns a single value. Check the Microsoft documentation for specifics on this topic.

Scenario 1

The default behavior of the window frames is dependent on whether the ORDER BY clause is included or not. There are two configurations to consider:

If the ORDER BY clause and the PARTITION BY clause are omitted and we do not include the window frame clause

If the ORDER BY clause is omitted but the PARTITION BY clause is included and we do not include the window frame clause

The default window frame behavior for both these cases is

```
ROWS BETWEEN UNBOUNDED PRECEDING AND UNBOUNDED FOLLOWING
```

Scenario 2

On the other hand, if we include an ORDER BY clause, the following two conditions also have a default window frame behavior:

If the ORDER BY clause is included but the PARTITION BY clause is omitted and we do not include the window frame clause

If the ORDER BY clause is included and the PARTITION BY clause is included and we do not include the window frame clause

The default window frame behavior for both these cases is

```
RANGE BETWEEN UNBOUNDED PRECEDING AND CURRENT ROW
```

Make sure you keep these default behaviors in mind as you start to develop queries that use the window functions. This is very important; otherwise, you will get some unexpected and possibly erroneous results. Your users will not be happy (is your resume up to date?).

Table A-3 might help you remember these default behaviors.

Table A-3. *Window Frame Default Behaviors*

ORDER BY	PARTITION BY	Default Frame
No	No	ROWS BETWEEN UNBOUNDED PRECEDING AND UNBOUNDED FOLLOWING
No	Yes	ROWS BETWEEN UNBOUNDED PRECEDING AND UNBOUNDED FOLLOWING
Yes	No	RANGE BETWEEN UNBOUNDED PRECEDING AND CURRENT ROW
Yes	Yes	RANGE BETWEEN UNBOUNDED PRECEDING AND CURRENT ROW

Remember you can override default behavior by including the ROWS or RANGE clause you need.

By the way, the following window frame clauses are not supported:

```
RANGE BETWEEN CURRENT ROW AND n FOLLOWING (will not work)
RANGE BETWEEN n PRECEDING AND CURRENT ROW (will not work)
RANGE BETWEEN n PRECEDING AND n FOLLOWING (will not work)
```

These do not work as the RANGE clause is a logical operation, so specifying a number for the rows to include in the window frame is not allowed. If you try to use these, you will get this wonderful error message:

RANGE is only supported with UNBOUNDED and CURRENT ROW window frame delimiters.

Pretty clear, don't you think?

One more time, here is a summary of the ROWS window frame specifications:

```
ROWS UNBOUNDED PRECEDING
ROWS BETWEEN UNBOUNDED PRECEDING AND CURRENT ROW
ROWS BETWEEN <row number> PRECEDING AND <row number> FOLLOWING
ROWS BETWEEN <row number> PRECEDING AND CURRENT ROW
ROWS BETWEEN CURRENT ROW AND UNBOUNDED FOLLOWING
ROWS BETWEEN CURRENT ROW AND <row number> FOLLOWING
```

And here is a summary of the RANGE window frame specifications:

```
RANGE UNBOUNDED PRECEDING
RANGE BETWEEN UNBOUNDED PRECEDING AND CURRENT ROW
RANGE BETWEEN CURRENT ROW AND UNBOUNDED FOLLOWING
```

Next, the syntax is presented for the three categories of functions.

The Aggregate Functions

We begin with the aggregate window functions. These were used in Chapters 2, 5, 8, and 11.

COUNT( )

Definition: Count the number of values in a table or the number of times a value appears in a column. An option exists to include or not include duplicates by using the ALL or DISTINCT keyword as a parameter.

Syntax

As an aggregate function in a SELECT statement:

```
COUNT(ALL | DISTINCT expression | *)
```

Using an OVER() clause:

```
COUNT(ALL expression | *))
OVER (
      PARTITION BY <column list or expression)
      ORDER BY <column list or expression>
      ROWS/RANGE <window frame specification >
      ) AS [Column Name]
      GO
```

The PARTITION BY clause is optional, the ORDER BY clause is optional, and so is the window frame clause.

Return value: INTEGER

Window frame ROWS/RANGE specifications are supported.

COUNT_BIG()

Definition: Same as COUNT() except it returns a BIGINT data type.

Syntax

As an aggregate function in a SELECT statement:

```
COUNT_BIG(ALL | DISTINCT expression | *)
```

Using an OVER() clause:

```
COUNT_BIG(ALL expression | *))
OVER (
      PARTITION BY <column list or expression)
      ORDER BY <column list or expression>
      ROWS/RANGE <window frame specification >
      ) AS [Column Name]
      GO
```

The PARTITION BY clause and the ORDER BY clause are optional, and so is the window frame clause.

Return value: BIGINT

Window frame ROWS/RANGE specifications are supported.

SUM( )

Definition: Totals values in a column. An option exists to include or not include duplicates by using the ALL or DISTINCT keyword as a parameter.

Syntax

As an aggregate function in a SELECT statement:

```
SUM(ALL | DISTINCT expression)
```

Using an OVER( ) clause:

```
SUM(ALL expression)OVER (
    PARTITION BY <column list or expression)
    ORDER BY <column list or expression>
    ROWS/RANGE <window frame specification>
    ) AS [Column Name]
    GO
```

The PARTITION BY clause and the ORDER BY clause are optional, and so is the window frame clause.

Return value: Depends on the data type of the expression

Window frame ROWS/RANGE specifications are supported.

MAX( )

Definition: Find the maximum column value in a set of rows. An option exists to include or not include duplicates by using the ALL or DISTINCT keyword as a parameter.

Syntax

As an aggregate function in a SELECT statement:

```
MAX(ALL | DISTINCT expression)
```

Using an OVER( ) clause:

```
MAX(ALL expression)OVER (
    PARTITION BY <column list or expression)
    ORDER BY <column list or expression>
```

```
ROWS/RANGE <window frame specification >
) AS [Column Name]
GO
```

The PARTITION BY clause and the ORDER BY clause are optional, and so is the window frame clause.

Return value: Depends on the data type of the expression

Window frame ROWS/RANGE specifications are supported.

MIN( )

Definition: Find the minimum column value in a set of rows. An option exists to include or not include duplicates by using the ALL or DISTINCT keyword as a parameter.

Syntax

As an aggregate function in a SELECT statement:

```
MIN(ALL | DISTINCT expression)
```

Using an OVER() clause:

```
MIN(ALL expression)OVER (
        PARTITION BY <column list or expression)
        ORDER BY <column list or expression>
        ROWS/RANGE <window frame specification >
    ) AS [Column Name]
    GO
```

The PARTITION BY clause and the ORDER BY clause are optional, and so is the window frame clause.

Return value: Depends on the data type of the expression

Window frame ROWS/RANGE specifications are supported.

AVG( )

Definition: Calculate the average column value from a set of rows. An option exists to include or not include duplicates by using the ALL or DISTINCT keyword as a parameter.

Syntax

As an aggregate function in a SELECT statement:

```
AVG(ALL | DISTINCT expression)
```

Using an OVER() clause:

```
AVG(ALL | DISTINCT expression)OVER (
        PARTITION BY <column list or expression)
        ORDER BY <column list or expression>
        ROWS/RANGE <window frame specification >
    ) AS [Column Name]
    GO
```

The PARTITION BY clause is optional, the ORDER BY clause is optional, and so is the window frame clause.

Return value: Depends on the data type of the expression

Window frame ROWS/RANGE specifications are supported.

GROUPING()

Definition: Groups rows from a query as defined by the GROUP BY clause in order to create rollup summaries. Used with ROLLUP, CUBE, or GROUPING sets in the GROUP BY clause.

Syntax

```
GROUPING(column expression)
```

The column or the expression needs to contain a column or columns in the GROUP BY clause of the query.

Here is a code snippet for a GROUP BY clause that will generate rollup values:

```
GROUP BY TransYear,
    TransQuarter,
    TransMonth,
    StoreNo,
    ProductNo,
    MonthlySales WITH ROLLUP
```

Return type: TINYINT

Does not support OVER() and of course ROWS/RANGE window frame specifications.

STRING_AGG()

Definition: Used to aggregate text strings or numerical values returned by a column in a query and delimit them with a specified delimiter, like a "," comma character.

Syntax

```
STRING_AGG(expression,delimiter)
```

Optionally uses

```
WITHIN GROUP (ORDER BY clause ASC | DESC)
```

Return type: TINYINT

Does not support OVER() and of course ROWS/RANGE window frame specifications. It does support the WITHIN GROUP clause as shown.

STDEV()

Definition: Returns the statistical standard deviation for when the entire population of the data set is not known. An option exists to include or not include duplicates by using the ALL or DISTINCT keyword as a parameter.

Syntax

As an aggregate function in a SELECT statement:

```
STDEV(ALL | DISTINCT expression)
```

Using an OVER() clause:

```
STDEV(ALL expression)
OVER (
          PARTITION BY <column list or expression)
          ORDER BY <column list or expression>
          ROWS/RANGE <window frame specification >
     ) AS [Column Name]
     GO
```

The PARTITION BY clause and the ORDER BY clause are optional, and so is the window frame clause.

Return value: FLOAT

Window frame ROWS/RANGE specifications are supported.

STDEVP( )

Definition: Returns the statistical standard deviation for when the entire population of the data set is known. An option exists to include or not include duplicates by using the ALL or DISTINCT keyword as a parameter.

Syntax

As an aggregate function in a SELECT statement:

```
STDEVP(ALL | DISTINCT expression)
```

Using an OVER() clause:

```
STDEVP(ALL expression)
OVER (
      PARTITION BY <column list or expression)
      ORDER BY <column list or expression>
      ROWS/RANGE <window frame specification >
      ) AS [Column Name]
      GO
```

The PARTITION BY and ORDER BY clauses are optional, and so is the window frame clause.

Return value: FLOAT

Window frame ROWS/RANGE specifications are supported.

VAR( )

Definition: Returns the statistical variance for when the entire population of the data set is not known. An option exists to include or not include duplicates by using the ALL or DISTINCT keyword as a parameter.

Syntax

As an aggregate function in a SELECT statement:

```
VAR(ALL | DISTINCT expression)
```

Using an OVER() clause:

```
VAR(ALL expression)
OVER (
     PARTITION BY <column list or expression)
     ORDER BY <column list or expression>
     ROWS/RANGE <window frame specification >
     ) AS [Column Name]
     GO
```

The PARTITION BY and ORDER BY clauses are optional, and so is the window frame clause.

Return value: FLOAT

Window frame ROWS/RANGE specifications are supported.

VARP()

Definition: Returns the statistical variance for when the entire population of the data set is known. An option exists to include or not include duplicates by using the ALL or DISTINCT keyword as a parameter.

Syntax

As an aggregate function in a SELECT statement:

```
VARP(ALL | DISTINCT expression)
```

Using an OVER() clause:

```
VARP(ALL expression)
OVER (
     PARTITION BY <column list or expression)
     ORDER BY <column list or expression>
     ROWS/RANGE <window frame specification >
     ) AS [Column Name]
     GO
```

The PARTITION BY clause is optional, the ORDER BY clause is optional, and so is the window frame clause.

Return value: FLOAT

Window frame ROWS/RANGE specifications are supported.

The Analytical Functions

The following is the syntax for the analytical window functions.

CUME_DIST()

Definition: This function calculates the relative position of a value within a data set like a table, partition, or table variable loaded with test data. An option exists to include or not include duplicates by using the ALL or DISTINCT keyword as a parameter.

Syntax:

```
CUME_DIST() OVER (
      PARTITION BY clause
      ORDER BY clause
) AS [Column Name]
```

Return type: float(53)

The PARTITION BY clause is optional; the ORDER BY clause is mandatory.

Window frame ROWS/RANGE specifications are not supported.

FIRST_VALUE()

Definition: Given a set of sorted values, the FIRST_VALUE() function will return the first value.

Syntax:

```
FIRST_VALUE(expression) OVER (
      PARTITION BY clause
      ORDER BY clause
```

```
     ROWS Window Frame Specification |
     RANGE Window Frame Specification
) AS [Column Name]
```

The PARTITION BY and ROWS/RANGE clauses are optional; the ORDER BY clause is mandatory.

Return type: Same as the expression

Window frame ROWS/RANGE specifications are supported.

LAST_VALUE()

Definition: Given a set of sorted values, the LAST_VALUE() function will return the last value.

Syntax:

```
LAST_VALUE(expression) OVER (
     PARTITION BY clause
     ORDER BY clause
     ROWS Window Frame Specification |
     RANGE Window Frame Specification
) AS [Column Name]
```

The PARTITION BY and ROWS/RANGE clauses are optional; the ORDER BY clause is mandatory.

Return type: Same as the expression

Window frame ROWS/RANGE specifications are supported.

LAG()

Definition: This function is used to retrieve the previous value of a column relative to the current row column value.

Syntax:

```
LAG( column or expression, optional offset, optional return default
value) OVER (
     PARTITION BY clause
```

```
      ORDER BY clause
) AS [Column Name]
```

The PARTITION BY clause is optional; the ORDER BY clause is mandatory.

Return type: NULL

Window frame ROWS/RANGE specifications are not supported.

LEAD( )

Definition: This function is used to retrieve the next value relative to the current row column.

Syntax:

```
LEAD( column or expression, optional offset, optional return default
value) OVER (
      PARTITION BY clause
      ORDER BY clause
) AS [Column Name]
```

The PARTITION BY clause is optional; the ORDER BY clause is mandatory.

Return type: NULL

Window frame ROWS/RANGE specifications are not supported.

PERCENT_RANK( )

Definition: Using a data set like results from a query, a partition, or a table variable, this function calculates the relative rank of each individual value relative to the entire data set (as a percentage).

Syntax:

```
PERCENT_RANK() OVER (
      PARTITION BY clause
      ORDER BY clause ASC|DESC
) AS [Column Name]
```

The PARTITION BY clause is optional; the ORDER BY clause is mandatory.

Return type: FLOAT(53)

Window frame ROWS/RANGE specifications are not supported.

PERCENTILE_CONT()

Definition: The PERCENTILE_CONT() function works on a continuous set of data based on a required percentile so as to return a value within the data set that satisfies the percentile (you need to supply a percentile as a parameter, like .25, .5, .75, etc.).

Continuous data is data measured over a period of time, like boiler temperatures in equipment over a 24-hour period or equipment failures over a period of time. You can use them in calculations like generating the average temperatures over 24 hours, but summing them up makes no sense.

Recall the questions you can ask yourself to differentiate between discrete and continuous data:

- Can I count it (over periods of time)? It is discrete data.

- Can I add (sum) it up? It is continuous data.

- Can I use it in calculations, like averages? It is continuous data.

Syntax:

```
PERCENT_CONT(value between 0.0 and 1.0)
      WITHIN GROUP (ORDER BY column or expression ASC|DESC)
      OVER (PARTITION BY clause)
) AS [Column Name]
```

The PARTITION BY clause is optional; the ORDER BY in the WITHIN GROUP clause is mandatory.

Return type: FLOAT(53)

Window frame ROWS/RANGE specifications are not supported.

PERCENTILE_DISC()

Definition: The PERCENTILE_DISC() function works on a discrete set of data based on a required percentile so as to return a value within the data set that satisfies the percentile. Discrete data values are numbers that have values that you can count over time, like investment account deposits over days, months, and years. You can use the SUM() function to calculate the total deposit amounts or withdrawal amounts in a day, but the account balance needs to be the result of adding the prior day's account balance plus all new deposits minus all new withdrawals.

Recall the questions you can ask yourself to differentiate between discrete and continuous data:

- Can I count it (over periods of time)? It is discrete data.

- Can I add (sum) it up? It is continuous data.

- Can I use it in calculations, like averages? It is continuous data.

Syntax:

```
PERCENT_DISC(value between 0,0 and 1.0)
     WITHIN GROUP (ORDER BY column or expression ASC|DESC)
     OVER (PARTITION BY clause)
) AS [Column Name]
```

The PARTITION BY clause is optional; the ORDER BY in the WITHIN GROUP clause is mandatory.

Return type: FLOAT(53)

Window frame ROWS/RANGE specifications are not supported.

The Window/Ranking Functions

We conclude with the syntax for the last category of functions, the window/ranking functions.

RANK( )

Definition: The RANK( ) function returns the number of rows before the current row plus the current row based on the value being ranked.

Syntax:

```
RANK() OVER (
      PARTITION BY clause
      ORDER BY clause ASC|DESC
) AS [Column Name]
```

The PARTITION BY clause is optional; the ORDER BY clause is mandatory.

Return type: BIGINT

Window frame ROWS/RANGE specifications are not supported.

DENSE_RANK( )

Definition: The DENSE_RANK( ) function works almost the same as the RANK( ) function. In case of ties, the dense rank value is the same for the ties, but the next value after the duplicate values is simply assigned the next rank number.

Syntax:

```
DENSE_RANK() OVER (
      PARTITION BY clause
      ORDER BY clause ASC|DESC
) AS [Column Name]
```

The PARTITION BY clause is optional; the ORDER BY clause is mandatory.

Return type: BIGINT

Window frame ROWS/RANGE specifications are not supported.

NTILE( )

Definition: The NTILE( ) function allows you to divide a set of rows in a data set into tiles or buckets (I like this term better). If you have a data set of 12 rows and you want to assign four tiles, each tile will have three rows. If you have a data set with 13 rows and want four tiles, the last tile will have more rows than the others.

Syntax:

```
NTILE() OVER (
      PARTITION BY clause
      ORDER BY clause ASC|DESC
) AS [Column Name]
```

The PARTITION BY clause is optional; the ORDER BY clause is mandatory.

Return type: BIGINT

Window frame ROWS/RANGE specifications are not supported.

ROW_NUMBER()

Definition: Given a set of rows in the data result set, it will assign a sequential number to each row in the data set (or partition) returned by a query.

Syntax:

```
ROW_NUMBER() OVER (
      PARTITION BY clause
      ORDER BY clause ASC|DESC
) AS [Column Name]
```

The PARTITION BY clause is optional; the ORDER BY clause is mandatory.

Return type: BIGINT

Window frame ROWS/RANGE specifications are not supported.

For more details on syntax, arguments, and return types plus examples, please refer to the online Microsoft documentation.

I suggest you practice all the different variations on small data sets so you can become familiar with how these work.

For example, here is a small practice data set you can use:

```
USE TEST
GO

DECLARE @TestTable TABLE (
      KeyValue SMALLINT IDENTITY NOT NULL,
      NumValue INT,
      TextValue VARCHAR(128)
      );
```

```
INSERT INTO @TestTable VALUES
(10,'TEXTA'),
(20,'TEXTA'),
(20,'TEXTB'),
(30,'TEXTC'),
(40,'TEXTC'),
(50,'TEXTC'),
(60,'TEXTD'),
(70,'TEXTE'),
(80,'TEXTE'),
(90,'TEXTF'),
(100,'TEXTG');

SELECT VAR(NumValue) OVER(ORDER BY NumValue) AS DistinctNumValue
FROM @TestTable;

SELECT VAR(NumValue) OVER(PARTITION BY TextValue ORDER BY NumValue) AS
DistinctNumValue
FROM @TestTable;

SELECT VAR(NumValue) OVER() AS DistinctNumValue
FROM @TestTable;
GO
```

APPENDIX B

Statistical Functions

The following is a short list of statistical functions you should be familiar with as you embark on your journey to become an advanced SQL Server developer or data architect.

Standard Deviation

The STDEV() function calculates the statistical standard deviation of a set of values when the entire population is not known or available.

The STDEVP() function calculates the statistical standard deviation when the entire population of the data set is known.

But what does this mean?

What this means is how close or how far apart the numbers are in the data set from the arithmetic **MEAN** (which is another word for average). In our context the average or **MEAN** is the sum of all data values in question divided by the number of data values. There are other types of **MEAN** like geometric, harmonic, and weighted **MEAN**, which are discussed in the following, but first a simple example.

Please refer to Listing B-1.

Listing B-1. Standard Deviation

```
DECLARE @X FLOAT;
DECLARE @Y FLOAT;

DECLARE @SampleData TABLE (
    KeyVal SMALLINT IDENTITY NOT NULL,
    ColVal FLOAT
    );
```

1009

© Angelo Bobak 2023
A. Bobak, *SQL Server Analytical Toolkit*, https://doi.org/10.1007/978-1-4842-8667-8

```
INSERT INTO @SampleData VALUES
      (10.00),(20.00),(30.00),(35.00),(40.00),(42.00),(44.00)
      ,(46.00),(48.00),(50.00),(50.00),(48.00),(46.00),(44.00)
      ,(42.00),(40.00),(35.00),(30.00),(20.00),(10.00);

SELECT ColVal,AVG(ColVal) OVER () AS MEAN,STDEVP(ColVal)
      OVER ()AS STANDARD_DEVIATION
FROM @SampleData
GO
```

Figure B-1 is a graph generated with Microsoft Excel from the data returned by this query.

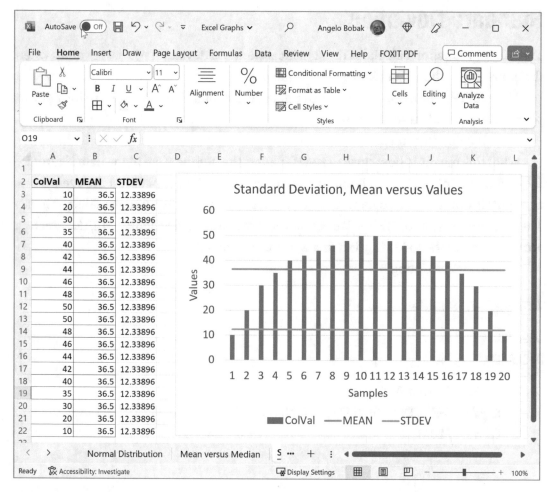

Figure B-1. *Standard deviation and mean by values*

Typically, we would use the values in a Microsoft Excel spreadsheet so we can visually see the results. We see the MEAN and standard deviation superimposed on the bar graph as a point of reference. We will see that standard deviation plays an important role in the generation of normal distribution curves called bell curves.

Standard deviation is one of the parameters used in the formula that generates the normal distribution values. Normal distribution will be discussed shortly.

Variance

The VAR() and VARP() functions are used to generate the variance for a set of values in a data sample. The same discussion related to the data sets used in the STDEV()/STDEVP() examples applies to the VAR() and VARP() functions relative to population of the data:

The VAR() function works on a partial set of the data when the entire data set is not known or available, while the VARP() function works on the whole data set population when it is known.

Informally, if you take all the differences of each value from the mean (average) and square the results and then add all the results and finally divide them by the number of data samples, you get the variance.

For our scenario, let's say we had some data that forecasted sales amounts for one of our products during a particular year. When sales data is finally generated for the product, we want to see how close or far the sales figures are to or from the forecast.

It answers the following questions:

- Did we miss our targets?

- Did we meet our targets?

- Did we exceed our targets?

Here is a simple example. Please refer to Listing B-2.

Listing B-2. Variance

```
DECLARE @SampleData TABLE (
    KeyVal SMALLINT IDENTITY NOT NULL,
    ColVal FLOAT
    );
```

```
INSERT INTO @SampleData VALUES
     (10.00),(20.00),(30.00),(35.00),(40.00),(42.00)
     ,(44.00),(46.00),(48.00),(50.00),50.00),(48.00),(46.00)
     ,(44.00),(42.00),(40.00),(35.00),(30.00),(20.00),(10.00);
SELECT ColVal
     ,VAR(ColVal)OVER () AS VARIANCE
     ,VARP(ColVal) OVER () AS VARIANCEP
FROM @SampleData;
GO
```

I am using a small generic data set to keep the example simple.

As can be seen an OVER() clause in its most basic form is used. There is no ORDER BY or PARTITION BY clause for both function calls. Let's check out the results.

Please refer to Figure B-2.

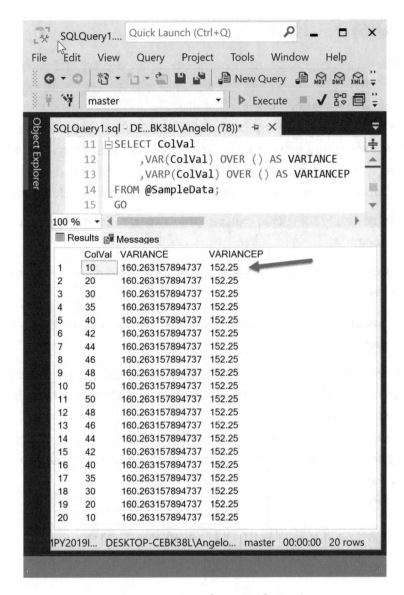

Figure B-2. *Comparing variance output for populations*

As can be seen, the results for VAR(P) are smaller as the entire population of the data is being used.

Normal Distribution

In this example we will create our own normal distribution query and compare it to the Microsoft Excel NDIST() function. Normal distribution is used to create a graph called a bell graph because it resembles an inverted bell.

When the graph takes on this shape, the MODE, MEDIAN, and MEAN values are the same, which makes sense if you think about it. This means the graph is symmetrical.

Depending on the data, the bell curve can lean to the right or to the left.

Standard deviation is used to describe the spread of the data within the distribution:

- Approximately 68% of the data is within 1 standard deviation from the mean.

- Approximately 95% of the data is within 2 standard deviations from the mean.

- Approximately 99.7% of the data is within 3 standard deviations from the mean.

An example of finding the normal distribution is when we dealt with equipment failure rates in a power plant. We wanted to know the spread of the failures across the same equipment type in all locations.

The following is a small example that we will use to generate normal distribution values for a generic data set and then compare them to the results generated by Microsoft Excel for the same data.

Here is the formula we use in our query:

$$(1/SQRT(2.0 * PI() * VARIANCE)) * EXP($$
$$(-1 * POWER((VALUE - MEAN),2))/(2 * VARIANCE)$$
$$)$$

Please refer to Listing B-3.

Listing B-3. Calculate the Normal Distribution

```
DECLARE @SampleData TABLE (
    KeyVal SMALLINT IDENTITY NOT NULL,
    ColVal FLOAT
    );
```

```
INSERT INTO @SampleData VALUES
     (10.00),(20.00),(30.00),(35.00),(40.00),(42.00),(44.00)
     ,(46.00),(48.00),(50.00),(50.00),(48.00),(46.00),(44.00)
     ,(42.00),(40.00),(35.00),(30.00),(20.00),(10.00);

DECLARE @StdDev   FLOAT;
DECLARE @Mean     FLOAT;
DECLARE @Variance FLOAT;

SELECT @StdDev = STDEV(ColVal) OVER()
FROM @SampleData;

SELECT @Mean = AVG(ColVal) OVER()
FROM @SampleData;

SELECT @Variance = VAR(ColVal)OVER ()
FROM @SampleData;

SELECT ColVal
     ,@StdDev AS STDEV
     ,@Mean AS MEAN
     ,(1/SQRT(2.0 * PI() * @Variance))
          * EXP((-1 * POWER((ColVal - @Mean),2))/(2 * @Variance)) AS NDIST
FROM @SampleData;
GO
```

We pretty much used functions available with SQL Server for this formula. If you are not sure what they are, check out the Microsoft documentation.

Let's copy our results to a Microsoft Excel spreadsheet and generate our bell curve. Then we use the same results and use the Excel-built in function NORM.DIST() to generate the normal distribution values to see if they match.

Please refer to Figure B-3a.

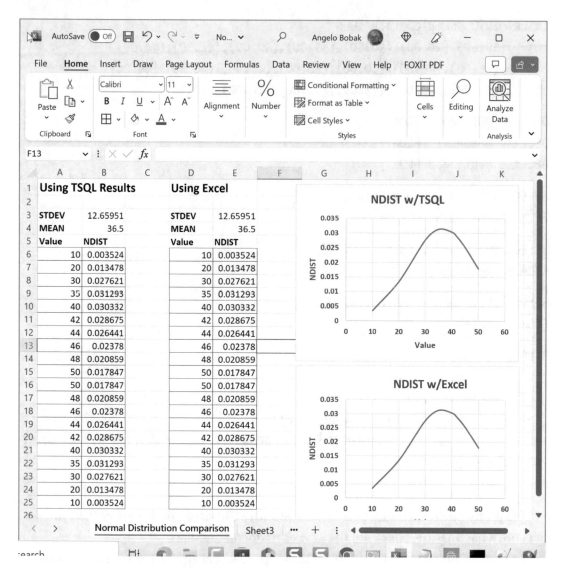

Figure B-3a. *Normal Distribution – SQL vs. Excel*

Success. It looks like both sets of results match. Well, it sort of looks like a bell but close enough.

Here are a few more bell curves based on various patterns of data.

Please refer to Figure B-3b

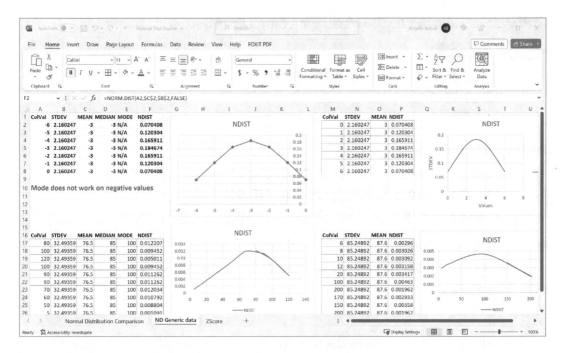

Figure B-3b. *Normal distribution – for various patterns*

Try experimenting on your own by generating some random patterns in Excel and then calculating the normal distribution with the NORM.DIST function in Excel:

```
=NORM.DIST(x,MEAN,STDEV,FALSE)
```

Just use the preceding parameters for each value in the data set. The variable X is the value, the MEAN and STDEV functions are built-in Excel functions, and FALSE/TRUE is one of two options you need to fill in:

> TRUE: Use the cumulative distribution function.

> FALSE: Use the probability density function.

Mean (Average)

Statistical MEAN is just another word for the average of a set of values. Given ten values, the MEAN is calculated by summing all ten values and then dividing them by the number of values to get the average. The following is a simple example.

Please refer to Listing B-4.

Listing B-4. Calculating the Mean for a Small Data Set

```
DECLARE @SampleData TABLE (
      KeyVal SMALLINT IDENTITY NOT NULL,
      ColVal FLOAT
      );

INSERT INTO @SampleData VALUES
      (10.00),(20.00),(30.00),(35.00),(40.00),(42.00),(44.00)
      ,(46.00),(48.00),(50.00),(50.00),(48.00),(46.00)
      ,(44.00),(42.00),(40.00),(35.00),(30.00),(20.00),(10.00);

SELECT KeyVal,ColVal,
      AVG(ColVal) OVER(ORDER BY KeyVal ASC) AS [Rolling Mean]
FROM @SampleData;
GO
```

The calculation performed by the function is easy. Sum up all the values and divide the results by the number of values to produce the average.

In this example, to make things interesting, an OVER() clause is used so we get a rolling average effect.

Please refer to Figure B-4 for the results.

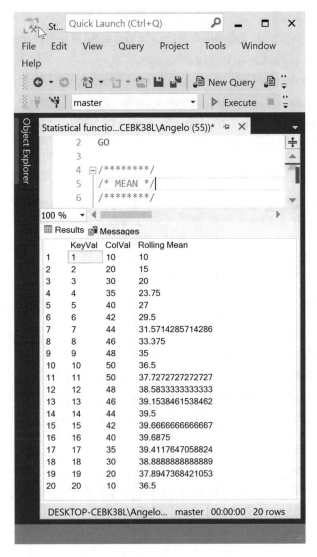

Figure B-4. *Statistical MEAN (average)*

As the rows are processed, the average is calculated for the current row and all prior row values. Try validating these results in Microsoft Excel.

Median

Median is used to find the numerical value of a data set that occurs in the middle of a data set. For example, given the following data set

{1,2,3,4,5,**6**,7,8,9,9,10}

the median is the value 6. Comparing this to the MEAN or average value in this case, we get the value 5.818182. Close but not the same.

Sometimes if the data set has the correct value spread, the MEAN and MEDIAN values are the same. Remember our normal distribution bell curve?

To calculate this value with TSQL, we use the `PERCENTILE_CONT(.50)` or `PERCENTILE_DISC(.50)` function.

Please refer to Listing B-5.

Listing B-5. Calculating the Median with PERCENTILE_CONT and PERCENTILE_DISC

```
DECLARE @SampleData TABLE (
     KeyVal SMALLINT IDENTITY NOT NULL,
     ColVal FLOAT
     );

INSERT INTO @SampleData VALUES
     (10.00),(20.00),(30.00),(35.00),(40.00),(42.00),(44.00)
     ,(46.00),(48.00),(50.00),(50.00),(48.00),(46.00),(44.00)
     ,(42.00),(40.00),(35.00),(30.00),(20.00),(10.00);

SELECT KeyVal,ColVal,
     PERCENTILE_CONT(.50)
     WITHIN GROUP(ORDER BY ColVal) OVER()  AS MEDIAN_PCT_CONT,
     PERCENTILE_DISC(.50)
     WITHIN GROUP(ORDER BY ColVal) OVER()  AS MEDIAN_PCT_DISC
FROM @SampleData;
GO
```

This time we use the `ORDER BY` clause in the `WITHIN GROUP()` clause and pass the value .50 to each function.

Please refer to Figure B-5.

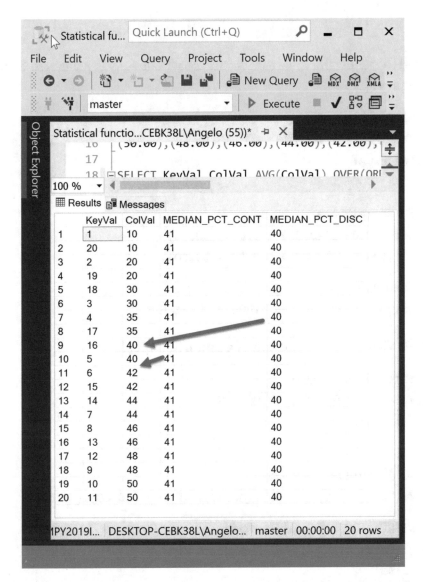

Figure B-5. *Calculating median values with the PERCENTILE_CONT() and PERCENTILE_DISC() functions*

The PERCENTILE_DISC() function will return a value that exists in the data set, while the PERCENTILE_CONT() will usually return an interpolated value that may or may not exist in the data set.

Mode

Mode is used to identify the value that occurs most frequently in a data set. Given the following data set

```
DECLARE @SampleData TABLE (
    KeyVal SMALLINT IDENTITY NOT NULL,
    ColVal FLOAT
    );

INSERT INTO @SampleData
    VALUES (10.00),(20.00),(30.00),(30.00)
        ,(40.00),(40.00),(40.00),(50.00);
```

we can see that the value 40 appears three times.

Here is a simple example. Please refer to Listing B-6.

Listing B-6. Calculating the Mode of a Small Set of Values

```
DECLARE @SampleData TABLE (
    KeyVal SMALLINT IDENTITY NOT NULL,
    ColVal FLOAT
    );

INSERT INTO @SampleData VALUES
(10.00),(20.00),(30.00),(30.00),(40.00),(40.00),(40.00),(50.00);

SELECT DISTINCT TOP 1 ColVal AS Mode
FROM @SampleData
GROUP BY ColVal
HAVING COUNT(*) > 1
ORDER BY ColVal DESC
GO
```

A few tricks are used like sorting the results in descending order and selecting the TOP 1 result. The value returned is 40.

Here is another version of the code using a CTE so we can compensate for ties. Please refer to Listing B-7.

Listing B-7. Mode for Ties

```
DECLARE @SampleData TABLE (
     KeyVal SMALLINT IDENTITY NOT NULL,
     ColVal FLOAT
     );

INSERT INTO @SampleData VALUES
(10.00),(30.00),(30.00),(30.00),(40.00),(40.00),(40.00),(50.00),
(60.00),(70.00),(80.00),(90.00),(90.00),(140.00),(240.00),(350.00);

;WITH ModeCTE(ColVal,Ties)
AS
(
SELECT DISTINCT ColVal AS Mode,COUNT(*) AS Ties
FROM @SampleData
GROUP BY ColVal
HAVING COUNT(*) > 1
)

SELECT ColVal AS Mode,MAX(Ties) AS MaxTies
FROM ModeCTE
WHERE Ties = (SELECT MAX(Ties) FROM ModeCTE)
GROUP BY ColVal,Ties
GO
```

A subquery is used so we can filter out the mode value and then find out if there is more than one set of values that satisfies the mode.

Please refer to Figure B-6 for the results.

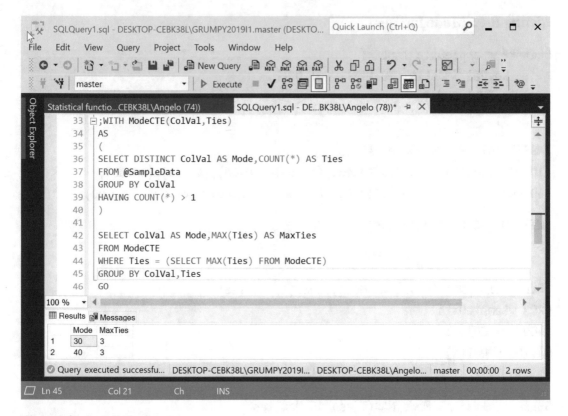

Figure B-6. *Mode for ties*

All ties identified.

Geometric Mean

This functions similar to the MEAN formula, but instead of summing up the values, the product of the values is taken (multiplication vs. addition).

Next, the nth root of the final product result is taken. The nth root represents the number of values in the data set. So, if you have two values, you take the square root; if you have three values, you take the cube root; and so on.

For example, if you have the following three numbers, 3, 3, and 3, we generate the product by multiplying each value against the next:

$$3 * 3 = 9$$

$$3 * 3 = 27$$

Next, we take the nth root, which in this case is the cube root (3), we get the answer 3.

Based on this logic, the following query generates the geometric mean for a small data sample. Please refer to Listing B-8.

Listing B-8. Calculating the Geometric Mean of a Small Set of Values

```
DECLARE @Rows             FLOAT;
DECLARE @ResultValue    VARCHAR(256);
DECLARE @ParmDefinition NVARCHAR(500);
DECLARE @Parms            NVARCHAR(500);
DECLARE @Formula          NVARCHAR(500);

DECLARE @SampleData TABLE (
    KeyVal SMALLINT IDENTITY NOT NULL,
    ColVal FLOAT
    );

INSERT INTO @SampleData VALUES
(2.00),(4.00),(6.00),(4.00),(6.00);

SELECT @Rows = COUNT(*) FROM @SampleData;

-- product of the reciprocals
SELECT @Formula = N'SELECT @ResultValueOUT = (' + STRING_
AGG(ColVal,'*') + ')'
FROM @SampleData;

SELECT @Formula

SET @Parms = N'@ResultValueOUT VARCHAR(256) OUTPUT';

EXEC sp_executesql @Formula,@Parms,@ResultValueOUT=@ResultValue OUTPUT;

-- take nth root of final product calculation
SELECT @ResultValue,POWER(@ResultValue,(1/@Rows)) AS GeoMean;
GO
```

Let's walk through the steps as the function is a bit tricky.

Step 1: Initialize a string so it contains the formula for the products of each value:

```
'SELECT @ResultValueOUT = (2*4*6*4*6)'
```

We want to execute this little query so as to get the value we will use for the geometric mean.

Step 2: Set up the string that will capture the output of the dynamic query when executed with the stored procedure EXEC sp_executesql.

Step 3: Write a query that will generate the reciprocal of the final product value by using the POWER() function. We raise the product value by the reciprocal of the number of data elements in the result set:

```
SELECT @ResultValue,POWER(@ResultValue,(1/@Rows)) AS GeoMean;
```

Walk in the park!

Let's take a look at the result, which also shows the dynamic SELECT query that contains the formula.

Please refer to Figure B-7.

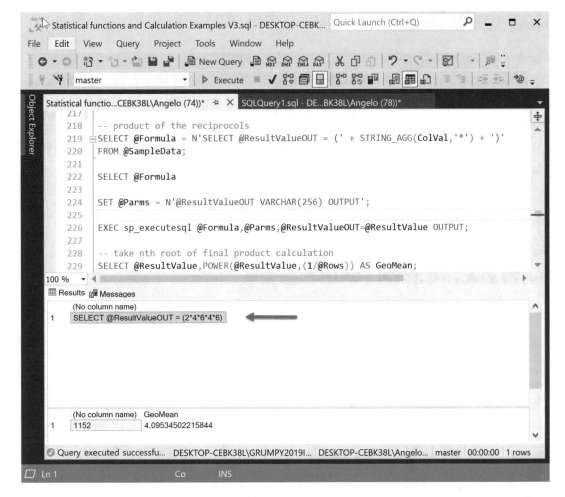

Figure B-7. *Generating the geometric mean*

The statement appears together with the results, which display the value generated by multiplying each value and the final geometric mean result. This was just one way of generating the geometric mean.

A word of warning: Be careful when using dynamic SQL to generate results as there could be security risks, like SQL injection where someone could append a malicious query with the intended query.

Try rewriting this query so it uses a `CTE` or `WHILE()` loop to calculate the products when multiplying each value in the data set against the next value.

Harmonic Mean

This is another type of average calculation. Given a set of values, the harmonic mean is calculated by generating the reciprocals of each value and adding them together to produce a sum. In the prior example for the geometric mean, we took the product of the values.

The average of the sum is then calculated by dividing it by the number of samples.

Finally, we take the reciprocal of the average.

Please refer to Listing B-9.

Listing B-9. Harmonic Mean

```
DECLARE @Rows FLOAT;
DECLARE @SampleReciprocol FLOAT;

DECLARE @SampleData TABLE (
    KeyVal SMALLINT IDENTITY NOT NULL,
    ColVal FLOAT
    );

INSERT INTO @SampleData VALUES
(2.00),(4.00),(6.00),(4.00),(6.00);

SELECT @Rows = COUNT(*) FROM @SampleData;

-- generate reciprocal
SELECT 1.0/ColVal
FROM @SampleData

-- sum the reciprocals
SELECT SUM(1.0/ColVal) AS SumRec
FROM @SampleData;

-- generate the average
SELECT (SUM(1.0/ColVal))/@Rows AS AvgSumRec
FROM @SampleData;
```

```
-- generate the final reciprocol
SELECT @Rows/(SUM(1.0/ColVal)) AS AvgSumRec
FROM @SampleData;
GO
```

The steps are simple:

Step 1: Generate the sum of the reciprocals.

Step 2: Calculate the average of the result in step 1 by dividing it with the value of the number of samples.

Step 3: Generate the reciprocal of step 2.

Another simple algorithm. The value returned is 3.75.

Notice how each step builds on the prior step so as to come up with the final query that includes all the steps.

Weighted Mean (Average)

Weighted mean is used when you want to assign weights to values so as to favor one over the other. For example, if we want to rate the categories of features for two sports cars, we can assign a value between 0 and 100 to each category and then assign weights to each category.

In this example, we have two car models, the SuperSport model and the SuperEuroSport model.

The categories we will use to rate them are fuel economy, color, horsepower, comfort, and speed.

The following is a table that identifies the ratings and weights we want to assign to each category for each car model.

Please refer to Table B-1.

Table B-1. *Car Model Features: Weights and Ratings*

Model	Category	Rating	Weight	Score
SuperSport	Fuel economy	20	0.125	2.5
SuperSport	Color	20	0.125	2.5
SuperSport	Horsepower	30	0.5	15
SuperSport	Comfort	10	0.125	1.25
SuperSport	Speed	20	0.125	2.5
SuperEuroSport	Fuel economy	10	0.125	1.25
SuperEuroSport	Color	10	0.125	1.25
SuperEuroSport	Horsepower	40	0.5	20
SuperEuroSport	Comfort	10	0.125	1.25
SuperEuroSport	Speed	30	0.125	3.75

Based on this data, let's create the weighted mean query to identify which car model wins the rating process.

Please refer to Listing B-10.

Listing B-10. Weighted Mean ala TSQL

```
DECLARE @CarRating TABLE (
      KeyVal      SMALLINT IDENTITY NOT NULL,
      CarModel    VARCHAR(64) NOT NULL,
      Category    VARCHAR(64) NOT NULL,
      Rating      FLOAT       NOT NULL,
      CategoryWeight FLOAT NOT NULL
      );

INSERT INTO @CarRating VALUES
('SuperSport','Fuel Economy',20.0,(1.0/8.0)),
('SuperSport','Color',20.0,(1.0/8.0)),
('SuperSport','Horse Power',30.0,(1.0/2.0)),
('SuperSport','Comfort',10.0,(1.0/8.0)),
('SuperSport','Speed',20.0,(1.0/8.0)),
```

```
('SuperEuroSport','Fuel Economy',10.0,(1.0/8.0)),
('SuperEuroSport','Color',10.0,(1.0/8.0)),
('SuperEuroSport','Horse Power',40.0,(1.0/2.0)),
('SuperEuroSport','Comfort',10.0,(1.0/8.0)),
('SuperEuroSport','Speed',30.0,(1.0/8.0));

SELECT CarModel,Category,Rating,CategoryWeight
FROM @CarRating;

SELECT CarModel,Category,(Rating * CategoryWeight) AS WeightedMean
FROM @CarRating;

SELECT DISTINCT CarModel
     ,SUM((Rating * CategoryWeight)) OVER (
          PARTITION BY CarModel
          ORDER BY CarModel
          ) AS WeightedMean
FROM @CarRating
GO
```

As can be seen we are using the SUM() function with an OVER() clause to generate the scores. The partition is defined by the CarModel column, and the partition data set is also ordered by the CarModel column.

A few queries are used to show you the results for each step of the process.

The following are the results. Please refer to Figure B-8.

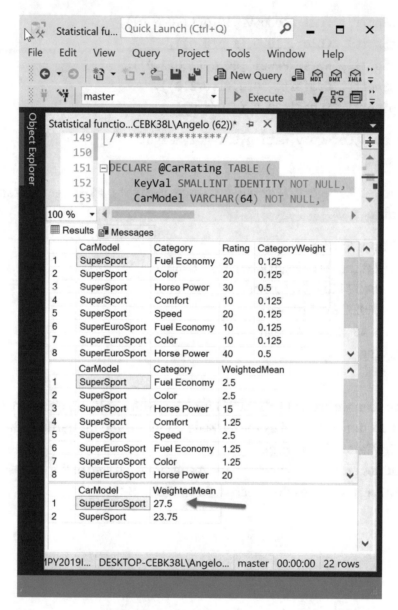

Figure B-8. *Weighted mean car model selection calculation*

Looks like the SuperEuroSport model is the winner.

Last but not least, here is a nice bar graph to present a visual of the analysis.

The following are the results. Please refer to Figure B-9.

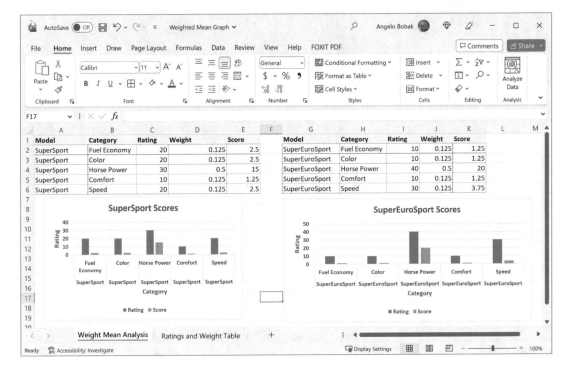

Figure B-9. *Weighted mean car model selection calculation graph*

Nothing like combining analysis reports and graphs to get a complete picture for analysis.

Summary

Now you know enough about statistics to make you dangerous.

All kidding aside, I hope my little tutorial on some basic statistics functions and formulas has awakened an interest in this fascinating and important field of study.

Consider these as more tools in your SQL Server toolkit. Try placing the code in stored procedures and functions so you can use them over and over in your scripts, queries, and reports.

Index

A

© Angelo Bobak 2023
A. Bobak, *SQL Server Analytical Toolkit*, https://doi.org/10.1007/978-1-4842-8667-8

O

Y, Z